PAINT AND SURFACE COATINGS

Theory and Practice

Second edition

Editors:

R LAMBOURNE and T A STRIVENS

WILLIAM ANDREW PUBLISHING

Published by Woodhead Publishing Ltd, Abington Hall,
Abington, Cambridge CB1 6AH, England

First published 1987, Ellis Horwood Ltd
Reprinted 1988
Reprinted in paperback 1993
Second edition 1999 Woodhead Publishing Ltd

Published in the United States of America by William Andrew Publishing

Library of Congress Cataloging-in-Publication Data
A catalog record for this book is available from the Library of Congress.

ISBN 1-884207-73-1

Printed and bound in England.

Contents

3 **Pigments for paint**
   A G Abel

4 **Solvents, thinners, and diluents**
   R Lambourne

5 **Additives for paint**
   R A Jeffs and W Jones

List of contributors

A G Abel	Industrial Manager (Technology), Clariant Specialties Ltd, Calverley Lane, Horsforth, Leeds LS18 4RP, UK
D A Ansdell	Formerly Technical Manager, Automotive Products, ICI Paints Division, Slough SL2 5DS, UK
J Bentley	Consultant, Paint Research Association, Teddington, Middlesex, UK
T R Bullett	Formerly Research Director, Paint Research Association, Teddington, Middlesex, UK
A Doroszkowski	Principal Research Scientist, Paint Research Association, Teddington, Middlesex, UK
F K Farkas	Consultant, FKF & Associates
J A Graystone	Principal Research Scientist, Paint Research Association, Teddington, Middlesex, UK
G R Hayward	Consultant, Hales End, Main Street, Laxton Village, E. Northants NN17 3AT, UK
R A Jeffs	Formerly Research Department, ICI Paints Division, Slough SL2 5DS, UK
W Jones	Formerly Research Department, ICI Paints Division, Slough SL2 5DS, UK
R Lambourne	Formerly Research Department, ICI Paints Division and Industrial Colloid Advisory Group, Bristol University, UK
A H Mawby	Director, Refinish Marketing, PPG Industries (UK) Ltd, Ladywood, Birmingham B116 0AD, UK
T A Strivens	Formerly Divisional Research Associate, ICI Paints Division, Slough SL2 5DS, UK
G P A Turner	Formerly Technical Manager, Industrial Coatings Research, ICI Paints Division, Slough SL2 5DS, UK

Preface to first edition

For many years I have felt that there has been a need for a book on the science and technology of paints and surface coatings that would provide science graduates entering the paint industry with a bridge between academia and the applied science and technology of paints. Whilst there have been many excellent books dealing with the technology there have not to my knowledge been any that have sought to provide a basic understanding of the chemistry and physics of coatings. Many of the one-time standard technological texts are now out of date (and out of print), so it seemed appropriate to attempt to produce a book that will, I hope, fill a gap. Nevertheless, it was with some trepidation that I undertook the task of editing a book covering such a diverse technology. The diversity of the technology is such that rarely will an acknowledged expert in one aspect of the technology feel confident to claim expertise in another. It therefore seemed to me that a work produced by a single author would not meet the objectives I had in mind, and I sought the help of friends and colleagues in the industry to contribute individual chapters on subjects where I knew them to have the requisite expertise. Fortunately, I was able to persuade sufficient contributions from individuals for whom I have the highest regard in respect of their knowledge and years of experience within the paint industry to satisfy myself of the ultimate authenticity of the book as a whole.

However, because of limitations of space it is impossible for a book of this kind to be completely comprehensive. Thus I have had to make decisions regarding content, and have adopted a framework which gives more space, for example, to the physics of paint and the physical chemistry of dispersions than most books of this kind. In doing so I have had to reduce the breadth (and in some cases the depth) of treatment of specific technologies. Thus, whilst the chapters on automotive painting and architectural paints are fairly detailed, the treatment of general industrial finishing is less an 'in depth' account of specific technologies, but is intended to illustrate the very wide range of requirements of manufacturing industry and the problems the paint technologist may encounter as a result of this.

In chapters dealing with the fundamental principles underlying the technology authors have been invited to provide critical accounts of the science and technol-

xii   Preface to first edition

ogy as it stands today. This is reflected in the extensive lists of references to original work mostly published within the last decade. It is hoped that readers wishing to delve further to increase their understanding will find these references a valuable source of information.

It is important to record that apart from the authors, a number of individuals have contributed to the production of this book. I would like to record my thanks to Dr Gordon Fettis, Research Manager of ICI Paints Division, for his support and encouragement from its inception, and for the use of many of the facilities of ICI in the production of the manuscript. Thanks are also due to Mrs Millie Cohen (of ICI) and Mrs Kate Slattery (of Bristol University) who between them typed the major part of the manuscript.

R Lambourne

Preface to second edition

When I was invited to edit the second edition of this book, I took the decision to retain as far as possible the original team of authors. In addition, valuable new chapters (20 and 21) on the use of computers in the paint industry and health and safety have been contributed by Mr J Bentley and Mr G R Hayward, respectively. Owing to the unfortunate and untimely death of Miss J F Rolinson, Chapter 3 on pigments has not been revised. Instead, thanks to Dr A G Abel, a completely new chapter on this subject has been provided.

I believe this resulting new edition will provide a useful text for those wishing to explore various aspects of paint technology and its underlying science, whilst its literature references will provide a useful start to the study of any particular aspect of that technology.

I would like to pay tribute to the team of authors, who have provided me with revised or new chapters, and for their support, help and encouragement in producing this second edition.

T A Strivens

1

Paint composition and applications — a general introduction

R Lambourne

1.1 A short history of paint

Primitive men are credited with making the first paints about 25 000 years ago. They were hunters and cave dwellers and were probably inspired by the rock formations of their cave walls to outline and colour the shapes of the animals they hunted. It is possible that by creating these images they thought their power over their prey would be increased.

Chemical analysis of cave paintings discovered at Altamira (Spain) and Lascaux (France) show that the main pigments used by Palaeolithic artists were based upon iron and manganese oxides. These provide the three fundamental colours found in most cave paintings, namely black, red, and yellow, together with intermediate tints. Carbon from burnt wood, yellow iron carbonate, and chalk may also have been used. Surprisingly, there is no trace of a white pigment (the commonest pigment in use today) at Lascaux, where the natural colour of the rock was used as a pale background. However, white pigments do occur in some prehistoric paintings in Africa.

These earth pigments were ground to a fine powder in a pestle and mortar. Naturally hollowed stones are thought to have been used as mortars and bones as pestles, following the finds of such articles stained with pigments. The powdered pigments were probably mixed with water, bone marrow, animal fats, egg white, or vegetable sugars to form paints. They were applied by finger 'dabbing', or crudely made pads or brushes from hair, animal fur, or moss. Cave paintings have survived because of their sheltered positions deep in caves which were subsequently sealed off. These paints have very poor durability, the binders serving merely to make the pigments stick to the cave walls.

The Egyptians developed the art of paint-making considerably during the period circa 3000–600 BC. They developed a wider colour range of pigments which included the blues, lapis lazuli (a sodium silicate–sodium sulphide mixed crystal), and azurite (chemically similar to malachite). Red and yellow ochres (iron oxide), yellow orpiment (arsenic trisulphide), malachite green (basic copper carbonate), lamp-black, and white pigment gypsum (calcium sulphate) all came into use during this period. The first synthetic pigment, known today as Egyptian Blue, was produced almost 5000 years ago. It was obtained by calcining lime, sodium carbonate malachite, and silica at a temperature above 830 °C. The Egyptians also developed the first lake pigments. These were prepared by precipitating soluble organic dyes onto an inorganic (mineral) base and 'fixing' them chemically to form an insoluble compound. A red dye obtained from the roots of the madder plant was used in the first instance. This is no longer used other than in artists' colours ('rose madder') because it fades rapidly on exposure to sunlight, and it has been replaced by alizarin. Lake pigments still, however, represent an important group of pigments today. Red lead was used in preservative paints for timber at this time, but was more extensively used by the Romans. The resins used were almost all naturally occurring gums; waxes which were applied molten as suitable solvents were unknown. Linseed and other drying oils were known, but there is no evidence that they were used in paints.

The Greeks and Romans in the period 600BC–AD400 almost certainly appreciated that paint could preserve as well as decorate objects. Varnishes incorporating drying oils were introduced during this period. However, it was not until the thirteenth century that the protective value of drying oils began to be recognized in Europe. During the Middle Ages much painting, especially on wood, was protected by varnishing. The varnish was made by dissolving suitable resins in hot linseed, hempseed, or walnut oil, all of which tend to darken with time.

By the late eighteenth century, demands for paints of all types had increased to such an extent that it became worthwhile for people to go into business to make paint and varnishes for others to use. In 1833, J W Neil advised varnish makers always to have an assistant present during the varnish making process, for safety. 'Never do anything in a hurry or a flutter . . . a nervous or timorous person is unfit either for a maker or assistant, and the greatest number of accidents occur either through hurry, fear or drunkenness.' This admonition is indicative of the increase in scale of manufacture and the dangers of use of open-pan varnish kettles.

The industrial revolution had a major effect on the development of the paint industry. The increasing use of iron and steel for construction and engineering purposes resulted in the need for anti-corrosive primers which would delay or prevent rusting and corrosion. Lead- and zinc-based paints were developed to fulfil these needs. It is interesting to note that one of the simplest paints based upon red lead dispersed in linseed oil is still probably one of the best anti-corrosive primers for structural steel. Lead-based paints are being superseded not because better products have been produced, but because of the recognition of their toxicity and the hazards attendant upon their use.

An acceleration of the rate of scientific discovery had a growing impact on the development of paints from the eighteenth century to the present day. Prussian blue, the first artifical pigment with a known chemistry, was discovered in 1704. The use of turpentine as a paint solvent was first described in 1740. Metal driers, for speeding up the drying of vegetable oils, came into use about 1840.

The basis of formaldehyde resin chemistry was laid down between 1850 and 1890 although it was not used in paints until the twentieth century. Likewise, it was discovered in 1877 that nitrocellulose could be made safe to use as a plastic or film, by plasticizing it with camphor, but it was not until after the First World War that it was used in any significant amount in paints. The necessary impetus for this to happen came with the mass production of the motor car. Vast quantities of nitro-cellulose were manufactured for explosives during the war. At the end of the war, with the decline in the need for explosives, alternative outlets for nitrocellulose needed to be found, and the mass production of motor cars provided the necessary market. The war had accelerated the exploitation of the discoveries of chemistry and the growth of the chemical industry. New coloured pigments and dyestuffs, manufactured synthetically, became available, and in 1918 a new white pigment, tita-nium dioxide, which was to replace white lead completely, was introduced. Titanium dioxide improved the whiteness and 'hiding' or obliterating power of paint, but when originally introduced it contributed to more rapid breakdown of paints in which it was used because of its photoactivity. Subsequent research has overcome this problem and ensured that the modern pigmentary forms of titanium dioxide can be used in any type of composition without suffering any disadvantage of this kind.

The most recent influences on coating developments are related to environ-mental considerations, and the need to conform to health and safety legislation. Cost/benefit relationships have also become more important in an increasingly com-petitive world market and have influenced formulation practice markedly.

Subsequent chapters of this book will be largely concerned with developments that have taken place in the twentieth century, of which most have occurred within the last fifty years.

1.2 Paint or surface coating?

The terms 'paint' and 'surface coating' are often used interchangeably. Surface coating is the more general description of any material that may be applied as a thin continuous layer to a surface. Purists regard the term 'surface coating' as tau-tological. However, it has been used widely in the UK and in North America to distinguish painting from other forms of surface treatment, such as electroplating, anodizing, and the lamination of polymer film onto a surface. Paint was tradition-ally used to describe pigmented materials as distinct from clear films which are more properly called lacquers or varnishes. We shall be most concerned with paint in the context of this book; but, as we shall see, modern painting processes may include composite systems in which a total paint system comprises several thin films, some, but not all, of which may be pigmented. We shall use both terms as appropriate to the context in which specific paint compositions are being discussed.

The purpose of paints and surface coatings is two-fold. They may be required to provide the solution to aesthetic or protective problems, or both. For example, in painting the motor car the paint will be expected to enhance the appearance of the car body in terms of colour and gloss, and if the body is fabricated out of mild steel it will be required to give protection against corrosion. If the body is formed from glass fibre reinforced plastic the paint will only be required for aesthetic purposes. There are obviously very sound economic reasons why it is attractive to colour only the outer surface of articles that might otherwise be self-coloured by using materials of fabrication, e.g. plastics that are pigmented, particularly if a wide choice of coloured effects is required. This topic will be developed in the chapters on paints for specific markets (Chapters 9–13).

In considering the nature of paints it will become abundantly clear that the relationship between the coating and the substrate is extremely important. The requirements for a paint that is to be applied to wood are different from those of a paint to be applied to a metal substrate. Moreover, the method by which the paint is applied and cured (or dried) is likely to be very different. In formulating a paint for a particular purpose it will be essential for the formulator to know the use to which the painted article is to be put, and physical or mechanical requirements are likely to be called for. He will also have to know how it is to be applied and cured. Thus, a paint for an item made from cast iron may call for good resistance to damage by impact (e.g. chipping), whilst a coating on a beer can will call for a high degree of flexibility. These different requirements will be described in Chapters 9–13 which will deal with specific areas of paint usage.

It has long been recognized that it is difficult, if not impossible, to meet the requirements of many painting processes by the use of a single coat of paint. If one lists the requirements of a typical paint system it is easy to see why. Many, if not all, of the following are likely to be required: opacity (obliteration); colour; sheen (gloss); smoothness (or texture); adhesion to substrate; specific mechanical or physical properties; chemical resistance; corrosion protection; and the all-embracing term 'durability'. Durability is an important area that we shall return to in many contexts. The number of different layers that comprise the paint system will depend on the type of substrate and in what context the coated object is used. A typical architectural (gloss) paint system might consist of a 'primer', an 'undercoat', and a 'topcoat'. All three are likely to be pigmented compositions, and it is probable that there will be more than one coat (or layer) of each of these paints. An architect may well specify one coat of primer, two coats of undercoat, and two coats of topcoat. The purpose of these individual layers and hence their composition is likely to be very different. The primer is designed largely to seal the substrate and provide a means of achieving good adhesion between substrate and undercoat. It may contribute to opacity, but this will not be its main purpose. The undercoat will be used for two purposes: to contribute significantly to the obliteration of the substrate and to provide a smooth surface upon which to apply the topcoat. The smooth surface is obtained by abrading the dried undercoat (after each coat has dried) with fine tungsten carbide paper. The topcoat is then applied to complete the process of obliteration and to provide the appropriate aesthetic effect (i.e. colour and sheen). The system as a whole would be required to give protection to the wood or metal substrate to which it is applied.

The interrelationship between these multilayers is worth considering. The mechanical and physical properties of the individual coatings will often be very different. The function of the primer in promoting adhesion has already been mentioned. It may also be required to relieve stresses that are built up within the coating system as a result of hardening and ultimately embrittlement of the topcoat (and undercoat) as a result of ageing, or to accommodate stresses imposed by the differential movement of the substrate. The softwoods used in the construction of window frames are known to expand and contract between the dry (summer) and wet (winter) conditions by at least 10% across the grain, but much smaller changes are observed in the direction of the grain. The undercoat will be formulated in a colour close to that of the topcoat, but it may serve this purpose to several closely related topcoat colours. It will normally be highly pigmented, in contrast to the topcoat which will not. The reason for the latter is the need, for example, to maximize gloss and extensibility. The use of the type and concentration of pigmentary material in the undercoat would not be conducive to maximizing these properties in the topcoat.

The primer will frequently be required to contribute to corrosion protection. Those formulated for use on steel are likely therefore to incorporate a chemically active anti-corrosive pigment. Corrosion protection may be achieved by yet another means, the chemical treatment of the substrate. Thus many industrial coating processes involve a chemical pretreatment of metal, mainly aluminium or ferrous substrates. The latter is most frequently treated with a phosphate solution that produces a crystalline phosphate layer. Subsequent paint application, i.e. priming, is therefore to a crystalline inorganic layer and not directly to an (uncoated) pure metal surface.

Surfaces are seldom what they seem. With the exception of the noble metals almost all surfaces that will be commonly regarded as 'metal' surfaces will present to the paint a surface that is not a metal but an oxide layer. Even so the purity or cleanliness of the surface may well be an unknown quantity. Since this surface will have an important effect on such properties as the adhesive performance of the paint system it is important to appreciate this point. Just as most surfaces will be 'dirty' and thus be ill-defined, it is necessary to produce paint systems that can accommodate the contamination and general variability of surfaces. These types of system must be 'tolerant' to all but excessive contamination and are often described as 'robust' if the required degree of tolerance can be achieved. This is not to say that industrial coating processes do not require certain pretreatments such as degreasing, and may involve the chemical pretreatments indicated above.

The deterioration of paints which occurs in many situations is largely due to changes in the chemical nature of the film former with consequent changes in its mechanical properties, and research continues unabated to formulate polymers and resins to improve the performance of paints in use. The development of new improved pigments may contribute to improvements in durability, but in most cases the weakest link in the system is the film former. One consequence of this is the development of systems for specific end uses. Such approaches are adopted when it is practicable to avoid the compromises that are otherwise likely to be required for a general-purpose product. For economic and marketing reasons the best product may not be available for a specific end use, and a compromise of cost and performance may be required. Indeed the cost-effectiveness of a particular coating composition will usually dominate other considerations particularly in the industrial paint markets.

1.3 The components of paint

The composition of a paint is indicated in Table 1.1 which also indicates the function of the main components.

Not all paints have every ingredient, as will be indicated in subsequent specialist chapters. For example gloss paints will not contain extenders which are coarse particle inorganic materials. These are used in matt paints such as the surfacers or primer surfacers used in the motor industry.

Major differences occur between the polymers or resins that are used in paints formulated for different purposes. This is because of differences between the methods of application and cure, the nature of the substrate, and the conditions of

Table 1.1 — The composition of paints

	Components	Typical function
Vehicle (continuous phase)	Polymer or resin (Binder)	Provides the basis of continuous film, sealing or otherwise protecting the surface to which the paint is applied. Varies in chemical composition according to the end use.
	Solvent or diluent	The means by which the paint may be applied. Avoided in a small number of compositions such as powder coatings and 100% polymerizable systems.
Pigment (discontinuous phase)	Additives	Minor components, wide in variety and effect, e.g. catalysts, driers, flow agents.
	Primary pigment (fine particle organic or inorganic)	Provides opacity, colour, and other optical or visual effects. Is thus most frequently used for aesthetic reasons. In primers the pigment may be included for anti-corrosive properties.
	Extender (coarse particle inorganic matter)	Used for a wide range of purposes including opacity/ obliteration (as an adjunct to primary pigment); to facilitate sanding, e.g. in primer surfacers.

use. Thus architectural ('decorative' or 'household') paints will be required to be applied *in situ* at ambient temperatures (which may be between 7 and 30 °C depending on climate and geographical location). They will 'dry' or 'cure' by one of two mechanisms: (i) atmospheric oxidation or (ii) the evaporation of diluent (water) accompanied by the coalescence of latex particles comprising the binder. Many industrial finishing processes will require the use of heat or other forms of radiation (UV, IR, electron beam) to induce chemical reactions, such as free radical or condensation polymerization, to convert liquid polymers to highly crosslinked solids. The most common of these processing methods uses 'thermosetting' polymers which will frequently be admixtures of two quite different chemical types of material, e.g. alkyd combined with amino resin. There is a similarity between both the oxidative drying and industrial thermosetting processes in so far that in both cases the polymers used are initially of low molecular weight and the curing process leads to crosslinking of these polymers to yield highly complex extremely high molecular weight products. In contrast, it is possible to produce coatings without the need for crosslinking, for use in both of these distinctive markets. In the case of decorative or architectural paints this is exemplified by the emulsion paint, in which the binder is in the form of high molecular weight polymer particles suspended in an aqueous medium. Lacquers used in the motor industry may be based upon high molecular weight polymer in solution. Both systems avoid the need for crosslinking to achieve satisfactory film properties, but recent technological developments suggest that improved film properties can be achieved by the introduction of crosslinking in some form, e.g. by the use of microgels.

1.3.1   Polymer or resin film formers
The organic chemistry of film formers is described in detail in Chapter 2. It will be useful, however, to indicate here some of the range of polymers and resins that have

Table 1.2

Low molecular weight	High molecular weight
Oleoresinous binders	Nitrocellulose
Alkyds	Solution vinyls
Polyurethanes	Solution acrylics
Urethane oils	Non-aqueous dispersion polymers (NADs)
Amino resins	Polylvinyl acetate (PVA) ⎫
Phenolic resins	Acrylic ⎬ latexes
Epoxide resins	Styrene/butadiene ⎭
Unsaturated polyesters	
Chlorinated rubber	

come to be used as film formers, and to indicate their general areas of application. Film formers or binders may be classified according to their molecular weight. Thus low molecular weight polymers that will not form solid films normally without further chemical reaction form one class. High molecular weight polymers that will form useful films without further chemical reaction form the second class. Examples of polymers and resins classified by this means are shown in Table 1.2.

1.3.1.1   Low molecular weight film formers

Oleoresinous binders
These are prepared by heating vegetable oils with naturally occurring resins such as wood rosin, fossil resins such as Congo Copal and Kauri gum. They would also include oil-modified phenolics. To a large extent these types of resinous binder have been displaced by alkyd resins and the like, but many are capable of providing products that in performance are the equal of their successors, particularly in specific applications such as in architectural undercoat paint. They are less reproducible than condensation polymers like alkyds and are less attractive in terms of modern manufacturing processes.

Alkyds
Alkyds are polyesters derived as the reaction products of vegetable oil triglycerides, polyols (e.g. glycerol), and dibasic acids or their anhydrides (e.g. phthalic anhydride). They are generally formulated with very different end uses in mind, and classified according to vegetable oil content (described by the term 'oil length') in three broad categories: short oil, medium oil, and long oil, roughly corresponding to <45%, 45–60%, >60% respectively. The variation in oil length is usually coincident with changes in the nature of the vegetable oil used and consequently the end use. Thus, for architectural gloss paint of maximum exterior durability, the alkyd binder will be a long oil alkyd based upon a drying oil such as linseed or soya bean oils (i.e. an unsaturated triglyceride oil). The drying oil provides the means by which the film former dries. In this case the conversion from a low molecular weight liquid polymer to a highly crosslinked solid arises as a result of oxidation. One characteristic of the long oil alkyd is that it is soluble in aliphatic hydrocarbons. In contrast the short oil alkyd is likely to be based on a saturated triglyceride (such as coconut oil). It will not be soluble in aliphatic hydrocarbons and will normally be dissolved in a high boiling aromatic. Although the short oil alkyd may be capable of forming a lacquer-

like film it would have a low softening point, and it is necessary to crosslink it in order to achieve a satisfactory film. In this case it is usually combined with an amino resin and crosslinked by condensation in a heat-curing process. It is debatable in this type of system whether the amino resin is crosslinking the alkyd or the alkyd is plasticizing the highly crosslinked amino resin. The former explanation is usually preferred because the proportion of the alkyd is invariably greater than that of the amino resin. The alkyd/amino resin ratio usually falls between 2:1 and 4:1 by weight. These types of system are used in industrial finishing. Alkyd/melamine formaldehyde compositions have found use in the automotive market for many years. Alkyd/urea formaldehyde compositions have found use in the domestic appliance market, although in both cases there have been new products developed that have displaced these materials to some extent, particularly when more stringent performance requirements have to be met.

Polyurethanes, urethane alkyds, and urethane oils
Structurally, these materials resemble alkyds in which polyester linkages are replaced partially or totally by urethane linkages.

$$-NH-\underset{\underset{O}{\|}}{C}-O-$$

Polyurethanes also include two-pack compositions for the industrial and refinish markets in which the curing is achieved by reaction between free isocyanate groups in one component and hydroxyl groups in the second component. The advantages of urethane oils and urethane alkyds are derived from the resistance of the urethane link to hydrolysis. In decorative (architectural) paints it is common practice to use binders which are a mixture of a long oil alkyd and urethane alkyd for maximum durability.

Amino resins
The most common types of amino resin are reaction products of urea or melamine (1:3:5 triamino triazine) and formaldehyde. The resins are prepared in alcoholic media, which enables the molecular weight and degree of branching to be controlled within practically determined limits related to the end use of the resin. The effect of this modification is shown in the solubility and reactivity of the resins so produced. The polymers produced are generally regarded as being derived from the hydroxymethyl derivatives of melamine and urea respectively; subsequent addition condensation and etherification reactions lead to complex, highly branched polymeric species. Curing or crosslinking to solid films (usually in combination with an alkyd or other polymer) can be achieved thermally (oven-curing) or at room temperature. In both cases the presence of an acid catalyst is essential if adequate and rapid cure is to be obtained.

   The crosslinking capability of amino resins may also be utilized as a means of curing acrylic resins. In this type of film former a minor proportion (usually a few percent) of a monomer such as N-butoxymethyl acrylamide is incorporated into the polymer. This provides reactive sites which enable the acrylic copolymer to be crosslinked. A suitable choice of monomers allows the acrylic resin to be plasticized internally, so that the use of added plasticizers is avoided. These resins are of particular interest where high levels of performance and particularly good adherence and flexibility are required.

Phenolic resins

The reaction of formaldehyde with a phenol gives rise to a range of resins that, in combination with other resins or drying oils, find use in industrial coatings. Broadly two main types of phenolic are produced, novolacs and resoles. Novolacs are low molecular weight linear condensation products of formaldehyde and phenols that are alkyl substituted in the *para* position. If the substituent alkyl group contains four or more carbon atoms (i.e. butyl or above — in the homologous series) the resin is likely to be oil-soluble. Resoles are products of the reaction of unsubstituted phenols with formaldehyde. Since the *para* position on the phenolic ring is available for reaction as well as the *ortho* position these resins are highly branched and can with continued reaction be converted into hard intractable glassy solids. Phenolics tend to confer chemical resistance to the compositions in which they are used. They are always used in combination with other resinous film formers. In some cases they may be prereacted with the other resin components, or they may be simply blended together. Thus a phenolic resin (e.g. a novolac) may be reacted with rosin or ester gum and then blended with a bodied (heat-treated) drying oil to form the binder in an architectural paint primer, or used simply in an unpigmented form as an oleoresinous varnish. Phenolic-based compositions find use in chemically resistant systems such as are required for pipes and tank linings.

Epoxide resins

The use of the epoxide or oxirane group

$$-CHCH_2$$
$$\diagdown \diagup$$
$$O$$

as a means of synthesizing resins and as a means of crosslinking binders is now well established. A large group of epoxide resins is based upon the reaction products of epichlorhydrin and bisphenol A (diphenylolpropane). These resins may be esterified with unsaturated fatty acids to give epoxy esters. These are film formers in their own right and resemble air-drying alkyds. They exhibit better chemical resistance than alkyds but are less durable than long oil alkyds in some circumstances. The way in which they break down may, however, be turned to good use, for example when used as 'self-cleaning' coatings. In this case the films may be pigmented with uncoated titania in the anatase form so that degradation in ultraviolet radiation causes erosion of the surface layers of the film, otherwise known as chalking. Such films gradually weather away, always presenting a white surface to the elements.

   Epoxide resins may also be used in conjunction with melamine formaldehyde (MF) or phenolics, or they may be formulated into epoxy-alkyds, i.e. where they are effectively being used as polyols in admixture with less functional polyols such as glycerol. The epoxide group offers great versatility in curing, and a wide range of two-pack compositions are possible. One of the most popular methods of crosslinking uses the reaction with polyamides. This is the same method of cure as used in epoxy adhesive compositions. The crosslinking occurs as an addition of terminal amino groups of the polyamide to the epoxy group. This reaction occurs slowly at room temperature. Crosslinking through the epoxide group can also arise from the use of polyamines or by means of the acid-catalysed polymerization to give ether crosslinks. It will be apparent that most of these products will be used in industrial applications.

Unsaturated polyesters
Unlike the previously described binders, unsaturated polyesters offer the benefit of totally polymerizable systems because the solvent in which they are dissolved is a polymerizable monomer. The simplest and most common polyesters are prepared from maleic anhydride/phthalic anhydride mixtures esterified with glycols such as propyleneglycol. The resins so produced are dissolved in styrene or vinyl toluene. The free radical copolymerization of the vinyl monomer and the maleic unsatura-tion in the polyester is usually initiated by a transition metal/organic hydroperox-ide system at ambient temperature or by the use of the thermal scission of a diacyl peroxide at higher temperatures. Unsaturated polyesters have found extensive use both pigmented and as clears in the wood finish market. They are capable of widely divergent uses depending on their composition. Chemically resistant finishes for tank linings, for example, can be formulated on polyesters derived from isophthalic and terephthalic acids. Another class of chemically resistant finish is based upon chlorinated polyesters. In this case the polyester incorporates chlorendic anhydride in place of the more common phthalic anhydride.

Chlorinated rubber
Chlorinated rubber is a film-forming resin that is available in a wide range of mol-ecular weights, from 3500 to about 20000. It is prepared by chlorinating rubber in solution, the commercial product containing about 65% of chlorine. It is used as the main binder in air-drying paints which are to be used in situations requiring a chem-ically resistant product of great durability. Because the polymer is a brittle solid, in paint applications chlorinated rubber requires plasticization. Chlorinated rubbers are also used in conjunction with other resins with which they are compatible, such as alkyds. Paints based on chlorinated rubber have been used for building, masonry, swimming pools, road marking, and marine purposes.

1.3.1.2   High molecular weight film formers
Almost all high molecular weight polymers are produced by the free radical initi-ated polymerization of mixtures of vinyl, acrylate, or methacrylate monomers. They may be polymerized in solution, in suspension or dispersion. Dispersion polymer-ization may be in hydrocarbon diluents (NAD) or in aqueous media ('emulsion polymers'). The reaction processes differ considerably between these systems as will become apparent when the subject is developed in Chapter 2.

   One major exception to the foregoing is nitrocellulose. This material is formed by the direct nitration of cellulose in the presence of sulphuric acid. It is available in grades determined by the degree of nitration which, in turn, determine its solu-bility in various solvents. The grades used in refinish paints, wood-finish lacquers, etc. require the molecular weight of the original cellulose to be reduced somewhat, to meet viscosity requirements in the solvents commonly used.

   In most cases the high molecular weight polymers do not need to be crosslinked in order to develop adequate film properties. A small number of solution polymers of moderately high molecular weight are, however, crosslinked through reactive groups incorporated into the polymer chain. The physical properties of films produced from high polymers may be only marginally affected by the manner in which they were prepared or their physical form at the stage of film application. Thus automotive finishes derived from solution acrylics and NADs are virtually indistinguishable, albeit that the methods of application, processing conditions, etc. may be significantly different. In most cases the product that is selected will be dependent on the economics of the process overall rather than the product cost alone. The need to meet end use specification is the main reason for the similarity in film properties and performance of alternative product formulations.

Aqueous latexes (emulsion polymers) as a group have been one of the fastest-growing sectors of the paint market. Starting with the use of homopolymer PVA as a binder in matt and low-sheen decorative wall paints we have seen the development of more robust systems using internal plasticization (i.e. the incorporation of a plasticizing comonomer) and, more recently, the use of acrylic and methacrylic copolymer latexes. Improved performance has enabled the introduction of these paints into exterior masonry applications.

The most recent developments in this field have entailed preparing latex particles with specific morphologies, such as core/shell particles in which the particle cores may be crosslinked. These physical changes in the structure of the latex particles enable significant improvements to be achieved in the mechanical properties of the derived films, and in film integrity as measured by 'scrub resistance'. For industrial paint applications, functionalized polymers that will undergo crosslinking after coalescence have been developed. These compositions have low volatile organic content (VOC) and are thus 'environmentally friendly'.

1.3.2   Pigments

1.3.2.1   Primary pigments

Primary pigments comprise solid particulate material which is dispersed in the binder or film former described previously. We shall distinguish between them and supplementary pigments, extenders, fillers, etc. in that they contribute one or more of the principal functions, namely colour, opacification, and anti-corrosive properties. The supplementary pigments, extenders, although important, do not in general contribute to these properties to a major extent. Their function is related to reductions in cost, but they can contribute properties to a paint that may be less obvious than colour or opacity. Thus they may enhance opacity, control surface sheen, and facilitate ease of sanding.

The principal pigment in use is titanium dioxide. This is primarily because of reasons of fashion, e.g. in the decorative paint market there has been a tendency for white and pastel shades to gain greater acceptance over strong colours in recent years. The same is true to a lesser extent in other fields, such as the motor car industry. In this case early motor cars were painted black, and it was difficult to obtain cars in other colours. Nowadays it is the black car that is a comparative rarity. Titanium dioxide in its pigmentary form is a highly developed pigment. The high refractive index rutile form is most commonly used, and the pigment is manufactured within very close limits with respect to particle size and particle size distribution. Although normally regarded as being chemically inert the surface of the titanium dioxide crystal is photoactive, and the pigmentary form is surface coated to minimize photochemical activity at the binder/pigment interface. Typical coatings contain silica and alumina in various proportions, and in addition to reducing photochemical activity they can improve the dispersibility of the pigment in the binder. Considerable research has been carried out into titanium dioxide pigments, and manufacturers are now able to offer a range of grades that are suitable for use in different types of paint media. The uses of titanium dioxide and other white pigments are detailed in Chapter 3, and its optical properties are discussed in Chapters 17 and 18.

Coloured pigments fall into two main groups, inorganic pigments and organic pigments. As a result of legislation governing the handling and use of toxic materials, many of the inorganic pigments traditionally used by the paint industry have been replaced by other less toxic materials. Thus except in a few industrial compositions lead pigments of all types (e.g. chromate, oxide, carbonate) have to a large extent been replaced, and it is to be expected that lead in all of its forms and chromates will ultimately be eliminated from all paints. Their removal from the scene gives rise to problems if a coloured pigment is used for the dual purposes of providing colour and contributing positively to the anti-corrosive properties of the paint film.

Some commonly used pigments are shown in Table 1.3. The selection and use of pigments is covered in detail in Chapter 3. The physics of colour are treated in Chapter 18 which includes a treatment of the principles and practice of colour matching. At one time colour matching was an art carried out visually by very skilful individuals. Whilst the skilled colour matcher has not been entirely replaced, there has been a gradual change to instrumental colour matching, a trend that has been accelerated by the development of sophisticated colorimeters and the use of the computer.

Table 1.3 — Some typical primary pigments

Colour	Inorganic	Organic
Black	Carbon black Copper carbonate Manganese dioxide	Aniline black
Yellow	Lead, zinc, and barium chromates Cadmium sulphide Iron oxides	Nickel azo yellow
Blue/violet	Ultramarine Prussian blue Cobalt blue	Phthalocyanin blue Indanthrone blue Carbazol violet
Green	Chromium oxide	Phthalocyanin green
Red	Red iron oxide Cadmium selenide Red lead Chrome red	Toluidine red Quinacridones
White	Titanium dioxide Zinc oxide Antimony oxide Lead carbonate (basic)	—

In automotive paints there has been a growth in the use of aluminium flake as a pigment to provide attractive metallic finishes. The dimensions of the flakes and their orientation within the paint film give rise to optical effects that cannot be achieved in other ways. Inevitably such paints are subject to problems in quality control, application, and colour matching that do not exist in solid colours. Nevertheless these problems are sufficiently well understood to be kept under control, and their continued use is amply justified aesthetically.

1.3.2.2   Extenders, fillers, and supplementary pigments

All three names have been applied to a wide range of materials that have been incorporated into paints for a variety of purposes. They tend to be relatively cheap materials, and for this reason may be used in conjunction with primary pigments to achieve a specific type of paint. For example it would be technically difficult and prohibitively expensive to produce satisfactory matt white emulsion paint using titanium dioxide as the only pigment. Titanium dioxide is not cost effective as a matting agent and indeed is not designed for this purpose. It is much more economic to use a coarse particle extender such as calcium carbonate in conjunction with TiO_2 to achieve whiteness and opacity in a matt or semi-matt product (e.g. a matt latex decorative paint, an undercoat or primer). Extenders do not normally contribute to colour, and in most cases it is essential that they be colourless. The particle sizes of extenders range from submicrometre to a few tens of micrometres; their refractive index is generally close to that of the organic binders in which they are used, and so they contribute little opacity from light scattering. Platelet type extenders such as wet-ground mica can influence the water permeability of films, and many therefore contribute to improved corrosion resistance. Talcs are often used (e.g. in automotive primer surfacers) to improve the sanding of the paint prior to the application of a topcoat. Many of the extenders in common use are naturally occurring materials that are refined to varying extents according to the use to which they are put. Whilst every attempt is made to ensure that they are reproducible they tend to be more variable and with a greater diversity of particle shape, size, and size distribution than primary pigments. A list of typical inorganic extenders is given in Table 1.4.

Table 1.4 — Some typical extenders

Chemical nature	Type
Barium sulphate	Barytes Blanc fixe
Calcium carbonate	Chalk Calcite Precipitated chalk
Calcium sulphate	Gypsum Anhydrite Precipitated calcium sulphate
Silicate	Silica Diatomaceous silica Clay Talc Mica

In recent years there have been several attempts to make synthetic (polymeric) extenders to meet special needs, in particular to replace some of the titanium dioxide in the paint film. One such material, 'Spindrift', originating from Australia, is in the form of polymer beads (spherical particles up to 30 μm in diameter which incorporate submicrometre air bubbles and a small proportion of pigmentary titanium dioxide). The introduction of air bubbles into the polymer that forms the beads influences the light-scattering power of the titanium dioxide in one of two ways. It can reduce the average effective refractive index of the polymer matrix and so enhance the light-scattering of the titanium dioxide, if the bubbles are very small (<0.1 μm). If the bubbles are large (~0.8 μm) they are able to scatter light in their own right. Other methods of introducing voids into paints in a systematic way to take advantage of the scattering of light by bubbles have included the emulsification of small droplets of a volatile fluid into aqueous latex paints which leave cavities within the film by evaporation after coalescence ('Pittment'); yet another is the use of non-coalescing latex particles in combination with coalescing latexes to give rise to cusp-like voids within the film (Glidden). All of these methods seek to achieve a cost-effective contribution to opacity without sacrifice of film integrity and other properties. These methods have received limited acceptance, mainly owing to economic reasons. Nevertheless, they do increase the opportunities open to the paint formulator to formulate a paint for a given purpose.

1.3.3   Solvents

Solvents are used in paint compositions for two main purposes. They enable the paint to be made, and they enable it to be applied to surfaces. This may seem to be stating the obvious, but it is important to appreciate that so far as the paint film performance is concerned the solvent plays no long-term role in this. This is not to say that in the early life of the film solvent retention does not affect hardness, flexibility, and other film properties.

The term solvent is used frequently to include liquids that do not dissolve the polymeric binder, and in these cases it is more properly called a diluent. The function of the diluent is the same as a solvent, as stated above. In water-based systems the water may act as a true solvent for some components, but be a non-solvent for the main film former. This is the case in decorative emulsion paints. More often in these cases it is common to refer to the 'aqueous phase' of the composition, acknowledging that the water present, although not a solvent for the film former, is present as the major component of the liquid-dispersing phase.

A wide range of organic liquids are used as paint solvents, the type of solvent depending on the nature of the film former. A detailed account of solvency, solvent selection, etc. is given in Chapter 4.

Considerable research effort has gone on into the study of the thermodynamics of solutions. This has provided the paint formulator with much more precise methods of solvent selection. The improved methods are based upon a better knowledge and understanding of molecular attractions in liquids and a recognition of the additivity of molecular attractions in mixed solvent systems. It is very rare for a single solvent to be acceptable in most situations, and the newer methods based upon solubility parameter concepts enable the more rational selection of solvent mixtures to meet a particular need.

Solvency alone is not the only criterion upon which solvent choice is made. Other important factors include evaporation rate, odour, toxicity, flammability, and cost.

These factors assume different degrees of importance depending on how the paint is used. If the paint is applied under industrial manufacturing conditions it is likely that problems associated with odour, toxicity, and flammability may be under control, but this is by no means certain. The need to install expensive extraction equipment or after-burners may preclude the use of some solvents and therefore some types of paint composition. Closely related to considerations of toxicity are those of pollution. Many countries have enacted legislation to protect the individual and the environment. This legislation has had a profound effect on the development of the paint industry and has influenced both the raw material supplier on the one hand and the paint user on the other. In North America there has been an enormous growth in water-borne systems at the expense of solvent-borne paints. This is a trend that is likely to continue.

Other alternatives to solvent-based paints are 100% polymerizable systems and powder coatings. In the former case a polymerizable monomer such as styrene fulfils the role of solvent for the composition, being converted into polymer in the curing reaction. With powder coatings, solvents may be used in the early stages of the paint-making process, but they are removed and recycled, so do not provide a hazard or problem for the user.

1.3.4   Paint additives

The simplest paint composition comprising a pigment dispersed in a binder, carried in a solvent (or non-solvent liquid phase) is rarely satisfactory in practice. Defects are readily observed in a number of characteristics of the liquid paint and in the dry film. These defects arise through a number of limitations both in chemical and physical terms, and they must be eliminated or at least mitigated in some way before the paint can be considered a satisfactory article of commerce.

Some of the main defects worth mentioning are settlement of pigment and skinning in the can; aeration and bubble retention on application; cissing, sagging, and shrivelling of the paint film; pigment flotation; and flooding. These defects represent only a small number of defects that can be observed in various paints. It is perhaps worth-while to describe cissing, shrivelling, sagging, flotation, and flooding here. They will be dealt with in more detail in Chapter 5, which describes some of the more common types of paint additives.

- 'cissing' is the appearance of small, saucer-like depressions in the surface of the film;
- 'shrivelling' is the development of a wrinkled surface in films that dry by oxidation;
- 'sagging' is the development of an uneven coating as the result of excessive flow of a paint on a vertical surface;
- 'floating' is the term used for the colour differences that can occur in a paint film because of the spontaneous separation of component pigments after application;
- 'flooding' (also known as 'brush disturbance') is the permanent colour change of a paint subject to shear after application.

To overcome these defects their cause requires to be understood and a remedy found. In some cases the defect may be overcome by minor reformulation. Shrivelling, for example, is commonly due to an imbalance between the surface oxidative crosslinking of a film and the rate of crosslinking within the film. This can usually be overcome by changing the drier combination, which consists of an active transition metal drier such as cobalt which promotes oxidation and a 'through' drier such as lead or zirconium which influences crosslinking, but does not *per se* catalyse the oxidation process. In other cases simple reformulation will not provide a remedy, and specific additives have been developed to help in these cases. Thus anti-settling agents, anti-skinning agents, flow agents, etc. are available from specialist manufacturers for most defects and for most paint systems.

The problems of 'floating' and 'flooding' are associated with colloidal stability of the pigment dispersion and may arise from a number of different causes. The differential separation of pigment illustrated by floating occurs as a result of the differences in particle size of the component pigment and may be overcome by coflocculation of the pigments in the system. Another method of curing the condition may be to introduce a small proportion of a very fine particle extender such as alumina, of opposite surface charge to the fine particle pigment, to coflocculate with the latter.

The flooding (or brush disturbance) problem is indicative of flocculation occurring as a film dries. Under shear, as the brush disturbs the paint, the pigment is redispersed and the paint becomes paler in shade. This is because an increase in the back-scattering of incident light occurs, owing to the white pigment becoming deflocculated.

Cissing and sagging are illustrative of other aspects of physical properties associated with surface chemistry and rheology. In the former case, the effect is caused by a localized change in the surface tension of the film. In extreme cases this can give rise to incomplete wetting of the substrate, often distinguished by the term 'crawling'. Sagging, on the other hand, is a bulk property of the film that may be influenced by the colloidal stability of the composition. Ideal, colloidally stable dispersions tend to exhibit Newtonian behavior, i.e. their viscosity is independent of shear rate. This means that on a vertical surface a Newtonian liquid that is of a suitable viscosity to be spread by a brush, i.e. with a viscosity of about $0.5\,N\,s\,m^{-2}$, will flow excessively unless the viscosity rises rapidly as a result of solvent loss. Alternatively, the paint formulator may aim to induce non-Newtonian behaviour such that the low shear viscosity of the product is very high. Thus, sagging may be avoided by either or a combination of these effects.

1.4   Paint making

Having considered briefly the most important constituents of paint, their function, and what they contribute to the final product, we need to consider the paint-making process. It is possible for a paint manufacturer to make almost all types of paint without any chemical processing. This is because each of the constituents can be purchased from a specialist manufacturer. Providing the paint manufacturer has certain basic plant capable of storing, mixing, dispersing, blending, and filling, he or she can be in business. Indeed some small volume or specialist manufacturers do just this, using basic formulations often provided by the resin or pigment supplier. However, the major paint manufacturers world-wide, seeking to capture a significant proportion of world markets, mount considerable research and development effort to produce products that are technically superior to those of their competitors or, if no better, can be produced more cheaply.

One of the most important parts of the paint-making process is the dispersion of pigment to prepare a stable and reproducible product. This is usually carried out in two stages, the dispersion of the pigment in part of the binder solution (or other dispersing medium) to form a 'millbase', followed by blending ('second staging') with the remaining binder solution. Finally, minor components of the composition, such as driers, flow agents, solvent (to adjust the viscosity), and tinters (according to the requirements of colour matching) are added.

The dispersion process involves the wetting of the pigment with the dispersing medium, the separation of particles from their aggregated state and their stabilization in suspension, either as individual particles or in a lightly flocculated condition. The millbase is prepared in one of a number of mills (depending on the type of paint to be prepared), ranging from ball mills to cavitation mixers and attritors. Details of the processes involved and the types of mill that are available are given in Chapter 8.

1.5   Methods of application

There are four main methods of applying paint:

- by spreading, e.g. by brush, roller, paint pad, or doctor blade;
- by spraying, e.g. air-fed spray, airless spray, hot spray, and electrostatic spray;
- by flow coating, e.g. dipping, curtain coating, roller coating, and reverse roller coating;
- by electrodeposition.

The methods adopted depend on the market in which the paint is used, each type of paint being formulated to meet the needs of the application method. Spreading by brush or hand-held roller is the main method for applying decorative/architectural paints and the maintenance of structural steelwork and buildings generally. It is also important in marine maintenance, although other methods (e.g. airless spray) may be used during the construction of a ship.

Application by spraying is the most widely applicable method. It is used for painting motor cars in the factory and by refinishers following accident damage; it is used in the wood-finishing industries (e.g. furniture) and in general industrial paints (e.g. domestic appliances). The various forms of spray painting make it a particularly versatile method of application. The flow coating methods are limited essentially to flat stock (e.g. chipboard) and coil coating (aluminium or steel coil) where they are much valued because of the high rates of finishing that can be achieved.

Electrodeposition has become established as the main method of priming the steel body shells of motor cars. The total process which involves degreasing, phosphate treatment, electrodeposition of primer, and then spray application of surfacer and finishing coats has raised the standards of corrosion resistance and general appearance considerably during this period. Electrodeposition may take place with the car body acting as either the anode or the cathode. In recent years it has been claimed that cathodic forms of electropaint give the better corrosion protection.

Mainly because of environmental considerations, powder coating has been a major growth area in industrial finishing, particularly in Western Europe, which produced 53% of the total world market for powder coatings in 1993. Powder coatings are normally applied by electrostatic spray, with minimal overspray losses and without the emission of VOCs.

Table 1.5

Market	%
Car manufacture	10.7
Car refinishing	6.3
Metal manufactured goods (inc. drum and can coatings, domestic appliances)	36.8
Woodfinishes	16.9
Marine and offshore	11.2
Industrial construction and maintenance	11.3
Other	6.8

Source: EC Report, *Reduction of VOC Emissions from Industrial Coating of Metal Surfaces*, Warren Spring Laboratory, September 1991.

The methods of application will be described in the appropriate chapters, discussing each of the main market areas.

1.6   Paint markets

It is convenient to discuss the markets for paint in terms of products for specific end uses. The end use determines the nature of the coating or system required to fulfil property requirements, such as mechanical properties, durability and chemical resistance properties related to the use of the article coated, which vary from an ocean-going liner to a beverage can. Thus, specialized products may be manufactured by small specialist paint companies; geographical considerations may be taken into account in some instances so that the user can exercise closer control over the continuity of supply and quality of the coatings purchased.

The principal markets for paints are the decorative (household or architectural) paints market and the widely divergent industrial paint markets. The requirements of the former are met by two types of composition based on autoxidative resin systems and latex paints; the technology of industrial coatings is more diverse in terms of film property requirements, methods of application and cure. In the majority of industrial coating operations the application methods and curing processes are thus carried out under factory conditions, under which greater control can be exercised.

Some idea of the scale of paint manufacture can be obtained from statistics published by a number of organizations world-wide. For example, the total EC industrial coatings market (1991) was 2 390 000 tonnes per annum (Table 1.5).

It is interesting to note that four countries within Europe consume almost 75% of the industrial coatings produced — Germany (28%), Italy (18%), the UK and France (about 13% each). The leading paint manufacturing countries in the world are all to be found in the Far East — Japan (2 090 000 tonnes), China (940 000 tonnes), South Korea (430 000 tonnes), Taiwan (280 000 tonnes), and the Philippines (230 000 tonnes) (source: National Statistics, 1992).

2

Organic film formers

J Bentley

2.1 Introduction

The first chapter has indicated the major types of surface-coating resins used, and this chapter will describe their chemistry in more detail, including their preparation.

In this introductory section an outline is presented of the theory of polymer formation and curing; the mechanism specific to each type of resin is then covered more fully in the subsequent sections of the chapter. The classes of resins available and the factors that decide the choice of resin for a particular use are also indicated. The properties and uses of each type of resin, again, are detailed more fully in the later relevant sections.

A number of terms are used interchangeably to describe the film-forming component of paint, as will already be apparent. 'Film former', 'vehicle', or 'binder' relates to the evident fact that this component carries and then binds any particulate components together, and that this provides the continuous film-forming portion of the coating. Resin or varnish are older terms relating to the previous more prevalent use of natural resins in solution or 'dissolved' in oils as the film former; they date from the time when the chemistry and composition of these components were far less well understood. Nowadays, with our better knowledge of the materials used, along with the wide application of the sophisticated polymers used also in the plastics and adhesives industries but tailored to our own use, it is strictly more correct to refer to this component as the polymeric film-forming component. The interchangeable use of old and new nomenclature is also found in the manufacture of film formers where 'kettle' refers to the polymerization reactor, and 'churn' to the thinning tank normally part of the manufacturing plant.

Film-forming polymers may or may not be made in the presence of solvent; however, since the polymers in solvent-free form generally range from highly viscous liquids to hard brittle solids, they are practically always handled in storage and in the paint-making process in solution (or in dispersion) with significant quantities of solvent or diluent included (and here the terms solvent and diluent include the full range of organic solvents and water). The exceptions are where unsaturated monomers and liquid oligomeric materials are used in place of solvent as diluents in high solids finishes. Solid resins are also used alone for the specialized application of powder coatings.

Many polymers used will be in true solution, with solvent being the other component. However, in other cases, either for reasons connected with the polymer preparation or with its final use, the polymer will exist in the form of a fine-particle dispersion in non-solvent (diluent); this is true for aqueous emulsions, non-aqueous dispersions, and for the emulsified materials used in electrodeposition and other water-borne applications. In some cases the system may be mixed solution/dispersion, for example a solution containing micellar polymer dispersion, micro-emulsion or microgel. A particular striking consequence of whether the polymer is dissolved or dispersed is the viscosity; dispersions are invariably more fluid at comparable solids contents than solutions. Most significantly while solution viscosity increases as molecular weight rises, the viscosity of emulsions or dispersions is independent of molecular weight. For any given polymer in solution or dispersion, viscosity will broadly increase as the solids increases (though water-borne solutions often exhibit unusual behaviour); the viscosity will decrease if the temperature rises.

As with any utility product, the paint user is concerned mostly with the ability of the material to provide final protective and decorative effects; he or she has little regards for composition, except in so far as it guides him or her to the ability of the material to satisfy those needs. Table 1.1 in Chapter 1 has listed the function of paint components and Table 2.1 shows the contribution that the three major components — resin, pigment, and solvent — make to the most important properties of a typical gloss paint. Informed readers will see limitations in the above, but it is primarily intended to highlight the broader binder/solvent contributions.

Generally all polymer types can provide a spectrum of compositions covering a span of properties at varying cost, and so given user criteria may be satisfied by selection from a number of resin types. The principal final choice for the user, whose application and cure conditions will probably have been determined by scale and now possibly environmental considerations, ultimately would appear to concern balancing performance and cost; true cost includes the total of paint cost, labour and equipment cost, and energy for cure.

The industry is constantly striving for higher performance and novel products. Factors influencing system design may well include a need to guarantee performance in such diverse applications as decorative maintenance paints and in automobile and coated coil products, and to apply total quality management concepts

Table 2.1 — Contribution of major paint components to final paint properties

Property	Film former	Pigment	Solvent
Application	Major	Minor	Major
Cure rate	Major	None	Significant
Cost	Major	Major	Minor
Mechanical properties	Major	Minor	—
Durability	Major	Major	—
Colour	Minor	Major	—

as enacted in the ISO 9002 standards required of suppliers. Other forces that increasingly apply are legislation to control usage and pollution, availability and cost of energy supplies, and periodic abundance or shortage of natural and oil-based materials. Most recent developments are the carrying out of life-cycle assessments (LCAs) on paints in connection with eco-labelling studies, along with needs to apply environmental management systems (EMSs) in manufacture [1, 2]. LCA carries through to ultimate disposal including consideration of recycling of paint, paint waste, and packaging. Eco-labelling is now a major issue and can drive the choice of all components of the paint system; compliant coatings are those fully meeting both legislation and voluntary agreements. These issues present many challenges to the resin formulator [3, 4].

The choice and amount of solvent or diluent used will be constrained by hazard and eco-labelling and then will depend on the nature of the polymer and the method of application; the quantity of solvent (solids) and final viscosity will then depend on the latter. The method of application generally imposes constraints regarding solvent boiling point and evaporation rate, for example, to ensure good spray or brush application. If the polymer can be prepared in the presence of little or no solvent, the solvent required by the method of paint application has little practical significance to the resin chemist, i.e. an alkyd or polyester may easily be thinned at end point with high or low boiling solvent. However, for an acrylic resin it is usually necessary to use a solvent or solvent blend of low chain transfer properties; the boiling point must be such that the reaction mixture can be refluxed to remove heat of polymerization and such that an initiator system is available at that temperature capable of efficient conversion of monomer to polymer. The enduring market trend is in reducing quantities of all organic solvent (particularly hydrocarbon solvent) used and in an increase in the use of water as a major part of the solvent/diluent system.

The mechanical properties of any given polymer 'improve' as molecular weight increases up to a value at which no change is seen with a further increase. In contrast the viscosity of the solution of a polymer continues to increase with molecular weight without a break. This imposes the constraint in designing a surface-coating system that if the polymer is to be made for optimum application and final properties, then molecular weight needs careful specification and control. Furthermore, since for many polymer systems the molecular weight necessary for good mechanical properties and durability will be high, considerable amounts of solvent will be needed to obtain good application properties, if the polymer is to be applied in solution at that molecular weight. The kind of system this describes is a lacquer, drying by solvent evaporation alone to leave a film of the polymer with useful properties, with no subsequent change of molecular weight or further reaction occurring. Practically this system, initially used with varnishes such as Shellac and French polish, continues with plasticized nitrocellulose and with thermoplastic acrylic lacquer systems used in automotive refinishing systems.

The simplest way of avoiding the molecular weight/viscosity conflict is the use of dispersion rather than solution systems. Examples are the decorative aqueous emulsions now made in high volume, the dispersed polymer used in water-borne systems, non-aqueous dispersion (NAD) systems, and organosols. The use of dispersions can also allow the use of cheaper, less polluting diluents, and was particularly useful when the Californian Rule 66 legislation controlling the VOC (volatile organic content) of factory exhaust emissions was first introduced. Dispersed polymers are, however, more complex to formulate, and more restricted in composition than solution polymers.

Most surface-coating systems avoid the molecular weight constraint by using 'crosslinking' or 'curing' reactions; the system is assembled with one or more reactive polymer components of relatively low molecular weight, capable of further reaction after application to high or infinite molecular weight. Crosslinking implies a multifunctionality such that each initial component molecule links to a number of other molecules, so that an infinite network is formed in the final coating.

Methods used for film formation with typical polymer systems are shown in Table 2.2. In every case except the lacquer system, the polymer system assembled includes free reactive groups in its architecture appropriate to the method of curing chosen. The curing reactions are discussed later in this chapter under the appropriate resin types. Apparent overlap of techniques may be found; for example, stoving may be used to accelerate solvent evaporation so blurring the distinction between thermoplastic and thermosetting acrylics; equally, stoving may accelerate curing of an alkyd capable of slower but ultimately satisfactory air-drying unaided.

Electrodeposited films normally require a subsequent heat cure; electrodeposition insolubilizes the polymer on deposition, but the film is soft and not fully resistant to other agents until crosslinking occurs. Electrostatic spraying of powder coating lightly adheres the powder to the substrate, but this may be brushed off until fused and possibly cured by stoving.

Polymers are conveniently categorized by their method of polymerization from monomers, which may be by stepwise or functional group polymerization, or by chain addition polymerization. These terms are more satisfactory than the older common terms, condensation and addition polymerization, being both more precise and more generally applicable.

Stepwise or *condensation* polymerization is identified where the monomer units initially present are nearly all incorporated into larger molecules at an early stage in the reaction; these larger molecules remain reactive and continue to join together so that the average molecular weight increases with time, but once the reaction is under way the yield of polymer species does not vary with reaction time. The reaction usually involves two separate monomer species with different but co-reactive groups. Most usually, though not invariably, some small molecule such as water will be eliminated as each step occurs. Alkyds and polyesters are typical stepwise polymers. The concept of functionality is particularly relevant to this type of polymerization in that a minimum requirement of each 'monomer' molecule is that it shall

Table 2.2 — Methods of film formation for typical polymer systems

Method	External agent	Typical polymer system
Solvent evaporation	None or heat	Lacquer systems
Environmental cure	Oxygen	Oil-modified alkyd
	Moisture	Moisture-curing urethane
Vapour phase curing	Amine	Hydroxy acrylic/isocyanate blend
2-pack	None or heat	2-pack epoxy/amine
Radiation	Infrared/ultraviolet/electron beam cure	Photocuring unsaturated polyester
Thermosetting	Stoving oven	Alkyd/nitrogen resin blend Thermosetting acrylic

Table 2.3 — Classification of polymer types used in coatings

Type	Subgroup
Natural resins	Resins, gums, rosin
Modified natural resins	Cellulose, starch, nitrocellulose
Stepwise (condensation) polymers	Polyester, alkyd resins
	Formaldehyde resins (urea, melamine, phenol)
	Epoxy resins
Chain addition polymers	Acrylic polymers and copolymers
	Vinyl polymers
Ether polymers	Polyethylene oxides and glycols

possess two reactive functional groups, if the product of each reaction step is to be able to participate in further reactions.

Chain addition polymerization has the characteristic that high molecular weight polymer is formed from the start by a chain reaction, and monomer concentration decreases steadily throughout the polymerization period; thus the yield of polymer increases with time, unlike stepwise polymerization. Polymerization is initiated by some active species capable of rupturing one of the bonds in the monomer, and may be radical, electrophilic, or nucleophilic in character. Only free radical chain addition polymerization is generally practised in production by surface-coating manufacturers though they may purchase polymers made by other techniques. Acrylic polymers are the best known chain addition polymers.

Group transfer polymerization is one special technique [5], enabling polymers of varied and controlled architecture and functionality to be prepared. Other investigations into structure control techniques [6] have provided star, hyperbranched, and dendritic structures [7, 8] finding application in high solids finishes (see Section 2.18). A further new technique, giving low molecular weight polymers of narrow distribution, also with potential for high solids applications, is catalytic chain transfer polymerization (CCTP) [9].

Table 2.3 shows the broad types of polymers used in coatings, partly in terms of the above classification. Ether polymers are separated into a group of their own, in part to acknowledge that materials such as polyethylene glycol can be prepared by both stepwise and addition polymerization, and also to avoid further confusion. Epoxy resins are formed by stepwise polymerization, but it should be noted that the epoxide group may take part in both stepwise and chain addition polymerization.

Finally, while the polymer phase has been referred to as the continuous phase, it should not be assumed that this phase is always homogeneous. Providing the bulk of the phase is continuous, there are many instances, a number described later in this chapter, where identifiable polymer inclusions are present, altering the properties of the final film. These inclusions are invariably of sub-micron size. Often these are rubbery, toughening the film, but there are also instances where these are hard affecting pigment distribution, or voided providing additional opacity.

2.2 Natural polymers

The vehicles used as surface coatings were originally based on natural oils, gums, and resins, giving, when combined, a range of both lacquer and autoxidatively drying products. The naturally occurring triglyceride oils still find considerable use in oil-modified alkyds and to a lesser extent in other fatty acid-modified products such as epoxy esters, and these are described under separate headings. The use of other natural products has now diminished to very minor amounts, in part owing to limited availability; some mention of those natural resins still used is made in the section on oleoresinous vehicles.

Under the description of modified natural products mention must be made of cellulose derivatives, particularly 'nitrocellulose', which is the major binder in cellulose lacquers which still find use. Nitrocellulose, more accurately named cellulose nitrate, is obtained by the nitration of cellulose under carefully specified conditions, which control the amount of chain degradation and the extent of nitration of the hydroxyl groups of the cellulose. The nitration reaction in essence is

$$HC-OH \xrightarrow{HNO_3} HC-ONO_2$$

Organic esters of cellulose can also be produced, and the mixed ester cellulose acetate butyrate is used as a modifying resin additive, particularly for acrylic lacquers. The mixed ester is used in preference to the softer higher esters, or the acetate which has poorer dimensional stability (see Section 2.7.4). Sucrose esters are now available, finding similar uses.

Cellulose derivatives such as hydroxy ethyl cellulose and the salts of carboxy methyl cellulose also find use as protective colloids in the emulsion polymerization of vinyl monomers (see Section 2.8).

2.3   Oils and fatty acids

Vegetable oils and their derived fatty acids still play an important a role in surface coatings today because of their availability as a renewable resource, their variety and their versatility.

Oils are mixed glycerol esters of the long chain (generally C_{18}) monocarboxylic acids known as fatty acids (Fig. 2.1); unrefined oils also contain free fatty acids, lecithins, and other constituents, the first two finding their own application in coatings.

Oils useful in coatings (Table 2.4) include linseed oil, soya bean oil, coconut oil, and 'tall oil'. Many countries will find some application for their own indigenous oils as well as those detailed here. When chemically combined into resins, oils contribute flexibility and, with many oils, oxidative crosslinking potential, one of the properties most exploited in paint.

It should be added that in searches for a synthetic alternative to drying oils, only allyl ethers have been found to be both accessible and possess similar and useful autoxidative abilities. These have found limited application in high solids finishes and unsaturated polyesters (see Sections 2.6.4 and 2.18).

Oils are classified as drying, semi-drying, or non-drying, and this is related to the behaviour of the unmodified oil, depending on whether it is, on its own, able to oxidize and crosslink to a dry film. This behaviour is directly related to the concentrations of the various fatty acids (Table 2.5) contained in the structure. These fatty

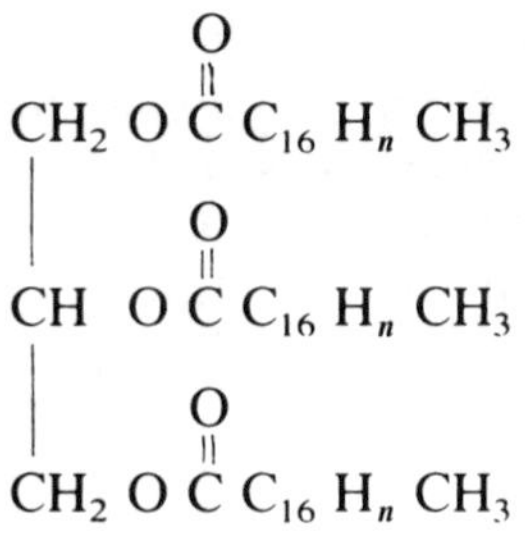

$$
\begin{array}{l}
\overset{\displaystyle O}{\underset{\displaystyle \|}{}} \\
CH_2\ O\ \overset{\|}{C}\ C_{16}\ H_n\ CH_3 \\
\quad |\qquad O \\
\quad\ \ \ \ \ \ \overset{\|}{} \\
CH\ \ O\ \overset{\|}{C}\ C_{16}\ H_n\ CH_3 \\
\quad |\qquad O \\
\quad\ \ \ \ \ \ \overset{\|}{} \\
CH_2\ O\ \overset{\|}{C}\ C_{16}\ H_n\ CH_3
\end{array}
$$

n may be 32, 30, 28, or 26

Fig. 2.1 — Representative structure of naturally occurring oil.

Table 2.4 — Composition of major oils used in surface coatings

	Saturated acids	Oleic acid	9,12- Linoleic acid	9,12,15- Linolenic acid	Conjugated acid
Tung	6	7	4	3	80[d]
Linseed	10	20–24	14–19	48–54	0
Soya bean	14	22–28	52–55	5–9	0
Castor oil	2–4	90–92[a]	3–6	0	0
Dehydrated castor oil	2–4	6–8	48–50	0	40–42[b]
Tall	3	30–35	35–40	2–5	10–15[c]
Coconut	89–94	6–8	0–2	0	0

[a] Principally ricinoleic acid, not oleic.
[b] Conjugated 9,11 linoleic acid.
[c] Conjugated linoleic acid and octadeca 5,9,12-trienoic (pinolenic) acid, proportions dependent on source and refinement.
[d] α-Eleostearic acid.

Table 2.5 — Structures of the commonly occurring unsaturated fatty acids

Octadeca-9-enoic (oleic) $CH_3\cdot(CH_2)_7\cdot CH{=}CH\cdot(CH_2)_7\cdot COOH$
Octadeca-12-hydroxy-9-enoic (ricinoleic)
$\qquad\qquad CH_3\cdot(CH_2)_4\cdot CH_2\cdot CHOH\cdot CH_2\cdot CH{=}CH\cdot(CH_2)_7\cdot COOH$
Octadeca-9,12-dienoic (linoleic)
$\qquad\qquad CH_3\cdot(CH_2)_4{-}CH{=}CH\cdot CH_2\cdot CH{=}CH\cdot(CH_2)_7\cdot COOH$
Octadeca-9,11-dienoic (conjugated linoleic)
$\qquad\qquad CH_3\cdot(CH_2)_5{-}CH{=}CH\cdot CH{=}CH\cdot(CH_2)_7\cdot COOH$
Octadeca-9,12,15-trienoic (linolenic)
$\qquad\qquad CH_3\cdot CH_2\cdot CH{=}CH\cdot CH_2\cdot CH{=}CH\cdot CH_2\cdot CH{=}CH\cdot(CH_2)_7\cdot COOH$
Octadeca-5,9,12-trienoic (pinolenic)
$\qquad\qquad CH_3\cdot(CH_2)_4\cdot CH{=}CH\cdot CH_2\cdot CH{=}CH\cdot(CH_2)_2\cdot CH{=}CH\cdot(CH_2)_3\cdot COOH$
Octadeca-9,11,13-trienoic (α-eleostearic)
$\qquad\qquad CH_3\cdot(CH_2)_3\cdot CH{=}CH\cdot CH{=}CH\cdot CH{=}CH\cdot(CH_2)_7\cdot COOH$

acids contain, with few exceptions, 18 carbon atoms including the terminal carboxyl. Fatty acids may be saturated (no double bonds), mono-unsaturated (one double bond) or polyunsaturated (two or more double bonds); to be considered 'drying' an oil must contain at least 50% of polyunsaturated acids. The fully saturated C_{18} fatty acid is stearic acid, of which some is found in all oils. The C_{12} saturated fatty acid, lauric acid, is the predominant component of coconut oil, where it constitutes 90% of its composition.

Thus linseed oil, a good drying oil contains over 60% of the polyunsaturated linoleic and linolenic acids, while soya bean oil, usually classified as semi-drying, contains just over 50% of linoleic acid. By contrast, coconut oil, a non-drying oil contains 90% saturated lauric acid and less than 10% unsaturated fatty acid.

Polyunsaturated fatty acids may be conjugated or non-conjugated, and tung oil, now in short supply and little used, contains nearly 80% of the conjugated α-eleostearic acid. By contrast 9,12,15-linolenic acid has three non-conjugated double bonds, while 9,12-linoleic acid has two non-conjugated double bonds. High conjugation is reflected in higher drying ability and greater tendency to heat body over that expected from the double bond content; thus, tung oil can in fact readily be gelled by heating, and for this reason — if incorporated into an alkyd resin — would not be processed by the monoglyceride process, this being considered too risky.

The drying oils, tung and linseed, may find application either unmodified or only heat bodied in coating compositions in simple blends with other resins. In contrast, most other oils find little application alone but only after chemical modification; soya bean oil, for example, finds its major application in chemically combined form in long oil alkyds. Oils such as coconut oil do not dry or heat body and have no application alone, but only find use when reacted into alkyds. In this case the oil component provides plasticization. Coconut alkyds have good colour and are used for stoving finishes and appliance enamels.

The two most important oils to the industry nowadays are probably soya bean oil and 'tall oil', the latter misnamed, as it is a fatty acid mixture containing over 50% polyunsaturated fatty acid, principally 9,11-linoleic acid. Tall oil is a by-product of the paper industry, and to be useful in coatings must be refined to reduce its rosin content to 4% or less from the higher concentration in the crude product. However, it is the fractionally distilled form [10, 11], which can now be interchanged with soya fatty acids (and in alkyds where soya bean oil is replaced with tall oil and polyol) depending on seasonal fluctuations of availability and price. Tall oil of Scandinavian origin may be more favoured than that from America owing to its higher unsaturated fatty acid content, which includes the triply unsaturated pinolenic acid. Tall oil is probably more variable in quality, dependent on source and refinement, than any other oil or fatty acid, and so needs careful evaluation and specification for any critical use.

An oil with a more distinctive fatty acid is castor oil which contains nearly 90% of ricinoleic acid, or 12-hydroxy oleic acid. It has some use as a polyol for polyurethane preparation; a major use exploiting the hydroxyl content is in the preparation of higher hydroxyl containing alkyds for crosslinking with nitrogen resins or isocyanates, and for use as plasticizing alkyds for nitrocellulose where high polarity is necessary for compatibility.

Oils are generally purified by acid or alkali refining for the coatings industry in order to remove materials such as free fatty acids and lecithins, to give a neutral clear product and to improve colour. When further reaction is required, oils may be separated into constituent fatty acids and glycerol, and though more expensive than oil, the free fatty acids of all the common oils are currently available to the coatings industry.

2.3.1 Modified oils

Drying oils may be heat-bodied at temperatures in the region of 290 °C to produce 'stand oils', where the viscosity increase is due principally to dimerization reactions through double bonds of unsaturated fatty acid moieties. This reaction may be followed through decrease in the iodine value. 'Stand oils' find application in oleoresinous vehicles and in alkyd resins. The separated dimer fatty acids also find specialized uses, for example in polyamide resin manufacture.

So-called boiled oils were once prepared by heating metal oxides in oil until fatty acids released during bodying solubilized the oxides as soaps. Nowadays a 'boiled oil' may be prepared more conveniently by blending commercial driers with a 'stand oil'.

Dehydration of castor oil is important, since the removal of the hydroxy group and an adjacent hydrogen atom creates an additional double bond, so increasing the residual unsaturation. Dehydrated castor oil (DCO) has distinctive characteristics since the second double bond generated is generally positioned relative to the original one to give both conjugated as well as non-conjugated polyunsaturation. Some bodying may occur on dehydration, although this can be minimized by use of a good vacuum and suitable catalysts, thus reducing the time necessary for the reaction. By virtue of its unsaturated fatty acid content DCO is classed as a drying oil. DCO finds major application in stoving alkyds and in alkyds required for further modification by vinylation; it is little used in air-drying finishes because of the incidence of surface 'nip' (tackiness) in dried films. DCO fatty acids are also available to resin formulators.

Castor oil may also be hydrogenated, and hydrogenated castor oil (HCO) has some application in alkyd resins for stoving application with MF resins; HCO fatty acid, more usually named 12-hydroxy stearic acid, finds application particularly for graft copolymers [12] for use in dispersants and surfactants (see Section 2.9).

Isomerization of oils and fatty acids is used to improve the drying properties particularly of medium oil length alkyds. These processes includes transformation of the double bonds of non-conjugated fatty acids (principally linoleic) to conjugated form, and in certain cases from the *cis* to the *trans* isomer; a number of processes exist for carrying this out [13]. Improvements from using these oils/fatty acids include better colour, initial dry, and resistance to water, acid, and alkali. Weathering and gloss retention are also said to be improved.

Oils and fatty acids containing unsaturation can be reacted with maleic anhydride. Where conjugated double bonds are present a Diels–Alder reaction (Fig. 2.2) is possible, and this proceeds exothermically. Where the bonds are non-conjugated (linolenic or linoleic) the first addition is by the Ene reaction; this has the consequence of moving the double bonds from a non-conjugated to a conjugated configuration. There is then the possible side effect that a sluggish first Ene reaction (Fig. 2.3) may be followed by a faster exothermic Diels–Alder second reaction.

Maleinization is carried out to increase acid functionality of fatty acids, for example for subsequent water solubilization (see Sections 2.5.7 and 2.16). It also occurs *in situ* in alkyd preparations where maleic anhydride is included in the for-

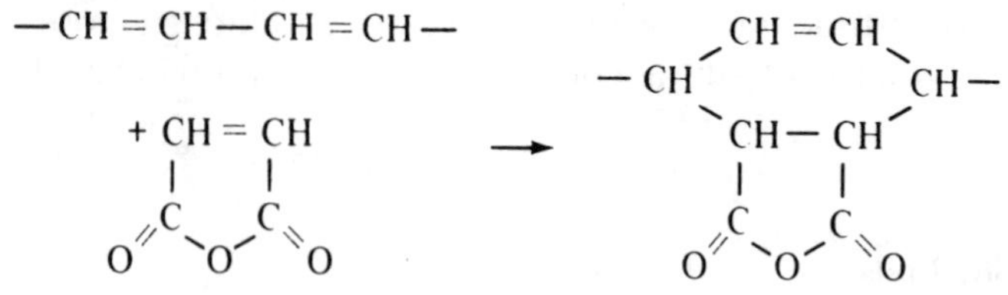

$$-CH = CH-CH = CH-$$

$$+ CH = CH$$

Fig. 2.2 — Diels–Alder reaction.

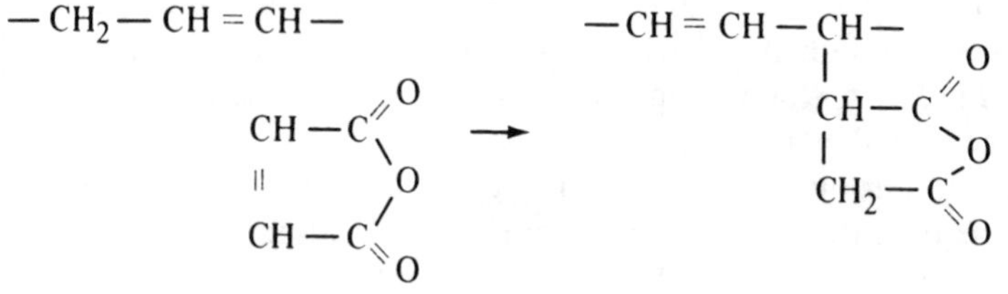

Fig. 2.3 — Ene reaction.

mulation, and its effect is to increase viscosity by increasing acid functionality. Care is needed in formulation since a small maleic addition can for this reason have an unexpectedly large effect.

True 'urethane oils' are obtained where a monoglyceride produced by reaction of an oil with glycerol (see Section 2.5.8) is reacted with a diisocyanate; most 'polyurethanes' of this type are, however, urethane alkyds where some condensation of monoglyceride with dibasic acid is first carried out prior to diisocyanate reaction.

Oils and fatty acids may be vinylated, but this alone has little application, though it may be a first step in the preparation of a water soluble or a vinylated alkyd (see Sections 2.5.7 and 2.5.8).

2.3.2 Drying oil polymerization

As previously stated, drying oil fatty acids may dimerize under heat treatment by Diels–Alder or Ene reaction of unsaturated groups. However, the most exploited reaction of these unsaturated groups is autoxidative crosslinking, accelerated by the presence of heavy metal driers and also with the aid of heat in stoving compositions.

The theories of autoxidative dry have been well reviewed [14], and the mechanisms recently further investigated [15, 16]; only an outline is given here. It is now accepted that the first steps in autoxidative drying are free radical in nature and involve hydroperoxide formation. It is believed that in the process double bonds may shift from the *cis* to the *trans* structure; for linoleates hydroperoxidation can lead to a movement in the double bond adjacent to the position of attachment of the hydroperoxide leading to a conjugated structure that stabilizes the product. While Khan [17] explained this by a non-radical reaction transition state, Fig. 2.4 shows a more likely mechanism with a delocalized radical transition state [18].

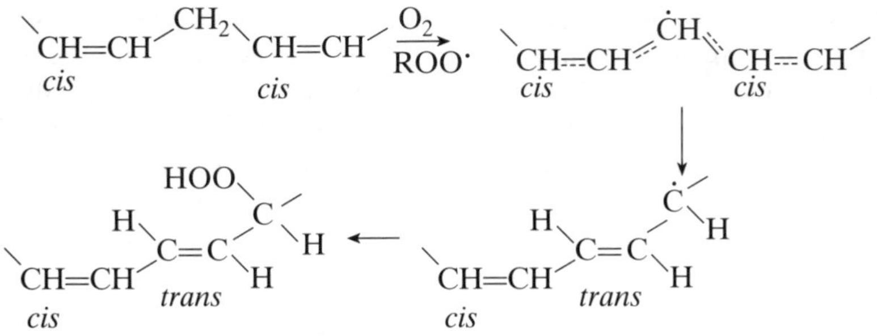

Fig. 2.4 — Oxidation of 9, 12-*cis*–*cis*-linoleate to conjugated *cis*–*trans*-linoleate hydroperoxide.

It is unclear whether added metal drier effects hydroperoxide formation. It is, however, well established that further decomposition and crosslinking reactions of hydroperoxides are accelerated by the presence of metallic driers (particularly cobalt). The use of a hydroperoxide with cobalt drier to initiate the cure of unsaturated polyesters is an example of this reaction being used elsewhere. Just as this latter reaction is identifiable as free radical in nature, so the autoxidative crosslinking reaction is also largely free radical in mechanism. This may be exploited by the copolymerization of high boiling or involatile vinyl monomers into an autoxidatively drying system to modify final properties [19, 20] (see also 2.18):

$$2 \text{ ROOH} \xrightarrow{\text{Co}^{2+} \text{ Soap}} \text{RO} + \text{RO}_2 + \text{H}_2\text{O}$$

The actual crosslinking reactions following radical formation are complex, and as well as the desired high molecular weight compounds, scission products with aldehydes prominent [21, 22] are also produced, giving characteristic drying smells. Yellowing is also a feature [23]. Polymerization reactions certainly involve both radical dimerizations and reactions further involving residual unsaturation. Recent evidence also indicates the transient presence of epoxides and endoperoxides [15].

Metallic catalysts or accelerators used are principally derivatives of heavy metals, particularly cobalt, with added calcium, zirconium and aluminium and other transition metals to enhance through drying properties in paint compositions; metals are normally added as soaps of long chain acids so that they are fully soluble in the media.

Synthetic branched acids have now displaced the use of naphthenates. Lead has been almost entirely eliminated because of its toxicity, and replaced with zirconium and aluminium, as above [24].

2.3.3   Characterizing oils

The iodine value of an oil is the number of grams of iodine absorbed by 100 g of oil; since iodine adds across double bonds this is a measure of the unsaturation present. The Wijs method is the most commonly used, though this can be inaccurate with conjugated systems.

The saponification value gives an indication of the molecular weight of the component fatty acid chains, assuming the original oil is intact and entirely triglyceride.

It must be noted that oils, being derived from annual crops of typically leguminous plants, can vary from season to season in quality (that is in fatty acid distribution); crop yields also vary, causing supply and price fluctuations. Geographical variations also occur as cited earlier in the difference between Scandinavian and American tall oils.

2.4   Oleoresinous media

Oleoresinous vehicles are those manufactured by heating together oils and either natural or certain preformed synthetic resins. In the process, the resin dissolves or disperses in the oil portion of the vehicle. The equipment for preparing this type of product may be essentially simple, and these vehicles were among the first used in the coatings industry after the exploitation of simple gum solutions and natural oils in either unmodified or bodied state. The manufacturing process involves heating oil and resin together until the product becomes clear (this may be tested by cooling a small sample 'pill' of material on a cold plate) and of the required viscosity.

The temperatures used are frequently those at which oil bodying occurs, that is around 240 °C or higher, and the best performance is often obtained when the resin is taken close to gelation point; for this reason great skill is required of the maker of this type of vehicle. Reaction near to end point may be rapid, and thus quick though subjective tests such as bodying to 'a short string' may have to be employed. Following attainment of end point the product may be cooled and thinned, often in one operation.

Oleoresinous vehicles continue in use for a number of applications, though they have been displaced by alkyds and other synthetic resins for many other uses. For example, clear varnishes such as can lacquers, primers and undercoats, aluminium paints, and marine coatings may still use oleoresinous vehicles to advantage. A high residual use persists in printing inks, though this is not a topic considered further in this chapter.

Oils used are exclusively drying and semi-drying oils, with the more highly unsaturated oils, i.e. tung and linseed, preferred. The resins used have included a number of natural 'fossil' resins such as rosins, copals, shellac, etc., of which only 'tall' rosin is now in good supply. (Rosin consists of a mixture of acids including abietic acid.) Bitumen is a natural resin, and black japan which contains bitumen is, in fact, an oleoresinous vehicle prepared as described above. Synthetic resins now used include rosin derivatives, and phenolic and epoxy resins. The latter two retain considerable importance for oleoresinous marine and insulating varnishes.

An important concept first used in oleoresinous vehicles and then used in defining alkyd formulations is that of oil length. In this context it refers to the oil content of the final resin; different nomenclatures may be used, either US gallons of oil per 100 lb of resin or in a more precise definition, the weight of oil % in the total non-volatile varnish produced. Short oil vehicles contain less than 67% oil and are fast drying, giving hard films but that lack flexibility; they are used for floor varnishes and gold size. Long oil vehicles (more than 67% oil) are slower drying but more flexible and are used for exterior varnishes and undercoats.

2.5   Alkyd resins

The alkyd resin was one of the first applications of synthetic polymers in surface coatings technology; it was successful in chemically combining oil or oil-derived fatty acids into a polyester polymer structure. Benefits were the enhancement of the mechanical properties, drying speed and durability of these vehicles over and above those of the oils themselves and the oleoresinous vehicles then available. Though now surpassed by more sophisticated polymers for the more exacting applications, the alkyd, because of its partial reliance on renewable natural resource, and especially because of the enormous variety of compositions possible, still accounts for a considerable volume of total surface-coating resins produced. The long oil alkyd, optionally blended with other modified alkyds to increase toughness or alter rheology, remains the major resin used in brush-applied solvent-based decorative coatings. Recent application has extended into higher solids and use in emulsified form, as ways of solvent reduction, in producing eco-labelled products. High solids compositions are covered later in this chapter (see Section 2.18). Short oil alkyds still have use in some automotive and general industrial stoving compositions where they are combined with melamine/formaldehyde hardening resins.

Oils are not directly reactive with polyester components and therefore cannot be incorporated into an alkyd structure without prior modification. We can, however, start with fatty acids saponified from the oil, though for many applications this will be an expensive route, especially if glycerol is intended to be included in the formulation. The method evolved to achieve oil modification is to carry out, as a first stage, a so-called monoglyceride preparation. In this process glycerol (or other polyol) and oil in the molar ratio of 2:1 are reacted at around 240°C in the presence of a basic catalyst (sodium hydroxide, lithium hydroxide, etc.) forming 'monoglyceride', the mono fatty acid ester of glycerol (Fig. 2.5).

Practically, the reaction does not go to completion, and an equilibrium distribution of species is present including oil, polyol, and mono- and diglycerides. If polyols other than glycerol are to be used, such as pentaerythritol (PE) in long oil alkyds, a 'monoglyceride' stage may still be carried out using the appropriate amount of polyol to give a notional dihydroxy functional product. The satisfactory attainment of sufficient randomization of structure may be tested for by an alcohol tolerance test, when no separation of free oil should be observed. Sometimes with polyols such as PE the alcohol test is not satisfactory. In that case, reaction is carried out on a small scale in a wide-necked tube with the correct proportions of 'monoglyceride' sample and polybasic acid to test for the ability to derive a clear final product without gelation resulting. With some oils it is important not to prolong the time at

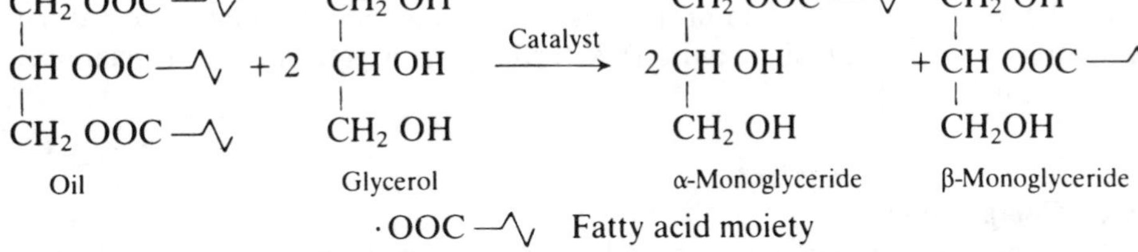

Fig. 2.5 — Schematic representation of monoglyceride formation.

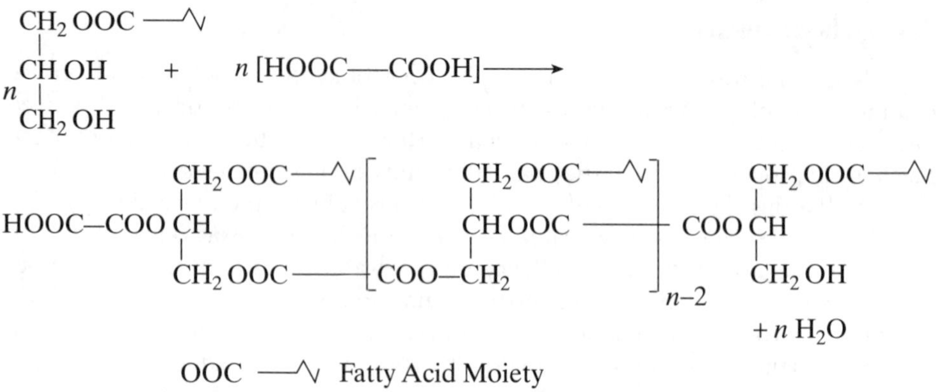

$$
\begin{array}{c}
\text{CH}_2\,\text{OOC} \longrightarrow \\
| \\
n \;\; \text{CH OH} \qquad + \qquad n\,[\text{HOOC—COOH}] \longrightarrow \\
| \\
\text{CH}_2\,\text{OH}
\end{array}
$$

$$
\begin{array}{ccc}
\text{CH}_2\,\text{OOC}\longrightarrow & \left[\begin{array}{c} \text{CH}_2\,\text{OOC}\longrightarrow \\ | \\ \text{CH OOC} \longrightarrow \\ | \\ \text{CH}_2\,\text{OOC}\longrightarrow \text{COO—CH}_2 \end{array} \right]_{n-2} & \text{CH}_2\,\text{OOC}\longrightarrow \\
\text{HOOC—COO CH} & \longrightarrow \text{COO} & \text{COO CH} \\
| & & | \\
\text{CH}_2\,\text{OOC} & & \text{CH}_2\,\text{OH} \\
& & + n\,\text{H}_2\text{O}
\end{array}
$$

OOC $\longrightarrow$ Fatty Acid Moiety

Fig. 2.6 — Schematic representation of glycerol alkyd polymer formation.

temperature for this stage in order to minimize oil bodying; for the same reason it is good practice to standardize on heat-up rate and time for the normal reaction, to ensure reproducible and predictable behaviour in the following stage.

Preparation of the final resin is carried out by the addition of further polyol and dibasic acid in the so-called bodying stage; in this stage, multiple esterification reactions occur (Fig. 2.6) and water formed as a by-product of reaction must be removed. Esterification results in the steady building-up of polymer structure and hence viscosity increases; the stage is taken to a degree of reaction at which the desired viscosity at a specified concentration (in solvent) and temperature has been attained.

While the monoglyceride method as described is used for long oil alkyds, the simpler so-called 'fatty acid process' is also practised; here, all ingredients including fatty acid (not oil) are charged together, and the process is frequently used for shorter oil alkyds where polyols other than glycerol are required. Other techniques are possible, for example an acidolysis route where oil is first equilibrated with polybasic acid prior to reaction with polyol in an apparent reversal of the monoglyceride process; this method is not widely favoured. Where it is necessary to incorporate tung oil into an alkyd, when to carry out a monoglyceride state would risk excessive bodying, a variant is the fatty acid/oil process. Here the tung oil is taken along with a good proportion of fatty acid, polyol, and dibasic acid, and a one-stage reaction carried out. Providing the oil concentration is not high, transesterification of sufficient magnitude takes place during processing to react the oil into the resin and ensure a clear final product. Another technique proposed for alkyd manufacture is the 'high polymer' technique [25, 26], where by stepwise addition of ingredients, more favourable molecular weight and reactive group distributions are said to be obtained and certain properties enhanced; this is claimed to be particularly effective with tall oil alkyds. This technique has recently been examined theoretically [27, 28].

2.5.1 Composition
Each of the main ingredients of an alkyd contributes to its properties in a predictable manner [29]. Oil content is expressed as 'oil length' where long oil alkyds

contain over 60% oil, medium oil between 40 and 60%, and short oil below 40%. Long oil alkyds are generally made with drying oils and are soluble in aliphatic solvents. They are low viscosity resins that air-dry slowly to give soft flexible films with poorer gloss retention and durability.

Shorter oil length generally results in the need for aromatic solvent, giving higher viscosity resins at lower solids. If formulated with drying or semi-drying oils, and of medium or long oil length, the alkyd can air-dry oxidatively at room temperature. Otherwise it will not do so, and to form hard films it must be stoved with a crosslinking resin. Medium and short oil alkyds give hard films with good gloss retention and chemical resistance; they are less flexible than long oil alkyds and are normally dried by stoving.

Aromatic acids such a phthalic anhydride or isophthalic acid, and maleic anhydride contribute hardness, chemical resistance, and durability to alkyds; in contrast long chain dibasic acids such as adipic, sebacic, and azelaic acids, are sometimes used to plasticize alkyds and provide flexibility, not hardness. Phthalic anhydride dominates in conventional alkyd formulations because of its low cost and ready availability. It should be noted that it has added economy in use, in that by virtue of its anhydride structure, water evolution is halved compared with a dibasic acid; the first stage of its reaction is exothermic, with so-called 'half ester' formation.

Polyols used are generally at least trifunctional to permit branching or crosslinking and can provide the alkyd with hydroxyl groups for further reaction. In general they contribute good colour and colour retention, but vary in their chemical and weathering resistance (see also Section 2.6). Most general-purpose alkyds are formulated using phthalic anhydride with properties modified by polyol variation. The effects of polyol change may be readily related to structure so that in a series where glycerol is replaced by trimethylol ethane and then by trimethylolpropane, aliphatic solubility increases, viscosity falls, and flexibility increases, contributed to by increased side chain length; otherwise varying the polyol component may have a significant effect, not least because with varying polyol molecular weight the phthalic content of the final resin is affected, and hence the hardness of the final alkyd.

Rosin was originally included in alkyds as a modifier because of its low cost. However, it may also be added, as may other monobasic acids such as benzoic acid, to enable oil length to be shortened without causing increased viscosity. Rosin itself contributes to faster drying and film hardness, but degrades weathering performance; its use is restricted to alkyds for primer applications.

Where short oil length and aliphatic solubility present conflicting requirements, trimethylol propane can be used as polyol, though with some penalty of increasing softness. *Para* tertiary butyl benzoic acid can also be used as a monobasic acid to modify solubility, and both are effective because of the contribution of their aliphatic side groups.

Isophthalic acid gives higher molecular weight resins than orthophthalic, with better drying characteristics and harder, more durable films. Its use can narrow molecular weight distribution and enable lower viscosity/higher solids resins to be made [30]. However, it is more difficult to incorporate, remaining solid in the reaction mixture until at least one of its acid groups has reacted. Like orthophthalic anhydride it can sublime, but unlike orthophthalic anhydride, it can lead to overhead blockages in manufacturing plant because of its lower solubility in xylene which is used as the entraining solvent for the removal of water of reaction. Special understanding is necessary to formulate high solids alkyds with longer oil length [31], and these principles are discussed further in section 2.18. Terephthalic acid has little use in alkyds because of its extremely low solubility in reaction mixtures and its very slow reactivity.

Mention must be made of Cardura E10 (Fig. 2.7; Shell Chemicals) which is the glycidyl ester of the branched Versatic Acid (Shell Chemicals), the latter being a 'tertiary' or trialkyl acetic acid with a total of ten carbon atoms. By virtue of the epoxide group, it reacts rapidly and completely with carboxyl groups at temperatures above 150°C; when incorporated into an alkyd, the effect is that which would be contributed by a non-drying oil or fatty acid component. Cardura E10 contributes special properties of good colour, gloss retention, and very good resistance to yellowing and staining [32–34]. Derived alkyds can be used in stoving enamels and in nitrocellulose lacquers. Care should be taken in the manufacture of alkyds which include both Cardura E10 and phthalic anhydride since the reaction is both rapid and exothermic (around 110kJ per epoxy equivalent). Special formulating techniques are available for this type of alkyd that also allow control of polymer architecture [35]. Cardura E10 has special use for reducing the acid value in resins as a final treatment, for example where zero acid value is required for resins for use with photocatalyst, in metallic paints or with certain drier systems:

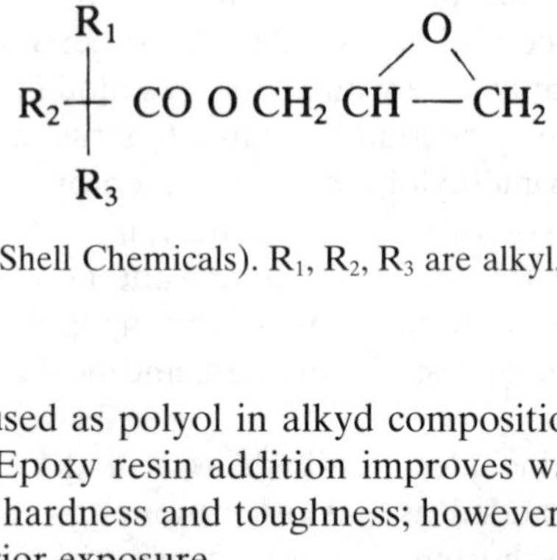

$$R_2 \overset{\displaystyle R_1}{\underset{\displaystyle R_3}{\vert\!\!\!\vert}} CO\ O\ CH_2\ CH \overset{O}{\overbrace{}} CH_2$$

Fig. 2.7 — Cardura E10 (Shell Chemicals). R_1, R_2, R_3 are alkyl, one of which is methyl.

Epoxy resin may be used as polyol in alkyd compositions in combination with the more usual polyols. Epoxy resin addition improves water and chemical resistance and contributes to hardness and toughness; however, it adds to cost and can lead to chalking on exterior exposure.

Since alkyds are soluble in either aromatic or aliphatic solvent dependent on their oil length, they are tolerant of additional stronger alcohol, ester, or ketone solvents for special applications.

2.5.2 Alkyd reactions and structure

The major polymer-forming reaction in alkyd preparation is the esterification reaction between acid and alcohol, and hence the mechanism is through a type of stepwise polymerization (Fig. 2.6). In alkyds this reaction is generally not catalysed. Side reactions that can occur include etherification of polyol especially in prolonged monoglyceride stages.

Pentaerythritol is particularly prone to etherification, and grades should be chosen with low dipentaerythritol content.

A reason for lower molecular weight with orthophthalic alkyds compared with those prepared with iso- or terephthalic acids is the tendency to form intramolecular ring structures in *o*-phthalate esters. Because of this, the effective functionality of phthalic anhydride is practically less than two.

It has been claimed that alkyds owe some of their special properties, including application and dry, to the presence of 'microgel' in the alkyd [36]. While this seems to be proven, it would appear difficult to exploit in formulating for a microgel content unless a 'microgel' prepared separately were to be deliberately blended with the composition [37]. It has been suggested that gelation of an alkyd occurs through separation of microgel particles, until particles become crowded enough to coalesce into a 'macrogel'.

2.5.3 Alkyd reaction control

Alkyd preparations must be controlled to a viscosity end point by sampling, with acid value also monitored. Viscosity measurement may be simply achieved by using Gardner bubble tubes with thinned samples; however, improved speed of measurement and higher accuracy may be achieved by using a heated cone and plate viscometer [38]. The sampling interval will often need to be reduced as the end point is approached, owing to an increased rate of rise in viscosity, especially with short oil alkyds.

Where behaviour has been characterized from previous batches, a 'flare path' guide may be provided to follow the viscosity and acid value change; if reference to this shows deviation from normal progress, adjustment may be made by a small addition of acid or polyol. Where solvent process and oil bodying is occurring, adjustment may be made by changing the reflux rate, providing this is carried out at an early enough stage in the process. It is also possible to make temperature changes during processing to modify the bodying rate.

On attainment of end point, cooling and thinning are necessary to halt the reaction; thinning is also essential since the alkyd resin, in the absence of thinning solvent, will either be extremely viscous or solid when cold.

For new formulations, process temperature will be established by consideration of the need for higher temperatures for oil bodying and faster reaction in long oil alkyds, or for lower temperature for easier control of end point in shorter oil alkyds. Acid value at gelation (AV gel) prediction may be possible as the preparation proceeds (see Section 2.5.4). The results of experiments will allow modification in both composition and processing conditions.

2.5.4 Describing alkyds

In specifications for alkyds, oil type, oil length, and solvent will be stated, and sometimes the dibasic acid and its % in the composition.

Oil length refers to the % weight of glyceride oil calculated on the theoretical yield of solid alkyd; fatty acid content for non-glycerol-containing compositions may alternatively be quoted. In both cases this is the major indication of the flexibility, solubility, and — where containing drying oil — drying potential to be expected for the resin.

Solids, viscosity, and acid value (AV) will be determined as for other resins, and hydroxyl value (OHV) may also be measured. Both the latter are quoted in units of weight of potassium hydroxide in milligrams, equivalent to 1 g of solid resin.

Excess hydroxyl content, in contrast to OHV, which can be found experimentally, is calculated from the molar formula and expressed as the excess hydroxyl % over that required to react completely with the polybasic acid moieties present. Excess hydroxyl % may be used as a formulating parameter to derive formulations with a required AV gel or average functionality.

Acid value at gelation (AV gel) may be calculated theoretically, and may also be determined practically by extrapolation of a graph of acid value against 1/viscosity as an alkyd preparation proceeds to its end point.

2.5.5   Alkyd formulating

Practical alkyd formulating involves a process of calculation, application of previous practical experience, and trial preparation to arrive at the desired product. Principally oil length and oil/fatty acid type will be specified for the application, and polyol and dibasic acid dictated by a combination of other considerations including solubility, availability, and cost.

The average functionality concept is one method of formulating as first proposed by Carothers:

$$F_{average} = 2 \times \frac{\text{total acid equivalents}}{\text{total mol}} \quad \text{(hydroxyl in excess)}$$

where for gelation to occur F should be two or more. This is the basis of the well-respected scheme by Patton, except that his constant $k = 2/F$. The basic functionality approach is recognized to require correction and Patton recommends appropriate factors for various ingredients.

An alternative method useful in arriving at a trial formulation involves calculating the extent of reaction at gelation, most usefully expressed as AV gel, which for example might be around five units below that desired in the final resin. It will be found necessary to vary the hydroxyl content, expressed as excess OH in order to alter the AV gel. Various equations have been produced as gelation theory has been developed, by for example Flory and Stockmeyer, and formulating guides are available which review the whole field of alkyd calculation [39, 40].

Frequently, one of the forms of the Stockmeyer equation will be used, such as:

$$p^2 = \frac{\left(\sum gB\right)^2}{\left(\sum f^2 A - \sum fA\right)\left(\sum g^2 - \sum gB\right)}$$

where p = degree of reaction of acid groups at gelation, f, g = functionality of carboxyl and hydroxyl moieties respectively, and A, B = mols of carboxyl and hydroxyl moieties present. Before applying this equation to oil-containing formulations, the mols of oil should be split into constituent fatty acids and glycerol.

It is recognized that to formulate accurately, the true picture invariably lies between the two approaches. While the Flory – Stockmeyer approach is statistically correct, its inaccuracy is due to the actual formation of a number of cyclic rather than long chain polyesters through intramolecular reaction. The effect is more pronounced with the anhydrides such as phthalic anhydride, a very common ingredient. While the simplest correction can be to consider phthalic anhydride as having a functionality of less than 2 (e.g. 1.85), Kilb developed the best theoretical solution, applied practically by Bernado and Bruins [41].

As a simpler alternative to some quite complicated calculations, those offered by Weiderhorn [42] may be found a simpler and equally practical alternative, being less complex to use in the circumstances to which they apply.

In formulating short oil alkyds, monobasic acid may need to be included, dictated by solubility and hydroxyl content requirements. The use of bodied oil in long oil alkyds will be determined by the need to increase molecular weight and obtain higher viscosity, when this is not attainable by other means; the effect will be to raise the practical AV gel above that calculated.

With microcomputers being generally available, people formulating alkyds (and polyesters) regularly are likely to use computer programs to calculate both stoichiometry and gelation parameters. Programs of high functionality have been written [43]. However, writing one's own program is quite feasible; alternatively simpler programs are available from raw material suppliers [44, 45].

Some further comments on alkyd and polyester formulating are included in Chapter 20 on modelling.

2.5.6   Alkyd preparation equipment

Alkyds may be prepared on both laboratory and production scales by both the so-called 'fusion' and 'solvent' process techniques. In the former, ingredients are heated in vented stirred pots; problems encountered are mainly due to variable losses of reactant, particularly phthalic anhydride which is volatilized along with the water of reaction. In the latter process, more sophisticated equipment is needed, namely reactors fitted with condensers, water separation equipment, and solvent return systems. Organic solvent, typically xylene, is added to the resin in the bodying stage (2–5%) and recycled via the overheads, water being removed by azeotropic distillation. It can be of advantage to fit a solvent process plant with a vacuum facility which, besides providing assistance with loading, can be used, for example, to reduce the residual solvent content at the end of processing a batch.

Problems of phthalic anhydride sublimation in the solvent process may be minimized by careful plant design or avoided by using bubble cap scrubbers or fractionating equipment [46]. Inert gas is normally supplied in all alkyd processing to improve colour and to increase safety. Process temperatures may be between 180°C and 260°C. Where consistent practice such as heating rate, gas flow, and ventilation can be achieved, it must be stressed that the fusion process can be satisfactory for almost all classes of alkyd, provided that environmental problems caused by fume can be controlled.

2.5.7   Water-soluble and dispersed alkyds

Water-soluble resins are discussed more generally in section 2.16. This section specifically concerns alkyd resins. One issue is the current inability of emulsion polymer systems (section 2.8) to provide traditional high gloss decorative coatings; for this reason, alkyd systems continue to be used. While high solids systems have been developed to satisfy eco-labelling requirements, emulsified alkyd systems have also evolved for glossy decorative applications.

The usual route to water solubility is to formulate to acid value in the region 40–60, followed by partial neutralization with alkali, ammonia, or amine. Since phthalate-based alkyds have poor hydrolysis resistance and hence poor storage stability, isophthalic acid (IPA) alkyds may be preferred. Alternatively, alkyds will be formulated using trimellitic anhydride (TMA) or dimethylol propionic acid (DMPA). The most common method in processing water-soluble alkyds containing TMA is first to react all the ingredients except the anhydride to form a polymer with an acid number below 10; and then to add the TMA at a reduced temperature of around 175 °C. Under these conditions the TMA is attached by a single ester link to hydroxyl groups on the preformed polymer, and the two remaining carboxyls are available for salt formation and subsequent water solubilization. This is termed the 'ring-opening' technique from the single stage reaction of the anhydride group of the TMA.

Water-soluble alkyds of this type find application in stoving compositions with water soluble melamine/formaldehyde resins as crosslinker. They may be supplied thinned in water-miscible solvent (e.g. butyl glycol) and water thinned at a late stage to gain maximum storage stability.

Water solubility can be conferred by incorporating water-soluble polymer. Thus, polyethylene glycols confer water solubility or dispersibility to the alkyd which may be exploited in the manufacture of gloss paints with easy brush-clean properties (as well as other use as surfactant) [47].

Alkyd emulsions may be prepared by emulsifying using surfactants [48]; polymeric non-ionic surfactants give advantages over normal anionic surfactants, especially when they contain autoxidizable reactivity [49]. All resin emulsification processes produce better results when using inversion, and in the case of alkyd emulsions, special techniques using the phase inversion temperature (PIT) have been developed [50]. The various problems additional to stability, found in the use of alkyd emulsions have been discussed in depth [51]. These include pigmentation, drier distribution, and system rheology. In the preparation of more stable emulsions, mixed ionic/non-ionic stabilization is now favoured.

To overcome deficiencies further, techniques have evolved to prepare alkyd emulsions with core/shell-like structures. One technique has been to use as a component, fatty acid grafted with copolymer rich in methacrylic acid [52]. Another is to use emulsified alkyd as the seed for growth of an acrylic polymer shell.

2.5.8 Modified alkyds

Urethane alkyds may be prepared by replacing part of the dibasic acid in the alkyd formulation with a diisocyanate such as toluene diisocyanate (TDI); the formulating calculations applicable are those used for alkyds. Long oil urethane alkyds are used in decorative paint formulations to impart greater toughness and quicker drying characteristics. While total replacement of dibasic acid is possible (thus producing a urethane oil), it is more normal to formulate an alkyd with a reduced dibasic acid content. This is processed to a low acid value, but a high residual hydroxyl content and the thinned resin is then treated with diisocyanate. The isocyanate groups react with hydroxyl groups to form urethane links, but, unlike acids in ester formation, do so exothermically at lower temperatures and without by-product evolution. It is essential to complete the reaction of all isocyanate groups present because of the high toxicity associated with free isocyanates. In-process testing may be by titration with amine, with final verification being by infrared spectroscopy. Small remaining residues of isocyanate may be removed by deliberate addition of low molecular weight alcohol. For the preparation of urethane alkyds, a monoglyceride catalyst should be chosen for the alkyd stage that does not promote unwanted isocyanate reactions, e.g. allophanate formation, and such catalysts as calcium oxide or a soluble calcium soap should be used.

Another modification of long oil alkyd is that with polyamide resin, used to impart thixotropy in the final resin solution. Polyamide resins used are typically those derived from dimer fatty acid and a diamine. The physical properties of these alkyds are very delicately balanced. Such characteristics as gel strength, gel recovery rate, and non-drip properties of derived paints are intimately connected with both the composition and the processing of the alkyd, along with the proportion and type of polyamide and degree of reaction [53]. If this type of alkyd is being prepared, or when formulating paint compositions containing such an alkyd, care should be taken to avoid the use of, or contamination with, polar, e.g. alcoholic, diluents; these destroy the rheological structure to which hydrogen bonding almost certainly provides a considerable contribution. Typical levels of polyamide modification are around 5%, and care in processing is necessary to stop the reaction in its final stage at such a point as to achieve clarity and the optimum and desired gel properties [54].

Silicone modification of long oil alkyds imparts much improved gloss retention for their use in structural maintenance and marine paints. The same hydroxy or methoxy functional silicone resins are applicable (Section 2.14) as used for modification of polyesters. In general, the degree of enhancement in properties is in proportion to the level of modification with these expensive materials, and formulators need to weigh carefully the balance of cost versus durability enhancement achieved.

Phenol formaldehyde resins of both reactive and non-reactive oil-soluble types can be used as alkyd modifiers. They increase film hardness and improve resistance to water and chemicals, but cause yellowing. Such modified alkyds are useful in primer systems as cheaper but less chemically resistant alternatives to epoxy resins.

Vinyl modification of alkyds is used to impart faster dry, increase hardness, and give better colour, water, and alkali resistance. However, decreased gloss retention and poorer resistance to solvents result, the latter making formulation for early recoatability difficult to achieve. Vinylated alkyds are typically used for metal finishes. Styrene and vinyl toluene are the usual modifiers, methyl methacrylate being much less common. The modification stage involves free radical chain addition polymerization of the monomer in the presence of the alkyd under conditions where grafting to the unsaturated oil component of the alkyd is encouraged. This is typically by using at least some dehydrated castor oil in the alkyd composition since the conjugated fatty acid moieties then present encourage the necessary co-reaction; the typical use of peroxide catalysts such as di-*tert*-butyl peroxide also encourage grafting by hydrogen abstraction.

A number of techniques for preparing vinylated alkyds are possible; most common is post-vinylation of the thinned alkyd by feeding monomer and initiator into the resin held at 150–160 °C. Less common alternatives are to vinylate the oil prior to alkyd manufacture or to vinylate the monoglyceride before esterification with dibasic acid. These latter techniques can allow final dilution in low boiling or reactive solvents in which the vinylation reaction would have been impossible. Where difficulties in achieving adequate grafting are experienced, resulting in cloudy resin preparations, possibilities exist of including maleic anhy-dride or methacrylic acid in the alkyd polyester structure to provide additional graft-ing sites.

To the list of modifications must be included the possibility of precondensation with melamine/formaldehyde (MF) resins. Normally for a coating composition that is multicomponent, viscosity will be minimized by keeping the components unre-acted until the curing stages; however, for some systems, advantages of increased stability, compatibility, or improvement in cure properties are seen by partial pre-reaction, even as here where the two polymers are compatible. For low bake curing alkyd/MF systems, precondensation of alkyd and MF can enhance final cure prop-erties, while reducing the tendency of the MF component to self-condense when highly acidic catalyst systems are used.

2.6 Polyester resins

In the chemical sense, the term polyester embraces saturated polyesters, unsaturated polyesters, and alkyds. However, the term alkyd is normally reserved for the oil-modified alkyd discussed above. Similarly, unsaturated polyesters containing vinyl unsaturation, which most typically are maleic-containing resins partially or totally thinned with a vinyl or acrylic monomer, are referred to as such; hence the term polyester is generally reserved for oil-free, acid, or hydroxy functional polyester resins. Polyester resins are typically composed mainly of co-reacted di- or polyhy-dric alcohols and di- or tri-basic acid or anhydride, and will be thinned with normal solvents. Much of the following applies nonetheless to both saturated and unsatu-rated polymers, the special features of unsaturated polyesters being referred to at the end of this section. The use of monobasic acids is not excluded from our definition of polyester. It is convenient to call a polyester containing lauric or stearic acid an alkyd, but to consider as polyesters those containing only shorter chain aliphatic acids, including lower molecular weight branched synthetic fatty acids, and monobasic aromatic acids, such as benzoic acid.

Polyesters can be formulated both at low molecular weight for use in high solids compositions, and at higher molecular weight, and can be both hydroxy and acid functional. Since they exclude the cheaper oil or fatty acid components of the alkyd and are so normally intrinsically more expensive, much effort has gone into under-standing their formulation, to exploit the higher performance of which they are capable; also, effort has gone into the development of new raw materials. The latter are in the main new polyols to suit high-durability applications such as automotive clearcoats and the painting of steel coil strip.

In choosing polyols, three factors affect durability. Both steric factors and the 'neighbouring group' or 'anchimeric' effects affect resistance to hydrolysis [55]. The absence of hydrogen atoms on the carbon atom beta to the hydroxyl group and sub-sequent ester group is a principal factor determining resistance of the ester link to break down under the influence of heat or radiation. These requirements are met by such polyols as 1,4-cyclohexane dimethanol (CHDM) and 2,2,4-trimethyl-1,3-pentane diol (TMPD) where the beta hydrogen content is reduced or hindered, and in materials such as trimethylol propane (TMP), neopentyl glycol (NPG), hydroxypivalyl hydroxy pivalate (HPHP), and 2-butyl-2-ethyl propane diol (BEPD) (Fig. 2.8).

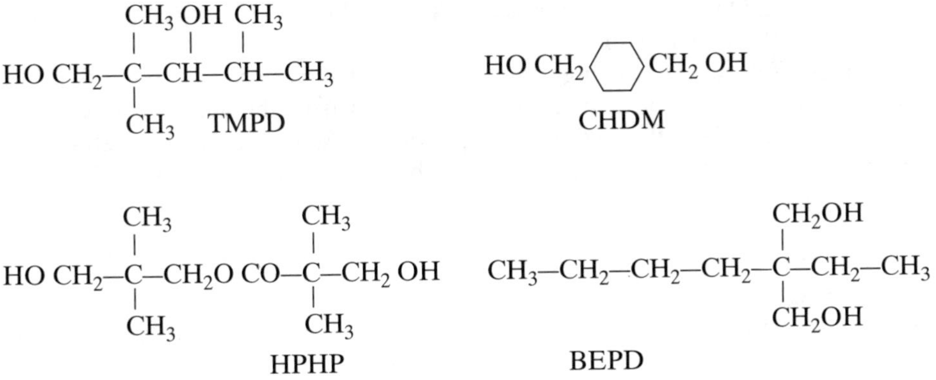

CH₃ OH CH₃ structures...

HO CH₂—C—CH—CH—CH₃   TMPD
       | | |
      CH₃   (with CH₃)

HO CH₂⟨ ⟩CH₂ OH   CHDM

HO CH₂—C—CH₂O CO—C—CH₂ OH   HPHP

CH₃—CH₂—CH₂—CH₂—C—CH₂—CH₃   BEPD

Fig. 2.8 — Typical polyol structures with hindered β positions relative to OH group.

2.6.1 Formulation

Polyesters are formulated in similar fashion to alkyds by making calculations of average functionality, supplemented by calculation of acid value of gelation by the Stockmeyer method [41]. In addition, for a theoretical hydroxy functional polyester, number average molecular weight can be calculated from the formula

$$M_n = \frac{\text{formula weight}}{\sum M - \sum E_a + \left(\dfrac{\text{formula weight} \times AV}{56100} \right)}$$

where M = mols of acid, hydroxy components; E_a = equivalents of acid.

The hardness/flexibility of polyesters may be adjusted by either blending 'softer' aliphatic dibasic acids with the 'harder' aromatic acids or by the inclusion in the polyol blend of the more rigid CHDM in place of 'softer' aliphatic polyols. With the general use of IPA rather than orthophthalic anhydride, crystallization problems can occur with the final polyester solutions. One technique of controlling this is to include a small proportion of terephthalic acid (TPA) to disrupt chain symmetry.

Solvent-borne polyesters for such applications as coil coating, automotive basecoat or clearcoat, and wood finish may be formulated from a blend of the 'hard' and 'soft' acids, IPA and adipic acid, with diols and triols such as NPG and TMP. Formulation variants can include phthalic anhydride for cheapness, and longer chain glycols added for extra flexibility. Monobasic acids such as pelargonic or benzoic acid may be included. These polyesters will generally be crosslinked by MF resin in stoving applications or in non-bake applications as a 2-pack formulation with poly-functional isocyanate adduct. The triol included can contribute to branching and to in-chain hydroxy/functionality depending on the formulation; for some applications such as coil coating where high molecular weight may be an advantage, triol content will be minimized.

With polyesters for high solids coatings, the total polyol in the composition is increased, and the resin processed to lower molecular weight and higher hydroxyl content. Water-soluble polyesters are formulated to a high acid value, often by a 2-stage technique, as described for the preparation of water-borne alkyds, where ring opening with trimellitic anhydride is carried out as a stage subsequent to polymerization. Dimethylol propionic acid has also been used to prepare high acid value water-soluble polyesters.

Polyesters for powder coatings, since they must have a softening point typically >40 °C, will generally not contain long chain plasticizing dibasic acids, but one or more aromatic acids with possibly a simple diol. They may be of high acid value for epoxy resin cure or low acid value and high hydroxyl content for melamine/formaldehyde or masked isocyanate resin curing.

2.6.2   Polyester preparation

For polyesters the general techniques used in alkyd manufacture are used, but with refinements appropriate to the materials used. All-solid initial charges will arise with many formulations, requiring careful initial melting. Difficulties in achieving clarity, even with complete reaction, can be encountered when isophthalic acid is present, and this may be exacerbated when the even less soluble and less reactive terephthalic acid is present. Since a number of the glycols used in polyester manufacture are volatile (e.g. ethylene glycol, neopentyl glycol) and are easily lost with the water of reaction, any serious attempts to prepare polyesters containing these components reproducibly demand a reactor fitted with a fractionating column [56, 57]. Pressure processing has been advocated as a means of raising reaction temperatures and hence reaction rates, without serious loss of glycol, particularly where ethylene and propylene glycols are included in the formulations. Where high molecular weight polyesters are desired, it can also be useful to have vacuum available for use later in the reaction to strip final residual water of reaction and achieve increased molecular weight (the technique is that used in the manufacture of polyesters of fibre-forming molecular weight [58]). For high molecular weight polyesters, particular care needs to be taken regarding glycol loss, since small losses can cause a sufficient imbalance in the formulation to restrict the molecular weight.

Catalysts are frequently used to increase reaction rate in polyester preparation, although the choice of catalyst needs care since colour may be adversely affected. Tin catalysts are particularly useful [59].

2.6.3   Modification

Silicone modification is used to enhance the properties of polyesters, particularly with respect to durability. Modification of both solvent-borne [60] and water-borne [61] polyesters is possible, and the former has become firmly established practice for coil application. In solvent-borne compositions, siliconization is carried out as a second stage following polyester preparation, using a silicone level of typically 20% to 50%; with high AV water-borne resins, the siliconization stage may be carried out after esterification and before the TMA addition stage. Silicone resins in common use are methoxy functional (Section 2.14), and methanol evolution occurs in reaction, which under some conditions may be used to monitor the reaction. The degree of reaction in terms of methoxy functionality may be of the order of 70–80%.

2.6.4   Unsaturated polyesters

The established unsaturated polyester resins are polyesters derived from polyols and dibasic acids, which also include some constituent containing a double bond, and thinned in a polymerizable monomer. By far the majority of these polyesters are linear and contain co-condensed maleic anhydride as the source of unsaturation. In surface coatings these resins have been used mainly in wood finish and refinish two-pack finishes, including a significant use in refinish surfacers and putties. In thin-film coatings, difficulties were found in controlling air inhibition of the curing reaction at the surface of the film, which limited their use. Expansion of the monomer range available, and increased use of radiation curing, has enabled this problem to be overcome to a large extent, and their use has risen.

For radiation curing, polyesters are of the post-acrylated type rather than the maleic type described below. For this application they are, however, in competition with acrylic and urethane oligomers (see Section 2.19).

Generally, unsaturated polyesters are formulated with ortho and isophthalic acids along with maleic anhydride, with appropriate aliphatic glycols, such as diethylene glycol, chosen as co-reactants to control flexibility. The actual amount of maleic in proportion to other acids present may be from 25% to 75% molar. During preparation, a degree of isomerization of *cis* maleate residues to *trans* fumarate occurs (Fig. 2.9), this being important to final curing, as copolymerization with styrene in particular is favoured with the *trans* configured fumarate. This isomerization can be virtually 100%, but varies with reaction mixture composition and reaction conditions. The post-acrylation method referred to above involves the preparation of a saturated polyester with excess hydroxyl groups which are esterified with acrylic acid as a final stage in the preparative process.

Preparation of unsaturated polyesters is straightforward and is carried out in a similar manner to saturated polyesters [57]. However, where the slower reacting iso and terephthalic acids are used rather then *o*-phthalic anhydride it is claimed that multistage processing, reacting the aromatic acid in a first stage with subsequent reaction of maleic anhydride, gives better performing products [62]. Since unsaturated polyesters are thinned in monomers capable of polymerizing thermally, care needs to be taken in this stage and as low a thinning temperature used as possible. For additional safety in thinning, and stability on storage, adding an inhibitor such

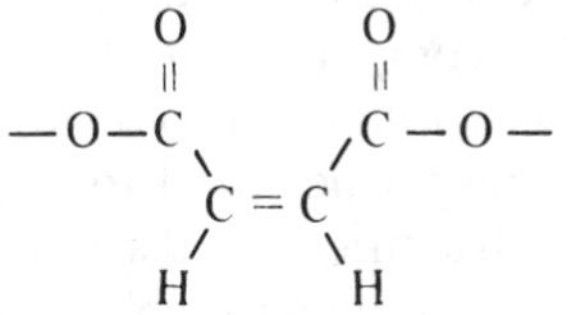

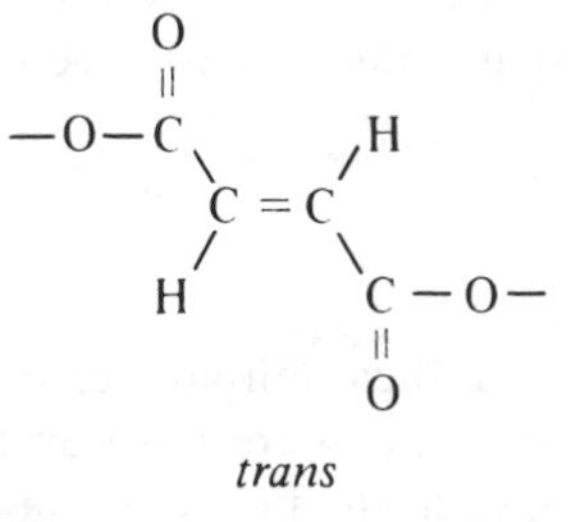

Fig. 2.9 — *Cis* and *trans* configurations of maleate and fumarate.

as *para-tert*-butyl catechol to the monomers in the thinning tank is necessary before resin dissolution. Alternatively, the resin may be allowed to solidify and the broken up solid subsequently dissolved in the monomer.

The monomer traditionally used is styrene, but vinyl toluene, methyl methacrylate, and some allyl ethers and esters such as diallyl phthalate are also used. The ultimate polymerization mechanism is most commonly a redox-initiated chain addition polymerization operating at ambient temperature, though thermal initiation is possible. The redox system is two-pack, where an organic peroxide or hydroperoxide is used as the second component, with the reducing component (amine or metal soap) mixed with the resin (see next section).

Oxygen is an inhibitor for styrene polymerizations, and exposed styrene thinned unsaturated polyester films suffer air inhibition which shows up as surface tack. This can be minimized by either the mechanical exclusion of oxygen by wax incorporation, or by the introduction into the composition of chemical groups which react with oxygen. Allyl ether groups have been used in such a role either through the addition of allylated monomers [63] or by reacting allylated materials such as allyl glycidyl ether or trimethyl propane diallyl ether into the backbone of the polyester [64]. (The allyl ether grouping can hydroperoxidize and then break down and oxidatively crosslink, in a similar manner to drying oils [65].)

Radiation curing unsaturated polyester compositions probably suffer less from air inhibition because the higher radical flux allows the reaction to proceed faster than the rate at which oxygen dissolves in the film. However, typical radiation curing compositions now use acrylated resins along with higher functionality and lower volatility monomers than styrene. The polyester resins used hence have acrylic acid attached at their end, replacing in-chain maleate/fumarate residues (see Section 2.19).

2.7   Acrylic polymers

Acrylic polymers are widely used for their excellent properties of clarity, strength, and chemical and weather resistance. The term acrylic has come to represent those polymers containing acrylate and methacrylate esters in their structure along with certain other vinyl unsaturated compounds. Both thermoplastic and thermosetting systems are possible, the latter formulated to include monomers possessing additional functional groups that can further react to give crosslinks following the formation of the initial polymer structure. Vinyl/acrylic polymerization is particularly versatile, in that possibilities are far wider than in condensation polymerization, of controlling polymer architecture and in introducing special features; for example, by using modification stages following the initial formation. Various references in the literature fully discuss the formulation and use of acrylic polymers in coatings [66, 67].

2.7.1   Radical polymerization

The kinetics and mechanism of radical-initiated chain addition polymerization [68] are important in all those processes where unsaturated vinyl monomers are used, either in the resin preparation or in the final cure mechanism. In resin preparation, processes discussed in this chapter include preparation of solution acrylic thermo-

Initiator breakdown $I:I \rightarrow I\cdot + I\cdot$
Initiation and propagation $I\cdot + M_n \rightarrow I(M)_n\cdot$
Termination $I(M)_n\cdot + \cdot(M)_mI \rightarrow I(M)_{m+n}I$ by combination
                                 $I(M)_n\cdot + \cdot(M)_mI \rightarrow I(M)_{n-1}(M-H) + I(M)_{m-1}(M+H)$
                                                                                                       by disproportionation
Transfer $I(M)_n\cdot + Polymer \rightarrow I(M)_nH + Polymer\cdot$ to Polymer
                                 $I(M)_n\cdot + RSH \quad \rightarrow I(M)_nH + RS\cdot$ to transfer agent
                                 $I(M)_n\cdot + Solvent \rightarrow I(M)_nH + Solvent\cdot$ to solvent where
                                                                                                      present

Fig. 2.10 — Representation of main reactions occurring in free radical chain addition polymerization.

plastic and thermosetting resins, emulsion and dispersion polymers, and the vinylation of alkyds. These reactions are also involved in the cure of unsaturated polyesters and of radiation-curable polymers. Autoxidative curing of oils and alkyds also occurs by a free radical mechanism, but not in the simple manner described below. Radical polymerization is the most widespread form of chain addition polymerization.

The main polymer-forming reaction is a chain propagation step which follows an initial initiation step (Fig. 2.10). A variety of chain transfer reactions are possible before chain growth ceases by a termination step.

Radicals produced by transfer, if sufficiently active, can initiate new polymer chains where a monomer is present which is readily polymerized. Radicals produced by so-called chain transfer agents are designed to initiate new polymer chains; these agents are introduced to control molecular weight and are usually low molecular weight mercaptans, e.g. primary octyl mercaptan.

Transfer to polymer is not normally a useful feature. At extremes such as with vinyl acetate polymerization, transfer at methyl groups on the acetate leads to grafted polymer side chains being present which are relatively easily hydrolysed, leading to undesirable structural breakdown of the polymer in use. Grafting to epoxy resin is used to prepare water-based can lacquers (see Section 2.16).

Transfer to solvent generally does not occur to a very significant extent and does not lead to radicals initiating new polymer chains. However, polymerization systems for high solids applications may be designed with, for example, alcohol solvent present where its chain-terminating effect can assist with molecular weight control. The nature of the solvents present can significantly affect the decomposition rates of most peroxy initiators.

Initiators (Table 2.6) used to give free radicals, are compounds typically breaking down under the influence of heat; two types predominate, those with an azo link ($-N=N-$) and those with a peroxy link ($-O-O-$). Initiator breakdown is characterized by 'half-life' which varies with temperature.

The 'half-life' of an initiator is the time within which 50% of the material has decomposed at a specified temperature. For many processes an initiator/temperature combination will be chosen so that the half-life under reaction conditions is in the region of 5–30 minutes. This ensures the steady generation of radicals at such a rate that the heat of reaction can be safely contained and high conversion of monomer to polymer ensured.

Despite the handling problems caused by its solid nature and poor solubility, azo-di-isobutyronitrile (AIBN) is often used for its non-grafting characteristics; di-

Table 2.6 — Typical initiators used in polymer preparation

Name	Formula	Half life 30 min
Azo-di-isobutyronitrile (AIBN)	CH$_3$—C—N=N—C—CH$_3$ with CH$_3$ and CN substituents	89 °C
t-Butyl-per-2-ethylhexanoate	t-BuOOCOCH(CH$_2$CH$_3$) (CH$_2$)$_3$CH$_3$	96 °C
Di-benzoyl peroxide	COOOOC (phenyl groups)	99 °C
t-Butyl perbenzoate	t-BuOOCO (phenyl group)	130 °C
Di-t-butyl peroxide	t-BuOOt-Bu	156 °C

benzoyl peroxide, though of only a slightly higher half-life temperature, conversely has a strong grafting tendency, a fact that may be deliberately exploited, for example to produce acrylic grafts onto other polymers such as epoxides. (Grafting is where the radical extracts a hydrogen atom from a site on the polymer chain, leaving a radical site onto which further polymer growth occurs.) This latter initiator is also a solid, normally supplied as a paste in plasticizer; for many uses it must first be dissolved and the solution assayed. It is therefore another less than ideal initiator. t-Butyl perbenzoate and t-butyl per-2-ethyl hexanoate are, however, liquid initiators which are more readily handled and may now be preferred. Many peroxide initiators produce acidic breakdown products that may have adverse effects in formulations. Dibenzoyl peroxide and perbenzoates can also leave benzene among the residues from side reactions.

It has been stated above that initiator decomposition may be affected by (solvent) environment. Reducing agents may in fact be deliberately added along with the initiator to produce *redox* (reduction–oxidation) systems where initiator decomposition is induced at far lower temperatures than would otherwise be the case. These systems have a number of specialist applications in both polymerization and curing reactions. Figure 2.11 shows the manner of decomposition of benzoyl peroxide by both thermal and amine-induced redox initiation routes to produce initiating benzoyl radicals. Redox initiation is frequently used in emulsion polymerization (see Section 2.8), and for cure of unsaturated polyesters.

2.7.2   Monomers and copolymerization

Monomers may be classified as 'hard', 'soft', or 'reactive', based on the properties they confer on the final polymer or copolymer. Hard monomers are, for example, methyl methacrylate, styrene, and vinyl acetate. The acrylates are 'softer' than methacrylates, and useful 'soft' monomers include ethyl acrylate and 2-ethyl hexyl acrylate, and also the long chain methacrylates. Reactive monomers may have hydroxy groups, for example hydroxy ethyl acrylate. Acrylamide and glycidyl methacrylate possess exploitable reactivity, the latter being particularly versatile. Acidic monomers such as methacrylic acid are also reactive and they are often included in small amounts so that the acid groups may enhance pigment dispersion and provide catalysis for cure of the derived polymer (Table 2.7).

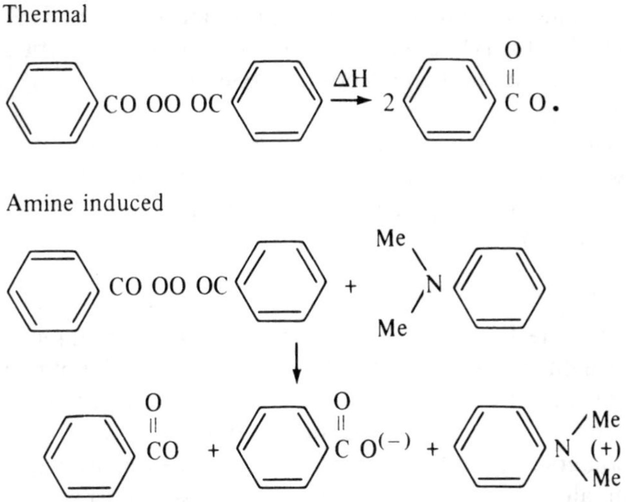

Thermal

Amine induced

Fig. 2.11 — Thermal and amine-induced decomposition of benzoyl peroxide.

Table 2.7 — Typical unsaturated monomers

Monomer	Structure	Polymer $T_g(°C)$
Butyl methacrylate	n Bu—O—C—C=CH$_2$ (O, CH$_3$)	22
Ethyl acrylate	CH_3—CH_2O—C—CH=CH$_2$	−22
2-Ethyl hexyl acrylate	$C_4 H_9$—CH—CH$_2$—O—C—CH=CH$_2$ (C_2H_5)	−70
2-Hydroxy propyl methacrylate	CH_3—CH—CH$_2$O—C—C=CH$_2$ (OH, CH$_3$)	76
Methacrylic acid	HO—C—C=CH$_2$ (CH$_3$)	210
Methyl methacrylate	CH_3—O—C—C=CH$_2$ (CH$_3$)	105
Styrene	CH_2=CH (phenyl)	100
Vinyl acetate	CH_2=CH O C CH$_3$	30

Methyl methacrylate as a hard monomer imparts resistance to petrol, UV resistance, and gloss retention. It is therefore used in copolymers for topcoats, particularly for automotive application. Butyl methacrylate is a softer monomer imparting excellent low-bake humidity resistance, but its plasticizing effect is limited. It gives good intercoat adhesion and solvent resistance and excellent UV resistance and gloss retention. Ethyl acrylate has good plasticizing properties but as a monomer has a highly unpleasant and toxic vapour. Its copolymers are fairly resistant to UV and give good gloss retention.

Practical coating polymer systems are rarely homopolymer but are copolymers of hard and soft monomers. Polymer hardness is characterized by glass transition temperature (T_g) and for any given copolymer the T_g of the copolymer may be estimated by the equation

$$1/Tg = W_1/Tg_1 + W_2/Tg_2 \text{ etc.}$$

where T_{g1}, etc. are the T_g's of homopolymers of the component monomers in K, and W_1, etc. their weight fraction present. (The calculated T_g will not be the final film T_g if the polymer is thermosetting since crosslinking will further raise the T_g, and this should be taken into account where appropriate.)

While monomers can combine during copolymerization in a variety of configurations (random, alternating, block, or graft), the vast majority of acrylic polymers actually used in coatings are random. This randomness also has the effect that the tacticity and crystallization phenomena so important in bulk polymer properties are generally not apparent or important in coating polymers. The most common structural effects actually experienced are phase separation and domain effects occurring either by accident or design.

The way different monomers react in copolymerization with other monomers depends on both their own structure and the nature of other monomers present [69]. Monomers of similar structure generally copolymerize readily and randomly. Thus all acrylates and methacrylates, including the parent acids, can be copolymerized together satisfactorily in almost any composition, though the longer chain monomers may be slower to polymerize and hinder attainment of complete conversion. Styrene can be included up to certain levels in acrylic polymers to reduce cost, and raise hardness and refractive index, but can be difficult to convert in higher concentrations. Maleic anhydride and maleate esters, though apparently able to copolymerize satisfactorily with other monomers when included in small amounts, do not homopolymerize but can form alternating structures with certain other monomers, notably styrene; they should be used with circumspection unless there is good understanding of this behaviour. Vinyl acetate and other vinyl esters are difficult to polymerize except in disperse systems.

When two monomers copolymerize, the rate may generally be slower than that of either polymerizing alone. The polymer radicals may react more readily with one monomer than another, and in fact four growth reactions can be envisaged, proceeding at different rates as shown by the following scheme:

$$R-M_1. + M_1 \xrightarrow{k_{11}} R-M_1.$$
$$R-M_1. + M_2 \xrightarrow{k_{12}} R-M_2.$$
$$R-M_2. + M_1 \xrightarrow{k_{21}} R-M_1.$$
$$R-M_2. + M_2 \xrightarrow{k_{22}} R-M_2.$$

where k_{11} and k_{22} are the reaction rate constants for chain ends with one monomer radical adding the same monomer, while k_{12} and k_{21} are for chain ends with one monomer radical adding the other monomer. The ratios k_{11}/k_{12} and k_{22}/k_{21} are referred to as the reactivity ratios, r_1 and r_2, and characterize the relative rates of reaction of the monomers with each other. Thus $r > 1$ indicates a polymer radical reacting more rapidly with its own monomer, and $r < 1$ indicates a polymer radical reacting selectively with the comonomer.

The product r_1r_2 has interest in that if, for example, $r_1r_2 = 1$ the monomers copolymerize truly randomly, while if r_1r_2 approaches zero the monomers trend strongly towards alternating in an equimolar manner. The kinetics of copolymerization has had considerable attention, and comprehensive tables of r_1 and r_2 values have been published [70]. Hence successful analytical and predictive modelling of simpler chain addition processes is now possible (see Section 20.5).

A more general scheme, which can be used to explore the copolymerisation behaviour of multi-monomer systems, is the Q–e scheme [71]. The Q parameter refers to the degree of resonance stabilization of the growing polymer chain, whereas the e value represents its polarity. Unfortunately, this scheme is rarely precise enough in practice, so its use is limited to gaining first approximations of reactivity. The recently disclosed Revised Patterns of Reactivity Scheme [72] is a four parameter scheme that is claimed to enable more reliable reactivity ratios to be calculated. Here parameters for monomer and radical reactivities and polarity are used.

Since monomers can polymerize thermally and by exposure to light, it is necessary to store monomers in opaque containers in cool conditions and to ensure that an adequate inhibitor level is maintained to prevent adventitious polymerization. Inhibitors are typically phenolic, and for some monomers checks on levels may be necessary on prolonged storage. Monomers also have toxicity ratings ranging from the mild to very severe, and many are highly flammable.

2.7.3   Formulation and preparation

When formulating acrylic resins, monomer composition may be determined by durability and cost requirements, and by functionality [66]. Hard and soft monomer ratios will be adjusted for T_g, which along with molecular weight may have a fairly rigid definition for the intended application of the coating. The major task of the formulator may be to achieve these and to obtain conversion. The major factors affecting molecular weight in preparation are the choice of the initiator and its concentration, temperature, and monomer concentration, and the concentration present of chain transfer species, if any. Normally polymerization will be carried out with up to 50% of solvent present and at reflux, the temperature thus defined by the solvents used. Initiator contents vary between 0.1 and 5% and polymerization temperatures for thermal initiator from 90 to 150 °C. The molecular weight (number average) is often found to vary inversely with the square root of initiator concentration at a given temperature.

Polymerization of acrylic monomers may be by bulk, suspension, emulsion, dispersion, or solution polymerization techniques; for surface coating applications solution, emulsion, and dispersion techniques are most common. The first of these is discussed below, and the others elsewhere in this chapter.

Solution polymerization has a number of possible modes, the simplest being the one-shot technique where solvent, monomer, and catalyst are heated together until conversion is achieved. However, vinyl polymerization is highly exothermic (50–70 kJ/mol) and this heat is generally removed in a refluxing system by the condenser. The one-shot process may give rise to dangerously high exotherm peaks. Thus processes where monomer and initiator, or initiator alone, are fed to the other refluxing reactants over 1–5 hours, moderate the reaction and the heat evolution rate, enabling the heat to be removed in a more controlled fashion. The process, where monomer and initiator are fed into refluxing solvent, also serves to give better control of molecular weight and molecular weight distribution, and is most favoured on technical as well as safety grounds. Mixed monomer and initiator can be fed together, or for ultimate safety, from separate vessels, provided good control of the two concurrent feeds can be achieved. While post-reaction thinning can be carried out, polymerization is always in the presence of at least 30–40% solvent so that the viscosity of the polymerizing mixture can be low enough to allow good mixing and heat transfer. Having a separator after the condenser in most processes is not essential, though useful to remove adventitious water. However, it is essential for resins where a condensation reaction proceeds concurrently with the addition polymerization, e.g. with some acrylamide-containing polymers.

The solvent present will depend on the exact nature of the acrylic polymer. Acrylic polymer solubility is affected by the nature of the side group, and shorter side chain polymers are relatively polar and require ketone, ester, or ether–alcohol solvents. As the side chain length increases, aromatic solvent can be introduced. In the choice of solvent composition it may also be necessary to consider the crosslinking method and any additional crosslinking agent which may be introduced, and any requirements dictated by the method of application.

Acrylic processes are usually followed by measuring the viscosity and solids of the reactor contents, principally to monitor conversion. In these processes no correction is possible once reaction has commenced, if molecular weight is found incorrect at any stage. For many formulations conversion does not continue unaided to completion after the end of the feed period, and one or more 'spikes' (or more preferably a feed) [73] of initiator will be necessary to complete the process. Accurate measurement of molecular weight may be by gel permeation chromatography (GPC) or may be deduced from reduced viscosity measurement in a U-tube viscometer using a dilute solution of the polymer in dichloroethane.

2.7.4 Thermoplastic acrylic resins

Thermoplastic acrylic resins have found application particularly for automotive topcoats both for factory application, and currently still for refinishing following accident damage repair. Molecular weight and molecular weight distribution need careful specification and control for lacquers, particularly with metallic formulations where rheological behaviour affects the orientation of the aluminium flakes in the paint film, so necessary for the flip tone effects required. Too high a molecular weight gives low spraying solids, and very high molecular weight 'tails' can result in cobwebbing on spraying; low molecular weight degrades film strength, mechanical properties, and durability. It has been found that the optimum molecular weight is in the region of 80000 for polymethyl methacrylate lacquers [74].

Polymethyl methacrylate homo- or copolymers for these applications can be prepared using a peroxide initiator in hydrocarbon/ketone blends, the reaction optionally being carried out at above atmospheric pressure by a one-shot process [75]. Polymethyl methacrylate films require plasticization to improve cold crack resistance, adhesion, and flexibility, and early compositions used such external plasticizers as butyl benzyl phthalate or a low molecular weight polyester. Phthalate plasticizers can be volatile, causing film embrittlement. Since problems of repair crazing and cracking are also found in externally plasticized films and post-polishing is necessary, further development became necessary. The use of cellulose acetate butyrate allowed the use of the bake–sand–bake process, but the most dramatic improvement was found when two-phase polymer films were designed [76]. These could be achieved, for example, by the blending of a methyl methacrylate/butyl acrylate (MMA/BA) copolymer with homopolymer polymethyl methacrylate solution when apparently clear but two-phase films with enhanced resistance to crazing resulted.

In lacquer formulations containing MMA, plasticizing monomers are introduced to tailor flexibility and hardness to application requirements. Also, small amounts of an acid monomer such as methacrylic acid may be included to introduce a degree of polarity to help pigment wetting and adhesion; for adhesion improvement amino monomers such as dimethyl amino methacrylate may alternatively be included, though these may introduce slight yellow coloration to otherwise colorless formulations.

Thermoplastic resins may be blended with nitrocellulose and plasticizing alkyd for use in low-bake automotive finishes, both for use by low volume motor manufacturers and for repair.

2.7.5 Thermosetting acrylic resins

Thermosetting acrylic resins were formulated to overcome the defects of thermoplastic compositions, particularly apparent in industrial applications. Advantages from thermosetting resins accrue in improved chemical and alkali resistance, higher application solids in cheaper solvents, and less softening at higher service temperatures. Thermosetting resins can be self-crosslinking or may require a co-reacting polymer or hardener to be blended with them; most fall into the latter category.

Normal thermosetting acrylics may have molecular weights of 10 000–30 000, with high solids formulations lower, and the hardeners and resins themselves will typically be polyfunctional (>2). The concentration of functional monomers used in the backbone will be in the range 5–25% by weight. The overriding principle in design of the complete curing system is that on crosslinking, an infinite network of crosslinks between polymer chains will be created. Thus, polymer of low molecular weight and low viscosity at the time of application is converted to infinite molecular weight and total insolubility after curing.

Hydroxy functional thermosetting acrylic resins are not self-reacting, but must be blended with, for example, nitrogen resins or blocked isocyanates for stoving applications or with isocyanate adducts for use as two-pack (2K) finishes for room temperature or low bake refinish paint application.

Hydroxy functionality is introduced by the incorporation of hydroxy alkyl (meth) acrylates such as hydroxy ethyl acrylate at up to 25% weight concentration in the total monomer blend [77]. A small percentage of an acidic monomer is usually introduced to give a final resin acid value of typically 5–10 in order to catalyse curing reactions with nitrogen resins. The constituents must be balanced to obtain the desired hardness, flexibility, and durability, and care must be taken to achieve the necessary compatibility and storage stability with the chosen amino resin. This type of resin along with a melamine formaldehyde (MF) resin is used, for example, in automotive topcoats and clearcoats, the ratio of acrylic to melamine/formaldehyde being typically between 80–20 and 70–30. The crosslinking reactions observed are those typical for amino resins with hydroxy-containing materials (see Section 2.10). Acrylic/MF compositions are superior to alkyd/MF for colour and exterior durability, but not necessarily for gloss. Acrylic/polyester/MF blends may be used for clearcoats. For repair of automotive topcoats in the factory, acid catalysts such as the half esters of dicarboxylic acid anhydrides or acid phosphates, are added to reduce the stoving temperature necessary, and acrylic/MFs are superior to alkyd/MFs under these conditions.

Hydroxy acrylic resins are generally prepared by the techniques previously described, typically in aromatic or ester solvent, feeding the monomer blend along with a peroxy initiator such as *t*-butyl perbenzoate into the refluxing solvent. Chain transfer agents may be used to achieve better molecular weight control.

Hydroxy acrylics can also be prepared by *in situ* hydroxylation of acid-containing polymers with ethylene or propylene oxides. Added flexibility can be given to the crosslinks by extending the hydroxyl group away from the acrylic backbone by reaction of existing hydroxy residues from hydroxy monomers with caprolactone.

The most distinctive class of self-crosslinking thermosetting compositions are those from acrylamide interpolymers reacted with formaldehyde and then alkoxylated [78,79]. The ultimate reactive group is the alkoxy methyl derivative of acrylamide, typically the butyl or iso-butyl ether. This group, which reacts in a near identical manner to the groups in nitrogen resins, also improves the compatibility of the polymer with alkyd and epoxy resins and nitrocellulose. These resins may hence also be blended with the acrylic resin, and co-reaction can then occur. Typical curing reactions are shown in Fig. 2.12.

Three methods of polymerization are possible in the manufacture of these acrylic polymers. The monomers, including the acrylamide, may be copolymerized and the amide groups subsequently reacted with formaldehyde and then alcohol. Concurrent reaction may be carried out when solvents, including alcohol, monomers, and formaldehyde, are charged with polymerization and methylolation taking place simultaneously. As a final possibility, *N*-alkoxy methyl acrylamide may be prepared as an intermediate prior to the polymerization stage. Formaldehyde in these compositions is most usually included in alcoholic rather than aqueous solution (e.g. 40% formaldehyde in butanol) or as paraformaldehyde. The amide/formaldehyde reaction may be acid or base catalysed. For the etherification reaction, it is necessary to remove water of reaction by azeotropic distillation. Excess of the alcohol, necessarily present for acrylamide solvation as well as being a reactant, should remain at the end as solvent to give stability to the resin. Polymerization may be initiated by a peroxide initiator, and a chain transfer agent will be present since the final polymer being crosslinking need only be of low molecular weight.

Polymer—CO—NH—CH$_2$ O Bu+HO—Polymer→

Polymer—CO NH CH$_2$ O—Polymer+BuOH

Polymer—CO—NH—CH$_2$ O Bu+BuO CH$_2$ NH CO—Polymer→

Polymer—CO NH CH$_2$ NH CO—Polymer+BuO CH$_2$ O Bu

Fig. 2.12 — Typical curing reactions of *N*-butoxymethyl acrylamide copolymers.

Thermosetting acrylamide polymers may contain styrene for alkali, detergent, and salt spray resistance, acrylate esters for flexibility, and acrylonitrile for improving toughness and solvent resistance. A wide range of compositions is therefore possible, for example, for domestic appliance, general industrial, or post-formed coil-coated strip application. An acid monomer is generally included to provide catalysis of curing. Epoxy addition has been used in appliance finishes to obtain extremely good detergent and stain resistance, and silicone modification can also be carried out to improve weathering performance. High AV variants are possible, which when amine-neutralized can be used in water-borne applications [80].

Carboxy functional thermosetting acrylic resins have been prepared which may be crosslinked with diepoxides. These were for a long time used for industrial finishes for appliances, coil coatings, and drums [81]. Their use has now been superseded in most of these applications by the hydroxy and acrylamide thermosetting acrylics.

Other functionalities may be introduced into acrylic resins (also emulsion polymers, section 2.8); for example, alkoxysilane functionality is provided by using trialkoxysilylalkyl methacrylate [82,83]. Coatings of this type find application in building exterior, automotive OEM and automotive refinishing. These resins cure through hydrolysis of the alkoxysilyl group, and exterior exposure properties are excellent. Acetoacetylated resins are made through copolymerization of acetoacetoxyethyl methacrylate [84,85]; a variety of subsequent curing reactions are possible, such as the Michael reaction shown in Fig. 2.13.

Acetoacetoxy ethyl methacrylate (AAEMA)

Trimethoxysilyl propyl methacrylate

Post-modification of acrylic resins is possible; for example acrylic resins with a high acid content may be condensed with the epoxy intermediate Cardura E10 (Shell Chemicals) to link plasticizing side chains to the polymer [32]. A reactive, though secondary hydroxyl group remains for possible crosslinking reactions. (The pre-prepared Cardura E10/unsaturated acid adducts may alternatively be used as monomers in preparation.) Saturated or unsaturated fatty acid modification has also been carried out on hydroxy functional acrylic resins for either plasticization purposes or to introduce air-drying functionality.

Water-soluble or dispersible acrylic resins will be prepared in a water-miscible solvent such as a glycol ether. They will be acid functional by the incorporation of, for example, acrylic acid; or amine functional by the incorporation of for example

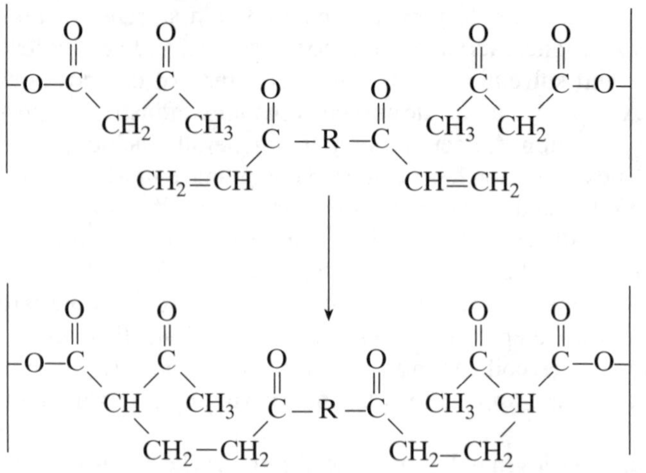

Fig. 2.13 — Michael addition reaction of resins containing incorporated AAEMA monomer.

dimethyl amino ethyl methacrylate, or glycidyl methacrylate which is reacted with amine (see Section 2.16).

2.8   Emulsion and dispersion polymers

This section will deal with fine particle addition polymers largely made by emulsion polymerization. Aqueous emulsion polymers are now probably the highest volume resins used by the coatings industry, principally because of the high usage of emulsion paints for home decoration; for this application the polymer is in the non-functional, room temperature coalescing, thermoplastic form. In parallel, other uses of more specialized emulsion formulations also exist, for example in industrial finishes where a hydroxy functional polymer is used, and in automotive and refinish basecoats where structurally engineered core/shell latexes are used [86,87].

It may be noted that high usage exists for emulsions in the adhesive and textile industries and emulsion polymerization is a route to poly (vinyl chloride) (PVC) and synthetic rubber polymer preparation.

The colloidal term 'emulsion' refers strictly to two-phase systems of immiscible liquids, where small droplets of one form the dispersed phase in the other, which is the continuous phase. In the terminology of the polymer industry, emulsion polymerization and emulsion polymer describe the process and end product of polymerizing addition monomers in water in the presence of surfactant, using water-soluble initiators to form fine-particle, stable dispersions. The term latex is also interchangeably used with emulsion for the final polymer dispersion. The polymer particles are typically sub-micrometre (0.1–0.5 µm) such that one litre of emulsion may contain 10^{16} individual particles of total surface area $2000 \, m^2$. The monomers and fundamental chemistry of polymer formation are those of acrylic polymerization described earlier.

Emulsion polymerization is a member of the polymerization class known as dispersion polymerization, of which the other major member used in surface coatings

is NAD polymerization. Certain other disperse polymer classes also find application in the coatings industry [88–90]. Latexes stabilized by non-ionic rather than
anionic surfactants may be mentioned as specialist emulsion polymers; the stabilizing groups may, for example, be polyethylene glycol chains [91, 92], providing
steric stabilization in a like manner to non-aqueous dispersions (see Section 2.9).
Polyurethane dispersions are mentioned later (Section 2.13.3).

While suspension polymerization produces particles too coarse for coatings
application, forms of micro suspension polymerization have been used to prepare
composite particles incorporating preformed polymer with additional polymer polymerized *in situ*, for use in coatings [93, 94].

A typical emulsion polymerization formulation will contain, besides water,
around 50% of monomers, blended for the required T_g, with surfactant, and often
colloid, initiator, and usually pH buffer and fungicide [95]. Hard monomers used in
emulsion polymerization may be vinyl acetate, methyl methacrylate, styrene, and
the gaseous vinyl chloride. Soft monomers used include butyl acrylate, 2-ethylhexyl
acrylate, Vinyl Versatate 10 or VeoVa 10 (Shell Chemicals), maleate esters, and the
gaseous monomers ethylene and vinylidene chloride (see Table 2.7 and Fig. 2.14).

The most suitable monomers are those with low, but not very low, water solubility; monomers of very low solubility can be difficult to use satisfactorily. To use any
of the gaseous monomers requires special plant; the techniques for handling these
are briefly mentioned at the end of this section. Other monomers may be included
in formulations, for example acids such as acrylic and methacrylic acids and
adhesion-promoting monomers. It is important that films coalesce as the diluent
evaporates, and the minimum film-forming temperature (MFFT) of the paint composition is a characteristic, closely related to T_g of the polymer. MFFT is also
affected by materials present such as surfactant and by inhomogeneity of polymer

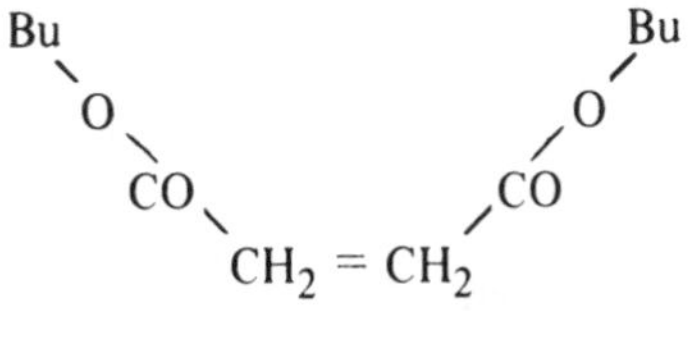

Dibutyl maleate

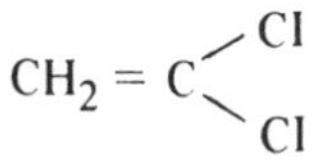

Vinylidene chloride

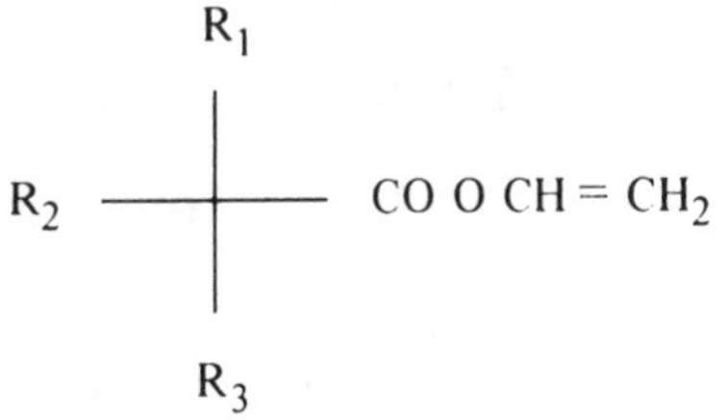

VeoVa 9 and 10 (Shell Chemicals)

where R_1, R_2, R_3 are alkyl,
one of which is methyl.

Fig. 2.14 — Monomers particularly used in emulsion polymerization.

composition at the particle surface [96]. Higher T_g polymers that would not otherwise coalesce at room temperature may be induced to do so by incorporating a transient plasticizer or 'coalescing agent' such as Texanol (Eastman) into the paint composition. MFFT is normally determined in preference to T_g for emulsion compositions, since it is difficult to allow for these deviating effects; for the usual decorative applications it may typically be in the range 0–20°C. Lower MFFTs are required with more highly pigmented and extended finishes.

Available surfactants are anionic, cationic, and non-ionic, and essential characteristics of these surface-active materials are that their molecules possess two dissimilar structural groups; one will be a water-soluble or hydrophilic group and the other a water-insoluble hydrophobic moiety (Fig 2.15). The composition, solubility properties, location, and relative sizes of the dissimilar groups in relation to the overall structure determine the surface activity of the compound. It will be noted that reactive stabilizers now find increased use both in general and in specific uses [97].

The role of surfactants [98] is conventionally considered to be firstly to provide a locus for the monomer to polymerize in surfactant micelles and, secondly, to stabilize the polymer particles as they form. Anionic and cationic surfactants function as electrostatic stabilizers, preventing coagulation by electrostatic repulsion arising from the charges located at the particle surface. Polymeric non-ionic stabilizers function through steric stabilization described in Section 2.9 (see also Fig. 2.18). When surfactants are dissolved in water at concentrations above a certain level, the molecules associate to form 'micelles' instead of being present in true solution; these micelles can solubilize monomer.

It is normal to use mixed anionic and non-ionic surfactants in emulsion polymerization; cationic surfactants are rarely used. The non-ionic nonyl phenol ethoxylates are being phased out on environmental toxicity grounds, with alkyl ethoxylates being found satisfactory replacements.

Anionic surfactants
Stearate soaps $\qquad$ $CH_3 \, C_{16} \, H_{32} \, COO^- \, Na^+$

Dodecyl benzene sulphonate $\qquad$ $CH_3 \, C_{11} \, H_{22} \, \langle\!\!\!\rangle\!\!-\!SO^-_3 \, Na^+$

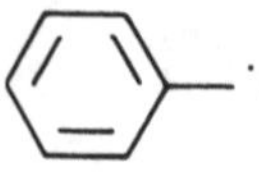

Sodium dioctyl sulpho succinate $\qquad$ $CH_3 \, C_7 \, H_{14} \, OOC \, CH_2$
$$CH_3 \, C_7 \, H_{14} \, OOC \, CH \, SO^-_3 \, Na^+$$

Cationic surfactant
Cetyl trimethyl ammonium bromide $\quad$ $(CH_3)_3 \, N^+ \, C_{15} \, H_{30} \, CH_3 \, Br^-$

Nonionic surfactants
Polyethoxylated nonyl phenol $\qquad$ $CH_3 \, C_8 \, H_{16} \, \langle\!\!\!\rangle\!\!\!\; O(C_2H_4O)_n \, H$

Polyethoxylated polypropylene glycol $\quad$ $HO(C_2H_4O)_n\!\!-\!\!(C_3H_6O)_m\!\!-\!\!(C_2H_4O)_p \, H$

Fig. 2.15 — Typical surfactants used in emulsion polymerization.

'Protective colloids' are water-soluble polymers such as poly(meth)acrylic acid or its copolymers, so-called polyvinyl alcohol (partly hydrolysed polyvinyl acetate), or substituted celluloses such as hydroxy ethyl cellulose. The properties of these colloids vary with molecular weight, degree of branching, and composition (proportion of water-soluble acid or hydroxy present). When used in emulsion polymerizations they may become grafted by growing chains of the polymer being formed, especially in the case of the celluloses, or may undergo chain scission. They assist in particle size control and in determining the rheology of the final paint, particularly in the derived gel structure and in the degree of thixotropy found.

A typical thermally initiated emulsion polymerization will involve two characteristic stages known as the seed stage and feed stage. In the seed stage, an aqueous charge of water, surfactant, and colloid will be raised to reaction temperature (80–90 °C) and 5–10% of the monomer mixture will be added along with a portion of the initiator (typically a water-soluble persulphate). In this stage the first polymer particles are formed. If redox initiation is used (typically water-soluble persulphate again, but with a reductant such as sodium metabisulphite), the temperature will be in the region of 50 °C or less.

The seed formulation contains monomer droplets stabilized by surfactant, initiator, and a small amount of monomer in solution, and surfactant, both in solution and micellar form. According to the micellar nucleation theory, radicals are formed in solution from initiator breakdown; these add monomer units dissolved in the water forming oligo-radicals, until the chain reaches such a size that the growing radical enters a micelle, where the monomer present within the micelle also polymerizes. Termination can occur when a further growing polymer radical enters the particle. Alternatively if coagulative nucleation is occurring, oligo-radicals coagulate together on their own to form nucleus particles which then become swollen with monomer and grow in like manner to micellar nuclii. In the seed stage, during which no reactants are added, initial monomer and initiator will largely be converted to polymer with the particle number roughly corresponding to the number of micelles of surfactant initially present (or until most of the surfactant present is mainly absorbed onto nuclear particles formed from coagulation processes). Monomer in droplets will diffuse through the aqueous phase into seed particles during this stage. At the end of the seed stage, monomer droplets will have disappeared. The concentration of surfactant remaining which is not associated with polymer particles will be small.

In the feed stage, remaining monomer and initiator solutions are fed together, the monomer being swollen into existing particles, or disolved in the aqueous phase. Droplets temporarily formed as neat monomer is fed in, quickly diffuse. Polymer formation proceeds as monomer in particles polymerizes, being replenished by monomer in solution. At the same time radicals enter the monomer-swollen particles, causing both termination and re-initiation of polymerization. As the particles grow, remaining surfactant from the water phase is absorbed onto the surface of the particles to stabilize the dispersion. In the overall process the entities shown in Fig. 2.16 are believed to be involved. Particles are stabilized from flocculation and coalescence by mutual repulsion of surface charges from the anionic surfactant.

The final stage of polymerization may include a further shot of initiator, probably redox, to complete conversion, followed by cooling and addition of biocide if required.

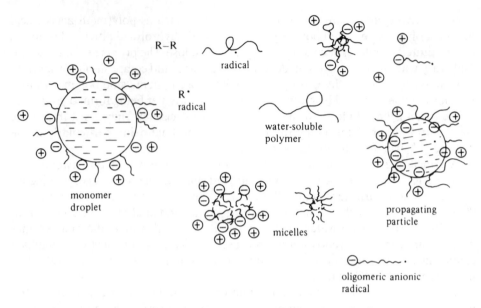

Fig. 2.16 — Entities involved in emulsion polymerization.

Various factors make emulsion polymerization more difficult to control than normal acrylic polymerizations. Agitation is critical in initial stages to get good disolution/emulsification of seed reactants, and in the later stages to get heat extraction through the reactor cooling surfaces and good incorporation of added monomer. At the same time formulations may be shear-sensitive, and excess agitation is to be avoided. Reflux in latex processes generally has an adverse effect, hence initial heating followed by progressive application of cooling is necessary to hold a steady temperature. Bit content and polymer buildup on reactor walls can cause problems.

The kinetics of emulsion polymerization are complex, and efforts to understand the overall mechanism have been extensive [99, 100]. With better understanding, particularly of the thermodynamics involved, techniques are now established to control the internal structure of particles in many ways, so that they may for example contain voids or interpenetrating networks, or have separately distinguishable core and shell structures [101]. Larger particles have also been made which include both pigment and air filled voids in order to provide improved paint opacity. Particles may be rubbery 'microgels' for mechanical property modification, or engineered for rheology control, as with solvent-borne microgels [102, 103]. Typical structures are shown in Fig. 2.17. They are generally prepared by changing the monomer feed composition during the course of the feed stage. A striking example of their use is in water-borne metallic basecoats for both automotive and refinish applications. The use of a crosslinked core, with a mantle of acrylic polyelectrolyte containing hydroxyl and carboxyl groups [86], gives paint with low viscosity when sprayed, but high viscosity once applied. This prevents sagging and ensures alignment of aluminium flakes.

Though it is possible to prepare emulsions with hard monomer only, when so-called external plasticizer must be added to the paint formulation, virtually all emul-

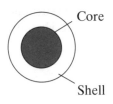

Core

Shell

Core/shell particle

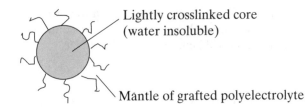

Lightly crosslinked core
(water insoluble)

Mantle of grafted polyelectrolyte

Aqueous acrylic microgel particle

Fig. 2.17 — Structured particles preparable by emulsion polymerization.

sions used nowadays are internally plasticized with copolymerized soft monomer. Styrene-containing emulsions find little use in coatings in the UK though they are used in continental Europe. General-purpose emulsions are often formulated with vinyl acetate plasticized with an acrylate such as butyl or 2-ethyl-hexyl acrylate, or a dialkyl maleate. Formulations of this type have good chalking resistance but poor alkali resistance and hydrolysis resistance; to improve these latter properties VeoVa 10 (Shell Chemicals) and analogues [104, 105] may be used as plasticizing monomer. All-acrylic formulations containing methyl methacrylate and an acrylate plasticizing monomer generally give higher quality emulsions except for poorer chalking performance. VeoVa 9 (Shell Chemicals) may upgrade water resistance of acrylic formulations.

In order to use gaseous monomers, equipment is needed that is capable of working under pressure; these monomers can be used to give cheaper emulsions by copolymerizing vinyl chloride and ethylene with vinyl acetate. Ethylene in particular cheapens formulations but is so soft that typically vinyl chloride is incorporated as a hard monomer along with vinyl acetate. Polymers containing vinyl chloride and vinylidene chloride find particular application for anti-corrosive primers. Vinyl chloride monomer is carcinogenic, and final emulsions must be carefully treated by steam stripping to remove all traces of free monomer from the product.

Emulsions are characterized by solids and viscosity, by MFFT [106] and by freeze–thaw stability. Film tests are additionally carried out to assess freedom from bits. pH may also need control particularly with vinyl acetate-containing emulsions in order to prevent hydrolysis.

As well as the established use of autoclaves for the preparation of 'pressure' polymers, a technique capable of handling gaseous as well as liquid monomers is the loop reactor [107, 108]. This is a form of continuous reactor whereby the reacting mixture is pumped around a heated/cooled loop, monomers and initiator being pumped in at one point in the loop, and at another point product 'overflows' from the loop at a similar rate to the incoming feeds. Claimed advantages include low installation, capital, and running costs, while disadvantages include inability to exploit, for example, techniques for making products with core/shell morphology.

The polymer composition may include acids to modify rheology; monomers to improve adhesion, or monomers to provide crosslinking opportunities may be included. For primers to adhere well to wood and to old gloss paint, ureido [109, 110] monomers are used. For room temperature crosslinking, methoxy silane [83] functional monomers are used. Acetoacetate functional monomers provide a number of reaction possibilities [85, 111].

Polymers for thermosetting applications will be formulated and prepared similarly to those described above, but will contain hydroxy or modified acrylamide
monomer and will generally be colloid-free.

2.9 Non-aqueous dispersion polymerization

Bearing many superficial similarities to emulsion polymers that exist in aqueous
media, the other type of polymer dispersion used in the coatings industry is the non-
aqueous dispersion (NAD). Here the polymer, normally acrylic, is dispersed in a
non-aqueous solvent, typically aliphatic hydrocarbon [112]. Acrylic polymer NADs
have found application chiefly in automotive systems and as modifiers in decorative
systems [113, 114], though a range of other applications has been described;
however, both condensation and addition polymer dispersions may be prepared,
with a variety of particle structures including layered, heterogeneous, and vesiculated [90].

NAD polymer dispersions may typically be submicrometre, of size 0.1–0.5 μm
similar to aqueous polymers, and the chain addition polymerization mechanisms are
again those earlier described under acrylic polymers. Unlike emulsion polymers,
however, where particles are stabilized by charge-repulsion mechanisms derived
from ionic surfactants, the only successful method available for stabilizing NADs is
steric stabilization involving the presence at the particle–medium interface of an
adsorbed solvated polymer layer. Charge stabilization has an inadequate effect in
the media used in NAD polymerization because of their dielectric constant, one or
two orders of magnitude lower than that for water. In steric stabilization, the repul-
sive forces that prevent flocculation when particles collide arise from the increase
in local solvated polymer concentration arising at the point of contact; the system
reacts to eliminate this local excess concentration by causing the particles to separate (Fig. 2.18). Steric stabilization is also effective in media of high dielectric constant where charge stabilization is normally used, this being the situation when
non-ionic stabilizers are used [91].

The most successful stabilizers used in dispersion polymerization have been
based on block or graft copolymers [115], and special techniques have been developed to prepare amphipathic graft copolymers suitable for use in NAD preparation
[116]. Preformed graft stabilizers based on poly-12-hydroxystearic acid (PHS) are
simple to prepare and effective in both addition and condensation NAD polymer-

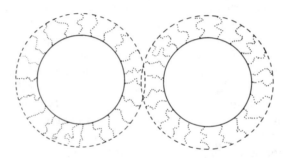

Fig. 2.18 — Representation of steric repulsion from soluble chains attached to dispersed
particles.

izations. PHS oligomer may be converted to a 'macromonomer' by reacting the carboxy group with glycidyl methacrylate (GMA). The 'macromonomer' is then, in turn, copolymerized with an equal weight of methyl methacrylate or a similar monomer to give a polymer of molecular weight 10–20000, which is a 'comb' stabilizer with five to ten soluble chains, pendant from a polymeric anchor backbone (Fig. 2.19).

The formation of the graft, and its method of stabilization of dispersed particles by adsorption of the insoluble anchor component, is shown diagrammatically (Fig. 2.19). Poly-12-hydroxystearic acid derivatives also gain application more generally as dispersion agents [12]. The special techniques to prepare the 'macromonomers' necessary for preparing graft dispersants of defined structure have been reviewed [117].

Virtually the full range of addition polymers may be used in NAD polymerization. It is of note that those found more difficult to polymerize in solution may be more readily polymerized in dispersion (as with aqueous emulsion polymerization). The major limitation for monomer composition is that the polymer produced shall be insoluble in the medium used. Thus, when in aliphatic media, the proportion of longer chain plasticizing monomers included, must be limited to satisfy this condition.

Excellent control of particle size is achieved in NAD preparations and it is possible to prepare dispersions of uniform particle size, far more so than with emulsion polymerization. Similarly, by using the control differently, different sizes of particles can be produced giving a more efficient packing arrangement enabling dispersions of high solids of up to 85% by volume to be obtained [118, 119].

NAD acrylic polymers found major application in automotive thermosetting polymers, and here hydroxy monomers were included in the monomer blend used. The polymer initially formed in dispersion may be blended with stronger solvent to dissolve a portion or all of the dispersed polymer. Using this, it is possible to

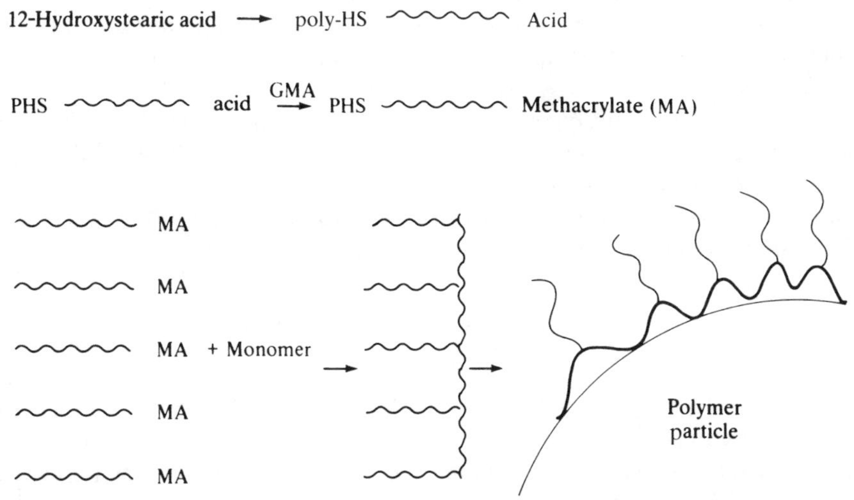

Fig. 2.19 — Schematic representation of poly-12-hydroxystearic acid 'comb' stabilizer preparation and use.

formulate particle compositions that range in state from the polymer being in true solution, through solvent swollen gel, to stable unswollen polymer particles. The reason for using this type of NAD is that the presence of swollen insoluble polymer has a profound effect on the evaporation rate of the solvents used and on the rate of increase of viscosity during evaporation. These effects give wide control over the formation of thicker coatings by spray application, with little sagging and good control of metallic pigment orientation. It was ultimately found most useful to prepare organic 'microgels' for blending with solution-prepared thermosetting hydroxy acrylic polymers to assist rheological control during application [102]. Monomers used include, for example, glycidyl methacrylate and methacrylic acid reacted *in situ* during the preparation, catalysed by added amine, such that the final particles become completely insoluble in organic solvents. These particles may be modified with an auxiliary polymer following the microgel core preparation, and finally diluted with a strong solvent blend. NADs continue in use in decorative systems as physical property modifiers [114] and for rheology control in high solids finishes.

2.10   Amino resins

Amino or nitrogen resins are the condensation products of certain compounds with two or more amine groups, particularly urea and melamine, with formaldehyde. These condensation products are generally alkylated and may also be partly polymerized. The family of acrylamide-derived thermosetting acrylic resins bear considerable resemblance to amino resins in both their formation and crosslinking reactions.

Nitrogen resins are used as hardening or crosslinking resins with a range of hydroxy functional resins in both stoving and two-pack room temperature curing systems; as such they are normally the minor component by weight. The special properties they confer are excellent colour and colour retention, hardness, and chemical resistance. Certain of the family of nitrogen resins are water-soluble. It may be noted that nitrogen resins find considerable use outside the coatings industry for bonding chipboard, as laminating adhesives and as a provider of hard laminate worksurfaces.

The major amino resin types are those derived from melamine (MF) and urea (UF), with benzoguanamine also used. Resins based on glycoluril [120] find some use for cure of powder coatings [121] (Fig. 2.20).

In all cases the first reaction in preparation is the condensation of formaldehyde with an amino group, as represented by:

$$\diagdown NH + HCHO \xrightarrow{\text{H}^+ \text{ or OH}^-} \diagdown NCH_2OH$$

With melamine two molecules of formaldehyde can add on to each amino group; with urea only one molecule condenses under normal circumstances. Basic conditions are used for this stage for both urea and melamine resins, though for melamine resins acidic conditions are also effective. In all cases the reaction is exothermic. The preparation of both dimethylol urea and the fully reacted hexa-methylol melamine is possible in a good state of purity.

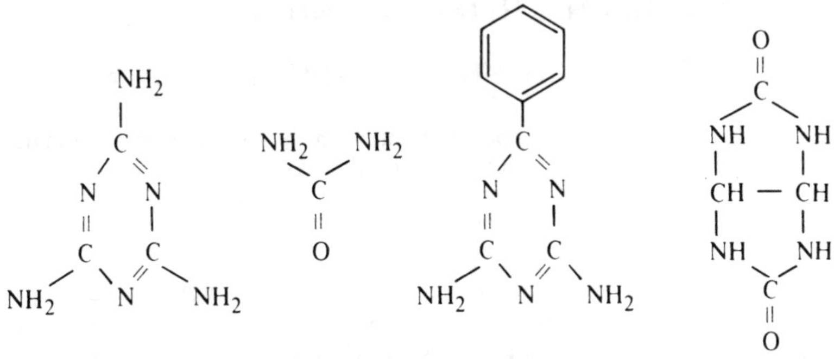

Fig. 2.20 — Typical amino compounds used in formaldehyde condensate resins.

The simple formaldehyde adducts that are water soluble, though finding applications in other industries, have no use in surface coatings. For coatings these adducts will be at least partly alkylated with lower alcohols, and this confers compatibility with other resins and allows solubility in organic solvents. The alcohols mainly used are propanol, normal or isobutanols, and, only in the case of melamine, methanol; the etherification reaction may be represented by:

$$\text{\textbackslash NCH}_2\text{OH} + \text{HO Alkyl} \rightarrow \text{\textbackslash NCH}_2\text{O Alkyl} + \text{H}_2\text{O}$$

The preparation of alkylated amino resins is normally three-stage, and pH adjustment may be carried out for each of the reaction stages. It is essential to have solvent process plant available to recycle and distil off solvent and to remove water of reaction, with the use of vacuum also necessary for some products. The detailed process will depend on the form of formaldehyde used which may be in either aqueous or alcoholic solution or as paraformaldehyde. Efficient removal of water (which is made more difficult by the presence of alcohol) is necessary to progress the esterification stage since the reaction is essentially an equilibrium. Good cooling of the condensate is necessary to allow effective separation of water from the solvent fraction. Employing a fractionating column can be advantageous, to aid the separation.

The first stage of reaction is the formaldehyde addition, which, if pH is favourable, will commence during heat-up. The melamine or urea will be dissolved in alcohol, and formaldehyde and some aromatic solvent may be present to modify the reflux temperature. This stage can be completed under total reflux. The second stage is the etherification stage during which water is removed. The third stage is the solvent distillation stage that may be completed using vacuum and is necessary to concentrate the product. With higher alcohols the resin is normally left in an alcohol-rich solvent blend in order to retain storage stability. Nevertheless storage life, especially in warmer conditions, may always be limited. Butylated MFs can be made with solids content of around 80%; above this reactivity loss is unacceptable for standard applications.

$$-NH\ CH_2\ OH + H_2\ N- \rightarrow -NH-CH_2- + H_2O$$

$$-NH\ CH_2\ OH + HO\ CH_2\ NH- \rightarrow -NH\ CH_2\ O\ CH_2\ NH- + H_2O$$

Fig. 2.21 — Reactions leading to polymerization in the preparation of formaldehyde condensation polymers.

The simple reaction scheme above is complicated in practice by a possible range of side reactions leading to a degree of polymerization (Fig. 2.21). Because of these and the impossibility generally of any meaningful in-process testing to control any particular stage, the key to successful nitrogen resin manufacture is strict control of pH, time and temperature regimes, distillation rates, and distillate quantities.

The degree of polymerization when it occurs is not high, though gelation in preparation is possible if a resin is misformulated. It should be noted that in polymerization reactions the secondary NH groups remaining after first additions onto primary amine groups are themselves reactive; with urea, for example, a mean functionality of three may be observed as the overall reactivity.

The effect of variation of the rates and extent of different reactions may show in a number of ways. For example, when excessive alkylation occurs at the expense of polymerization more water is formed, the tolerance of the final resin for mineral spirits will increase, the solids content will be high, and the viscosity will be decreased.

2.10.1   Characterization
Mineral spirits or petroleum ether tolerance, which indicates the degree of alkylation and the compatibility with other resins, is expressed as the volume compatibility to a given weight of resin, determined as the threshold at which haze develops in the blend. Solids are normally measured at a longer time and lower temperature than for other resins to obtain more reproducible and meaningful results; more than with any other resin, though, the figure obtained is not a true figure since a degree of decomposition and crosslinking is unavoidable during solids determination. Water contents are normally measured as an assurance of correct processing, and pH is characterized since this may affect storage stability.

2.10.2   Formulating
The condensation reactions involved, particularly the main formaldehyde addition and etherification reactions, are reversible, and so excess formaldehyde and alcohol concentrations are necessary to push equilibria towards the desired products. When formulating resins the ratio of formaldehyde to melamine or urea determines the degree of methylolation. With urea charging 2.4 mol of formaldehyde to attain the normally required 2 mol addition may be necessary; with melamine resins, where reactivities up to 6 are both possible and useful, a wide range of ratios may be used. The quantity of alcohol included, along with the processing conditions, then determines the degree of alkylation and polymerization, and the characteristics of the final resin. Again, a considerable excess of alcohol is normally required to be present, though its presence in the final resin helps storage stability. Reactivity of

—NH CH$_2$ O Bu + HO – Polymer→—NH CH$_2$ O Polymer + BuOH

—NH CH$_2$ O H + HO – Polymer→—NH CH$_2$ O Polymer + H$_2$O

—NH CH$_2$ O Bu +
        HOOC – Polymer→—NH CH$_2$ O CO Polymer + BuOH

Fig. 2.22 — Crosslinking reactions possible between formaldehyde condensate resins and other polymers.

butylated melamine resins, for example, increases with decreasing formaldehyde to melamine ratio, and with decreasing degree of etherification.

Unlike other resins, specifying with accuracy the ratio of reactants actually incorporated into the resin is not possible; rather, specifying the ratio of reactants charged initially may be more meaningful.

2.10.3 Curing

Amino resins react on heating both with themselves, and with any other resin present containing functional groups, which will normally be hydroxy but equally possibly amino or carboxyl. The reactions are similar to those shown for formation, involving condensation and elimination reactions, where water, alcohol, or formaldehyde may be the products evolved (Fig. 2.22); the reactions leading to polymerization in preparation also occur.

The factors affecting reactivity are complex [122, 123]. As with formation, the curing rate and the relative importance of the above reactions are affected strongly by pH as altered by the presence of (typically) acid catalysts [124–126].

2.10.4 Uses

Melamine formaldehyde resins, because of their greater functionality, have significantly different properties from urea formaldehyde resins, having better chemical resistance, better colour retention at elevated temperatures, better exterior durability, and shorter baking schedules when in combination with hydroxy resins. It should be noted that their reactivity may be less than that expected for steric reasons. They are also more expensive. UF resins are very satisfactory, however, for general-purpose industrial finishes, and are extensively used in acid-catalysed room temperature cured wood finishes. Benzoguanamine resins find use for container coatings.

While strong acid catalysis of blends containing UF resin proceeds satisfactorily, with MF resins it may lead to self-reaction rather than co-reaction. Consequently for low bake or room temperature curing with MF resin blends, where high acid catalyst levels may be imperative with alkyd resins, for example, it is sometimes advantageous to prepare a precondensate with the MF resin.

Alkylated melamine resins will be chosen by their reactivity where high reactivity may mean high viscosity and rapid curing, but low mineral spirit tolerance and so poorer compatibility with alkyd resins. While the generally used resins are butylated, other alcohols are used. Secondary and tertiary alcohol-modified resins are generally slower curing but may result in harder films.

Special mention must be made of hexamethoxymethyl melamine (HMMM) resins that have secured their own place in coating compositions. Though available as the pure material as a waxy solid, they are most widely used in slightly modified or condensed form when they are more easily handled viscous liquids. They are soluble in water and in all common organic solvents except aliphatic hydrocarbons, and are compatible with practically all resin media. They are usually available at 100% solids. Unlike other alkylated MFs they are considerably more stable. As with other alkylated MF their preparation requires a high alcohol excess during reaction to ensure complete reaction; the problems of preparation include the removal of methanol containing distillate.

The superior adhesion and flexibility obtainable may be attributed to the lower tendency to self-condense and to a more satisfactory film structure [122, 123]. There is, however, a sharper optimum to the amount of HMMM required in any composition for the best properties. The resin finds particular use in high solids and waterborne compositions. It should be noted that HMMM is also more effective in that there is only an 18% loss of solids on stoving compared with 29% with a butylated equivalent.

HMMM has the disadvantage of increasing surface tension in the paint, resulting in application and film defects, and mixed methyl/butyl ether resins overcome these defects. Adequate water miscibility is retained if the methyl/butyl balance is correct, while hydrolytic stability and water resistance are improved. These mixed ether resins are finding increasing application.

2.11   Phenol formaldehyde resins

In the early years of the coatings industry, only naturally occurring resins were available to enhance the properties of natural oils. Early in this century, the availability of the hard oil-soluble phenolic formaldehyde (PF) resin allowed a more scientific approach to varnish making. The development of heat-reactive, soluble phenolic resins in turn enabled the development of baking finishes with excellent solvent and corrosion resistance; these resins still find application alone or in blends with alkyd or epoxy resins for can coatings, and tank and drum linings [127]. However, they have in turn been supplanted as crosslinking resins by amino resins because of the latter's far superior colour for most applications.

Unmodified phenolics may be oil-insoluble or soluble and may be heat-reactive or non-reactive. The initial reaction with formaldehyde and the subsequent reaction under acid conditions are shown in Fig. 2.23. The dimer shown can lose additional water to form a resin. This kind of resin is a *Novolac*, and in simple form, made with unsubstituted phenol alone, finds little use in coatings.

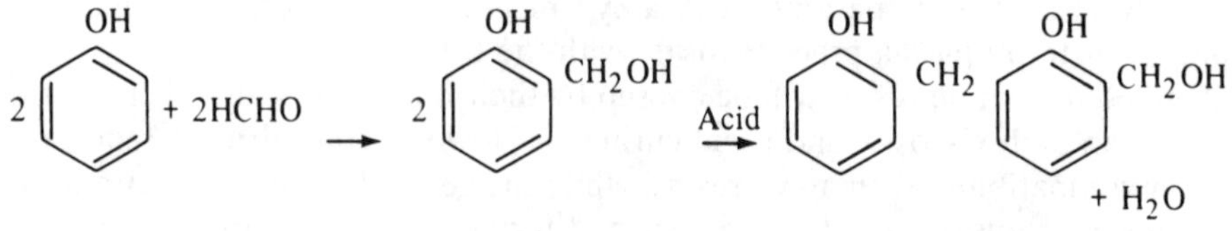

Fig. 2.23 — Novolac type phenolic resin formation.

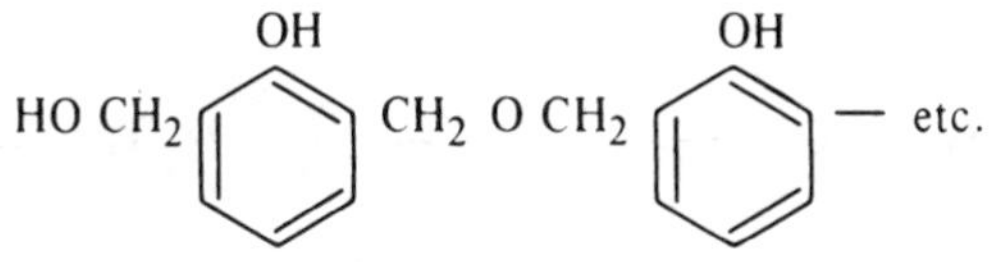

Fig. 2.24 — Resole type phenolic resin structure.

Alkali-catalysed phenolic resins are most common for coatings since these condense by ether formation and hence provide softer resins with some residual methylol groups for further reaction (Fig. 2.24). This class of phenolic resin is known as a *Resole*, and when made with phenol is only alcohol-soluble, but is thermo-hardening, unlike a Novolac. It can form a crosslinked structure without any additional curing agent.

Other phenols may be used including cresol, 2,4-xylenol, and the *para*-substituted phenols such as *p*-phenyl phenol, *p*-*tert* butyl phenol, and diphenylol propane.

The more useful PF resins of both classes are made with *p*-substituted phenols. Oil-soluble and oil-reactive resins are obtained with *p*-*t*-butyl and *p*-phenyl phenols, where acid-catalysed Novolacs are non-reactive, and alkaline-catalysed Resoles are oil-reactive. It is also possible to etherify methylol groups with alcohols such as butanol, so improving oil solubility.

Rosin-modified phenolics have found considerable application for oleoresinous vehicles, and are prepared by heating a Resole-type PF with rosin when combination of unsaturated double bonds in the rosin and hydroxyl groups of the Resole occurs. The product is then esterified with a polyhydric alcohol such as glycerol or pentaerythritol to reduce the acid value. This type of resin can be either dissolved or cooked into oil for varnish preparation.

2.12   Epoxy resins

The epoxide or oxirane group has many reactions useful in resin chemistry, in particular those with carboxyl, hydroxyl, phenol, and amine (Fig. 2.25).

These reactions, which do not require high temperatures, are exothermic (70–80 kJ/epoxy equivalent) and are often readily catalysed. They can be exploited in the assembly of polymers and in curing reactions; in certain circumstances novel multistage reaction routes can be devised. The amine–epoxide reaction is particularly used in the cure of epoxy resins (see Section 2.12.2). The reaction with dicarboxylic anhydrides, as well as occurring in the manufacture of alkyds containing Cardura E10 (Shell Chemicals), is also used for curing powder coatings; in this reaction, requiring both initiation and catalysis, polyesters are produced both rapidly and exothermically without any water of reaction being evolved [128]. Since hydroxyl groups may always be present when epoxy resins are being cured, even when other reactive groups predominate, it should be noted that conditions of temperature and catalysis may enhance or suppress the possible etherification reaction with hydroxyls in relation to the other epoxy group reactions taking place [129]. For example, in the acid–epoxy reaction, base catalysis suppresses the etherification reactions that would otherwise occur.

The compound glycidyl methacrylate that has both vinyl unsaturation and an epoxy group is particularly useful as a bridge between condensation and addition

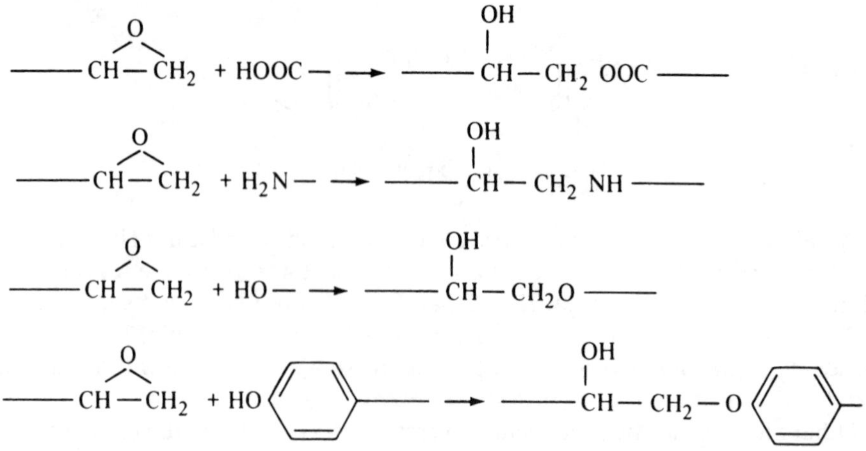

$$
\begin{array}{ccc}
\overset{O}{\overset{\triangle}{-\!\!-\!\!-CH-CH_2}} + HOOC- & \longrightarrow & -\!\!-\!\!-\overset{OH}{\underset{|}{CH}}-CH_2\, OOC-\!\!-\!\!- \\
\end{array}
$$

$$
\begin{array}{ccc}
\overset{O}{\overset{\triangle}{-CH-CH_2}} + H_2N- & \longrightarrow & -\overset{OH}{\underset{|}{CH}}-CH_2\, NH-\!\!-\!\!- \\
\end{array}
$$

$$
\begin{array}{ccc}
\overset{O}{\overset{\triangle}{-\!\!-\!\!-CH-CH_2}} + HO- & \longrightarrow & -\!\!-\!\!-\overset{OH}{\underset{|}{CH}}-CH_2O-\!\!-\!\!- \\
\end{array}
$$

$$
\overset{O}{\overset{\triangle}{-\!\!-\!\!-CH-CH_2}} + HO\text{—}\bigcirc\text{—} \longrightarrow -\!\!-\!\!-\overset{OH}{\underset{|}{CH}}-CH_2-O\text{—}\bigcirc\text{—}
$$

Fig. 2.25 — Typical reactions of the epoxide group.

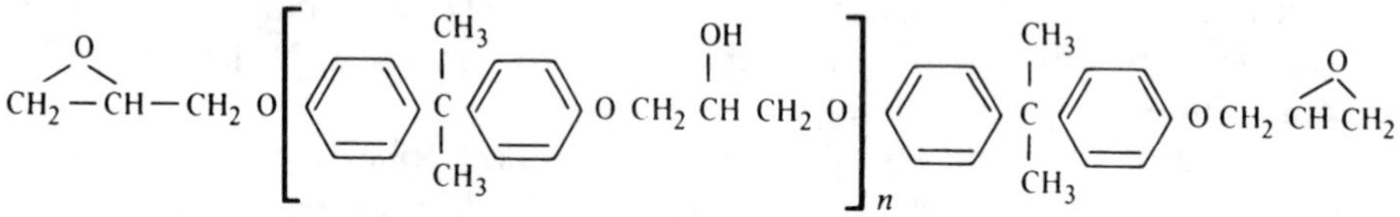

$$
\overset{O}{\overset{\triangle}{CH_2-CH-CH_2}}\, O\left[\bigcirc\,\overset{CH_3}{\underset{CH_3}{\overset{|}{\underset{|}{C}}}}\,\bigcirc\, O\, CH_2\,\overset{OH}{\underset{|}{CH}}\, CH_2\, O\right]_n \bigcirc\,\overset{CH_3}{\underset{CH_3}{\overset{|}{\underset{|}{C}}}}\,\bigcirc\, O\, CH_2\,\overset{O}{\overset{\triangle}{CH\, CH_2}}
$$

Fig. 2.26 — Generalized structure of bisphenol epoxide resins.

stages in polymerizations. Possibilities occur to react either group initially, the second then reacting in a subsequent stage.

The best-known epoxide containing materials are the range of preformed epoxy resins based on the reaction between diphenylol propane (bisphenol A) and epichlorhydrin [130]; their general structure is shown in Fig. 2.26.

Where n approaches 0, the resin approximates to the diglycidyl ether of diphenylol propane, and the product is liquid. As n increases from 2 up to about 13, solid forms with increasing melting points are encountered (Table 2.8). The lower-melting grades are often modified by pre-reaction before use in coatings, though the higher melting grades can be used in unmodified form for can or drum lining applications.

Because of problems including handling toxic epichlorhydrin, it is normal for coatings manufacturers to purchase epoxy resins of this type; it is, however, now common practice to 'chain-extend' liquid grades to higher molecular weight by reaction with diphenylol propane [131, 132]. Precatalysed liquid epoxy resin is available for use with this technique, which can reduce stocking of a range of grades and can provide access to intermediate solid grades; cost savings are also possible.

The formula above would indicate that all epoxy resins possess two terminal epoxide groups and all, except where $n = 0$, possess in-chain hydroxy groups. Higher molecular weight epoxy resins frequently depart from this linear structure owing to the incidence of side reactions in their preparation.

Considerable care is now taken in epoxy resin manufacture to eliminate free epichlorhydrin from the product because of its carcinogenic nature. Low molecular weight di-epoxides may also be carcinogenic, and consequently the available range

Table 2.8 — Typical grades of bisphenol epoxide resins and properties

Number of repeat units (n)	Melting point (°C)	Epoxide equivalent
0.5	Viscous liquid	225–290
2	64–76	450–525
4	95–105	850–950
9	125–132	1650–2050
12	140–155	2400–4000

of low molecular weight epoxies is now limited, with certain low molecular weight aliphatic di-epoxides now no longer available.

The outstanding properties of cured epoxy resins may be explained by their structure. The very stable carbon–carbon and ether links in the backbone contribute to chemical resistance, while a factor in their toughness is the wide spacing between the reactive epoxide groups and in turn the hydroxyl groups. The polar hydroxy groups, some of which may always remain, also help adhesion by hydrogen bonding. The aromatic ring structure enhances thermal stability and rigidity. Though these properties are attractive, aromatic epoxy resins yellow, and for this reason their major application is in primer and undercoat compositions where adhesion and corrosion resistance are particularly valuable. They are also used in can coatings for their good one-coat performance.

Epoxy resins may be formulated from phenols other than bisphenol A, for example diphenylol methane (bisphenol F). Epoxy-type materials may also be glycidyl ethers of other resins such as PF novolacs [133].

Epoxide compounds are characterized by melting point and by their epoxide group content or epoxy equivalent which may be expressed in a number of ways.

2.12.1   Epoxy esters

Both the terminal epoxide groups, and the secondary hydroxy groups of solid epoxy resins, can be reacted with fatty acids to produce the so-called epoxy ester. In esterification reactions each epoxide group is equivalent to two hydroxy groups. Epoxy esters are prepared by heating the fatty acid and epoxy resin in an inert atmosphere, preferably under azeotropic conditions to remove water of reaction, with temperatures of between 240 and 260 °C normally being used. The reaction may optionally be accelerated by the addition of, for example, calcium or zinc soaps at 0.1 or 0.2% weight on total charge.

Typically the epoxy resin where $n = 4$ is used, and fatty acid content chosen to esterify between 40 and 80% of the available groups including hydroxyl. Medium (50 ~ 70% modified) and long (over 70%) oil epoxy esters of drying oil fatty acids are used in air-drying brushing finishes. Short (30–50%) oil-drying or non-drying fatty acid esters are used in industrial stoving primers and finishes. Stoved epoxy resin ester films, especially when cured with MF resin, are harder and of superior adhesion, flexibility, and chemical resistance than similar alkyd/MF formulations. Increased fatty acid content, as would be expected, imparts better aliphatic solubility, better exterior durability, and decreased hardness, gloss, and chemical resistance. Linseed, tall, and DCO esters are most usual, though all fatty acids and rosin can be used.

2.12.2   Other epoxy applications

Epoxy resins may be cooked into alkyd formulations, replacing part of the polyol. They also find a place unmodified as a third component in alkyd/MF compositions to upgrade the resistance properties of the films.

Epoxy resins react with PF resins to form insoluble coatings, and well-formulated high molecular weight epoxy/PF coatings meet the highest standards of chemical resistance. These products are suitable for the linings of food cans and collapsible tubes, coatings for steel and aluminium containers, and wire enamels. Curing probably involves the formation of polyether links between the hydroxyl groups of the epoxy, and methylol groups present in PF resins of the resole type; the epoxy also reacts with phenolic hydroxyl groups on the PF. With some PF resins, compatibility problems on cold blending may be solved by pre-condensation, involving refluxing the epoxy and PF resins together in solution, when some reactive groups combine, leaving the remainder free to react in the curing process.

Urea/formaldehyde resins or melamine/formaldehyde resins may be used to cure epoxy resins, giving stoved films of paler colour, but with a reduced level of chemical resistance compared with phenol/formaldehyde resins. Again, the higher molecular weight epoxy resins are preferred.

Two-pack epoxy/isocyanate finishes require separate solutions of high molecular weight epoxy resin and polyisocyanate adduct as the two components; the epoxy resin must be in alcohol-free solvent since the curing reaction is predominantly with in-chain hydroxyl groups on the epoxy resin. One-pack finishes can be formulated with blocked isocyanates.

Epoxy resins may also be used in two-pack compositions with polyamines or polyamides [134]. The films obtained possess outstanding chemical resistance, hardness, abrasion resistance, flexibility, and adhesion. Low molecular weight solid epoxy resins are used most. Though primary or secondary amines such as triethylenetetramine may be used, in order to avoid the toxic hazards involved in handling amines, amine adducts with low molecular weight solid epoxy resin will nowadays be used as hardeners. These adducts are prepared by the reaction of excess of an amine such as diethylene triamine with an epoxy resin to produce fully amine-terminated adduct. Reactive polyamide resins formed from dimerized fatty acids with diamine are also used.

Reactive coal tar pitches may be incorporated into an epoxy resin base for curing with amine, amine adduct, or with a polyamide resin. The derived coatings have excellent chemical resistance and are not brittle; they hence find use as pipeline, tank and marine coatings.

Epoxy resin modification with a silicone resin is possible to enhance water resistance; epoxy/silicone combinations are used in blends with other polymers with the cure mechanisms mentioned above.

2.13   Isocyanates

The isocyanate group (Table 2.9) is highly reactive with itself, and a number of other groups, and may be catalysed in its reactions. It reacts exothermically (40 kJ/mol) with many groups with active hydrogen atoms, in particular alcohols, amines, phenols, and water at room or moderately elevated temperatures [135]. Amine groups are most reactive, so that hydroxy groups are the most common reactive

Table 2.9 — Typical reactions and reaction conditions for isocyanates

Reaction	Temperature (°C)	Catalysts
RNCO + Alkyl OH → RNHCOO Alkyl	25–50	Varied
RNCO + Alkyl NH$_2$ → RNHCONH Alkyl	10–35	Not required
RNCO + Aryl OH → RNHCONH Aryl	50–75	Tertiary amine

groups on resins cured by isocyanates. However amine groups are used with, for example, cathodic electrocoat resins cured with blocked isocyanates. Secondary alcohols and secondary amines react in a similarly way to primary but with less vigour. Both aromatic and aliphatic polyisocyanates are available, with the former generally more reactive than the latter.

Water reacts with the isocyanate group at similar rates to secondary alcohols; the initial product is an unstable carbamic acid (RNHCOOH) which decomposes to a primary amine and carbon dioxide, and this amine group is of course then available to react with additional isocyanate. This reaction is normally to be avoided in coating formulations, though it is exploited in the production of solid and flexible foams. For this reason, systems to be reacted with isocyanates should be dehydrated, and low moisture content urethane grade solvents used.

Isocyanates also react with themselves under certain conditions and in the presence of certain catalysts [135]; these reactions are generally undesirable. Tin compounds are strong catalysts for isocyanate reactions and are also generally 'clean' in not promoting side reactions. It will be evident that all reaction products of isocyanates contain active hydrogen groups attached to the nitrogen atom of the urethane or substituted urea reaction product; hence, steric factors permitting, reaction can continue with these products if additional isocyanate is present, to give allophanates and biurets. All of the groups mentioned above may be present on simple molecules or attached to polymer molecules.

Low molecular weight diisocyanates are used in coatings manufacture in preparing urethane alkyds and blocked isocyanate crosslinkers. Toluene diisocyanate (TDI) is the main diisocyanate used for urethane alkyds (normally the mixed 2,4 and 2,6 isomers). Products derived from aromatic isocyanates such as TDI and diphenyl methane diisocyanate (MDI) tend to yellow; hence aliphatic isocyanates such as hexamethylene diisocyanate (HDI) and isophorone diisocyanate (IPDI) now find higher usage (Fig. 2.27). Trimethyl hexamethylene diisocyanate (TMDI) and hydrogenated diphenyl methane diisocyanate (H$_{12}$MDI), not illustrated, also find application. Great care is necessary in the handling of all volatile isocyanates since their vapours are irritant and they are sensitizers at extremely low concentrations.

2.13.1   Isocyanate adducts and blocked isocyanates
Owing to the toxicity of low molecular weight (volatile) isocyanates, polyfunctional (3+) isocyanate adducts of higher molecular weight are invariably used in two-pack applications of isocyanates. Three different adduct types have been used, namely the triol adduct, the biuret and the isocyanurate. An example of the first is the reaction product of 3 mol of TDI with 1 mol of trimethylol propane. The biuret is

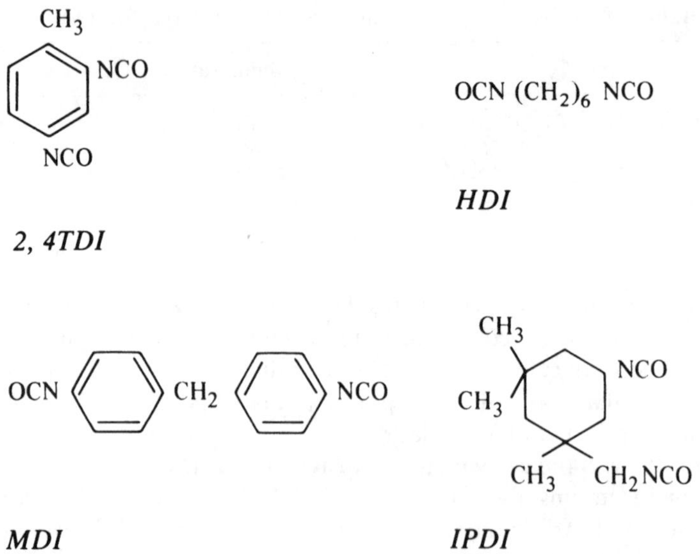

CH$_3$

NCO

NCO

OCN (CH$_2$)$_6$ NCO

HDI

2, 4TDI

OCN ⟨ ⟩ CH$_2$ ⟨ ⟩ NCO

CH$_3$

NCO

CH$_3$

CH$_3$ CH$_2$NCO

MDI

IPDI

Fig. 2.27 — Typical diisocyanates used in coatings resins.

prepared by reaction of 3 mol of diisocyanate with 1 mol of water, and the isocyanurate is the trimer formed by the reaction together of three isocyanate groups. Isocyanurates from HDI and IPDI are those most frequently used in 2K refinish paints. Isocyanurates may also be mixed with uretdione (isocyanate dimer). The structures of the HDI biuret, uretdione, and isocyanurate are illustrated in Fig. 2.28.

The reaction of the isocyanate group with low molecular weight alcohols, phenols, and certain other compounds such as lactams, may be considered reversible. If an isocyanate that has been 'blocked' or masked in this way is heated with a polymer with reactive groups, the low molecular weight blocking material is released. The urethane or substituted urea bond is then remade with the reactive group on the polymer. Whether free isocyanate groups are fully released in the reaction with a polymer is a matter of conjecture. Blocked isocyanate curing agents are used in automotive and various other solvent applied finishes, and powder coatings. Depending on the blocking agent, curing temperatures may be between 100 and 175 °C [136–138] Automotive cathodic electrocoat formulations cure using blocked isocyanate; for powder coatings caprolactam blocked IPDI based crosslinker is used.

2.13.2   Polyurethane coatings

Two-pack (2K) finishes based on hydroxy functional resins and isocyanate adducts find application for tough high solvent-resistant coatings curing at atmospheric temperature or under moderate stoving conditions. Resins that may be used are alkyd, polyester, polyether, epoxy, and acrylic. Alkyds used are frequently castor oil-based. Two-pack alkyd finishes are used for wood finishing. Two-pack polyester finishes find application for high-durability transport finishes including marine and aircraft and by automotive manufacturers for painting plastic parts fitted after the main metal body has been painted and stoved at high temperature.

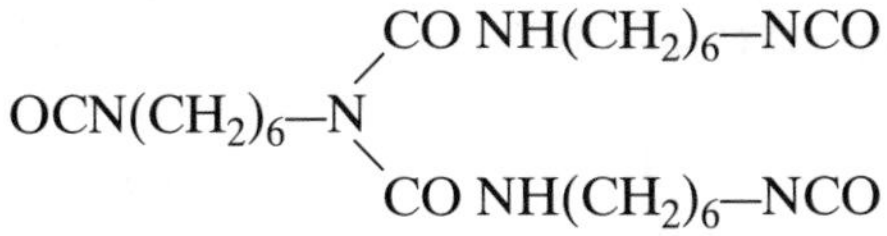

$(CH_2)_6$ NCO

HDI biuret

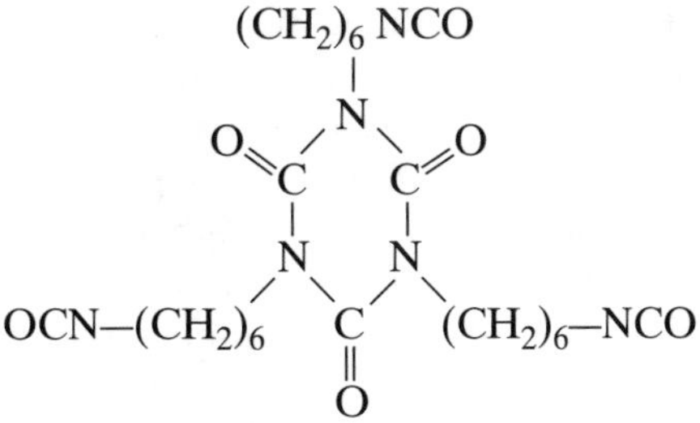

HDI isocyanurate

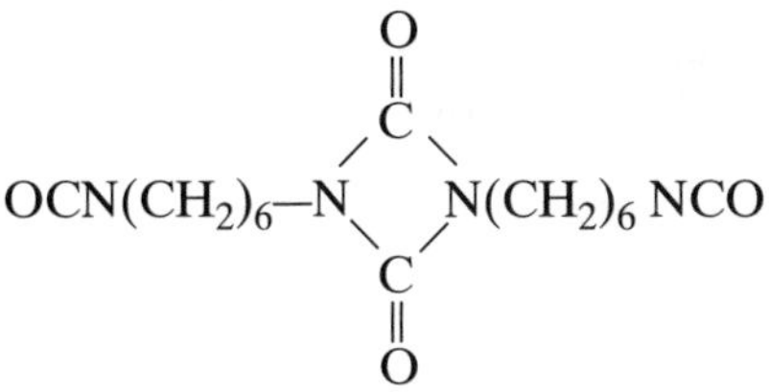

HDI uretdione

Fig. 2.28 — HDI biuret HDI isocyanurate HDI uretdione.

Two-pack systems with hydroxy functional acrylic and polyester resins (2K) have high usage in car refinishing, and in most European countries this type of finish now predominates. However, because of the toxicity of isocyanates and the danger of inhaling droplets while spraying, producers have in the past striven to develop systems avoiding the use of isocyanate, but with similar properties. For these, the isocyanate component could be replaced by a crosslinker such as a melamine/formaldehyde resin. Masked isocyanate polyurethane (PU) curing systems, sometimes mixed curing with MF, dominate in automotive OEM primers, and both solvent and water-borne systems are in use [139].

Two-pack systems such as the hydroxy polymer/isocyanate one may be difficult to formulate because of 'pot-life' difficulties, where conflict occurs between the need to accelerate curing but to retard bulk solidification to protect the application equipment. A development has been to accelerate curing of this type of system after application, by exposure to an atmosphere containing volatile amine catalyst (vapour phase curing) [135, 140].

Moisture-curing products may also be used, based on stable isocyanate terminated polymers obtained from the reaction of excess low molecular weight di- or polyisocyanate with a polyfunctional hydroxyl-terminated polyether. The pigmentation of moisture-curing finishes is difficult, though dehydrating agents are available to assist; the most frequent use of this type of product is for clear wood varnishes.

A special class of PU block copolymers incorporating both hydrophobic and hydrophilic segments is used in aqueous systems as one of the group of additives known as associative thickeners. HEUR (hydrophobically modified ethylene oxide urethane) associative thickeners consist of polyethylene glycol blocks linked via urethane segments. The end blocks are hydrophobic long aliphatic chains.

2.13.3   Polyurethane dispersions

Fully reacted, aqueous polyurethane dispersions have long been available, mainly used in applications other than traditional coatings, but their use in coatings is now increasing. They are typically prepared from DMPA (see Section 2.5.7), polyester-polyol and an excess of aliphatic isocyanate. These are chain extended with short chain diamines and solubilised with t-amine. [141, 142]. Crosslinking is possible with added aziridine or carbodiimide crosslinkers, autoxidatively by fatty acid modification or even by radiation curing via attachment of acrylate.

A recent development is of 2K reactive PU systems, where HDI adduct is homogenized into a pre-prepared aqueous millbase, and the hydroxy functional co-reactants can function as emulsifier for the isocyanate component. These protect the isocyanate groups from reacting with water, and while some reaction with hydroxyls does commence in the droplets, pot life extends to several hours.

2.14   Silicone resins

Silicone–oxygen and silicon–carbon bonds are particularly stable, and this has a beneficial influence on the behaviour of the semi-organic silicone resins, so that they are exceptionally resistant to thermal decomposition and oxidation.

For the surface coatings formulator, a range of reactive silicone resins is available for use in the preparation of silicone-modified polymers, and these may be either hydroxy or methoxy functional. Typical structures are shown in Fig. 2.29.

R may be phenyl or methyl, and both alkyl and aryl organo-silicones are currently available for resin modification. Reaction into a resin structure occurs through its available hydroxyl groups, when either water or methanol will be eliminated (Fig. 2.30); catalysis of these reactions is possible but not essential.

Polymers that may be modified include alkyd, polyester, acrylic, and epoxy. Silicone modification may typically be from 15% to 40%, though higher levels of modification of alkyds, for example, are possible for special heat-resistant applica-

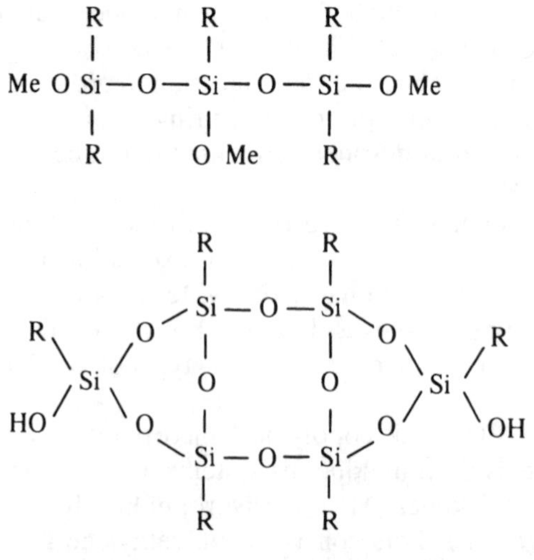

Fig. 2.29 — Typical silicone resins for polymer modification.

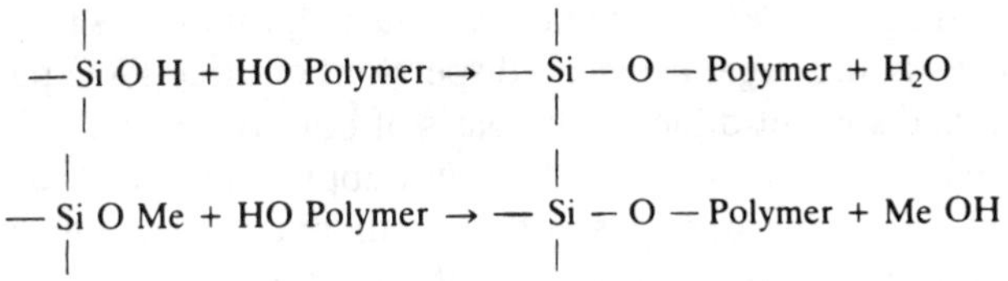

$$\begin{array}{ccc}
| & & | \\
-\text{Si O H} + \text{HO Polymer} \rightarrow & -\text{Si} - \text{O} - \text{Polymer} + \text{H}_2\text{O} \\
| & & | \\
\end{array}$$

$$\begin{array}{ccc}
| & & | \\
-\text{Si O Me} + \text{HO Polymer} \rightarrow & -\text{Si} - \text{O} - \text{Polymer} + \text{Me OH} \\
| & & | \\
\end{array}$$

Fig. 2.30 — Silicone resin reactions with hydroxy polymer.

tions. The single largest use of silicone-modified resins is for coated steel coil for building structure facing. Silicone modification always considerably enhances durability (see Sections 2.5.8, 2.6.3, and 2.16).

Pure silicone surface coatings are available for special application where very high heat resistance properties are required. These resins cure by the same mechanisms as for modification, but here will be catalysed:

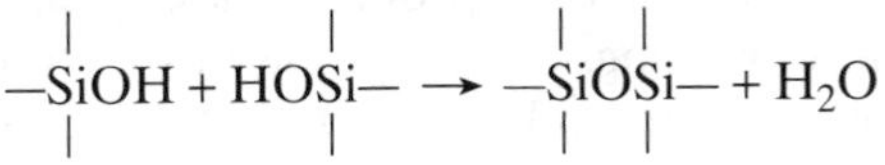

$$\begin{array}{ccc}
| & | & | \quad | \\
-\text{SiOH} + \text{HOSi}- & \rightarrow & -\text{SiOSi}- + \text{H}_2\text{O} \\
| & | & | \quad |
\end{array}$$

Alkoxy–silyl functional acrylic resins and emulsions were mentioned in Sections 2.7.5 and 2.8 respectively. These are moisture curing resins. Non-reactive linear silicones, e.g. poly dimethyl siloxanes of low viscosity, find frequent use in very low concentrations as flow control or marr aids in paint formulations:

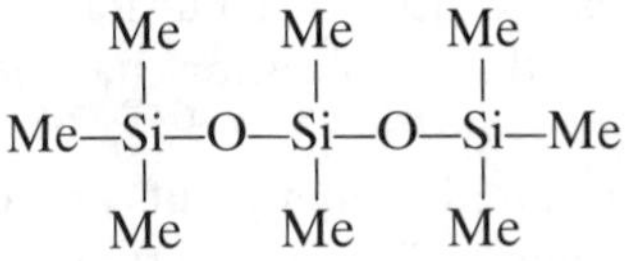

$$\begin{array}{ccccc}
\text{Me} & & \text{Me} & & \text{Me} \\
| & & | & & | \\
\text{Me}-\text{Si}-\text{O}-\text{Si}-\text{O}-\text{Si}-\text{Me} \\
| & & | & & | \\
\text{Me} & & \text{Me} & & \text{Me}
\end{array}$$

These particular silicones are generally used in solvent based coatings. For water based systems, a range of block copolymers of poly dimethyl siloxane with polyalkylene oxides are available. Both poly ethylene oxide and poly ethylene oxide/propylene oxide di-block copolymers are used.

2.15 Vinyl resins

The term 'vinyl resin' commonly refers to polymers and copolymers of vinyl chloride, though the term has more general meaning. Vinyl chloride is a cheap monomer whose polymers possess good colour, flexibility, and chemical resistance. Vinyl chloride is, however, gaseous and a recognized carcinogen; hence, though it is readily polymerized by emulsion or suspension polymerization techniques in an autoclave, the polymer will be purchased by most surface-coatings manufacturers. For this reason polymer preparation is not further discussed. The polymers are nevertheless quite widely used in coatings.

Both homopolymers, and copolymers usually with vinyl acetate, are used; small amounts of an acidic monomer may also be present. A serious problem with vinyl chloride-containing polymers, if heated, is dehydrochlorination, and stabilization is often necessary with, for example, organo-tin compounds or group 2 metal carboxylates.

A major advantage of PVC connected with its polar nature is its ability to accept plasticizer. Plastisols and organosols are dispersions of dried PVC particles in either plasticizer or mixed solvent/diluent. The basis of both types of coating composition is that the particulate dispersion is stable until applied and heated, when the plasticizer or solvent swells the particles, softening them and allowing coalescence; this is resisted at room temperature because of the crystalline nature of the PVC particles. With plastisols, plasticizer is retained in the coating after coalescence; with organosols the solvent and diluent are lost through evaporation.

The formulation of an organosol requires much care and a fine balance between diluent and solvent. If the mixture is too rich in diluent, the particles flocculate and give a high-viscosity dispersion; if too rich in solvent the viscosity is too high, this time owing to solvent swelling of the particles.

Polyvinyl acetate/vinyl chloride copolymers are soluble in ketones or ketone/aromatic solvent blends, and this allows their use in lacquer formulations.

Vinyl resins are used in coatings for strip for Venetian blinds and for bottle tops where extreme flexibility and extrusion properties are required. They are also used in heavy-duty and marine coatings where properties of toughness, elasticity, and water resistance are paramount, and have application as thick coatings on coil-coated strip for building facings. Polyvinylidene fluoride dispersion is now used in the highest durability coil coating finishes in blends with acrylic resin containing up to 80% vinyl polymer. Compositions containing lower concentrations, for example 30–50% of vinyl polymer, may also be used where the acrylic resin will be a thermosetting resin of, for example, the self-reacting acrylamide-containing type.

PVC copolymer emulsions are used in emulsion paint compositions, and brief mention of formulations for this application has been made under emulsion polymerization.

Other resins find use; chlorinated rubber paints are formulated using chlorinated paraffins as plasticizing component. Chlorinated rubber may also be used as an additive with resins such as alkyds and acrylics. Polyvinyl acetals (formal and butyrals) are also used, the latter particularly in resistive or active primers.

2.16   Water-borne systems

The four main technologies providing solvent reduction or elimination as so-called compliant coatings are waterborne, high solids, radiation-curing and powder coatings. The replacement of solvent in a coating by water also improves safety in application from the point of view of flammability and toxicity. Though water-borne systems have always found some application, legislation has accelerated the switch to waterborne systems. Electrodepositable coatings are a special class of water-borne system.

Water is however very different to organic solvents (Table 2.10), in ways which create many problems for the formulator. To reformulate a resin system to replace organic solvent by water (Table 2.10), requires an increase in the polarity of the whole polymer system either by the incorporation of some water soluble groups, or by the inclusion of surfactants, or both [142]. The polymer may finally be either in solution or in dispersed form. Unless very carefully formulated so that hydrophilic groups are destroyed or deactivated in curing, water sensitivity may be a problem in the final coating.

Table 2.10 — Comparison of the Properties of Water and Xylene

	Water	Xylene
Molecular weight	18	106
Melting point, boiling point (°C)	0, 100	−25, 139
Vapour pressure (mm) 20 °C	23	8
Latent heat of vaporisation (J/g)	2259	395
Surface tension (mN/m) 25 °C	72	27.5
Solubility parameter δ_D	15.5	20.1
δ_P	16	1.8
δ_H	42.3	2.5

The high surface tension creates problems with flow and wetting, and the increased likelihood of surface defects. Replacement of solvent by water need not affect the time to dry, though a consequence of the higher heat of vaporization could be increased energy requirements for stoving. However air-flows in ovens can generally be appreciably lower. Difficulties with control of rheology can be a reason to retain some solvent in an otherwise waterborne composition [143].

Polymers fully water-soluble without the addition of salt-forming additives exist; examples are polyethylene oxide or glycol, polyvinyl pyrrolidone, polyacrylamide, and copolymers containing a high proportion of these materials. Polyethylene glycol, typically having one or two terminal hydroxyl groups, has a major use as a reactant to make alkyds water-soluble or dispersible; polyethylene oxide occurs as the water-soluble portion of many surfactants.

Other water-soluble polymers, including polyvinyl alcohol, acrylic polymers with a very high acid content, and modified celluloses, find application as colloid in emulsion polymerisations. Some formaldehyde condensate resins, including hexamethoxymethyl melamine and certain phenolics, are water soluble, and therefore are used in preference to other resins as crosslinking resins in water-borne systems [144].

It will be recognized that polymers not containing soluble groups may be emulsified by added surfactant; this is not, however, common practice. Special cases of water-borne products prepared solely using surfactants were discussed in the section on emulsion polymerization (Section 2.8), and alkyd emulsions were also covered earlier. There are cases where resins are blended and one resin acts as emulsifier for other polymer species present; this is the case, for example, where the final cure requires the presence of an insoluble curing agent. The following is concerned with systems solubilized or dispersed with the aid of hydrophilic groups on the polymer backbone. Here the polymer will normally be prepared in the absence of water, and only when at its terminal MW is it transferred into water.

The major route to water solubility/dispersibility, is by the preparation of polymers with acid or amine groups on their backbone, which may be solubilized by salt formation by the addition of an alkali/amine or acid respectively. When the resins dissolve in water, these salt groups ionize, the counter-ions being carried in the bulk solution. There is a particular advantage in using volatile amine or acid in that water sensitivity in the final film is reduced, owing to the loss of the salt-forming agent; this sensitivity can be further reduced or eliminated if the crosslinking reaction can then take place with these groups. As a general point it should be noted that many successful formulations are dispersions (colloidal or micellar), not solutions, simply because this can realize the highest solids and lowest viscosity. Resins may be prepared with a higher acid or amine content than necessary, the 'degree of neutralization' with salt-forming acid or amine then being about 60% to 85%. This may be fine tuned to give the optimum viscosity, stability, freedom from settlement, and application properties.

Early water-borne vehicles were alkali neutralized maleic adducts, the simplest being maleinized oils. Maleinized fatty acids reacted with polymer backbones still play a role in water-borne systems. Epoxy resins esterified with maleinized linseed oil fatty acids alone or in combination with other fatty acids followed by neutralization, can provide vehicles for both normal spray or dip and for electrodeposited application. Polybutadiene may be maleinized and can then similarly form the basis of water-borne systems. Both oil and polybutadiene-based systems may be cured by oxidation, usually stoved without the addition of further crosslinker. Water-borne alkyd and polyester systems have already been mentioned (Sections 2.5.7 and 2.6.1), specially formulated with a high acid value to give water solubility or to be water dispersible. Acidic acrylic systems are also preparable where, for example, maleic anhydride or acrylic acid are copolymerized with the other unsaturated monomers, and such thermosetting systems have already been described. Epoxy systems may be given acidic functionality by the ring-opening half ester reaction with a dicarboxylic anhydride, or trimellitic anhydride as described for alkyds and polyesters [145].

Water-borne resins may be self-crosslinking (alkyds, thermosetting acrylics) or may be crosslinked by added soluble or co-emulsified water-insoluble crosslinking resins. Fully soluble crosslinking resins may be blended into the system at any stage, and these include MF, UF, and phenolic resins, reacting as previously described for solvent-borne systems.

Carboxyl groups of waterborne anionic polymers can participate in reactions with a number of added crosslinkers. β hydroxyalkylamides were originally proposed for waterborne systems [146], though they now find use in the cure of powder coatings (see Section 2.20). Two other specialized crosslinkers are polycarbodiimides and aziridines, where use is practically restricted to curing polyurethane dispersions (see Section 2.13.3). The structure of these is shown in Fig. 2.31.

Epoxy resins can be formulated to be amine functional by reaction with secondary amines that will then be solubilized with, for example, acetic acid (Fig. 2.32).

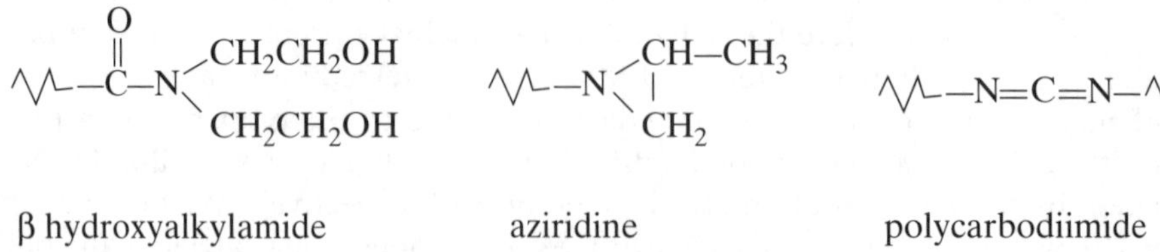

β hydroxyalkylamide   aziridine   polycarbodiimide

Fig. 2.31 — Functional group of β hydroxyalkylamide, aziridine and polycarbodiimide crosslinkers.

Crosslinking resins must be added to these systems, for example MF, phenolic, or blocked isocyanate. A method of increasing the number of amine groups is by reacting with a diketimine, as shown later (Fig. 2.35). The ketimine can be reacted with epoxy resin, without any risk of gelation, and subsequently hydrolyzed to release two primary amine groups.

Water-borne epoxy resins find particular application for metal primers and for can coatings [147], and for the latter usage water reducible acrylic grafted epoxide resins find extensive application [148, 149]. These are bisphenol epoxy resins grafted with styrene/acrylic acid copolymers, using a grafting initiator such as benzoyl peroxide (Fig. 2.33).

Specially formulated acrylic emulsions are also used in thermosetting water-borne systems, and these are prepared in the manner described in section 2.8 on emulsion polymerization. The monomers will include a hydroxy functional monomer, and the preparation will generally be carried out without the inclusion of colloid, only surfactant being present. These are usually called 'colloid-free' latexes.

Aqueous polyurethane dispersions have already been discussed (Section 2.13.3).

Water-soluble resins may be silicone modified, and the advantages of this have been demonstrated for alkyds and polyesters where siliconization is carried out by pre-reacting the silicone intermediate with polyol. For acrylic latex products siliconization of the latex may be carried out as a subsequent step to the emulsion polymerization [61].

Certain problems are found specific to water-borne products, and require additional care in formulation. Condensation products can be prone to hydrolysis. Curing may be adversely affected by the solubilizing agent either by pH effects or,

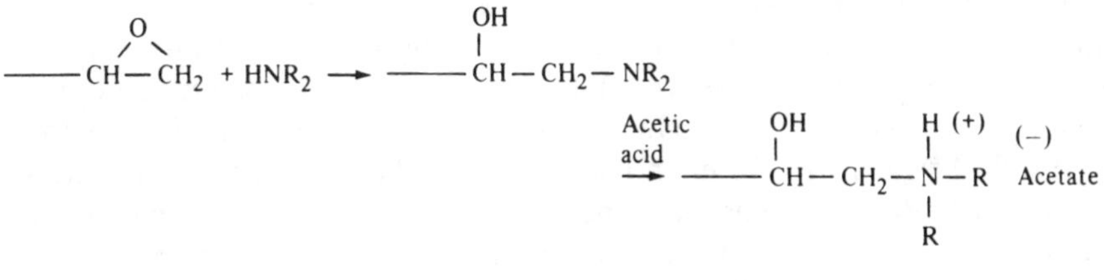

Fig. 2.32 — Epoxy resin/amine reaction and subsequent water solubilization.

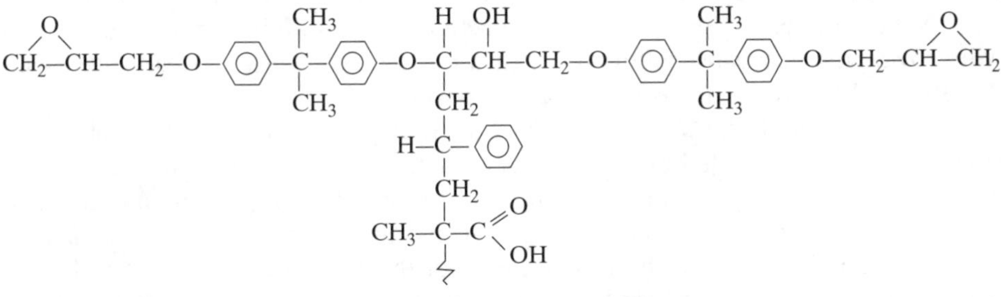

Fig. 2.33 — Epoxy acrylic graft structure.

with air-drying alkyds, by neutralizing amine retarding oxidative crosslinking [150]. The presence of amines can also adversely affect colour retention.

Most water-soluble compositions contain a proportion of water-miscible co-solvent [151] which may have been a necessary component in the resin prepara-tion state prior to water addition. However, its presence may also allow the paint formulator greater latitude to control properties such as stability, drying and particularly rheology [152], and still meet any required VOC levels required by pollution legislation. Disperse compositions may need to contain organic sol-vent to aid coalescence. The presence of water-miscible solvents may also aid emulsification in disperse systems as with the techniques to prepare so-called 'microemulsions'.

2.17   Resins for electrodeposition

Electrodepositable resins are a special class of water-borne resin [153, 154]. The polymer is carried in an aqueous medium and, on application of current via suit-able electrodes, polymer in the vicinity of one of the electrodes is destabilized and deposited on the electrode. The deposited polymer builds up to form an insulating layer that ultimately limits further deposition. The electrodeposition process has been called electrophoretic deposition, though it is now recognized that elec-trophororesis plays little part in the process.

In anodic electrodeposition negatively charged polymer is deposited on the anode, and in cathodic electrodeposition positively charged polymer is deposited on the cathode. Compared with conventional painting, very uniform coverage of external surfaces can be obtained, and deposition of paint inside partially closed areas can be achieved ('throwing'). Paint utilization is high, and almost complete automation of the process is possible. These advantages are not achieved with simpler dipping processes. Since the films can only be built up on metal surfaces and to a limiting thickness, electropaints are either primers or for some industrial applications, one-coat finishes. Unpigmented clear systems are in use for coat-ing metallic brightwork. Both anodic and cathodic commercial processes exist [155, 156]. Cathodic systems have now supplanted anodic systems particularly for automotive applications. Electrocoating is important in the application of can lacquers.

Almost all polymer types described in the previous section under water-borne coatings can be adapted to electrodeposition. In most cases, to be suitable for the electrodeposition process, the resin will be held in stable particulate or micellar dis-persion, but not complete solution, by the action of hydrophilic ionic groups, which will provide the necessary colloidal stabilization. Non-ionically sterically stabilized dispersions, however, may also be electrodeposited [157, 158].

The system should be designed such that the deposited film has high electrical resistance so that shielded areas can receive adequate coverage. The system may contain organic solvent to aid the dispersion process and dispersion stability, and act as a flow promoter during coating and curing. The role of the neutralizing acid or base is fundamental both to stability of the dispersion and to the electro-coating process. In practice, for anionic systems either alkali or amines may be used, and for cathodic systems lactic and acetic acids are usual choices for the neutraliz-ing agent.

The earliest anodic electrodeposition vehicles were based on maleinized oils and oil derivatives, and the chemistry of these developed through vinylated and alkyd type condensates to the use of epoxy esters based on maleinized fatty acids. Since epoxy resin-based systems exhibit such good performance as metal primers, solubilized epoxy vehicles play a major role in both anodic and cathodic systems, especially for automotive applications.

Alkyd and acrylic systems have been developed for electrodepositing and have found application in industrial systems. Anodic alkyd systems are based on high acid value resins, particularly those that are trimellitic anhydride-derived, and may be drying oil-modified. They may therefore be autoxidative/stoving, or if non-drying oil-containing or oil-free, cured by co-emulsified or soluble melamine/formaldehyde resin. Acrylic systems have been proposed for both anodic and cathodic formulations, in the former case by inclusion of higher than normal concentrations, of, for example, acrylic acid [104, 159]. In the latter case, they have been formed by the inclusion of a copolymerized amino monomer, such as dimethylaminoethyl methacrylate [160] or by including a glycidyl monomer that can be subsequently reacted with an amine [161].

Cathodic epoxy systems may be either primary or secondary amino functional, reacting epoxy resins with amines or quaternary amine salts can be used to give adducts with terminal secondary, tertiary, or quaternary amines or their salts. The relatively weak nature of these amines can result in poor dispersibility of these systems, and it is more desirable to incorporate primary amine groups. This can be difficult to achieve. However, useful methods have been developed; these include reacting an excess of a di-primary amine with an epoxy resin [162], or blocking primary amine groups on molecules with ketones, before reaction with epoxides through other functionalities [163]. These are illustrated in Fig. 2.34 and 2.35.

Because of their extremely high film alkalinity these systems crosslink sluggishly with melamine and phenolic crosslinkers. They can be effectively crosslinked with blocked isocyanates, which are designed to be stable at bath temperatures but

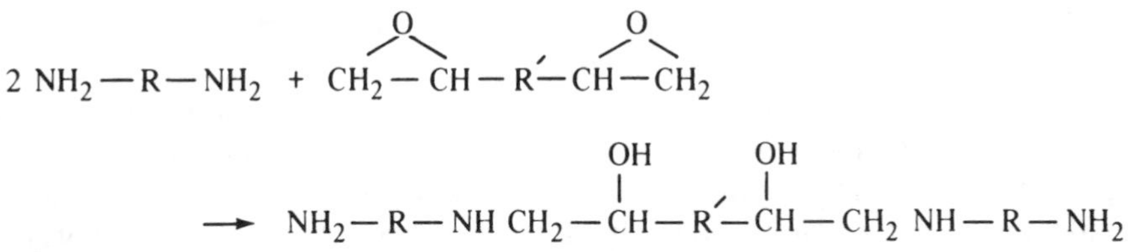

$$2\ NH_2-R-NH_2\ +\ CH_2-CH-R-CH-CH_2$$

$$\longrightarrow\ NH_2-R-NH\ CH_2-\overset{OH}{CH}-R-\overset{OH}{CH}-CH_2\ NH-R-NH_2$$

Fig. 2.34 — Reaction of epoxy resin with excess diamine.

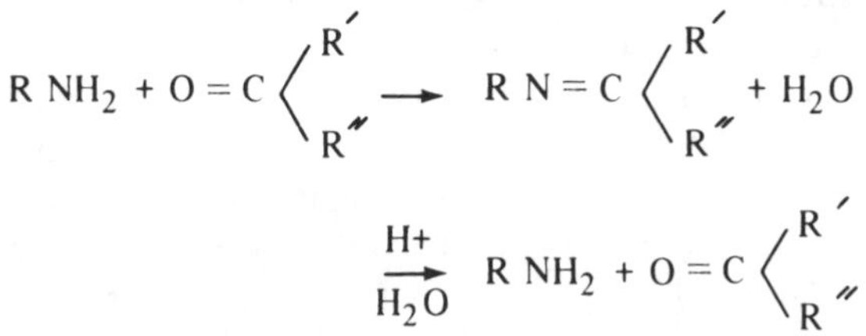

$$R\ NH_2 + O = C\begin{array}{c} R' \\ R'' \end{array} \longrightarrow R\ N = C\begin{array}{c} R' \\ R'' \end{array} + H_2O$$

$$\xrightarrow[H_2O]{H+}\ R\ NH_2 + O = C\begin{array}{c} R' \\ R'' \end{array}$$

Fig. 2.35 — Ketimine formation and hydrolysis.

unblock at reasonable stoving temperatures [164]. Acrylic cationic systems may also be cured with these curing agents for one-coat systems where good colour requirements are more important than corrosion resistance.

2.18   High solids coatings

Changing one's paint formulation to higher solids is the simplest means available to reduce actual organic solvent content to achieve lower VOC. Current expectation will be for high solids spray paint to be applied at >75% solids, and for a brush-applied paint to be >85% solids. The trend to higher solids and hence lower solvent content is seen in all areas of paint application.

A number of approaches to achieving higher solids are possible. Powder coatings are of course a 100% solids system. 100% solids can be achieved where reactive diluents (e.g. unsaturated monomers) are used in place of solvent, as is the practice with unsaturated polyesters, and with radiation-curing polymers. The use of a reactive diluents as a partial replacement for solvent is possible with most systems. With epoxy systems, low molecular weight or monofunctional epoxides can be used. With PU systems, oxazolidines and low molecular weight polyols can be used. As described earlier (Section 2.3.2), with autoxidatively drying alkyds, some solvent may be replaced by diluent which is a vinyl monomer, which copolymerizes into the structure through the free radical nature of the autoxidative drying process [19]. Allyl ether-based oligomers have been examined as components of high solids alkyd systems [165] (though formation of acrolein limits their use) and specially formulated unsaturated melamine resins have also been used.

On the face of it, with all high solids systems, savings should be made in that less solvent is needed, and less energy used in the evaporation of solvent; however, the approach generally requires special formulating techniques and invariably the use of more expensive ingredients.

It is possible to obtain small gains in application solids with any paint by applying at higher viscosity or by heating the paint as a means of reducing viscosity. However, to achieve a significant increase in solids and to retain good application characteristics, it is necessary to redesign the overall system using polymer of lower molecular weight than before; the result of lower molecular weight is that less diluent is required to achieve application viscosity. However, lower molecular weight polymer always requires more crosslinking reaction in order to achieve final film properties, and lowered molecular weight has the consequence that special care is necessary with the concentration of reactive functional groups and their distribution.

Polymers for high solids systems are often referred to as 'oligomeric'. This term is used for low molecular weight polymers whose molecular weight is in that part of the molecular weight spectrum where polymer properties are only just starting to become apparent; this is in particular below the molecular weight where chain entanglement is significantly affecting viscosity. An oligomer may be of molecular weight 1000–5000, where the molecular weight for an older conventional thermosetting polymer might be 10000–40000, and a thermoplastic polymer not requiring crosslinking 80000–100000 [166, 167]. Telechelic refers to linear or branched polymer molecules possessing terminal functional groups; the design of these for high solids systems has been described [168].

To achieve low viscosity/high solids, good understanding is necessary of the relationship between molecular weight and viscosity for the resin, and for the interaction between all the components (oligomer, crosslinker, solvent, and pigment) [169]. Molecular weight and molecular weight distribution both require careful control. Narrowing of the molecular weight distribution is desirable; for alkyd systems narrower molecular weight for high solids systems has been achieved using iso-phthalic acid rather than *o*-phthalic anhydride.

In conventional coatings relatively few functional groups need react to yield crosslinked films. In high solids systems, a substantially larger number of groups must react to reach the same final crosslinked structure. This increase in functional group content does, however, tend to increase viscosity since these groups are generally polar; this in part offsets the viscosity decrease achieved through molecular weight reduction. The formulator must nevertheless try and ensure that each oligomer molecule has at least two functional groups attached.

Polyester formulation for high solids is relatively simple, using similar ingredients as for higher molecular weight polyester. Polyesters are usually certain to have two terminal functional groups, more if branched. For alkyds, higher functionality in reactants may be required. Thus high solids alkyds may be formulated using six functional dipentaerythritol and trifunctional trimellitic anhydride or tetrafunctional pyromellitic anhydride [170]. Higher solids alkyds will be of very long oil length in order to ensure a high overall content of oxidisable groupings and a sufficient number on each individual alkyd polymer molecule.

In the case of acrylic resins additional problems are experienced in preparation over normal formulations. High concentrations of expensive initiator may be required to reduce molecular weight; alternatively, high chain transfer agent concentrations will be required, leading to residual odour problems, difficult to tolerate for some applications. The use of high solids in preparation, which may be essential, results in higher degrees of chain transfer to polymer, leading to broadening of molecular weight distribution and consequent adverse effects on rheology. Unlike polyesters, in acrylics functional group content from a functional monomer tends to be random (see Section 20.6.1); it can hence be beneficial to include functional transfer agents such as mercaptoethanol to ensure that as many polymer molecules as possible carry at least one terminal reactive group.

Also in acrylics, the use of such bulky monomers as isobornyl methacrylate and cyclohexyl methacrylate has been found to give lower viscosity at comparable molecular weights [171, 172]. Higher molecular density is also achieved through branching of the polymer, so that a hyperbranched polymer has a substantially lower viscosity at comparable molecular weight or solution concentration, compared to linear polymer. Formulating practice has developed techniques to provide branched, hyperbranched, star and dendritic structures. Acrylic resins can be made more highly branched using ethylene glycol dimethacrylate [173] or using multifunctional transfer agents [174]. Techniques aiming for star or dendritic structures have been applied to obtaining both high solids alkyds and acrylics [172, 173, 175].

High solids acrylic and polyester resins are normally hydroxy functional, and hexamethoxymethyl melamine resins have found application as crosslinkers in high solids thermosetting finishes. The reasons for their suitability have been discussed [176]. With corresponding higher concentrations of functional groups in the oligomer part of the formulation, an amino resin may constitute as much as 50% or more of the composition, compared with 10–40% in more conventional finishes. Blocked isocyanate are also used.

High solids 2K finishes with isocyanate adducts use the same acrylic and polyester resins described above.

2.19   Radiation-curing polymers

Radiation that can assist or control curing includes electron beam cure (EBC), ultraviolet (UV), and infrared (IR), and all these methods are practised commercially. Reduced energy consumption, improved environment protection, and, in particular, suitability for wood, paper, and plastic substrates have contributed to recent growth. For UV cure, photoinitiators are required, and for IR cure thermal initiators, in both cases normally to initiate chain addition polymerization. For EBC no added initiator is required, the electron beam producing radicals in the polymerizing layer through its own high energy. The systems employed are normally high solids, and typically the resin system is a blend of prepolymer and reactive diluent. Unsaturated polyesters are typical prepolymers and are used in combination with suitable vinyl or acrylic monomer diluents. The systems and mechanisms for UV, IR, and EBC have been reviewed [177]. Owing to the opacity of most pigments to UV radiation, the majority of UV curing finishes are clearcoats.

In addition to the unsaturated polyesters described earlier in this chapter, for UV cure, epoxy and polyurethane resins are also used. Epoxy acrylates are prepared by the reaction of bisphenol epoxy resins with acrylic acid. Polyurethane resins for photocuring are obtained by the complete reaction of di-functional isocyanates or polyfunctional isocyanate adducts with hydroxy ethyl acrylate. Rather than unsaturated polyesters with fumarate unsaturation, for UV curing, polyester acrylates are preferred, prepared by the esterification of hydroxy functional polyether or polyester with acrylic acid. In the preparation of all acrylate resins careful control of inhibitor levels is necessary to prevent polymerization, and final dilution with monomer may be necessary to retain fluidity. A major advantage of this type of prepolymer is low viscosity.

Monomers are necessary as reactive diluents especially for viscosity control, and apart from their viscosity and solvating ability other factors needing consideration include their volatility and photo-response. Crosslinking mechanisms are complex, and for any resin/monomer system there may be an optimum resin/monomer ratio for a given dosage of radiation to attain the required film characteristics. Acrylates give the fastest cure speed, and polyfunctionality is normally necessary; hence such monomers as hexanediol diacrylate and trimethylol propane triacrylate find considerable application. A broad range of such multifunctional monomers is now available commercially. The order of reactivity for unsaturated groups in radiation curing systems is acrylate > methacrylate > allyl = vinyl. UV/visible photo-initiators for unsaturated systems will be free radical generators, for example aromatic ketone; ketone/amine combinations are also used [178, 179]. Curing reactions are those described earlier for acrylic polymerization.

The cationic curing of epoxy resins using photo-initiators which release Brønsted acids on irradiation is a development in this field; systems using this method are based mainly on cycloaliphatic epoxy resins and aliphatic diepoxides; bisphenol epoxides degrade the curing efficiency of this system.

The most recent developments are in UV curing of water-borne finishes and of powder coatings. For the latter, heating is still necessary to fuse the coating, but further raising of temperature to effect cure is not.

2.20   Powder-coating compositions

Powder coatings possess advantages over conventional coatings in that no polluting solvent loss occurs on application, and owing to the use of electrostatic spray, little material is lost. Electrostatic spraying is also employed with solvent-borne coatings and is a method particularly suited to automation. Powders have three distinct temperature regimes to consider and this imposes special constraints on their formulation [180, 181]. The ingredients, and particularly the final composition, have to be solid and glassy at room temperature in order to be ground to size and for the powder to stay free-flowing on storage, and this requires that the blend must not soften or sinter below 40 °C. Most powders are produced by comminuting an extruded melt which means that the final curing composition has to withstand this melting process. (Pigmentation is also achieved in the course of the extrusion.) Finally good flowout and then cure must occur at higher temperatures still after application. In formulation and production, additional factors such as particle size control, melt flow characteristics and resistivity need careful attention.

Epoxy compositions are widely used and are based on bisphenol resins cured with amines (dicyandiamide derivatives in particular) or phenolic crosslinkers. Optionally in so-called hybrid systems, saturated polyester may be mixed with the epoxy. Dimerized fatty acid modification may also be carried out to improve film levelling and adhesion, and epoxy resins containing branched species derived from epoxidized novolacs may be introduced to increase hardness. Dicarboxylic acid anhydrides and acidic polyesters are also used to cure epoxy powders, and these compositions have better performance than amine cured epoxies, being non-yellowing.

Polyester powder coatings may be hydroxy functional, and use melamine/formaldehyde or masked isocyanate curing agents [136], or may be formulated from acidic polyesters cured with added epoxy resin or triglycidyl isocyanurate (TGIC); the latter offers excellent durability and colour stability. Caprolactam is used as a blocking agent for masked isocyanates for curing powders, though release of this

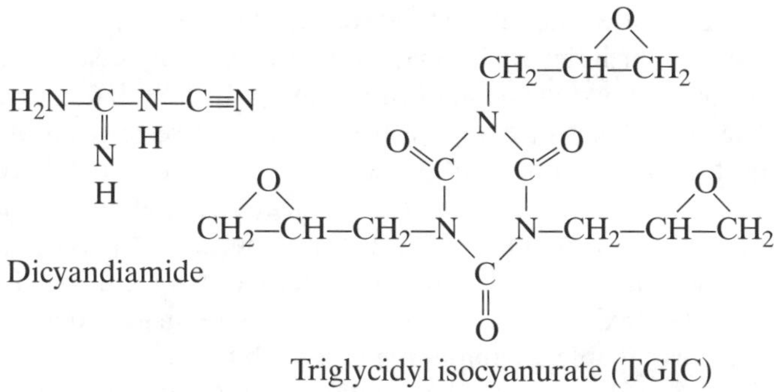

Fig. 2.36 — Typical curing agents for powder coatings.

agent from the curing film may be troublesome. Typical curing agents for powder coatings are shown in Fig. 2.36. The use of TGIC is now under threat on toxicity grounds, but β-hydroxylkylamides (Fig. 2.31) are finding some use as a replacement [182].

Acrylic systems are available but to date have found limited application; possible use in automotive clearcoats may increase use dramatically.

Thermoplastic powders are also used containing, for example, nylon 11 or nylon 12, polyvinyl chloride, cellulose acetate and butyrate, or polyethylene. These are used as thicker coatings. Redispersed powders may be applied from water as the so-called slurry coatings or aqueous powder suspensions, and may also be electrodeposited (electropowder coatings).

2.21 Resin manufacture

In the preceding sections of this chapter, the manufacturing methods normally used for each resin type have been indicated. This brief section summarizes these, highlights certain general points, and indicates future trends. Plant design and operations require advanced engineering considerations and interested readers should consult elsewhere [183].

Resin processes divide mainly between high temperature (over 180 °C) and low temperature processes where the latter are generally more fiercely exothermic. For exothermic processes, heat balance and safety, including considering the prevention of reaction runaway, are important in scale-up. Resins may be made in the melt or in solution, but will always finally be used in liquid form, either as solution or emulsified to a liquid dispersion form. Otherwise, resins will be produced in dispersed form by emulsion or dispersion polymerization.

Resin manufacturing plant has now moved a long way from the simple portable stirred and heated pot, satisfactory only for oleoresinous vehicles and step growth polymers such as alkyds, to the almost universal use of fixed plant. The reactor of this plant will have efficient agitation, be fitted with means of heating and cooling with sophisticated control, be equipped with vapour-condensing systems capable of azeotropic distillation, and will normally be attached to a thinning and cooling vessel. The plant may be multipurpose, but it is more usual to segregate high temperature, low temperature, and aqueous emulsion resins to different plants. One reason is the need of the latter two types of product for feed vessels for both monomers and initiator. Emulsion reactors, because of the inherent sensitivity of emulsions to shear, often need more carefully designed agitation systems and require other special facilities such as aqueous preparation vessels.

Computers now find extensive application in controlling larger scale resin and paint plant. The advantages of computerized control of resin plant are claimed to include stricter timing and control of process phases and improved accuracy [46].

Resin manufacturing processes for paint application are almost exclusively batch processes, and only few examples of large-scale continuous processes exist [107]. Most resin processes are carried out at atmospheric pressure, although examples of acrylic resin processes exist where reaction is carried out in sealed pressurized reactors [75]. Autoclaves capable of containing very high pressures are used in the manufacture of vinyl chloride and ethylene containing aqueous emulsions (see Section 2.8). Resin reactors may frequently have vacuum facilities attached, both for fume control and for use in vacuum distillation. Automation of testing remains an ultimate goal, and to date only a few working examples exist [184].

Bibliography

ALLEN G & BEVINGTON J C (eds), *Comprehensive Polymer Science*, 7 volumes, Pergamon (1987), ISBN 0 08 032515 7.
MUNK P, *Introduction to Macromolecular Science*, Wiley (1992). ISBN 0 471 157389 2.
PAUL S, *Surface Coatings Science and Technology*, Wiley (1996). ISBN 0 471 61406 8.
STOYE D & FREITAG W (eds), *Resins for Coatings*, Hanser (1996). ISBN 3 446 17475 3.
Surface Coatings Association of Australia, *Surface Coatings*, Volume 1, *Raw Materials and their Usage*, Chapman & Hall (1993). ISBN 0 412 5521 0.
WICKS Z W, JONES F N & PAPPAS S P, *Organic Coatings Science and Technology*, 2nd ed. Wiley (1999). ISBN 0 471 24507 0.

References

[1] SARVIMÄKI I, *Surface Coatings Internat* **77** 339 (1994).
[2] GEURINK P J A & BANCKEN E L J, In: *PRA 12th International Conference Water-borne Coatings*, Milan (1992).
[3] SCHUMACHER H, *Europ Coatings J* **1991**(12) 852.
[4] HAZEL N J, *J Coatings Technol* **68** (861) 117 (1996).
[5] HERTLER W R, *Macromol Symp* **88** 55 (1994).
[6] CORNER T, *J Oil Col Chem Assoc* **75** 307 (1992).
[7] TOMALIA D A, *et al*, *ACS PMSE* **73** 75 (1995); Turner S R, *ACS PMSE* **73** 77 (1995).
[8] PETTERSSON B, *Pigment & Resin Tech* **25** (4) 4 (1996).
[9] DUPONT DE NEMOURS E I, United States Patent 4,680,357 (1987); United States Patent 5,028,677 (1991).
[10] ENNOR K S & OXLEY J, *J Oil Col Chem Assoc* **50** 577 (1967).
[11] HASE A & PAJAKKALA S, *Lipid Technol* **6** (5) 110 (1994).
[12] LOWER E S, *Manufacturing Chemist* **52** (12) 50 (1981).
[13] VICTOR WOLFF, British Patent 1,141,690 (1969); Pacific Vegetable Oil Corporation, United States Patent 3,278,567 (1966).
[14] WEXLER H, *Chem Rev* **64** 591 (1964).
[15] MUIZEBELT W J, HUBERT J C & VENDERBOSCH R A M, *Progr Org Coat* **24** 263 (1994); *J Coatings Tech* **70** (876) 83 (1998).
[16] CARR C, DRING I S & FALLA N A R, in: *Proc XVI International Conference Organic Coatings Science and Technology*, Athens (1990).
[17] KHAN N A, *Can J Chem* **37** 1029 (1959).
[18] PORTER N A & WUJEK D G, *JACS* **106** 2626 (1984).
[19] BARRETT K E J, *J Oil Col Chem Assoc* **49** 443 (1966).
[20] KUO C-P, CHEN Z, LATHIA N & DIRLIKOV S, *21st Waterborne, High Solids and Powder Coatings Symposium*, New Orleans, 1994, Pt. 2, 797.
[21] HANCOCK R A, LEEVES N J & NICKS P F, *Progr Org Coat* **17** 321, 337 (1989).
[22] LEEVES N J, *Autoxidative Degradation of Unsaturated Fatty Acid Esters*, Thesis, Royal Holloway College, Egham, UK (1985).
[23] DAVISON S E L K, *Chemical Processes Accompanying the Autoxidation of Paint Films*, Thesis, Royal Holloway & Bedford New College, Egham, UK (1989).
[24] LOVE D J, *Polym Paint Col J* **175** 436 (1985).
[25] KRAFT W M, *Amer Paint J* **41** (28) 96 (1957).
[26] Heydon Newport Chemical Corporation, United States Patent 2,973,331 (1961).
[27] YANG Y, ZHANG H & HE J, *Macromol Theory Simul* **4** 953 (1995).
[28] TOBITA H & OHTANI Y, *Polymer* **33** 801 (1992); *Polymer* **33** 2194 (1992).
[29] ROONEY J F, *Official Digest* **36** (475 Part 2) 32 (1964).
[30] HAZAN A F, *Paint & Resin* **Oct/Nov 1993**, 5.
[31] HOLMBERG K, *High Solids Alkyd Resins*, Marcel Dekker (1987). ISBN 0 8247 7778 6.
[32] SHELL CHEMICALS, *Cardura E10*, Technical Manual CA 1.1; Henry N *et al*, *Coatings World* **1998**(4) 49.

[33] HERZBERG S, *J Paint Technol* **41** 222 (1969).
[34] SHELL CHEMICALS, *The Manufacture of 'Cardura' Resins*, Technical Manual CA 2.1.
[35] HOLMBERG K & JOHANSSON L, *Proc. 8th International Conference Organic Coatings Science and Technology*, Athens, 255 (1982).
[36] BOBALEK E G, MOORE E R, LEVY S S, & LEE C C, *J Appl Polym Sci* **8** 625 (1964).
[37] Imperial Chemical Industries, British Patent, 1,242,054 (1971); British Patent, 1,594,123 (1978).
[38] POND P S & MONK C J H, *J Oil Col Chem Assoc* **53** 876 (1970).
[39] TYSALL L A, *Calculation Techniques in the Formulation of Alkyd and Related Resins*, Paint Research Association, Teddington (1982).
[40] PATTON T C, *Alkyd Resin Technology (Formulating Techniques and Allied Calculations)*, Interscience (1962).
[41] BERNARDO J J & BRUINS P, *J Paint Technol* **40** 558 (1968).
[42] WEIDERHORN N M, *Amer Paint J* **41**(2) 106 (1956).
[43] NELEN P J C, *17th FATIPEC Congress* 283 (1984).
[44] *Technical Report No GTSR-74B*, AMOCO Chemicals.
[45] *Technical Information Leaflet no. 0005 T10*, Perstorp Polyols.
[46] KASKA J & LESEK F, *Progr Org Coat* **19** 283 (1991).
[47] Imperial Chemical Industries, British Patent 1,370,914 (1972); Baker A S, *Paint Resin* **52**(2) 37 (1982).
[48] LANE K R, *PRA 4th Asia-Pacific Conference Hong-Kong* (1993); *Technical Reports GTSR-103A and GTSR-104*, AMOCO Chemicals.
[49] ÖSTBERG G et al, *Progr Org Coat* **24** 281 (1994); Akzo-Nobel, European Patent 593487 (1996).
[50] ÖSTBERG G & BERGENSTÅHL B, *J Coatings Technol* **68**(858) 39 (1996); Feustel D, *Surface Coatings Austral* **1994 (Jan/Feb.)** 6.
[51] HOFLAND A, *J Coatings Tech* **67**(848) 113 (1995); Beetsma J, *Pigment & Resin Tech* **27**(1) 12 (1998).
[52] GOBEC M, *Färg och Lack Scandinavia* **1997**(5)5; Vianova, European Patents 0267562 (1987) and 0758365 (1997).
[53] SHAKLEFORD W E & GLASER D W, *J Paint Technol* **38** 293 (1966).
[54] GARBETT I, *PRA 2nd Middle East Conference Bahrein* (1995) Paper 3.
[55] TURPIN E T, *J Paint Technol* **47**(602) 40 (1975).
[56] AMOCO Chemicals Corporation, *How to Process Better Coating Resins with AMOCO PIA and TMA*, Amoco Chemicals Bulletin IP-65d.
[57] AMOCO Chemicals Corporation, *Processing Unsaturated Polyesters Based on AMOCO Isophthalic Acid (IPA)*.
[58] Imperial Chemical Industries, United States Patent 3,050,533 (1962).
[59] FRADET A & MARÉCHAL E, *Adv Polym Sci* **43** 51 (1982).
[60] EARHART K A, *Paint & Varnish Prodn*, **1972**(1) 35; ibid **1972** (2) 37.
[61] PRICE J G, *Silicone Resins as Modifiers for Waterborne Coatings*, Dow Corning, UK (1978).
[62] Standard Oil, United States Patent 3,252,941 (1964).
[63] Imperial Chemical Industries, British Patent 784,611 (1957).
[64] FARBENFABRIKEN BAYER, British Patent 810,222 (1956).
[65] JOHANSSON M and HULT A, *J Polym Sci A Polym Chem Ed* **29** 1639 (1991); Figure 7 in *Färg och Lack Scandinavia* **1992**(10) 190.
[66] BROWN W H & MIRANDA T L, *Official Digest* **36** (475 Part 2), 92 (1964).
[67] OLDRING P K T & LAM P K H (eds), *Waterborne and Solvent Based Acrylics and Their End User Applications*, Wiley/SITA, London (1996).
[68] RUDIN A, *Elements of Polymer Science and Engineering*, Chapters 6 & 8. Academic Press (1999).
[69] ALLEN G and BERINGTON J C (ed.), *Comprehensive Polymer Science*, Volume 3. Chapters 2 & 15. Pergamon.
[70] GREENLEY R Z, in: Brandrup J and Immergut E H (eds) *Polymer Handbook* 4th/II edn, p 181, Wiley Interscience (1999).
[71] GREENLEY R Z, II p 309, ibid.
[72] JENKINS A D & JENKINS J, p II 321, *ibid*; *Polymer Int.* **44** 391 (1997).
[73] O'DRISCOLL K F, PONUSWAMY S R, & PENLIDIS A, *ACS Symposium Series* **404** 321 (1988); O'Driscoll K F, Ponuswamy S R & Penlidis A, *J Appl Polym Sci* **39** 1299 (1990).
[74] DUPONT DE NEMOURS E I, British Patent 763,158 (1956).
[75] DUPONT DE NEMOURS E I, British Patent 807,895 (1959).
[76] ZIMMT W S, *Ind Eng Chem Prod Res Dev* **18** 91 (1979).
[77] TAYLOR J R & PRICE T I, *J Oil Col Chem Assoc* **50** 139 (1967).
[78] CHRISTENSON R M & HART D P, *Official Digest* **33** 684 (1961).
[79] VOGEL H A & BITTLE H G, *Official Digest* **33** 699 (1961).
[80] PPG Industries, British Patent 1,556,025 (1975); British Patent 1,559,284 (1976).
[81] MURDOCK J D & SEGALL G M, *Official Digest* **33** 709 (1961).
[82] FURUKAWA H et al, *Progr Org Coat* **24** 81 (1994).
[83] CHEN M J et al, *Surface Coatings Internat* **1996**(12), 539; *J Coatings Tech* **68** (870) 43 (1997).

[84] DEL RECTOR F, BLOUNT W W & LEONARD D R, *J Coatings Technol* **61**(771) 31 (1989).
[85] FENG J, *et al*, *J Coatings Tech* **70** (881) 57 (1998).
[86] UK Patent 2206591A (1989); Pritchett R J, *PRA 12th International Conference Waterborne Paints*, Paper 11 (1990).
[87] NAKAMURA H & TACHI K, ASC PMSE **76** 126 (1997); Imperial Chemical Industries, United States Patent 4,403,003 (1983).
[88] YANAGIHARA T, *Progr Org Coat* **11** 205 (1983).
[89] THOMPSON M W, in: *Polymer Colloids*, Buscall R, Corner T & Stageman J F (eds), Elsevier Applied Science (1985).
[90] HUNKELER D *et al*, *Adv Polym Sci* **112** 116 (1994).
[91] BROMLEY C W A, *Colloids and Surfaces*, **17** 1 (1986).
[92] PALLUEL A L L, WESTBY M J, BROMLEY C W A, DAVIES S P & BACKHOUSE A J, *Makromol Chem Macromol Symp* **35/36** 509 (1990).
[93] LYONS C, *Surface Coatings Austral* **1987**(12) 14.
[94] RITCHIE P, *Surface Coatings Austral* **1989**(9) 17.
[95] EL AASSER M S in *Polymer Latexes: Preparation, Characteristics and Application*, ACS Symposium Series **492** (1992).
[96] LLEWELLYN I & PEARCE M F, *J Oil Col Chem Assoc* **49** 1032 (1966).
[97] GUYOT A & TAUER K, *Adv Polym Sci* **111** 43 (1994).
[98] BONDY C, *J Oil Col Chem Assoc* **49** 1045 (1966).
[99] LOVELL P A & EL-AASSER M S (eds), *Emulsion Polymerisation and Emulsion Polymers*, Wiley (1997).
[100] GILBERT R G, *Emulsion Polymerisation — A mechanistic approach*, Academic Press (1995).
[101] LEE S & RUDIN A, in: *Polymer Latexes: Preparation, Characteristics and Application*, ACS Symposium Series **492** (1992), p. 234.
[102] Imperial Chemical Industries, United States Patent 4,403,003 (1983).
[103] BACKHOUSE A J, *J Coatings Technol* **54** (693) 83 (1982).
[104] SLINKX M, *Surface Coatings Aust*, **1997**(5) 24; Decocq F *et al*, *Paint & Coat Ind* **1998**(4) 56.
[105] SMITH O W, COLLINS M J, MARTIN P S & BASSETT D R, *Progr Org Coat* **22** 19 (1993).
[106] GORDON P G, DAVIES M A S & WATERS J A, *J Oil Col Chem Assoc* **67** 197 (1984).
[107] GEDDES K R, *Surface Coatings Internat.* **76** 330 (1993).
[108] KHAN B, *European Coatings J* **1991**(5) 307; ibid. **1991**(12) 886.
[109] ROHM & HAAS, British Patent 1,072,894 (1967).
[110] Imperial Chemical Industries, British Patent 2,112,773 (1983).
[111] GEURINK P J A *et al*, *Progr Org Coat* **27** 73 (1996).
[112] WALBRIDGE D J, in: *Science and Technology of Polymer Colloids*, Volume 1, Poehlein G W, Ottewill R H & Goodwin J W (eds), Martinus Nijhoff (1983).
[113] BROMLEY C W A B, *J Coatings Technol* **61** (768) 39 (1989).
[114] Imperial Chemical Industries Ltd., British Patent 2164050A (1986).
[115] WALBRIDGE D J, in: *Dispersion Polymerisation in Organic Media*, Barrett K E J (ed.), Wiley (1975).
[116] WAITE F A, *J Oil Col Chem Assoc* **54** 342 (1971).
[117] REMPP P F & FRANTA E, *Adv Polym Sci* **58** 1 (1984).
[118] BARRETT K E J & THOMPSON M W, in: *Dispersion Polymerisation in Organic Media*, Barrett K E J (ed.), Wiley (1975).
[119] Imperial Chemical Industries, British Patent 1,157,630 (1969).
[120] PAREKH G G, *J Coatings Technol* **51** (658) 101 (1979).
[121] FAWER B, *Coatings World* **1** (1) 57 (1996).
[122] ZUYLEN J VAN, *J Oil Col Chem Assoc* **52** 861 (1969).
[123] KORAL J N & PETROPOULOS J C, *J Paint Technol* **38** 600 (1966).
[124] BAUER D R & DICKIE R A, *J Polym Sci Polym Phys Ed* **18** 1997 (1980); *ibid* 18 2015 (1980).
[125] OLDRING P K T (Ed.), *The Chemistry and Application of Amino Crosslinking Agents on Aminoplasts*, Wiley/SITA, London (1999).
[126] HOLMBERG K, *Proc 10th International Conference Organic Coatings Science and Technology, Athens*, 71 (1984).
[127] OLDRING P K T (Ed.), *The Chemistry and Application of Phenolic Resins or Phenoplasts*, Wiley/SITA, London (1998).
[128] MAURI A N *et al*, *Macromolecules* **30** 1616 (1997).
[129] SCHECHTER J, WYNSTRA J, & KURKJY R P, *Ind Eng Chem* **48** 94 (1956).
[130] ELLIS B, (ed), *Chemistry and Technology of Epoxy Resins*, Blackie (1993); Oldring P K T (ed.), *Waterborne and Solvent Based Epoxies and their End User Applications*, Wiley/SITA, London (1996).
[131] SOMERVILLE G R & PARRY M L, *J Paint Technol* **42** 42 (1970).
[132] GUTHRIE J T, MARTEN A & NIELD E, *J Oil Col Chem Assoc* **75** 213 (1992).
[133] MOSS N S, *Polym Paint Col J* **174** 265 (1984).
[134] HULL C G & SINCLAIR J H, in: *Handbook of Coatings Additives* Volume 2, Calbo L J (Ed.), Chapter 10, Marcel Dekker (1987).

[135] THOMAS P (Ed.), *Waterborne and Solvent Based Polyurethanes and Their End User Applications*, Wiley/SITA, London (1998).
[136] WICKS Z W, *Progr Org Coat* **3** 73 (1975).
[137] KORDOMENOS P I *et al*, *J Coatings Technol* **54**(687) 43 (1982).
[138] MCBRIDE P, *J Oil Col Chem Assoc* **65** 257 (1982).
[139] BOCK M & CASSELMANN H, *21st Waterborne High Solids and Powder Coatings Symposium*, New Orleans, p 65, (1993).
[140] BLUNDELL D & BRYAN H H, *J Oil Col Chem Assoc* **61**(769) 39 (1989).
[141] BECHARA I, *Europ Coatings J* **1998** (4) 236; Tennebroek R, *et al*, *Europ Coatings J* **1997** (11) 1016.
[142] PADGETT J C, *J Coatings Technol* **66**(12) 89 (1994).
[143] ZHENG-ZHENG, J *et al*, *J Coatings Technol* **60**(757) 31 (1988).
[144] BLANK W J & HENSLEY W L, *J Paint Technol* **46**(593) 46 (1974).
[145] KRISHNAMURTI K, *Progr Org Coat* **11** 167 (1983).
[146] LOMAX J & SWIFT G, *J Coatings Technol* **50**(643) 49 (1978).
[147] MOSS N S & DEMMER C G, *J Oil Col Chem Assoc* **65** 249 (1982).
[148] ROBINSON P V, *J Coatings Technol* **53**(674) 23 (1981).
[149] WOO J T K & TOMAN A, *Progr Org Coat* **21** 371 (1993).
[150] HILL L W & WICKS Z W, *Progr Org Coat* **8** 161 (1980).
[151] GRANT P M, *J Coatings Technol* **53**(677) 33 (1981).
[152] DOREN K, FREITAG W & STOYE D, *Waterborne Coatings*, Hansa, (1994).
[153] SCHENK H U, SPOOR H, & MARX M, *Progr Org Coat* **7** 1 (1979).
[154] DUFFY J I (ed.), *Electrodeposition Processes, Equipment and Compositions*, Noyes Data Corporaton, (1982).
[155] WISMER M, *et al*, *J Coatings Technol* **54**(688) 35 (1982).
[156] KORDOMENOS P I & NORDSTROM J D, *J Coatings Technol* **54**(686) 33 (1982).
[157] Imperial Chemical Industries, European Patent, 15,655; European Patent, 109,760 A2 (1984).
[158] DOROSZKOWSKI A, *Colloids Surfaces* **17**(1) 13 (1986).
[159] PPG Industries, United States Patent 3,947,339 (1976).
[160] BASF, United States Patent 3,455,806 (1969).
[161] SCM, United States Patent 3,975,251 (1976).
[162] Imperial Chemical Industries, British Patent 2,050,381 (1981).
[163] PPG Industries, United States Patent 4,017,438 (1973).
[164] PPG Industries, United States Patent 3,799,854 (1974); United States Patent 4,101,486 (1978).
[165] STAMICARBON B V, European Patent 0 357 128 A1 (1989); Coady C J, *J Oil Col Chem Ass* **1993**(1) 17.
[166] GIBSON D & LEARY B, *J Coatings Technol* **49** 53 (1977).
[167] NOREN G K, *Polymer News* **10** 39 (1984).
[168] ATHEY R D, *Progr Org Coat* **7** 289 (1979).
[169] HILL L W & WICKS Z W, *Progr Org Coat* **10** 55 (1982).
[170] BASF, European Patent 87/00589 (1987); Hofland A & Reker W, *Proc 17th Waterborne and High Solids Coatings Symposium*, New Orleans (1990).
[171] ZEZZA C A & TALMO K D, *J Coatings Technol* **68**(856) 49 (1996).
[172] HUYBRECHTS J & DUSEK K, *Surface Coatings Internat* **81**(3) 117 (1998); ibid **81**(4) 172 (1998); ibid **81**(5) 234 (1998).
[173] SIMMS J A, *Progr Org Coat* **22** 367 (1993).
[174] ROHM & HAAS, UK Patent 2,172,598 (1986); Imperial Chemical Industries, European Patent 0448224 (1992).
[175] JOHANSSON M, MALMSTRÖM E, & HULT A, *TRIP* **4**(12) 398 (1996).
[176] BLANK W J, *J Coatings Technol* **54**(687) 26 (1982).
[177] PAPPAS S P (ed.) *Radiation Curing: Science and Technology*, Plenum Press, New York (1992).
[178] GREEN P N, *Polym Paint Col J* **175** 246 (1985).
[179] HAGEMAN H J, *Progr Org Coat* **13** 123 (1985).
[180] MISEV T A, Chapter 9.3 in: Paul S, *Surface Coatings, Science and Technology*.
[181] GRIBBLE P R, *Int Waterborne High Solids & Powder Coatings Symposium*, New Orleans, 1998, p. 218.
[182] PLEDGER A, Waterborne, *High Solids and Powder Coatings Symposium*, p. 480, New Orleans (1996); KAPLAN A, *Europ Coatings J* 1998 (6) 448.
[183] GREAVY C MC (ed.), *Polymer Reactor Engineering*, Chapman & Hall (1993).
[184] HÄNDEL E, *Farbe Lack* **102**(11) 135 (1996).

3

Pigments for paint

A G Abel*

3.1 Introduction

This chapter introduces pigment technology and is designed for the newcomer to the paint industry, who has some chemical knowledge. It is intended to be comprehensive, but can only provide an introduction to the aspiring paint technologist, or someone closely connected to the paint industry. Other publications, including suppliers' literature, go into greater detail.

In this chapter, pigments will be defined and their required qualities detailed. The methods of classification and the terminology used in the paint industry are explained. Pigment manufacturing methods are outlined and reference is made to their environmental and toxicological properties, with implications for the paint industry. A description is given of the application properties of pigments, with advice on the rules to be adopted in order to select the correct pigments for use in paint.

Finally, there are brief descriptions of the more important pigment types used in paint, grouped in colours, with special mention of other pigments which are used to obtain special properties such as corrosion inhibitors, extenders, aluminium, and pearlescent pigments.

3.2 Definition

The word 'pigment' derives from the Latin *pigmentum* — *pingere*, to paint. An internationally accepted [1, 2] definition has been provided by the Color Pigments Manufacturers Association of America (CPMA, but formerly known as Dry Color Manufacturers Association of America, DCMA):

colored, black, white or fluorescent particulate organic or inorganic solids, which are insoluble in, and essentially physically and chemically unaffected by, the vehicle or substrate in

---

*Based on original chapter by Miss J F Rollinson.

which they are incorporated. They alter appearance by selective absorption and/or scattering of light. Pigments are usually dispersed in vehicles or substrates for application, as for instance in the manufacture of inks, paints, plastics, or other polymeric materials. Pigments retain a crystal or particulate structure throughout the coloration process.

This definition concentrates on the products used to give an aesthetic effect, such as colour or opacity, but there are other types of pigment that are used for specific purposes, such as 'extenders'. Extenders do not usually affect the colour or opacity of a paint film, but may play an important role in the film properties such as reinforcement, gloss (usually to make the film less glossy), hardness, etc. They can also affect the properties of the paint itself, such as its rheological properties, settling characteristics, or simply its cost. Extenders play an important part in paint formulation so some of the important types are described in Section 3.12.11.

Finally there are pigments that may affect the appearance of the film, but are primarily used for other properties such as fire retardance, corrosion protection, or to improve the weatherability of the substrate, mainly by absorbing ultraviolet light or other harmful radiation.

Just to confuse, the term pigment is also used for substances that give colour to animal and vegetable tissue. These biological pigments have no connection to the pigments described in this chapter; many are in fact dyes.

If we break down this definition we can recognize the essential requirements of a pigment.

3.3 Required qualities of pigments

3.3.1 Appearance
Most pigments are used to provide a visual effect, mainly colour and opacity. Having achieved the required colour and opacity, it is also important to ensure the pigment will remain essentially insoluble in the system in which it is used and will give the required physical properties, such as light fastness, weatherability, and resistance to chemicals. The paint technologist must meet these criteria and is usually expected to do so within the economic restraints imposed by the end application.

3.3.2 Colour of pigments
The colour of a pigment is mainly dependent on its chemical structure. The selective absorption and reflection of various wavelengths of light that impinge on the pigmented surface determines its hue (i.e. whether it is red or yellow, etc.). A blue pigment appears so because it *reflects* the blue wavelengths of the incident white light that falls upon it and absorbs all of the other wavelengths. Hence, a blue car in orange sodium light looks black, because sodium light contains virtually no blue component. Black pigments absorb almost all the light falling upon them, whereas white pigments scatter and reflect virtually all the visible light falling on their surfaces. Fluorescent pigments have an interesting characteristic. As well as having high reflection in specific areas of the visible spectrum, they also absorb light in areas outside the visible spectrum (the ultraviolet part where our eyes are unable to detect), splitting the energy up, and re-emitting it in the visible spectrum. Hence, they appear to emit more light than actually falls upon them, producing their brilliant colour.

The differing absorption and reflection characteristics of compounds are attributed to the arrangements of the electrons within their molecules and to their energy and frequency vibration. A molecule will absorb electromagnetic radiation from part of the spectrum. This absorption excites the electrons, resulting in the promotion of an electron from its ground state, E_1, to an orbital of higher energy E_2. The wavelength of the light absorbed is determined by the difference in energy E between the two orbitals concerned:

$$E = E_2 - E_1 = hc/\gamma$$

where h is Planck's constant, c is the velocity of light, and γ is the wavelength of light.

A given molecule has a limited number of orbitals, each with its own characteristic energy. This means that the energy difference E described above has certain definitive values. The pigment molecule is therefore only able to absorb light at particular wavelengths, determined by the energy difference E — that is characteristic of the given molecule. As the electronic excitation is accompanied by a multitude of rotational and vibrational transitions, the absorption is not at a single wavelength, but over a wide band. The wavelengths not involved in the absorption are reflected and determine the colour of the molecule, and therefore produce the complementary colour to the wavelengths absorbed.

3.3.2.1   Fluorescent pigments

Most pigments described as fluorescent do not fit our definition. They are fluorescent dyes that have been dissolved into a resin, often achieving good migration fastness on account of their reacting with the resin. The dyes themselves are often basic dyes, with poor light fastness, and while as 'pigments' they have improved durability, nevertheless they are still considered poor and are limited to interior applications. They also have poor opacity and are quite expensive.

There is an exception, azomethine (CI Pigment Yellow 101), which is a true pigment and has been known since the end of the nineteenth century. It finds little use in the paint industry, but its structure is interesting, especially when one considers the importance of conjugation (*qv*) in a molecule to develop colour.

CI Pigment Yellow 101 reference CI 48052 'Azomethine'

3.3.2.2   Coloured organic compounds

For organic compounds it has been found that certain groups of atoms are essential for colour. These groups are known as chromophores. Colour is more or less limited to aromatic compounds. An exception is the important coloured aliphatic compound, lycopene, which is responsible for the red colour of tomatoes. The molecule consists of a long hydrocarbon 'conjugated' chain; this is, consisting of alternate single and double bonds. This allows free movement of the electrons giving rise to the absorption of light as previously described and hence colour [3].

Obviously, the nature of the benzene ring gives many more opportunities for conjugation, therefore most coloured organic compounds contain benzene rings and are thus aromatic in nature. Even nitrobenzene is slightly coloured. The azo group, $-N=N-$ (and its tautomers), also provides a very good linkage, maintaining a conjugated system, hence azobenzene is coloured, especially in solution.

Azo Benzene

Additional substituents on the molecule, known as auxochromes, can modify the colour by donating or accepting electrons. Substituents that donate electrons usually have lone electron pairs, e.g. alkoxy, hydroxy, alkyl and arylamino groups, whereas groups that can accept electrons include NO_2, $COOH$, $COOR$, SO_2 [4, pp. 12–15].

More details regarding the colour of organic compounds can be found in reference [5].

3.3.2.3   Coloured inorganic compounds

Colour in inorganic pigments is more difficult to explain, partly because no single theory fully accounts for what is observed. The main cause of colour is due to charge transfer spectra and/or d–d transition spectra, especially of the 3d transition metals. However, this does not explain all inorganic pigments. It is well known that molecules containing a metal in two valency states can be intensely coloured, e.g. Prussian blue (CI Pigment Blue 27). It is thought that the deep colour of such compounds is due to transfer of an electron from ions in the lower valency state to a higher one.

Another cause of colour is the trapping of a cation in the three-dimensional lattice carrying a negative charge. Therefore, within these cavities there is room for positive and negative ions and neutral molecules. Ultramarine (CI Pigment Blue 29) is an example of such a molecule.

A very detailed account of colour in inorganic compounds appears in reference [6].

3.3.2.4   Other factors influencing colour

It is not only the pigment's chemical nature that determines its colour. The crystal structure is now recognized as playing an important role. Several pigments can exist in more than one crystal form, a property known as polymorphism, and these forms can be of very different colours. The clearest examples are linear *trans*-quinacridone, CI Pigment Violet 19, which exists in three different crystal forms (see Section 3.12.6.9), and copper phthalocyanine, which exists in several different forms (see Section 3.12.8.4). Polymorphism is not limited to organic pigments. Titanium dioxide exists in three crystal forms and other polymorphic pigments such as lead chromates and lead molybdates also have more than one crystal form.

Particle size also influences colour. Smaller particles are usually brighter in shade and change the hue of a pigment. As a general rule, smaller particles give:

- greener yellows;
- yellower oranges;
- yellower reds up to mid red;
- bluer reds from mid reds;
- redder violets;
- greener blues;
- yellower greens.

The effect of the crystal structure and particle size on colour is well covered in reference [7].

Pigment manufacturers have become very skilled in producing pigments with the desired crystal form and even with a narrow particle size distribution in order to impart the desired colour, physical properties, and hence performance.

3.3.3   Tinctorial strength

As well as colour, tinctorial strength must be considered when choosing a pigment. The higher the tinctorial strength, the less pigment is required to achieve a standard depth of shade, therefore it has economic implications. The chemical structure is fundamental to the tinctorial strength of the pigment molecule. In the case of organic pigments the colour and tinctorial strength are dependent on the ability to absorb certain wavelengths of light and in turn this is related to the conjugation in the molecule. Highly conjugated molecules show increased tinctorial strength. Therefore, increasing the aromaticity of the molecule, promoting a planar molecule and the addition of certain substituents into the molecule, mainly electron donors as described earlier (see Section 3.3.2) all encourage high tinctorial strength.

Inorganic pigments that are coloured because of their having metals in two valency states, show high tinctorial strength, whereas those that have a cation trapped in a crystal lattice are weakly coloured.

Particle size can make a dramatic difference to the tinctorial strength of a pigment, smaller particles yielding higher tinctorial strength. Manufacturing conditions are the main factor in influencing the particle size of pigment crystals, so pigment manufacturers play a crucial role. They can reduce the size of the particles by preventing the growth of crystals during synthesis. Rosin is often used for this purpose, particularly in pigments intended for the printing ink industry, but more sophisticated techniques have to be used for higher-quality pigments. The paint manufacturer can increase tinctorial strength by efficient dispersion, which is particularly important for pigments with a very small particle size. However, it is not usually possible to reduce the size of the primary crystal particles during the dispersion process.

3.3.4   Particulate nature of pigments

Pigments can be either crystalline or non-crystalline (amorphous), i.e. the atoms within each pigment molecule are arranged in either an orderly or random fashion respectively. From our definition of *pigment* we have described them as maintaining a particulate structure throughout the coloration process and as being insoluble. Therefore the following colorants *cannot* be considered pigments.

3.3.4.1   Solvent soluble dyes

Some colorants do dissolve during their incorporation into a binder and then return to an insoluble form, or are held within the crystalline structure of the binder. These colorants cannot be strictly considered pigments. Most are considered to be solvent soluble dyes. They are frequently used for the coloration of plastics. Other solvent soluble dyes, usually metal complexes, are used for the coloration of solvent-based packaging inks and clear lacquers. Their high transparency, high-colour strength, and brilliant, strong shades are particularly useful when printing on aluminium foil. They have to be dissolved in specific solvents, usually alcohols, ketones, or esters. It is this limited solubility and sometimes their low fastness properties that often limit their application. There are also some relatively simple 'solvent soluble dyes' that are soluble in hydrocarbons, including some plastics and waxes. They have poor light fastness and cannot be overcoated, as they dissolve in the top coating, thus severely limiting their application.

In the textile industry, a number of colorants would be found to be insoluble within the fibre.

3.3.4.2   Vat dyes

Vat dyes are highly insoluble, but to apply them to a fibre they are dissolved by reduction with powerful reducing agents, forming the so-called soluble leuco form, which is then used to dye the fibre. Oxidation follows, converting the dye to the insoluble form within the fibre. Because the colorant has gone through a soluble stage it cannot be considered a pigment. However, several compounds that are used as vat 'dyes' can also be used directly as pigments, by eliminating this solubilizing process, for example CI Vat Orange 7, which is also CI Pigment Orange 43.

3.3.4.3   Disperse dyes

Disperse dyes are mainly used for the coloration of polyester fibres. They are *almost* insoluble, but have to have a slight solubility. When put into a dyebath, they have to be dispersed, usually with the help of surface-active agents. A small proportion of the dye goes into solution. However, the dye prefers to be in the polymer phase, and therefore enters the fibre. Equilibrium is re-established by more dye going into solution, once again entering the fibre. In this way the depth of the dyeing builds up, usually at temperatures above 100 °C obtained by dyeing under pressure. Again, as they go through a soluble stage, they cannot be considered pigments. Several disperse dyes can also be used as solvent soluble dyes.

3.3.4.4   Azoic dyes

These include some of the oldest known dyes, owing their origin to the textile industry. In 1880, Read Holliday in Huddersfield (later to become part of ICI, now Zeneca) patented its 'developing dyes'. It found that cotton impregnated with naphthol (2-naphthol) and dried, could then be passed through a liquor of an ice-cold diazonium compound. It was in fact synthesizing a pigment inside the fibre and consequently produced a cloth with very good fastness properties. But again the colorant is not a pigment as it does not remain insoluble throughout the coloration process.

3.3.5 Insolubility

In the definition of a pigment, it was stated that it must be 'insoluble in, and essentially physically and chemically unaffected by the vehicle or substrate'. Therefore, when a pigment is incorporated into our paint formulation, it must not significantly dissolve in either the vehicle or solvents. Nor must it react with any of the components, such as crosslinking agents. Pigments are required to retain these properties even when the paint is being dried, which is frequently carried out at elevated temperatures. Once in the dried film the pigment must also remain unaffected by the substrate and to agents with which it comes into contact, including water, which may simply be in the form of condensation, or acidic industrial atmospheres.

Although we have defined pigments as being insoluble, in practice they may dissolve under certain conditions leading to application problems. Organic pigments may dissolve to a limited extent in organic solvents, and although inorganic pigments are completely insoluble in such solvents, other chemicals may still affect them. The application problems associated with a pigment being soluble in a system include the following.

3.3.5.1 Blooming

If the pigment dissolves in the solvent, as the paint dries, the solvent comes to the surface and evaporates, leaving crystals of the pigment on the surface in the form of a fine powder. This deposit looks very much like the bloom seen on the surface of soft fruit — hence its name. As solubility increases with temperature, this phenomenon is made worse at elevated temperatures, such as in stoving paints. As the pigment has also dissolved in the binder, usually in the form of a supersaturated solution, it continues to migrate even when the solvent has evaporated and continues to appear on the surface even after being wiped clear.

3.3.5.2 Plate out

The effect of plate out looks similar to blooming, but occurs in plastics and powder coatings. However, it is not due to the pigment dissolving, but rather to the surface of the pigment not being properly wetted out. It usually occurs mainly with complex pigments and once wiped from the surface does not reappear.

3.3.5.3 Bleeding

Pigments in a dried paint film may dissolve in the solvent contained in a new coat of paint applied on top of the original film. If this topcoat is the same colour this will be of little consequence. However, if the topcoat is a different colour, particularly a white or pale colour, the result can be disastrous. It is most noticeable when a white paint is applied over a red paint. The red pigment bleeds into the new white paint film, staining it pink. Subsequent coats will not cover the original, as the pigment continues to dissolve. Again elevated temperatures exacerbate the problem. Although usually associated with red pigments, it is possible with any pigment possessing only poor to moderate solvent fastness.

3.3.5.4 Recrystallization

This phenomenon was almost unknown until the introduction of beadmills. During the milling stage heat is generated, which dissolves a portion of the pigment. The paint often goes through quality control without a problem, appearing strong and bright, because the pigment is acting as a dye in a supersaturated solution. However, over a period of time the dissolved 'pigment' starts to precipitate out, losing brilliance and colour strength. This becomes especially noticeable in the case of paints containing two differently coloured pigments, e.g. a blue and a yellow, that have different solubility characteristics. The more soluble yellow pigment dissolves and then as it comes out of solution and precipitates, the paint will go bluer in shade. Recrystallization can even take place in aqueous systems. It can be avoided by using less soluble pigments and/or by controlling the temperature during the dispersion process.

3.3.5.5 Imparting insolubility into an organic molecule

Insolubility can be imparted to a molecule in a number of ways; however, most of them increase costs, and therefore it is a property that is critical for the optimum economic performance:

- *Increasing molecular size* decreases solubility, therefore in general the more complex the pigment, the better is its resistance to solvents. However, just building up the size of the molecule is likely to have other effects, such as reducing tinting strength and increasing costs, so it would be unlikely to be an efficient method on its own.
- *Adding certain substituents* can either increase or decrease a pigment's insolubility. Long chain alkyls, alkoxy, alkylamino, and acids such as sulfonic acids groups all increase solubility, whereas insolubilizing groups include carbonamide (—CONH—), nitro, and halogens. These groups have proved particularly effective for the so-called 'azo' pigments.
- *Introducing a metal group* reduces solubility. The formation of metal salt with certain water-soluble dyes that contain carboxy acid or sulpho groups can make the colorant insoluble. The more acid groups in the dye, the better the solvent fastness obtained. This is a relatively cheap method, but such pigments, known in Europe as *toners*, have a limited application in paint because of their poor resistance to chemicals. Several new and more complex toners have been recently introduced for the coloration of plastics. The introduction of metal groups by the *formation of metal complexes*, where the bond is covalent also produces pigments that are highly insoluble.
- *Intermolecular bonding* can often result in pigments having much better resistance to solvents than one would expect from their molecular structure. The best example of such a pigment is linear *trans*-quinacridone (CI Pigment Violet 19). If the shape of the crystal does not allow the formation of secondary bonding (i.e. the α crystal modification), the molecule does not have good resistance to solvents, whereas the β and γ modifications have excellent solvent fastness.

3.3.5.6 Testing solvent resistance

The bleed resistance can be tested by overcoating the pigmented paint with a white paint. Normally, either a nitro-cellulose lacquer or stoving paint is used for this test. Fastness to individual solvents can be tested by incorporating a small amount of pigment into an envelope of filter paper, then immersing in solvent for a given amount of time. The staining of the solvent is compared with a standard iodine/potassium permanganate solution. Rating 5 indicates no staining, whereas severe staining is rated 1. For quality control purposes a representative solvent blend can be used.

3.3.6 Opacity

The terms 'hiding power' and 'opacity' are often interchanged, but there is a subtle and important difference. Hiding power is the ability of a pigmented coating to obliterate the surface. It is dependent on the ability of the film to absorb and scatter light. Naturally, the thickness of the film and the concentration of the pigment play a fundamental role. The colour is also important: dark, saturated colours absorb most light falling upon them, so blacks and deep blues cover up the surface, whereas yellows do not. Yet carbon black and most organic blue pigments are quite transparent, because they do not scatter the light that falls on them. At the other extreme, titanium dioxide absorbs almost no light, yet its ability to scatter light ensures that at a high enough concentration it will cover the substrate being coated. In practice a combination of pigments is often used in order to obtain optimum results.

In most coating systems we want to obliterate the surface, but there are certain applications where this is not the case. This is particularly true of printing ink applications, where high transparency is often a necessity. Even certain types of paint may require transparency, especially wood finishes, certain can coatings and in metallic and pearlescent finishes, now so popular in automotive finishes.

A key factor in the opacity of a pigment is its *refractive index* (RI), which measures the ability of a substance to bend light. The opacifying effect is proportional to the difference between the refractive index of the pigment and that of the medium in which it is dispersed. This is one of the main reasons why titanium dioxide is now almost universally used as the white pigment in paint. In Table 3.1 the refractive index is given for white pigments, important extender pigments, and for media in which they may be incorporated. It explains why titanium dioxide gives such good opacity and why extender pigments are transparent or nearly so.

The refractive index is fundamental to a given compound providing it retains the same crystal structure. As most inorganic pigments have a high refractive index and organic pigments have much lower values, it follows that most inorganic pigments

Table 3.1

Medium	RI	Pigment or extender	RI
Air	1.0	Calcium carbonate	1.58
		China clay (aluminium silicate)	1.56
Water	1.33	Talc (magnesium silicate)	1.55
		Barytes (barium sulphate)	1.64
		Lithopone 30% (zinc sulphide/barium sulphate)	1.84
Film formers	1.4–1.6	Zinc oxide	2.01
		Zinc sulphide	2.37
		Titanium dioxide: Anatase	2.55
		Rutile	2.76

Scatter

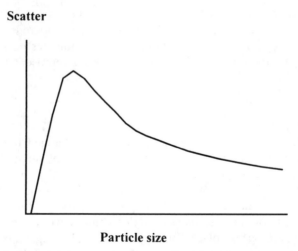

Particle size

Fig. 3.1 — The effect of particle size on the ability of a pigment to scatter light.

are opaque, whereas organic pigments are transparent. If the refractive index was the only factor affecting opacity this would always be the case, but there is another factor that also plays an important role, the particle size of the pigment — or more correctly the particle size distribution. As the particle size increases, the ability of the particle to scatter light increases, up to a maximum (see Fig. 3.1). It then starts to decrease. This ability to scatter light increases the hiding power of the pigment, and therefore the hiding power also reaches a maximum and then decreases as the particle size increases.

Whereas the refractive index of a compound cannot be altered, the pigment manufacturer can influence the particle size of pigments; consequently particle size selection has become one of the principal developments in pigment technology in recent years.

The particle size that gives optimum opacity can be calculated for any given wavelength [4, pp. 129–133]. For maximum opacity, the desired particle size is approximately half the dominant wavelength. A whole series of organic pigments is now available in which the particle size has been made as near as possible to this optimum point. As these particles are considerable larger than conventional pigments, they also have a lower surface area and therefore less binder demand. So not only are such pigments inherently more opaque, but it is possible to use them at higher concentrations without adversely affecting flow properties, gloss, or weatherability. Indeed the lightfastness and weatherability are much improved. The disadvantage of such pigments is that they have much reduced tinctorial strength, but as they are usually intended for full and near full shade paints, where one cannot use the even more opaque titanium dioxide, this is not such a problem.

There has even been the development of organic pigments that have even larger particles, moving them beyond this point of maximum opacity, in order to combine transparency with the higher light fastness one sees from larger particles. This phenomenon has been called 'ultra transparency', but has the disadvantage that such pigments have low tinctorial strength.

The techniques used to manufacture pigments in as narrow a particle size distribution as possible are published and depend on dissolving the pigment in various solvents, at elevated temperatures and pressures, then recrystallizing under carefully controlled conditions [8].

As titanium dioxide is used for its covering power, many studies have been undertaken to determine how this property can be optimized. Whereas organic pigments have a particle size of between 0.1 and 0.5 μm, inorganic pigment can be much larger. Therefore titanium dioxide manufacturers often try to reduce the size of their pigment particles, down to around 0.25 μm, thereby enhancing opacity. Crystal size also plays a role, as does the state of dispersion and, needless to say, pigment concentration [9].

3.3.7 Transparency

Usually, transparency is obtained by producing as small a pigment particle as possible. This is achieved by surrounding the particles as soon as they are formed with a coating, which prevents the growth of crystals. The most common products used for this coating are rosin or rosin derivatives [10]. Rosin is a natural product, whose main constituent is abietic acid. It is often reacted with alkalis, such as sodium hydroxide, to form alkaline earth metal rosinates. This is particularly useful for printing ink pigments that are required to have high transparency and it has the added advantage that such pigments are more easily dispersed. For higher-quality pigments that are required to have better mechanical stability, alternative coatings such as more complex rosin derivatives with higher melting points and greater resistance to oxidation or other surface treatments are used.

There is an important group of transparent inorganic pigments. Although iron oxide pigments are normally opaque, they can be made in a transparent form. The transparent form is widely used for metallic finishes, where its high transparency not only gives an attractive finish, but its excellent weatherability can actually improve the weatherability of pigments with which they can be combined. This is known as a synergistic effect. Transparent iron oxides depend on the particles being unusually small, and also having a specific crystal shape.

The dispersion process can influence the transparency obtained, as it consists of breaking up clusters of particles to individual primary particles. However, dispersion does not break down these primary particles, which are the result of the pigment's manufacturing process, so all one can do is to make maximum use of the pigment's inherent particle size. Good dispersion will make maximum use of a small particle's transparency.

3.3.7.1   Measurement of transparency and opacity

Transparency is simply assessed by applying the coating over a black and white contrast chart and measuring the colour difference. The greater the colour difference, the higher the transparency.

Covering power is measured in a similar way, but is normally expressed in terms of what film thickness achieves complete obliteration of the surface. Therefore, the coating is applied in a wedge shape over a contrast chart. The film thickness is built up over the length of the chart, which is attached to a metal panel. The point at which complete obliteration is observed is noted and the film thickness at that point measured.

3.3.8   Durability

For most paint applications, pigments are required to be durable. They should neither fade nor darken to any extent, a function not just of the pigment but of the whole system. The pigment concentration and other pigments used in the paint play a role. Durability must also be considered in terms of the end use. Clearly an automotive paint must have very much better light and weather fastness than the paint on a gardening or kitchen tool.

Pigments must not weaken the film formed by the binder. Indeed, in a well-formulated paint it is likely to assist in producing a cohesive, hard, elastic film, protecting the substrate and increasing the life expectancy of the paint film. This will also be dependent on whether the paint film is sufficiently thick, another property that can be influenced by correct pigmentation.

For corrosion-resistant paints, the pigment is a fundamental component, often providing chemical protection to the substrate. Strictly speaking this contradicts the definition of a pigment. It would be impossible to do justice to such pigments in a single chapter; indeed corrosion inhibition is a technology in its own right.

3.3.8.1   Light fastness

When considering light fastness one must look at the whole pigmented system, not just the pigment. The binder imparts a varying degree of protection to the pigment, so the same pigment will tend to have better light fastness in a polymer than it will in paint. Even within paints, one generally finds that a pigment's light fastness improves from an aqueous emulsion system, through an air-drying alkyd to an acrylic/melamine formaldehyde system. Pigments will nearly always have a much poorer light fastness in a printing ink system, where there is less resin to protect the pigment and there is a double effect of light passing through the pigmented layer, being reflected by the substrate and back through the pigmented layer. Similarly, metallic finishes are more prone to fading than solid colours and hence are often overcoated with a clear resin containing an UV absorber.

Other pigments may also play a role in a pigmented system. Titanium dioxide promotes the photodegradation of most organic pigments, so the higher the ratio of titanium dioxide (i.e. the paler the shade), the worse the light fastness. On the other hand iron oxide appears to improve the light fastness of organic pigments, probably because it is an effective absorber of UV light. If two pigments together give a better light fastness than one would expect it is called a synergistic effect, where they are lower, it is called an antagonistic effect.

A few inorganic pigments are unchanged by exposure to light, but most pigments and all organics are changed in some way. This may show itself as darkening on the one hand or complete fading on the other. The energy responsible for fading is sometimes mistakenly believed to be exclusively UV light. This is not always the case: certain pigments may appear to have good fastness in UV, but fade on outdoor exposure, and vice versa. While the chemical constitution plays a fundamental role in a pigment's ability to resist light, other factors such as the pigment concentration, the crystal modification, and the particle size distribution have an effect. Additionally, other factors in the environment can dramatically affect results, such as the presence of water and chemicals in the atmosphere or in the paint system. Hence the light fastness of a pigmented system can only truly be tested in the final formulation and application. Light fastness tests must be carried out only under carefully controlled test conditions. Therefore, much effort has gone into perfecting a test system that will give as meaningful results as possible when evaluating the light fastness of a pigmented system.

3.3.8.2   Light source

Although natural daylight would be the ideal, it is normal to try to accelerate the test by using an intense source of light that corresponds as closely as possible to natural daylight. One of the most commonly used light sources is the xenon arc lamp, which can be further improved by filters, which screen out certain parts of the UV spectrum. As the xenon lamp ages, the lamp's energy distribution changes, so it is important to follow manufacturers' instructions regarding the life of the lamp. Control of humidity is essential when testing light fastness.

3.3.8.3   Blue Wool Scale

Light fastness is quantified by exposing the sample alongside a Blue Wool Scale and comparing the fading obtained with the sample with the strip of wool showing similar fading. The Blue Wool Scale consists of eight strips of wool all dyed with standard dyes. In theory, strip 1 fades twice as fast as 2, which fades twice as fast as 3, etc. Therefore, a difference of two points is a four-fold difference and three points is an eight-fold difference, etc. In practice this is not quite the case especially at the top end of the scale, where most paints are needed to be, but it is a good guide. It can also be difficult to assess the light fastness of a glossy paint film with a textile substrate, which can be of a completely different colour and depth of colour. Work is going on to try to develop the use of printed strips, but there are problems, particularly at the lower end of the scale. It is also agreed that certain modern paint systems, such as automotive and coil coating systems, need a light fastness in excess of eight, and therefore require an extended scale.

Ratings are based on the 1–8 scale, where 1 is poor and 8 is excellent. Interim steps can be indicated, e.g. 5–6 and darkening can be denoted by 'd' after the figures.

3.3.8.4   Weatherability

Although closely related to light fastness, weatherability adds the extra dimension of atmospheric conditions, including salt from the sea, waste gases from industrial areas, or very low humidity from desert conditions. It is therefore necessary to include in any weatherability data the exact conditions of exposure. Machines are available which in addition to a xenon lamp, include wet cycles interspersed between longer dry cycles, depending on the exposure anticipated.

Weatherability is designated in terms of the 1–5 Grey Scale, where 5 is no change, and 1 is a severe change. To repeat, as well as the figure it is necessarily to denote the place and period of exposure, if necessary with a note about the prevailing conditions.

3.3.8.5   Effects of chemistry on light and weatherfastness

Light fastness in organic molecules, especially azo pigments, usually increases with increasing molecular weight. A planar molecular structure also promotes light fastness, probably due to intermolecular bonding, as exemplified by linear *trans*-quinacridone, where the β and γ modifications have much better fastness properties than the α form. Polycyclic pigments with a symmetrical structure generally have high fastness, e.g. phthalocyanine, qunacridones, perylenes, and carbazole violet.

Additional groups can be introduced into organic, especially azo pigment molecules, which increase the light fastness. Chlorine groups invariably improve light fastness. Other groups that have a positive effect include $-NO_2$, $-CH_3$, $-OCH_3$, $-CONH-$, $-SO_2$, and NH$-$.

3.3.9   Chemical stability

The requirements of a pigment to withstand any chemicals with which it may come into contact may well be a critical factor in the selection process. One has to consider two aspects of chemical attack. First it can come from within the paint itself. The resin, crosslinking agents, UV-initiators, and any other additive may react with the pigment and change its performance. In the early days of UV-cured coatings, additives on the pigment drastically reduced storage stability, resulting in the coating gelling in the can. One has to be very careful when selecting pigments for powder coatings, as the initiator can change their shade and reduce fastness properties. Reputable pigment manufacturers publish data on such systems and can usually help in case of difficulties.

The second possible attack can come from chemicals with which the coating comes into contact. Water, in the form of condensation, can seriously affect a paint film, including domestic situations such as bathrooms and kitchens. Many of the detergents used for cleaning paintwork are harsh and aggressive towards the pigment. If the coating is to come into contact with food it is essential that not only the coating is unaffected, but the food must also remain completely unchanged. Before one dismisses the possibility of food affecting a paint film, remember how many foods contain acid (vinegar), alcohol, and fatty acids.

Most test methods associated with chemical stability consist of applying the chemical to the surface of the coating, keeping them in contact for a given time, then measuring the discoloration of the coating and/or the staining of the chemical concerned. By the very nature of the wide demands made on a coating, published data will be of limited value; therefore it may be necessary to devise special tests to reproduce conditions found in practice.

3.3.10   Heat stability

Few pigments degrade at temperatures normally associated with coatings. However, several pigments will change shade because they become more soluble as the temperature is increased. Therefore for organic pigments, heat stability is closely related to solvent fastness. Nevertheless, pigments that may prove entirely satisfactory at one stoving temperature may prove quite unsatisfactory at an application requiring a 10°C higher stoving temperature.

Chemical stability is also more likely to be critical at elevated temperatures; this is often very relevant in powder coating systems. Another critical area is in coil coatings, where metal complex pigments may react with stabilizers at elevated temperatures, producing distinct changes in shade. Elevated temperatures can also result in a modification in the crystal structure of pigments.

Pigments with a highly crystalline structure are usually more heat stable than polymorphic pigments, where the different crystal modifications may respond differently to heat. Generally, inorganic pigments have superior heat stability, but an exception is yellow iron oxide, which loses water from the crystal at elevated temperatures.

Heat stability must always be regarded as system dependent and this must be reflected in any tests. All tests work on the principle of assessing the colour at a series of different temperatures and measuring the colour difference between the sample in question and a standard that has been processed at the minimum temperature.

3.4 Pigment classification

Trying to place all pigments into a logical and meaningful classification system is not really possible. This is especially true for pigments that are not primarily used for their colour, such as white and extender pigments. Coloured pigments have traditionally been classified as either organic or inorganic. While there are characteristics associated with each of these two classes of pigment, modern manufacturing techniques are developing new products, which impart properties not previously associated with their chemical type, e.g. organic pigments have traditionally been transparent; however, there are now organic pigments with high opacity.

Inorganic pigments have a long history, going back to the early cave paintings that are about 30000 years old. They can occur naturally, but for the manufacture of paint they usually need refining. However, they are increasingly synthesized. All the white pigments are inorganic as are a broad range of coloured pigments.

Organic pigments are relatively new. Although natural dyes have been precipitated on to inorganic bases (known as lakes) and used in artists' colours since the middle ages (e.g. madder lake and crimson lake), true organic pigments have only been known since the early years of the twentieth century. They are derived from fossil fuels, usually oil.

The properties that have traditionally been associated with inorganic and organic pigments are summarized in Table 3.2. A more detailed classification of inorganic and organic pigments into chemical types is shown in Fig. 3.2 and typical properties associated with chemical types are shown in Table 3.3.

Table 3.2 — Traditional properties associated with organic and inorganic pigments

Pigment property	Inorganic	Organic
Colour	Often dull	Usually bright
Opacity	Normally high	Normally transparent
Colour strength	Usually low	Normally high
Fastness to solvents — bleed resistance	Good	Varies widely from good to poor
Resistance to chemicals	Varies	Varies
Heat resistance	Mostly good	Varies
Durability	Usually good	Varies
Price	Often inexpensive	Varies, but some are expensive

Table 3.3 — General properties of pigment chemical types

Properties in paint	Titanium dioxide (coated)	Iron oxide (not *trans*)	Prussian blue	Lead chromate	Carbon black	Monoazo	Disazo	Disazo condensation	Metal azo toner	Phthalocyanine
Colour	E	F	E	E	E	E	E	E	E	E
Opacity	E	E	E	E	E	F-G	F-G	F-G	F-G	P
Tinting strength	E	P-F	G	F	E	G-E	E	F-G	G-E	E
Heat stability	E	G-E	G	G	G	P-F	G	E	G	E
Fastness to solvents	E	E	E	E	E	P-F	G	E	G	E
Resistance to										
Acids	E	E	F	F-G	E	E	E	E	P	E
Alkalis	E	E	P	P	E	E	E	E	P	E
Chemicals	E	E	P	G	E	E	E	E	P	E
Light fastness										
Full/near full shades	E	E	G	G	E	G-E	F-G	E	F-G	E
Pale reductions	E	E	F	P	E	F	P-F	F-G	P-F	E
Dispersibility	E	E	F	G	F-G	G	G	G	G	F
Flow properties	E	E	G	E	F	G	F	G	F	P
Flocculation resistance	G	G	G	E	P	G	G	G	G	F
Price	Low	Low	Low	Low	Low	Low–high	Low–medium	Medium–high	Low–medium	Low–Medium

E, excellent; G, good; F, fair; P, poor (however, within most categories there can be notable exceptions).

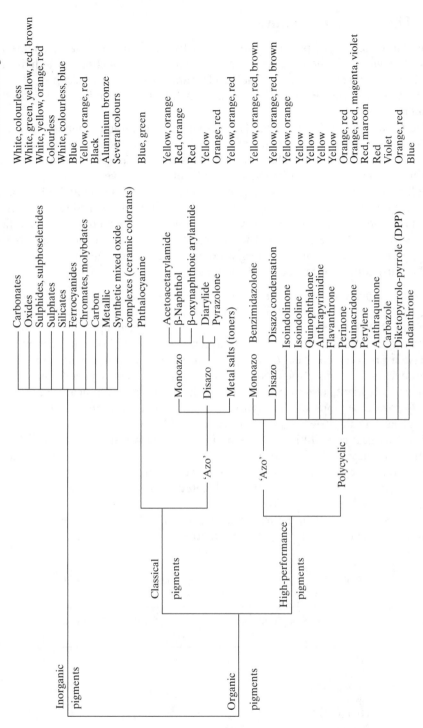

Fig. 3.2 — Pigments for paint, white, coloured and extenders (classification by chemical type).

3.5   Pigment nomenclature

Although a number of pigments are known by their common names, sometimes associated with their chemical type (e.g. toluidine red), connections with their origin (e.g. Prussian blue), or their original trade name (e.g. Hansa® Yellow), the most common system used for identifying pigments are Colour Index (CI) names and numbers.

The *Colour Index International* [11] provides a logical and comprehensive way of identifying the essential chemical nature of dyes, pigments, intermediates and optical brightening agents. However, there is also a *Pigment and Solvent Dyes* version [12]. This only lists pigments and solvent soluble dyes.

Colorants are designated by a generic name, e.g. CI Pigment Yellow 3. This generic name is used for all pigments (or dyes) that have an identical chemical structure. The chemical structure, where known, is indicated by the five-figure *constitution number*, e.g. 11710. The chemical class to which the colour belongs (e.g. monoazo) and the hue (e.g. bright greenish yellow) are given.

The third edition has three main sections:

1   Tables indicate the pigments' physical properties, such as solvent fastness and light fastness. However, this can be misleading as products designated CI Pigment Yellow 3 do not necessarily have identical application properties. Particle size distribution, surface treatments, and method of manufacture can all have a pronounced effect on the pigment's colour, colour strength and fastness properties. Therefore, pigments registered under such a generic name while chemically identical may not necessarily have the same physical properties.

Separate sections of the complete *Colour Index International* list dyes by application categories, e.g. CI Acid Blue 1, CI Vat Orange 7.

2   The chemical constitution, where published is shown in order of their CI constitution number. An indication of the method of synthesis, the name of the first known maker and references are often given where these are known. There are a few products that have not been allocated a constitution number, as the constitution has not been disclosed.

3   This section is split into three sub-sections. The first section lists the names, addresses, and code letters attributed to colorant manufacturers. The second

Table 3.4 — Number and chemical types of pigments listed in the 1997 *Colour Index*

Colour Index pigment	No. of different chemical types listed
White	15
Black	24
Brown	23
Blue	36
Green	19
Yellow	122
Orange	36
Red	146
Violet	26
Metal	3

Table 3.5 — Main pigments by volume

Pigment	Type
CI Pigment White 6	Titanium dioxide
CI Pigment Red 101	Red iron oxide
CI Pigment Yellow 42	Yellow iron oxide
CI Pigment Blue 15, 15:1, 15:2, 15:3, 15:4	Copper phthalocyanine
CI Pigment Yellow 13	Diarylide yellow (AAMX type)
CI Pigment Red 57:1	Monoazo 'Lithol® Rubine' toner

section lists all the colorants in order of generic name and lists all the commercial names that have been registered as being that generic name with the manufacturer's code. Finally the commercial names of all registered products are given with their appropriate manufacturer's code, generic name and constitution number.

The complete *Colour Index International*, is now available in the form of a CD-ROM. The 'Pigment and Solvent Dye' edition [12] is simpler to use, as it combines the chemical structure and associated information with all the commercial names of products with that structure. The main application is listed and there is an opportunity for manufacturers to distinguish their commercial product. A second section lists commercial products under their trade name, giving their CI name and constitution number. A breakdown of the entries in this edition is given in Table 3.5. This is also available in a CD-ROM format.

Of all these pigments, excluding extenders, about 30 account for 80% volume of the total manufacture and consumption. In Table 3.5 some of these main pigments are listed. It should be noted that in the case of the coloured pigments, the influence of the printing ink industry, which has a higher consumption of coloured pigment than the paint industry, predominates.

3.6   Further types of pigments and terms used

3.6.1   Inorganic pigments

The most important pigment in coatings is without doubt titanium dioxide — or more correctly titanium (IV) oxide. It combines a pure white colour with a high refractive index, making it opaque. Efforts to improve its surface have resulted in pigments that are easy to disperse. Early grades were prone to a defect known as chalking. This was caused by a photochemical reaction in which radicals broke down the binder, leaving a chalky effect on the surface of paints, owing to particles of pigment becoming exposed on the surface. This reduces the gloss of the finish and in the case of coloured finishes gives a whitening effect on the surface that develops over a period of time. However, surface treatment of the pigment can now minimize this effect.

Titanium dioxide is now universally used as the white pigment in paint, but like many organic pigments it exists in more than one crystal modification [13]. Brookite is not used as a pigment. The anatase grade is a slightly bluer white and is softer in texture. However, it has two major disadvantages. It is far more likely to chalk than the rutile crystal form and it has a lower refractive index (2.55) and is therefore less opaque than the rutile grade (refractive index of 2.76). Consequently, for surface coatings the rutile grade is usually preferred.

If we trap a foreign ion into the rutile crystal lattice, we can obtain some weakly coloured pigments that are known as mixed metal oxides, or less correctly as 'titanates'. Such pigments were developed for the ceramic industry, but are now used for either pale shades with excellent fastness properties or in place of titanium dioxide in near full shades. Nickel (greenish yellow) and chromium (reddish yellows) ions are the most commonly used to obtain such pigments.

3.6.2   Types of organic pigment

Many of the organic pigments we now use derive from dyes that were developed for the textile industry. In Table 3.5, the last pigment is described as a 'toner'. This term can be confusing because it has different meanings in different countries. Another term that causes confusion is 'lake', again because usage varies in different parts of the world. In this chapter we will adopt the UK usage, as it is the most precise [11].

In the USA the term 'toner' is used to describe all organic pigments, as they are used to 'tone' duller inorganic pigments, whereas the term 'lake' is used to describe both lakes and toners in UK usage.

3.6.2.1   Pigment toners

Toners are essentially metal salts of dyes, mainly acid dyes, in which the anion is the source of the colour. Most acid dyes are lithium, potassium, sodium, or ammonium salts. However, salts produced with metals such as calcium, manganese, barium, and strontium are mainly insoluble. Therefore, this is an easy method of producing relatively simple and economic pigments that find significant usage, especially in printing inks.

CI Pigment Red 57:1 (Lithol® Rubine or Rubine toner), which is normally used as the standard magenta in three and four colour printing processes, is the most important pigment of this type. CI Pigment Red 53:1 (Lake Red C — which in spite of its name is in fact a toner) and CI Pigment Red 49:1 (Lithol® Red) are the most important in inks. CI Pigment Red 48:4 (2B toner) is frequently used in industrial paints. However, this method of producing insolubility in the molecule is now being increasingly adopted in the development of pigments for plastics, where much more complex anions are being utilized, providing toners with a significantly better light fastness. CI Pigment Yellow 183, Pigment Yellow 191 and Pigment Red 257 are good examples.

The metal ion used for the toner does have an influence on both the colour and the fastness properties of the pigment:

$$Na \rightarrow Ba \rightarrow Ca \rightarrow Sr \rightarrow Mn$$
$$\rightarrow \text{Improving lightfastness} \rightarrow$$
$$\text{Orange} \rightarrow \qquad \rightarrow \text{Bordeaux}$$

It is also possible to produce salts from basic dyes, in which it is the cation that produces the colour, but this is a little more difficult. Cationic dyes (or basic dyes) are usually in the form of quarternary ammonium salts, so in order to make the molecule insoluble one has to use complex acids, producing phosphomolybdate, phosphotungstomolybdate, phosphotungstate, silicomolybdate, or ferrocyanide complexes. These pigments are often known as Fanal® pigments. Their use is mainly limited to printing inks, usually for packaging. They have intense, very bright colours, but limited fastness properties especially to light and chemicals.

3.6.2.2 Lakes

The first lakes were natural dyes which were then absorbed on to an absorbent inorganic base such as blanc fixe or alumina hydrate. Following the discovery of the first synthetic dye by Perkin in 1856 and subsequent discoveries, countless new dyes became available and many of these could be similarly precipitated. Such lakes are rarely used today except occasionally in inks and food dyes (tartrazine lake, etc.). Some of the last to survive are quinizarin lake (CI Pigment Violet 5), which uses alumina as a base, and pigment scarlet 3B lake (CI Pigment Red 6O:1), which uses aluminium hydroxide and zinc oxide as a base.

Just in case anyone is tempted to feel the need to go back to nature, beware! The consequences of producing lakes from natural dyes appear to have had a much more damaging impact on the environment than current manufacturing methods. In the Middle Ages the production of blue dyes and lakes from woad was severely curtailed because of the catastrophic effect it was having on the environment. The woad plant requires large quantities of potassium, so the soil on which it was cultivated quickly became devoid of the mineral and thus completely infertile. Another step in the production process involves fermentation, during which obnoxious gases are given off, making life intolerable for those living within 3 km (two miles). Finally, the waste produced contaminated watercourses, reeking havoc on town and countryside alike.

To reiterate, the term 'lake' is widely used in the USA and elsewhere to describe what are known as 'toners' in the UK.

3.6.2.3 Organic pigments

Organic molecules that are insoluble, without having to precipitate on to a base or being made insoluble by conversion into a salt, used to be called pigment dyestuffs. However, this confusing term is no longer encouraged. The term 'organic pigment' will suffice. They can be conveniently sub-divided into two, azo pigments and polycyclics.

3.6.3 Azo pigments

This is by far the most important sector and includes about 70% of all organic pigments, especially in the yellow, orange, and red parts of the spectrum. As well as including many pure organic pigments, the azo types also include many of the toners. The traditional structure has always been assumed to be the azo structure, characterized by the —N=N— group, e.g. CI Pigment Red 3 has always been designated by its assumed constitution based on an azo structure:

Traditional structure assigned to CI Pigment Red 3

However, work [14] on the measurement of bond lengths and bond angles convincingly shows that this pigment does not actually contain the azo group, but has a hydrazone structure. This would allow a more planar structure and in part explains the insolubility of the molecule. CI Pigment Red 3 is now considered to have a hydrazone structure:

More recent structure assigned to CI Pigment Red 3

Many other pigment structures have been investigated and the evidence suggests that all had the hydrazone, rather than the previously assumed azo structures [15, 16]. While the term 'azo' is unlikely to be changed, like the *Colour Index* we will adopt the 'hydrazone' structure, as the evidence that this is the correct structure is very convincing.

Azo pigments can be subdivided into monoazo and disazo pigments.

3.6.3.1 Monoazo pigments

Monoazo pigments have the simpler structure and tend to have poorer solvent stability. The simplest types are the β-naphthol pigments represented by CI Pigment Red 3 (toluidine red). In the yellow part of the spectrum the simplest pigments are the arylamide yellow pigments, such as CI Pigment Yellow 1. These pigments (with the exception of CI Pigment Yellow 97 — see section 3.12.4) can only be used for water-based media or for media that are soluble in weak organic solvents such as white spirits. They have poor heat stability, whereas light fastness varies from poor to very good.

β-Naphthol colorants are among the oldest synthetic colours still being manufactured. They owe their origin to the textile industry, being introduced in 1880 as 'developing dyes'. Cotton impregnated with β-naphthol (2-naphthol) and dried, is then passed through a liquor of ice-cold diazonium compound, forming a coloured pigment inside the fibre. Five years later, Gallois and Ullrich produced para red (CI Pigment Red 1) in its own right and it can therefore be considered the first true synthetic organic pigment. Although para red is rarely used today, several pigments with a similar structure are still important, especially for decorative paints.

More complex are the *arylamide red* pigments (also known as *BON arylamides* from their chemical basis–Beta Oxy Naphthoic acid, or *naphthol AS* ® pigments). They have an additional benzene ring attached to the β-naphthol by a carbonamide linkage (—CONH—). The higher molecular weight and this carbonamide linkage improve the solvent resistance of these pigments. The extra benzene ring also provides more sites for additional substituents, thus extending the colour range from orange through to violet and allows the possibility of improving solvent resistance even further. Consequently, these pigments have a wide range of properties from moderate to very good.

Solvent fastness can be increased even further by the addition of the carbonamide group in a cyclical form. This replaces a benzene ring with 5-aminocarbonyl benzimidazolone, giving rise to the name given to such pigments — *benzimidazolone*. The range of colours goes from very green shade yellow, through orange, red right through to violet. They have good to excellent solvent fastness and good to excellent light fastness.

3.6.3.2   Disazo pigments

Disazo pigments can be divided into two groups. The *diarylide* pigments form the larger and older group. They are mostly based on 3,3'-dichlorobenzidine, but their low light fastness means they find most use in the printing ink industry. Nevertheless, their high colour strength and good resistance to solvents and heat does mean they can be used for some industrial paint applications. Of particular interest is the opaque form of CI Pigment Yellow 83 on account of its superior light fastness and can therefore be used for more demanding applications. The colour range can be extended into the orange and even red part of the spectrum by using aryl-substituted pyrazolones.

The other group contains the high-performance *disazo condensation* pigments, that consist of two monoazo pigment molecules containing an acid group joined together by condensing with a diamine. The higher molecular weight thus obtained, plus the two carbonamide groups formed by the condensation reaction, give such molecules very high resistance to solvents and heat, coupled with good light fastness. The range of colours obtained goes all through the spectrum from greenish yellow to bluish shade red and also includes brown.

3.6.4   Polycyclic pigments

This is a convenient term used to link most of the organic pigments that are not classified as azo. They are defined as compounds with more than one five or six-membered ring in their chemical structure. Many of the pigments in this category were originally developed as vat dyes. They only came into consideration as pigments when work in the USA modified their surface by a process called conditioning, allowing the products to be dispersed into a fine particle size. They are expensive to manufacture and can only be economically justified where their good heat stability chemical and solvent resistance and good durability, even in pale shades, are needed. A few of these pigments do have a tendency to darken on exposure to light, especially when used in higher concentrations.

Within this classification are also included *phthalocyanine* pigments, which are of such importance they constitute a group in their own right, having various crystal forms and substituted derivatives. A number of other polycyclic pigments have been developed. *Quinacridone* pigments also exist in two crystal forms and include various substituted grades, the best known of which is quinacridone magenta (CI Pigment Red 122). *Dioxazine violets* are most useful pigments, allowing one to redden off blue and blue off reds without losing brightness.

In many ways *isoindolinone* pigments can be considered to be the ultimate in chemical 'architecture'. The molecule contains two —CONH— groups and all the spare sites on the outside benzene rings are chlorinated. As one could predict, these pigments have excellent fastness to light and excellent solvent fastness, but have relatively low colour strength. A simpler pigment, with slightly lower fastness but higher colour strength is *isoindoline* (CI Pigment Yellow 139).

An important group of pigments for high-quality coatings are the *perylenes*, which range from a mid red (CI Pigment Red 149) through a bluish red grade (CI Pigment Red 224) to black (CI Pigment Black 31 and Black 32). Some of these pigments are available in high-opacity grades, ideal for full shades, or in high strength, transparent grades that are used for metallics.

In the yellow/orange part of the spectrum there have been a number of metal complex pigments based on *azomethine*. These pigments give very interesting effects when used in metallic finishes but are less interesting for solid shades as they are rather dull.

A significant new group of pigments has been the *diketopyrrolo-pyrrole* (or DPP) which combine high fastness properties with a wide range of shades. Again they lend themselves to careful selection of particle size, so the same chemical pigment may be in more than one form, one having high opacity and the other high strength and transparency. They are mainly used in automotive and other high-fastness demand finishes.

3.7   Particulate nature of pigments and the dispersion process

3.7.1   Crystal structure

Pigments can be crystalline or non-crystalline (i.e. amorphous). For pigments to be crystalline, the atoms within each molecule are arranged in an orderly fashion, but in amorphous pigments the atoms are random. Some materials are capable of existing in several different crystalline forms, a phenomenon called polymorphism.

Application properties, not least colour, are not only dependent on the chemical constitution but also on these different structures. Certain pigments may have chemically identical entities in different crystal forms, which are not suitable for use as a pigment. Examples of polymorphic pigments include titanium dioxide, phthalocyanine blue and linear trans quinacridone.

Pigment manufacturers are developing techniques that influence the formation of a desired crystal form and particle size distribution, in order to optimize the commercial product for end applications.

3.7.2   Particle shape

The shape of a pigment particle is determined by its chemical structure, its crystalline structure (or lack of it) or how the pigment is synthesized. The primary particles of a pigment may be:

- Spherical — unlikely in practice;
- Nodular — irregular shaped;

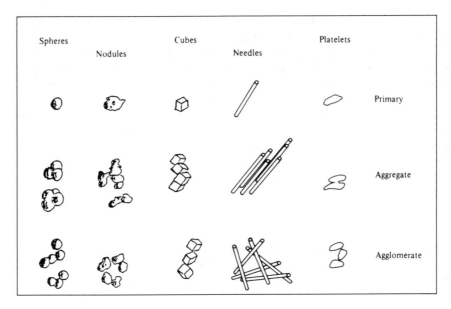

Fig. 3.3 — Particle shapes.

- Prismatic — cubic;
- Acicular — needle-like;
- Lamellar — plate-like.

These shapes are illustrated in Fig. 3.3.

Primary particles consist of single particles. The smaller these particles, the greater their surface energy and therefore there is an increased tendency to clump together during the manufacturing process. However, even if it were possible to supply pigments in the form of primary particles, it would be quite impractical, as they would be more like smoke than a powder. In practice, they only exist as the pigment is synthesized. When the particles clump together during the manufacturing process they form either aggregates or agglomerates.

Aggregates are joined along crystal boundaries during synthesis or drying. They are difficult to separate and therefore the pigment manufacturer tries to avoid their formation during the pigment's production.

Agglomerates are loose clusters of primary particles, and can be broken down by efficient dispersion processes.

The states of aggregation and agglomeration are key factors in the pigment's dispersibility. Once dispersed it is still possible for particles to re-agglomerate, into loosely held groups, known as flocculates. It frequently occurs when there is a rapid change of state, such as too rapid a dilution, or the addition of an incompatible substance. Flocculation results in a loss of tinctorial strength. However, flocculates are usually easier to separate than true agglomerates, and even normal shear such as brushing out is sufficient. This results in an uneven increase in tinctorial strength, depending on how much shear has been developed during brushing out. Small particles are more prone to flocculation than larger ones, so pigments most at risk are grades of carbon black and conventional fine organic pigments, such as phthalocyanine and dioxazine violet pigments. However, there are an increasing number of flocculation-stable grades coming on the market.

The crystal shape of a pigment, particularly in the case of larger particles, does influence the way they pack in the paint film and consequently can affect the properties of the paint. Thus, properly dispersed acicular particles are said to reinforce the paint film like the fibres in glass fibre reinforced plastics. Lamellar particles such as aluminium and mica, form an overlapping laminar structure, not unlike roof tiles, and like roof tiles they offer resistance to the passage of water, allowing the paint film to impart good protection properties to the substrate.

Particle shape can also influence the shade of a pigment. It has been shown that β-copper phthalocyanines (CI Pigment Blue 15:3) give different hues, depending on the deviation from isometry to acicularity [17].

3.7.3   Particle size and particle size distribution

It is important to distinguish between the particle size distribution of the powder pigment and that of a pigment dispersed in the binder system. Particle size is normally expressed as an average diameter of the predominant primary particles. However, pigment particles are not usually spherical, and therefore may have different dimensions depending on whether one measures the length, width or height. This requires making a number of assumptions and simplifications. Typical ranges are:

- carbon black — 0.01 to 0.08 μm;
- organics — 0.01 to 1.00 μm;
- inorganics — 0.10 to 5.00 μm;
- titanium dioxide — 0.22 to 0.24 μm.

Extender pigments can be among the coarsest pigment particles, up to 50 μm, but other types can be exceptionally fine (e.g. the precipitated silicas).

The particle size of small particles is difficult to measure, and is made even more difficult by the fact that one needs a dilute suspension, and dilution can often result in flocculation, thus changing the size of the particles. Sedimentation methods were employed [18] but increasingly techniques involving the use of lasers are employed [19].

3.7.4   Surface area

Although measuring the surface area does not give any information about the shape of crystals, it is still most useful as it is closely linked to the pigment's demand for binder. It is usually defined in terms of the surface area of 1 gram of pigment and has the unit of square metre (m^2). Typical values for organic pigments are between 10 and 130 m^2. Most manufacturers give the surface area of their pigments in technical literature. Measurement is usually made by calculating the adsorption properties of the pigment, using the Brunauer, Emmett, and Teller (BET) equation with either nitrogen or argon as the gas [20].

Larger particles have a smaller surface area and therefore a lower demand for binder. Measurement of the surface area has largely replaced the former measurement 'oil absorption' which tended to be too subjective. A pigment's oil absorption is expressed as the weight of an acid-refined linseed oil in grams, required to produce a smooth paste with 100 grams of pigment. Even with standard oils the method is dependent on how the test procedure is conducted, but can give useful advice on formulating efficient mill bases for the dispersion process. The ease with which a pigment can be wetted out also influences the oil absorption.

Care must be taken when using heavily surface-treated pigments, as these can give very strange results, but such pigments are not widely used in the paint industry.

3.7.5   Dispersion

Efficient and effective pigment dispersion is necessary in order to obtain optimum tinctorial strength, cleanliness of shade and good gloss from the final coating. Most organic pigments also show better transparency as dispersion improves, while in the case of the larger particle size inorganic pigments, opacity is improved by good dispersion.

The dispersion process consists of the permanent breaking down of agglomerates into, as far as possible, primary particles. The smaller the primary particle, the more difficult this process becomes, although the finishing treatments given to the pigment, including the addition of surface-active agents, can bring about tremendous improvements. Consequently, organic pigments tend to require more dispersion than inorganic pigments. It is possible to overdisperse some pigments, especially those with needle-like crystals, such as bluish red iron oxides, and inorganic pigments which have a protective coating, such as molybdate red. If this coating is worn down, the fastness properties of the pigment can be significantly reduced. Although the process is often described as 'grinding', this is rarely accurate. There are four aspects to the dispersion process, which can to some extent take place alongside one another.

1   *Deagglomeration* is the breaking down of the agglomerates by the shear forces of the equipment being employed. A mixture of crushing action and mechanical shearing force provides the means of breaking down agglomerates. A higher viscosity leads to greater shearing forces, but the specific equipment being used will put practical limits on the viscosity of the millbase.
2   *Wetting out* takes place at the pigment surface and requires a binder or surface active agent to anchor itself on to the pigment's surface and act as a bridge between the pigment and the binder itself. The wetting process can be calculated [4, p. 77]. The calculation reveals that the time to wet out is directly proportional to the viscosity. The heat generated by the mechanical shearing process results in an increase in the temperature of the mixture, thus assisting in the wetting out process. However, this temperature rise also reduces the viscosity, thus reducing the effectiveness of deagglomeration, a phenomenon supported by experience and research [21].
3   *Distribution* requires the pigment to be evenly spread out throughout the binder system. It tends to be favoured by a lower viscosity.
4   *Stabilization* prevents the pigments from re-agglomerating. It is often neglected in the manufacturing process. While the theories that try to explain stabilization are complex, in practice it is found that it is during the dilution process that prob-

lems are most likely to occur. These problems can be minimized by diluting as slowly as possible, so that the pigment can be stabilized at each stage before further dilution takes place.

3.7.6 Dispersion equipment

The choice of dispersion equipment must be made with the end application clearly in mind. One of the most effective machines for dispersion is the three-roll mill, which can deal with high-viscosity systems and coarse pigments. However, it is rather slow and labour intensive. On the other hand, ball mills require very little labour, but are not very effective when dealing with coarse pigments and higher viscosity systems. Between these two pieces of equipment there have been some very significant developments based on sand and bead mills. Every machine has its own specific requirements and therefore the experience of those employed to run them efficiently must be considered [22].

No matter how proficient modern mills are, all operate more efficiently when used with a good premix prior to passing through the mill. An effective premix should contain sufficient wetting agent or binder to coat the pigment particles. However, it must also be of low enough viscosity to produce a millbase that will proceed through the mill. It must be remembered that the dispersion process itself will inevitable increase viscosity as binder is absorbed by the new pigment surfaces that are produced, owing to the resultant increased dispersed particle phase volume. This effect may, however, be counteracted by the viscosity decrease resulting from the temperature increase during the milling process.

The beads (or sand) in the mill also play an important role. The nearer to a perfect sphere the better, hence Ottawa sand is considered to be the best natural milling material. The higher the density of the beads, the better the dispersion that can be achieved. Glass beads are frequently used but steatite, zirconium oxide, and the more resistant to wear zirconium silicate are more dense and therefore have advantages. In spite of its price, even yttrium-stabilized zirconium beads provide a higher density and greater hardness. However, it is important to ensure the beads used are suitable for the mills being employed as they may result in wearing out the lining of the mill's chamber. Economics can also play an important role, some of the more expensive beads such as the yttrium-stabilized zirconia are reputed to wear out at a much slower rate and retain their spherical shape. They may also remain much smoother, thus producing less heat and wear on the impeller discs and the chamber walls.

While smaller beads can also improve dispersion, when mixed with the millbase these smaller beads will result in a higher viscosity that can strain or even block the mill. The less dense the milling media, the more critical this effect, so it is possible to use smaller steatite beads than glass under the same conditions. The technology of separating beads from the millbase is constantly developing and seals and screens are now much more efficient and open up new options.

The dispersion process is accompanied by an increase in temperature of the millbase. This can be very significant and some pigments will be adversely affected, especially pigments with poor resistance to solvents. The pigments can dissolve, leading to recrystallization and subsequent colour strength and shade stability problems. Therefore, when dispersing such pigments it is important to ensure there is sufficient mill cooling to prevent the millbase becoming too hot, by using as cold a chilling medium as possible and by having a sufficient cooling surface area to millbase volume. Note that as mills increase in size, the ratio of surface area to volume reduces proportionally to the diameter.

3.8 Manufacture of pigments

Many of the inorganic pigments can be obtained from natural sources, but paint industry's demand for consistency mean that even these natural pigments, such as iron oxides, are now being made synthetically. The main pigments used in their natural form are the so-called extenders, which include china clay, barytes, and chalk. Commercial organic pigments are all made synthetically, originally from coal tar but now from petroleum distillates.

3.8.1 Synthetic inorganic pigments

Each group of inorganic pigments has its own manufacturing route. Some are the result of simple chemical reactions, such as lead chromates and molybdates. Chrome yellow (CI Pigment Yellow 34) can be manufactured by mixing solutions of lead nitrate with sodium chromate.

Other inorganic pigments such as titanium dioxide (CI Pigment White 6) depend on the extraction of the desired compound from ores, either from ilmenite ($FeTiO_3$) by the sulphate process or from rutile (impure TiO_2) by the chloride process.

In recent years the chloride route has developed, as it lends itself to continuous production, produces a slightly purer white, and has advantages with respect to environmental concerns. The rutile ore is treated with chlorine in the presence of coke at a temperature of around 800–1000 °C. Titanium tetrachloride gas is formed. It is condensed and purified by fractional distillation. The purified titanium tetrachloride is then superheated and passed into a reaction chamber with oxygen at about 1500 °C. After cooling the reaction products are separated and the chlorine gas recycled [23].

Ultramarine blue, which is a synthetic version of the semiprecious mineral lapis lazuli, can be manufactured by the calcination of china clay, silica, sulphur, and a carbonaceous material such as coal tar pitch. Temperatures of around 800 °C for two to four days are required, followed by cooling in an oxidising atmosphere for several days. For other pigments, calcination can sometimes be used to change or modify the crystal structure.

Iron oxide pigments do occur naturally, but for the paint industry synthetic grades predominate. They cover a wide area of the spectrum from yellow ochre (CI Pigment Yellow 42), through red ochre (CI Pigment Red 101), brown (CI Pigment Brown 6) to black (CI Pigment Black 11). They are manufactured by various methods [24].

Carbon black (CI Pigment Black 7) is made by the partial combustion of natural gas. The process most frequently used is the furnace process [25].

Following any synthesis, pigments have to be washed, filtered, dried, milled, and finally classified. Additives to improve the manufacturing procedures and also to modify the final pigment properties are included at various stages of the production process.

3.8.2   Synthetic organic pigments

The synthesis of organic pigments from oil is a multistage process involving many critical steps. In the case of some of the more complex pigments, which can involve over 20 syntheses, it may take many months, and until the final step one is never sure that the process has been satisfactory.

The pigment manufacturers are unlikely to carry out all these steps themselves, so that materials from each stage may be sold between manufacturers to achieve such benefits as cost savings derived from economies of scale and greater utilization of specialist plant.

The materials at each stage are given the terms:

1   primaries;
2   substituted primaries;
3   intermediates;
4   crude.

Primaries are obtained directly from petroleum distillation. The most important are the aromatic hydrocarbons:

Benzene 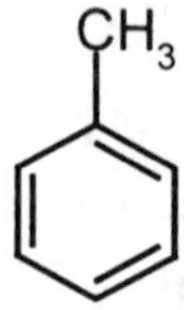   Toluene

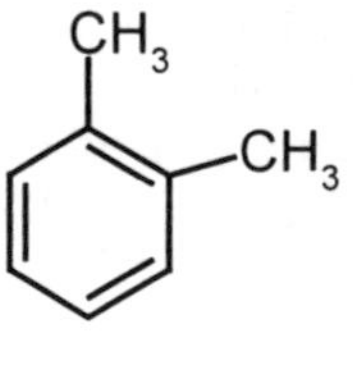

Ortho-xylene 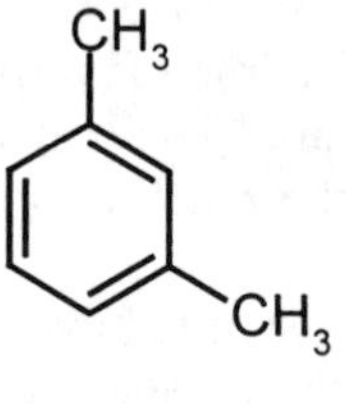   *Meta*-xylene 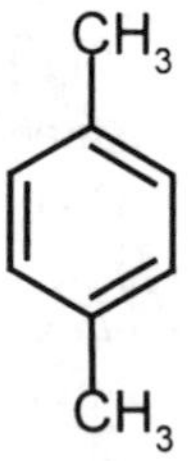   *Para*-xylene

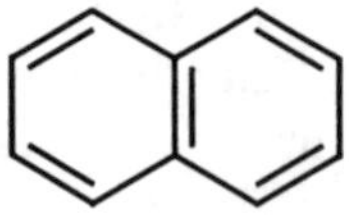

Naphthalene 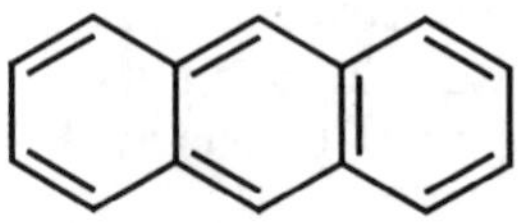   Anthracene

These primaries are converted into 'substituted primaries' and then into intermediates by a variety of well-known classical chemical reactions. Hundreds of

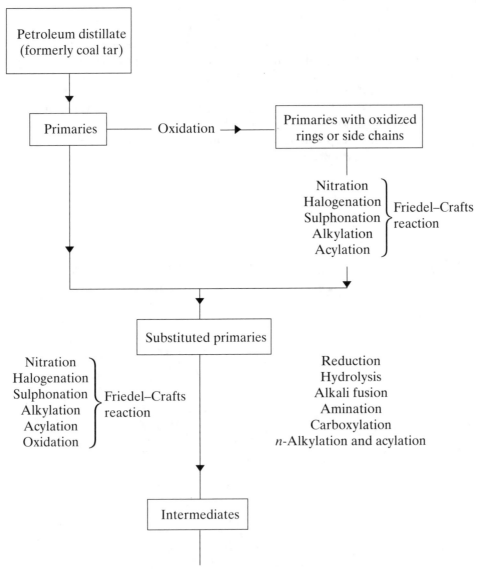

Fig. 3.4 — Typical production processes for organic pigments.

intermediates are involved. Some may be manufactured as specialist products from one or two manufacturers; others may be commodity type products bought on the open market.

A schematic production process is shown in Fig. 3.4.

As typical examples of pigment manufacture we will look at two very important classes of pigment. Since around 80% of all organic pigments are classified as azo pigments we will use the preparation of toluidine red (CI Pigment Red 3) as typical to illustrate a simple azo process:

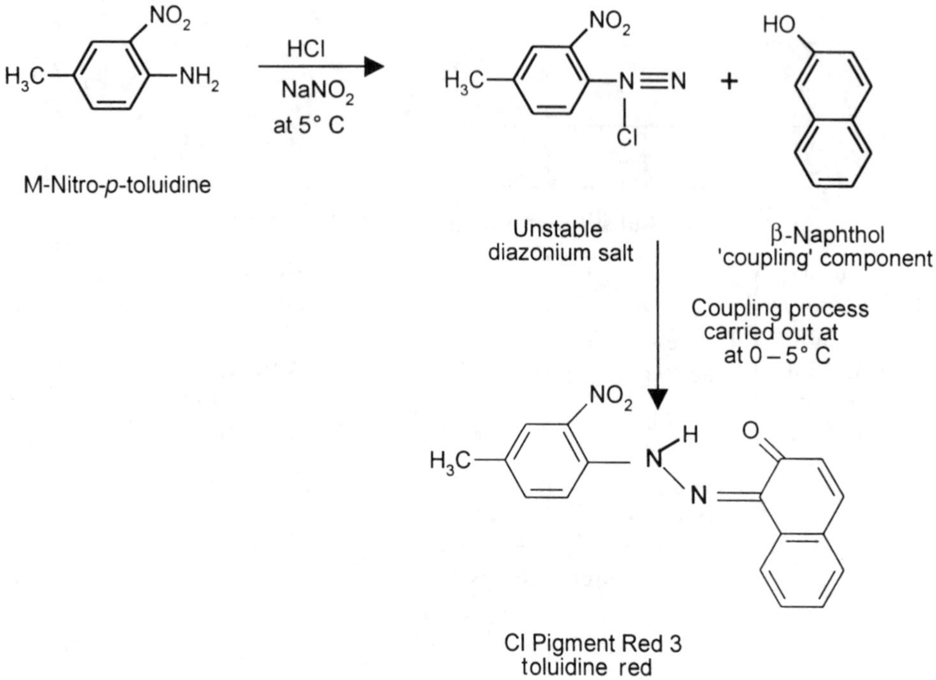

Fig. 3.5 — Production of typical azo pigment, toluidine red.

There are several methods of manufacturing copper phthalocyanine, but two predominate. The most frequently used and the cheaper route starts from phthalic anhydride and urea and has been used in the UK and the USA. It takes place in either an organic solvent or in the vapour phase. The vapour phase is becoming more popular as it produces less undesirable by-products. Phthalic anhydride is melted with urea above 150°C in a rotating drum in the presence of a catalyst. A copper salt, such as copper (II) chloride is added and the temperature increased to about 200°C. The reaction mixture is cooled, washed and steam distilled to remove impurities. The resulting product is known as crude copper phthalocyanine and is of little use unless further processed. Some pigment manufacturers produce their own 'crude' while others purchase the crude and condition themselves.

Production of a phthalocyanine pigment from phthalic anhydride

However, a technically superior product can be obtained from phthalonitrile, using a process developed in Europe and traditionally favoured by German manufacturers.

3.8.3 Pigment conditioning

The chemical synthesis of both organic and inorganic pigments is by no means the final stage. The chemical compound formed has to go through further processing before becoming a commercial product, and it is often these stages that allow the experienced pigment manufacturer to offer commercial products that have application advantages. These stages usually include many washings, filtering, drying, and classification.

For a typical *azo* pigment, synthesis would result in the formation of tiny insoluble particles (primary particles). The manufacturing conditions such as temperature, pH, and rates of addition will all influence the crystal shape and crystal size of the product. The solid particles have to be separated from the liquor by filtration, which is achieved by means of a filter press. The slurry is squeezed under intense hydraulic pressure, which results in the so-called 'presscake'. Normal presscakes would contain about 30% pigment, but modern equipment can achieve over 50% pigment. This looks like a damp powder. The drying process results in the primary particles joining together to form agglomerates and aggregates. Agglomerates are loose associates of primary particles and can be broken down by the dispersion process, whereas aggregates are joined along crystal boundaries. The resulting hard-grained pigment is virtually impossible to disperse, so if it were used in paint the paint would develop poor colour strength.

Even if one could obtain pigments in the form of primary particles, it would be quite impractical. They are so small and light that they would resemble smoke. However, by careful control of the conditions and after treatment one can develop optimum application properties. The first stage is to wash out the salts and impurities from the presscake with water. Various methods are then used, but traditionally heating the purified pigment suspension in water removes some of the very fine particles that are mainly responsible for forming harder agglomerates. In the case of the more complex pigments with lower solubility, it is necessary to use solvents or solvent/water mixtures at elevated temperatures. This treatment has become very well developed. In addition certain auxiliaries may be added. The most common is rosin or rosin derivatives. They are able to coat the particles as they are formed and prevent the growth of particles. Later, during the dispersion process it facilitates improved wetting of the particles, assisting their incorporation into the binder. Other auxiliaries are sometimes used which resemble the pigment molecule, but contain groups that have an affinity for the application media, again improving wetting and facilitating easier dispersion.

The first stage in the drying process consists of applying hydraulic pressure and squeezing out most of the water. This produces presscakes, but still contains between 45–70% water, so has limited application. Therefore, it is normal to dry the press-cake. This drying process is almost a technology in its own right. It can be a long, slow process lasting as long as two days. The presscake is spread out as thinly as possible and dried on trays or on conveyer belts, using large steam-heated ovens. Pigments that are heat sensitive may be dried under partial vacuum.

Another method of drying involves the use of spray dryers, where the pigment paste is sprayed into a heated cone shaped chamber.

The dried pigment then has to be pulverized. Each pigment will have its own characteristics and too heavy a pulverization can lead to re-agglomeration.

Azo pigments can be manufactured as a batch process or on a continuous oper-ation [4, p. 208].

Crude phthalocyanine, which is in the β form, can be converted to a pigment grade by a number of methods, depending on the crystal modification required. An important stage in conditioning is to dissolve some of the pigment in concentrated sulphuric acid. When added to cold water the α crystal form is produced. The pres-ence of surface-active agents improves the quality.

The pigmentary grade can also be prepared by dry milling of the crude with a salt such as sodium chloride or sodium sulphate. The crude β grade slowly changes into the α form, but a complete conversion is not usually possible and commercial grades may contain around 10% of the β form. However, if a small amount of polar solvent is present, the β form is retained.

3.9   Toxicity and the environment

Pigment suppliers and paint chemists have a responsibility to protect their own health and the health of those who use their products, and the environment. However, ideas on how this can best be achieved with respect to people, animals, and plants are constantly changing. Suppliers have a responsibility to provide full infor-mation on their products, including their physiological properties and their effect on the environment. Details on how to handle pigments safely in the workplace and how to dispose of wastes should be given. This allows users to assess the hazards associ-ated with such raw materials and formulate suitable handling procedures.

Compared with many other chemicals, most pigments are of low hazard, pri-marily because of their low solubility, which limits their bioavailability and hence their toxicity. However, as this is such a complex subject, the industry is vulnerable to scare stories, badly thought-out legislation, and misinformation. In an effort to overcome such problems the pigment manufacturers through their relevant indus-try associations, have made a great effort to be proactive in going beyond their legal obligations and communicate the hazards associated with pigments and to explain in clear language some of the issues.

In 1993 the Color Pigments Manufacturers Association, Inc. (CPMA) published *Safe Handling of Color Pigments* [26]. In addition to providing details of most of the hazards associated with pigments, it explains many of the terms used in conveying safety information. It also details the legislation affecting pigments and details of the various agencies charged with the enforcement of such legislation in North America.

Outside North America, the industry recognized the contribution made by this publication and four organizations co-operated to produce a similar publication but with an emphasis on European legislation. In 1995 ETAD (representing organic pigment manufacturers), BCMA (representing the British pigment manufacturers), VdMI (representing inorganic pigment manufacturers and many users in Germany) and EPSOM (representing French pigment manufacturers) published *Safe Handling of Pigments* [2].

Both publications have been widely circulated throughout the pigment-using industry and provide an excellent starting point for anyone trying to achieve a better understanding of issues and concerns associated with the use of both organic and inorganic pigments and are therefore essential reading.

3.10   Choosing pigments

Before selecting a pigment for a specific paint the formulator needs to consider the type of paint being manufactured and its end use. The markets in which they are to be used may classify paints. The usual classifications are those used for buildings, either architectural or decorative, automotive finishes, either OEM (original equipment manufacturers) or VR (vehicle refinishes), and those used for industrial finishes. In addition there are powder coatings and coil coatings, both of which cover a range of demands.

3.10.1   Paint types

The binder in the paint system plays an important role in the selection of the pigment, in particular the solvent in which it is dissolved. Water is increasingly being used as a solvent and this makes relatively little demand on pigments, except for some toners. Long oil alkyd paints, as used in many decorative gloss paints, usually use white spirit as their solvent. Most pigments are insoluble, or almost insoluble in white spirit, so this rarely restricts the pigments one can select. General industrial finishes can be based on a whole range of solvents: many such as xylene, ketones, and esters are very powerful solvents and can dissolve pigments with poor or only moderate resistance to solvents. If such paints are dried by stoving, the elevated temperature makes even greater demands on the pigments used. One also has to consider if the coating is likely to be overcoated. It is safe to assume that most cars will at some time have to be repaired, so any pigment used in the original finish will have to be fast to overcoating.

In powder coatings many of the crosslinking agents can affect the pigment, so it is important to ensure the pigments are stable to these agents, at the temperatures likely to be employed during application.

The type of resin used and the solvent used for the resin therefore remain key factors in the choice of pigment.

3.10.2   Paint performance

Consideration has to be given to the end use of the paint, especially with respect to its durability requirements, any resistance to chemicals and the price that can be tolerated for improved performance. Clearly it would be as futile to use cheap, low-quality pigments in an automotive finish as it would to use a high-quality pigment in a paint that was to be used in a gardening implement! The difference between the price of a cheap organic pigment and one giving a high performance can be a factor of around 20. The speciality pigments developed for high performance are often very complex in nature and have limited application, hence their high price.

3.10.3   Opacity and transparency

In pale shades, the inorganic pigment used, usually titanium dioxide, normally provides the covering power. The influence of the coloured pigment is insignificant. However, deeper shades have a much lower proportion of titanium dioxide, hence one needs more opaque coloured pigments. Such shades are known as solid shades.

Certain special effects can only be achieved using transparent finishes. These include the metallic and pearlescent paints frequently used in automotive finishes. The aluminium or pearlescent pigments provide the covering power, but the coloured pigments used are required to have high transparency in order to allow the sparkle of the aluminium or pearlescent pigment to show through, giving clarity to the finish. This results in the paint appearing to have a different colour depending from which angle it is viewed, and hence the length of the light path through the coating. This is the so-called flip tone. Although dyes can be used, they rarely provide sufficient resistance properties. As the light passes through the film twice, this can make extra demands on light fastness, so only pigments with the highest standards of light fastness can be employed for such applications. Transparent pigments are also required in some clear lacquers such as flamboyant finishes that are used for bicycles, can coatings, or as wood stains.

The restrictions placed on certain inorganic pigments based on metals restricted by legislation have resulted in the search for safer inorganic pigments, or for more opaque organic pigments. Several organic pigments have been introduced that are characterized by high opacity, especially in the yellow, orange, and red parts of the spectrum. Conversely, iron oxide pigments can be prepared in a very fine particle form, resulting in their having high transparency. Transparent iron oxide pigments are frequently used in automotive finishes where they tend to lift the performance of brighter organic pigments as they are able to absorb UV radiation.

3.10.4   Colour and blending of pigments

Few colours, with the exception of black and white, can be matched with a single pigment. Instrumental methods for matching colours have simplified the process, but a working knowledge of pigment technology is of great advantage before one chooses the products to be used, as few pigments find universal application.

Trying to keep the formulation simple and thereby using as few pigments as possible is a good principle. In a pigment blend, solvent resistance depends on the worst pigment present. Likewise, the pigment with the worst light fastness has a disproportionate effect on the light fastness of the total system, e.g. a pale orange made from CI Pigment Yellow 74 and CI Pigment Red 3 will rapidly fade towards the yellow side because of the inadequate fastness of the red.

Purity of shade is also a factor one must consider. It is normal to use dull pigment, such as iron oxide, and tint to the required shade with a brighter organic pigment, thus utilizing the economy and the fastness of the iron oxide to the maximum. When bright colours are mixed together, the resulting shade becomes duller the further they are from each other in the spectrum, hence blue and red give dull purple (or maroon) shades, whereas red and violet gives a much brighter purple shade.

Another consideration when blending pigments is the avoidance of using mixtures of pigments with wide differences in particle size, as this can give rise to flotation and flooding. This effect can often be seen in grey paints based on carbon black (as small as $0.01 \mu m$) and titanium dioxide ($0.25 \mu m$).

3.10.5 Quantity of pigments in a paint
The amount of pigment used in a paint is determined by:

- its intensity and tinctorial strength;
- the required opacity;
- the gloss required;
- the resistance and durability specified.

The paint technologist works on one of the two main concepts, either pigment volume concentration (PVC) or pigment to binder ratio (P:B). The PVC is of fundamental concern when formulating paints that are required to have optimum performance with respect to durability. It is known that there is a critical point that represents the densest packing of the pigment particles commensurate with the degree of dispersion of the system [27]. It is a complex calculation but essential for paints that have to meet the highest performance standards with respect to durability.

The P:B ratio, by weight or occasionally by volume, is a much simpler calculation, often used to assist in formulating a good millbase and for balancing a formulation for gloss and opacity.

For systems requiring high gloss, a low PVC is required, whereas primers and undercoats can have a much higher PVC — up to 90%.

3.11 Physical forms of pigment

Most pigments are supplied as free-flowing powders, although some are supplied as low-dusting granules. However, there are several other forms available depending on the requirements of the paint manufacturer.

3.11.1   Pigment preparations

As pigments have become more complex in nature and the systems into which they are incorporated have become more numerous and complex, traditional dispersion techniques have not always proved satisfactory. The best example is in water-based emulsion paints. Although one can disperse titanium dioxide and extenders into emulsion systems, it is much more difficult to disperse organic or even coloured inorganic pigments into the resin without affecting the properties of the resin itself. Therefore, such pigments are normally dispersed into a non-ionic and/or anionic surfactant system and added to the paint as a paste. The pigment manufacturer, the paint manufacturer, or a third party specializing in such products can produce these pastes.

Such aqueous preparations are required to have a high degree of compatibility and should be resistant to drying out. Most organic pastes will contain around 40–55% pigment content, iron oxide pastes can have a much higher pigment content, but carbon black pastes are usually much lower.

Multi-purpose pastes are a natural progression from such products. As well as having good compatibility with emulsion paints, they can also be used for the coloration of air-drying alkyd paints. This is achieved by using surfactants that not only act as dispersing agents and aid compatibility in emulsion paint, but also act as emulsifying agents in solvent-based systems. Again they can be supplied by certain pigment producers, made by the paint manufacturer or supplied from a third party. These colorant pastes are the basis of the tinting systems frequently seen in large paint retailers, providing a much greater selection of shades than would be available in a ready mixed range.

Certain paint systems require very high degrees of dispersion. This is frequently necessary in automotive metallic shades and specialist companies are able to provide such a service. Also in the general industrial area, there is an increasing demand for pigments that offer a wide range of compatibility, allowing the paint manufacturer to concentrate on producing high volumes of base paints and tinting to shade with stainers.

3.11.2   Presscake

Most pigments are synthesized in an aqueous system. The pigment then has to be dried, a costly operation. However, many applications are water based. Earlier presscakes contained around 70% water and felt like damp powders, but it is now possible to extract a much higher proportion of water using hydraulic and vacuum techniques, making presscakes with as little as 40% water available. Such products are primarily of interest to aqueous preparation manufacturers.

3.11.3   Flush pastes

Organic pigments in the form of flush pastes are now widely used for the production of offset inks. If the pigment presscake is mixed with a binder, such as linseed oil, and thoroughly mixed in the presence of selected surfactants, the pigment eventually transfers from the aqueous phase to the oil phase. Such pigment pastes have undergone little agglomeration and aggregation, so they are relatively easy to disperse. As the end user has little control on the binder, flush pastes have very limited use in the paint industry.

3.11.4   Masterbatch

Most plastics, especially polyolefins and engineering polymers, are coloured with masterbatch. These are either single pigments or blends to a certain standard shade, which are highly dispersed in a suitable medium, allowing consistent and easy coloration of polymers. This approach is finding favour with manufacturers of powder coatings and can also include the incorporation of other ingredients, such as stabilisers. However, to date most manufacturers continue to start from powder pigments.

3.12   Notes on families of pigments

In this section some of the more important pigment types are briefly described and compared. The information cannot be complete, and should only be used in conjunction with manufacturers' data and tests in the system being used. The pigments are grouped in colours, but there can be some overlap, so for instance, while most orange pigments are detailed under orange, some may appear in the red or yellow sections, if their chemistry better fits these colour areas.

3.12.1   White pigments

All pigments in this sector are inorganic and it is totally dominated by titanium dioxide, or more correctly titanium (IV) oxide. It was first manufactured in 1916, developed through the 1930s, but only became the dominant white pigment after the Second World War. It has largely displaced other white pigments. Whereas coloured pigments are compared by their tinctorial strength, in the case of white pigments the reducing power is used as the comparison. This is the amount of white pigment needed to produce an equal depth of shade when used with a standard amount of coloured pigment.

- Titanium dioxide
- White lead
- Zinc oxide
- Zinc sulphide
- Lithopone
- Antimony oxide

3.12.1.1   Titanium dioxide
Colour Index — CI Pigment White 6.
Formula — TiO_2.

Titanium dioxide exists in three crystal forms: *brookite* is not used as a pigment, *anatase* is only used occasionally and is softer in texture and has a slightly better white colour than the most commonly used crystal form *rutile*. However, rutile has a higher refractive index (around 2.76 compared with around 2.55 for anatase) and is therefore more opaque and it has much less tendency to chalk (see section 3.6.1).

Modern grades have much better chalk resistance imparted by treating the photoreactive sites on the surface of the crystals with inorganic oxides, such as those of aluminium, silicon and zinc.

Two different processes are used to manufacture the rutile grade. The sulphate method is the older and starts from the ore ilmenite. Ferrous sulphate and dilute sulphuric acid are by-products of the process, and it is their disposal that has caused environmental problems. Pigments produced by this method used to be slightly stained by the iron salt, giving it a slightly yellow colour, but this is no longer necessarily the case.

The more modern chloride process commences with the ore of titanium dioxide called rutile. The ore is mixed with coke ground to a powder. It is then reacted with chlorine at temperatures around 900 °C, to form titanium tetrachloride, along with oxides of carbon. Titanium tetrachloride is a liquid and can be condensed, separating off impurities. The purified titanium tetrachloride is vaporized, super-heated, and reacted with oxygen at around 1500 °C. The chlorine that is given off is recycled.

Titanium dioxide possesses many of the ideal properties demanded of a white pigment. It has a good white colour and very high resistance to organic solvents, most chemicals, and heat. However, it is its high refractive index which makes it such an attractive pigment, allowing the production of highly opaque coatings. It has good durability and is unaffected by industrial atmospheres. Its photoreactivity does reduce the light fastness of some coloured pigments, including almost all organics.

Early grades suffered from poor dispersibility, but again coating the pigment particles can produce much-improved commercial products. The balance between dispersibility and durability generally requires products that are specific to a range of applications, such as water-based paints or solvent-based stoving finishes. Being inert, it has favourable physiological properties and can be used for food packaging, toys, and other sensitive applications, providing it meets purity criteria [28].

In spite of its excellent properties, titanium dioxide is reasonably priced. Nevertheless attempts are made to replace or partially replace it with cheaper pigments or semi-opaque extenders.

3.12.1.2   White lead
Colour Index — CI Pigment White 1.
Formula — $2PbCO_3Pb(OH)_2$.

This pigment reacts with acidic binders giving tough elastic films, which have good durability, especially on woodwork. However, the pigment also reacts with the sulphurous gases in industrial atmospheres and goes black. Concerns regarding the toxicity of lead compounds severely limit its use.

3.12.1.3   Zinc oxide
Colour Index — CI Pigment White 4.
Formula — ZnO.

Zinc oxide can also react with binders by producing zinc soaps in oleoresinous paints, which confer hardness and resistance to abrasion and moisture, but can cause severe thickening in the can. It gives some mildew resistance in paints and reduces the breakdown in binders, which results in chalking, so can be a minor component of titanium dioxide. It was widely used before the introduction of titanium dioxide, but its use has now declined.

3.12.1.4 Zinc sulphide

Colour Index — CI Pigment White 7.

Formula — ZnS.

This pigment produces a good, strong white, good opacity and a high degree of chemical inertness. However, it chalks badly.

3.12.1.5 Lithopone

Colour Index — CI Pigment White 5.

Formula — either ZnS 30% / $BaSO_4$ 70% or ZnS 60% / $BaSO_4$ 40%.

These pigments were available before zinc sulphide. Their properties follow those of zinc sulphide, but are correspondingly weaker owing to the barium sulphate content. They also have poor resistance to chalking. They are relatively cheap and sometimes partially replace titanium dioxide.

3.12.1.6 Antimony oxide

Colour Index — CI Pigment White 11.

Formula — Sb_2O_3.

Originally used to reduce the chalking of anatase titanium dioxide, this pigment is inert and moderately opaque. Its main use today is in fire-retardant paints.

3.12.2 Black pigments

Carbon black is the most common black pigment, although there are a couple of organic blacks, one based on anthraquinone, and also a black iron oxide. Within the term carbon black there is a wide variety of types available:

- Carbon black
- Graphite
- Black iron oxide
- Micaceous iron oxide

There are also a couple of organic black pigments:

- Aniline black — CI Pigment Black 1
- Anthraquinone black — CI Pigment Black 20

3.12.2.1 Carbon black

Colour Index — CI Pigment Black 6, 7, 8.

Formula — C (Carbon).

Carbon black is one of the oldest of pigments, being used by early man in cave paintings 27000 years ago, by the Egyptians as far back as 2500 BC in inks, then through the Middle Ages when the printing process was first developed. They are manufactured by thermal oxidative dissociation of hydrocarbon oils or gas. Although carbon blacks are inorganic, they have many of the properties of organic pigments. They consist of elemental carbon, but there is a wide variety of products available, dependent on the starting materials and method of manufacture. The grade of carbon black selected must be matched to its particular field of application, as commercial grades are manufactured to carefully defined requirements. The paint industry is only a modest user of these pigments. By far the biggest user is the rubber industry, where carbon black not only imparts colour but also increases the strength and is therefore an essential component in tyres. The printing inks industry also uses significant quantities.

These pigments have excellent light fastness and good resistance to solvents, although in some of the cheaper pigment grades, solvents can extract some coloured impurities. Chemical stability and heat stability are excellent. The particle size determines the intensity of blackness, known as jetness, smaller particles giving higher values. There is an internationally recognized scale used to classify carbon blacks, signifying the manufacturing process and the jetness.

The first two letters designate the colour strength of the pigment; high colour (HC), medium colour (MC), regular colour (RC), and low colour (LC). The final letter describes the manufacturing process; furnace (F) and channel/gas process (C):

				Range of average particle sizes
HCC	High colour channel	HCF	High colour furnace	0.010–0.015 μm
MCC	Medium colour channel	MCF	Medium colour furnace	0.016–0.024 μm
RCC	Regular colour channel	RCF	Regular colour furnace	0.025–0.035 μm
		LCF	Low colour furnace	>0.036 μm

The manufacturing process used to produce channel blacks consists of burning petroleum gases from several small jets in an atmosphere deficient in oxygen. The flame plays on rollers and the deposit is scraped off. This process is now almost defunct because it had disastrous consequences on the environment, producing severe air pollution and also because it is less economic.

Furnace blacks are produced by the thermal oxidative process, which involves burning hydrocarbon oils with a high aromatic content, using a single flame, in a limited supply of air. The carbon is collected by electrostatic precipitation. Particles produced by this process lie between 0.01 and 0.08 μm.

The finer particle blacks are used for high-quality finishes such as in automotive paints on account of their higher jetness. Medium size blacks are used for intermediate quality paints, whereas the coarser pigments are used for decorative paints. For tinting purposes coarser grades are usually used as their lower strength means they are easier to control. As the carbon black is in the form of a light dusty powder, it is often converted to an easy to disperse pellet form by the dry pelleting process.

Carbon black pigments can cause problems in paint as over a period of time they absorb active ingredients in the formulation, such as the metal soaps used as driers in air-drying alkyds. It is normal to double the drier content to compensate. Another problem is that being such small particles they have a tendency to flocculate after dispersion. This can be minimized by taking great care when diluting paints. Only small stepwise additions should be made with thorough homogenization between additions.

Carbon black, having very small particles with a corresponding large surface area, can prove difficult to disperse. The dispersibility can be improved by slightly oxidizing the surface of the pigment (this can be shown in the designation by the suffix 'o', e.g. HCC[o]) or by the addition of small quantities of organic groups such as carboxylic acids. Such groups reduce the pH of the aqueous extract and aid the wetting of the particles. Another problem associated with paints having a high proportion of carbon black is structure. The particles of carbon appear to form clusters, often compared to a cluster of grapes, held together by forces ranging from weak physical attraction to chemical bonds. This structure affects the dispersibility, the jetness, gloss, and, most of all, the viscosity of the paint.

Although the price of carbon black varies considerably, generally increasing as the particle size decreases, it is a relatively cheap pigment largely because of the very large quantities manufactured for the rubber industry. Speciality grades made for specific applications are more expensive because there is less economy of scale.

3.12.2.2 Graphite
Colour Index — CI Pigment Black 10.
Formula — C, elemental carbon crystallized in a hexagonal system.

Graphite is the traditional 'lead' in pencils. It is a soft pigment consisting of inert plate-like particles. These lamellar plates form layers in a paint film, which prevent the penetration of water and are therefore used to reinforce other pigments in anti-corrosive paints. As it has low tinctorial strength and its colour is of low intensity, it is not used for its colour, but it does confer high spreading rates owing to the slippery nature of the particles.

Graphite can be obtained from natural sources, in varying levels of purity from 40 to 70% in combination with silica. It can also be manufactured by the Acheson process, which consists of heating anthracite in electric furnaces at high temperatures.

3.12.2.3 Black iron oxides
Colour Index — CI Pigment Black 11.
Formula — ferroso-ferric oxide FeO, Fe_2O_3.

This pigment occurs naturally as magnetite and can be used in its natural form. However, like most iron oxide pigments used in paint, it is now usually manufactured synthetically. It is a cheap, inert pigment with excellent solvent and chemical resistance, and excellent durability. It has limited heat stability and low tinctorial strength.

Its main use is for applications where the tendency of carbon black to float, especially in grey tones in combination with titanium dioxide, cannot be tolerated. Its lower tinting strength can also be an advantage when used as a tinter, as it allows more control.

3.12.2.4 Black micaceous iron oxides
Colour Index — no listings.
Formula — lamellar shaped particles of Fe_2O_3.

Its main use is in heavy-duty coatings to protect structural steelwork, where its plate-like structure impedes the passage of oxygen and moisture. It is inert and its ability to absorb to UV radiation also protects the polymers used as binders. It has a greyish appearance and a shiny surface, not unlike mica, hence its name. Care must be taken not to over-disperse because the platelets can be broken and rendered ineffective.

3.12.2.5 Aniline black
Colour Index — CI Pigment Black 1 (however, the structure given is not considered adequate).
Formula — azine structure:

Aniline black is probably the oldest synthetic organic pigment. It was discovered around 1860 and is not dissimilar to a product with the same name discovered by Perkin, who is regarded as the father of organic dyes and pigments. It is used in some speciality coatings where very deep blacks are required, which it achieves because of its low scattering power and very strong absorption of light. It has a very high binder demand so produces matt effects in paint, often described as velvety in appearance. Its fastness properties are quite good, but its chromium (VI) content limits its application where physiological properties have to be considered.

3.12.2.6   Anthraquinone black
Colour Index — CI Pigment Black 20.

This pigment is used only in camouflage paints, because its infrared spectra satisfies various standards. It has moderately good light fastness and only moderate solvent resistance.

3.12.3   Brown pigments
This area is dominated by iron oxide, but a few organic pigments are used for speciality applications.

- Iron oxides — CI Pigment Brown 6 (synthetic), 7 (natural)
- Metal complex — CI Pigment Brown 33
- Benzimidazolone — CI Pigment Brown 25
- Azo condensation — CI Pigment Brown 23, 41, 42

The azo browns will be considered in the red section (section 3.12.6).

3.12.3.1   Iron oxide brown
Colour Index — CI Pigment Brown 6, 7.
Formula — Fe_2O_3, ferric oxide.

When this pigment is derived from its natural form it is known as burnt sienna or burnt umber. Pigments are made from naturally occurring ores that are then heated. The various shades obtained are dependent on the impurities, especially the manganese oxide content. The names are familiar more by their use for artists' colours than from their use in commercial paints. They have low tinting strength and are not opaque.

The synthetic grades give a rich brown colour with excellent fastness properties, but do command a premium over yellow and red oxides. They are not used to any extent in paint formulations, as similar shades are more economically obtained using mixtures of cheaper pigments.

3.12.3.2   Metal complex brown
Colour Index — CI Pigment Brown 33.
Formula — (Zn, Fe) (FeCr)$_2$O$_4$ in a spinel structure.

This pigment is used mainly in ceramics, but can be used in coil coatings, where its high heat stability and excellent fastness properties make it a useful pigment.

3.12.4   Yellow pigments
There are a large number of pigments in this part of the spectrum, both organic and inorganic. The choice depends on the end application with respect to brightness of shade, opacity, fastness requirements, physiological properties, and economic considerations. As well as being used in yellow paints, yellow pigments are also used in oranges, greens and browns, so it is a very important sector for the paint technologist.

Inorganic:

* Lead chromate
* Cadmium yellow
* Yellow oxides
* Complex inorganic colour pigments (mixed phase metal oxides)
* Bismuth vanadate

Organic:

* Arylamide
* Diarylide
* Benzimidazolone
* Disazo condensation
* Organic metal complexes
* Isoindolinone
* Isoindoline
* Quinophthalone
* Anthrapyrimidine
* Flavanthrone

3.12.4.1   Lead chromate
Colour Index — CI Pigment Yellow 34, Orange 21.
Formula — PbCrO$_4$/PbSO$_4$; the orange is PbCrO$_4$. xPb(OH)$_2$.

These pigments can be pure lead chromate or mixed phase lead chromate and lead sulphate. They can range in shade from a greenish yellow through to orange. The most important pigments are:

* Primrose chrome — a pale, greenish yellow (containing between 45% and 55% lead sulphate), where the pigment has a metastable orthorhombic structure crystal structure, which needs to be stabilized by the addition of additives.
* Lemon chromes — still a greenish yellow, but somewhat redder than primrose, depending on the sulphate content (it contains between 20% and 40% lead sulphate, the more sulphate, the greener the shade). It has a monoclinic crystal structure.

- Middle chromes — reddish yellow, mainly pure lead chromate precipitated in the monoclinic structure.
- Orange chrome — chemically this pigment is basic lead chromate and contains no sulphate. It has a tetragonal crystal structure. Varying the proportion of 'x' in the formulation can alter the shade, but the crystal structure also plays a role.

Lead chromes are noted for their excellent opacity, low oil absorption, very bright shades, and high chroma (i.e. they give deep or saturated shades), making them ideal for full shade yellow paints. They also possess excellent solvent resistance and moderately good heat stability, which can be further improved by chemical stabilization. They are sensitive to alkalis and acids, including industrial atmospheres, which cause them to fade. Their light fastness is usually satisfactory in full shade, but they darken on exposure to light. All these deficiencies can be much improved by surface-treating commercial grades with such material as silica, antimony, alumina, or various metals.

The one drawback that severely limits the wider use of these pigments is that they contain both lead and chromium (VI). This effectively limits their application to certain industrial finishes [2, Section 3.7]. In Europe, paints containing these pigments are required to carry warning labels. When being used for paint manufacture, careful attention must be paid to the handling instructions contained in the safety data sheets. Disposal of waste can also be problematic and local regulations carefully followed.

They are relatively cheap to use and replacing them especially in full shade formulations cannot be achieved without a considerable increase in cost.

3.12.4.2   Cadmium yellow
Colour Index — CI Pigment Yellow 37, 35 (lithopone type).
Formula — CdS, cadmium sulphide, containing some zinc sulphide. The lithopone type contains barium sulphate.

The relatively high price of cadmium pigments has always limited their use in paints, but they do possess excellent heat stability, solvent fastness, and alkali stability. They have good light fastness in full shades, but fade in reductions and discolour in industrial atmospheres. They have excellent opacity and offer bright colours with high chroma.

In Europe, their use in almost all coatings is prohibited, owing to the environmental problems associated with cadmium.

3.12.4.3   Yellow iron oxide
Colour Index — CI Pigment Yellow 42 (synthetic), 43 (natural).
Formula — $Fe_2O_3 \cdot H_2O$, hydrated ferric oxide (or more correctly $FeO(OH)$).

Yellow iron oxide is by far the most important coloured inorganic pigment used in the paint industry. It occurs naturally as limonite, but is better known as raw sienna or yellow ochre. It ranges in shade from a bright yellow to very dull yellow depending on the purity. Natural grades have been used as pigments since humans lived in caves and are still widely used for colouring concrete, but in the paint industry the synthetic grades predominate.

There are various routes for its synthesis. It has a needle-like shape (acicular) and the final colour is dependent on the dimensions of the crystal. The longer the needle, the greener and purer the colour. The greener grades can be overdispersed, which results in the crystals breaking and changing colour. The crystals can exist in three forms, α, β, and γ, but it is the α form that predominates.

These pigments are almost perfect for the paint manufacturer, except for one limitation. Unfortunately, this limitation is their colour. They are very much duller than either lead chromes or organic alternatives. Nevertheless, it is normal to utilize their excellent application properties with their relatively low price as far as possible and then tinting to shade using brighter alternatives. They are the basis of most 'magnolia' decorative shades, through to high-quality beige industrial colours.

They have excellent resistance to solvents and chemicals. They have moderate tinting strength, but this can be an advantage, as it requires relatively little organic pigment to brighten them. Their light fastness and durability is excellent; in addition they even improve the light fastness of other pigments with which they may be combined and the polymers in which they may be incorporated, as they absorb damaging UV radiation. They have a high refractive index so normally have good covering power, but can be manufactured in a transparent form. They are among the easiest of pigments to disperse. The only technical property, which limits their application, is their heat stability. Above 105 °C they start to lose water of crystallization, giving a shade shift towards red, and this accelerates as the temperature and/or time of exposure are increased.

Transparency is achieved by preparing the pigment in as fine a particle size as possible by:

- controlling precipitation under special conditions; and
- attrition in high-energy mills.

Transparent grades are much more difficult to disperse, so are often sent to dispersion houses, where special techniques are employed to maximize transparency. They are much more expensive than conventional grades, but are widely used for automotive metallic shades and transparent wood finishes.

3.12.4.4 Complex inorganic colour pigments (formerly known as mixed phase metal oxides)

Colour Index — CI Pigment Yellow 53, 119, 164, Brown 24.

Formula — three of these pigments have a rutile titanium dioxide structure, with another metal oxide trapped in the crystal lattice.

These pigments were formerly known as 'mixed phase metal oxide', but this name implies that these pigments are mixtures of metal oxides. This is not the case; they are in a homogeneous chemical phase, in which a host lattice traps a 'foreign' cation. CI Pigment Yellow 53 is derived from the rutile titanium dioxide crystal lattice, in which there is a partial replacement of the Ti^{4+} ions with Ni^{2+} ions, at the same time incorporating antimony (V) oxide such that the Ni/Sb ratio is $1:2$. A typical composition of this pigment might be $(Ti_{0.85}\ Sb_{0105}\ Ni_{0.05})\ O_8$.

These pigments were formerly developed for the coloration of ceramics and are often referred to, incorrectly, as titanates. They have excellent solvent fastness and heat stability. Their light fastness is good to excellent and they have good chemical stability (although zinc ferrite brown is less stable). The earlier grades were difficult to disperse, but more recent commercial products are much improved. Their opacity is good but much lower than titanium dioxide. They have quite low tinctorial strength and are relatively expensive. These pigments are being used for high-quality applications such as automotive and coil coatings, often in combination with organic pigments, to provide medium to full shade alternatives to chrome pigments. They can also be used alone for pale shades. Although they may contain heavy metals such as nickel, chrome, and antimony, which may pose threats to the environment, these pigments are so inert they can be regarded as physiologically and environmentally harmless.

3.12.4.5 Bismuth vanadate
Colour Index — CI Pigment Yellow 184.
Formula — between $BiVO_4$ and mixed crystals of $BiVO_4$ and Bi_2MoO_6.

This pigment, which has only been on the market for a relatively short time, could arguably be included in the above category, but its manufacture and application characteristics are somewhat different. It is a bright, intense green shade yellow pigment. It possesses high opacity, high light fastness, and excellent solvent and heat resistance. It can be used for high-quality applications such as automotive coatings, powder coating, and coil coatings. In combination with organic pigments it produces very bright shades with high chroma and high covering power. Relatively small proportions of organic pigment seem to have a dramatic effect on the resulting hue.

Early grades suffered from a serious defect in that they were photochromic, that is they changed shade in the presence of light, gradually returning to their original shade when the light source was removed. However, this problem has now been overcome by introducing inorganic stabilizers. They can be used for most high-quality applications, including automotive finishes, coil coatings, and powder coatings. They are relatively expensive, but economic in use. Toxicity studies indicate that the pigment can only be harmful at high concentrations, so low dusting grades minimize this risk. Current information suggests that they pose low risk to the environment.

3.12.4.6 Arylamide yellow (Hansa® yellows)
Colour Index — CI pigment Yellow 1, 3, 65, 73, 74, 75, 97, 111 (there are several other pigments in this group, but these are the more important grades).
Formula — as represented by CI Pigment Yellow 1, 'Yellow G'.

Arylamide pigments were discovered at the beginning of the twentieth century. Although there are many pigments in the group, they are characterized by moderate to very good light fastness, moderate tinctorial strength, poor to moderate solvent resistance and poor heat stability. Consequently, they are mainly used in both water-based and white spirit-based decorative paints. Their nomenclature usually indicates their hue, 10G being very green, 5G being on the green side, G being virtually mid shade, whereas R and 3R (no longer commercially significant) are on the red side.

CI Pigment Yellow 3 (Yellow 10G) is one of the greenest in the series. The presence of two chlorine groups in the molecule gives it good light fastness, but it has low tinctorial strength. It is frequently offered in an easy to disperse form, which also possesses significantly higher colour strength than conventional grades.

CI Pigment Yellow 1 (Yellow G) is a mid shade yellow with moderate light fastness and moderate colour strength. Again it is frequently offered as an easily dispersible grade, with additional colour strength.

CI Pigment Yellow 73 (Yellow 4GX) is a mid yellow shade with similar colour and strength to Yellow G. Its main advantage is that it has significantly better resistance to recrystallization, an important consideration when dispersing in modern bead mills that can generate heat during the dispersion process, causing the pigment partially to dissolve.

CI Pigment Yellow 74 (Yellow 5GX) has grown in importance in recent years at the expense of Yellow G and 10G. It is available in various qualities. The conventional grade is approximately half way between G and 10G, but with much higher tinctorial strength. Its light fastness is close to Yellow G. As it is only slightly more expensive than Yellow G it is much more economic to use. Alternatively, it can be manufactured with a much larger particle size, resulting in a correspondingly higher opacity, making it ideal for full and near full shades. Although the colour strength is much reduced, it has much higher light fastness, and can therefore be used for pale shades.

CI Pigment Yellow 97 is not typical of these pigments:

As one would expect from the extra benzene ring attached to the remaining molecule by a sulphonamide ($-SO_2NH-$), it possesses much better solvent fastness than other grades, and can be used in stoving enamels. Its light fastness is also very good and for a time it was used in automotive finishes. It commands a significantly higher price than the other arylamide yellows, which are relatively cheap.

3.12.4.7   Diarylide yellows (Disazo structure)
Colour Index — CI Pigment Yellow 12, 13, 14, 17, 81, 83 (these are the most important members of the series but there are others).

Formula — as represented by CI Pigment Yellow 12:

As these products are manufactured from 3, 3'-dichlorobenzidine, they used to be referred to as 'benzidine' yellows. This should be avoided as it confuses them with benzidine dyes, over which there are legitimate safety concerns. Providing they are not used above 200 °C these pigments can be used. Diarylide pigments are the most important class of organic yellow pigments. However, this is largely because of their use in the printing ink industry, where they are the basis of the yellow process ink, used in three and four colour printing processes. They are characterized by high colour strength, good resistance to solvents and chemicals but poor light fastness. As they start to dissociate above 200 °C they should not be used above this temperature. Their use in paints is limited by their poor light fastness, but some modern grades, especially CI Pigment Yellow 83, do find use in industrial finishes and powder coatings. In recent years a new CI Pigment Yellow 83 has been introduced in a larger particle size grade, which has high opacity and much improved light fastness. It is used as a replacement for middle chrome even for vehicle refinishing paints and commercial vehicle finishes.

Diarylide yellow pigments are relatively inexpensive and their high strength makes them very economic, but the more specialist grades are significantly more expensive.

3.12.4.8   Benzimidazolone yellows

Colour Index — CI Pigment Yellow 120, 151, 154, 175, 181, 194, Orange 36, 60, 62.
Formula — Represented by CI Pigment Yellow 154:

These pigments range in shade from a green shade yellow through to orange. They are mainly monoazo in structure, but when compared with other monoazo yellow pigments have much improved solvent resistance. This is due to the benzimidazolone group on the molecule, which is in effect a cyclical carbonamide group.

They are characterized by good to excellent light fastness, good solvent and chemical resistance, and good heat stability. Several grades are available in higher particle size forms, which improves light fastness even further and allows them to be used in demanding applications such as automotive finishes and powder coatings. Although significantly more expensive than other monoazo pigments, they are often more economic than other high-quality pigments in this part of the spectrum.

3.12.4.9 Disazo condensation pigments
Colour Index — CI Pigment Yellow 93, 94, 95, 128, 166.
Formula — Represented by CI Pigment Yellow 95:

When discussing how it is possible to build solvent resistance to a molecule, two of the methods included building up the molecular size and introducing various insolubilizing groups such as the carbonamide group (—CONH—). The disazo condensation pigments utilize both approaches. This is achieved by putting together two acid derivatives of arylamide molecules, and condensing them by means of a diamine. This is not quite as easy as it sounds, but pigments produced in this way cover the whole of the spectrum from yellow, through orange and bluish red to brown.

As one would expect, such pigments have very good solvent resistance and excellent heat stability. They have reasonably good tinctorial strength and excellent chemical stability. They are relatively difficult to manufacture and therefore moderately expensive. Their main application is for the coloration of plastics, but individual products such as CI Pigment Yellow 128, are used in paints, including automotive original finishes.

3.12.4.10 Organic metal complexes
Colour Index — CI Pigment Yellow 129, 153, Orange 65, and 68. All azo methine
    complexes, and CI Pigment Yellow 150, Green 8 and 10. All azo complexes.
Formula — represented by CI Pigment Yellow 153:

The introduction of a metal group into the molecule imparts solvent resistance into the pigment. This metal group can be the centre of the molecule as in azo methine complexes. The metals used can include nickel, cobalt (II), or copper. They are mostly rather dull in shade but transparent. However, one member of the group, CI Pigment Orange 68, is available in both transparent and highly opaque forms. Their main application areas vary, but several were used for automotive metallic finishes. In spite of the great expectations from this group of pigments, few have made a great impact on the paint industry, and few of the above-mentioned products remain commercially available.

3.12.4.11   Isoindolinone yellow pigments
Colour Index — CI Pigment Yellow 109, 110, 173, Orange 61.
Formula — represented by CI Pigment Yellow 110:

Isoindolinone pigments are chemically related to the *azo methine* types. Having already discussed how chlorine groups impart insolubility and promote light fastness in a molecule, it can readily be seen that these pigments are going to have excellent solvent resistance and good light fastness. They also have excellent chemical resistance and heat stability, making them ideal for plastics and high-quality paint systems, including automotive OEM, but also for pale decorative paints and tinting systems. They have only moderate tinctorial strength.

CI Pigment Yellow 173 still has an isoindolinone structure, but only has chlorine groups on the middle benzene ring, and therefore is lower in fastness properties. Nevertheless, it is still used in automotive metallic finishes, often in combination with transparent iron oxides. It is duller in shade than the other two yellows. CI Pigment Orange 61 is also less fast than the yellows. They are moderately expensive.

3.12.4.12   Isoindoline yellows
Colour Index — CI Pigment Yellow 139, 185, Orange 69.

Isoindoline pigments are chemically related to the *methine* types. They have good general fastness properties especially for higher-quality industrial finishes. The yellows are not stable to alkalis. In general they are a little cheaper than the isoindolinone pigments and have higher tinctorial strength.

3.12.4.13   Quinophthalone
Colour Index — CI Pigment Yellow 138.

This pigment can be prepared in a highly opaque form that has a reddish yellow shade, possessing excellent light fastness and good resistance to solvents and heat. It is used as a chrome replacement in high-quality finishes. In its more transparent form it is much greener and has higher tinctorial strength, but with lower fastness properties.

3.12.4.14   Anthrapyrimidine
Colour Index — CI Pigment Yellow 108.

Anthrapyrimidine yellow was originally introduced as a vat dye (CI Vat Yellow 20), but as knowledge became available on how to treat the surface of these dyes in order to improve dispersibility, a process known as 'conditioning', a commercial grade of pigment was developed. For a period of time it was considered to be of the highest quality, but it has only moderately good resistance to solvents and its durability is not of the highest order, tending to darken when exposed to light. It is quite a dull colour and, although transparent, its light fastness is not good enough for today's automotive paints. Its complex nature makes it expensive.

3.12.4.15   Flavanthrone
Colour Index — CI Pigment Yellow 24.

This pigment was also known as a vat dye, but needs to be in a pure state and conditioned in order to obtain a commercial pigment. It is reddish yellow, stronger than anthrapyrimidine and with better heat stability. It is transparent and therefore is useful for metallic finishes where it shows good durability, although strong shades do darken on exposure to light.

3.12.5   Orange pigments

The borders between yellow/orange and orange/red are often difficult to define. Many orange pigments can be obtained following similar chemistry to the inorganic yellow chrome pigments, and to the arylamide, isoindolinone, and isoindoline yellow organic pigments. Others tend to possess chemistry more related to red pigments, such as molybdate and cadmium reds, β-naphthol and BON arylamide pigments. However, the following pigments stand in their own right as predominately orange pigments:

- Pyrazolone orange
- Perinone orange

3.12.5.1   Pyrazolone Orange
Colour Index — CI Pigment Orange 13, 34.
Formula — Represented by CI Pigment Orange 34:

These pigments are closely related to the diarylide yellow pigments, and their properties are also similar. They are bright and have high tinctorial strength. Their solvent resistance and heat stability are moderate to good, but light fastness is poor. Therefore, their main use is for printing inks. The light fastness of CI Pigment Orange 34 is the better of the two and can be used in industrial finishes, especially when produced in its opaque form, making it a useful replacement for chrome pigments. They are reasonably cheap, so combined with their high strength, they are economic in use.

3.12.5.2   Perinone orange
Colour Index — CI Pigment Orange 43.

Perinone orange was developed as a vat dye (CI Vat Orange 7) and still finds a use for this application. It has a very bright shade and high tinctorial strength. It has excellent fastness to most solvents and has very high heat stability. Its light fastness is very good, but strong shades darken. It is a useful pigment for tinting systems on account of its brightness, strength and good properties. However, it is expensive to manufacture and its correspondingly high price limits its application.

Colour Index — CI Pigment Orange 43.

There is also a perinone red. The orange is the *trans* isomer whereas the red is the *cis* isomer. It is rather dull in shade and its fastness properties are somewhat lower. Although cheaper, it has limited application.

3.12.6   Red pigments
This is by far the largest colour group of pigments, and no one type predominates, perhaps reflecting that fact that all the chemical types in this part of the spectrum have major limitations. It is therefore essential that the paint technologist has a good knowledge of the products available, their strengths and limitations, as a well-informed selection will provide technical and commercial advantages. The most important types are as follows.

Inorganic:

- Lead molybdate
- Cadmium red
- Red iron oxide

Organic:

- β-naphthol
- BON arylamides

- Toners
- Benzimidazolone
- Disazo condensation
- Quinacridone
- Perylene
- Anthraquinone
- Dibromanthrone
- Pyranthrone
- Diketopyrrolo-pyrrole pigments (DPP)

3.12.6.1 Molybdate red (or orange), scarlet chrome
Colour Index — CI Pigment Red 104.
Formula — $Pb(Cr, S, Mo) O_4$, mixed-phase crystals of lead chromate, lead sulphate, lead molybdate in a tetragonal structure.

The shades vary from orange to intense scarlet. They are bright and have good opacity, excellent solvent resistance, and good durability, especially grades that have been stabilized. Heat stability is also good, but more demanding applications require stabilized grades. Alkali stability is only moderate. They disperse easily; indeed over-dispersion can damage their surface treatment and reduce their resistance properties. Molybdate red pigments are frequently used in combination with organic pigments to extend their colour range considerably, while maintaining excellent opacity. They are relatively cheap and would be the ideal pigment if it were not for their lead and hexavalent chromium content, which in Europe restricts their use to mainly industrial finishes.

It is necessary to comply with strict labelling regulations for both the pigment and paints made from these pigments.

3.12.6.2 Cadmium red
Colour Index — CI Pigment Red 108, Orange 20.
Formula — CdS/CdSe.

These pigments are bright and opaque and have excellent solvent resistance and heat stability. Their light fastness is good but they are not stable to acids, or industrial atmospheres. Cadmium pigments are relatively expensive and have only moderate tinctorial strength. Their use in most coatings is prohibited on physiological and environmental grounds.

3.12.6.3 Red iron oxide
Colour Index — CI Pigment Red 101 (synthetic), 102 (natural).
Formula — Fe_2O_3 Ferric oxide.

This pigment in the form of the mineral 'haematite' was used in the cave paintings found in Spain and France. It can be manufactured from the yellow 'raw sienna' by calcining and driving off the water of hydration. However, most red oxides used in paint are made synthetically by a number of manufacturing processes, each imparting its own specific characteristics on the resulting pigment. The shade is dependent on the particle size and shape, smaller particles giving a paler shade. The shape of the crystal can be influenced by seeding with appropriate crystals and by the degree of calcination.

Red iron oxides have excellent application properties. They have excellent heat stability, resistance to chemicals and solvents. Heat stability is excellent and they are normally opaque. They have quite low tinctorial strength, but in spite of this their low price makes them very economic in use. Providing they are purified, they appear to be physiologically harmless and have good environmental properties. It is possible to over-disperse them, which can make them darker in shade but as they are relatively easy to disperse, this should not be necessary. They do tend to settle in low-viscosity systems. Their only real limitation is their colour—they have a dull brownish red shade.

3.12.6.4 β-Naphthol reds
Colour Index — CI Pigment Red 3, 4, and Orange 5.
Formula — represented by CI Pigment Red 3:

β-Naphthol red pigments have a limited colour range from orange to mid red. For paint, two products predominate, toluidine red (CI Pigment Red 3) and dinitroaniline orange (CI Pigment Orange 5). Both pigments can be prepared in a range of shades depending on the particle size distribution of the commercial products, which in turn depends on the manufacturing conditions.

The nitro group plays an important part in the molecule, imparting both insolubility and light fastness. However, the improvement is relative, the molecule is a very simple one and consequently has limited solvent resistance and heat stability. Light fastness is reasonably good in full, or near full, shades, but falls off rapidly in paler shades. Chemical stability is generally good. These pigments are relatively cheap and therefore find application in both solvent-based and water-based decorative paints as well as cheap industrial paints, where good gloss retention is not essential.

For more demanding applications, they have been replaced with Naphthol AS® pigments.

3.12.6.5 Naphthol AS® pigments (also known as BON arylamide pigments)
Colour Index — CI Pigment Red 2, 5, 12, 23, 112, 146, 170, Orange 38. These are the most important pigments used in paint; there are many other pigments in the series.
Formula — as represented by CI Pigment Red 112:

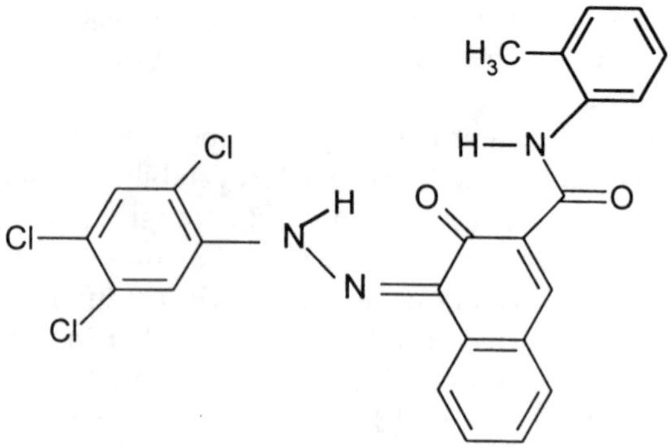

The addition of an extra benzene ring on this molecule has three effects on the pigment's properties. It increases the molecular weight, thus improving solvent resistance (the addition of a carbonamide group also improves insolubility). Finally, the extra benzene ring provides additional sites for further substitution that can alter the shade, improve solvent resistance and/or improve light fastness.

Consequently, Naphthol AS® pigments cover the spectrum from orange, through scarlet to bordeaux. Their properties can also vary considerably. They possess good chemical stability, but some of the pigments have only limited light fastness and solvent resistance, whereas others attain good general fastness properties. To some extent, properties can often be predicted from the structure. The pigment represented above, CI Pigment Red 112, has good light fastness and solvent resistance. It is used in decorative paints, including tinter systems, and cheaper industrial paints. It is available in a high-strength version or as a highly opaque grade. However, for industrial paints, including vehicle refinish systems, CI Pigment Red 170 has largely replaced it:

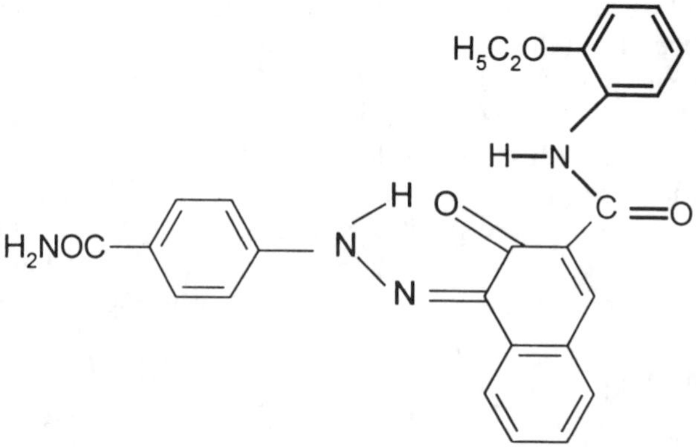

It has a similar shade to CI Pigment Red 112, but is available in a strong bluish shade or in a yellower, opaque form. The extra carbonamide group imparts much higher solvent resistance; it can be used in combination with higher quality pigments, for automotive finishes, especially in its opaque form.

It is possible to add another carbonamide group, e.g. CI Pigment Orange 38:

This pigment finds little use in paint as it only has moderate light fastness, but is worthy of inclusion to illustrate the effect of the carbonamide group, as this pigment can be used for the coloration of polymers such as PVC and polyolefins.

These pigments vary enormously in price, but usually offer the most economic colorations, providing their fastness properties are adequate.

3.12.6.6   Toner pigments

Colour Index — CI Pigment Red 48, 57, 60, 68 (an additional colon and number signifies the metal group). There are several other pigments of this type, but have little application in paint, usually due to poor light fastness.
Formula — CI Pigment Red 48:4:

One of the simplest ways of improving a coloured molecule's insolubility in organic solvents is to introduce a metal into the molecule. This can most readily be achieved by adding one or more acid groups to the molecule, then reacting this acid with an alkaline earth metal (or manganese) salt. The metal used has a significant effect on the shade of the pigment and on its fastness properties. This class of pigment is very important to the printing ink industry, where CI Pigment Red 57:1 is the normal standard for the magenta process colour used in three and four colour sets.

The base molecule can be simple β-naphthol azo molecules or β-naphthol azo molecules with a carboxylic acid on the naphthol (known as BONA toners, **B**eta **O**xy **N**aphthoic **A**cid). As well as being very economic, they cover a wide range of the spectrum from yellow (although not very important) through red and magenta to violet. They have reasonably good solvent fastness, much better than similar pigments without the metal group, and several are very heat stable. Light fastness varies and partially depends on the metal, manganese giving the best results. Their biggest drawback is their poor resistance to acids and alkalis; even water can affect them. Poor chemical stability limits their use in paints to industrial finishes where this property is not necessary.

There have been some developments in recent years, where the more complex Naphthol AS® molecules have had an acid group added. These pigments provide economic, heat-stable pigments for polymer coloration.

3.12.6.7　Benzimidazolone red and brown pigments
Colour Index — CI Pigment Red 171, 175, 176, 185, 208, Violet 32, Brown 25.
Formula — as represented by CI Pigment Red 185:

The introduction of the carbonamide groups via the benzimidazolone group provides this range of pigments with their high solvent resistance and good heat stability. However, several of the pigments in the series are rather dull in shade and others, although they have good light fastness, do not meet the highest requirements. Their main application tends to be in plastics, but CI Pigment Red 171, Red 175 and Brown 25 are used in metallic automotive and vehicle refinishing paints. They are relatively economic when compared with other high-quality pigments.

3.12.6.8　Disazo condensation pigments
Colour Index — CI Pigment Orange 31, Red 144, 166, 214, 220, 221, 242, Brown 23.
Formula — as represented by CI Pigment Red 242:

As in the yellow and orange series, the red disazo condensation pigments are characterized by good solvent resistance and excellent heat stability, because of the linking of two azo red molecules to a diamine by means of carbonamide groups. They range in shade from scarlet through bluish red to violet and brown. Their main use is in plastics, but several are used for industrial and occasionally vehicle refinishing paints and as the basis of tinting systems. They generally have bright shades.

CI Pigment Brown 23 is used for OEM automotive metallics and finds wide use in PVC coil coatings. Their price generally lies between the benzimidazolone pigments and the more complex polycyclic pigments described later.

3.12.6.9   Quinacridone pigments
Colour Index — CI Pigment Red 122, 192, 202, 207, 209, Violet 19.
Formula — as represented by CI Pigment Violet 19:

Few classes of pigment have seen the sustained growth shown by the quinacridone series. They are distinguished by their bright colours, which have excellent solvent stability, heat stability, chemical resistance, and high light fastness.

CI Pigment Violet 19 is unsubstituted and is available in two crystal forms, one having a bright mid-red shade and the other being red violet. A third crystal form is known, but it does not have favourable pigmentary properties, due to the crystal shape not allowing the formation of secondary bonds.

They are relatively expensive, but find application in most pigment using industries. In paint they are used in automotive OEM, vehicle refinishing, high-quality industrial finishes, and decorative tinting systems. Their high fastness properties are attributed to the formation of secondary bonds between the molecules. This can only exist in the β and γ crystal modifications.

There is a similar class of pigments known as 'quinacridone quinone' pigments, which are less bright and have poorer fastness properties, but are little used outside the USA.

3.12.6.10   Perylene pigments
Colour Index — CI Pigment Red 123, 149, 178, 179, 190, 224, Violet 29, Black 31, 32.
Formula — as represented by CI Pigment Red 224:

Although discovered in the early years of the twentieth century and used as vat dyes, these colorants were not developed as commercial pigments for nearly half a century. They have excellent solvent resistance, heat stability and light fastness. Chemical resistance is normally excellent but CI Pigment Red 224 lacks alkali stability. Some grades are available in both high-opacity and high-strength forms. The transparent forms are used for OEM automotive finishes whereas the opaque forms are used in full shades, where their high cost can be justified by their excellent properties.

3.12.6.11   Anthraquinone
Colour Index — CI Pigment Red 177.

This pigment has a unique bright, strong shade that is frequently used in automotive metallics, often in combination with transparent iron oxides, and formerly in solid shades with molybdate reds. It has good resistance to solvents and heat stability. Its light fastness is good, but only attains the standards required by automotive finishes when used in combination with pigments that improve performance. In paler white reductions, light fastness rapidly declines. It is relatively expensive.

3.12.6.12   Dibromanthrone
Colour Index — CI Pigment Red 168.

This pigment is also a vat dye. It has a bright scarlet shade with very good resistance to solvents. Its heat stability is sufficient for most paint systems and it has exceptionally good light fastness. Its high price limits its use mainly to automotive OEM and vehicle refinishing paints, coil coatings, and occasionally to tinting systems, where even in pale tints it retains excellent light fastness.

3.12.6.13   Pyranthrone red
Colour Index — CI Pigment Red 216, 226, Orange 51.
Formula — as represented by CI Pigment Red 216:

These pigments are chemically related to flavanthrone and are mainly used for their transparency. They have a rather dull shade (except CI Pigment Orange 51) and good fastness to solvents, but the heat stability of CI Pigment 216 is somewhat limited. Good, light fastness deteriorates rapidly in combinations with titanium dioxide.

3.12.6.14   Diketopyrrolo-pyrrole pigments (DPP)
Colour Index — CI Pigment Red 254, 255, 264, 270, 272, Orange 71, 73.
Formula — as represented by CI Pigment Red 254:

These pigments have only recently been introduced, but immediately found application on account of their bright shade, good opacity, excellent heat stability and resistance to solvents. They meet the requirements of the most demanding applications, including automotive OEM, although often used in combination with other more economic pigments, especially high opacity grades of CI Pigment Red 170 and CI Pigment 48:4 (2B manganese toner).

3.12.7   Violet pigments
Although violet paints are not often seen, violet pigments attract considerable usage, because they bridge the gap between red and blue. If one mixes blue and red,

the result is a dull maroon colour or even brown, whereas a violet pigment allows one to blue off red paints and redden off blue paints, without going dull. Violet can also be used for tinting white to counter the basic yellow tint of titanium dioxide. For the paint industry, one chemical type dominates—dioxazine violet.

3.12.7.1   Dioxazine violet
Colour Index — CI Pigment Violet 23, 37.
Formula — As represented by CI Pigment Violet 23:

These pigments are widely used in a variety of paint systems from water-based emulsions to automotive OEM and coil coatings. They possess excellent solvent fastness and heat stability, and very good light fastness. Although relatively expensive, their very high strength makes them reasonably economic for most applications. CI Pigment Violet 23 (carbazole violet) is more transparent and bluer than CI Pigment Violet 37, and is therefore preferred for metallic paints.

Their very small particle size makes them vulnerable to flocculation and can also lead to plate out in some powder coating systems.

3.12.8   Blue pigments
There are relatively few blue pigments, because one chemical type totally dominates this part of the spectrum. Phthalocyanine provides us with what must be considered almost the ideal pigment. There are a large number of manufacturers, many just specializing in phthalocyanines. In the paint industry, a vat pigment, indanthrone, is used for some special high quality applications, and two inorganic pigments, ultramarine and prussian blue are occasionally used. The printing ink industry does use some cationic toners (phospho tungsto molybdic acid, ferrocyanide and alkali blue pigments), but their poor solvent and chemical resistance coupled with poor light fastness means they have virtually no use in paint.

Inorganic:

- Prussian blue
- Ultramarine
- Cobalt blue

Organic:

- Copper and copper-free phthalocyanine
- Indanthrone

3.12.8.1   Prussian blue
Colour Index — CI Pigment Blue 27.
Formula — $FeK_4Fe(CN)_6$ ferric potassium ferrocyanide, or the ammonium salt.

This pigment is principally used in printing inks, where it is called milori or iron blue. It gives an intense, strong colour that improves the jetness of black inks. It is solvent-fast and has reasonably good heat stability. However, it has poor alkali stability, making it unsuitable for colouring emulsion paints that are to be applied to plaster.

3.12.8.2   Ultramarine
Colour Index — CI Pigment Blue 29, Violet 15.
Formula — approximately $Na_7Al_6Si_6O_{24}S_3$.

Ultramarine is a synthetic form of lapis lazuli, a naturally occurring mineral, prized by medieval artists. It is used for its brilliant colour and clean shade, ranging from pink, through violet to green. It has excellent heat stability and solvent resistance. It has good light fastness and alkali stability, but is unstable to acids, making it unsuitable for paints being used in industrial atmospheres.

3.12.8.3   Cobalt blue/green
Colour Index — CI Pigment Blue 36, Green 50.
Formula — $Co(Al, Cr)_2O_4$ (blue), $(Co, Ni, Zn)_2 TiO_4$.

These colorants belong to the 'complex inorganic colour pigment'. They are highly heat stable, have excellent solvent fastness, chemical stability, and light fastness. They can be used for the most demanding applications, but they are most used in very pale shades where high light fastness is required.

3.12.8.4   Copper phthalocyanine
Colour Index — CI Pigment Blue 15, 15:1, 15:2, 15:3, 15:4, 15:6 and Blue 16.

The phthalocyanine molecule, complexed around iron, was discovered in 1928 by Scottish Dyes (later to become ICI and now Zeneca); however, commercial grades were not introduced until around 1940. In many ways they are ideal pigments, having high colour strength, excellent heat stability, and good light fastness. Most grades have excellent resistance to solvents. While the chemical synthesis of phthalocyanine is relatively easy, converting it into a commercial product is more difficult.

The pigment is polymorphic, existing in at least five crystal forms, but only three are commercially produced:

- α   phthalocyanine, reddish shade but unstable to aromatic hydrocarbons;
- β   phthalocyanine, greenish blue;
- γ   phthalocyanine does not exist as a commercial pigment;
- δ   phthalocyanine does not exist as a commercial pigment;
- ε   phthalocyanine, slightly redder than the α form.

The α modification (CI Pigment Blue 15) has a reddish shade but is its unstable form and when used in aromatic solvents converts to the β grade. It can be stabilized (CI Pigment Blue 15:1) by slightly chlorinating (about half an atom of chlorine per phthalocyanine molecule). In doing so one also makes the pigment greener and tinctorially weaker. It can be treated to make it stable to flocculation (CI Pigment Blue 15:2).

CI Pigment Blue 15:3 is the β modification and has a greener, purer shade thus giving the cyan standard for three and four colour printing inks. If treated to make it flocculation stable, it is designated CI Pigment Blue 15:4.

CI Pigment Blue 15:6 is the δ modification, but is considered something of a speciality and little used. CI Pigment Blue 16 is the metal-free grade. It is a β crystal modification and possesses high transparency and even slightly superior durability, and was widely used for metallic shades, where it gave attractive flip tones. However, its use has declined in recent years, largely because of improved grades of the conventional β modification. In addition there is also a trichlorinated phthalocyanine which is also used for metallics and a grade which replaces the copper with an aluminium compound, in order to reduce copper content.

In spite of their excellent performance characteristics, copper phthalocyanine pigments are among the cheapest organic pigments on the market, their very high tinctorial strength making them very economic in use. Many companies manufacture copper phthalocyanine pigments.

Copper phthalocyanine pigments are quite transparent and can be used in solid, reduced, and metallic automotive coatings. Because the primary particles of phthalocyanine are so small, they have a tendency to flocculate and the loss of colour strength in such a tinctorially strong pigment is very noticeable. The flocculation stable grades are prepared by coating the particles with products that will prevent the particles from re-aggregating. A number of products can be used:

- sulphonated phthalocyanine;
- long chain amine derivatives of sulphonated phthalocyanine;
- aluminium benzoate;
- acidic resins, e.g. rosin;
- chloromethyl derivatives of phthalocyanine.

These stabilized phthalocyanines often show a greater purity of shade, easier dispersibility and better gloss than untreated grades, but are significantly more expensive.

3.12.8.5   *Indanthrone*
Colour Index — CI Pigment Blue 60.

Indanthrone blue is also a vat dye (CI Vat Blue 4). As a pigment it is polymorphic, with four known crystal forms, but the α modification provides us with the most stable pigment. It has a deep reddish blue shade, possessing much better stability to flocculation and even higher durability than phthalocyanine blues. It has excellent chemical and solvent resistance, very good heat stability and is highly transparent. It is used in high-performance paints, including automotive OEM finishes, especially pale shades and metallics. A chlorinated version (CI Pigment Blue 64) was used, but there are currently no known manufacturers.

3.12.9   Green pigments

As in the blue part of the spectrum, copper phthalocyanine pigments dominate the green sector. In practice, greens are often obtained by mixing yellows and blues, the brightness required and economics determining the best approach. Inorganic pigments play a relatively unimportant role.

Inorganics:

- Chrome green (Brunswick green)
- Chromium oxide
- Hydrated chromium oxide

Organic:

- Halogenated copper phthalocyanine

3.12.9.1   Chrome green (Brunswick green)
Colour Index — CI Pigment Green 15.
Formula — A co-precipitate or dry blend of chrome yellow and Prussian blue, $PbCrO_4. \times PbSO_4. FeNH_4Fe (CN)_6$.

These pigments inherit the strengths and weaknesses of the parent pigments. They have an additional disadvantage because they wet out at different rates, leading to floating and flooding. Some grades substitute phthalocyanine blue for Prussian blue.

3.12.9.2   Chromium oxide green
Colour Index — CI Pigment Green 17.
Formula — Cr_2O_3 chromium sesquioxide.

This pigment has a dull shade, good opacity but low tinctorial strength. It has excellent heat stability, excellent resistance to solvents and chemicals, and excellent light fastness. It should not be confused with other 'chrome' pigments, as the chrome is in the trivalent form, which is inert and appears to be free from toxicological and adverse environmental effects. It is a very hard abrasive pigment and can cause excessive wear on dispersion equipment.

It is used mainly on account of its excellent fastness properties and its ability to reflect infrared radiation, allowing its use in camouflage coatings.

3.12.9.3   Hydrated chromium oxide
Colour Index — CI Pigment Green 18.
Formula — hydrated chromium oxide, $Cr_2O(OH)_4$.

This pigment is chemically similar to chromium oxide, but the water of hydration leads to a much brighter and transparent pigment. Its light fastness is excellent but it is less heat stable and has slightly less stability to acids.

3.12.9.4   Phthalocyanine green
Colour Index — CI Pigment Green 7, 36.
Formula — As represented by CI Pigment Green 7:

The colour range of phthalocyanine green is dependent on the degree and type of halogenation. CI Pigment Green 7 has a bluish green tone, with excellent properties, even better than the blue grades. It has excellent solvent and chemical resistance, high heat stability, and excellent light fastness. It has a high tinctorial strength, and while the early products and less sophisticated grades are very difficult to disperse and have a tendency to flocculate, most modern commercial products are reasonably easy to disperse and are less prone to flocculation. It is normally transparent and therefore this pigment can be used for virtually all paint systems including all types of automotive OEM finishes, including metallics, powder coatings, industrial finishes and decorative paints.

CI Pigment Green 36 is chloro-brominated and has similar properties, but is much yellower in shade and is very bright. Its significantly higher price and much lower tinctorial strength tend to limit its application to high-quality finishes.

3.12.10   Special effect pigments

Certain pigments are used to give special effects, rather than for the colour. The best-known products to the paint technologists are the pigments used in metallic and flamboyant finishes to give a sparkle effect. They can give this effect in two quite different ways, as represented by:

- aluminium flake pigments;
- pearlescent pigments.

3.12.10.1   Aluminium flake pigments
Colour Index — CI Pigment Metal 1.
Formula — aluminium (Al).

There are two types of aluminium pigment:

- leafing;
- non-leafing.

The classification is based more on the effect they give, rather than any fundamental difference in the aluminium. These effects are achieved by the use of surface-active agents used on the aluminium.

Adding stearic acid to the aluminium particles produces the leafing types, which give the appearance of a continuous film of metal, almost mirror-like. The aluminium particles align themselves on the surface during the paint's drying process. They can be used with transparent pigments to produce coloured effects, but the aluminium must be added after the dispersion process as over-dispersion damages their stearate coating. Because the pigment aligns itself as plates, it prevents water penetrating the paint film, and provides corrosion protection to the surface of the substrate.

The non-leafing types remain within the film, producing a 'polychromatic' effect, so called because the colour looks different depending on the angle of view and the angle of the incident light. The effect is often described as 'flop' and is usually associated with automotive metallic paints and hammer finishes. They are often coloured with transparent or semi-transparent pigments. Lauric or oleic acid is used on the surface during the production of the pigment.

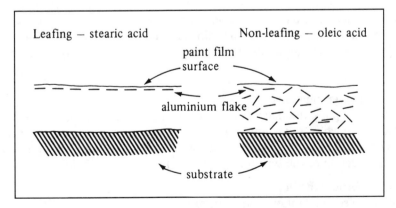

Fig. 3.5 — Aluminium flake pigments in a paint film (cross-section).

Aluminium pigments are produced by ball milling finely divided aluminium powder, down to a particle size of anything between 1.2 µm and 80 µm, in the presence of the appropriate stearate. Filtering and sieving follow to provide the various grades supplied to the paint industry.

Although aluminium shows no acute toxicity, care has to be taken when using it in acidic media, as both the leafing properties and sheen can be adversely affected, and hydrogen gas can be evolved. The stearic acid in leafing grades can also react with naphthanates and rosinates used as driers, which results in the deleafing of the aluminium. Most commercial grades are available as dust-free pastes with white spirit or other hydrocarbon solvents, as aluminium powder can cause explosive mixtures in air.

3.12.10.2   Pearlescent pigments

Colour Index — Not designated.

Formula — Principally mica coated with metal compounds, especially titanium dioxide.

A number of natural products and minerals give a pearlescent effect, including:

- natural platelets composed of guanine with hypoxanthine, commonly called 'fish silver', as they are derived from the scales of fish;
- basic lead carbonate, $Pb_3(OH)_2 (CO_3)_2$;
- bismuth oxychloride, $BiOCl$;
- mica flakes coated with titanium, iron or chrome oxides.

The pigment particles are transparent, thin platelets of a high refractive index, which partially reflect and partially transmit incident light. These platelets can stack up in layers, creating a third dimension and giving rise to the typical pearlescent lustre. The iridescent or 'interference' pigments are those where at a particular optical thickness, interference occurs between the partly reflected and partly transmitted light rays, cancelling out some wavelengths and reinforcing others, giving rise to a play of subtle colours or 'iridescence' as well as the pearly lustre.

Unlike normal pigments, where colours are subtractive (i.e. blue and yellow gives green), mixing interference pigments leads to 'additive' colour mixing (i.e. blue and yellow produces white). The smaller particles give a satin effect, whereas coarser particles produce a sparkle, similar to aluminium.

Pearlescent pigments of the titanium-coated mica type are now widely used in automotive finishes, usually in combination with highly transparent pigments to produce brilliant effects. The titanium dioxide used in the pigment should be of the rutile type, otherwise durability of the paint film can be adversely affected.

3.12.11   Extender pigments

As the name implies, the main purpose of such pigments is simply to 'extend' the more expensive white and coloured pigments, so cheapening the paint formulation. However, the use of such pigments contributes much more to the paint's properties so careful selection of the type and quantity can affect such features as:

- flow properties (rheology);
- stability to sedimentation (anti-settling);
- film strength.

Most extenders are white in colour and have a refractive index close to those of commonly used binders (1.4–1.7). Hence, unlike titanium dioxide, which has a refractive index of around 2.7, they give relatively little opacifying effect. Most are naturally occurring, but may require purifying. Others are produced synthetically.

- Calcium carbonate (synthetic and natural)
- Aluminium silicate (china clay)
- Magnesium silicate (talc)
- Barium sulphate (natural — barytes; synthetic — blanc fixe)
- Silica

3.12.11.1   Calcium carbonate (chalk, whiting, Paris white)
Colour Index — CI Pigment White 18.
Formula— $CaCO_3$.

Naturally, calcium carbonate is obtained from chalk or limestone. It is often known as whiting and is obtained in its natural state with a purity of around 96%.

Synthetic grades are often known as 'precipitated calcium carbonate', being produced by calcining limestone, followed by slaking and purifying. The calcium hydroxide suspension is then converted to the carbonate by passing carbon dioxide through the solution. As such, they are often made by using by-products of other industrial processes and from water softening processes.

They are cheap and are widely used in water- and solvent-based paints, for interior and exterior decoration. Their low oil absorption allows them to be used at relatively high levels imparting some structure to the wet paint, improving the stability to sedimentation of other heavier pigments and imparting sag resistance in paint films. Calcium carbonate reacts with acids, but in other respects is stable.

3.12.11.2   Aluminium silicate (china clay, kaolin, calcined clay)
Colour Index — CI Pigment White 19.
Formula — $Al_2O_3 \cdot 2SiO_2 \cdot 2H_2O$.

China clay is inert and has a good colour. It occurs naturally through the breakdown of granite, along with silica and mica. It exists as plate-like particles, with a particle size between 0.5 and 50 μm. It imparts some thixotropy in paints and is used as a cheap flatting agent and produces a structure that improves the suspension of other pigments. A major use is in water-borne decorative paints.

Fine particle size grades (between 0.2 and 0.4 μm), especially the calcined versions, can be used partially to replace the more expensive opaque pigments such as titanium dioxide, while retaining some opacity.

3.12.11.3   Magnesium silicate (talc, French chalk, asbestine)
Colour Index — CI Pigment White 26.
Formula — typically hydrated magnesium silicate $3Mg$ 0.4 $SiO_2 \cdot H_2O$.

The term 'talc' covers a number of minerals with varying chemical composition, but based on magnesia and silica, with calcium and aluminium as impurities. Rocks containing talc are crushed and then milled to give lamella or fibrous extender pigments, which are then classified.

Talc is inert and hydrophobic. Its plate-like form helps to prevent pigment settlement in the wet paint, improves flow behaviour, especially when brushing out and provides some resistance to humidity. It also enhances the sanding properties of paint films. It is used in both solvent- and water-based decorative paints as well as in undercoats and industrial finishes.

3.12.11.4   Barium sulphate (natural — barytes; synthetic — blanc fixe)
Colour Index — Barytes, CI Pigment White 22; blanc fixe, CI Pigment White 21.
Formula — $BaSO_4$.

Barium sulphate is very inert, insoluble and stable to light and heat. The natural form is obtained as the mineral 'heavy spar'. After being crushed, washed and dried it is usually micronized, reducing its particle size from $25\,\mu m$ to 2–$10\,\mu m$, thus aiding its dispersibility. The synthetic version is made by reacting available barium compounds with sulphuric acid or soluble sulphate salts, and has a finer texture than the natural grades, giving it a higher oil absorption.

Its refractive index (1.64) is higher than other extenders, which gives it some pigmentary properties. Its high density is also useful for paints sold by weight. It is used in primers, undercoats, and industrial finishes, where it hardens the film. Its high density leads to it having a tendency to settle. In spite of it being a barium salt, its insolubility ensures it is non-toxic.

3.12.11.5   Silica
Colour Index — CI Pigment White 27.
Formula — SiO_2.

Silica is available from both natural and synthetic sources. Natural sources include:

* diatomaceous earth, e.g. kieselguhr, Celite®.
* 'amorphous' silica (actually cryptocrystalline).
* crystalline silica — quartz, can cause silicosis, and therefore a health hazard.

Diatomaceous earth such as Kieselguhr consists of soft siliceous remains of diatoms (a type of fossilised algae) deposited over long periods of time. The synthetic grades include precipitated silica with extremely small particle size, which can even be colloidal.

The finer grades of these extenders are used as matting agents, reducing the gloss of paint films. They improve intercoat adhesion and the sanding properties of the paint film.

3.12.12   Corrosion-inhibiting pigments
Corrosion is the destruction of metal by chemical attack, such as rusting of iron or tarnishing of copper. It requires:

* a thermodynamically unstable metal;
* an electrolytic conductor of ions;
* an electrical conductor;
* an electron acceptor.

Pigments can help prevent corrosion by:

* helping to provide a physical barrier to the passage of water and oxygen;
* being sacrificially destroyed as an anode, thus protecting the anodic sites that have become pitted;

- providing soluble passivating ions to protect the metal; or
- producing an insoluble film, which prevents active corrosion.

Traditionally the main corrosion-inhibiting pigments were:

- Red lead, CI Pigment Red 105
- Basic lead silicochromate, not listed
- Zinc chromate, CI Pigment Yellow 36
- Strontium chromate, CI Pigment Yellow 32
- Calcium, strontium and zinc molybdate, not listed
- Calcium plumbate, CI Pigment Brown 10
- Zinc phosphate, CI Pigment White 32
- Zinc dust, CI Pigment Metal 6
- Micaceous iron oxide, not listed

However, concerns regarding the toxicity of many of these pigments, mainly those containing lead or chrome VI have resulted in new approaches. The alternatives are less universal and need carefully selecting for the precise application required. So before choosing an anti-corrosive pigment it is absolutely essential to read the safety data and the technical information, in order to obtain optimum results.

3.12.12.1   Red Lead
Colour Index — CI Pigment Red 105.
Formula — Pb_3O_4.

Red lead is mainly used in primers for metal protection, where it reacts with the acidic groups in the resin to give lead soaps that passivate iron and steel surfaces.

3.12.12.2   Basic lead silico chromate (Oncur)
Colour Index — core SiO_2 coating $PbSiO_3 \cdot 3PbO \cdot PbCrO_4 \cdot PbO$.

The lead silicate–lead chromate coating is firmly bound to the silica core, producing a pigment that is easy to disperse and gives excellent metal protection in structural steel and automotive paints. Finer grades can be used in electrocoat paints.

3.12.12.3   Zinc chromates: basic zinc chromate, zinc tetroxy chromate
Colour Index — CI Pigment Yellow 36.
Formula — $ZnCrO_4$, $K_2CrO_4 \cdot 3ZnCrO_4Zn(OH)_2$, $ZnCrO_4 \cdot 4Zn(OH)_2$.

These pigments liberate chromate ions, which passivate metal surfaces, producing a protective film at the anodes, thus inhibiting the anodic reaction. They were used for iron, steel, and aluminium protection, but their physiological properties make them unsuitable.

3.12.12.4   Calcium, strontium, and zinc molybdates
Colour Index — not listed.
Formula — $CaMoO_4$, $SrMoO_4$, $ZnMoO_4$.

These three pigments, all of which are white, passivate the anode. Their use has grown considerably in recent years on account of their more favourable physiological properties.

3.12.12.5   Calcium plumbate
Colour Index — CI Pigment Brown 10.
Formula — Ca_2PbO_4.

This pigment is a strong oxidizing agent and also reacts with the acid groups in binders with fatty acid groups such as linseed oil; to give lead and calcium soaps. This promotes adhesion in the paint film and confers toughness. The corrosion-inhibiting effects are due to the pigment's ability to oxidize soluble iron compounds formed in anodic areas, which can then form an insoluble film of iron compounds at the anode, thus neutralizing that element of the 'corrosion cell' and preventing further corrosion. Meanwhile calcium carbonate is generated at the cathodic region of the 'corrosion cell'.

3.12.12.6   Zinc phosphate
Colour Index — CI Pigment White 32.
Formula — $Zn_3 (PO_4)_2 \cdot 2H_2O$.

More specifically, it is zinc orthophosphate dihydrate that has the superior corrosion resistance properties. While this pigment was recognized as conferring good durability, excellent intercoat adhesion, and good flow properties in paint systems, its anti-corrosive properties were not fully understood. It is now recognized that in industrial atmospheres it reacts with ammonium sulphate to form complex hetero acids, which inhibit corrosion. Although it is a white powder, the pigment has very low reducing power.

A number of modifications are also offered, including aluminium zinc phosphate, zinc molybdenum phosphate and zinc silicophosphate hydrate. Zinc hydroxyphosphite also plays a role.

3.12.12.7   Zinc dust
Colour Index — CI Pigment Metal 6 and CI Pigment Black 16.
Formula — Zn.

Zinc dust is a fine bluish grey powder that reacts with oils to produce zinc soaps and with alkalis to give zincates. Corrosion resistance is imparted by a sacrificial chemical reaction of the pigment rather than the steel substrate. It also absorbs UV radiation, thus protecting the film formers in exterior coatings. Zinc-rich primers can be used for steel construction even where there will be subsequent welding.

The search to find more physiologically acceptable corrosion-inhibiting pigments has uncovered a number of possible inhibitors. Zinc, calcium, and strontium molybdates have been described, as has zinc phosphate. Other phosphates include basic molybdenum phosphate, aluminium triphosphate, and various polyphosphates. Borates can also be used, such as barium metaborate and zinc borate. Silicates such as calcium borosilicate, calcium barium phosphosilicate, calcium strontium phosphosilicate, and calcium strontium zinc phosphosilicate are all offered.

Mention has been made of **micaceous iron oxide** and aluminium acting as a barrier pigment; steel flake can offer similar protection.

As knowledge on the toxicity properties of these types of pigments is constantly developing, it is essential to keep up to date with published safety data. It is also necessary to give more attention to formulation details than was perhaps necessary with traditional pigments, as economics becomes a much more critical factor in the consideration of performance. The pigment loading and the pigment volume concentration that give optimum results are likely to have a much smaller 'window' than has previously been the case.

References

[1] DCMA, *Amer Inkmaker*, **June** (1989).
[2] ETAD/BCMA/VDMI/EPSOM, *Safe Handling of Pigments*, European Edition (1995).
[3] SANDERS J D, *Pigments for Inkmakers*, SITA Technology, London p. 19 (1989).
[4] HERBST W & HUNGER K, *Industrial Organic Pigments*, VCH, Weinheim (1993).
[5] JONES F, in *Pigments — An Introduction to Their Physical Chemistry*, Patterson D, Editor, Elsevier (1967), pp. 12–15.
[6] FEITKNECHST W, in *Pigments — An Introduction to Their Physical Chemistry*, Patterson D, Editor, Elsevier (1967), pp. 1–25.
[7] PATTERSON D, in *Pigments — An Introduction to Their Physical Chemistry*, Patterson D, Editor, Elsevier (1967), pp. 50–65.
[8] HOECHST, BP 1455653 Disazo Pigment (1974).
[9] BALFOUR J G, *J Oil Colour Chem Assoc JOCCA*, **June** (1990).
[10] SMITH J L, *Amer Ink Maker*, **September** (1994), pp. 41–48.
[11] *Colour Index International*, Third edition, Society of Dyers and Colourists American Association of Textile Chemists and Colorists (1971).
[12] *Colour Index International*, Third edition, *Pigment and Solvent Dyes*, Society of Dyers and Colourists American Association of Textile Chemists and Colorists (1982, updated 1987, 1997).
[13] CHRISTIE R, PIGMENTS, OCCA, pp. 1–3.
[14] WHITAKER A, *Z Kristallogr* **147** 99–112 (1978).
[15] WHITAKER A, *J Soc Dyers and Colourists* **94** 431–435 (1978).
[16] WHITAKER A, *J Soc Dyers and Colourists* **104** 294–300 (1988).
[17] SAPPOK R, *J Oil Colour Chem Assoc* **61** 299–308 (1978).
[18] ALLEN T, in *Pigments — An Introduction to Their Physical Chemistry*, Patterson D, Editor, Elsevier (1967), pp. 102–133.
[19] WASHINGTON C, *Particle size analysis — Pharmaceutics and other industries. Theory and Practice*, Ellis Horwood (1992).
[20] BRUNAUER S, EMMET P H, and TELLER E, *J Amer Chem Soc*, **60** 309 (1938).
[21] HERBST W, *The Dispersing of Pigment*, Farbe u Lack, **76** (12) 1190–1208 (1970).
[22] BERG I, in *Principles of Paint Formulation*, Woodbridge R, Editor, Blackie, Glasgow (1991).
[23] ROTHE U et al, *Paint Colour J*, **182** S10–S16 (1992).
[24] FULLER C W, in *Pigment Handbook*, Patton T C, Editor, John Wiley & Sons, New York, pp. 333–347 (1987).
[25] KUHNER G, *What is Carbon Black?*, Degussa, Frankfurt, PT 17-15-3-0992 Be.
[26] CPMA, *Safe Handling of Color Pigments*, Color Pigments Manufacturers Association, Alexandria, USA (1993).
[27] STEIG F, *Principles of Paint Formulation*. Woodbridge R, Blackie, Glasgow (1991).
[28] VALANTIN H & SCHALLER K H, *Human Biological Monitoring of Industrial Chemicals*, vol. 3, Titanium Commission of the European Community EUR 6609 (1980).

4

Solvents, thinners, and diluents

R. Lambourne

4.1 Introduction

With the notable exception of water, all solvents, thinners, and diluents used in surface coatings are organic compounds of low molecular weight. There are two main types of these materials, hydrocarbons and oxygenated compounds. Hydrocarbons include both aliphatic and aromatic types. The range of oxygenated solvents is much broader in the chemical sense, embracing ethers, ketones, esters, ether-alcohols, and simple alcohols. Less common in use are chlorinated hydrocarbons and nitro-paraffins.

The purpose of these low molecular weight materials is implied by their names. Thus a solvent dissolves the binder in a paint. Thinners and diluents may be added to paints to reduce the viscosity of the paint to meet application requirements. These statements do not, however, go far enough to describe the complex requirements for solvents, thinners, and diluents to achieve their practical purpose. This purpose is to enable a paint system to be applied in the first instance, according to a prese-lected method of application; to control the flow of wet paint on the substrate and to achieve a satisfactory, smooth, even thin film, which dries in a predetermined time. They are rarely used singly, since the dual requirements of solvency and evapora-tion rate usually call for properties of the solvent that cannot be obtained from the use of one solvent alone.

In the foregoing we have been largely concerned with the use of solvents or solvent mixtures that dissolve the binder. There is a growing class of binders (see Chapter 2) that are not in solution, but nevertheless do require the use of a diluent, which is a non-solvent for the binder, for application. Water is thus a diluent for aqueous latex paints, and aliphatic hydrocarbons are examples of diluents for some non-aqueous polymer dispersions. Even these materials may require a latent solvent or coalescing agent to ensure that the dispersed binder particles coalesce to form a continuous film on evaporation of the diluent.

Factors that affect the choice of solvents and the use of solvent mixtures are their solvency, viscosity, boiling point (or range), evaporation rate, flash point, chemical nature, odour, toxicity, and cost. Solvency is not an independent property of the liquid since it depends on the nature of the binder or film former. Odour, toxicity, and cost have become of increasing importance in recent years.

4.2   The market for solvents in the paint industry

Solvents are not produced for the paint industry alone. Nevertheless, the paint industry represents the largest outlet for organic solvents. The second largest sector is probably that for metal cleaning, i.e. principally degreasing, but this market is heavily dependent on chlorinated solvents. Other major industrial users of solvents are the adhesives and pharmaceutical industries.

The production of solvents has changed dramatically since the late 1960s because of two main factors: (i) environmental considerations that have led to considerable legislation particularly in the USA, and (ii) the economic recession. Solvent consumption dropped by almost 20% in 1975 and took almost ten years to recover, but with a different balance of solvents.

The main technical changes have been the decrease in the usage of aromatic solvents as a result of environmental legislation, notably the introduction of rule 66 in Los Angeles in 1967 and subsequently rules 442 and 443. This legislation was introduced to limit the emission of ultraviolet radiation absorbing materials into the atmosphere in an attempt to eliminate or minimize the well-known Los Angeles photochemical smog. It is important to note that legislation does not prohibit the use of solvents, only their emission into the atmosphere. This has meant that some users have not discarded aromatics and other atmospheric pollutants, but have had to seek alternative means of dealing with them. Two methods are commonly used: the introduction of solvent recovery plant enabling the re-use of solvents, for example the recovery of toluene and xylene in large printing works; and the use of 'after burners' on the fume extraction plant of industrial paint users. Both of these methods must be able to justify the capital expenditure. The savings resulting from recycling are obvious. In the case of the installation of after burners, the cost is recovered at least to some extent by utilizing the energy obtained by the combustion of the solvent.

Reliable data on the production and use of hydrocarbons are not available, largely because most hydrocarbons are produced for fuel and the solvent fractions represent only a minor proportion of the total. Data on oxygenated solvents are more reliable, but in some cases the solvent may be used as a raw material in chemical synthesis as well as a solvent, e.g. acetone, methanol. Ball [1] reviewed solvent markets and trends in the 1970s. Solvent consumption in Western Europe and the USA are compared in Table 4.1. Ten solvents commonly used in the paint industry are listed. It is interesting to note that although there continued to be a small increase in the use of organic solvents in Western Europe, overall the USA market shows a decline. This is due to the effects of the legislation referred to previously which has led to the development of alternative technology.

The use of aliphatic hydrocarbons was estimated to be at the rate of about two million tonnes per year in Western Europe in 1978. This compares with about 2.5 million tonnes in the USA for the same period, appearing to show a decline from a peak in 1987.

It is difficult to make direct comparisons between Ball's figures for solvent con-

Table 4.1 — Solvent consumption in Western Europe and the USA (000 tonnes) [1]

Solvent	Western Europe		USA	
	1974	1978	1974	1978
Methyl ethyl ketone	140	130	238	280
Methyl isobutyl ketone	65	50	86	70
Ethyl acetate	165	173	70	81
n-Isobutyl acetates	90	123	32	44
Perchlorethylene	350	375	292	315
Trichlorethylene	270	255	157	115
Isopropanol	250	250	427	360
Acetone	238	244	325	335
Methanol	250	265	270	240
Glycol ethers	142	150	121	140
Totals	1960	2015	2018	1980

Table 4.2 — Total solvent use in the USA (million tonnes) [2]

Solvent type	1987	1992	1997 (predicted)
Hydrocarbons	2.76	2.15	1.81
Alcohols/esters/ethers	1.64	1.90	2.07
Chlorinated solvents	0.91	0.63	0.31
Ketones	0.59	0.53	0.52
Glycols/esters/ethers	0.42	0.37	0.39
Other	0.10	0.12	0.15
Totals	7.42	5.70	5.25
Of which coatings applications employed:			
	4.58	3.10	2.77
	(61.%)	(54%)	(52.7%)

sumption (excluding hydrocarbons in 1974/78) and the more recent figures of Kirschner (Table 4.2). The latter includes data for 1987 and 1992 (with predictions for solvent usage in 1997), but the assumptions made and the methods of obtaining the data may be very different. However, it is probably safe to conclude that the use of organic solvents in paints has diminished over the past two decades and will continue to do so as a result of technological change stimulated largely by environmental considerations and more stringent legislation.

4.3   Solvent power or solvency

The dissolution of a polymer or a resin in a liquid is governed by the magnitude of the intermolecular forces that exist between the molecules of the liquid and the molecules of the polymer or resin. It is these same forces that govern the miscibility of liquids and the attraction between colloidal particles suspended in a liquid. Intermolecular forces have such an important bearing on so many of the factors that control practical paint properties that it is worth considering how they arise. There are three kinds of intermolecular force: dispersion or London forces, polar forces, and hydrogen bonding. Dispersion forces arise as a result of molecular perturbation such that at any instant there is a transient but finite dipolar effect leading to an attraction between the molecules. In the case of hydrocarbons this is the only source of molecular attraction. Polar forces originate from the interaction between permanent dipoles (Keesom effect) or between dipoles and induced dipoles (Debye effect). Hydrogen bonds, largely responsible for the high anomalous boiling point of water, are well known to the organic chemist. When a hydrogen atom is attached to an electronegative atom such as oxygen or nitrogen, there is a shift in electron density from the hydrogen to the electronegative atom. Thus the hydrogen acquires a small net positive charge and the electronegative atom becomes slightly negatively charged. The hydrogen can then interact with another electronegative atom within a second molecule to form a so-called hydrogen bond. These forces are responsible for such fundamental properties as the latent heat of vaporization, boiling point, surface tension, miscibility, and of course solvency.

4.3.1 Solubility parameters

The concept of solubility parameters was introduced by Hildebrand and Soolt [3] who was concerned with its application to mixtures of non-polar liquids. It was derived from considerations of the energy of association of molecules in the liquid phase, in terms of 'cohesive energy density', which is the ratio of the energy required to vaporize 1 cm^3 of liquid to its molar volume. The solubility parameter is the square root of the cohesive energy density of the liquid,

$$\delta = \sqrt{\frac{\Delta E_v}{V_m}}$$

where ΔE_v = energy of vaporization and V_m = molar volume. For liquids having similar values of δ, it is to be expected that they should be miscible in all proportions. Where δ is significantly different for two liquids it is unlikely that they will be miscible. Whilst this was an important step forward in an attempt to put solvency on a sound footing, it was of little immediate help to the paint technologists of the day who were dealing with a wide range of solvents, both polar and non-polar, and almost invariably used as mixtures. Moreover, since volatilization of polymers without degradation is impossible, the availability of data on polymers was not forthcoming at that time. Burrell [4] made the next major step forward in recognizing that a single parameter as defined by Hildebrand was insufficient to characterize the range of solvents in use in the paint industry. He did, however, recognize the potential of the solubility parameter concept and carried out some of the early experiments to determine solubility parameters of polymers by swelling. He saw the need to classify the hydrogen bonding capacity of solvents and proposed a system in which solvents were classified into one of three categories, poorly, moderately, and strongly hydrogen bonding. In doing so he established an empirical approach to the determination of the solubility parameters of polymers using a 'solvent spectrum'. A typical solvent spectrum is shown in Table 4.3. The solubility parameters of liquids can be calculated from experimentally determined energies of vaporization and their molar volumes. Solubility parameters of polymers can be determined empirically by contacting them with a range of solvents (e.g. the 'spectrum' in Table 4.3) and observing whether or not dissolution occurs. The solubility parameter is

Table 4.3 — A solvent 'spectrum' for determining the δ values of polymers

Group I δ (weakly H-bonded)		Group II (moderately H-bonded)		Group III (strongly H-bonded)	
6.9	Low odour mineral spirits	7.4	Diethyl ether	9.5	2-ethyl hexanol
7.3	n-hexane	8.0	Methylamyl acetate	10.3	n-Octanol
8.2	Cyclo-hexane	8.5	Butyl acetate	10.9	n-Amyl alcohol
8.5	Dipentene	8.9	Butyl 'Carbitol'	11.4	n-Butanol
8.9	Toluene	9.3	Dibutyl phthalate	11.9	n-Propanol
9.2	Benzene	9.9	'Cellosolve	12.7	Ethanol
9.5	Tetralin	10.4	Cyclo-pentanone	14.5	Methanol
10.0	Nitrobenzene	10.8	Methyl 'Cellosolve'		
10.7	1-Nitropropane	12.1	2,3-butylene carbonate		
11.1	Nitroethane	13.3	Propylene carbonate		
11.9	Acetonitrile	14.7	Ethylene carbonate		
12.7	Nitromethane				

Note: Several of these solvents are highly toxic. They will not be used in practical situations as solvents for binders. The purpose of the 'spectrum' is to determine δ values for polymers only.

taken as the centre value of the range of δ values for solvents which appear to dissolve the polymer. An alternative method is to use a lightly crosslinked form of a particular polymer and to identify the solvent that is capable of causing the maximum swelling. The δ value for the polymer is then equated with that of the solvent.

Small [5] proposed a method for calculating the δ value for a polymer from the simple equation,

$$\delta = \frac{\rho \sum G}{M}$$

where ρ is the density of the polymer and M its molecular weight. G is the 'molar attraction constant' for the constituent parts of the molecule which are summated. Although this is only in most cases a crude approximation it serves to give a value which is a guide to the choice of the range of solvents that one might select for either of the experimental methods mentioned above. Small's approach is not satisfactory for many of the polymers that are currently used in the paint industry. In the case of those polymers that are of low molecular weight and are converted at a later stage in the film-forming process into higher molecular weight products, the reactive group in the molecule may have a disproportionate effect on the solubility of the polymer. With high molecular weight polymers there has been a growth in the use of amphipathic polymers for many purposes. These may be block or graft copolymers in which different parts of the molecules have very different solubility characteristics. The implications of these structural considerations, e.g. in the context of non-aqueous dispersion polymerization (see Chapter 2), illustrate this point. Molar attraction constants (G values) are given in Table 4.4.

The need to use solvent mixtures arises for a number of reasons. We have already mentioned the need to control evaporation rate. On the other hand, a suitable single solvent may not be available. If available it may be too expensive or too toxic, or it may be particularly malodorous. It is therefore useful to predict as accurately as possible the solvency of mixtures having known individual δ values. A reasonable

Table 4.4 — Molar attraction constants at 25°C according to Small [5]

Group	G	Group	G
—CH$_3$	214	H	80–100
—CH$_2$—	133	O (ether or hydroxyl)	70
— CH —	28	CO (carbonyl)	275
— C —	–93	COO (ester)	310
CH$_2$=	190	CN (nitrile)	410
—CH=	111	Cl (mean)	260
>C=	19	Cl (single)	270
Ch ≡ C—	285	Cl (twinned as in > CCl$_2$)	260
—C ≡ C—	222	Cl (triple as in — CCl$_3$)	250
Phenyl	735		
Phenylene	658		
Naphthyl	1146		
5-membered ring	105–115		
6-membered ring	95–105		
Conjugation	20–30		

Example: Epikote 1004

Equivalent molecular mass of repeating unit,
$C_{18}H_{20}O_3 = 284$; $\rho = 1.15\,g/cm^3$

ΣG;   2 ether O   @   70 = 140
1 hydroxyl O   @   70 = 70
1 hydroxyl H   @   100 = 100
2 phenylene   @   658 = 1316
2 methyl   @   214 = 428
2 methylene   @   133 = 266
1 tert C   @   –93 –93
                    $\Sigma G = 2227$

Thus, $\dfrac{\Sigma G}{M} = \dfrac{1.15 \times 2227}{284} = 9.02$

approximation is to calculate the δ value for the mixture simply from the volume fraction of the constituent solvents. For example, a polyester resin that is readily soluble in ethyl acetate (δ = 9.1, Group II) is not soluble in either *n*-butanol (δ = 11.4, Group III) or xylene (δ = 8.8, Group I). However, it is soluble in a mixture of xylene and butanol in the ratio 4/1, by volume. The average δ value of the mixture is 9.3.

Interesting as these observations may be it is clear that the single value Hildebrand solubility parameter, even when used in conjunction with a broadly defined hydrogen bonding capacity, falls short of what is required to define solvent power in the paint industry. Nevertheless it did have sufficient promise as a method for predicting solubility for subsequent workers in this field to attempt to refine the concept and to improve its validity.

4.3.2   Three-dimensional solubility parameters

All subsequent developments in this field have been based upon the resolution of the single 'solubility parameter' into three constituent parts related to the three types of inter molecular forces described above, namely dispersion forces, polar forces, and hydrogen bonding. Crowley, Teague, & Lowe [6] were the first to put forward a scheme which used a solubility parameter δ (equivalent to that of Hildebrand), the dipole moment of the molecule, μ, and a hydrogen bonding parameter, γ. The hydrogen bonding parameter, γ, was derived from the spectroscopic data published by Gordy [7]. Although an improvement on the original Hildebrand solubility parameter as applied by Burrell to paint polymer and resin systems, the Crowley method has not gained the acceptance within the industry as have two subsequent treatments, due to Hansen [8] and Nelson, Hemwall, & Edwards [9].

More recently, work on the prediction of solubility using the solubility parameter concept has been directed at the behaviour of solvent blends, particularly during the evaporation stage of coating formation [10].

4.3.2.1   The Hansen solubility parameter system

Hansen considered that the cohesive energy density of a solvent results from the summation of energies of volatilization from all of the intermolecular attractions present in the liquid,

$$\frac{\Delta E_t^v}{V_m} = \frac{\Delta E_d^v}{V_m} + \frac{\Delta E_p^v}{V_m} + \frac{\Delta E_h^v}{V_m}$$

where ΔE^v, subscripts T, d, p, and h respectively represent the energies per mole of solvent, and energy contributions arising from dispersion, polar, and hydrogen bonding respectively. V_m is the molar volume. Alternatively, this may be written in terms of the solubility parameter δ, in the form

$$\delta_s^2 = \delta_d^2 + \delta_p^2 + \delta_h^2$$

where δ_s, δ_d, δ_p, and δ_h are the solubility parameters corresponding to the solvent, dispersion, polar, and hydrogen bonding respectively. Hansen used a homomorph concept to estimate the value of δ_d for a given solvent. This involves the assumption that a hydrocarbon having approximately the same size and shape as the solvent molecule will have the same energy of volatilization at the same reduced temperature (the ratio of the absolute temperature to the critical temperature) as the solvent, i.e. the dispersion force contributions are likely to be the same.

The difference between the energies of volatilization of the homomorph and the solvent will thus give a measure of the contributions from polar and hydrogen bonding forces. In solubility parameter terms this means,

$$\delta^d \text{ for solvent} = \delta \text{ for homomorph}$$

therefore

$$\delta_s^2 - \delta_d^2 = \delta_p^2 + \delta_h^2$$

Hansen determined the values of δ_p and δ_h initially by an empirical method involving a large number of solubility experiments for several polymers and then selecting values of δ_p and δ_h which gave the best 'volumes' of solubility for the system. Later he used another method of calculating δ_p and δ_h independently and found generally good agreement with his previous empirical values. Because the concept uses three numerical values it is convenient to plot them on three axes, hence the term 'volume' of solubility which is simply a reference to the volume on a three-dimensional plot representing the three parameters, into which the parameters of a given solvent would have to fall if the solvent is to dissolve a given polymer or resin. In Hansen's case this would be an oblate spheroid. However, for convenience the scale for the dispersion contribution is doubled to convert the volume of solubility into a sphere. In this form the sphere of solubility of a polymer or resin can be defined in terms of its position relative to the three axes, i.e. by the coordinates of its centre and its radius, R_p. An alternative way of treating solubility data is to use two-dimensional plots, where one is in effect using the projection of the sphere of solubility onto one or other of the two-dimensional axes as indicated in Fig. 4.1. The third parameter defines the circle which should enclose solvents for the polymer or resin in question. There is, however, no sound theoretical basis for the assumption that all solvents for a given polymer will fall within the sphere of solubility, particularly as the sphere is only arbitrarily defined and without resorting to the artifice of doubling the scale for one of the parameters would not be a

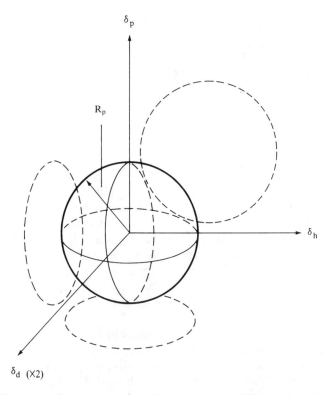

Fig. 4.1 — Diagrammatic representation of a polymer solubility plot based on Hansen's parameter system. (The dotted circles represent projections of the 'sphere of solubility'.)

sphere at all. Since the sphere is determined empirically the concept does have a certain simplicity which makes it easy to apply to practical systems, and many workers have adopted Hansen's method because of this.

The distance, R_s, between the three-dimensional coordinates of the solubility parameters of the solvent and the coordinates of the centre of the sphere of solubility of the polymer determines whether or not the polymer will be soluble in the solvent. That is, if $R_p > R_s$ the polymer will be soluble in the solvent because the coordinates for the solvent fall within the sphere of solubility for the polymer.

As an illustration of this point it is worthwhile seeing how the Hansen method compares with the Burrell method for predicting the solubility of the polyester of our previous example (Section 4.3.1) in the 4/1 mixture of xylene and n-butanol.

$$R_s = \left[4(\delta_{do} - \delta_d)^2 + (\delta_{po} - \delta_p)^2 + (\delta_{ho} + \delta_h)^2\right]^{\frac{1}{2}}$$

From Tables 4.5 and 4.6 the Hansen solubility parameters are as follows:

	δ_d	δ_p	δ_h	
n-butanol	7.8	2.8	7.7	
xylene	8.7	0.5	1.5	
saturated polyester (Desmophen 850)	10.53	7.30	6.0	$R_p = 8.2$

Since the solvents are in the proportion 1/4, butanol/xylene, the equivalent solubility parameters for the mixture are:

	δ_d	δ_p	δ_h
20% butanol	1.56	0.56	1.54
80% xylene	6.96	0.40	1.20
mixture	8.52	0.96	2.74

therefore

$$R_s = \left[4(10.53 - 8.52)^2 + (7.30 - 0.96)^2 + (6 - 2.74)^2\right]^{\frac{1}{2}}$$
$$= \left[16.16 + 40.20 + 10.63\right]^{\frac{1}{2}}$$
$$= (66.99)^{\frac{1}{2}} = 8.18$$

Since R_s is less than R_p this predicts that the polyester will be soluble in this solvent mixture. But since the values are almost equal it suggests that the polyester is only just soluble. It can be argued, therefore, that Hansen's method can give more information than the Burrell method, since the latter does not give any indication of the influence of degree of disparity between the δ values for polymer and solvent. A selection of the Hansen δ values for solvents is given in Table 4.4, a selection of δ and R_p values for polymers in Table 4.5 and δ and R_p values for a selection of pigments in Table 4.7. A more extensive range of values will be found in reference [11].

As has been indicated in the foregoing discussion of the three-dimensional solubility parameter concept, early work was concerned mainly with aspects of solvent miscibility and the dissolution of polymers in solvents and mixtures thereof. It is now recognized that it has even wider applications, such as in polymer compatibility, interactions at pigment surfaces (wetting and adsorption properties), and in predicting the chemical and solvent resistance of the coatings [12].

Table 4.5 — Hansen's solubility parameters for a selection
of solvents

Solvent	δ_d	δ_p	δ_h
n-hexane	7.3	0	0
Cyclohexane	8.2	0	0.1
Toluene	8.8	0.7	1.0
Xylene	8.7	0.5	1.5
Acetone	7.6	5.1	3.4
Methyl ethyl ketone	7.8	4.4	2.5
Methyl isobutyl ketone	7.5	3.0	2.0
Ethyl acetate	7.7	2.6	3.5
n-Butyl acetate	7.7	1.8	3.1
Ethanol	7.7	4.3	9.5
n-Propanol	7.8	3.3	8.5
n-Butanol	7.8	2.8	7.7
Ethoxyethanol	7.9	4.5	7.0
Tetrahydrofuran	8.2	2.8	3.9
Ethylene glycol	8.3	5.4	12.7
Glycerol	8.5	5.9	14.3
Dichloromethane	8.9	3.1	3.0
Carbon tetrachloride	8.7	0	0.3
Water	7.6	7.8	20.7

Table 4.6 — Hansen's solubility parameters for a selection of polymers

Polymer type	Trade name	Supplier	Computed parameters			
			δ_d	δ_p	δ_h	R_p
Short oil alkyd (34% oil length, OL)	Plexal C34	Polyplex	904	4.50	2.40	5.20
Long oil alkyd (66% OL)	Plexal P65	Polyplex	9.98	1.68	2.23	6.70
Epoxy	Epikote 1001	Shell	9.95	5.88	5.61	6.20
Saturated polyester	Desmophen	Bayer	10.53	7.30	6.00	8.20
Polyamide	Versamid 930	General Mills	8.52	−0.94	7.28	4.70
Isocyanate (Phenol blocked)	Suprasec F5100	ICI	9.87	6.43	6.39	5.70
Urea/formaldehyde	Plastopal H	BASF	10.17	4.05	7.31	6.20
Melamine/formaldehyde	Cymel 300	American Cyanamid	9.95	4.17	5.20	7.20
Poly(styrene)	Polystyrene LG	BASF	10.40	2.81	2.10	6.20
Poly(methyl methacrylate)	Perspex	ICI	9.11	5.14	3.67	4.20
Poly(vinyl chloride)	Vipla KR	Montecatini	8.91	3.68	4.08	1.70
Poly(vinyl acetate)	Mowilith 50	Hoechst	10.23	5.51	4.72	6.70
Nitrocellulose	H23 ($\frac{1}{2}$ sec)	Hagedorm	7.53	7.20	4.32	5.60

Table 4.7 — Hansen's solubility parameters for a selection of pigments [12]

Pigment	δ	δ	δ	R_p
Zinc silicate	23.5	17.5	16.8	15.6
TiO$_2$ (Kronos RN 57)	24.1	14.9	19.4	17.2
Heliogen blue LG	21.9	10.2	13.3	12.3
Aluminium powder	19.0	6.1	7.2	4.9
Red iron oxide	20.7	12.3	14.3	11.5

With respect to the use of solubility parameters in describing the wetting and de-wetting of surfaces in general, the surface tension of liquids has been correlated quite successfully with the three Hansen parameters according to the equation.

$$\gamma = 0.0715V^{1/3}\left[\delta_D^2 + 0.632(\delta_P^2 + \delta_H^3)\right]$$

It has been shown that in pigmented systems colour development is at a maximum when the solvent quality for the binder is only moderate. Too poor a solvent leads to excess binder on the surface of the pigment, in a flatter configuration than for a slightly better solvent which allows for polymer chain segment extension into liquid phase. Too good a solvent can cause the binder to be desorbed and the colloidal stability of the system deteriorates.

An efficient aid to the reformulation of solvent systems for paints has been developed by Shell Chemicals in the form of a series of computer programs, under the name 'Shell Solvent Computer Programs' (SSCP), that assist the paint technologist by reducing the number of formulations that have to be evaluated experimentally. The procedure is not limited to a simple calculation of the properties of solvent blends, but also includes data on properties that relate directly to the performance of the paint formulation. For example, the changes in the solvent parameters as a result of changes in the solvent composition that accompany evaporation can be examined for quite complicated systems, thus reducing the need for extensive experimental work [10].

4.3.2.2 The Nelson, Hemwall, and Edwards solubility parameter system

Nelson and co-workers developed their system almost contemporaneously with Hansen, but quite independently. They proposed a system which uses the Hildebrand solubility parameter δ, a fractional polarity term, and a 'net hydrogen bond accepting index', θ_A. The latter allows for differences in hydrogen bonding characteristics of the various solvent types. It thus attempts to quantify what Burrell had recognized as an important factor (cf. his 'solvent spectrum').

$\theta_A = K_\gamma$, where γ is the spectroscopic value for hydrogen bonding [6]. The coefficient K has three values: −1 for simple alcohols, 0 for glycol–ethers, and +1 for all other solvents.

The system is commonly used in the form of 'solubility maps' an example of which is shown in Fig. 4.2. This illustrates the application of the method to Epikote 1004, an epoxy resin (ex Shell Chemicals plc). A and B on the map represent the solubility parameters of xylene and n-butanol respectively, neither of which is a solvent for the resin. However, the line joining these two points intersects the region

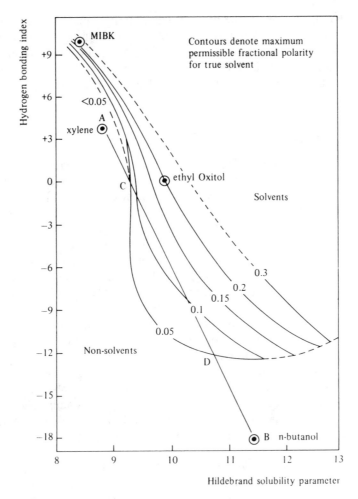

Fig. 4.2 — Solubility map for Epikote 1004 (solids > c. 30% W).

of solubility at the two points C and D, so that it would be expected that Epikote 1004 will be dissolved by mixtures of xylene and *n*-butanol for all compositions lying between C and D. Since the solubility parameters combine linearly, the composition of the mixtures at C and D can be calculated. They turn out to be 85% xylene, 15% *n*-butanol at C, and 25% xylene and 75% *n*-butanol at D. These blends now have to be checked against the fractional polarity contours on the map. Neither mixture may have a fractional polarity of greater than 0.05 if they are to be true solvents. The fractional polarity for xylene is 0.001 and for *n*-butanol is 0.096. Thus the average fractional polarities for the two mixtures are

Blend C 85/15 xylene/butanol = 0.00085 + 0.0144 = 0.01525

Blend D 25/75 xylene/butanol = 0.00025 + 0.072 = 0.07225

These results indicate that Blend C would be a solvent for the resin, but that Blend D would not. Clearly one would only have to calculate the proportions of these two solvents to give a value of fractional polarity equal to or less than 0.05 in order to achieve solubility. A number of solubility maps of this sort have been published, and for more detailed information the reader is directed to reference [13].

Both the Hansen method and the Nelson, Hemwall, and Edwards method lend themselves readily to computerization, e.g. in Rocklin & Edwards [14].

4.4   Solvent effects on viscosity

Establishing that a given solvent or mixture of solvents is capable of dissolving the resinous or polymeric components of a paint is only the first step in solvent selection. The solubility parameter concept may aid in choosing solvents for a given system, but it does not tell us anything about the condition of the polymer or resin in solution; i.e. the intermolecular forces described previously also control the disentanglement and extension of the polymer chains. A 'good' solvent will tend to maximize (at relatively low polymer concentration) the viscosity of the solution at a given concentration. For solvents of similar solvent power the viscosities of a particular polymer at the same concentration in each will be in the ratio of the solvent viscosities. This implies that the solvent/polymer interactions in both solutions are equal. It is not generally understood that good solvents tend to give more viscous solutions, although it is recognized that the addition of a non-solvent for a polymer to a solution of that polymer in a good solvent can reduce the viscosity of that solution. There is a point at which precipitation will take place when the polymer will have collapsed back on itself, no longer being solvated by the 'good' solvent. This effect is used to characterize polymers such as aminoplasts which are usually dissolved in aromatic hydrocarbon/butanol mixtures. Their tolerance to a non-solvent such as hexane is a measure of the degree of butylation and molecular weight.

Solvents are Newtonian fluids. That is, their viscosity is independent of shear rate. Likewise, most polymer solutions are Newtonian, although some may be non-Newtonian at high concentrations. A few polymers such as polyamides are designed to produce non-Newtonian effects, but it is debatable if they are in true solution, and it is believed that the non-Newtonian behaviour (in this case shear thickening or thixotropy) is due to structure developing through a dispersed micro-particulate phase.

It is convenient to compare the viscosities of polymer solutions under standard conditions, since this enables direct comparisons to be made between solvents with respect to solvent power, or comparisons to be made between the same polymer type, but with differing molecular weights. The following terms have come into use:

$$\text{Relative viscosity} = \frac{\text{Solution viscosity}}{\text{Solvent viscosity}}$$

$$\text{Specific viscosity} = \frac{\text{Solution viscosity} - \text{Solvent viscosity}}{\text{Solvent viscosity}}$$
$$= \text{Relative viscosity} - 1$$

$$\text{Reduced viscosity} = \frac{\text{Specific viscosity}}{\text{Concentration (g per decilitre)}}$$

Relative viscosity and specific viscosity are dimensionless quantities; reduced viscosity has the dimension decilitres per gram. If reduced viscosities are measured for a range of dilute solutions, e.g. <1% concentration, and they are plotted against concentration, a quantity known as the intrinsic viscosity of the polymer can be obtained by extrapolation to zero concentration. The intrinsic viscosity of a polymer is related to its molecular size and thus its molecular weight.

The foregoing is applicable to linear flexible homopolymers or random copolymers in solution. With block or graft copolymers it is possible for the polymer to be dissolved by a solvent for one component block or graft, the remaining component remaining unsolvated and collapsed into a tight coil. In these cases the polymer is likely to be in a colloidal state, in micellar solution. Such a solvent is described as a 'hemi-solvent' for the polymer. The effects of solvent on polymer configuration will be described in more detail in connection with the use of polymer stabilizers for dispersion (see Chapter 8).

4.5   Evaporation of solvents from coatings

In arriving at a suitable solvent blend for a surface coating, one has to consider the rate of evaporation of the solvent from the paint in relation to the conditions of application and cure. For example, thinners supplied for motor refinishing may be formulated for rapid or slow evaporation according to the ambient conditions. Also, the rate of solvent loss from a thin film governs the propensity of the coating to flow, so that it may be important to use a solvent blend of two or more solvents. This enables a rapid increase in viscosity to occur by the evaporation of a volatile component, and flow to be controlled by a less volatile one.

Evaporation takes place in essentially two stages. Initially the solvent loss is dependent on the vapour pressure of the solvent and is not markedly affected by the presence of dissolved polymer. As the polymer film is formed, solvent is retained within the film and is lost subsequently by a diffusion-controlled, slow process. This latter stage may be dominant when as much as 20% of the solvent is retained in the film.

The initial rates of evaporation can be predicted from an application of Raoult's Law, from a knowledge of activity coefficients of the components of a mixture of solvents and its composition. The measurement of evaporation rates is done very simply by a gravimetric method. Shell have developed an automatic evaporometer which is now accepted as an ASTM Standard, No. D 3539.76 [15].

Two methods are commonly used, evaporation from a thin film of the liquid, or from a filter paper, using equal volumes of liquids in each case. The results obtained are comparable for all except the most volatile of solvents. The filter paper method is preferred by many since the evaporating surface remains constant throughout the determination. Newman and Nunn [16] quote relative evaporation rates for some common solvents and tabulate their order of retention. It is common to use n-butyl acetate as a standard for evaporation and to quote all other solvents in comparison with butyl acetate, taken as unity. It is interesting to note from Table 4.8 that the retention of a given solvent is not related to its relative initial evaporation rate or its molar volume. However, within a homologous series the expected order of retention seems to apply, e.g. acetone, methyl ethyl ketone, and methyl isobutyl ketone.

Table 4.8 — Order of solvent retention (data from reference [16])

Solvent	Relative evaporation rate	Molar volume (cm^3/mole)	
Methanol	3.2	40	
Acetone	7.8	73	
2-Methoxyethanol	0.51	79	
Methyl ethyl ketone	4.6	90	
Ethyl acetate	0.35	97	
2 Ethoxy ethanol	4.3	97	
n-Heptane	3.3	146	
2 n-Butoxyethanol	0.076	130	
n-Butyl acetate	1.0	132	Increasing order of retention
2 Methoxyethyl acetate	0.35	117	
2 Ethoxyethyl acetate	0.19	135	
Toluene	2.3	106	
Methyl isobutyl ketone	1.4	124	
Isobutyl acetate	1.7	133	
Cyclohexane	4.5	108	
Methyl cyclohexane	3.5	126	
Cyclohexanone	0.25	103	
Methyl cyclohexanone	0.18	122	▼

4.6   Flashpoint

The flashpoint of a solvent mixture is important to those concerned in the manufacture, handling, storage, transport, and use of solvent-containing products. A considerable amount of legislation has been enacted to control the use of flammable liquids because of the fire hazards they represent, and standard methods have been established for the determination of flashpoint.

The flashpoint of a liquid may be defined as the lowest temperature at which the liquid, in contact with air, is capable of being ignited by a spark or flame under specified conditions. Two methods of determining flashpoint are in common use, the Abel method (Institute of Petroleum Test Method 170) and the Pensky-Marten closed cup method (ASTM D93). Each calls for a specific piece of equipment designed for the test. The apparatus and its method of use is described in each of the test methods referred to.

There have been many attempts to predict the flashpoints of paints containing solvent mixtures [17, 18]. They are based broadly on the assumption that the flashpoint is related inversely to the vapour pressure of the solvent and its activity coefficient. These predictive approaches have gained little support, however, since to meet most regulations related to transportation and storage an experimentally determined value by a specified method is almost always required. In practice it is necessary to know whether a product has a flashpoint above a specified limit to satisfy regulations regarding shipment. For example in the UK a product is not classed as 'highly flammable' unless it has a flashpoint below 32 °C and is combustible. Flashpoints of a number of common solvent constituents of paints are given in Table 4.9.

As a general guide the flashpoint of a mixture of solvents can be assumed to be that of the solvent with the lowest flashpoint. However, this is not always the case, and lower values are sometimes obtained experimentally.

Table 4.9 — Boiling points (BP), flashpoints (FP), and threshold limit values (TLV) for some common solvents

Solvent	BP (°C)	FP (°C)	TLV (ppm)
Ketones			
Acetone	57	–17	750
Methyl ethyl ketone	80	–4	200
Methyl isobutyl ketone	116	13	100
Diacetone alcohol	168	54[†]	50
Di isobutyl ketone		47	
Esters			
Ethyl acetate	77	–3	400
Isobutyl acetate	118	16	
n-Butyl acetate	127	25	150
2-Ethoxyethyl acetate	156	52[†]	50
2-butoxyethyl acetate	1	115[†]	
Alcohols			
Methanol	65	10	200
Ethanol	78	12	1000
n-Propanol	97	22	200
Isopropanol	82	10	400
n-butanol	118	33	50
Isobutanol	108	27	50
Sec butanol	100	14	
Hydrocarbons			
Toluene	111	5	100
Xylene	138–144	24	100
White spirit (LAWS)	155–195	39	500
SBP 6	138–165	28	—

FP by Abel method, except [†] which were by the Pensky-Marten closed cup method. TLV quoted is time weighted average (TWA).

4.7 Toxicity and environmental pollution

In recent years it has become evident that many of the solvents in common use represent a health hazard, and legislation has been introduced to control their use. In addition, even with solvents of low toxicity their odour may be unacceptable both in the working environment and in the vicinity of manufacturing and user plants. As has been mentioned previously, solvent emission into the atmosphere can be controlled by the use of afterburners on extraction systems or by recycling.

Studies of the toxicity of solvents, as with any chemical compound, cover a comprehensive range of standard test procedures, including animal testing for toxic effects arising from skin contract with the liquid, exposure to vapour at a range of well-defined concentrations, and by ingestion. Data have also been collected over many years on the effects of exposure of workers in most industries to the chemicals which they use. These data, which are constantly being revised as more information is obtained, are used to define acceptable working conditions for those exposed to the solvents in their daily work.

The most commonly applied limit on solvent concentration in the atmosphere is the 'threshold limit value' (TLV). This is usually applied as a time weighted average value which takes into account the period that an individual may be exposed during his or her working day, e.g. eight hours. Thus the TLV gives the average concentration (ppm) that should not be exceeded within the working day. Such exposure is regarded as being insufficient to be a health hazard. Clearly the concentration may be exceeded for a short period if most of the time an individual is working in an atmosphere in which the concentration of the solvent is less than the TLV. An exception to this is if a 'ceiling' value for a given material has been defined. In this case the concentration of the material must not exceed the 'ceiling' value if a health hazard is to be avoided. Fortunately, few solvents come into this category. Some TLVs are given in Table 4.9.

4.7.1 Volatile organic compounds and environmental legislation

The presence of toxic solvents in the atmosphere of the work-place is not the only problem giving concern in recent years. It is now well established that the emission of volatile organic compounds (VOCs) into the atmosphere can have serious environmental implications. Thus, effects on the concentration of ozone in the stratosphere (resulting in the depletion of ozone (particularly in the polar regions) have been much publicized as 'holes in the ozone layer', which give rise to increased ultraviolet radiation penetration of the earth's atmosphere. The mechanism of this process, in which chlorinated hydrocarbons play an important part, is well understood. A consequence of the increased intensity of UV light in the troposphere is an increase in the incidence of skin cancers. The formation of summer smogs (associated with large conurbations such as Los Angeles and Athens) is a result of ozone creation in the troposphere which is also encouraged by VOC emissions. It has also been observed that a smog can be formed at substantial distances from the original emission source. Smogs can cause eye irritation and respiratory problems in human beings and animals, and can be harmful to crops and trees. The World Health Organization guideline levels are an 8 hour average exposure of 50–60 p.p.b., but higher levels are frequently recorded in the UK and continental Europe. The 'greenhouse effect' which is caused mainly by increases in the carbon dioxide concentration in the atmosphere is also exacerbated by some VOCs. 'Acid rain', although caused principally by sulphur dioxide and nitrogen oxides in the atmosphere, can contain small amounts of hydrochloric acid that is derived from chlorofluorocarbons (CFCs) [19].

All of these effects of the release of chemicals in the atmosphere were recognized as major issues during the 1970s and 1980s as requiring global action to achieve significant improvement. This realization resulted in the United Nations taking an increased role in setting up Conventions to address these issues. Various agencies of the United Nations organization are involved, including the UN Environment Programme (UNEP), UN Economic Commission for Europe (UNECE), and the UN Conference on Environment and Development (UNCED). These bodies provide a forum for discussion and cooperation, make recommendations, and publish detailed Convention statements and Protocols. However, ultimately it is for national governments to introduce and enforce the implementation of any legislation introduced to their statute books.

The paint industry in the USA, for example, is attempting to minimize (or to eliminate) the use of solvents to reduce the costs of dealing with hazardous wastes as well as other chemical emissions for which mandatory limits have been laid down according to the 'Toxics Release Inventory' (TRI), published by the US Environmental Protection Agency. The TRI is applied in conjunction with the '33–50' programme which calls for voluntary emission reductions of 17 chemicals, including 6 commonly used solvents. The 33–50 programme is so-called because companies agreed to reduce releases of the 17 listed chemicals by 33% by 1992 and by 50% by 1995, as measured against 1988 data. The EPA reported that its 33% target was achieved, a reduction of 40% being reported in 1992.

The representatives of the member states of the Montreal Protocol met in September 1997 and some optimism was expressed to the effect that the tightening of global controls on ozone-depleting substances had resulted in a thickening of the ozone layer by 15–20% in the previous year [20]. Optimistically, it is stated that 'the ozone layer is expected to start to recover within a few years, and to recover fully by the middle of next century'. This is in spite of considerable in-fighting between Canada and the USA on the one hand and the EC on the phasing out of HCFCs.

Of the three main types of solvent that are subject to legislation (hydrocarbons, oxygen-containing solvents, and chlorinated solvents), the replacement of chlorinated solvents provides the biggest challenge. Ozone-depleters, such as 1,1,1-trichloroethane and CFC-113, are subject to a production and import ban in the USA with effect from 1 January 1996 under the Montreal Protocol on 'Substances that Deplete the Ozone Layer'. The Montreal Protocol was signed by the representatives of 60 governments in September 1987, committing those governments to actions the end of the millennium [2].

In the USA the Clean Air Act Amendments 1990 extended existing law by adding over 700 pages of new and amended requirements, and sets an agenda that will extend well into the twenty-first century. The major provisions of this document which affect the paint industry have been discussed by Wigglesworth [19].

Acknowledgement

Acknowledgement is due to Mr P. Kershaw and the Shell Chemical Company for permission to produce Fig. 4.1 and 4.2 from the *Shell Bulletin* cited.

References

[1] BALL T M D, *Polymc Paint J* **170** (4029) 595 *et seq.* (1980).
[2] KIRSCHNER E M, *Chem Eng News*, **June 20 1994** 13–20.
[3] HILDEBRAND J H & SCOTT R, *The Solubility of Non-Electrolytes*, 3rd edn, Rheinhold (1950); *Regular Solutions*, Prentice Hall (1962).
[4] BURRELL H, *Official Digest*, **27** (369) 726 (1955).
[5] SMALL P A, *J Appl Chem* **3** 71 (1953).
[6] CROWLEY J D, TEAGUE G S & LOWE J W, *J Paint Technol* **38** (496) 269 (1966); *ibid.* **39** (504) 19 (1967).
[7] GORDY W, *J Chem Phys* **7** 93 (1939); *ibid.* **8**, 170 (1940); *ibid.* **9** 204 (1941).
[8] HANSEN C M, *J Paint Technol* **39** (505) 104 (1967); *ibid.* **39** (511) 505 (1967).
[9] NELSON R C, HEMWALL R W & EDWARDS G D, *J Paint Technol* **42** (550) 636 (1970).
[10] BEERS N C M, Proc XXI FATIPEC Congress, Amsterdam, June (1992), Vol. 4, 157–171; *EuroCoat J* **1993** (3) 147–154, 320.
[11] HANSEN C M & BEERBOWER A 'Solubility Parameters', in *Kirk Othmer Encyclopedia of Chemical Technology*, Supplementary Vol., 2nd edn, Interscience, 889–910 (1971).
[12] HANSEN C M, *EuroCoat J* **1994** (5) 305–317.
[13] Shell Chemicals Technical Bulletin, *Solubility Parameters*, ICS(X) 78/1.
[14] ROCKLIN A L & EDWARDS G D, *J Coating Technol* **48** (620) 68 (1976).
[15] Shell Chemicals Technical Bulletin, *Evaporation of Organic Solvents from Surface Coatings*, ICS/77/4.

[16] NEWMAN D J & NUNN C J, *Prog. Organic Coatings*, **3** 221 (1975).
[17] WALSHAM J G, in *Solvents, Theory and Practice*, ed R W Tess, Advances in Chemistry Series, no. 124 (1973).
[18] Shell Technical Bulletin, *Predicting the Flash Point of Solvent Mixtures*, ICS/77/5.
[19] WIGGLESWORTH D J in *Chemistry and Physics of Coatings*, ed. A R Marrion, Royal Society of Chemistry, London, 8–20 (1994).
[20] ENOS Report 271, August 1997, p. 42.

5

Additives for paint

R A Jeffs and W Jones

5.1 Introduction

In this chapter an additive is counted as one of those substances included in a paint formulation at a low level which, nevertheless, has a marked effect on the properties of the paint. Generally we do not include ingredients that we regard as part of the fundamental formulation. For example, textured coatings and fire-retardant paints use special formulating ingredients. Although these are used in relatively small quantities one does not usually convert a standard paint to products in these classes by a simple addition, therefore they remain outside the scope of this chapter.

The brief remarks made under each heading will, we hope, prove helpful, but the following must be remembered:

- Any listing of this type can never be complete or up to date with new developments.
- Sometimes an additive is very specific to a particular formulation, being effective in some types of paint and valueless in others.
- The use of additives very frequently gives rise to undesirable secondary effects, especially if used unwisely or to excess.
- Care must be taken that additives are not the cause of 'spiral formulation' where, for example, a surfactant is used to aid surface wetting but gives rise to foaming which is corrected by silicone solution, thus causing cissing, in turn overcome by a surfactant.
- In any formulation which has been developed over a long period, perhaps at the hands of several formulators, expediency may have caused the inclusion of a variety of additives. When further troubles arise it may be sounder practice to *take out* additives rather that seek yet others. The present authors have often stripped-back a formulation to its known essential, and only returned additives to the formulation as an up-to-date essential, and only returned additives to the formulation as an up-to-date review has found them truly necessary. Some had become a permanent feature, although originally added to overcome the eccentricity of a particular batch.

- Additives are invariably related to curing paint's shortcomings. The description of paint faults is often imprecise. Make sure that the fault an additive claims to overcome is the one you are interested in. Foaming and biological protection are two areas with pitfalls for the unwary.

Some semblance of order is attempted into the sub-headings. The classes of product are in alphabetical order.

In an effort to make the entries helpful to the chemist in trouble we list, after each section, raw materials and products mentioned in the text or others that may be useful. The numbers that follow in brackets refer to the listing at the end if this chapter, where you will find the names of suppliers and, where relevant, their UK agent. This is not a comprehensive list of additive suppliers — detailed addresses are available from trade publications such as *Polymers Paint Colour Year Book* (published by Fuel and Metallurgical Journals Ltd, Redhill, Surrey).

Many of the products listed are registered trade names.

5.2   Anti-corrosive pigment enhancers

Various proprietary materials are offered to enhance the corrosion protection properties afforded by conventional anticorrosive pigments. Some of these are tannic acid derived (Kelate). Albarex is a treated extender used to part-replace true anticorrosive pigments such as zinc phosphate. Alcophor 827, described as a zinc salt of an organ nitrogen compound, also augments prime anticorrosive pigments.

Alcophor 827 [18/19]	Albaex [27/9]
Kelate [29/24]	Ferrophos
Anticor 70[4]	Molywhite MZAP [24]

5.3   Antifoams

Latex paints are stabilized with surfactants and colloids which, unfortunately, also help to stabilize air introduced during manufacture or during application, and thus form a stable foam. Non-aqueous paints (indeed any liquid other than a pure one) may also show bubbling. The antifoams on the market may be directed to a particular class of paint or offered for general use. Sometimes two antifoam additions are made, one at an early stage of manufacture and the other just prior to filling-out.

Usually antifoams are of high surface activity and good mobility while not being actually soluble in the foaming liquid. Commonly they work by lowering the surface tension in the neighbourhood of the bubble, causing them to coalesce to larger, less stable bubbles which then break. At their simplest, these additives may be solutions of single substances such as pine oil, dibutyl phosphate, or short chain (C_6–C_{10}) alcohols. On the other hand, they may be complex undisclosed compositions comprising mineral or silicone oils carried on fine-particle silica in the presence of surfactants. Many work by providing an element of incompatibility, and thus create centres from which bubble collapse can start. The slow wetting-out or emulsification of antifoam agents is the reason why they so often lose their effectiveness during prolonged processing or on long storage.

Testing a candidate for effectiveness can be difficult, for rarely do the shaking or stirring tests give more than an indication of what happens in real production conditions. As so often, an additive that works excellently in one composition will be worthless in another. Nevertheless the following are given as a few examples of the host on the market:

BYK 023	[6]
BYK 031 Mineral oil-based, stable to over 100°C	
Pyrenol 20/22 Non-silicone type for epoxy systems	[18/20]
Defoamer L401 Silicone based for stoving alkyds	[12]
Nopco range to latexes	[20]
EFKA range	[15]

5.4 Antisettling agents

The prevention of settling of pigment when a solvent-based paint is stored is often a matter of compromise. Good dispersing and deflocculants aid gloss and opacity but are not favourable for settling. Controlling the final rheological properties of the paint is also a careful balance between resisting the settlement and harming the gloss. As an additive to the dispersion stage the well-established but nonetheless valuable soya-lecithin can have profound antisettling effects. It is possible, however, to exceed the optimum level and induce very heavy vehicle separation.

To impart a small measure of thixotropy, proprietary products abound. Some years age it was not uncommon to include a fraction of a percent of aluminium stearate. This proved difficult on plant scale. More handleable are the stearate-coated calcium carbonates (e.g. Winnofil) where calcium carbonate is a permissible extender. Another family used to prevent settling are the modified hydrogenerated castor oils. These are usually purchased as gels in solvent and are chosen to suit the paint and the milling temperature if 'seeding' on storage is to be avoided (Crayvallac).

Grades of modified montmorillonite clay also impart some structure. Grades such as Bentone SDI and Easigel need no polar-solvent for activator; older grades such as Benetone 34 and Perdem 44 did and, depending on how effectively this was done, variations in structure could result. Byk Anti-terra 203 is claimed more reliable as a gellent. It should be borne in mind that often antisettling aids are used with each other to enhance the effect. The fine-particle pyrogenic silicas (e.g. Aerosil and Cab-o-sil) are useful on their own or with, for example, Byk Anti-terra 203.

Since so many of the above additives affect viscosity at low shear rates (thus holding pigments in suspension despite gravitation forces) they are also used for the other low shear effects, principally antisagging, flow control, and colour flotation.

In latex paint, structure is so commonly formulated into the system that settling is not a problem. For other types of aqueous paints the choice of colloid is important, and judicious use of bentonite or pyrogenic silicas can be added to the pigmentation, or recourse may be made to cellulosic or synthetic polymeric thickeners. Guidance on antisetting may be sought from suppliers of:

Crayvallac	
MPA	[33]

Bentones	[33]
Perchem	[26]
Easigel	[26]
Winofil	[14]
BYK Anti-Terra 203	[6]
Aerosil	[10]
Primal RM5/RM8	[34]
Aluminium stearate	[17]

5.5 Antiskinning agents

The driers in autoxidative air-drying paints, of course, are essential for the proper balance of surface and through-drying characteristics. Unfortunately they may also cause the formation of a skin on the surface of stored paint. Originally a percent or two of pine oil or dipentene was used to alleviate the problem, with phenolic antioxidants such as guaiacol (less than 0.1%) held in reserve for the more stubborn cases. They have largely been displaced by the more easily used oximes. Butyraldoxime, and especially methyl ethyl ketoxime, bought as such, or under one of the proprietary names, are now widely used at about 0.2% on the paint. Cyclohexanone oxime is a powder and finds use because of its mild odour, Being volatile the oximes are lost from the film at an early stage and therefore do not significantly retard the drying. This volatility can, however, be a disadvantage in that a container, once opened, may now skin on being re-lidded and stored.

Some antioxidants in certain systems can cause loss of drying potential on storage, so do check drying as well as skinning performance.

Butyraloxime	[36/22]
Methyl ethyl ketoxime	[36/22]
Cyclohexanone oxime	[36/22]
Exkins 1, 2 and 3	[36/22]
Pine oil	[16]
Dipentene	[16]

5.6 Can-corrosion inhibitors

Aqueous paints of all types, even when packaged in lacquered tinplate containers, tend to cause corrosion, especially at internal seams and handle stud, etc.

Sodium nitrite and sodium benzoate, often in conjunction with each other at about 1% levels, are long-established inhibitors deriving from car antifreeze experience. A possible disadvantage of using salts of this type is their adverse effect on the water sensitivity of the paint film.

Ser-Ad FA179 at less than 0.3% is claimed to be an effective inhibitor of this type of corrosion, and has the advantage of insolubilizing during film formation.

SER-AD FA179	[36/22]
Sandocorin 8132B	
Borchigen Anti Rust D	[5/7]

5.7   Dehydrators/antigassing additives

With some paints it is important to keep the moisture level low for storage stability reasons. Moisture curing polyurethane paints can only use pigments from which adsorbed moisture has been largely removed, if crosslinking and gassing in the can on storage is to be avoided.

Additive TI is a monomeric isocyanate which reacts with water avidly and is therefore used to scavenge water from pigments at the dispersion stage. Additive OF (a triethyl orthoformate) is added to the paint to control the effects of residual moisture during storage.

Another group of products in which water can cause problems on storage are the aluminium paints and zinc-dust primers where reaction of the metal with moisture liberates hydrogen, causing pressure in a closed can. Sylosiv A1 and ZN1 are commonly used here, although they will not cope if grossly water-contaminated solvents or binders have been used.

Additive TI   [4]
Additive OF   [4]
Sylsiv

5.8   Dispersion aids

One of the main aims of the paint formulator is the optimization of the dispersion of pigments in the binder. Wherever possible he or she will do this without recourse to additives, but some binders are poor wetters, and some pigments are difficult to wet (see Chapters 6, 7 and 8). If long dispersion times, poor hiding, inferior colour development, and unsatisfactory gloss are to be avoided, then dispersion aids may well be required. The choice of dispersants, in the broadest sense, is very wide, and they are often specific to the binders, pigments, and solvents in the formulation. In fact the earliest ones used were the metal soaps, normally added last in the preparation of paint as through-driers, which were found to improve dispersion markedly if added to the millbase. Calcium and zinc octoates still find use as additives to the dispersion stage and should not be overlooked.

Additive manufacturers often publish charts showing which dispersant to use to best effect. Some are suitable for both water and solvent-carried systems. The offers may be anionic, cationic, non-ionic, or amphoteric. Usually they are characterized by possessing anchor groups which are attracted to the pigment surface and are necessarily rather specific to particular groups of pigments, and commonly a polymeric, or at any rate a higher molecular weight component that is compatible with the binder and solvents in use.

Few general rules on their use can be given, except that it is usual to incorporate a wetting aid at an early stage so that it has the best opportunity to meet with the pigment surface and not have to complete with — or worse, to displace — other liquid components of the composition. Secondly, it is common to make later liquid additions to the millbase carefully, and with stirring if the dispersing aids themselves are not to be displaced from the pigment surface.

Of the variety of these products on the market, one range, the Solsperse 'hyperdispersants' highlights how effective, but equally how specific, such additives can be.

This range usually needs first a bridging molecular chosen to be of similar nature of the pigments being dispersed, which then attracts strongly to stabilizing molecule suited to the polarity of the solvent. When this is achieved, pigment loading of a millbase can be many times greater than dispersing in conventional binders. A short selection from the products on the market includes:

Disperbyk cationic polymeric solvent systems   [6]
Solsperse range 48/15 polymeric surfactants solvent systems [37/14]
Colorol E modified soya lecithin emulsion paints   [9]
Dispex A40 ammonium polyacrylate aqueous paints   [3]
Tamol 731 di-isobutylene-maleic dispersion emulsion paints   [34]

5.9   Driers

Although it might be argued that the use of driers is so fundamentally a part of the paint formulation that it ought not to be considered an additive, it is, by convention, included here.

The oxidation and polymerization reactions that occur as a paint drier will take place without a catalyst, but are speeded greatly by the presence of certain metal organic compounds. The catalytic activity depends on the ability of the metal cation to be readily oxidized from a stable lower valency to a less stable higher valency. All driers are therefore multivalent. In practice, usually a mixture of metal driers is used. Whereas inorganic salts, say cobalt nitrate, can be used to some effect in an autoxidative aqueous paint, by far the greater number of driers are metal compounds of organic acids ('soaps'), which will emulsify into aqueous systems.

In earlier days lead oxide (PbO), for example, was reacted with linseed oil fatty acids or rosin to make respectively the so-called lead linoleate and lead resinate. These worked but were unstable on storage, and of uncertain metal content. Most paint manufacturers buy-in carefully controlled, stable metal driers, usually based on octoic acid (of which the most readily available isomer is 2-ethyl hexanoic acid). The acid radical provides solubility; the metal cation is the all-important part of the drier. The literature shows that over 40 transition metals have been examined, and ten or so showed worthwhile activity.

5.9.1   Notes on driers

Cobalt
Cobalt is by far the most powerful drier. When used alone it gives a pronounced surface dry, leaving the underfilm mobile, which gives rise to wrinkling in thick films. It is used on its own occasionally, e.g. in aluminium air-dry paints (where very thin films are applied and where lead would dull the aluminium), and in some stoving paints.

Lead
Lead is traditionally the most commonly used metal, ever since pigments were seen to aid drying. It is a 'through' drier. It is still used in combination with other metals unless toxicity or sulphide staining of the dry film rules it out. Toxicity and hazard labelling requirements have much restricted its use.

Manganese

Manganese is the second most powerful drier, but is not in the class of cobalt. Surface and through-drying is contributed. Its limitation is its colour; manganous, light brown goes predominantly manganic, dark brown. With most driers there is no benefit in adding excessive quantities — with manganese, drying markedly falls off beyond an optimum level.

Iron

Iron is rather an old fashioned drier of poor colour. It was used in cheap stoving enamels in dark colours. It has no real use in air-drying paints.

Zinc

Zinc is used as the 'soap' or added as zinc oxide pigment in drier quantities. It slows the initial surface dry and accelerates the through dry. The resulting film is harder when zinc is used.

Calcium

Calcium is not a prime drier, in that little effect can be seen when used on its own. As an auxiliary drier to a lead/cobalt drier system it helps prevent precipitation of the lead on storage, and drying at low temperature is improved. Some years ago when 'blooming' of dry alkyd was a problem then the presence of calcium drier helped.

Cerium

Cerium was promoted some years ago as a substitute for lead (through drier). Although effective, the pronounced yellowing of the film discouraged its use.

Vanadium

Comments rather as cerium apply to vanadium; discoloration of the film and loss on dry storage. It has no real usage.

Barium

Barium is an auxiliary drier, as is calcium. There is no general legislation on toxic hazard at the moment but the level of barium allowed in paints for toys, etc. is strictly controlled. It finds use in some lead-free drier combinations, but concern exists for its long-term future.

Zirconium

Zirconium is finding wide favour as a through-drier to replace lead, although the chemistry of its activity is unrelated. It is sold also in combination with cobalt to make lead drier substitution easy. Its toxicity clearance seems satisfactory.

Aluminium

Aluminium comes up for review every few years as a rather unusual non-toxic through-drier for a paint system that is lead-free. It is not a straightforward substitution for lead — generally the vehicle needs to be tailor-made to suit this drier. Good colour, good through-drier, and hard films are the potential benefits.

5.9.2   Drier combinations

Most resin manufacturers will provide recommended drier combination and levels to use with their products. Technical service is also available from the drier manufacturers themselves.

For drier manufacturers, see listing at end of chapter [5/7].

5.10   Electrical properties

It may be necessary to modify the electrical properties of a liquid paint for two main reasons.

- Antistatic additives may be added to the hydrocarbon solvents to improve conductivity and therefore help avoid electrostatic build-up and the associated spark and fire risk when storing and transporting.
- To lower the electrical resistance of paints based on non-polar solvents, to enable them to be applied satisfactorily by electrostatic spray.

For the first use, it is not uncommon for some companies to make an addition automatically of, say, Stadis 425 at 0.3% to all incoming hydrocarbon solvents. For electrostatic spraying the formulator will choose his or her solvents for optimum resistivity. If conductivity needs increasing, then minor additions of products such as Byk E80 or Ransprep may be made.

Oxygenated solvents (e.g. the glycol ethers), although often desirable for their solvency characteristics polar, show too high conductivity for good electrostatic spraying and good 'wrap-around'.

Stadis 450   [23]
Stadis 425   [23]
Ramsprep   [31]
Byk ES80   [6]

5.11   Flash corrosion inhibitors

Where aqueous paints find use directly onto ferrous metals, e.g. water-borne primers, rust staining may mar the film, usually in the form of a scatter of light brown spots. This so-called 'flash corrosion' may be overcome by the use of inhibitors. Generally, the same additives listed from protection against in-can corrosion find use here but at slightly higher additions.

Serad FA 279   [36/22]
Drewgard 432   [12]
Ray BO 60   [32]
Flash Rust Inhibitor 179   [1]
Prestantil 448   [25/9]

5.12   Floating and flooding additives

Most coloured paints change their colour slightly during drying, owing to the migration of some of the pigment to the surface. If it always took place in a regular manner there would be little problem.

Floating is the term usually reserved for colour striations or mottled effects, whereas flooding is the marked development of colour which, although at first uniform, is lightened again if a part-dried area is disturbed (e.g. by lapping-in by brush). Because of the high viscosity of the setting paint-film the area does not then re-develop its colour. It follows that both effects may manifest themselves when two or more pigments are presented in a formulation. Flooding is often considered to be a severe form of floating. Both can be alleviated by dispersing the pigments intimately together, so that any flocculation that takes place is co-flocculation. Fine particle-size extenders such as aluminium oxide and modified precipitated calcium carbonates included at the dispersion stage are often of value. Disperbyk 181 is claimed to even out the electrical charges of organic pigment surfaces, although the level required to do so is greater than that normally expected of an additive. At the other end of the scale, introducing some controlled flocculation, e.g. with Anti-terra P, may prove a remedy.

Corrective additions to a tinted paint include the various silicone-containing products, e.g. Dow Corning PA3 or non-silicone additives such as Henkel's Product 963.

Suiting the additive to the formulation is particularly relevant in this area. In checking the control of floating, the authors have found that observing the lack of Bénard cell formation on small pools of paint dropped on to a glass panel, especially useful in judging both effectiveness and the right level of addition of these additives.

Aluminium oxide	[10]
Disperbyk 181	[6]
Dow Corning PA3	[11]
Efka 64/66	[13/14]
Borchigen ND	[5/7]

5.13   In-can preservatives

Latex and other aqueous paints are particularly prone to spoilage by microorganisms. Good housekeeping is vital to prevent major infection occurring at the production unit, but preservatives will still be needed to prevent deterioration of colloids, etc. The evil-smelling breakdown products of glue-size and casein are now rarely met, but bacterial infection can still cause gassing in the can, and enzyme breakdown of cellulosic thickeners will cause loss of viscosity.

Ideally, the biocide is added to the first charge of water, thus ensuring its presence from the earliest stage. It follows that in-can preservatives are readily water soluble. Although organo-mercurials and organo-tin compounds are effective and still find use, there has been a marked swing to metal-free organic biocides, for low mammalian toxicity is becoming increasingly important. End use will determine the importance of this and other properties. Where the product finds household use, then low irritancy and low sensitization will also rate high; pale colours will need a biocide that does not cause discoloration.

Generally, useful products available include the benzisothiazolinone derivatives, e.g. members of the proxel range, substituted oxazolidines, e.g. Nuosept 95, and proprietary blends of 'heterocylics' with or without formaldehyde release agents, e.g. Mergal K6N and Mergal K7.

Proxel range   [37/14]
Mergal range   [21]
Preventol range   [4]
Acticide

5.14   In-film preservatives

Given the right environment, most classes of paint film will support mould growth, causing, typically, black stains at the junction of walls and ceiling in bathrooms. Similar growth may be seen on the outside of buildings, and the role of greenish algae in defacing exterior surfaces is increasingly recognized. Biocides incorporated into paints can help to keep the films free from such growth, but do not expect too much from them if the surface is obliterated by nutrients. The present authors recall the difficulties of keeping painted walls clean when they were heavily coated with syrup, in a Barbados sugar refinery.

Biocides are not usually added as an afterthought; neither are they planned into the formulation, but generally the level used is below 5% and therefore they are often considered as additives. Unlike biocides, used as in-can preservatives for emulsion paints which need to be highly soluble, those used for film protection need to be almost insoluble if they are not to be leached quickly from the film and the effectiveness lost with condensation or the first showers of rain.

Proprietary film preservatives are often blends of biocides to give best protection against a broad spectrum of fungi and algae. Testing for fungicidal and algicidal efficacy is a specialist activity, but fortunately many of the suppliers offer an evaluation service to help choose type and level of biocide to suit a particular paint composition and the environment it must meet.

Suppliers include:

Proxel range   [37/14]
Mergal 588   [21]
Metatin range   [1]

5.15   Insecticidal additives

The control of household flies by the use of an insecticide in a coating seems to be worth while. Two main problems arise. Firstly, to render the active ingredient available to a settled insect the level of additive needs to be high and is formulated into a flat paint. Secondly, and more fundamental, is the desirability of the end effect. Where flying insects are a problem the public prefer them on-the-wing rather than dropping lifeless into food. There are special applications where such paints find use, e.g. in ships' holds to kill cockroaches. Nor must one overlook the insecticides used in penetrating compositions for the protection of wood. Organo-metallic compounds (e.g. tributyl tin oxide, zinc octoate, and copper naphthenate) have found use in woodcare compositions.

Chlorinated aromatic compounds, e.g. 6-chloro epoxy hydroxy naphthalene and 1-dichloro 2,2'bis-(p-chlorophenyl) ethane, our records show, have been offered as insecticides for addition to paint.

Zinc octoate	[17]
Copper napththenate preventol	[1, 17]
Priem insecticide	[30]
Tributyltin oxide	[1]
Preventol	[4]

5.16 Optical whiteners

These materials absorb ultraviolet wavelengths and re-emit the energy in the visible waveband. If they are chosen to emit in the blue-violet region they can give a boost to the colour of whites, overcoming any tendency to yellowness. Although widely used in the paper and detergent industries they have not found a significant outlet in the surface-coating industry because of their short-lived effectiveness and the cost premium incurred. An example of a product of this type is Uvitex OB.

The use of optical whiteners is not to be confused with the 'improvement' of a white by the addition of, say Colour Index (CI) Pigment Violet 23 to counter yellowness. This it does, but the resulting undesirable greyness often rules out this use.

Uvitex OB

5.17 Reodorants

Most paints smell when drying. The mild odour of latex paints is quite acceptable to most people and receives little attention, but that from solvent-borne systems is more noticeable. It stems, in the first place, from the evaporation of the solvents used, but in oxidative paints a secondary and more persistent odour from the drying reactions follows. Still sought is an additive that will truly remove these odours.

Generally, care can be taken in the initial formulation to avoid particularly offensive-smelling components, e.g. by choosing lower odour solvents. However, the odour from both sources can be modified or masked by the use of industrial perfumes, such as find use in polishes and household sprays. It is the writers' experience that the majority of people prefer the original paint odour to the 'muddy' combination-odour that so often results from the use of reodorants in paints.

Reodorant suppliers are listed at the end of the chapter [16, 28].

5.18 Ultraviolet absorbers

Many pigments fade and many binders degrade owing to the effect of incident radiation, especially ultraviolet. The use of a coat of varnish was shown to slow down the fading of a fugitive paint many years ago, but unfortunately unpigmented varnish films themselves degrade quickly. Recently ultraviolet absorbers (akin to those used in sun-screen cosmetic creams) have been shown to improve the performance of such a protective clearcoat. A further advancement in this technology has been achieved by combining a light-stabilizer with the UV absorber. The resultant protection is achieved in two stages. The UV absorber (about 1%) converts the undesirable short wavelengths to heat energy, and the light-stabilizer (a sterically hindered amine) (HALS) captures the free radicals generated that would cause film degradation. This technology has made possible the use of 'basecoat plus clearcoat' automotive finishing systems and overcomes the early problems of under-film chalking, delamination, and cracking of the top clear coat.

Tinuvin 1130 UV absorber
Tinuvin 1123 HALS
Tinuvin 1144 HALS
Sanduvor VSU UV absorber
Sanduvor 3058 HALS
Sanduvor 3212 HALS/UV absorber

5.19   Additive supplies

All addresses are in the UK unless otherwise specified.
[1] ACIMA Chemical Industries Ltd, PO Box CH9470, Buchs SG, Switzerland.
[2] Albright & Wilson Ltd, PO Box 3, Oldbury B68 ONN.
[3] Allied Colloids, PO Box 38, Low Moor, Bradford.
[4] Bayer (UK) Ltd, Bayer House, Strawberry Hill, Newbury RG13 1JA.
[5] Borchers (Gerb) AG, Elsabethstrasse 14, Postfach 1812, D-4000 Dusseldorf 1, Germany.
[6] Byk CERA GmbH, Abelstrasse 14, D4230 Wesel, Germany.
[7] Chemitrade Ltd, Station House, 81–83 Fulham High Street, London SW6 3JW.
[8] Cray Valley Produce, Machen, Newport, Gwent.
[9] Croxton & Garry Ltd, Curtis Road, Dorking RH4 1XA.
[10] Degussa Ltd, Winterton House, Winterton Way, Macclesfield SK11 OLP.
[11] Don Corning Ltd, 185 Kings Court, Reading RG1 4EX.
[12] Drew Ameroid International, PO Box 6, 4930 AA Geerthudenberg, Netherlands.
[13] Efka Chemicals BV, PO Box 358, 2180 AJ Hill, Egom, Netherlands.
[14] Ellis & Everard Industrial Specialities, Caspian House, 61 East Parade, Bradford BD1 5EP.
[15] Efka Chemicals BV, PO Box 358, 2180 AJ Hillegom, The Netherlands.
[16] Ferguson & Menzies Ltd, 312 Broomloam Road, Glasgow, Strathclyde G5 2JN.
[17] Harcross Durham Chemicals Ltd, Birtley, Chester-le-Street DH3 1QX.
[18] Henkel Chemicals GmbH, Dusseldorf, Germany.
[19] Henkel UK Ltd, Merit House, The Hyde, Edgware Road, London NW9.
[20] Henkel Performance Chemicals, Norco House, Kirkstall Road, Leeds LS3.
[21] Hoechst (UK) Ltd, Hoechst House, 50 Salisbury Road, Hounslow, Middlesex.
[22] Huls (UK) Ltd, Featherstone Road, Wolverton Mill South, Milton Keynes MK12 3TB.
[23] Imstara Ltd, 3 Greenhill, Winksworth, Derby DE4 4EN.
[24] K & K Greeff, Arglyle House, Epson Avenue, Stanley Green, Hanforth, Wilmslow SK9 3RN.
[25] Fratelli Lamberti Spa, 21041 Alb Izzate, Italy.
[26] Laporte Surface Treatments Ltd, Spring Road, Smethwick, Warley B66 1OT.
[27] Pluss Stauffer, CH-L665 Otrigen, Switzerland.

[28] PMC Specialities Group, 65B Wigmore Street, London W1H 9LG.

[29] PR BSA, Avenue de Brogneville 12, B1150 Brussels, Belgium.

[30] Priem Wilhelm GmbH & Co., D4800 Bielefield 1, Osningstrasse 12, Germany.

[31] Ramsburg (UK) Ltd, Ramsburg House, Hamm Moor Lane, Weybridge K15 2RH.

[32] Raybo Chemical Corporation, USA.

[33] Rheox Inc, rue de L'Hopital 31, B1000, Brussels, Belgium. Rheox UK Ltd, Barons Court, Manchester Road, Wilmslow SK9 1BQ.

[34] Rohm & Haas (UK) Ltd, Lennig House, 2 Masons Avenue, Croydon CR9 3NB.

[35] Schwegmann Bernd Kg, Buchenwag 1, D5300, Bonn 3, PO Box 300, 860 Germany.

[36] Servo BV, PO Box 1, 7490 AA Delden, The Netherlands.

[37] Zeneca, PO Box 42, Blackley, Manchester.

6

The physical chemistry of dispersion

A Doroszkowski

6.1 Introduction

The physical chemistry of dispersion may range from producing a polymer dispersion, such as latex, where the particle dispersion is formed *in situ*, to pigment suspension.

In preparing a latex, there is a strong element of polymerization kinetics which does not apply to pigment dispersion. On the other hand, in pigment dispersion there may be certain interfacial considerations which do not apply to latex formation. However, from a colloidal point of view, once the dispersion is produced, the physical chemistry of the dispersion is the same, whether polymer or pigment.

In this chapter the emphasis is on examining the physical chemistry of the dispersion where the disperse phase is preformed, i.e. from a pigment standpoint, although the same considerations apply to polymer particles if they constitute the disperse phase.

Our objective is therefore to look at the physical chemistry involved in forming a dispersion; to examine the factors influencing dispersion stability; and to consider how they can be measured or assessed.

For simplicity, we can define a paint as a colloidal dispersion of a pigment (the 'disperse' phase) in a polymer solution (the 'continuous' phase). Emulsion paints have both the polymer and pigment as the disperse phase. While the chemical structure of the polymer is important in determining the properties of a paint, the state of dispersion of the pigment in the polymer is no less important.

In practice it is fair to say that most of the problems that arise in a paint come back to the state of the pigment dispersion. The state of pigment dispersion can affect the:

- optical properties, e.g. colour [1];
- flow properties [2];
- durability [3];
- opacity [4];
- gloss [5];
- storage stability [6].

To produce a 'good' dispersion of colloidal particles from a dry powder we have to go through a number of processes which can be subdivided, arbitrarily, as separate operations, since in practice they may occur simultaneously. They are:

- immersion and wetting of the pigment;
- distribution and colloidal stabilization of the pigment.

These simple titles may mask many complex processes but are well worth consideration, as is the 'simple' practical question 'Is it better to disperse TiO_2 pigment in an alkyd solution using an aromatic or aliphatic solvent?'

6.2   Immersion and wetting of the pigment

Suppose we consider a single solid particle of pigment which will become immersed in a liquid. This can be represented schematically as a three-stage process:

Stage 1	*Stage* 2	*Stage* 3
Particle in air	Creation of a hole in the liquid	Filling of the hole by the particle

If the particle has an area A, and γ_{SV} is the free energy per unit area, then the total energy of the three stages can be written as:

Stage 1: The surface energy of the particle is simply $A\,\gamma_{SV}$.

Stage 2: The energy of the particle in the vapour plus the work done to create a hole in the liquid identical to the volume and area of the pigment is $A\,\gamma_{SV} + A\,\gamma_{LV}$.

Stage 3: The work done to plug the hole in the liquid with the particle is $A\,\gamma_{LS} - A\,\gamma_{SV} - A\,\gamma_{LV}$.

Then the total energy change on immersing a particle is the sum of stages 2 + 3 − 1:

i.e. $\quad A\,\gamma_{LS} - A\,\gamma_{SV}$ $\hfill (6.1)$

From Young's equation we have $\gamma_{SV} = \gamma_{LS} + \gamma_{LV}\cos\theta$:
  The energy change on substitution in equation (6.1) is

$\quad -A\,\gamma_{LV}\cos\theta$ $\hfill (6.2)$

If we examine Young's equation in terms of wetting, or contact angle (θ), that is:

$$\cos \theta = \frac{\gamma_{SV} - \gamma_{SL}}{\gamma_{LV}} \qquad (6.3)$$

then, provided that $\theta < 90°$, a decrease in γ_{LV} will reduce θ and improve wetting, hence an aliphatic hydrocarbon is preferable to an aromatic solvent, since γ_{LV}(aliphatic) $< \gamma_{LV}$(aromatic).

If, however, we add a surface active agent, and it adsorbs at the air interface it will reduce γ_{LV}, and if it adsorbs on the particle surface it will decrease γ_{SL}. Both these effects will lead to better wetting.

However, if $\theta = 0$, as in a high-energy surface such as one might expect of a TiO_2 surface, it would be better to have γ_{LV} maximized, i.e. the aromatic solvent should be preferable to the aliphatic solvent as the dispersion medium.

6.2.1 Penetration of agglomerates

If we consider the spaces between the powder particles as simple capillaries of apparent radius r, then the surface pressure (P) required to force a liquid into a capillary is:

$$p = \frac{-2\gamma_{LV} \cos \theta}{r} \qquad (6.4)$$

Force $= -2\pi r \, \gamma_{LV} \cos \theta$
Area $= \pi r^2$

Hence penetration will occur spontaneously (ignoring gravitational effects) only if $\theta < 90°$ and if the pressure within the capillaries does not build up to counter the ingress of liquid.

Thus, to enhance liquid penetration of agglomerates it is desirable to maximize γ_{LV} and decrease θ. But since changes in γ_{LV} go hand in hand with θ this is difficult to realize. The addition of surface active agents will tend to decrease both γ_{LV} and θ, especially in aqueous media, hence the assessment of which is the dominant effect is best obtained by trial.

The argument used above is rather a simplistic one which uses the surface energy of a solid in equilibrium with the vapour of the liquid. Heertjes & Witvoet [7] have examined the wetting of agglomerates and have shown that only when $\theta = 0$ can complete wetting of the powder aglomerate be obtained.

The rate of liquid penetration into an agglomerate was derived by Washburn [8] as:

$$\frac{dl}{dt} = \frac{r\gamma_{LV} \cos \theta}{4\eta} \qquad (6.5)$$

where dl/dt is the rate of penetration of the liquid of viscosity (η) in a capillary of radius r and length l.

In a packed bed of powder it is customary to employ an 'effective pore radius' or a 'tortuosity factor'. Thus $r/4$ can be replaced by a factor K, which is assumed to be constant for a particular packing of particles. Then equation (6.5) becomes

$$l^2 = \frac{Kt \, \gamma_{LV} \cos\theta}{\eta} \tag{6.5a}$$

Thus by inspection of equation (6.5a) we can see that to facilitate penetration of the powder we want to:

- maximize $\gamma_{LV} \cos\theta$;
- minimize the viscosity (η);
- have K as large as possible, e.g. loosely packed agglomerates of pigment.

The Washburn equation describes a system in which the walls of the tube are covered with a duplex film (i.e. one where the surface energy of the film is the same as that of the surface of the bulk material). Good [9] generalized the Washburn equation to cover the case where the surface is free from adsorbed vapour. That is,

$$l^2 = Kt \left[\frac{\gamma_{LV} \cos\theta + \Pi_e - \Pi_{(t=0)}}{\eta} \right] \tag{6.5b}$$

where Π_e is the spreading pressure of the adsorbed film that is in equilibrium with the saturated vapour, and $\Pi_{(t=0)}$ is the spreading pressure for the film that exists at zero time.

If we consider the question of whether it is better to disperse TiO_2 in aliphatic or aromatic hydrocarbon, then while it is difficult to assign a specific surface energy to the TiO_2, because of its varied surface coating, which consists of mixed hydroxides of alumina, silica and titania, it is nevertheless a high surface energy material and can be likened to a water surface for simplicity [10], therefore $\cos\theta = 1$ for both the aromatic and aliphatic hydrocarbons, unless one of the liquids is autophobic [11]. Hence, in order to maximize $\gamma_{LV} \cos\theta$ it is better to use the aromatic hydrocarbon.

Crowl [12] has demonstrated the effect of $\gamma_{LV} \cos\theta$ on grind time, as shown in Table 6.1.

Table 6.1 — Adhesion tension ($\gamma_{LV} \cos\theta$) and rate of milling: laboratory ballmilling of rutile titanium dioxide

Medium	Adhesion tension ($\gamma_{LV} \cos\theta$)	Time (hr) to reach Hegman 3[a]	Hegman 5[a]
5% isomerized rubber	21.9	0.9	2.0
5% alkyd	16.3	1.0	1.7
10% alkyd	14.4	1.7	2.5
10% isomerized rubber	12.3	2.0	4.0

[a] Hegman readings on the 0–8 scale, where 0 = coarse or poor dispersion, 8 = fine or best dispersion.

6.2.2 Solid surface characterization

The surface and interfacial tension forces can be subdivided into polar interactions (γ^{AB}) which are the (Lewis) acid parameter of surface tension, γ^+, and (Lewis) base parameter, γ^- (for hydrogen bonding γ^+ is the contribution of the proton donor (Brønsted acid) and γ^-, that of electron acceptor (Brønsted base). Thus H-bonding is a special case of Lewis acid–base interactions) and apolar Lifshitz–van der Waals interactions, γ^{LW} (comprising the London–van der Waals, or dispersion forces; van der Waals–Keesom, orientation; and van der Waals–Debye, induction, forces) as described by van Oss and co-workers) [13]. Hence the total surface tension of a material (i) is

$$\gamma_i = \gamma_i^{LW} + \gamma_i^{AB}$$

The total acid–base free energy of interaction between two materials i and j is not additive, unlike the apolar components because of the nature of the independent (asymmetric) interactions, hence the total interfacial free energy between substances i and j is:

$$\gamma_{ij}^{tot} = \left(\gamma_i^{LW} + \gamma_j^{LW} - 2(\gamma_i^{LW}\gamma_j^{LW})^{\frac{1}{2}}\right)$$
$$+2\left[(\gamma_i^+\gamma_i^-)^{\frac{1}{2}} + (\gamma_j^+\gamma_j^-)^{\frac{1}{2}} - (\gamma_i^+\gamma_j^-)^{\frac{1}{2}} - (\gamma_i^-\gamma_i^+)^{\frac{1}{2}}\right]$$

Young's equation, i.e. $(1 + \cos\Theta)\gamma_l^{tot} = -(\gamma_{sl} - \gamma_s - \gamma_l)$ can be written in the form:

$$(1 + \cos\Theta)\gamma_l^{tot} = 2(\gamma_s^{LW}\gamma_l^{LW})^{\frac{1}{2}} + 2(\gamma_s^+\gamma_l^-)^{\frac{1}{2}} + 2(\gamma_s^-\gamma_l^+)^{\frac{1}{2}}$$

where i = s (solid) and j = l (liquid).

By measuring the contact angle at the solid–liquid interface, using at least three different test liquids of known polar and apolar components (see Table 6.2) it is possible to obtain three simple equations in three unknowns. Hence the solid surface's polar and apolar surface tension (surface free energy) components can be determined, thereby characterizing the solid surface in energy terms ($mJ\,m^{-2}$) (or in $mN\,m^{-1}$ as tension). For example poly(methyl methacrylate) was found to have surface free energy components of [14]

$$\gamma^{LW} = 43.1; \gamma^+ = 0; \gamma^- = 11.3 \text{ (in } mJ\,m^{-2}).$$

Table 6.2 — Test liquids [13]

Test liquid	γ_l	γ^{LW}	γ^+	γ^-	η
Decane	23.8	23.8	0.0	0.0	0.00092
α-Bromonaphthaline	44.4	44.4	0.0	0.0	0.00489
Diiodomethane	50.8	50.8	0.0	0.0	0.0028
Water	72.8	21.8	25.5	25.5	0.0010
Formamide	58.0	39.0	2.28	39.6	0.00455
Ethylene glycol	48.0	29.0	1.92	47.0	0.0199
Glycerol	64.0	34.0	3.92	57.4	1.490

Surface free energies in $mJ\,m^{-2}$; viscosity (η) in Pa s.

The measurement of the contact angle between a powder and a liquid frequently presents problems, when the capillary rise technique is used. This is frequently due the difficulty in obtaining a reproducible powder-packing in a capillary tube, especially when pigments constitute the powder. To overcome the problems associated with the capillary rise in a column of fine powder, van Oss and co-workers used a 'thin-layer wicking' (TLW) technique [13, 15]. The technique essentially consists of making reproducible, thin-layer chromatography substrates on suitable microscope glass slides; measuring the liquid (solvent) rise as one would in standard thin layer chromatography (TLC), to determine the apparent pore radius (r) of the substrate, using an alkane ($\cos\Theta = 1$); then repeating the process with other test liquids on similar plates, using the experimentaly determined value of r.

Although the TLW technique appears to work reasonably well in characterizing pigment and other powder surfaces, it should be noted that when applied to titanium dioxide pigment characterization, the 'solvent front' moves very slowly (about 1 cm in an hour, at the most) and great care has to be taken to prevent solvent evaporation, so as not to make the measurements meaningless. However, it is possible to differentiate between various grades of commercial titanium dioxide pigments using this approach.

There are some criticisms associated with the TLW technique as discussed by Chibowski and Gonzalez-Caballero [14]. The original form of the Washburn equation describes the process correctly only if the liquid completely wets the powder (zero contact angle) and a duplex film of the liquid has been formed well ahead of the penetrating front, hence the solid surface free energy has no influence on the rate of wicking. Chibowski and co-workers, however, have shown that by using equations based on Good's derivation of the 'Washburn' equation, this argument can be negated and that the TLW technique is a useful tool in characterizing solid surface free energies, provided care is exercised to ensure that there is no solvent loss from the moving front, and preferably horizontal sandwich plates are used [16–18].

6.3   Deagglomeration (mechanical breakdown of agglomerates)

Pigmentary titanium dioxide exists in powder form as loose agglomerates of about 30–50 μm in diameter. The surface coating of the pigment, by the manufacturer, has a large effect on reducing the cohesive forces of the powder and thus assists in the disintegration (or deaggregation) process [19]. It is difficult to define the 'grind' or dispersion stage where the loose agglomerates are broken down into finer particles after all the available surface has been wetted out, and the process is generally treated empirically, as described in Chapter 8.

The surface coating applied by the pigment manufacturer, for example onto TiO_2, is proprietary information and probably not completely understood, though there are many publications attempting to describe it [20, 21]. The chemical analysis of the pigment and surface coating is generally in terms of 'equivalent to' Al_2O_3, SiO_2, ZnO, TiO_2 etc. even though the coating consists of mixed hydroxides which are probably more correctly described by $M_x(OH)_y$, where M can be a mixture of Al, Si, Ti, Zr, etc. (as well as organic treatments with polyols, to facilitate stages of paint manufacture). The main purpose of the surface coating is to deactivate the surface of

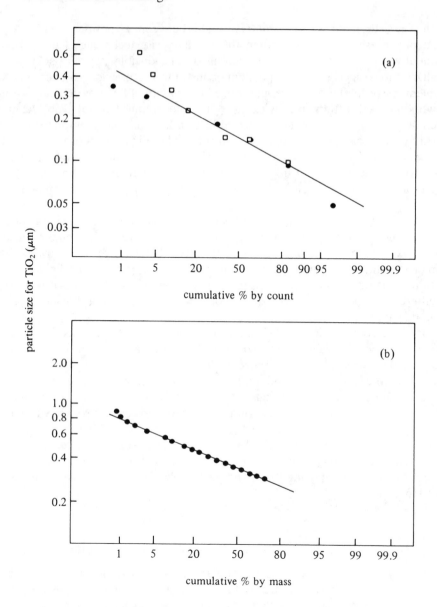

Log — probability plots of particle size
(a) by electron microscopy, prior to milling; $d_{gc}=0.16\,\mu m$
    $\sigma=1.52$, therefore $d_{dm}=0.3$
    □ counting all particles (even when obvious aggregates)
    ● counting only single crystals (provided more than half the perimeter was visible)
(b) by x-ray sedimentation, after milling; $d_{gm}=0.36$, $=1.5$

Fig. 6.1 — Particle size analysis of TiO$_2$ (log-probability plots).

the rutile pigment, which would otherwise accelerate the degradation of the resin on weathering. The pigment coating is also there to aid pigment dispersion by the paint manufacturer. The 'grinding stage' in millbase manufacture is not a comminution stage but a dispersion stage of the pigment to the primary particle size as made by the pigment manufacturer. Some of the 'primary' particles consist of sinters of TiO_2 crystals produced during the 'surface coating' stage in pigment manufacture and remain intact *per se* on completion of the 'grind stage', as can be shown by particle size analysis before and after incorporation of the pigment in a paint (see Fig. 6.1); that is, the median particle size by count ($d_{gc} = 0.16\,\mu m$) and distribution ($\sigma = 1.52$) of the TiO_2 before 'grinding' was found to be the same as that obtained by sedimentation analysis after 'grinding' in a ballmill (using the appropriate Hatch–Choate equation to convert particle size by mass to that of size by count; $d_{gc} = 0.16\,\mu m = d_{gm} = 0.3\,\mu m$ when $\sigma = 1.52$ which is in good agreement with the measured size $d_{gm} = 0.36\,\mu m$, $\sigma = 1.5$).

6.4   Dispersion — colloid stabilization

It is not enough to 'wet out' particles by the continuous phase to produce a stable colloidal dispersion. It is important to realize that attractive, interparticle forces are always present in pigment dispersions. These are the London, van der Waals (or surface) forces. These attraction forces are a consequence of the attractive interatomic forces amongst the atoms which constitute all particles.

Polar materials exert electrostatic forces on other dipoles (Keesom forces [22]), and polar molecules can attract non-polar molecules by inducing dipoles (Debye forces [23]). The attraction between non-polar atoms or molecules was not understood until it was realized that the electron cloud surrounding the nucleus could show local fluctuations of charge density. This produces a dipole moment, the direction of the dipole fluctuating with the frequency of the charge fluctuations. If there is another atom in the vicinity, then it becomes polarized and it interacts with the first atom.

London [24] showed how this treatment could be used to calculate the interaction between two similar atoms, and he extended it to dissimilar atoms. London's analysis is based on instantaneous dipoles which fluctuate over periods of 10^{-15} to 10^{-16} s. If two atoms are further apart than a certain distance, by the time the electric field from one dipole has reached and polarized another, the first atom will have changed. There will be poor correlation between the two dipoles, and the two atoms will experience what are known as 'retarded' van der Waals forces.

In general, we can expect strong non-retarded forces at distances less than $100\,\text{Å}$ ($10\,nm$) and retarded forces at distances greater than this.

6.4.1   Forces between macroscopic bodies

The forces between two bodies, due to dispersion effects, are usually referred to as surface forces. They are not only due to the atoms on the surface, but also to the atoms within the bulk of the material. Hamaker [25] computed these forces by pairwise addition. They may be summarized as:

$$V_A = \frac{-A}{48\pi}\left[\frac{1}{d^2} + \frac{1}{(d+\delta)^2} - \frac{2}{\left(d+\dfrac{\delta}{2}\right)^2}\right] \tag{6.6}$$

if

$$d \gg \delta \text{ then } V_A = \frac{-\delta^2 A}{32\pi d^4} \tag{6.6a}$$

$$d < \delta \quad V_A = \frac{-A}{48\pi}\left[\frac{1}{d^2} - \frac{7}{\delta^2}\right] \tag{6.6b}$$

$$d \ll \delta \quad V_A = \frac{-A}{48\pi d^2} \tag{6.6c}$$

where V_A is the attraction energy for two plates of thickness δ at a distance of separation of $2d$ from each other, and A is the Hamaker constant for the substance comprising the plates.

Hamaker [26] also showed that the attraction energy (V_A) between two like spheres of radius a separated by a surface to surface distance d is:

$$V_A = \frac{-A}{6}\left[\frac{2}{S^2-4} + \frac{2}{S^2} + \ln\left(\frac{S^2-4}{S^2}\right)\right] \tag{6.7}$$

where

$$S = 2 + \frac{d}{a}$$

If $d \ll a$ then $V_A \simeq \dfrac{-Aa}{12d}$ \hfill (6.7a)

and for spheres of radii a_1, a_2:

$$V_A \simeq \frac{-Aa_1 a_2}{6d(a_1 + a_2)} \tag{6.7b}$$

We can compute these attraction forces in terms of particle size, medium, and distance of separation, and express the result as an energy of attraction (V_A). However, the Hamaker constant A is the value *in vacuo*, and if one is considering the attraction energy between two particles in a fluid, it has to be modified according to the environment. Hamaker showed that particles of substance A_1 in a medium of substance A_2 had a net Hamaker constant of $A = A_1 + A_2 - 2\sqrt{(A_1 A_2)}$. Thus if we have two different materials — material 1 and material 2 in medium 3 — the net Hamaker A_{132} is modified according to

$$A_{132} = \left(A_{11}^{\frac{1}{2}} - A_{33}^{\frac{1}{2}}\right)\left(A_{22}^{\frac{1}{2}} - A_{33}^{\frac{1}{2}}\right) \tag{6.8}$$

For a list of Hamaker constant values for common substances see Visser [27]. Gregory [28] showed how the Hamaker constant could be estimated from simple experimental measurements.

The point to note is that these forces are always attractive; they may be modified to varying extents depending on one's wish for accuracy, e.g. Vold's correction due to adsorbed layers [29] or Casimir and Polders' correction due to retardation of the attraction forces [30, 31]. However, simple examination of the published values of the Hamaker constants reveals that they may vary by an order of magnitude and therefore rarely justify laborious correction for varying subtle effects such as that of Vold [29] and Vincent and co-workers [32].

Although there is always an attraction force between two like particles, Visser [33] points out that in certain circumstances it is possible to have a negative Hamaker constant in a three-component system such as when $A_{11} < A_{33} < A_{22}$ or $A_{11} > A_{33} > A_{22}$ where A_{33} is the individual Hamaker constant of the medium. It is important to note that there is still an attractive potential between like particles of A_{11} in A_{33} as well as A_{22} particles attracting each other in medium A_{33}, but there will be a repulsion between A_{11} particles with A_{22} particles. A suggested example of this type of non-association is thought to be exemplified by poly(tetra fluoroethane) (PTFE) and graphite particles in water.

To obtain a 'colloidal dispersion' one must somehow overcome the omnipresent attraction energy by generating some kind of repulsion energy (V_R) between the particles, such that when the repulsion energy and attraction energy are added, there is still a significant net repulsion energy. Since the attraction and repulsion energies are dependent on interparticle distance, it is important to know something of particle spacing, especially in paint systems which, in conventional colloidal terms, are considered to be 'concentrated'. It can be shown that in a hexagonally close-packed array of spheres, the ratio of surface-to-surface separation (S) to centre-to-centre separation (C) is related to total volume by

$$\frac{S}{C} = 1 - \left(\frac{PVC}{0.74}\right)^{\frac{1}{3}} \tag{6.9}$$

where PVC is pigment volume concentration (%).

The surface-to-surface distance in relation to pigment volume concentration is given in Table 6.3, where the distance S is expressed in terms of sphere diameter. Thus for any particle size of mono-disperse spheres the average separation is equal to one particle diameter at a PVC of 9.25%. Therefore the average

Table 6.3 — Relationship between pigment volume concentration (PVC) and surface-to-surface spacing (S)

PVC	S^a
5	1.43d
10	0.95d
15	0.71d
20	0.55d
25	0.44d
30	0.35d
35	0.28d
40	0.23d
45	0.18d

[a] S is expressed in terms of particle diameter (d).

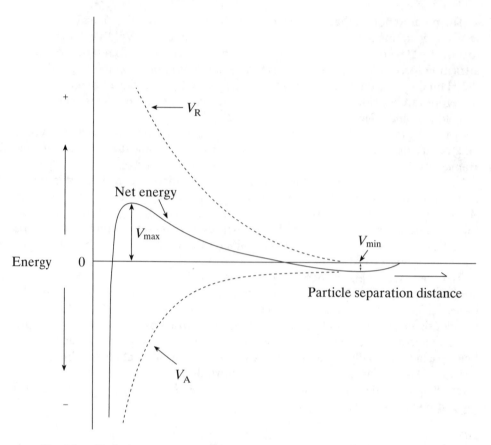

Fig. 6.2 — Typical energy v.s. particle separation curves for the approach of two like particles: V_R = repulsion energy; V_A = attraction energy, where the total interaction energy (V_{net}) is $V_{net} = V_R + V_A$. Note that weak flocculation, or secondary minimum flocculation (V_{min}), occurs if $V_A > V_R$ at large particle separations, even though there is strong repulsion (V_{max}) on closer approach; on very close approach there is very strong primary flocculation due to $V_A \gg V_R$.

interparticle distance for TiO_2 at this concentration is about 2000 Å (200 nm), and for smaller particles at the same volume concentration the distance is proportionally smaller.

If we compare repulsion energies and attraction energies, and remember that if particles are in Brownian motion, which at room temperature have an average translational energy of $\frac{3}{2}kT$, then to stabilize them we must have an energy barrier (V_{max}) which is significantly greater than this (see Fig. 6.2).

Depending on the nature of the net energy curve we can have a 'secondary minimum' or an energy trough which may bring about flocculation which is weak. This is called 'weak flocculation' (V_{min}) or 'secondary minimum flocculation' and is used to differentiate it from the stronger flocculation which occurs on closer approach of particles, sometimes called 'primary flocculation'. It is worth noting that sometimes in the literature [34] the terminology is different from that used in the paint industry, e.g. the term 'coagulation' in the literature is used to mean flocculation, when normally that term is reserved by industrial chemists for the irreversible association of latex or emulsion particles.

6.4.2   Dispersion stabilization by charge

To produce a stable colloid dispersion, a source of repulsion energy must be induced onto particles so that within the interparticle distances (at the desired particle concentration, i.e. PVC) there is sufficient net energy to prevent flocculation.

This repulsion energy may arise from coulombic forces as described by the DLVO theory (see below) or from 'steric stabilization'. Charge may be generated at a surface in different ways:

* preferential adsorption of ions;
* dissociation of surface groups;
* isomorphic substitution;
* adsorption of polyelectrolytes;
* accumulation of electrons.

Preferential adsorption of ions is the most common way of obtaining a charged particle surface, e.g. the adsorption of Ag^+ or I^- ion on Ag halide sols, or ionic surfactants on pigments. The dissociation of surface groups is a common feature with latexes which become charged by the dissociation of sulphate or carboxyl groups. The theory of stabilization of colloidal particles by charge has evolved over the years. It culminated in what is known as the DLVO theory, and is expounded by Verwey and Overbeek (the VO of DLVO) in their book [35]. There are many excellent reviews [36, 37] and papers on the subject, and the present author does not intend to go into the theory. In outline the double layer theory at a flat surface consists of an innermost adsorbed layer of ions, called the 'Stern layer'. The plane (in the Stern layer) going through the centre of the hydrated ions (when no specific adsorption takes place) is known as the 'outer Helmholtz plane' (or Stern plane), OHP, and is directly related to the hydration radius of the adsorbed ions.

If some specific adsorption takes place (usually the dehydration of the ion is a prerequisite) then the plane going through the centre of these ions is known as the 'inner Helmholtz plane', IHP. The distinction between the OHP and IHP is generally necessary since the specifically adsorbed ions allow a closer approach to the surface than the hydrated ions at the OHP and thus increase the potential decay.

Beyond the Stern (or compact) layer there is a diffuse layer, known as the 'Gouy' (or Gouy–Chapman) layer.

When the charge in the Stern layer, diffuse layer, and the surface is summed the total $= 0$ because of electroneutrality. The potentials just at the solid surface (ψ_0), at the IHP (ψ_s) and OHP (ψ_d) are, as Lyklema [36] points out, abstractions of reality, and are not the zeta potential, which is the experimentally measureable potential defined as that occurring at the 'slipping plane' in the ionic atmosphere around the particle.

However, while the zeta potential and the potential at the Helmholtz plane are not the same, for simple systems such as micelles, surfactant monolayers on particles, etc., there is considerable evidence to equate them [38].

6.4.3   Zeta potentials

Zeta potentials can be estimated from experimentally determined electrophoretic mobilities of particles. The equation used to convert observed mobilities into zeta potentials depends on the ratio of particle radius (a) to the thickness of the double layer $(1/\kappa)$, e.g. for 10^{-3} M aqueous solution at $25\,^\circ$C with a $1:1$ electrolyte, $1/\kappa = 1.10^{-6}$ cm.

Values for other electrolyte types or other concentrations modify κ on a simple proportionality basis, since for an aqueous solution of a symmetrical electrolyte the double layer thickness is

$$\frac{1}{\kappa} = \frac{3 \times 10^{-8}}{zc^{\frac{1}{2}}}\text{ cm}$$

where z is the valency, and c is expressed as molarity.

If $\kappa a > 200$ then the mobility (μ) is related by the Smoluchowski equation to the zeta potential (ζ) by

$$\mu = \frac{\varepsilon \zeta}{4\pi\eta}$$

where ε is the permittivity of the medium and η = viscosity.

If the mobility is measured in μm s^{-1}/V cm^{-1} then $\zeta' = 12.8\,\mu$mV$_{\text{per mobility unit}}$ for aqueous solutions.

When $\kappa a < 0.1$ then the Huckel equation

$$\mu = \frac{e\xi}{6\pi\eta}$$

applies.

For $200 < \kappa a < 0.1$ it is necessary to use the computations of O'Brien and White [39] which are an update of the earlier Wiersema *et al.* [40] computations relating mobility to zeta potential.

The conversion of electrophoretic mobility to zeta potential is based on the assumption that the particles are approximately spherical. If they are not, then unless κa is large everywhere, there is obviously doubt as to the value of the zeta potential obtained from the mobility calculation.

If the particle consists of a floccule comprising small spheres and if κa (for a small sphere) is large, the Smoluchowski equation applies.

Hence the zeta potential has frequently a small uncertainty attached to its value, and many workers prefer to quote just the experimentally determined electrophoretic mobility.

6.4.4   Measurement of electrophoretic mobility

The electrophoretic mobility of small particles can be measured by microelectrophoresis, where the time taken for small particles to traverse a known distance is measured, or alternatively by a moving-boundary method. The microelectrophoresis method has many advantages over the moving-boundary method, and it is the more frequently adopted method, although there are some circumstances which favour the moving-boundary method [41].

An alternative method based on the electrodeposition of particles has been devised by Franklin [42]. While the quoted values appear to be in good agreement with microelectrophoresis measurements, the method is fundamentally unsound because electrodeposition of a dispersion is based on electrocoagulation and does not depend on electrophoresis [43].

If the mobility of charged particles is examined in a microelectrophoresis cell then it will be noticed that there is a whole range of velocities, with some particles even moving in the opposite direction. This effect is due to electro-osmotic flow within the cell. The true electrophoretic mobility can only be determined at the 'stationary levels' where the electro-osmotic flow is balanced by the hydrodynamic flow. The position of the stationary levels depends on the shape of the microelectrophoresis cell, i.e. whether it is circular or rectangular. There are many varieties of microelectrophoretic cell [44], and there are many refinements to enable rapid and accurate measurements to be made, such as Rank Bros's microelectrophoresis apparatus [45] equipped with laser illumination, rotating prism, and video camera and monitor. For details on how to measure electrophoretic mobilities the reader is referred to Smith [38], James [46], and Hunter [47].

In extremis one can construct a simple 'flat cell' suitable for viewing under an ordinary microscope, just using a microscope glass slide and two sizes of glass cover slips, as shown in Fig. 6.3.

6.4.5   Application to colloid stability
When two charged surfaces approach each other they start to influence each other electrostatically as soon as the double layers overlap. For surfaces of the same sign the ensuing interaction is repulsion. In a qualitative interpretation a number of points have to be taken into account.

The essence of DLVO theory is that interparticle attraction falls off as an inverse power of interparticle distance and is independent of the electrolyte content of the continuous phase, while the coulombic (or electrostatic) repulsion falls off exponentially with a range equal to the Debye–Hückel thickness $1/\kappa$ of the ionic atmosphere.

The Stern layer does not take a direct role in the interaction, but its indirect role in dictating the value of ψ_d is enormous. In the DLVO theory, the double layers are considered as if they were purely diffuse, but in reality the theory applies to the two diffuse parts of the two interacting double layers. The surface potential ψ_0 should be replaced by Stern potential ψ_d, which in turn is replaced by the zeta potential for non-porous substances.

Typical double-layer thicknesses at varying electrolyte concentrations are given in Table 6.4, illustrated in Fig. 6.4 and calculated from

$$\kappa^2 = \frac{F^2 \Sigma c_i z_i}{\varepsilon_r \varepsilon_0 RT} \tag{6.10}$$

where F = Faraday constant,

$RT = 2.5 \times 10^{10}$ ergs at 298 K,

$\varepsilon_r \varepsilon_0$ are relative permittivity and permittivity in a vacuum and,

z_i; c_i are the charge and number of all ions in solution.

For an aqueous solution of a symmetrical electrolyte the double-layer thickness is

$$\frac{1}{\kappa} = \frac{3 \times 10^{-8}}{z c^{\frac{1}{2}}} \text{ cm}$$

where z is the valency and c is expressed as molarity.

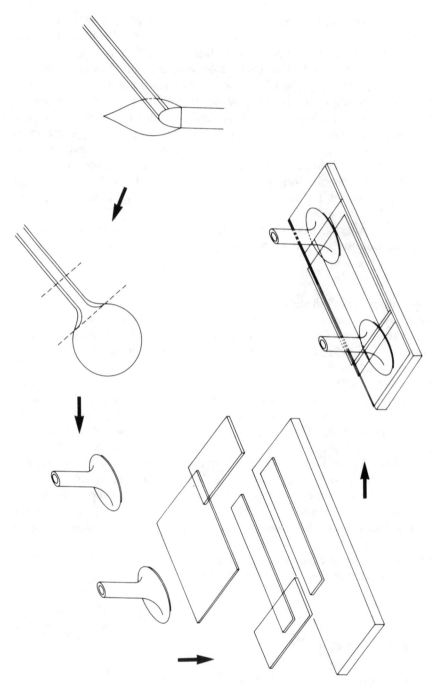

Fig. 6.3 — Construction diagram of a simple 'flat' electrophoretic cell made from a microscope slide and cover slips, cut and glued together with an epoxy adhesive. The electrode compartment is made by blowing a piece of glass tubing to flare the ends so that it can seal the opening to the cell as shown.

Table 6.4 — Effect of electrolyte on double layer thickness
Concentration of 1:1 electrolyte in water at 25 °C

Molar concentration	Double layer thickness $\frac{1}{\kappa}$ in cm × 10^{-8}
10^{-5}	1000
10^{-4}	300
10^{-3}	100
10^{-2}	30
10^{-1}	10

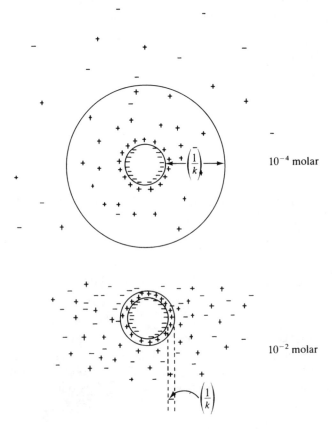

Fig. 6.4 — Schematic effect of electrolyte concentration on double layer thickness $1/\kappa$.

6.4.5.1 The dynamics of interaction

Much depends on the timescale of particle interaction, e.g. diffusion relative to adjustments of the double layers on overlap. If the rate of approach is fast, the double layers have little or no time to adjust.

If there is slow approach, then the double layers are continually at equilibrium. The DLVO approach, in this latter case, is that the energy of repulsion, V_R, can be calculated by reversible thermodynamics as the isothermal reversible work to bring the particle from infinity to distance H.

The underlying idea is that in an equilibrium encounter the potential should remain constant, since it is determined by adsorption of the charge-determining ion, the chemical potentials being fixed by their values in the bulk. This is what is known as 'constant potential' interaction.

'Constant charge' interaction is expected to occur if the interaction proceeds so fast that there is no readjustment of ionic adsorption/desorption, and hence the surface potential, ψ_0, is not constant but rises during overlap.

When all is said and done, there is not a great difference in V_R for the two types of interaction, but the constant charge case tends to give larger V_R than constant potential.

For two small spherical particles of radius a at constant charge the energy of repulsion is given by Verwey and Overbeek [35] as

$$V_R = \varepsilon a \psi_0^2 \left(\frac{e^{-\tau(s-2)}}{s} \right) \gamma \tag{6.11}$$

while for constant potential

$$V_R = \varepsilon a \psi_0^2 \left(\frac{e^{-\tau(s-2)}}{s} \right) \beta \tag{6.12}$$

where

$$\gamma = \frac{1+\alpha}{1 - \dfrac{e^{-\tau(s-2)}}{2s\tau} \left(\dfrac{\tau-1}{\tau+1} + e^{-2\tau} \right)(1+\alpha)}$$

and

$$\beta = \frac{1+\alpha}{1 + \left(\dfrac{e^{-\tau(s-2)}}{2s\tau} \right)(1 - e^{-2\tau})(1+\alpha)}$$

$s = R/a$; R is the centre-to-centre distance between two particles of radius a. Since γ and β are always between 0.6 and 1.0 we may neglect their influence in many cases [35, p. 152]. Hence equation (6.13) is a good approximation for the free energy of electrostatic repulsion. (Overbeek points out that there is no exact equation [36].)

$$V_R = 2\pi\varepsilon_r\varepsilon_0 a \left[\frac{4RT}{zF} \gamma' \right]^2 \ln\left[1 + \exp(-\kappa H)\right] \tag{6.13}$$

$$\simeq 2\pi\varepsilon_r\varepsilon_0 a \left(\frac{4RT}{zF} \gamma' \right)^2 \exp(-\kappa H) \tag{6.13a}$$

where $\gamma' = \tanh(zF\psi_0/4RT)$
ε_r = relative permittivity of medium
ε_0 = permittivity of a vacuum
R, T, F, z have their usual meanings

$$\kappa^2 = \frac{F^2 \Sigma c_i z_i}{\varepsilon_r \varepsilon_0 RT} \tag{6.13b}$$

For the interaction of two particles of different surface potential (ψ_{01}, ψ_{02}) and different radii, a_1 and a_2 (provided that $\kappa H > 10$ and $\psi_0 < 50\,mV$), is according to Hogg et al. [48],

$$V_R = \frac{\varepsilon a_1 a_2 (\psi_{01}^2 + \psi_{02}^2)}{4(a_1 + a_2)} B \tag{6.14}$$

where

$$B = \frac{2\psi_{01}\psi_{02}}{(\psi_{01}^2 + \psi_{02}^2)} \ln\left(\frac{1 + 2\exp - \kappa H}{1 - \exp - \kappa H}\right) + \ln\left[1 - \exp(-2\kappa H)\right]$$

Figures 6.5–6.7 illustrate the evaluation of V_{net} ($= V_R + V_A$) in terms of surface-to-surface distance (H) when the particle size and electrolyte content are kept constant, but the surface potential is varied; the surface potential and electrolyte content are kept constant, but the particle size is varied, and when the particle size and surface potential are kept constant, and the electrolyte content is varied.

6.4.5.2 Flocculation by electrolyte

From the evaluation of $V_{net} = V_R + V_A$ shown in Fig. 6.7 it is seen that the addition of an electrolyte will make a charge-stabilized dispersion flocculate. This property was recognized at the end of the nineteenth century when Schulze found that the

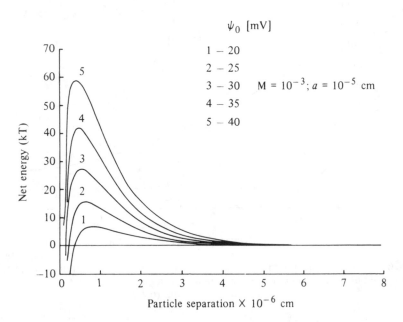

Fig. 6.5 — Net interaction energy ($V_{net} = V_R + V_A$)–particle separation curves, where particle size and electrolyte content are constant and the surface potential is varied (using equations (6.13) and (6.7).

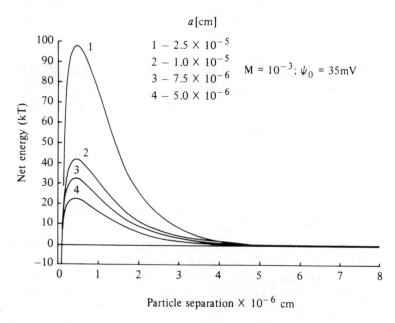

Fig. 6.6 — Net interaction energy ($V_{net} = V_R + V_A$)–particle separation curves, where electrolyte content and surface potential are constant and particle radius is varied (using equations (6.13) and (6.7)).

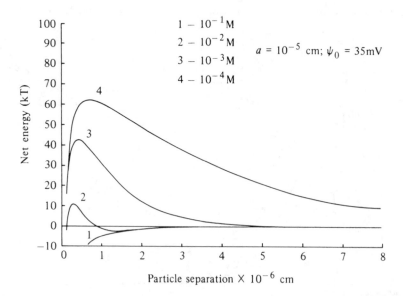

Fig. 6.7 — Net interaction energy ($V_{net} = V_R + V_A$)–particle separation curves, where particle size and surface potential are constant and the electrolyte content is varied (using equations (6.13) and (6.7)).

flocculating power of a counterion is greater, the higher its valency. It was confirmed by Hardy, and is now known as the Schultz–Hardy rule.

We can define the onset of dispersion instability using Overbeek's approach [37] as the condition when

$$V_{net} = V_R + V_A = 0 \qquad (6.14a)$$

and $\partial V_{net}/\partial H = 0$ are satisfied for the same value of H. That is,

$$\frac{\partial V_{net}}{\partial H} = \frac{\partial V_R}{\partial H} + \frac{\partial V_A}{\partial H} = 0,$$

using equations (6.13a) and (6.7a), where $d = H =$ surface-to-surface separation distance, we obtain

$$\frac{\partial V_{net}}{\partial H} = -\kappa V_R - \left(\frac{1}{H}\right) V_A = 0,$$

therefore the critical condition is when $\kappa H = 1$.

Substituting κ for H in equation (6.14a), we obtain the critical condition as:

$$\kappa_{crit} = 24\pi \varepsilon_r \varepsilon_0 (4RT/F)^2 \, e^{-1} (\gamma')^2 / A z^2$$
$$= 2.03 \times 10^{-5} (\gamma')^2 / z^2 A \qquad \text{in c.g.s. units,}$$

where: $\varepsilon_r =$ dielectric constant (78.5) for water at 25 °C.

$\varepsilon_0 = \varepsilon_{i/4\pi} = 1/4\pi$ ($\varepsilon_i = 1$ and dimensionless in c.g.s.–e.s.u. units, but not in SI units)

$e = 2.718$

$R = 8.314 \times 10^7$ erg T^{-1} mol^{-1}

$T = 298$ °K

$F = 2.892 \times 10^{14}$ e.s.u. mol^{-1},

and for water at 25 °C we can (for symmetrical z–z electrolyte) write equation (6.13b) as:

$$\kappa = \frac{z M^{\frac{1}{2}}}{3 \times 10^{-8}} \, \text{cm}^{-1}$$

and substituting for κ_{crit} we obtain the critical flocculation concentration (M_{crit}) as

$$M_{crit} = \frac{3.7 \times 10^{-25} (\gamma')^4}{z^6 A^2} \, \text{moles/litre} \qquad (6.14b)$$

If the surface potential (ψ_0) is high ($\psi_0 \gg 100 \, \text{mV}$) then $\gamma' = \tanh(zF\psi_0/4RT) \to 1$. Hence the DLVO theory predicts that the flocculation efficiency of indifferent lectrolytes should be inversely proportional to the sixth power of the valency (that is, $1/z^6$; hence the relative concentration to bring about flocculation by monovalent:divalent:trivalent ions is $1/1^6 : 1/2^6 : 1/3^6$ or $729 : 11.4 : 1$.

For low surface potentials ($\psi_0 < 25 \, \text{mV}$) we can equate the surface potential to the zeta potential and simplify equation (6.14b) to

$$M_{crit} = \frac{3.31 \times 10^{-33} \zeta^4}{A^2 z^2} \, \text{moles/litre}$$

where ζ is in mV since $(4RT/F) = 102.8\,\text{mV}$ at 25 °C and $\tanh x \simeq x$, for low values of x. Hence the ratio of flocculating concentrations for monovalent:divalent:trivalent ions becomes $1/1^2 : 1/2^2 : 1/3^2$ or $100:25:11$.

In practice there has been found to be good general agreement with the predictions of the DLVO theory, but since flocculation studies are dependent not only on equilibrium conditions, but also on kinetic factors and specific ion effects, care has to be exercised in using this simplified approach, for even ions of the same valency form a series of varying flocculating effectiveness such as the lyotropic (Hofmeister) series where the concentration to induce flocculation of a dispersion decreases with ion hydration according to $[Li^+] > [Na^+] > [K^+] > [Rb^+] > [Cs^+]$.

In emulsion paints the presence of electrolytes is thus seen to be a very important factor. Pigment millbases formulated on the use of charge alone, e.g. Calgon (sodium hexametaphosphate), are very prone to flocculation; sometimes even by the addition of latex, since the latex may carry a considerable concentration of electrolyte. Such dispersions, once made, are usually further stabilized by the addition of a water-soluble colloid (polymer) such as sodium carboxy methyl cellulose, which may be labelled only as a 'thickener'; nevertheless it increases the dispersion stability to the presence of electrolyte by its 'protective colloid' action.

6.4.5.3 Charge stabilization in media of low dielectric constant

Charge stabilization is of great importance in media of high dielectric constant, i.e. aqueous solution, but in non-aqueous and particularly in non-polar, systems (of low dielectric constant) repulsion between particles by charge is usually of minor importance [49].

There have been attempts to use charge to explain colloid stability in media of low dielectric constant such as p-xylene [50].

Osmond [51] points out that the attempted application of double layer theory, with all the corrections necessary to apply it to non-polar, non-aqueous media, reduces the DLVO approach to simple coulombic repulsion between two charged spheres in an inert dielectric (thus ignoring the existence of ions in solution).

Examination of Lyklema's calculation [52] of the total interaction between two spherical particles of radius 1000 Å (0.1 µm) (e.g. TiO_2 particles) in a low dielectric medium shown in Fig. 6.8, reveals that if the particles had a zeta potential of 45 mV and were at 9% PVC — i.e. average spacing of 2000 Å (0.2 µm) — then the maximum energy barrier to prevent close approach is only about 4 kT. While at 20% PVC — average spacing 1000 Å (0.1 nm) — only 2 kT is available to prevent flocculation, which is clearly insufficient.

Thus, because the repulsion energy–distance profile in non-polar media is very 'flat' (see Fig. 6.8); concentrated dispersions (their definition implying small interparticle distances) cannot be readily stabilized by charge. In very dilute concentration where interparticle distances are very large, charge may be a source of colloidal stabilization. While charge is not considered to be important in stabilizing concentrated non-aqueous dispersions in media of low dielectric constant, however, it can be a source of flocculation in such systems if two dispersions of opposite charge are mixed.

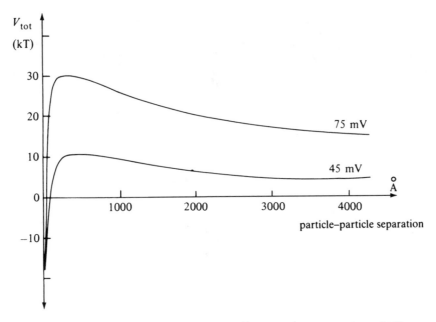

Fig. 6.8 — Net energy–distance plots in non-aqueous medium (from [52]).

6.5 Steric (or polymer) stabilization

An alternative source of repulsion energy to stabilize colloidal particles in both aqueous and non-aqueous (including non-polar) media is what is now known as 'steric' or 'entropic' stabilization. In 1966 Overbeek said 'The theory of protective (entropic) action is still in rather a primitive state' [53]. Since then great progress in the understanding of steric stabilization has been made with many comprehensive reviews on the subject [54–56].

Napper [57] presents the latest 'state-of-knowledge' in a very readable form. The important feature of steric stabilization is that it is not a steric hindrance as normally considered in organic chemistry terms, where molecular configuration prevents a reaction, but a source of energy change due to loss of entropy. It is sometimes known as 'polymer stabilization', but since non-polymeric material can also produce stabilization, this name too is wanting, as are the many alternatives that have been suggested, e.g. non-ionic, entropic.

In its simplest form, steric stabilization can be visualized as that due to a solvated layer of oligomeric or polymeric chains irreversibly attached to a particle surface of uniform concentration producing a 'solvated' sheath of thickness δ.

If two such particles approach each other, so that the solvated sheaths overlap, or redistribute their segment density in the overlap zone, there is a localized increase in polymer concentration (see Fig. 6.9). This localized increase in polymer concentration leads to the generation of an osmotic pressure from the solvent in the system. Hence the source of the repulsion energy is equivalent to the non-ideal component of the free energy of dilution.

It is possible to estimate the size of the adsorbed layer of polymer around a particle by measuring the viscosity of the dispersion at varying shear rates [58]. From these measurements it is also possible to obtain some idea of the effectiveness of

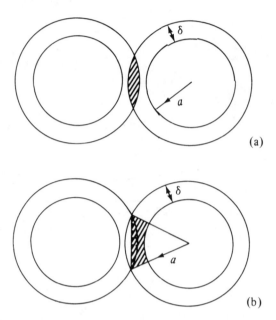

Fig. 6.9 — Schematic representation of steric-barrier overlap: (a) Ottewill and Walker [82], (b) Doroszkowski and Lambourne [75].

the adsorbed layer by examining the slope in a plot of log (dispersion viscosity) against the square root of the reciprocal shear rate, as suggested by Asbeck and Van Loo [59]. The greater the slope, the larger the degree of flocculation. If the viscosity is independent of shear then there is no detectable flocculation under these specific conditions.

A check on the validity of measuring adsorbed layer thickness was made by synthesizing a series of oligoesters of known chain length which could be adsorbed only through their terminal carboxyl group onto the basic pigment surface. The measured 'barrier thicknesses' were found to be in very good agreement with terminal oligoester adsorption [60], confirming the effectiveness of the method.

From measurements such as these, it is clear that the size of the adsorbed polymer is not all-important as has been suggested by some [61]. Smaller adsorbed layers can be more effective than larger ones [62]. Chain branching, and where it occurs can have a marked influence on the effectiveness of the stabilizing layer [60]. For example, poly-12-hydroxy stearic acid, which is used as the soluble moeity in a graft copolymer as described in [63], is very effective.

The theory of steric stabilization, like most, has undergone evolution. Its origins are irrelevant, except maybe to historians, whether it is founded in the concept of 'protective colloids' as discussed by Zigmondy [64] with his gold sols in 1901, or the early attempts to calculate the effect of entropy loss by adsorbed molecules such as envisaged by Mackor [65]. Nevertheless, steric stabilization was practised by crafts, including paintmakers, before scientists classified the various stages by nomenclature and explanation. Fischer [66] probably contributed most to the understanding of steric stabilization, since he envisaged the important influence of the solvent environment on steric stabilization, which was not appreciated even by later workers [67].

At present steric stabilization is generally considered to arise from two factors (which are not necessarily equal):

- A volume restriction term (i.e. the configuration of the adsorbed molecules is reduced by the presence of a restrictive plane V_{VR}).
- A mixing or osmotic term (i.e. an interpenetration or compression of adsorbed layers resulting in an increase in polymer concentration V_M).

These two contributions are regarded as being additive. That is, $V_{tot} = V_{VR} + V_M$.

It is difficult to evaluate the contribution of these two terms since it entails a knowledge of the polymer configuration of the adsorbed polymer, and more importantly the segment density distribution normal to the surface of the particle.

Meier [68] and Hesselink [69, 70] have attempted to calculate this segment distribution and hence the relative contribution of the two sources of repulsion energy. However, there is dispute in that by subdividing the repulsion energy in this way, there is double counting [71].

There have been, unfortunately, only a few attempts to measure the energy of repulsion by direct experiment [72–77], and it would appear that the theoretical and mathematically complex calculations of Meier and Hesselink grossly overestimate the repulsion energy measured by experiment [75]. The more simple estimates of energy of repulsion (derived from purely the 'osmotic' term V_{RM}) appear to be in better accord with practice. Napper, in many investigations [78–80], has shown the importance of the solvent environment in contributing to steric stabilization, and for all intents and purposes the osmotic contribution to steric stabilization is all that needs to be considered. Therefore the repulsion energy on the interaction of two solvated sheaths of stabilizing molecules may be simply written as:

$$V_R = \frac{c^2 RT}{v_1 \rho^2} (\psi_1 - \kappa_1) V, \tag{6.15}$$

where c = polymer concentration in solvated barrier,
ρ = polymer density,
v_1 = partial molar volume of solvent,
ψ_1, κ_1 = Flory's entropy and enthalpy parameters [81],
V = increase in segment concentration in terms of barrier thickness δ,
a = particle size and surface-to-surface separation (h).

Which in the Ottewill and Walker model [82] is

$$V = \left(\delta - \frac{h}{2} \right) \left(3a + 2\delta + \frac{h}{2} \right)$$

and the Doroszkowski and Lambourne model [75]

$$V = \frac{\pi x^2 R_1 (3R_1 - x)(2R_1 - x) \, 4(R_1^3 - r^3) + x R_1 (x - 3R_1)}{2(R_1^3 - r^3) + x R_1 (x - 3R_1)}$$

where $R_1 = a + \delta$; $r = a$; and $x = \delta - h/2$.

One has a more useful relationship if one rewrites equation (6.15) and substitutes $(\psi_1 - \kappa_1)$ by $\psi_1(1 - \theta/T)$ where θ is the theta temperature as defined by Flory [81], i.e. where the second virial coefficient is zero, meaning that the molecules behave in their ideal state and do not 'see or influence each other'.

Thus one can immediately realize that at the theta temperature, $[1 - (\theta/T)]$ reduces to zero ($\theta = T$), and the repulsion energy disappears, allowing the attractive energy to cause flocculation, as shown by Napper. Furthermore, if $T < \theta$, the repulsion energy not only does not contribute to the repulsion, but it actually becomes an attraction energy.

An important point to bear in mind is the c^2 term, which is the polymer concentration term in the 'steric barrier'. If this is small then flocculation can occur, even when the solvency is better than 'theta', which has been observed in a few instances.

Napper [83] has subdivided steric stabilization into 'entropic stabilization' and 'enthalpic stabilization'. Depending on whether the entropy parameter is positive or negative, e.g. in the case of steric stabilization by polyhydroxy stearic chains in aliphatic hydrocarbon, flocculation is obtained by cooling; and in enthalpic stabilization, flocculation is achieved by heating (as with polyethyleneglycol chains in water). A purist [84] might correctly insist that this is entropic or enthalpic flocculation, and not stabilization. Nevertheless, in simple practice it is a useful distinction.

Another simple way of achieving the theta point is by changing the degree of solvency of the continuous phase, i.e. by the addition of a non-solvent. Napper [85] has illustrated the utility of varying the solvency to achieve incipient flocculation at constant temperature, and has thus again demonstrated the overall importance of the 'osmotic' term in steric stabilization.

If we examine the requirement for steric stabilization by an adsorbed layer of molecules, then we can intuitively see that the effectiveness of the stabilization is dependent on the strength of the attachment of the 'polymeric' (or oligomeric) sheath to the particle surface. One can envisage three degrees of effectiveness of steric stabilization by the same molecule:

1   Suppose the stabilizing layer comprises an adsorbed layer of molecules (e.g. and oligomer of 12-hydroxy-stearic acid as shown in Fig. 6.10. Let us suppose that the molecules are covalently attached to the surface of a particle. Then if two such layers overlap to give an overlap zone of increased segment density, then the free energy of dilution (due to increased localized segment density) can only be relieved by separation of the two particles.

2   Suppose the stabilizing layer of the oligomer is strongly adsorbed on the surface of the particle, but not anchored to a specific site on the surface, i.e. the oligomer is free to move over the particle surface but cannot be desorbed. On overlap of the stabilizing sheaths (as two particles approach), the increase in segmental concentration can be redistributed over the whole of the surface layer by lateral compression of the molecules and thus the increase in surface concentration of the attached molecules is reduced, hence a lower free energy of dilution to repel (separate) the two particles. This situation is clearly less effective than case 1.

3   Suppose the oligomer molecules are weakly adsorbed on the particle surface, i.e. they are free to move over the particle surface, as in case 2, but they are also free to 'pop-off (desorb from) the particle surface if the compression forces become too great. This clearly will give rise to the weakest form of steric stabilization, for on overlap of the solvated sheaths, the surface molecules can relieve the increase in surface concentration by a combination of redistribution, desorption and particle separation.

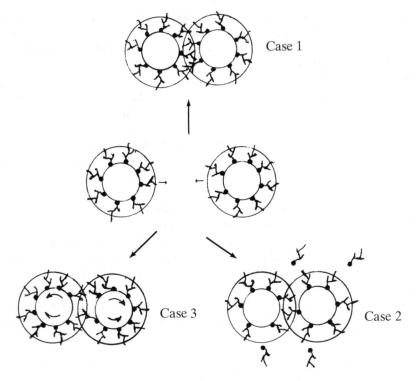

Fig. 6.10 — Degrees of effectiveness in stabilisation by steric barriers on close approach of like particles: (1) stabilishing molecules bound to site on particle surface (large localised increase in segmental concentration); (2) molecules free to move over particle surface (no desorption), hence reduced increase in segmental concentration, leading to weaker particle-particle repulsion; (3) molecules free to move over surface and desorb, hence increase in segmental concentration reduced by partial desorption, ie, steric stabilisation effectiveness is a > b > c.

Thus the effectiveness of steric stabilization is (1) > (2) > (3) because of the dependence on the nature of the molecules' attachment to the surface of the particles, as well as, the nature of the continuous phase, etc.

The molecular weight of the adsorbed polymer, which determines the size of the steric barrier, is not an overriding criterion. For example it has been suggested [86] that if one uses the best estimate of the Hamakar constant for TiO_2 in aliphatic hydrocarbon media as $A_{12} = 5 \times 10^{-20}$ J [27] then one is left with a value of 51 nm for the thickness of the solvated polymer sheath around the TiO_2 pigment to produce effective steric stabilization. This in turn must mean that the molecular weight of the polymer molecules (adsorbed in chains and loops) must be at least many tens of thousands, if not more! Yet pigment dispersion experience has shown that small molecules, about 500 in molecular weight, can give effective colloid stability [87–89].

There are many variations of the theory of steric stabilization. Napper (57, Ch. 11, 12], for example divides them into pragmatic' and 'ab initio' theories, Pragmatic theories assume some form of segment distribution in the adsorbed layer, e.g. a uniform segment distribution, or an exponential segment density distribution. Ab *initio* theories attempt to calculate segment distributions from first principles: these can be further subdivided into classical theories which use Flory's mixing and elastic (entropic) contributions to the steric repulsion energy, and the mean-field theories, which effectively combine these effects, treating the segments as having a real volume. While the latter may be more sophisticated in their approach, they are more difficult to use. Dividing the repulsion energy into separate mixing and elastic terms might be more 'synthetic' it is, however, conceptually more useful. While many theories on steric stabilization differ in complexity, and the simplifying assumptions made, their qualitative description of steric stabilization is similar. For a discussion of various attempts to estimate the steric repulsion energy the reader is referred to the review by Vincent and Whittington [90].

The study of the nature of colloid stability in the absence of solvent in a fluid–polymer matrix (polymer-melt condition) has somehow been neglected in the scientific literature, yet it is of the utmost importance in paint technology. This situation occurs during the final stages of solvent evaporation in air-drying films, re-flow in stoving (baking) finishes and in the dispersion stage, as well as 'cure' of powder coatings to mention but a few. The mechanism of colloid stabilization in the 'polymer-melt' condition is ascribed to the 'volume restriction' component of steric stabilization. For the total steric repulsion free energy (V_{tot}) comprises the mixing (V_M) and the volume restriction (V_{VR}) terms, i.e. $V_{tot} = V_M + V_{VR}$; which for spherical particles is [89, 91]:

$$V_{tot} = \frac{2\pi a k T}{V_1} \cdot \Phi_2 \left(\frac{1}{2} - X \right) S_M + \frac{2\pi a k T}{V_2} \cdot \Phi_2 S_E$$

where S_M and S_E are geometric functions of barrier thickness and particle separation; V_1 and V_2 are the solvent and polymer segment volumes respectively, Φ_2 is the average volume fraction of segments in the barrier; X is Flory's interaction parameter, and a, k, and T have their usual meaning.

Under 'melt' conditions the solvent can be regarded as a polymer, hence V_1 is very large. The above equation shows that the consequence of making V_1 large is that V_M becomes very small, leaving V_{VR} to be the source of free energy change to provide particle stabilization, as for example in making a dispersion of particles in molten polyethylene oxide [92].

6.6   Depletion flocculation and stabilization

Steric stabilization arises as a consequence of polymer adsorption onto the surface of particles. The effects of free polymer in solution on colloid stability give rise to what was coined by Napper and Feigin 'depletion flocculation' [93] and 'depletion stabilization'. (For a review on the topic see [57, Ch. 17].)

The concept of depletion flocculation may be traced back to Asakura and Oosawa [94, 95]. However, it was Li-in-on *et al.* [96] who showed that at higher concentration of free polymer in solution, the flocculating effect of added polymer disappeared.

There are variations on the theme of depletion flocculation [97–99], but basically the concept arises from consideration of non-adsorbing surfaces and polymer molecules in solution. When surfaces of colloidal particles approach each other to separations less than the diameter of the polymer molecule in solution, the polymer is then effectively excluded from the interparticle space, leaving only solvent. Thermodynamically, this is not favourable, leading to osmosis of the solvent into the polymer solution and thereby drawing the two particles even closer together, i.e. the particles flocculate.

Napper and Feigin [93] extend the depletion flocculation mechanism to depletion stabilization by considering what happens under equilibrium conditions to two inert, flat plates immersed in non-adsorbing polymer solution, and by examining the segmental distribution of a polymer molecule adjacent to the surfaces in the bulk solution from a statistical point of view. They showed that the polymer segment concentration adjacent to the flat, non-adsorbing surface was lower than in the bulk. By estimating the 'depletion-free energy' as a function of plate separation in terms of polymer molecule diameter, they showed that as the two plates were brought together there was first a repulsion as the polymer molecules began to be constrained, followed by an attraction on closer approach of the two surfaces (see Fig. 6.11).

They applied Derjaguin's method of converting flat plate potential energy curves into those for curved surfaces (i.e. sphere–sphere) interactions, by the equivalent of mathematical 'terracing' of a curve as shown in Fig. 6.12.

Vincent et al. studied the flocculating effect of added polymer in sterically stabilized latexes [100] and explained the findings with a semi-quantitative theory that is schematically represented in Fig. 6.13. They showed that the strength of the repulsion energy on overlap of the adsorbed polymer sheaths (G^{LL}) is reduced by the presence of polymer in the continuous phase by $G^{LL} = G^{LL,O} - n/2\, G^{PP}$ where $G^{LL,O}$ is repulsion energy of the polymer sheaths in the absence of polymer in solution and G^{PP} is the interaction between two polymer coils in solution; n is the number of polymer coils displaced into solution.

The theory predicts that flocculation by added polymer occurs between the polymer concentration Φ_{PS}^* when the expanded polymer coils in solution just touch one another (see Fig. 6.14); and Φ_{PS}^{**}, the concentration when uniform segment density, occurs as the polymer coils contract to their theta configuration (Φ_{PS}^{**}) on the further addition of polymer. That is,

$$\Phi_{PS}^* = \frac{M}{b*\langle S^2\rangle^{3/2} N\rho} \quad \text{and} \quad \Phi_{PS}^{**} = \frac{M}{b**\langle S^2\rangle_0^{3/2} N\rho}$$

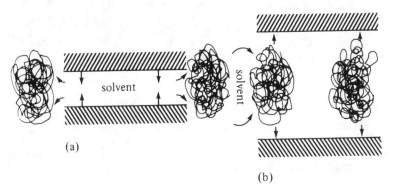

Fig. 6.11 — Schematic representation of depletion flocculation (a) and stabilization (b).

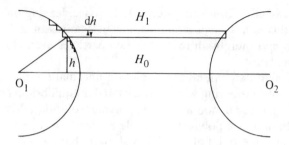

Fig. 6.12 — Derjaguin's flat plate-sphere conversion.

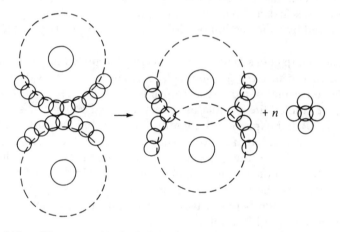

Fig. 6.13 — Vincent *et al.*'s depletion flocculation mechanism (from [100]).

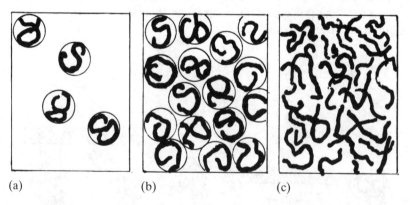

(a)   (b)   (c)

Fig. 6.14 — Schematic representation of polymer molecules in a good solvent (adapted from [102]); (a) the dilute region, where concentration ($c < c^*$) polymer coils are expanded and well separated from one another; (b) the intermediate region (semi-dilute) polymer chains start to overlap, but segment density is not large ($c^* < c < c^{**}$); (c) concentrated region where every segment is in contact with other segments ($c > c^{**}$) and polymer coils contracted (effectively in theta condition).

where M is the molecular weight of the polymer of density ρ; N is Avogadro's number; $\langle S^2 \rangle^{\frac{1}{2}}$ is the radius of gyration; b^* and b^{**} are packing factors (8 and 2.52 for cubic close packing, and 5.6 and 1.36 for hexagonal close packing respectively).

When applied, their theory gave a reasonable, broad co-relation with experiment for both aqueous and non-aqueous systems [102].

Scheutjens and Fleer [103] proposed a similar depletion flocculation–stabilization mechanism to Napper, also based on theoretical conformations of polymer molecules.

Both Napper and Feigin [93] and Scheutjens and Fleer [103] predict large depletion stabilization energies available to colloidal particles at suitable free polymer content, but the two theories differ in several respects. For example, Scheutjens and Fleer predict greater stabilization effects in poor solvency conditions (i.e. theta conditions), while Napper and Feigin favour good solvency.

One criticism of depletion stabilization theory is that it appears to teach that it is only necessary to have a sufficiently concentrated polymer solution, preferably of high molecular weight, to make a good pigment dispersion. In practice this is clearly found to be an inadequate requirement. It does not necessarily mean that depletion stabilization theory is wrong, but only incomplete, maybe because concentrated polymer solution theories are still poor. Fleer et al. [104], unlike Napper, have revised their estimates of depletion stabilization energy downwards to be more in line with Vincent et al., who strictly speaking do not have a depletion stabilization energy, but only an absence of depletion flocculation at high polymer concentrations.

The concept of depletion flocculation has an important bearing on paint properties. It teaches, for example, that even if good pigment dispersion is achieved in dilute solution, then the addition of even compatible non-adsorbing polymer may cause flocculation. (Note the care necessary in 'letting-down' a ballmill on completion of the pigment dispersion stage.) It also suggests the possibility of pigment flocculation occurring on the addition of solvent, e.g. dilution with thinners prior to spraying.

It also suggests that different methods of paint application which rely on varying degrees of solvent addition may cause different degrees of pigment flocculation, leading to variation in colour development, even when the dispersion is well deflocculated at low polymer content in the continuous phase. Depletion flocculation may also occur with charge-stabilized systems such as latexes on the addition of an 'inert' thickening agent, as shown by Sperry [105] which may produce different types of sedimentation phenomena.

6.7 Adsorption

To obtain good dispersion stability by steric stabilization it is important to fix the stabilizer molecules to the surface of the particle. The more firmly they are held, provided that they are also well solvated and free to adopt varying configurations, the better; for example, terminally adsorbed polymeric molecules such as polyhydroxystearic acid, have been found to be excellent stabilizers [63]. If, for example, the adsorbed molecules are attached to the surface of a particle and are able to move about on the surface, then on overlap of the solvated layers the increased concentration could be partly accommodated by a surface redistribution which will result in a lower repulsion energy than if surface redistribution is not possible. Furthermore, if the adsorption of the stabilizing molecules is weak, then not only could the increase in osmotic pressure be countered by surface redistribution, but also by stabilizer desorption, leaving little, if any, force for approaching particle repulsion, as discussed earlier.

But, if adsorption of the stabilizer molecules is so strong that the stabilizer molecules are 'nailed flat' to the surface of the particle, effectively producing a pseudo, non-solvated layer, then on the close approach of another similar layer there would be no entropy loss of the polymer chains, and all that would be achieved would be the extension of the particle surface by some small distance. The attraction potential would then be from the new surface, and the particles would have a composite Hamaker constant producing an attraction potential with no source of repulsion energy.

In the design of dispersion stabilizers it is important to have something with which to attach the solvated stabilizing molecule firmly to the surface, preferably at a point, leaving the rest of the molecule to be freely solvated in its best solvent, e.g. chemically bonding the stabilizer to the particle surface either covalently, as with reactive stabilizers such as those described by Thompson et al. [106], or by acid–base interaction in non-polar media such as reacting carboxyl terminated fatty chains, e.g. poly-12 hydroxy stearic acid with an 'alumina' coated pigment.

Sometimes it is not possible to react a stabilizer to the surface of the particle, and the demand for strong anchoring and maximum solvation can be achieved only by physical adsorption using amphipathic copolymers in hemi-solvents, such as block or graft copolymers, where one portion of the molecule is 'precipitated' onto the surface of the particle (i.e. the anchor component is non-solvated) leaving the solvated portion to produce the stabilizing barrier as described in [63]. For a comprehensive discussion on polymers at interfaces the reader is referred to Fleer et al. [107].

6.7.1 Adsorption isotherms

Adsorption isotherms are very useful in obtaining an understanding of what may be happening in a pigment dispersion. For example, it may be expected that when TiO_2 pigment is dispersed in butyl acetate, xylol, and white spirit, using the same dispersant [60], then the attraction energy because of the Hamaker constant values of the respective solvents should be of the order.

$$V_A^1 > V_A^2 > V_A^3$$

where V_A^1 = attraction energy in white spirit,
V_A^2 = attraction energy in xylol,
V_A^3 = attraction energy in butyl acetate.

Since the stabilizer is terminally adsorbed (i.e. the barrier thickness is the same), it should give the best stability in butyl acetate solution. However, viscometric measurements indicate that the reverse sequence holds, i.e. the dispersion with the least flocculation was in white spirit, followed by that in xylol, and the worst was in butyl acetate. Inspection of the adsorption isotherms immediately revealed that the surface concentration was the least in butyl acetate (see Fig. 6.15), hence the stabilizing barriers were the weakest in butyl acetate solution.

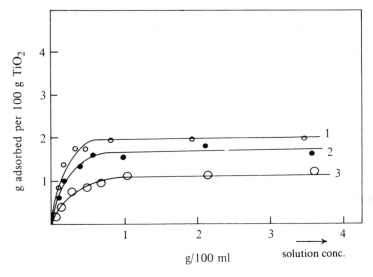

Fig. 6.15 — Adsorption isotherm of dimer 12-hydroxy-stearic acid in (1) white spirit; (2) xylol; (3)butyl acetate onto TiO_2 [60].

Adsorption isotherms are most readily measured by placing a fixed amount of pigment in a known amount of polymer solution, in glass jars, along with small (e.g. 8 mm) glass beads as a milling aid, and leaving the jars for at least 24 hours on rollers, so that constant agitation is obtained. Centrifugation will then enable the separation of the continuous phase, so that the amount of material adsorbed may be determined, as described in [60].

An alternative method to obtain an adsorption isotherm is to place the pigment in a chromatographic column and to pass a dilute solution of the adsorbate down the pigment column, as described by Crowl [108].

The advantage of the chromatographic method is that it enables preferential adsorption and adsorption reversibility to be easily determined. The method generally works well with organic pigments; however, with TiO_2 pigments flow rates through the pigment column are in practice very slow, making the method very much less attractive than might appear at first sight.

6.7.2   Free energy of adsorption and adsorption isotherms

6.7.2.1   Estimation of free energy of adsorption (ΔG_{ads}) from adsorption isotherms
Suppose that we examine a particle immersed in a liquid, e.g. water, and we imagine the surface of the particle to consist of a 'mosaic' of adsorption sites where 'S' represents such a site occupied by the solvent. If we introduce a solute molecule, X, into the liquid, then if it were to adsorb on the surface, it will displace the water molecule occupying the site. Let SX represent the adsorption of the solute at a site, and S(H$_2$O) the water molecule adsorption. We can express this as an equilibrium reaction by:

$$S(H_2O) + X \underset{k_2}{\overset{k_1}{\rightleftharpoons}} SX + H_2O \qquad \text{and} \qquad K = \frac{k_1}{k_2}$$

i.e. K is the partition, or equilibrium, constant for the solute between the surface of the particle and the solvent, and is related to the free energy of adsorption (ΔG_{ads}) by

$$\Delta G_{ads} = -RT \ln K$$

Using Glasstone's notation [109], where possible, we can express K as:

$$K = \frac{\text{(mole fraction solute on surface) (mole fraction solvent in bulk)}}{\text{(mole fraction solvent on surface) (mole fraction solute in bulk)}}$$

Suppose the surface has a sites per g adsorbent which are occupied by the solvent molecules in the absence of solute and that x_1 of these are occupied by the solute when dissolved in the solvent, then

$$K = \frac{(x_1/a)\,\text{(mole fraction solvent in bulk)}}{\left(\dfrac{a-x_1}{a}\right)\text{(mole fraction solute in bulk)}}$$

If we restrict our considerations to low (i.e. initial) adsorption, i.e. when $x_1 \ll a$, and have dilute solute concentration, then

$$K \approx \frac{(x_1/a)\,(1)}{(1)\,(N)}$$

$$\text{hence } \Delta G_{ads} = -RT \ln\left(\frac{x_1/a}{N}\right)$$

where N is the mole fraction of solute of concentration, c, g moles in solution of a liquid of molecular weight M, and density 'ρ' so that $N = Mc/1000\rho$ and hence for aqueous solution $N = c/55.5$.

Let x/m be the number of g moles adsorbed solute per g adsorbent. This is proportional to the fraction of sites occupied by solute, i.e. $x/m = p(x_1/a)$ where p is a proportionality constant.

If the surface is completely covered by a monolayer of solute, then $x_1/a = 1$; and $x/m = p = $ maximum adsorption (or plateau value). Thus for aqueous solution,

$$\Delta G_{ads} = -RT \ln\left(\frac{x/m \times 55.5}{c \times p}\right)$$

At low solution concentration, $(x/m)/c = $ initial slope (IS) of an adsorption isotherm, hence

$$\Delta G_{ads} = -RT \ln\left(\frac{\text{IS} \times 55.5}{\text{plateau value, } p}\right) \tag{6.16}$$

By plotting an adsorption isotherm in terms of g moles adsorbed per g adsorbate against molar equilibrium concentration it is thus possible to estimate the free energy of adsorption in a monolayer or Langmuir-type adsorption isotherm from the initial slope and saturation value.

The usefulness of equation (6.16) is that it gives a 'feel' for the strength of adsorption and especially its 'converse' use, i.e. in estimating the adsorption of a surfactant by particle surfaces such as those of pigment or emulsion droplets, and ensuring that sufficient material is added to satisfy the adsorption requirement of the dispersion. For if the free energy of adsorption is known, and for simple physical adsorption it is of the order of 5 Kcal/g mole, the cross-sectional area of a surfactant molecule is known (obtainable in the literature, or it can be estimated using molecular models). It is thus possible to construct the initial slope and plateau portions of the adsorption isotherm. By joining the initial slope to the plateau level, by drawing a gentle curve in the 'near-saturation' zone (as shown by the broken line in Fig. 6.16), the adsorption isotherm can be completed with a little 'artistic licence', hence the minimum surfactant requirement estimated.

The derivation of ΔG_{ads} as outlined above is based on a simple model of an adsorption surface composed of adsorption sites which are either occupied by solvent molecules or solute, such as a simple surfactant.

This is useful in giving an insight into the interpretation of an adsorption isotherm and illustrates the importance of the initial slope, as well as the saturation value

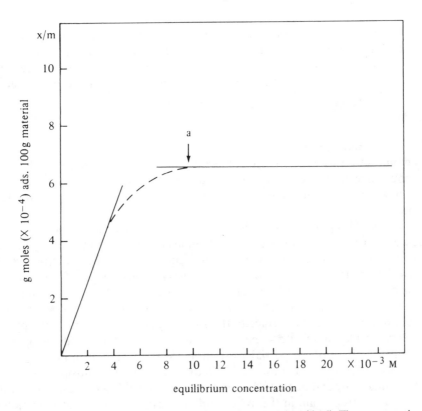

Adsorption

Fig. 6.16 — Prediction of surfactant adsorption using equation (6.16). The construction of the isotherm is based on the physical adsorption of surfactant (aqueous) at 5 Kcal per mole at the rate of 200 Å^2 per molecule onto a powder of density 1.5 g/cm^3, with average particle diameter of 0.5 μm. a is the required surfactant concentration which comprises 6.7×10^{-4} g moles surfactant per 100 g powder in equilibrium with a minimum surfactant solution concentration of 9.5×10^{-3} M.

(which also applies to polymer adsorption); but the calculation of ΔG_{ads} for a polymer molecule, which may occupy a varying number of sites depending on its conformation, should not be made unless refinements, or constraints to polymer chain configurations, are taken into account. A similar proviso is as made by Koral et al. [110] who determined the 'isosteric' heats of polymer adsorption from adsorption isotherms. That is, if the adsorption isotherm is measured at different temperatures, then it is possible to estimate the enthalpy of adsorption (ΔH) using an analogue of the Clausius–Clapeyron equation (Huckel equation):

$$\frac{\Delta H}{R} = \frac{H_s - H_L}{R} = \frac{d \ln(a_L/a_s)}{d(1/T)} \tag{6.17}$$

where H_s and H_L are the partial molar enthalpies of the polymer on the adsorbate and in solution: a_s and a_L are the corresponding activities of the polymer at the surface and in solution. Under very dilute condition $a_L = c$ and, if it is assumed that the activity of the polymer at the surface does not change if the amount adsorbed is kept constant, the above equation becomes:

$$\frac{\Delta H}{R} = \frac{d \ln c}{d(1/T)}. \tag{6.17a}$$

Hence by measuring the adsorption isotherm at different temperatures, taking equal amounts of the adsorbed material at different temperatures, and plotting the corresponding equilibrium solution concentration on a logarithmic scale against the reciprocal of the absolute temperature, a straight line is obtained whose slope is equal to $\Delta H/R$.

6.7.3 Adsorption and temperature

The amount of polymer adsorbed from solution frequently increases with temperature, thus indicating that the process is endothermic. However, there are also cases when the adsorption will decrease with temperature [111] or it may be unaffected by temperature [112].

Koral et al. [110] point out that the adsorption process could not be endothermic in the case of physical adsorption of simple molecules on a clean surface, since then the entropy lost on adsorption will be negative, necessitating the enthalpy to be also negative (to ensure a decrease in the free energy of adsorption). However, in the case of polymers one must consider the system as a whole. The adsorption of a polymer molecule at several sites on a surface requires that several solvent molecules are released from the surface to the solution. The translational entropy of the polymer molecule, along with some of its rotational and vibrational entropy, is lost on adsorption, because of partial restriction to its segmental mobility. Thus the solvent molecules which are desorbed gain their translational entropy, which cumulatively is much larger than that of the polymer molecule. The net result is that there is an overall entropy gain in the system on the adsorption of a polymer molecule which displaces solvent molecules, even if the process is endothermic.

The quality of the polymer solvent also has an effect on the amount of polymer adsorbed by the substrate. Generally the poorer the solvent the greater the amount adsorbed [110], and adsorption can be related to the solubility parameter [113]. However, the adsorption of polymer does not depend on polymer–solvent interaction alone, but also on substrate–solvent interaction as well [114]. It is also possible to obtain less polymer adsorption by a substrate on the addition of small amounts of non-solvent to a polymer solution, if the solvent has a strong interaction with the substrate [115].

6.7.4 Rate of adsorption and equilibration

It is difficult to define how quickly adsorption takes place since it depends on many factors, especially concentration [116], but it is generally considered to be fast [117, 118]. For a review of polymer adsorption see Lipatov and Sergeeva [119]. However, in most practical instances one is more interested in the state of equilibrium adsorption in a mixed component system; and with polymer adsorption from solution, the simple principle of first come, first adsorbed applies, i.e. adsorption is diffusion controlled [114, 120, 121]. The smaller molecules arrive at a surface first and are adsorbed only to be displaced on the later arrival of the larger molecules which are preferentially adsorbed [122]. This phenomenon of preferential adsorption of the higher molecular weight analogues frequently gives rise to the observed 'maxima' in adsorption isotherms of polymers [62]. In dispersing pigments, e.g. making a millbase, it is important to ensure that the preferentially adsorbed polymer is in sufficient quantity at the very start, since displacement may be very slow.

This is sometimes vividly demonstrated in practice, as for example when pigment is initially dispersed in a polyethylene glycol (PEG) modified alkyd, and the other polymeric constituents are added afterwards: then good pigment flushing into the aqueous phase occurs immediately, as shown when the freshly prepared oil-based paint adhering to a paint brush is washed in detergent solution (i.e. when the modified PEG alkyd is adsorbed on the pigment surface). However, if the pigment is milled in the other polymeric constituents first, and then the PEG alkyd is added, it then takes about four weeks for the 'brush wash' properties to appear, signalling that the PEG alkyd has displaced the other constituents from the pigment surface [123].

Another example of the slow re-equilibration of adsorbed molecules sometimes manifests itself as a 'colour drift' (a gradual shift in colour on storage), for example when the 'colour' is made by blending various (colour stable) pigment dispersions (tinters), especially if the tinters are based on different resins. One way of solving this colour drift problem is simply to pass the finished paint through the dispersion mill (e.g. a Sussmeyer). The shearing in the mill, as well as the temperature rise, accelerate the re-equilibration of the adsorbed species, just as agitation is found to accelerate adsorption of polymer [121] and thereby stops the colour change.

Stereo regularity is also an important factor in polymer adsorption, as shown by Miyamoto *et al.* [124] who found that isotactic polmethyl methacrylate is more strongly adsorbed from the same solvent than syndiotactic polymer of the same molecular weight. Syndiotactic PMMA (polylmethy (methacrylate)), in turn, is more strongly bound than atactic polymer, as shown by TLC development with ethyl acetate [125].

6.7.5 Specific adsorption

Adsorption of polymer from solution onto a surface depends on the nature of the solvent and the 'activity' of the surface.

While in the simple physical adsorption of polymer it can be seen intuitively that higher molecular weight species might be preferentially adsorbed in a homologous series, to the low molecular weight species, specific adsorption may override molecular weight considerations. A typical example is the preferential adsorption of the low molecular weight, more polar species, such as the phthalic half esters present in long oil alkyds. Walbridge et al. [126] showed that the adsorption of these materials could be explained in terms of acid–base interactions, where the base was the TiO_2 pigment coating and the metal of the 'driers'. They were able to show that the complex flocculation–deflocculation behaviour depended on the order of addition of drier and dimer fatty acid to a simple white, long oil alkyd-paint. They found that it also depended on the relative acid strength and the formation of irreversible carboxyl–pigment coating bonds, which in turn were a function of the temperature of dispersion. Just as ion exchange resins can bind acids (or bases), so can certain pigment surfaces behave in a similar manner. Solomon et al. [127] studied the effect of surface acid sites on mineral fillers and concluded that they were comparable in strength to, but fewer than, those found in 'cracking' catalysts. The presence of these sites, which were developed by heat, profoundly affected chemical reactions in polymer compounds, particularly in non-polar media.

6.8   Rate of flocculation

When considering dispersions and flocculation phenomena it is useful to have some idea of the rate of flocculation, at least in order of magnitude terms. Smoluchowski [128] showed that for uniform spheres which have neither attraction nor repulsion forces present and adhere to each other on collision, the rate of decrease in the number of particles can be expressed as: $-dN/dt = KN^2$ where N = number of particles, and K is the rate constant. On integration, this becomes

$$1/N = 1/N_0 + Kt \qquad (6.18)$$

where N_0 is the initial number of particles; i.e. $N = N_0$ at time t_0. If we define a 'half-life' time $t_{\frac{1}{2}}$ where the number of particles is one half the original number. i.e. $2N = N_0$ at time $t_{\frac{1}{2}}$ then on substitution in equation (6.18) we have $2/N_0 = 1/N_0 + Kt_{\frac{1}{2}}$, that is,

$$t_{\frac{1}{2}} = 1/KN_0 \qquad (6.18a)$$

If the rate constant is diffusion controlled, then $K = K_D = 4kT/3\eta$ where k = Boltzman constant; T = temperature, η = viscosity of the medium.

Table 6.5 gives the time taken for different particle sizes at various pigment volumes to reduce their number by one half, according to Smoluchowski's rapid flocculation kinetics based on Brownian motion when the medium is at room temperature and has a typical paint viscosity (1 poise, 0.1 Pa s). If flocculated, a typical gloss paint (TiO_2 at 15% PV) will revert to its flocculated state in half a second after having been subjected to disturbance, e.g. by brushing. Even if the vehicle viscosity is very much greater, the time taken for reflocculation to occur after shear is still very short. For a similar reason when Aerosil (fumed silica) is used as a structuring agent, the quantity required is small compared with TiO_2 volume, and the structure developed through flocculation is almost instantaneous. There have been a number of attempts to verify and improve on Smoluchowski's rapid flocculation kinetics by

Table 6.5 — The half life of particles of different particle size undergoing rapid flocculation at 25 °C in a medium of $0.1\,\mathrm{N\,s\,m^{-2}}$

Half-life time $(t_\frac{1}{2})$ (s)	% pigment volume (e.g. TiO_2) (~0.2 μm diam.)	Half-life time $(t_\frac{1}{2})$ (μs)	% vol (Aerosil) (~0.014 μm diam.)
7.6	1	2400	1
1.5	5	480	5
0.76	10	240	10
0.5	15	160	15
0.38	20	120	20

taking into account interparticular forces and hydrodynamic interactions [129–133]. While these findings suggest that the rate constant should be approximately half of Smoluchowski's value, they do not significantly alter the concept that in practical systems the flocculation of pigment is very fast.

Gedan et al. [134] have measured the rapid flocculation of polystyrene particles. While they are in contention with Smoluchowski's approximation that the rate constant for doublet, triplet, and quadruplet formation is the same, nevertheless Smoluchowski's theory is still considered to give a good insight as to the rate of dispersion flocculation and has been well confirmed by experiment [129, 130].

6.8.1 Sedimentation and flocculation

Pigments generally have a density greater than the resin solution constituting the paint, and under the influence of gravity they will tend to settle out, according to Stokes law. This may be expressed as

$$d = \left(\frac{18\eta h}{\Delta\rho g t}\right)^{\frac{1}{2}}, \tag{6.19}$$

where a particle of apparent diameter d (Stokes diameter) falls a distance h in a medium of viscosity η for a time t under the influence of gravity g when there is a difference in density of $\Delta\rho$ between the particle and medium.

If well deflocculated, the particle dispersion will sediment according to the above relationship to give a hard, compact deposit which is difficult to redisperse. However, a sedimentation velocity of up to circa $10^{-6}\,\mathrm{cm\,s^{-1}}$ is generally countered by diffusion and convection, and sedimentation may seem to be absent. If flocculation is present, pigment sedimentation is rapid and the deposited material is soft, voluminous, and readily reincorporated into suspension by stirring.

When sedimentation takes place the volume of the final sediment depends upon the degree of flocculation; the greater the flocculation, the larger the sediment volume. In the extreme, the sedimentation volume may equal the total volume and should not be confused with the deflocculated state when no sedimentation may occur. The sediment volume is not only dependent on the flocculation forces, but it also depends on particle shape and size distribution [135]. Care should be taken to ensure that sedimentation volumes are used only as a relative measure of the degree of flocculation in similar systems.

Plate 6.1 shows how sedimentation tubes may be used to determine the optimum

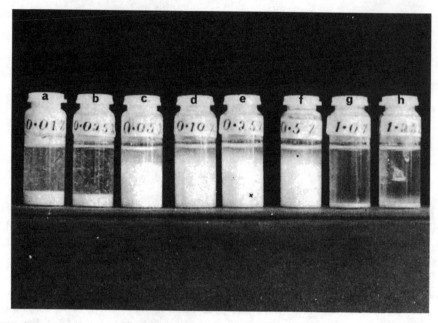

Plate 6.1 — Sedimentation tubes used to determine optimum concentration of Calgon to produce TiO$_2$ dispersion in water. Tube 'e' is marginally the best — note the lower sedimentation volume of tubes 'g' and 'h', indicating that although flocculated they are not as intensely flocculated as 'a' and 'b'.

concentration of dispersing agent (Calgon) to produce a well-deflocculated suspension of TiO$_2$ in water.

A simple alternative way of assessing the degree of pigment flocculation is to spot the dispersion onto filter paper and measure the ratio of pigment stain to solvent stain, in an anlogous manner to R_f values in chromatography; the higher the ratio the more deflocculated the dispersion.

Plate 6.2 illustrates the respective 'spot' testing of the dispersions shown in the sedimentation test. Note how tube and spot tests both show dispersion 'e' to be the best dispersion.

Smith and Pemberton [136] have described a variant of the 'spot' method of assessing flocculation which they claim gives better results.

Generally, in order to assess flocculation by observing sedimentation behaviour or measuring the R_f flocculation value, it is necessary to reduce the viscosity of a paint, by the addition of solvent, otherwise sedimentation is too slow to be of experimental value; though on dilution, care must be exercised, since this may introduce a new source of flocculation due to desorption, or depletion flocculation, or both.

Flocculation is more readily assessed in the liquid state where it can be measured by viscometry [58]. For example, if the viscosity of the continuous phase (η_0) is Newtonian, then if there is no particle–particle interaction, the dispersion viscosity (η) should also be Newtonian. If the dispersion is not Newtonian, then the degree of deviation from Newtonian behaviour is a measure of particle–particle interaction, i.e. of flocculation, provided that the disperse phase volume is <30% +, otherwise inertial effects may be observed [137]. By plotting log η against the reciprocal square

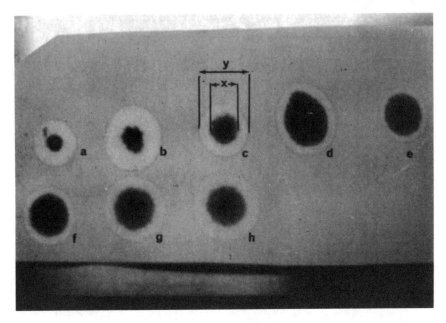

Plate 6.2 — 'Spot' tests on filter paper of the dispersions used in Plate 6.1 to determine optimum Calgon concentration for TiO_2 dispersion. $R_f = x/y$. The larger the value the better the dispersion. Spot 'e' is the best.

root of the shear rate, straight-line plots are obtained. The steeper the gradient the greater the degree of flocculation, as shown in Fig. 6.17.

This method of assessing flocculation has been found in practice to be very reliable and applicable to typical pigment concentrations in paint formulations. The comparison of flocculation by this method is strictly for similar volume concentrations, since the slope in the above plot will increase with disperse phase concentration. However, the effect of dissimilar disperse phase volumes can be estimated, and corrections can be applied to enable comparison.

It is worth noting that the rheological assessment of flocculation is a function of the total interaction of the system which is measured. Therefore, if there is a large difference in particle size between two dispersions, even though the two dispersions are compared at the same disperse phase volume, the dispersion with the smaller particle size and weaker interparticle interactions will appear to be more non-Newtonian (i.e. more flocculated). This is due simply to the interparticular forces being proportional to the first order of particle radius, while the number of particles per unit volume is proportional to the cube of the radius; hence the total 'strength' of a large number of weak bonds amongst small particles is greater than a few strong bonds between larger particles.

There have been attempts to estimate interparticle forces from rheological measurement by determining a 'critical shear rate' [138], but these estimates are very questionable [139].

The degree of pigment flocculation can also be assessed by colour measurement [140], anti-sag properties [126], or by direct visual observation through a microscope, as shown in Plate 6.3. Flocculation in a dry paint is more difficult to assess since microtoming thin sections of paint film for electron micrographs may alter the

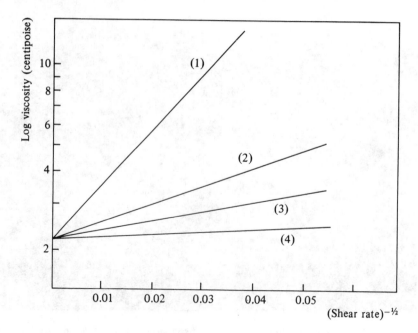

Fig. 6.17 — Log (dispersion) viscosity–(shear rate)$^{-\frac{1}{2}}$ plots as a measure of varying degrees of pigment flocculation [60] (1) most flocculated to (4) least flocculated.

apparent state of pigment dispersion [141]. Freeze fracture, followed by oxygen etching methods [142], has been developed to overcome this shortcoming; but at concentrations around 12% or more, visual interpretation becomes more difficult.

An alternative approach to the assessment of pigment flocculation has been developed by Balfour and Hird [143] who used the measurement of scatter coefficients determined in the IR region which are more sensitive to large particles. However, none of these methods can measure the state of pigment flocculation directly in both the liquid and solid state. Doroszkousk and Armitage [144] have described a method based on proton straggling which can measure the state of pigment flocculation directly, irrespective of the nature of the phase (i.e. dry or liquid paint).

In conclusion, the apparently simple question asked at the beginning of this chapter, as to whether it is better to disperse a pigment in aliphatic or aromatic hydrocarbon, is in reality extremely complex, for it contains conflicting considerations, viz:

1   It is necessary to maximize $\gamma \cos\theta$ for the system, and this may depend on the nature of the substrate and the concentration of alkyd. For TiO_2, which has a high surface energy, it would probably be better to use the aromatic solvent, since $\cos\theta = 1$ for both solvents.

2   From a consideration of attraction energies it would appear that the aromatic solvent might be the better choice when TiO_2 is the pigment, since this would produce a better match of Hamaker constants.

3   When considering the nature of the adsorbed layer and its contribution to steric

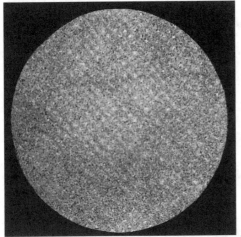

6.3a — A well-dispersed TiO_2. When viewed through a microscope ($\times 400$) the particles appear to shimmer, owing to Brownian motion. Although the particles seem to be visible, in fact they are not resolved, and only their diffraction pattern is seen as a dot.

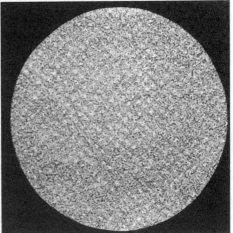

6.3b — Lightly flocculated TiO_2 dispersion. The particles have lost their Brownian motion and are seen as clusters ($\times 400$).

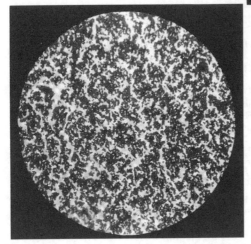

6.3c — Highly flocculated TiO_2 dispersion. The particles are flocculated into large clusters and are reminiscent of the mudflats of a river bed ($\times 400$).

Plate 6.3 — Photomicrographs of TiO_2 pigment dispersion showing varying degress of flocculation.

repulsion it would probably be better to use the aliphatic solvent (white spirit) since a greater concentration of resin might be adsorbed, and this may give a more robust dispersion at only a small trade-off in solubility of the solvated layer than using aromatic solvent (toluene).

Finally, consideration (3) will probably outweigh considerations (1) and (2) combined.

References

[1] BALFOUR J G, OCCA Conference 39–49 (1977).
[2] ASBECK W K, *Off Digest* **33** 65–83 (1961).
[3] BALFOUR J G, *JOCCA* **60** No. 9 365–76 (1977).
[4] PEACOCK J, *XI Fatipec Congress* 193–9 (1972).
[5] ELM A C, *Off Digest* **33** (433) 163.
[6] PATTON T C, p. 146 in *Paint Flow and Pigment Dispersion*, Interscience Publ. (1964).
[7] HEERTJES P M & WITVOET W C, *Powder Technol* **3** 339 (1970).
[8] WASHBURN E O, *Phys Rev* **17** 374 (1921).
[9] GOOD R J, *Chem Ind* 600 (1971).
[10] DOROSZKOWSKI A, LAMBOURNE R, & WALTON J, *Colloid Polym Sci* **255** 896–901 (1977).
[11] HARE E F & ZISMAN W A, *J Phys Chem* **59** 335 (1955).
[12] CROWL V T, *JOCCA* **55** (5) 388–420 (1972).
[13] CONSTANZO P M, GIESE R F, & VAN OSS C J, *J Adhesion Sci Technol* **4** 267–275 (1990).
[14] CHIBOWSKI E and GONZALEZ-CABALLERO F, *Langmuir* **9** 330–340 (1993).
[15] COSTANZO P M, GIESE R F, and VAN OSS C J, pp. 223–232 in *Advances in Measurement Control of Colloidal Processes*, Eds Williams R A & de Jaeger N C, Butterworth-Heinemann (1991).
[16] CHIBOWSKI E and HOLYSZ L, *Langmuir* **8** 710–716 (1992).
[17] CHIBOWSKI E and HOLYSZ L, *Langmuir* **8** 717–721 (1992).
[18] CHIBOWSKI E, *J Adhesion Sci Technol* **6** (9) 1069–1090 (1992).
[19] PARFITT G D, *XIV Fatipec Congress* 107 (1978).
[20] HUGHES W, *Fatipec Congress*, 67–82 (1970).
[21] HOWARD P B & PARFITT G D, *Croatica Chemica Acta* **45** 189–194 (1973); ibid **50** 15–30 (1977), see also *JOCCA* **54** 356–362 (1971).
[22] KEESOM W H, *Phys Zeit* **22** 129 (1921)
[23] DEBYE P, *Phys Zeit* **21** 178 (1920).
[24] LONDON F, *Zeit Phys* **63** 245 (1930).
[25] HAMAKER H C, *Rec Trav Chim* **55** 1015 (1936); ibid **56** 3727 (1937).
[26] HAMAKER H C, *Physica* **4** 1058 (1937).
[27] VISSER J, *Adv Colloid Interface Sci* **3** 331 (1972). see also BUSCALL R *Colloid Surf* **75** 269–272 (1993).
[28] GREGORY J, *Adv Colloid Interface Sci* **2** 396 (1970).
[29] VOLD M, *J Colloid Sci* **16** 1 (1961).
[30] CASIMIR H B G and POLDER D, *Phys Rev* **73** 360–372 (1948).
[31] SCHENKEL J H & KITCHENER J A, *Trans Farad Soc* **56** 161 (1960); see also TABOR D p 23–45 in: *Colloid Dispersions* Ed J W Goodwin Special Publ/Royal Soc of Chem (1981).
[32] OSMOND D W J, VINCENT B, & WAITE F A, *J Colloid Interface Sci* **42** (2) 262–269 (1973); ibid **42** (2) 270–285 (1973).
[33] VISSER J, *Adv Colloid Interface Sci* **15** 157–69 (1981).
[34] LA MER K, *J Colloid Sci* **19** 291 (1969).
[35] VERWEY E J W & OVERBEEK J TH G, *Theory of Lyophobic Colloids*, Elsevier (1948).
[36] LYKLEMA H, in *Colloid Dispersions*, Ed J W Goodwin, Special Publ/Royal Soc of Chem (1981).
[37] OVERBEEK TH G, *Adv Colloid Interface Sci* **16** 17–30 (1982).
[38] SMITH A L, in *Dispersion of Powders in Liquids*, p. 127, Chapter 3, 3rd edn, Ed G D Parfitt, Applied Sci Publ (1981).
[39] O'BRIEN R W & WHITE L R, *J Chem Soc Faraday* II **2** 1607 (1978).
[40] WIERSEMA P H, LOEB A L, & OVERBEEK TH G, *J Colloid Interface Sci* **22** 78 (1966).
[41] VAN OSS C J, *J Colloid Interface Sci* **21** 117 (1966).
[42] FRANKLIN M J B, *JOCCA* **51** 499–523 (1968).
[43] BECK F, *Prog Organic Coatings* **4** 1–60 (1976).
[44] SCHENKEL J H & KITCHENER J A, *Experientia* **14** 425 (1958); Mackor E L, *Rev Trav Chim* **70** 747–62 (1951).
[45] RANK BROS, Bottisham, Cambridge, UK.

[46] JAMES A M, in *Surface and Colloid Science*, vol 11, Chapter 4, pp. 121–185, Eds R J Good & R R Stromberg, Plenum Press (1979).

[47] HUNTER R, in *Zeta Potential in Colloid Science*, Chapter 4, pp. 125–178, Academic Press (1981).

[48] HOGG R, HEALY T W, & FUERSTENAN D W, *Trans Faraday Soc* **62** 1638 (1966).

[49] OVERBEEK TH G, *Disc Faraday Soc* **42** 10 (1966).

[50] MCGOWAN D N L & PARFITT G D, *Disc Faraday Soc* **42** 225 (1966).

[51] OSMOND D J W, ibid p 247.

[52] LYKLEMA H, *Adv Colloid Interface Sci* **2** 94 (1968).

[53] OVERBEEK TH G, *Disc Faraday Soc* **42** 10 (1966).

[54] VINCENT B, *Adv Colloid Interface Sci* **4** 193–277 (1974).

[55] PARFITT G D & PEACOCK J, in *Surface and Colloid Science*, Vol 10, Chapter 4, Ed E Matijevic Plenum Press (1978).

[56] OSMOND D J W, *Science and Technology of Polymer Colloids*, Vol 2369 NATO ASI series G W Poehlein, R H Otteweill, & J W Goodwin (eds) (1983).

[57] NAPPER D H, *Polymeric Stabilisation of Colloidal Dispersions*, Academic Press (1983).

[58] DOROSZKOWSKI A & LAMBOURNE R, *J Colloid Interface Sci* **26** 214–221 (1968).

[59] ASBECK W E & VAN LOO M, *Ind Eng Chem* **46** 1291 (1954).

[60] DOROSZKOWSKI A & LAMBOURNE R, *Disc Faraday Soc* **65** 253–262 (1978).

[61] GARVEY M J, *J Colloid Interface Sci* **61** (1) 194–196 (1972).

[62] DOROSZKOWSKI A & LAMBOURNE R, *CID Congress Chemie Physique et Applications Protiques* Vol 2, part 1 73–81, Barcelona (1968).

[63] WALBRIDGE D J, In: *Dispersion Polymerisation in Organic Liquids*, Chapter 3, Ed Barrett K E J, Wiley-Interscience Publ (1974).

[64] ZIGMONDY R, *Z Anal Chem* **40** 697 (1901).

[65] MACKOR E L, *J Colloid Sci* **6** 492 (1951).

[66] FISCHER E W, *Kolloid-Z* **160** 120 (1958).

[67] CLAYFIELD E J & LUMB E C, *J Colloid Sci* **22** 269 (1966).

[68] MEIER D J, *J Phys Chem* **71** 1861 (1967).

[69] HESSELINK F TH, *J Phys Chem* **73** 3488 (1969).

[70] HESSELINK F TH, *J Phys Chem* **75** 65 (1971).

[71] EVANS R & NAPPER D H, *Kolloid Z Z Polym* **251** 409–414 (1973); ibid 329.

[72] OTTEWILL R H, & BARCLAY L, & HARRINGTON A, *Kolloid Z Z Polym* **250** 655 (1972).

[73] OTTEWILL R H, CAIRNS R J R, OSMOND D W J, & WAGSTAFF I, *J Colloid Interface Sci* **54** 45 (1976).

[74] HOMOLA A M & ROBERTSON A A, *J Colloid Interface Sci* **54** 286 (1976).

[75] DOROSZKOWSKI A & LAMBOURNE R, *J Polym Sci* **C34** 253 (1971).

[76] DOROSZKOWSKI A & LAMBOURNE R, *J Colloid Interface Sci* **43** 97 (1973).

[77] KLEIN J, *Nature* **288** 248 (1980). see *Adv Colloid Interface Sci* **16** 101 (1982).

[78] NAPPER D H, *Trans Faraday Soc* **64** 1701 (1968).

[79] NAPPER D H, *J Colloid Interface Sci* **32** 106–114 (1970).

[80] EVEND R, DAVISON J B & NAPPER D H, *J Polym Sci* **B10** 449–453 (1972).

[81] FLORY P J, in *Principles of Polymer Chemistry*, p 523, Cornell Uni Press (1953).

[82] OTTEWILL R H & WALKER T, *Kolloid Z* **227** (1/2) 108–116 (1968).

[83] NAPPER D H, *Kolloid Z-Z Polym* **234** 1149 (1969).

[84] WAITE F A, OSMOND D J W, & VINCENT B, *Kolloid Z Z Polym* **253** 676–682 (1975).

[85] NAPPER D H, *Trans Faraday Soc* **64** 1701 (1968).

[86] WU D T, *Paint Ink Int* **5** (2) 36 (1992).

[87] RUCKSTEIN E, *Colloids and Surf* **69** 271–275 (1993).

[88] CORNER T & GERRARD J, *Colloids Surf* **5** 187 (1982).

[89] BUSCALL R & OTTEWILL R H, in *Polymer Colloids*, p 185 and p. 196 Eds Buscall R, Corner T & Stageman J F, Elsevier (1982).

[90] VINCENT B & WHITTINGTON S G, in *Surface and Colloid Science*, vol 12, Ed Matijevic E, Plenum New York (1982).

[91] SMITHAM J B, EVANS R & NAPPER D H, *J Chem Soc Faraday Trans 1* **71** 285 (1975).

[92] SMITHAM J B & NAPPER D H, *J Colloid Interface Sci* **54** (3) 467 (1976).

[93] NAPPER D H & FEIGIN R I, *J Colloid Interface Sci* **75** 525 (1980).

[94] ASAKARA S & OOSAWA F, *J Chem Phys* **22** 1255 (1954).

[95] ASAKARA S & OOSAWA F, *J Polymer Sci* **33** 183 (1958).

[96] LI-IN-ON F K, VINCENT B, & WAITE F A, *ACS Symposium Series* **9** 165 (1975).

[97] VRIJ A, *Pure Appl Chem* **48** 471 (1976).

[98] JOANNY J F & DE GENNES P G, *J Polym Sci Polym Phys* **17** 1073 (1979).

[99] PATHMAMANOHARAN C, HEK H DE & VRIJ A, *Colloid Polymer Sci* **259** 769 (1981).

[100] VINCENT B, LUCKHAM P F & WAITE F A, *J Colloid Interface Sci* **73** No 2 (1980).

[101] DE GENNES P G, '*Scaling Concepts in Polymer Physics*', 3rd edn, p. 77, Cornell University Press (1988).

[102] CLARKE J, VINCENT B, *J Chem Soc Faraday Trans 1* **77** 1831–43 (1981).

[103] SCHEUTJENS J M H M & FLEER G J, *Adv Coll Interface Sci* **16** 361 (1982).
[104] FLEER G, SCHEUTJENS J H M H, & VINCENT B, in *Polymer Adsorption and Dispersion Stability* pp. 245–263, Eds Vincent B & Goddard E D ACS Symp Ser (1984) **240**.
[105] SPERRY P R, *J Colloid Interface Sci* **99** 97–108 (1984).
[106] FLEER G J, COHEN STUART M A, SCHEUTJENS J M H M, COSGROVE T, & VINCENT B, *Polymers at Interfaces*, Chapman & Hall (1993).
[107] THOMPSON M W, GRAETZ C, WAITE F A, WATERS J A, EP13478.
[108] CROWL V T, *JOCCA* **46** 169–205 (1961).
[109] GLASSTONE S, *Textbook of Physical Chemistry*, 2nd edn, p 822, MacMillan (1956).
[110] KORAL L, ULLMAN R, & EIRICH F, *J Phys Chem* **62** 541 (1958).
[111] GILLILand E R & GUTOFF E B, *J Appl Polym Sci* **3** (7) 26–42 (1960).
[112] ELLERSTEIN S & ULLMAN R, *J Polym Sci* **55** 161, 129 (1961).
[113] MIZUHARA K, HARA K, & IMOTO T, *Kolloid-Z Z Polym* **229** 17–21 (1969).
[114] HOWARD G J & MCCONNELL P, *J Phys Chem* **71** 2974, 2981–2991 (1967).
[115] PERKEL R & ULLMAN R, *J Polym Sci* **54** 127–148 (1961).
[116] KISELEV A V, LYGIN V I *et al*, *Colloid J USSR* **30** 291–295 (1968).
[117] THIES C, *J Phys Chem* **70** (12) 3783–3789 (1966).
[118] EIRICH F R, *Effects of Polymers on Dispersion Properties Symposium* 125–143 (1983).
[119] LIPATOV YU S & SERGEEVA L U, *Adsorption of Polymers*, Halsted Press (1974).
[120] HOBDEN J F & JELLINEK H H G, *J Polym Sci* **11** 365 (1953).
[121] PATAT VON FRANZ & SCHLIEBEUER C, *Macromol Chem* **44-6** 643–668 (1961).
[122] FELTER R E, MOYER E S, & RAY L N, *J Polym Sci* **B7** 529–533 (1969).
[123] BAKER A S, JONES G M, & NICKS P F, BP 1370914 (1974).
[124] MIYAMOTO T, TOMOSHIGE S, & INAGAKI H, *Polym J* **6** (6) 564–570 (1974).
[125] BUTTER R, TAN Y, & CHALLA CR, *Polymer* **14** (4) 171–2 (1973).
[126] WALBRIDGE D J, SCOTT E I, & YOUNG C H, *Vth Conference in Organic Coatings Sci Technol* 526–546 (1979) Athens.
[127] SOLOMON D H, SWIFT J D, & MURPHY A J, *J Macromol Sci — Chem* **A5** (3) 587–601 (1971).
[128] SMOLUCHOWSKI M, *Z Phys* **17** 557 (1916); *Z Phys Chem* **92** 129 (1917).
[129] LICHTENFELD J W, PATHMAMANOHARAN C, & WIERSEMA P H, *J Colloid Interface Sci* **49** 281 (1974).
[130] LIPS A & WILLS E, *Trans Faraday Soc* **69** 1226 (1973).
[131] OVERBEEK J TH G, in *Colloid Science*, Vol 1, p. 282 Ed H R Kruyt Elsevier (1952).
[132] SPIELMAN L A, *J Colloid Interface Sci* **33** 562 (1970).
[133] HONIG E P, ROEBERSEN G J, & WIERSEMA P H, *J Colloid Interface Sci* **36** 97 (1971).
[134] GEDAN H, LICHTENFELD H, & SONNTAG H, *Coll Polym* Sci **260** 1151 (1982).
[135] VOLD M J, *J Coll Sci* **14** 168–174 (1959).
[136] SMITH A E & PEMBERTON E W, *Pig Resin Technol* **9** (11) 8 (1980).
[137] STRIVENS T A, *J Colloid Interface Sci* **57** (3) 476–87 (1976).
[138] ALBERS W & OVERBEEK J TH G, *J Colloid Sci* **15** 480–502 (1960).
[139] DOROSZKOWSKI A & LAMBOURNE R, *J Colloid Sci* **26** 128–130 (1968).
[140] KALWZA U, *Pig Resin Technol* **9** (4–7) (1980).
[141] HORNBY M R & MURLEY D R, *Prog Organic Coatings* **3** 261 (1975).
[142] MENOLD R, LUTTGE B, & KAISER W, *Adv Colloid Interface Sci* **5** 281–335 (1976).
[143] BALFOUR J G & HIRD M J, *JOCCA* **58** 331, (1975).
[144] DOROSZKOWSKI A & ARMITAGE B H, *XVII Fatipec Congress* **2** 311–330 (1984).

7

Particle size and size measurement

A Doroszkowski

7.1 Introduction

To understand the behaviour and properties of particle suspensions such as pigments in paint it is necessary to know something about the size and the size distribution of the particles. Particulate material may constitute pigment, extenders, emulsion droplets, or latex particles.

The question of 'What is a particle?' and hence 'What size is it?' is not always easy to answer, as shown in Plate 7.1. The answer to the question resides in another question: 'What property is to be investigated?' Particle size measurement is usually a means to an end and rarely justified in itself. In any particle size measurement it is important to use a method most suited to the problem in hand. When surfactant adsorption estimates are required, sizing by surface area should be carried out; if sedimentation considerations are required then particle sizing should be by a volume (mass) method in preference to a count by number method.

When the particles are perfect spheres and are all of the same size, then it is only necessary to know the particle diameter to be able to describe fully the size and size distribution. However, in practice such an occurrence is extremely rare. Particles are frequently irregular in shape and usually have a range of sizes, and what might be an adequate definition of 'particle size' for one set of conditions will be inadequate for another. It is therefore important to use a definition of particle size which is relevant to the property being considered. For example the variable floccule size shown in Plate 7.1 has no relation to the size of the monodisperse primary particles constituting the floccules.

7.2 Definitions

Spherical particles are readily described by their diameter, but non-spherical particles may be measured in many different ways. Some of the ways to express the size of irregular particles are illustrated in Fig. 7.1 [1].

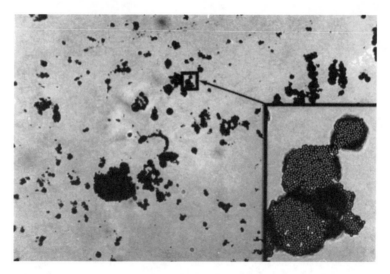

Plate 7.1 — What is a particle? (Insert at ×12 increase magnification).

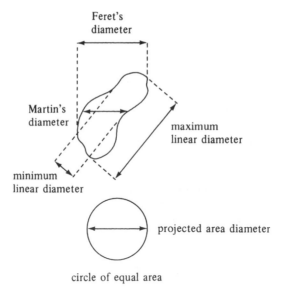

circle of equal area

Fig. 7.1 — Non-spherical particle diameter.

Feret's diameter is the distance between two tangents on opposite sides of the particle, perpendicular to the direction of scan.

Martin's diameter is the length of a line parallel to the direction of the scan that divides the particle profile into two equal areas.

Maximum and minimum linear diameters are two obvious linear measurements that may be used. These values can be amalgamated to give a single value in the form of the square root of their products, which is more representative of size than either value alone. However, the process is rather laborious, and frequently special scales consisting of a series of circles of different diameters are placed on photographs (or in the eyepiece of a microscope) where the irregular particles can be equated to a circle of equivalent area (or equivalent perimeter), i.e. the *projected area diameter* is the diameter of a circle having an area equal to the projected area of particle. Typical examples are the Patterson–Cawood graticule where the series of circles is graduated in an arithmetic series, and the Porton graticule which is graduated in a series based on $2^{\frac{1}{2}}$ [2].

7.2.1   Counting requirements

The question of how many particles should be counted to obtain a representative particle size distribution is frequently encountered. Intuitively, one can see that this must depend on the range of sizes. If the particle size distribution is monodisperse then a small number of measurements will suffice. Likewise the more polydisperse, the larger the number of measurements that will be required. Figure 7.2 is a simple practical demonstration of 'average' particle size measured against number of particles counted; the two types of 'average' being number and weight averages (described later). It is seen that somewhere after counting 350 particles there appears to be little change in the 'average' size from counting 1000 particles.

Time or fatigue is the usual criterion for limiting the number of particles measured. In practice the American Society of Testing and Materials [3] recommends that not fewer than 25 particles in the modal class should be measured, and that at least 10 particles should be present in each size class.

Sichel *et al.* [4] devised a technique called 'truncated multiple traversing' to minimize the number of measurements, yet maintaining reliability. The method is an adaptation of 'stratified sampling' [5]. The concept is that at least 10 particles must be observed in every size class that has a significant influence on the size curve. The method is exemplified in Table 7.1. In the first traverse, an area was searched and

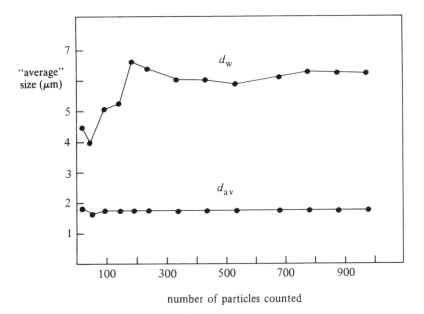

Fig. 7.2 — Average particle size vs number of particles.

Table 7.1 — Example of particle sizing using Sichel's 'truncated multiple traversing' [4]

Traverse number	Number of particles in size range												
	0–1	1–2	2–3	3–4	4–5	5–6	6–7	7–8	8–9	9–10	10–11	11–12	Totals
1st	0	5	11	34	41	24	5	3	1	0	1	0	125
2nd	1	7					4	2	1	1	0	0	16
3rd	0						5	5	3	1	0	0	14
4th	0								2	2	1	0	5
5th	1								3	2	1	1	8
6th	0									1	2	0	3
7th	0									2	1	0	3
8th	0									1	1	0	2
9th	0										1	0	1
10th	0										0	0	0
Column total	2	12	11	34	41	24	14	10	10	10	8	1	177
Nos. per traverse	0.2	6	11	34	41	24	4.7	3.3	1.3	1.3	0.8	0.1	128.4
Nos. per cent (frequency)	0.16	4.67	8.57	26.40	31.9	18.6	3.66	2.57	1.01	1.01	0.62	0.08	100

Total number counted = 177 particles (equivalent to counting 1250 particles)

sufficient measurements were obtained in sizes 2–3, 3–4, 4–5, and 5–6 units. A second traverse of a similar, but not the same area as the first, was made, except that only particles greater than 5–6 and less than 3–4 units were counted. The process was repeated until Sichel's criteria were satisfied by all major size classes, with the result that only 180 particles were counted, yet the reliability was equivalent to the counting of 1250 particles not using this method (ten traverses in each field giving an equivalent count of about 1250 particles, i.e. 125×10).

The overriding criteria in particle size counting are that the sample being measured is representative and that the sample preparation technique does not introduce bias. Therefore it is insufficient to measure a large number of particles from, say, one electron micrograph; more accurate results may be obtained by measuring fewer particles but from many different electron micrographs. The particles being counted must represent the total population from which they were obtained, must be dispersed randomly without reference to shape or size, and agglomerates are deflocculated. A frequent requirement for single particle counting is that more than half the particle perimeter is visible before it is counted. This, however, denies the existence of particle sinters which may occur, e.g. TiO_2 pigment (see Fig. 6.1 in Chapter 6 for this effect).

7.2.2 Average particle size — ways of expressing size

If a particle size plot is made of size against the number of particles in that size (frequency), a particle size distribution graph is obtained, or a histogram, depending on the method of presentation. While this is very informative it is cumbersome to use, hence the need of a simple expression to summarize the range of sizes, i.e. an 'average'. There are many ways of describing an 'average', depending on the property being emphasized (weighting).

Table 7.2 — Definitions of average diameters

Mode	The diameter that occurs most frequently.
Median diameter	The diameter for which 50% of the particles measured are less than the stated size.
Average diameter (d_{av})	The sum of all the diameters divided by the total number of particles $\dfrac{\sum (nd)}{\sum n}$
Geometric mean (d_g)	The nth root of the product of the diameters of n particles measured $d_g = n\sqrt{d_1 d_2 \cdots d_n}$ usually determined as $$\log d_g = \frac{\sum n(\log d)}{\sum n}$$
Harmonic mean	The reciprocal of the diameters measured $$\frac{1}{d_n} = \sum (n/d) \Big/ \sum n$$
Mean length diameter	(d_1) Measured $d_1 = \dfrac{\sum nd^2}{\sum nd}$ as an average it is comparable to arithmetic and geometric means; it represents the summation of surface areas divided by summation of diameters.
Surface mean (d_s)	$d_s = \sqrt{\sum nd^2 \Big/ \sum n}$
Volume mean (d_v)	$d_v = \sqrt[3]{\sum nd^2 \Big/ \sum n}$ The median value of this frequency is often called mass median diameter.
Volume–surface mean (d_{vs}) (or Sauter mean)	$d_{vs} = \sum nd^3 \Big/ \sum nd^2$ Average size based on specific surface per unit volume.
Weight mean (d_w) (or De Broucker mean)	$d_w = \sum nd^4 \Big/ \sum nd^3$ Average size based on unit weight of particle.

Some of the definitions frequently encountered are given in Table 7.2 and their centre of gravity on a curve is represented in Fig. 7.3.

7.2.3 Size-frequency curves

For most particle size determinations, the size–frequency curves obtained follow the probability law. The usual normal probability equation applies to distributions which are symmetrical about a vertical axis, sometimes called a Gaussian distribution. Since size distributions are frequently 'skewed' or asymmetrical, the normal law does not apply (see Fig. 7.4).

Fortunately, the asymmetrical curves can be made symmetrical in most cases, if the size is plotted on a logarithmic scale (the frequency remaining linear). Such distributions are known as log-normal distributions. Hatch and Choate [6] showed the importance of this property.

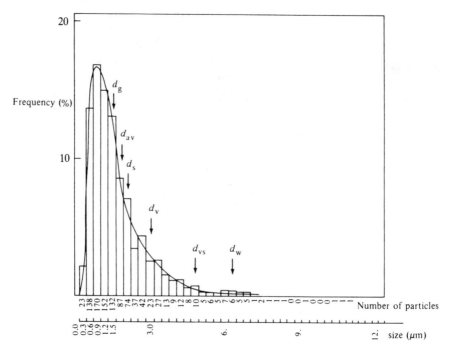

Fig. 7.3 — Effect of definition on weighting of 'average'.

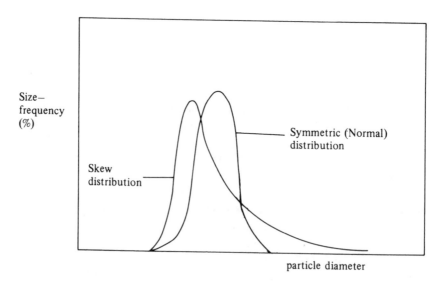

Fig. 7.4 — Distribution curves.

The equation of the normal probability curve (Fig. 7.5) as applied to size–frequency distribution is;

$$F(d) = \frac{\Sigma n}{\sigma \sqrt{2\pi}} \exp\left[\frac{-(d - d_{av})^2}{2\sigma^2}\right]$$

(7.1)

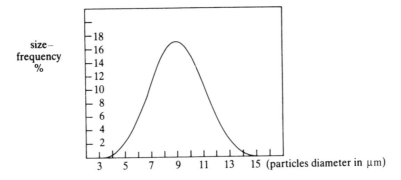

Fig. 7.5 — Typical symmetrical size distribution (Gaussian or normal law) plot.

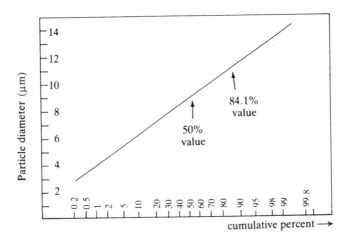

Fig. 7.6 — Gaussian curve (of Fig. 7.5) plotted on arithmetic-probability chart.

where $F(d)$ is the frequency of observations of diameter d; n is the total number of observations, d_{av} is the arithmetic average of the observations $[\Sigma(nd)/\Sigma n]$, and σ is the standard deviation given by $\sigma = \sqrt{[\Sigma n(d - d_{av})^2/\Sigma n]}$. The constants d_{av} and σ completely define the frequency distribution of a series of observations. Thus if the particle sizes are plotted on an 'arithmetic-probability' grid, the summation curve is a straight line where the mean value (the 50% value) is the simple arithmetic average d_{av}: the standard deviation (σ) is given by $\sigma = 84.1\%$ size -50% size $= 50\%$ size -15.9% size (if plotted with negative slope) (Fig. 7.6).

The asymmetrical distribution curve, where particle size is plotted on a linear scale (Fig. 7.3), can be converted into a symmetrical one if the particle diameters are fitted onto a log scale as in Fig. 7.7; i.e. equation (7.1) becomes

$$F(d) = \frac{\Sigma n}{\log(\sigma_g)\sqrt{2\pi}} \exp \frac{-(\log d - \log d_g)^2}{2\log^2 \sigma_g} \tag{7.2}$$

where d_g now refers to the geometric mean ($d_g = \sqrt[n]{[d_1 d_2 \ldots d_n]}$) and σ_g is obtained from equation (7.2a)

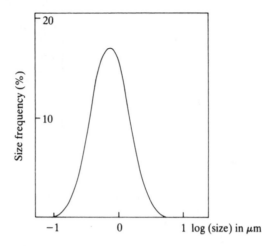

Fig. 7.7 — Transformation of skew distribution (of Fig. 7.3) into symmetrical plot using log scale for particle size (μm).

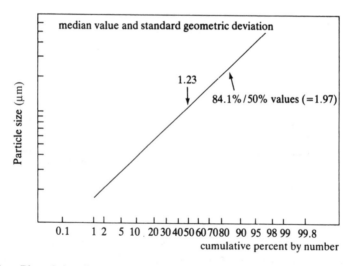

Fig. 7.8 — Plot of size distribution (shown in Figs. 7.3 and 7.7) represented on log-probability chart showing derivation.

$$\log \sigma_g = \sqrt{\frac{\sum \left[n(\log d - \log d_g)^2\right]}{\sum n}} \qquad (7.2a)$$

The terms $\log d_{gl}$ and $\log \sigma_g$ are called the log-geometric mean diameter and the log-geometric standard deviation respectively. These two values are very important since they completely define a log-normal size distribution which is typical of a dispersion process [7].

A simple way of plotting a log-normal distribution is to use log-probability graph paper (see Fig. 7.8, cf. arithmetic-probability with a normal distribution), where the particle size is plotted as the ordinate and cumulative percent by weight (or number) is plotted along the abcissa. The geometric median diameter (d_g) is the 50% value of the distribution and the geometric standard deviation (σ_g) is the 84.1% value divided by the 50% value (or the 15.9% value divided by the 50% value if the plot is with a negative slope).

Sometimes, a plot of the particle size distribution is not a straight line on log probability paper, as for example in Fig. 7.9. It can be shown that this is due to the absence of particles less than seven units (the curve asymptotes to this value). If we re-plot the results but use ($x - 7$) instead of x, a straight line is obtained, and again we can completely represent the distribution by two numbers.

There are many other types of deviation from a straight line in this type of plot, such as curves with points of inflection which are due to bimodal distributions. These different cases, how to resolve them, and the mathematical basis for the procedures are discussed by Irani and Calliss [8].

The geometric standard deviation is always the same in a log-normal particle size distribution, whether the sizes are plotted as cumulative percent by count or by weight. However the median values are different and hence care must be exercised to denote if the geometric median value is by weight (d_{gw}) or by count (d_{gc}).

The Hatch-Choate transformation equations [6] enable us to convert the geometric median by weight to that of the geometric median by count. They also enable us to convert one type of 'average' to that of another and hence are most useful in comparing distribution size measurements carried out by one method with that of another (see Table 7.3).

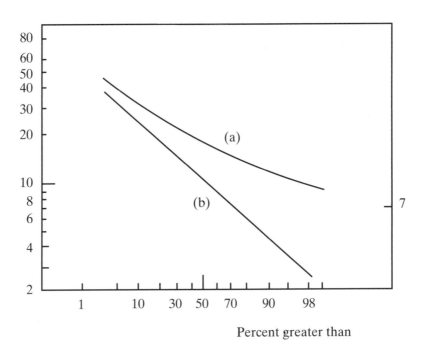

Fig. 7.9 — Abnormal log-normal distribution. Plot (a) gives curvature when particle size, x, is plotted on log-probability paper with the smallest size asymptoting to seven size units. Plot (b) is the same size distribution but re-plotting the size on the basis of ($x - 7$) which now gives a straight line plot (standard log-normal plot) as shown in [8].

Table 7.3 — Hatch–Choate transformation equations

To convert from	To		Use
d_{gm}, the geometric median mass	$d_{gc} = \text{antilog}\left(\dfrac{\sum n \log d}{\sum n}\right)$		
	$d_{av} = \Sigma nd/\Sigma n$	geometric median by count	$\log d_{gc} = \log d_{gm} - 6.908\log^2\sigma_g$
	$d_s = \sqrt{\Sigma nd^2/\Sigma n}$	arithmetic mean	$\log d_{av} = \log d_{gm} - 5.757\log^2\sigma_g$
	$d_v = \sqrt[3]{\Sigma nd^3/\Sigma n}$	surface mean	$\log d_s = \log d_{gm} - 4.605\log^2\sigma_g$
	$d_{vs} = \Sigma nd^3/\Sigma nd^2$	volume mean	$\log d_v = \log d_{gm} - 3.454\log^2\sigma_g$
	$d_w = \Sigma nd^4/\Sigma nd^3$	volume-surface	$d_{vs} = \log d_{gm} - 1.151\log^2\sigma_g$
		weight mean	$\log d_w = \log d_{gm} + 1.151\log^2\sigma_g$
d_{gc}, the geometric median by count	d_{av}		$\log d_{av} = \log d_{gc} + 1.151\log^2\sigma_g$
	d_s		$\log d_s = \log d_{gc} + 2.303\log^2\sigma_g$
	d_v		$\log d_v = \log d_{gc} + 3.454\log^2\sigma_g$
	d_{vs}		$\log d_{vs} = \log d_{gc} + 5.757\log^2\sigma_g$
	d_{gm}		$\log d_{gm} = \log d_{gc} + 6.908\log^2\sigma_g$
	d_w		$\log d_w = \log d_{gc} + 8.023\log^2\sigma_g$

In the transformation of particle size distributions, by number to that of weight, errors may be introduced since the largest and heaviest particles are frequently present in statistically small numbers. Jackson *et al.* [9] have calculated the errors which are likely to be encountered when converting count by number to that of mass, and say what steps have to be taken to ensure that the errors be small.

7.3   Sampling

Before any particle size measurements are made it is important that a truly representative sample is taken for the size measurement. If this is not done, then even if the measurement carried out is precise, it will be inaccurate with respect to the particle size representation. For example, suppose one has 25 kg bag, of a powder such as a pigment (or a powder coating); however, only a few mg of sample are required for the measurement. It is therefore important that the mg-sample is representative of the whole, and that size segregation has not taken place before one starts a particle size distribution analysis.

The minimum sample size required for a measurement is dependent on the mass fraction of the coarsest size and the desired accuracy. The exact relationship is given in graph form in British Standard Methods, Appendix B [10]. For example if the mass fraction of the largest-size class is 0.01, the mean size is 0.2 mm (specific gravity, SG = 1.5), and the coefficient of variation is 0.05, then 0.24 g is the minimum sample size. If, however, we had the same requirements in terms of accuracy, but were considering a 'powder coating' which had a mean size of 70 μm and a SG = 1.5, then the minimum sample must be 10.3 mg.

The most important segregation-causing property is particle size itself, and particularly so in free-flowing powders. When powders are poured into a heap, the fines

Plate 7.2 — Cross-section of a heap of a bimodal mixture of 2 mm (white) and 0.2 mm (black) particles, at 1:1 by wt. concentration and of similar density. Note how the larger particles segregate to the surface and periphery, while the smaller particles are concentrated in the centre.

tend to collect at the centre of the heap with the coarse particles on the surface as shown in Plate 7.2. In vibrating containers the larger particles collect at the top, even if the density of the larger particles is greater than that of the fines. This can occur by either a percolation of fines to the bottom, or by the larger particles 'walking' to the surface. 'Walking' takes place when a larger particle tilts under vibration so that the smaller particles near its edge pour underneath and consolidate the new position. Further random elevation of the other side of the large particle results in it rising relative to its original position.

It is best to assume that powder samples are heterogeneous, unless it is known to the contrary, whence 'scooping' an appropriate amount is permissible. Powders can be classed as free-flowing, non-free-flowing and non-flowing. Free-flowing powders show the greatest tendency to segregate particle sizes, and rotary sampling is the preferred method of obtaining a representative sample. Non-free-flowing and non-flowing powders can be reduced to sample measurement size by coning and quartering.

7.3.1 Sampling from large containers

There are a number of ways of sampling from large containers such as taking scoop samples at different depths and positions within the container, but not from the surface layer or at the container walls. Alternatively, by using special sampling

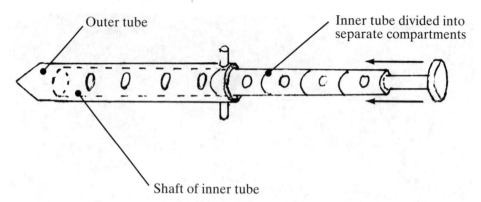

Fig. 7.10 — 'Sampling thief' for sampling powder. The inner tube is divided into separate compartments so that when the holes in the inner tube are aligned with the outer tube, powder can enter. The holes in the tube are then closed off and the samples of powder from different depths can be removed, when plunged into a heap (or container) of powder.

devices which sample at different positions simultaneously in a single operation such as the 'sampling thief' shown schematically in Fig. 7.10.

7.3.2   Sampling from heaps

Plate 7.2 is a photograph of a cross-section of a heap of a bimodal mixture of 2 mm (white) and 0.2 mm (black) spheres (1:1 by weight and of similar densities). It shows how the larger beads form a surface and peripheral layer, while the finer particles concentrate in the centre. For while segregation may occur to a lesser extent with finer particles, it does demonstrate the problem of sampling a heap. The best sampling technique would be to take samples from the flowing powder while the heap was being made, otherwise incremental samples taken from different portions of the heap have to be compounded, and a less representative sample is to be expected.

The heap may be 'coned and quartered'. This relies on radial symmetry when the heap is flattened and then carefully quartered, to give four identical samples; however, this approach does require operator skill, and may lead to bias if not carried out properly, see Fig. 7.11.

The best method of powder sampling is by the use of a spinning riffler as shown schematically in Fig. 7.12. There are many commercial instruments available which are based on this principle. The powder is loaded into a hopper and set in motion to form a stream. It is then collected incrementally, into a series of containers fixed to a rotating table. The rate of flow and the speed of rotation of the boxes is adjusted so that there are at least 30 'box-collections' during the total flow of the powder. Every second box may then be discarded, thereby reducing the sample by one-half. The process is repeated until the desired sample size is obtained (a quicker collection time, but with a lower degree of accuracy, is by collecting from a smaller number of boxes, e.g. every sixth box).

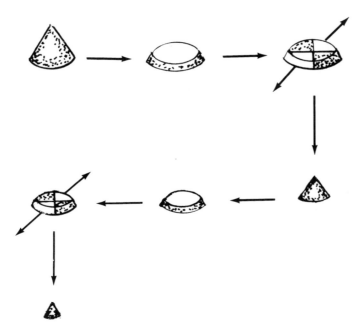

Fig. 7.11 — Coning and quartering. A conical heap is formed using a powder scoop (top left-hand side); the process is then repeated twice (not shown). The cone is carefully flattened to be of uniform thickness and radius (top centre): the cone is quartered, and the two opposing quarters are discarded (top right-hand side). The other two quarters are combined, and the process is repeated until the required sample size is obtained (bottom of diagram).

7.3.3   Reduction from laboratory to sample size

The spinning riffler can be used to reduce the sample size to about 1 g. For further reduction, the sample may be dispersed to form an aqueous (or non-aqueous) suspension by use of a surfactant and then collected at the appropriate dilution with a syringe. A small drop can then be placed on a microscope slide or on a formvar film, prior to making a grid specimen for viewing with an electron microscope. Care, however, has to be taken for size-segregation on the microscale can occur (see Plate 7.3) if a dispersion is 'spotted' when it is very fluid (or a solvent is dropped onto the 'spot') the smallest particles can be swept away from the larger ones by the spreading surface film (this is a form of fractionation by hydrodynamic chromatography — see later). Alternatively the sample may be dispersed in a viscous liquid such as a nitro-cellulose/plasticizer solution by mulling, using a palette knife and a glass plate (or a pestle and mortar). The resultant dispersion is then thinly spread on a microscope slide, using a glass rod, for specimen preparation/measurement.

7.3.4   Sampling from dispersion

The sampling of colloidal dispersions by aliquot removal is usually simpler (and more homogeneous) than that of free-flowing powders, even though the particle size

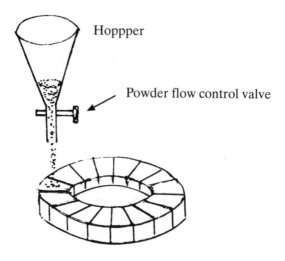

Hoppper

Powder flow control valve

Side view

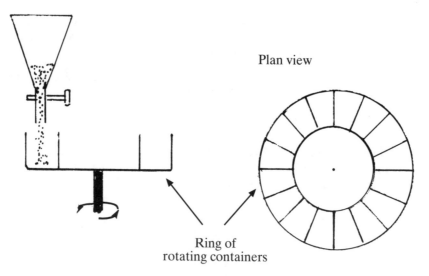

Plan view

Ring of
rotating containers

Fig. 7.12 — Spinning riffler.

ratios of the components in each may be comparable. However, the dispersion must be agitated before sampling to ensure there is no sediment or cream layer (note the uniformity of a coloured paint when stirred, which comprises different-sized coloured pigments). Further details on sampling of liquids and pastes is given in British Standard procedures [12].

For further reading on sampling the reader is referred to Kaye [13] and Allen [14].

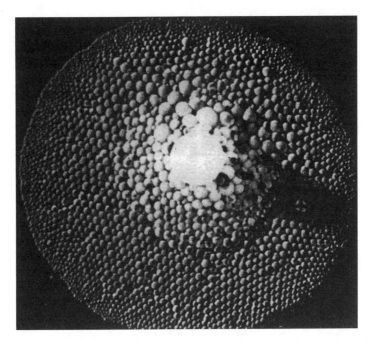

Plate 7.3 — Electron micrograph of a single latex droplet from [11]. Note how the smaller latex particles have segregated, by moving to the periphery of the latex droplet, when placed on a microscope slide.

7.4   Methods of particle sizing

It is a truism to say that the difficulty in measuring particle size, and especially particle size distribution, is inversely proportional to the size itself, the submicrometre particle size range being particularly difficult, yet most frequently encountered in the paint industry.

The basic method of particle sizing is still, probably, by visual inspection through a microscope, or an electron micrograph for fine particles. Calibration standards such as mono-disperse latexes used for other instruments are still determined this way. The advent of modern image analysers has taken the labour and tedium away from these measurements.

Particle size measurement techniques can be divided into two basic approaches, 'fractionating' and 'non-fractionating.' These two divisions can then be subdivided into sedimentation and classification approaches or imaging, field disturbance, and adsorption approaches. This can then be further subdivided into variations on a theme, as shown in Table 7.4, which presents an overview of the various methods of particle sizing, as well as size range; whether the method only gives an average particle size, an average size and an indication of size distribution, or a full particle size analysis.

It is not possible to discuss the various techniques in a single chapter such as this, hence only the more common techniques are described along with some simple approaches not requiring expensive equipment. A list of some 358 references on particle sizing techniques is presented by Barth and Sun [15] who appear to publish such lists every two years in *Analytical Chemistry.*

Table 7.4 — An overview of various methods of particle sizing

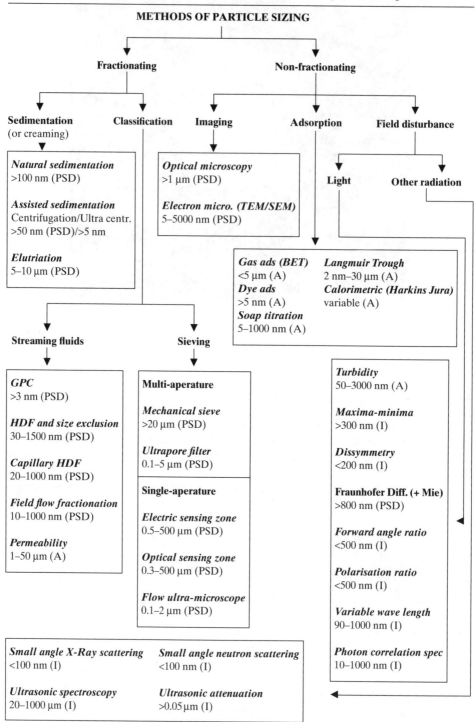

PSD = particle size distribution obtained; I = average size and indication of size distribution; A = average size only.

7.4.1   Direct measurement

Microscopy or electron microscopy is probably the definitive sizing technique producing full-size distribution analysis. Monodisperse particles such as latexes or gold sols, used as standards for other techniques, are certified by this means for use as calibrants in other particle sizing methods. There are well-established counting procedures such as those described at the beginning of this chapter, and many ways of simplifying the counting procedures such as the use of modern image analysers [16, 17] like the Quantimet 920 [18], Magiscan 2 [19], or the Zeiss Kontron sem-ips. These instruments can be directly coupled to an optical or electron microscope, or they can operate using just photographs of particles to be measured.

Image analysers consist of two components, a unit for converting the optical image into an electrical signal, and a computer that analyses the electric signal to give quantitative image information. The image scanner can enhance the image definition, or contrast, and be made to assess the particle in a variety of ways such as particle area, perimeter, Feret's or Martin's diameter, radius of equivalent disc area or perimeter; in fact the only limitation appears in the 'software' or computer programmes available. Light pens can be used to discard or separate 'touching' particles. Unfortunately, image analysers are very expensive, costing more than some types of electron microscope. Their accuracy is nevertheless dependent on proper, representative sample preparation.

When preparing a sample for direct measurement by microscopy and especially electron microscopy, great care has to be taken not to introduce instrumental artefacts, such as aggregation of the particles during slide preparation.

Soft or liquid particles are particularly difficult to prepare for electron microscope examination. They require crash-cooling techniques followed by freeze fracture [20] to produce replicas suitable for electron microscopy. If crash-cooled at high concentration (e.g. 15%) soft particles may be forced to coalesce even when cooling rates are in excess of $1000\,K/s^{-1}$ [21] to give unrepresentative emulsions (see Plate 7.4) for size analysis. Under these conditions one has to dilute emulsions to at least a few percent before crash-cooling, using Arcton/liquid nitrogen in order to obtain samples suitable for making replicas for electron microscope particle sizing. Even when successfully freeze-fractured, special counting techniques [22] have to be employed to obtain the true size of particles, since not all fractures will be through the diameter of the particle.

7.4.2   Sedimentation methods

Particle sizing by sedimentation will give full particle size distribution analysis whether it is due to natural sedimentation, i.e. under gravity alone, or enhanced sedimentation due to centrifugation. The particle diameter that is measured is 'Stokes diameter', which is the diameter of a sphere of the same density and free-falling speed as the particle being assessed.

Particle size measurements are always made at high dilution, and great care must be taken to ensure that the particles are well dispersed (deflocculated).

The simplest approach to obtain a particle size distribution with readily available, inexpensive equipment with an accuracy of between 2 and 5%, is to use the Andreasen pipette [23] (see Fig. 7.13) which consists of a cylindrical sedimentation vessel (about 550 ml) with a 10 ml pipette fitted with a two-way stopcock to enable sampling at a predetermined depth of usually 20 cm. A well-dispersed suspension

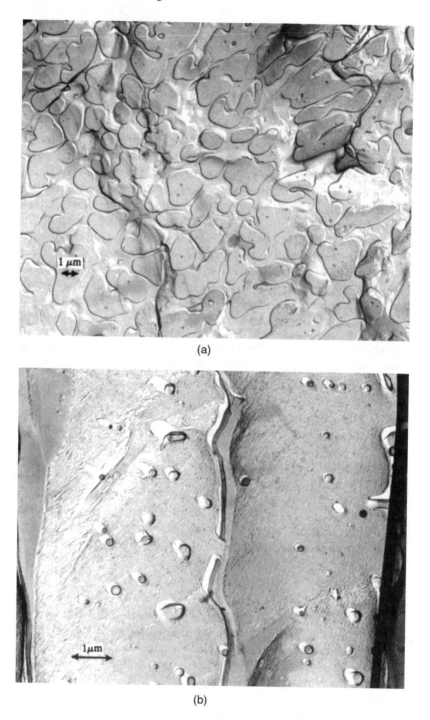

(a)

(b)

Plate 7.4 — Electron micrographs of replicas of freeze-fractured emulsion. (a) Effect of crash-cooling at too high a disperse phase volume with soft particles (15% DPV) — particles have coalesced to form larger particles about 2 μm in size. (b) The same particles crash cooled at high dilution ($\frac{1}{2}$% DPV) (note that (b) is ×2 magnification of (a).)

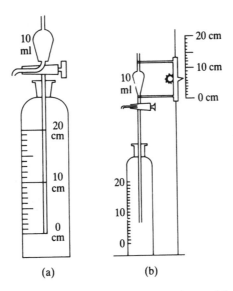

Fig. 7.13 — Schematic representation of Andreasen pipette (a) and variable depth modification (b).

of not more than 1% disperse phase volume is placed inside the vessel which is maintained at a constant temperature by means of a water bath. Samples of the dispersion are withdrawn slowly at regular time intervals (usually in a 2:1 time interval progression), and their solids per unit volume are determined.

The particle size calculations are based on determining the initial concentration of particles (zero time) and the concentration of the sample after a lapse of time t.

The calculation of particle size is dependent on Stokes law, that is,

$$d = \left(\frac{18\eta h}{(\sigma - \rho)gt} \right)^{\frac{1}{2}} \times 10^4 \qquad (7.3)$$

where d = Stokes diameter in µm;

σ = apparent density of particle $(g\,cm^{-3})$ (equal to true density for non-porous particles only);

ρ = density of medium $(g\,cm^{-3})$;
η = absolute viscosity of medium (poise);
h = distance (cm) through which particle falls in time t (s);
g = acceleration due to gravity $(cm\,s^{-2})$.

One thus obtains a cumulative percent by weight versus particle diameter plot. The Andreasen pipette method is a British Standard test method and is fully described in [24]. The particle size range is usually between 1 and 100 µm, but it can be used to determine the particle size distribution of dense submicrometre particles such as TiO₂ pigment if suitable precautions are taken. However, because of the lengthy sedimentation times for TiO₂ particles of about 0.2 µm the Andreasen pipette is frequently modified to sample at varying depths in order to speed up the measurements — see Plate 7.5 [25]. An analysis of the errors that can occur in size measurement by sedimentation has been given by Svarovsky and Allen [26].

A variation on particle size distribution by sedimentation under gravity is that obtained using a hydrometer [27]. In this procedure the suspension density is estimated at the effective distance beneath the suspension surface at the centre of gravity of a floating hydrometer, known as a soil hydrometer, having a density range from 0.995 to 1.038. The hydrometer is placed in a sedimentation cylinder containing a well-dispersed suspension of about 50 g pigment in 200 ml of dispersion medium; the density of the dispersion is read off at time intervals increasing exponentially, along with that of a control cylinder containing the dispersion medium (e.g. Calgon solution); the temperature is maintained to $\pm 0.1\,°C$. The percent pigment remaining in suspension after t minutes, at the level which the hydrometer measures the density of the suspension, is the ratio of $(R - B)$ to $(R_0 - B)$ multiplied by 100, where R = hydrometer reading in pigment suspension at time t, B = hydrometer reading in blank solution, and R_0 = initial hydrometer reading in pigment suspension.

Stokes' diameter is determined from equation (7.3), which can be related to the percent pigment remaining in suspension — the method is applicable generally to particles ranging in size from 1 to 15 μm, and even smaller if the pigment density is greater.

The use of a soil hydrometer assumes that the difference in density between the suspension and the suspension medium is proportional to the pigment concentration, and that the hydrometer measures the density at a given level somewhere near the centre of gravity of the hydrometer. The position of this level is somewhat uncertain. Furthermore, the hydrometer sinks as sedimentation progresses, hence the

Plate 7.5 — Adjustable height Andreasen pipette, and racking mechanism placed in thermostatted bath ready for use.

sampling depth increases. Removal and re-insertion of the hydrometer into the suspension also distrubs the settling of the suspension. To overcome these problems one can use small glass divers known as 'Berg's divers' of different densities which are immersed in the suspension [28, 29]. These take up an equilibrium position at the level of equal suspension density, which is assumed to be proportional to concentration. Since the divers are about 1 cm in length the uncertainty in height is small. The disadvantage is that the diver is not usually visible and must be moved laterally to the side of the vessel with a magnet, and might therefore disturb the suspension.

An alternative approach to size determination by simple sedimentation based on Stokes Law is to use a sedimentation balance as described by Oden [30]. Modern techniques employ micro-balances such as those built around the Cahn or Sartorius [31, 32] microbalances, thus enabling automatic recording and size calculation using a simple computer program. It is claimed that this method is more accurate, automatic, and uses a more rigorous analysis than that applicable to the Andreasen pipette. However, it is also very much more expensive, and like the pipette method is limited by the need to have a reasonable density difference between the particles and suspension medium, as well as a deflocculated dispersion.

When determining particle size by sedimentation the requirement of an analytical method to determine sample concentration is solved, in the Andreasen pipette method, by careful sample removal for analysis by non-volatile assay, and, in the sedimentation balance, by *in situ* weighing. Another convenient method of determining concentration continuously is to measure the attenuation of a beam of radiation passing through the sample. For this approach the attenuation must be proportional to the mass of sample lying in the beam. Visible light absorption can be used for coloured pigments which have negligible light-scattering properties. However, for particles which strongly scatter light, such as TiO_2, X-rays have to be used. Murley [33] describes such an apparatus (shown schematically in Fig. 7.14). A commercially available instrument working on this principle is built by Micromeritics in the form of the 'Sedigraph 5000ET'.

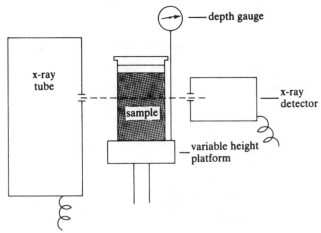

Fig. 7.14 — Schematic representation of static X-ray apparatus [33].

All the aforementioned methods rely on sedimentation due to gravity and hence are time-consuming for small particle sizes. Centrifugation can be applied to speed up sedimentation, as with the Joyce–Loebl disc centrifuge; and for finer particle sizes an ultra centrifuge is used [34]. The Joyce–Loebl disc centrifuge [35] has a hollow transparent disc-like cylinder which is rotated at a preselected speed. The particles to be measured are injected into the disc, using a buffered line start technique to prevent particle streaming and 'settle' through the spin fluid. A modified relationship of Stokes law is used to determine the particle size:

$$t = 9\eta/2w^2d^2\Delta\rho \ln(r_2/r_1)$$

where t = time elapsed after sample injection;
   η = viscosity of spin fluid;
   w = rotational speed of disc;
   d = equivalent particle diameter;
   $\Delta\rho$ = density difference between particle and spin fluid;
   r_1 = initial radius of spin fluid;
   r_2 = radius of sampling depth.

A stroboscope is necessary to detect that particle streaming does not occur, which otherwise invalidates the measurement. The early J–L instruments used Perspex for the construction of the centrifuge disc, which greatly restricted the type of spin fluid that could be used. The most recent model from the Joyce–Loebl stable of disc-centrifuges is the 'Disc Centrifuge 4' produced by J–L Automation Ltd, who have taken over from Joyce–Loebl, see Plate 7.6. Like previous models, it has a solvent-resistant centrifuge disc, and it is claimed by the manufacturers to produce com-

Plate 7.6 — Disc centrifuge (photo courtesy J-L Automation).

parative size distribution curves of particles in the range 0.01 to 60 µm (depending on density) by continuously monitoring the progress of particles past a light beam. True particle size distribution curves can be obtained from the instrument by calibration with standard samples, but careful interpretation of the data is required [41]. There are two ways of measuring particle size distributions with a disc centrifuge. They are the 'line-start' and 'homogeneous-start' methods. A comparison of the line-start approach with the homogeneous-start was studied by Coll and Searles [36]. They concluded that good agreement between modal diameters was obtained by the two methods. However, the homogeneous-start method (where the disc is filled with the, homogeneous, dispersion and changes in concentration are determined with spin-time) gave broader and sometimes distorted size distribution peaks. The only advantage in using the homogeneous-start method was that it enabled materials to be measured where the disperse phase was less dense than the continuous phase.

The difficulty in using a line-start (or two-layer) technique is that it tends to introduce unstable sedimentation (streaming). It consists of laying a thin layer of the dispersion onto the surface of the (clear) spin fluid that has a higher density than the continuous phase of the dispersion, but a lower density than the disperse phase. To overcome the problem of streaming, the 'buffered line-start' was developed by Jones [37]. It consists of eliminating the sharp interface, by establishing a region of gradually increasing density and viscosity between the dispersion and the spin fluid. This is achieved by placing a thin (buffer) layer of the continuous phase of the dispersion onto the surface of the spin fluid, and subjecting the disc to transient acceleration and de-acceleration, to produce the diffuse region before the dispersion itself is injected, as advocated by the makers of the J–L instrument. Makers of a similar disc centrifuge, Brookhaven Instruments Inc., advocate an 'external gradient' method. Here the buffer layer is prepared externally (in a syringe) and then injected, as opposed to creating it *in situ*. In essence, the two approaches are the same.

Data analysis considerations with respect to the measurement of particle size distributions using a disc centrifuge have been examined by Devon *et al.* [38]. The effect of detector slit width error in measurement is discussed by Rudin and co-workers [39], while the effect of the extinction coefficient is examined by Weiner *et al.* [40].

The major difficulty in determining the precise particle size using a disc centrifuge is in ascribing the correct density value for the particles, especially when the particles are very small and there is only a small difference in density between the particles and the spin fluid.

The problem of light-scattering particles where the attenuation of light per unit weight of sample is strongly particle size dependent, was overcome by Hornby and Tunstall [42] using X-rays, as mentioned earlier. They applied this principle to the design of a disc centrifuge which readily enabled them to measure fine particle size, and used this method to compare the particle size distribution of TiO_2 pigment milled for varying times in different grinding mills [33].

7.4.2.1   Elutriation [43]

The corollary of particles moving through a fluid is that a fluid moves through the particles, which is the basis of elutriation and permeametry. Elutriation (the converse of sedimentation) is where a fluid (usually a gas) is forced through a powder bed, and can thus be used to determine particle size distributions. The basis of the method is the ability of fluid velocities to support particles smaller than a given size. It is based on the following equations:

1   for streamline motion $V_m = K_s \left(\dfrac{\rho - \rho_0}{\rho_0} \right) d^2 V^{-1}$

2   *turbulent motion* $V_m = K_t \left(\dfrac{\rho - \rho_0}{\rho_0} \right)^{\frac{1}{2}} d^{\frac{1}{2}}$

where (c.g.s. units)

ρ = density of particle
ρ_0 = density of fluid
d = particle size (diameter)
V = fluid velocity
V_m = maximum particle velocity
K_s and K_t are constants dependent on particle shape.

Terminal-velocity constants for differently shaped particles are [44]:

Shape	K_s (c.g.s. units)	K_t
Sphere	54.5	24.5
Irregular (quartz)	36.0	50.0

The particles can be fractionated according to size by varying the quantity of air passing through the air jets or by altering the size of the elutriation chamber. One can thus obtain a cumulative size distribution by collecting the various fractions in settling jars. Elutriation methods of particle sizing are nowadays not favoured because the process is generally difficult to govern [45].

7.4.2.2   Permeametry

The use of gas permeability through a packed powder bed as proposed by Carman [46] was used by Lea and Nurse [47] to build the earliest permeability apparatus for routine particle sizing. Gooden and Smith [48] produced a variation of the Lea and Nurse apparatus; they incorporated a nomograph to simplify particle size calculations, and the instrument was commercialized in the form of the 'Fischer sub-sieve sizer'.

The apparatus provides good results in the 1 to 50 μm range, but should not be used beyond this range, because the flow mechanism in beds of submicrometre particles is largely molecular and not streamline, which is the basis of the sub-sieve analyser's size calculation.

Rigden [49] showed how to account for molecular flow and particle clusters which can exhibit a 'natural' porosity at low flow rates in which the constituent particles act as individuals. Pechukas and Gage [50] and Carman and Malherbe [51] made modifications to the apparatus, as did Hutto and Davies [52] who were able to show that their results were in good agreement with BET surface area measurements.

7.4.3   Chromatographic methods

Gel permeation chromatography (GPC) has become widely adopted as a means of sizing polymer molecules. The principle of sizing is based on treating a polymer molecule as a 'particle' of a certain size, and molecular weight distributions are calibrated on an equivalence to a polystyrene molecular weight standard which corresponds to a given volume. The method is now widely used to determine molecular weights via molecular volume [53, 54], and it is not proposed to discuss this approach, since the coiled molecular chain constituting the 'particle' is not a particle in the context of this chapter. There are many excellent commercial instruments and many articles and textbooks dealing with the subject. Although GPC instrumentation is expensive, it need not necessarily be so for the determination of molecular weight distribution, since there are inexpensive thin-layer chromatography GPC adaptations for molecular weight determinations [55, 56].

The extension of GPC to larger, solid particles is known as 'size exclusion chromatography' or 'hydrodynamic chromatography'. Although the two types of chromatography are different in principle they can occur simultaneously in a chromatographic column. The two types may be schematically differentiated as in Fig. 7.15.

In hydrodynamic chromatography (HDC), because larger particles project further from the wall of a capillary they will be subjected to faster flow towards the centre of the channel than smaller particles, whose centres can approach closer to the stiller flow at the channel wall. In size exclusion chromatography, a small particle may find refuge from the velocity gradient in a pore which is unavailable to a larger particle. In both instances the larger particles pass through the column faster than the small particles, and are sized according to their retention time as a parameter of particle size, as in GPC analysis. Although there are many papers dealing with particle size determination by chromatographic means [57–61], this sizing approach is as yet not of general, non-research application to pigments. Micromeritics Corporation, however, has marketed an instrument suitable for routine latex analysis in the form of a 'Flow Sizer HDC5600' which provides a full particle size distribution analysis of latex emulsions ranging from 30 to 1500 nm on this principle, with a claimed resolution equal to 5% of particle size using a column packed with a cation ion-exchange resin [62], thereby restricting analyses to anionic latexes. The column is stabilized by continuous circulation of eluent (initially eluent with latex), and once stabilized the instrument is quick and easy to use. Although the fractionating principle for sizing is simple, the design of the instrument is complex, requiring a fairly powerful small computer to deconvolute and process the detector signal to give the particle size distribution results. The 'HDC 5600 Flowsizer' has,

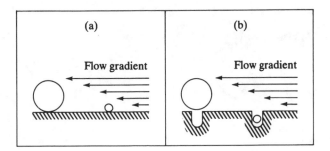

Fig. 7.15 — Schematic representation of (a) hydrodynamic and (b) size exclusion chromatography.

however, been taken off the market because of much adverse criticism, since frequently not all the sample injected was eluted from the column. The makers say they withdrew the HCD 5600 because of column stability problems, stating; 'In actual use the columns deteriorated too rapidly, resulting in the need for frequent recalibration. Maintenance costs just did not justify continuing the sale of the product.'

The analytical separation of submicrometre particles by capillary hydrodynamic fractionation (CHDF) was first documented by DosRamos and Silebi in 1989 [63]. However, size separation based on laminar flow was proposed by DiMarzio and Guttman in 1970 as the mechanism in size exclusion chromatography. According to them the mechanism of separation by flow through conduits is due to two effects:

- the laminar velocity profile of the liquid in the conduit;
- the steric exclusion of the particles from the slower velocity streams at the wall of the capillary, see Fig. 7.15.

A commercially available CHDF instrument is available in the form of the 'Matec CHDF' (Plate 7.7). It has a capillary tube about 10 m long, and an internal diameter of 5 μm, coiled in a thermostatted chamber. While the instrument is not an 'absolute instrument', i.e. it requires calibration because of 'inertial spreading', it is simple to operate, has a fast throughput with a measurement time of about 7–10 min. It has stable calibration, and gives good reproducibility. It has simple controlling factors — flow rate and ionic strength. DosRamos and Silebi claim that CHDF has significantly better resolution than SFFF [64] at normal operating con-

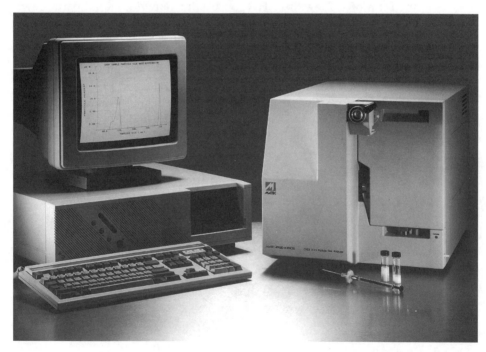

Plate 7.7 — MATEC CHDF-1100 submicrometre particle sizer (photo courtesy MATEC Applied Sciences).

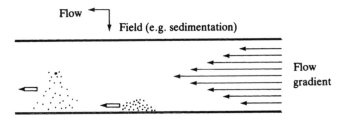

Fig. 7.16 — Principle of field flow fractionation.

ditions. They say that because of its speed, accuracy, and relatively low cost, CHDF is ideally suited as a routine characterization tool for small particles, i.e. less than 1 μm in size. At present the instrument can only handle aqueous dispersions, it is independent of particle density, and, in principle, it is not limited by the nature of the media.

Field flow fractionation [65] is a method for particle size separation, and hence sizing, based on applying a concentration-disturbing lateral field to a suspension flowing unidirectionally in a narrow tube as shown in Fig. 7.16. The applied field may be a thermal gradient [66] or an electrostatic or gravitational field (sedimentation field flow fractionation). Kirkland *et al.* [67, 68] describe a technique using an apparatus with a changing gravitational field developed by them called TDE–SFFF (time-delayed exponential sedimentation field flow fractionation) which has enabled them to obtain particle size distribution analysis in the range <0.01–1.0 μm of various pigments such as TiO_2, carbon black, and phthalocyanine blue as well as various latexes. Relative to constant-field SFFF, TDE–SFFF has the advantage of faster analysis time and enhances detection sensitivity while maintaining resolution. Retention time correlates with the logarithm of particle size (the largest particles have the longest retention time) provided that the particles have the same density. In the case of a sample containing several components with different densities it is not possible to attach an accurate particle size scale to retention time. Nevertheless, qualitative variations in the constituents enable the 'fingerprinting' of the sample.

7.4.4 Aperture methods

7.4.4.1 *Sieving*
Sieving is an old, well-established method of grading particles according to size; however, it is not as simple a method as it might appear. Although micro-sieves with an aperture width down to 5 μm have been developed [69], the method is generally used for particle fractionation from about 20 μm to 125 mm particles using standard woven wire test sieves.

The weight of particles collected between two sieves at the end of a sieving process can be readily determined. However, the nominal aperture width cannot be taken to represent the 'cut size' of the sieve. Sieving is a non-ideal classification process, and cut size has to be measured independently by calibration, using a known material for accurate work.

Sieves used for sieve analysis are standardized. Their nominal aperture widths follow a progression series. They may be woven wire screens as in BS410: 1969, or electroformed micromesh sieves (ASTM E161–607). The aperture width of a standard sieve itself may vary, and it should therefore be characterized by its aperture width distribution.

The particle size classification of a specific sieve depends on a number of factors which are independent of the mesh, e.g. solids loading, shaking action, tendency for solids to agglomerate, amount of near-mesh particles, etc. Leschowski [70] suggests that it would be better to plot sieving weight results against the median aperture width of a sieve rather than its mesh size. A sieve should be characterized by its median aperature, its confidence interval, and the standard deviation as obtained by calibration.

Great care has to be exercised in cleaning fine sieves; one should avoid brushing a fine sieve, e.g. below 200 µm, as this can readily damage the mesh, and it should be cleaned in an ultrasonic bath. Extremely fine micromesh sieves demand special wet sieving techniques as described by Daeschner [71] and Crawley [72].

7.4.4.2   Principle of sieving

When sieving, a powder is separated into two fractions, one that passes through the apertures in the mesh and a residue that is retained. The process is complicated in that non-spherical particles will only pass through the holes in the sieve when presented in a favourable orientation.

The sieving process may be divided into:

- the elimination of particles considerably smaller than the sieve apertures, which occurs fairly rapidly;
- the separation of 'near-mesh' particles, which is a gradual process, never reaching final completion.

The general approach to sieving is to eliminate the fines, and define an 'end point' to the test, when the elimination of 'near-mesh' particles has attained a practical limit. Whether sieving is performed in the dry or wet state depends on the characteristics of the material. Very fine particles are more quickly eliminated by wet sieving; it may also reduce the breakage of friable materials, but it may give results different from dry sieving.

There are two alternative recommended methods to define the 'end point' which marks the completion of 'near-mesh' particles:

1   To sieve until the rate at which particles passing through the mesh is reduced to a specified weight or percentage weight per minute.
2   To sieve for a specified time.

Method (1) is the better approach, but to simplify the procedure for routine tests with fairly consistent materials, method (2) may be adopted, provided that there is evidence to show that the time selected is adequate.

Both the dry and wet methods of sieving, along with the construction of a simple mechanical vibrator for sieves, are described in BS1796:1952 (now superseded by BS1796:1976, but without the design of the simple vibrator).

For a review of the sieving process, along with a discussion of the theory and practice of sieving, the reader is referred to Whitby [73] and Daeschner et al. [74].

Arietti et al. [75] have extended the sieving approach to particle size distribution analysis in the submicrometre size range, using a micro-pore filtration approach which appears to give good agreement with electron microscope sizing.

7.4.4.3   Electrical resistivity (electric zone — the Coulter principle)

The Coulter principle of particle sizing enables the measurement of particle sizes from 1 μm to a few hundred μm. With great care it is claimed to be able to measure down to 0.4 μm.

It was originally used to count blood cells [76–78]. Modifications were suggested by Kubitschek [79], which enabled the principle to be applied to the measurement of cell volume as well as number. The principle of the system is that a constant electric current (DC) is established between two electrodes placed in two separate containers (see Fig. 7.17) and linked by a glass sensing zone (orifice). A mercury siphon is made temporarily unbalanced by the application of a vacuum. Closing the tap (T) allows the suspension to be drawn through the aperture, as the mercury returns to its equilibrium position. Electrical contacts triggered by the mercury allow a precise, reproducible volume of suspension to pass through the orifice, altering the resistance by displacing their own volume of electrolyte, creating voltage pulses essentially proportional to the volume of the particle. These particle-generated pulses are amplified and counted in a 'discriminating' circuit. By systematically changing the sensitivity of the pulse detection (or counting) a cumulative count of particles (pulses) larger than a given size can be made, as the predetermined unit volume of suspension is transferred from one chamber to the other.

Calibration of the instrument is carried out for the specific electrolyte solution being used, usually with standard monodispersed latexes of known diameter supplied by the manufacturer.

The calibration constant (k) is determined from

$$d_c = kt_c^{\frac{1}{3}}$$

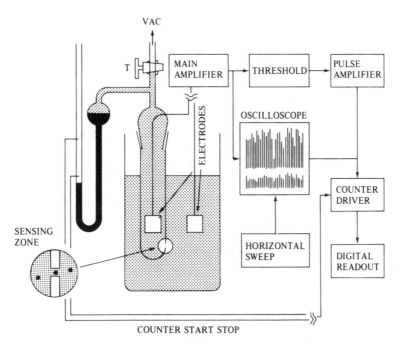

Fig. 7.17 — Schematic representation of the Coulter principle.

where d_c is the known number-volume diameter; t_c is the threshold level for the average of the pulses generated with the known particle size suspension.

Having thus calibrated the instrument with a standard latex, the constant (k) should hold for all other threshold sizes using that specific electrolyte.

7.4.4.4 Coincidence correction

The measurement of pulses is based on the assumption that each pulse is caused by a single particle. However, two particles may enter the sensing zone together, giving rise to a disproportionate signal. By making the measurements within certain counting rates, the effects of coincidence can be minimized from a statistical point of view, and they may be estimated experimentally by counting increasingly dilute suspension, as described in the Coulter *Users' manual*.

In conclusion, the Coulter principle for particle size distribution has been found to be a very useful, reliable, and an easy method to operate, as testified by over a thousand references on the subject [80], for the determination of particle sizes from about 0.6 μm upwards, and the reader is referred to Allen's review on the subject for further reading [81]. Micromeritics [82] manufactures a similar instrument to the Coulter Counter, making use of the electric sensing zone technique. The minimum particle size that can be measured by this method (~2% orifice diameter) is restricted by the inability to produce a suitable orifice, i.e. sensing zone, smaller than about 12–15 μm in diameter, not by the electronics of the instrument.

7.4.5 Optical (light-scattering) methods

A good review of light-scattering theory, as applied to particle size, is given by Kerker [83] and common methods used in size measurement are discussed by Collins *et al.* [84].

The source of scattered light is the re-radiation of light due to oscillating dipoles in polarizable particles induced by the oscillating electric field of a beam of light. The polarizability per unit volume at any position and time can be considered as the sum of a constant portion (giving rise to refraction) and a fluctuating part which produces scattering.

If we limit our considerations to spherical, non-adsorbing, and non-interacting particles, the light-scattering behaviour is mainly determined by two factors: (i) the ratio of particle size to the wavelength of the incident light beam in the medium (d/λ), and (ii) the relative refractive index, $m = n_1/n_2$ where n_1 and n_2 are the refractive indexes of the particles and their suspending medium respectively. In practice a dilute dispersion is illuminated by a narrow, intense beam of monochromatic light, and the intensity of the scattered light is measured at some angle Θ from the incident beam.

The three most common approaches to the measurement of particle size are:

1   By turbidity (transmission) where Θ is fixed at 180° and the light intensity is measured.
2   The light intensity is measured at some fixed angle (usually 90°).
3   The intensity of the scattered light is measured as a function of the angle.

The measurements usually have to be carried out at infinite dilution so that Rayleigh (when $d/\lambda \ll 1$), Rayleight–Debye (when $(n_1 - n_2)\, d/\lambda \ll 1$) or Lorenz–Mie (when $(n_1 - n_2)\, d/\lambda \lesssim 1$) theories can be applied, which are for single scattering centres.

For a Rayleigh scatterer the intensity of the scattered light I is angle-independent and given by:

$$I = \frac{16\pi^4 d^6}{x^2 \lambda^4} \left(\frac{n_1^2 - n_2^2}{n_1^2 + n_2^2} \right)^2$$

where x is the distance between the sample and the detector.

Scattering intensity is angle-dependent with a Rayleigh–Debye scatterer, and even more complex angle-dependency occurs with a Lorenz–Mie scatterer. Concentrated solutions produce multiple scattering, and the multiple scattering light theories become indeterminate in practice except on a semi-empirical basis. The advantages and disadvantages of the transmission, dissymmetry, maximum–minimum techniques, forward angle ratio, and polarization ratio, are all discussed by Collins *et al.* [84]. An extension of the turbidimetric approach has enabled the particle size distribution to be determined as well [85]. In addition to these methods of particle sizing is a method of variable wavelength patented by Tioxide International Limited [86] which entails the measurement of transmitted light at three different wavelengths, using a spectrophotometer. This enables the measurement of a mean particle size and its standard deviation which is useful in determining the efficiency of grinding. Nobbs [87] recently described a method using a thin-film technique which is a modification of conventional light-scattering measurements, but enables the measurement of apparent particle size to be made at high concentrations. Also of interest is a refractive index measurement of a dispersion method to obtain a particle size diameter [88].

7.4.5.1 *Fraunhofer diffraction*

As particle size increases and approaches the wavelength of the light source λ, the amount of light scattered at forward angles increases and becomes very much greater than in other directions. When the particle size d is much greater than λ then Fraunhofer diffraction (FD) theory describes the forward-scattering properties of a particle in a beam of light which can be considered as a limiting case of Lorenz–Mie theory. FD theory shows that the intensity of the scattering angle (diffraction pattern) is proportional to the particle size, and that the size of the scattering angle is inversely proportional to particle size, as shown in Fig. 7.18. A Fourier transform lens — a lens positioned between the particle field and the detector, such that the detector lies in the focal plane of the lens — is used.

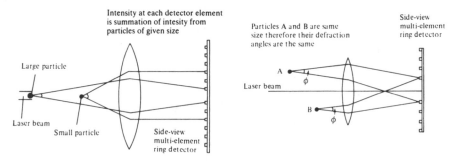

Fig. 7.18 — Diagrammatic representation of sizing by the FD principle.

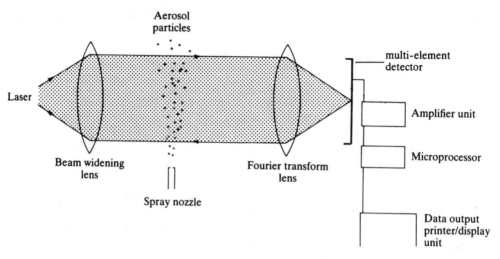

Fig. 7.19 — Schematic representation of typical FD instrument for measuring aerosol particle sizes.

A typical FD instrument used for particle size analysis is shown schematically in Fig. 7.19.

Because the geometry and positioning of the lens in the instrument are so arranged as to meet the requirements of FD theory [89], the diffraction pattern of a moving particle will be stationary when the detector is in the focal plane of the lens.

The detector, consisting of an array of light sensors, analyses the light energy distribution (that of a low-powered laser beam) over a finite area, and a microprocessor computes the particle size distribution.

Instruments of this type, such as Particle Measuring Systems Inc. PDPS 11–C which can be used with different probes, are suitable for accurately measuring particle size distributions from a few micrometres upwards. They are useful in obtaining the particle sizes of aerosols such as those generated when paint is applied by spray-gun applicators. This type of instrument can also be used to determine the particle size distribution in liquids as with the Malvern 3600E Laser Particle Sizer (Plate 7.8). Leeds & Northrup have similar instruments in their Microtrac range. Royco Instruments Division market HIAC instruments which can also measure the particle sizes of Aerosols using what they term a 'shadowing technique'.

For a correlation of particle size analysis using HIAC, Coulter Counter, Sedigraph, Quantimet 720, and Microtrac instruments the reader is referred to Johnston and Swanson [90].

7.4.5.2   *Photon correlation spectroscopy*

A different approach to the use of light scattering for particle size determination is 'photon correlation spectroscopy' (PCS), also known as quasi elastic light scattering or dynamic light scattering.

It was the advent of lasers in the early 1960s — which produce an intense, coherent monochromatic light — that enabled the use of time-correlation functions to be applied to the measurement of particle size. Time correlation functions are a way

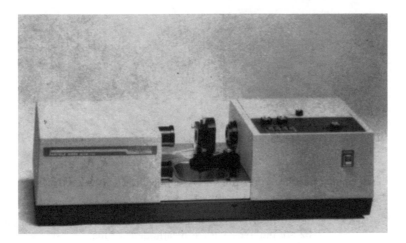

Plate 7.8 – The Malvern 3600E FD sizer (photo by courtesy of Malvern Instruments).

of describing by means of statistical mechanics the fluctuations of some property (in this case the number of photons emitted) on a time basis. This type of analysis requires a coherent, monochromatic incident radiation so that phase relationships in a beam are maintained, and examines the fluctuation of the radiation due to the random motion of the light scattering centres in a small volume which gives information as to the diffusion coefficient of the light-scattering centres.

When we examine scattered light intensity on a time basis, it will be found to fluctuate about an average value, if the light-scattering particles are undergoing random (Brownian) motion. The scattered electric field is a function of particle position and is therefore constantly changing. The intensity (proportional to the square of the electric field) is also fluctuating with time. By measuring these fluctuations it is possible to determine how these fluctuations decay over longer time-averaged periods, using auto-correlation theory to determine a diffusion coefficient for the particle. This in turn can be related to a particle diameter if certain assumptions as to particle shape and the viscosity of the medium are known, through the Stokes–Einstein equation.

A typical experimental set-up is shown in Fig. 7.20. The intensity of scattered light is an average effect and is a function of the individual fluctuations, just as the pressure of a gas is the average of the individual bombardment of gas molecules on the container wall.

For dilute suspensions with particles smaller than λ (the wavelength of the light beam)

$$\langle I_s(q) \rangle = KNM^2\, P(\Theta)\, B(C) \tag{7.4}$$

where $\langle I_s(q) \rangle$ is the time-averaged scattered intensity from the particles;

q is the wave vector amplitude of the scattering fluctuation;
K is an optical constant;
M is the mass of the particle;
N is the number of particles contributing to the scattering;
P is the particle form factor;
Θ is the scattering angle;

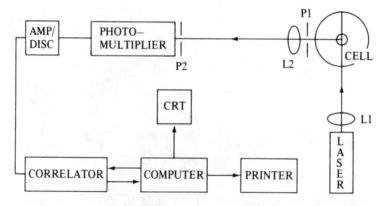

Fig. 7.20 — Block diagram of the BI–90 particle sizer. L1, L2 are focusing leneses and P1, P2 are pinholes; CRT is a cathode ray tube.

B is the concentration factor; and
C is the particle concentration.

The fluctuations due to thermal excitation may be resolved into various frequencies, and at any angle the scattering, due to a particular fluctuation, may be stated as

$$q = \frac{4\pi n}{\lambda} \sin \Theta/2$$

where n = refractive index.

The mass of a spherical particle is proportional to d^3, hence the M^2 term in equation (7.4) leads to a d^6 factor in scattering. The particle form factor, $P(\Theta)$, is known for simple shapes and for size less than λ. However, in the limit as Θ goes to zero, the particle form factors tend to unity. Hence angular PCS measurements with extrapolation to zero angle are required for polydisperse samples with large particles. When the particles are macromolecules and in true solution, then equation (7.4) can be used to determine molecular weight.

The PCS technique has been found to be very reliable and equal to electron microscopy for determining the size of monodisperse particles [91]. However, for polydisperse systems the method is much more problematical since distribution information is obtained from an analysis of the sums of exponentials contributing to the measured auto-correlation function. There are various mathematical approaches to this problem. The most frequently adopted approach is that of 'cumulant analysis', which gives two size parameters: the average diameter, which is related to the z-average diameter, and a polydispersity factor which is a function of the variance of the z-average diameter [92]. Unless measurements are extrapolated to zero angle and concentration, the apparent size is angle- and concentration-dependent.

PCS measurements, except for monodisperse systems, are relatively insensitive to size distribution. Separation into two peaks appears to be possible only if the size ratio is more than about 2:1, although improvements in this respect are continually being made.

Frequently the instruments display histograms which give size classification along the abscissa; however, the manufacturers are usually careful not to name the ordinate as percent size frequency as might be expected (which it is not), but use terms such as 'percent relative light scatter' or 'particle size distribution (arbitrary units)'.

The particle size distribution is dependent on the value of the second cumulant which is very sensitive to the presence of dust particles. In some instruments the effect of dust particles has been reduced by a subtractive process using a delayed-time baseline approach which effectively ignores light scatters larger than a predetermined size.

In PCS size determinations there is an underlying assumption (as with all light scattering) that all particles are of a homogeneous composition, even though their refractive index need not be known (i.e. only that of the continuous phase is required). This means in practice that one cannot measure the average size of a mixture of two different latex particles of different polymers, e.g. a mixture of a polystyrene and an acrylic latex.

There are many commercial instruments on the market for measuring particle size by PCS using the fixed angle (usually 90°) approach, such as Coulter Electronics Nanosizer [93] or its N4 series of instruments which can measure at a number of different angles. Brookhaven Instruments Corp. (USA) produce a B1–90 particle sizer which has an electronic 'dust filter' and is a fixed-angle instrument (Plate 7.9). It has now been superseded by the Zetaplus, which can measure both particle size and the zeta potential of particles. Malvern Instruments produces a 'System 4600' variable-angle Photon Correlation Particle Analyser and a fixed-angle instrument called the 'Autosizer', and Nicomp Instruments (part of Hiac/Royco Instruments)

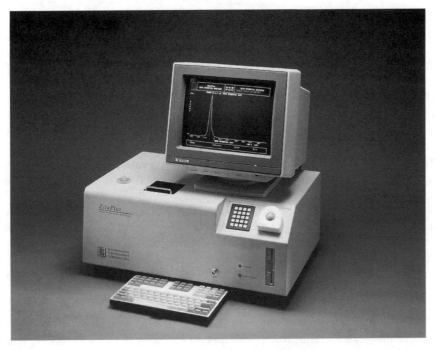

Plate 7.9 — Brookhaven's ZetaPlus PCS particle sizer which also can measure the zeta potential (courtesy Brookhaven Instruments Inc.).

also produces a number of models for particle-sizing by PCS, such as the Laser Particle Sizer Model 200D. All these instruments give an 'average' particle diameter and an indication of the particle size distribution. New PCS instruments using enabling-fibre optic techniques are capable of handling moderately concentrated dispersions, up to 10% disperse phase volume (although higher disperse phase volumes are claimed), and are now available such as the Malvern's AutoSizer Hi-C.

7.4.5.3   Ultramicroscopy

A classic method [94] to determine the size of colloidal particles beyond the resolution of an optical microscope was to employ dark-field vision (ultramicroscope) to count the number of particles per unit volume, and by knowing their mass per unit volume their average size was readily estimated.

Derjaguin *et al.* [95] improved on the method by introducing a flow technique to simplify the counting procedure. Since then other workers have extended the procedure [96].

Walsh *et al.* [97] built a flow ultramicroscope capable of sizing particles between 0.1 and 2.0 μm. By detecting and measuring the intensity of laser light pulses using a photo multiplier, they have been able to extend the technique to particle size distribution measurements as well. Their detector set at 20° was suitable for materials with lowish refractive index such as clays, polymer, or oil particles. However, it was not suitable for sizing materials with a high refractive index such as TiO_2.

7.4.6   Surface area methods

7.4.6.1   Gas adsorption

One can determine the average particle size of a solid from a knowledge of its surface area per unit weight and its density; for if the particles are considered to be uniform spheres, then the ratio of volume to area is $r/3$, where r is the radius of the equivalent sphere.

Gas adsorption [98] is the most frequently used method for determining the surface area of a solid. In principle, gas adsorption techniques can be applied to any gas–solid system, but in practice the method is restricted and is limited to the type of adsorption encountered.

To determine surface area it is necessary to have a suitable value for the area of the adsorbing molecule and a knowledge of the number of molecules for monolayer coverage. Equations to determine these values depend upon the nature of the forces between the gas and the solid. If they are purely non-specific, i.e. physical adsorption, then monolayer coverage can be calculated by using the semi-empirical BET equation. If, on the other hand, chemical adsorption occurs, monolayer coverage can be obtained, but the area occupied by the adsorbing molecule with depend on the lattice spacing of the atoms in the solid. Many gas–solid systems are intermediate in adsorption type and cannot be readily defined, e.g. active carbons or clays.

Classical equipment for surface area measurement used vacuum glassware. Modern commercial equipment now available is generally made of metal and uses electronic gauges instead of traditional mercury manometers and Macleod gauges.

Gas adsorption surface areas are determined from the full adsorption isotherm plot with automatic instruments such as Carlo Erba's Sorptomatic instrument or

Micromeritics Digisorb 2600. While automation may reduce other time factors it does not alter the time required to reach equilibrium conditions. To speed up measurements, instruments have been designed to make certain assumptions concerning the nature of the BET equation allowing surface areas to be determined from a single experimental point, such as the Micromeritics 2200 surface area analyser which uses a static method, while Perkin–Elmer's Sorptometer uses a dynamic adsorption (flow) method. Both instruments enable the surface area to be determined in less than an hour, once the sample has been 'conditioned'.

'Conditioning' or sample pretreatment can have a large effect on the measured surface area, and care has to be taken to ensure that the nature of the surface to be measured is not altered by the pretreatment.

One of the advantages of surface area measurement by gas adsorption is that it can indicate the presence of porosity. At liquid nitrogen temperatures, nitrogen condenses in the pores, according to the Kelvin equation; and the shape of the isotherm [99] reflects pore sizes between 1.5 and 30 nm in the solid, and it is most successfully analysed in the type IV isotherm. (Pore sizes in the range 8 nm upwards are usually determined by using mercury porosimetry. Typical commercial porosimeters can operate automatically down to 7.5 nm corresponding to pressures of about 2000 atmospheres.)

7.4.6.2   Solute adsorption

An alternative way of measuring surface areas is to use the adsorption of a solute from solution such as a fatty acid [100, 101]. While the cross-sectional area of a fatty acid is 20.4 Å^2 per molecule in a vertical orientation, it will not necessarily adopt this form of adsorption on all surfaces and in all solvents. It is therefore necessary to establish the nature of the adsorption and the molecules' effective area before using it for surface area determination (unless the measurements are relative ones on the same substrate using the same solvent).

Dyestuff adsorption has also been used o determine surface areas since its concentration is readily and accurately determined colorimetrically [102]. Again, care must be used in assigning an area for molecule occupancy as cautioned by Kipling and Wilson [103] who assigned an area of 102–108 Å^2 per molecule for methylene blue. Linge et al. [104, 105], however, found that methylene blue tended to absorb at between 69.6 and 76 Å^2 per molecule corresponding approximately to dimer formation.

In measuring surface area by dye adsorption Padday [106] suggests that one should precalibrate the method against some more reliable method such as gas adsorption.

Gregg and Sing [107] suggest that in using a dyestuff approach to the measurement of surface area it must be limited to cases where:

- the dyestuff is sufficiently soluble and a clear plateau is obtained in the isotherm;
- the orientation of the molecule must be known; and
- the number of molecular layers must be known.

Nevertheless, when applicable, surface areas by dyestuff adsorption can be readily determined even without the use of a spectrophotometer to determine dyestuff concentration in solution, which is the usual approach.

For example, to determine the concentration of dyestuff in a mother liquor after adsorption, it is only necessary to make a dye solution weaker in colour than the unknown solution but of known concentration, e.g. prepared by dilution of a standard solution. It is then placed in a test-tube and used as a reference colour; by adding a known volume of the unknown colour solution to a second similar tube and diluting the unknown sample with clear solvent (for the dyestuff) until the two tubes appear identical in colour, and recording how much solvent was required to bring the unknown concentration to the colour of the reference, then by simple arithmetic the concentration of the unknown sample can be determined. This makes use of the human eye as a powerful colour comparator, which it is.

7.4.6.3 'Surface' methods

Calorimetry to determine the heats of immersion also offers a means of measuring surface areas, as for example the Harkins and Jura method [108]. Heats of immersion have been advocated for determination of the specific surface of solids. However, the advantage of a single experimental determination is frequently outweighed by the small quantity of heat evolved and the care and instrumentation required to make a measurement [109].

Surface balance techniques have been used as methods for measuring the particle sizes of aluminium flake [110]. The method can be traced back to Edwards and Mason in 1934 [111].

The flakes are spread on a water surface, and successive expansions and contractions on the Langmuir balance-like equipment are carried out until a constant, compact, but planar area is obtained. From a knowledge of the sample weight and the two-dimensional area occupied, it is possible to calculate the average flat diameter of the particles as well as their thickness. Capes and Coleman [112] have used this approach to determine the size of mica flakes in the size range of 2–30 μm as well as sand particles in the range of 50–200 μm. Submicrometre latex particles have also been spread at the air–water interface [113] and the oil–water interface [114] where similar calculations can be made.

7.5   The best method?

The question as to what is the best method for particle sizing cannot be answered unless the question is qualified with a reason for the sizing. Even then, two independent methods are better than one (carried out twice), since in conjunction they will be more informative. For example, if the average TiO_2 pigment particle size is determined from specific surface areas using nitrogen adsorption as well as by sedimentation (or electron microscopy), it will be found that the 'average' values do not necessarily correlate even when all corrections are made. This is due to the surface coating on the pigment, and hence more can be learnt about the nature of the particle, and there is less possibility of being misled from the results of one type of measurement alone.

It is also worthwhile examining particle size distributions from a number and mass point of view. Figure 7.21, based on HDC measurement by count, gives quite a different impression of size distribution from that of Fig. 7.22, which is the same distribution on a mass frequency basis.

Figure 7.23 shows the results of particle size distribution measurements made, using the same sample of latex and plotted on a mass–frequency basis, using elec-

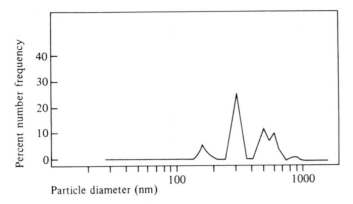

Fig. 7.21 — Particle size distribution by count.

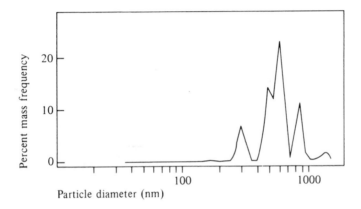

Fig. 7.22 — Particle size distribution by mass.

tron microscopy (over 4000 particles counted); disc centrifuge; four different PCS instruments; and hydrodynamic chromatography.

It has not been possible to describe in a single chapter all the 400 or so methods for particle sizing [115]. The author's classification of methods in a 'generic' form may not comply with that of another. However, by using the format chosen it has been possible to give some idea of the diversity of basic techniques available, while not citing all the variations.

The methods described here may be divided into two categories. The first relies on the most recent instrumentation where nearly all the calculations are carried out by microprocessors. The instruments are generally expensive to buy or else require sophisticated instrument-making facilities. In this instance only the underlying principle has been described along with examples of the instruments.

The second category, which enables measurements to be made without resort to a large expenditure of money, has in general been described in more detail to enable one to measure particle sizes on a do-it-yourself principle with readily available materials. In general these tend to be more classical methods, and are certainly no less valid than modern automated techniques.

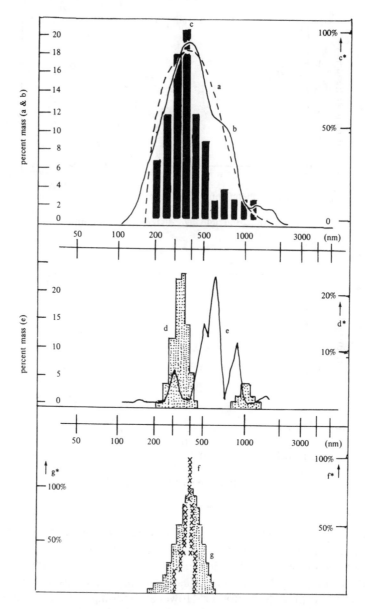

* The y-axis for the four PCS graphs is linear percent, and purely arbitrary.
The Nanosizer N4 has a maximum scale of 40% (SDP: differential intensity),
whereas the rest have their peaks normalized to 100% ('relative mass' – the
Autosizer does not label the ordinate). Only the Nanosizer correctly identifies
the ordinate as 'Size Differential Processor (results): Differential Intensity,
which gives an indication of relative mass.

Fig. 7.23 — Comparison of size distribution measurements, on the same sample of latex,
using: (a) Joyce-Lobel disc centrifuge; (b) electron microscopy (~4000 particles counted);
(c) PCS, Brookhaven BI-90; (d) PCS, Nanosizer N4; (e) HDC, Micromeritics HDC 5600;
(f) PCS, Nicomp 200; (g) PCS, Malvern Autosizer.

References

[1] KAYE B H, *Chem Eng* **73** 239 (1966).
[2] DELLY J, *The Particle Analyst: Reticles and Graticules*, Vol 1, No 77, Ann Arbor Sci Publ (1965).
[3] ASTM, E20–68 (1974) section 14.02.
[4] SICHEL H, SILVERMAN L, BILLINGS C, & FIRST M, *Particle Size Analysis in Industrial Hygiene*, p. 248, Academic Press (1971).
[5] HOEL P G, *Introduction to Mathematics and Statistics* 3rd edn, J Wiley & Sons, (1962).
[6] HATCH T & CHOATE S P, *J Franklin Inst* **207** 369–387 (1929).
[7] JELINEK Z K, in *Particle Size Analysis*, p. 14, Ellis Horwood (1974).
[8] IRANI R R & CALLISS C F, *Particle Size Measurement, Interpretation & Application*, J Wiley & Sons (1963).
[9] JACKSON M R, IGLARSH H, & SALKOWSKI M, *Powder Technol* **3** 317–322 (1969/70).
[10] BS 3406:Part 1:1986.
[11] HOY K L and PETERSON R H, *J Coat Technol* **64** (806) 62 (1992).
[12] BS 3900:Part A1:1992, ISO 1512:1991.
[13] KAYE B H, '*Direct Characterisation of Fine Particles*' Vol 61, Ch 2 Chemical Analysis series, Wiley-Interscience, (1981).
[14] ALLEN T, *Particle Size Measurement*, 3rd edn, Ch 1, Chapman & Hall (1981).
[15] BART H G & SUN S-T, *Anal Chem* **65** 55R–66R (1993).
[16] SWENSON R A & ATTLE J R, Counting, Measuring and Classifying with Image Analysis *Amer Lab* **11** 4 (1978).
[17] ATTLE J R, ONEG D, & SWENSON R A, *International Lab* 35–48 Oct 1980.
[18] Quantimet 920, Cambridge Instruments Ltd, Rustat Rd, Cambridge CB1 3QH, UK.
[19] Magiscan 2, Joyce-Loebl, Gateshead NE11 0QW, UK.
[20] MENOLD R, LUTTGE B, & KAISER W, *Adv Colloid Interface Sci* **5** 281–335 (1976).
[21] MENOLD R, LUTTGE B, & KAISER W, *Adv Colloid Interface Sci* **5** 306 (1976).
[22] CRUZ-ORIVE L M, *J Microscopy* **131** (3) 265–290 (1983); also Schwartz H A & Saltykov (1958), see Saltykov *Stereometric Metalography*, 2nd edn, Metallurgizdat (1958).
[23] Andreassen Pipette, *Technico Andreassen Pipette Apparatus PBW-200-W* technical brochure ex Gallenkamp.
[24] BS3406: Part 2: 1963.
[25] JELINEK Z K, in Particle Size Analysis p. 77, Ellis Horwood (1974).
[26] SVAROVSKY L & ALLEN C J, *Particle Size Analysis*, pp. 442–450, Ed M J Groves, Heyden & Son (1978).
[27] ASTM D3360–06.02.
[28] BERG S, ASTM Special Tech Publ No 234 p. 143–171 (1958).
[29] JARRETT B A & HEYWOOD H, in 'A Comparison of Methods for Particle Size Analysis' *British J Applied Sci Suppl* **3** 21 (1954).
[30] ODEN S, *Alexanders Colloid Chemistry* Vol 1, p. 877–882, Chem Catalogue Co (1926).
[31] SIEBERT P C, in *Particle Size Analysis*, p. 52, Eds Stockham J D, & Fochtman C G, Ann Arbor Sci (1977).
[32] SCOTT K J & MUMFORD D, *Powder Technol* **5** 321–328 (1971/2).
[33] MURLEY R D, *XII Fatipec Congress* 377–383 (1974).
[34] SVEDBURG T, *Ind Eng Chem (Anal ed)* **10** 113–127 (1938).
[35] Joyce-Loebl Centrifuge, Vickers Company, Adv-Information brochure 8:83. Now, DCF4, Centrifuge, JL Automation Ltd, Unit 2C, Hylton Park, Sunderland SR5 3NX, UK.
[36] COLL H & SEARLES C G, *J Colloid Interface Sci* **115** (1) 121–129, (1987).
[37] JONES M H, *Proc Soc Anal Chem* **3**, 116 (1966).
[38] DEVON M J, PROVDER T, & RUDIN A, *Particle Size Distribution II, Assessment & Characterisation*, ACS Symposium series 472, pp. 134–153, Ed T Provder (1991).
[39] DEVON M J, MEYER E, PROVDER T, RUDIN A, WEINER B, *Particle Size Distribution II, Assessment & Characterisation*, ACS Symposium series 472, Ch 10, pp. 154–168, Ed T Provder (1991).
[40] WEINER B B, FAIRHURST D, & TSCHARNUTER W W, *Particle Size Distribution II, Assessment & Characterisation*, ACS Symposium series 472, Ch 12 , pp. 184–195, Ed T Provder (1991).
[41] OPPENHEIMER L E, *J Colloid Interface Sci* **92** (2) 350–357 (1983).
[42] HORNBY M R & TUNSTALL D F, BP 1,387,442.
[43] SILVERMAN L, BILLINGS C E, & FIRST M W, *Particle Size Analysis in Industrial Hygiene*, pp. 137–146, Academic Press (1971).
[44] DALLAVALLE J H, *Micromeritics*, 2nd edn, p. 20, Isaac Pitman & Son Ltd (1948).
[45] HERDAN G, *Small Particle Statistics*, 2nd revised edn p. 360, Butterworth (1960).
[46] CARMAN P C, *J Soc Chem Ind* **57** 225–239 (1938).
[47] LEA F M & NURSE R W, *J Soc Chem Ind London* **58** 277 (1939).
[48] GOODEN E C & SMITH C M, *Ind Eng Chem* (Anal ed) **12** 497 (1940).
[49] RIGDEN P J, *J Soc Chem Ind* **66** 130–136 (1947).

[50] PECHUKAS A & GAGE F W, *Ind Eng Chem* (Anal ed) **18** 370–373 (1946); also *J App Chem* (*London*) **1** 105 (1951).

[51] CARMEN P C & MALHERBE P LE R, *J Soc Chem Ind* **69** 139–193 (1950) and *J Appl Chem* **1** 105–108 (1951).

[52] HUTTO F B & DAVIES D W, *Off Dig* **31** 429 (1959).

[53] ALTGELT K H & SEGAL L (Eds) *Gel Permeation Chromatography*, Marcel Dekker Inc (1971).

[54] YAU W W, KIRKLAND J J, & BLY D D, *Modern Size Exclusion Liquid Chromatography*, J Wiley & Sons (1979).

[55] OTOCKA E P, HELLMAN M Y, & MUGLIA P M, *Macromolecules* **5** 1273 (1972).

[56] BELENKII B G & GANKINA E S, *J Chromatog* **141** 13–90 (1977).

[57] SMALL H, SAUNDERS F L & SOLC J, *Adv Colloid Interface Sci* **6** 237–266 (1976).

[58] SMALL H, *Anal Chem* **54** (8) 892A–8A (1982).

[59] MCGOWAN G R, *J Colloid Interface Sci* **89** (1) 94–106 (1982); also Nagy D J, Silebi C A, & McHugh A J, *J Appl Polym Sci* **26** (5) 1555–1578 (1981).

[60] GIDDINGS J C & CALDWELL K D, *Anal Chem* **56** (12) 2093–9 (1984).

[61] RUDIN A & FRICK C D, *Polymer Latex 11*, preprints of Conference (item 12), London, May 1985.

[62] THORNTON T & MALEY R, *Polymer Latex 11*, preprints of Conference (item 14), London, May 1985.

[63] DOSRAMOS J G & SILEBI C A, *J Colloid Interface Sci* **130** 14 (1989).

[64] DOSRAMOS J G & SILEBI C A, *Particle Size Distribution II*, ACS series 472, p. 306 (1991); see also *Polym Internat* **30** 445–450 (1993).

[65] GIDDINGS J C, *J Chem Phys* **49** (1) 81–85 (1968).

[66] HOVINGH M E, THOMPSON G H, & GIDDINGS J C, *Anal Chem* **42** 195–203 (1970).

[67] KIRKLAND J J, REMENTER S W, & YAU W W, *Anal Chem* **53** 1730–1736 (1981).

[68] KIRKLAND J J & YAU W W, *Anal Chem* **55** (13) 2165–2170 (1983).

[69] HEIDENREICH E, in *Particle Size Analysis*, Ed M J Groves, Heyden (1978).

[70] LESCHOWSKI K, *Powder Technol* **24** 115–124 (1979).

[71] DAESCHNER H W, *Powder Technol* **2** 349 (1968/9).

[72] CRAWLEY D F C, *J Sci Instruments Ser 2* **1** 576 (1968).

[73] WHITBY K T, *ASTM Special Tech Publ* No 234 pp. 3–25 (1958).

[74] DAESCHNER H W, SEIBERT E E, & PETERS E D, *ASTM Special Tech Publ* No 234 26–50 (1958).

[75] ARIETTI R, TENCA F, & SCARPONE A, *XVI FATIPEC Congress* 277–300 (1978).

[76] COULTER W H, USP 2,656,508 (1953).

[77] COULTER W H, *Proc National Electronic Conf* **12** 1034 (1956).

[78] MORGAN B B, *Research, London* **10** 271 (1957).

[79] KUBITSCHEK H E, *Nature* **182** 234–5 (1958). See also *Research* **13** 128 (1960).

[80] *Industrial Bibliography* Oct 1982, produced by Coulter Electronics Ltd Luton UK.

[81] ALLEN T, in *Particle size measurement*, 2nd edn, Chapter 13, Chapman & Hall, London (1974).

[82] Micromeritics Inst Corp, One Micromeritics Drive, Norcross, Georgia 30093–1877, USA.

[83] KERKER M, *The Scattering of Light and Other Electromagnetic Radiation*, Academic Press (1969).

[84] COLLINS E A, DAVIDSON J A, & DANIELS C A, *J Paint Technol* **47** (604) 35–56 (1975).

[85] MELIK D H & FOGLER H S, *J Colloid Interface Sci* **92** (1) 1983.

[86] TUNSTALL D F, UKP 2,046,898.

[87] NOBBS J H, Paint Research Assoc Progress Report No 6 33–37 (1984).

[88] MEETEN G, *J Colloid Interface Sci* **72** 471 (1979).

[89] WEINER B B, in *Modern Methods of Particle Size Analysis*, Chapter 5, Ed Barth H G, Wiley Interscience Publ (1984); also Plantz P E, *ibid* Chapter 6.

[90] JOHNSTON P R & SWANSON R, *Powder Technol* **32** 119–124 (1982).

[91] LEE S P, TSCHARNUTER W, & CHU B, *J Polym Sci* **10** 2453 (1972).

[92] WEINER B B, Chapter 3 in *Modern Methods of Particle Size Analysis*, Chapter 3, Ed Barth H G, Wiley Interscience Publ (1984).

[93] LINES R W & MILLER B V, *Powder Technol* **24** 91–96 (1979).

[94] SIEDENTOPF H & ZSIGMONDY R, *Ann Physic* **10** 1 (1903).

[95] DERJAGUIN B V, VLASENKO G JA, STOROZHILOVA A I, & KUDRJAVTEEVA N M, *J Colloid Sci* **17** 605–627 (1962).

[96] MCFADYEN P & SMITH A L, *J Colloid Interface Sci* **45** 573 (1973).

[97] WALSH D J, ANDERSON J, PARKER A, & DIX M J, *Colloid Polym Sci* **259** 1003–9 (1981).

[98] GREGG S J & ALLEN K S W, *Adsorption, Surface Area and Porosity* Academic Press (1967).

[99] GREGG S J & SING K S W, *Adsorption, Surface Area and Porosity*, p. 7, Academic Press (1967).

[100] DE BOER J H, HOUBEN G M M, LIPPENS B C, & MEIJS, *J Catalysis* **1** 1 (1962).

[101] KIPLING J J & WRIGHT E H M, *J Chem Soc* 855–860 (1962).

[102] HERZ A H, DANUER R P, & JANUSONIS G A, in *Adsorption From Aqueous Solution*, pp. 173–197, eds Webb W W & Matijevic E, Advances in Chemistry Series 79 (1968).

[103] KIPLING J J & WILSON R B, *J Appl Chem* **10** 109–113 (1960).

[104] LINGE H G & TYLER R S, *Proc Australas Inst Min Metall* **271** 27–33 (1979).

[105] BARKER N W & LINGE H G, *Hydrometallurgy* **6** (3–4), 311–326 (1981).

[106] PADDAY J F, *Surface Area Determination Proc Int Symp* 1969, pp. 331–340 Ed Everett D H. Butterworth.

[107] GREGG S J & SING K S, *Adsorption Surface Area and Porosity*, p. 294, Academic Press (1967).

[108] HARKINS W O & JURA G, *J Amer Chem Soc* **66** 1362 (1944).

[109] GREGG S J, *The surface chemistry of solids*, 2nd edn, p. 280, Chapman & Hall (1961).

[110] EDWARDS J D & WRAY R I, *Aluminium Paint and Powder*, 3rd edn, p. 18, Rheinhold Publ Corp (1955).

[111] EDWARDS J D & MASON R B, *Ind Eng Chem Anal Ed* **6** 1951 (1934).

[112] CAPES C E & COLEMAN R D, *Ind Eng Chem Fundamentals* **12** (1) 1246 (1973).

[113] SHEPARD E & TCHEUREKEDJIAN N, *J Colloid Interface Sci* **28** 481 (1968).

[114] DOROSZKOWSKI A and LAMBOURNE R, *J Polym Sci Part* **C34** 253 (1971).

[115] SCARLET B, in *Particle Size Analysis*, p. 219, Eds Stanley-Wood N G, & Allen T, Wiley Heyden (1982).

8

The industrial paint-making process

F K Farkas

8.1 Introduction

The theory of wetting and the stabilization of pigment particles in paint media is discussed in Chapter 6. It is essential, however, to restate that the purpose of the pigment dispersion process is the wetting and separation of primary pigment particles from aggregates and agglomerates and their subsequent stabilization in suitable paint media, i.e. resin or dispersant solutions, during the process of dispersion.

All stages of this process are important and effect considerably the utilization of pigment, productivity, and the properties of the final product. The process is summarized diagrammatically in Fig. 8.1.

To prevent reaggregation during and after the dispersion it is important to select the correct ratios of pigments, resins, and solvents. In addition, the second stage of adding further amounts of resin solutions or solvents should be carried out in the dispersion equipment to eliminate the possibility of 'colloidal shock' (flocculation) on the final make up to paint. The process is carried out in various types of milling equipment where shear forces are applied to the pigment aggregates, in order to separate the primary pigment particles. This stage is often called 'grinding'.

Intermolecular forces also play an important part in influencing the wetting of the surface of the pigment and in bringing about some degree of spontaneous dispersion. The maximization of the effect of intermolecular forces within any paint system to achieve rapid and stable dispersion of pigments with the minimum application of shear forces is a desirable objective.

The cleanliness and the strength of the colour, the durability of a number of a shear-sensitive pigments, and the actual time and energy expended during the process are affected by the wetting process. The conditions by which the physical (shearing) forces are applied in the various dispersing machinery have similar influence.

8.2 The use of dispersants

Wetting plays an important part in the dispersion of pigments and therefore in the production of paints. All pigments have contaminants, e.g. air, moisture, and gases,

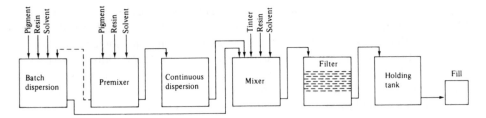

Fig. 8.1 — Flow diagram of the paint-making process.

adsorbed onto their surfaces. To wet the particles these contaminants must in most cases be displaced by the dispersing medium. It is essential therefore that the wetting efficiency of the dispersant is strong enough to overcome, or at least reduce, the cohesive forces within the liquid and the surface tension between the solid/liquid interface leading to adhesion of the wetting groups of the dispersant onto the surface of the pigment.

The majority of the vehicles used for paint-making can be considered as dispersants, their wetting efficiency depending on their molecular weight, structure, and the presence of substituent groups, e.g. carboxyl, hydroxyl, amine, and ester. Dispersants are now available especially formulated to be more efficient than the majority of film-forming media in paint, and to a large degree these can be multipurpose in nature, having a wide range of compatibility with a variety of paint systems.

Since the electrical charges in the molecules of liquid and pigment are responsible for the wetting process, the polarity of ingredients, resins, solvents, and solid particles plays an important part in the dispersion stage. Polarity is defined by the shape of the molecule and therefore by the arrangement of the electrical charges and whether they are symmetrical or asymmetrical.

Symmetrical non-polar pigment

Thio indigo red Y

Unsymmetrical polar pigment

Toluidine red

Symmetrical non polar solvent

Benzene

Unsymmetrical structures have increased polarity and are soluble in polar solvents

Chlorobenzene
Dipole moment 2.25 Debye units.
(Sum of electrical charges ×
distance between them)

Polar solvent
$n\,C_4\,H_9\,OH$ n-butyl alcohol

Dispersant oleic acid: $C_{17}\,H_{35}\,COOH$

| non-polar radical | polar reactive group |

Dispersants and paint media adsorb onto pigment surfaces by means of a wetting group (anchor group), leaving the non-polar radical which is soluble in the liquid phase extended. This mode is called 'steric stabilization'. The solvent balance, the compatibility of subsequent resin constituents, and the order of addition are of great importance to maintain stability. If the stabilizing chains partly or wholly collapse as a result of solvency changes, reaggregation and flocculation can take place to the detriment of the final product which, in most cases, is impossible or very expensive to recover.

A large variety of wetting agents (surfactants) are being used in the paint industry to reduce surface tension at the solid/liquid interface. Thus, new surfaces are created onto which the adsorption of paint media becomes easier during the dispersion process. The surfactants may be cationic where the adsorbable ion is positively charged, anionic where the adsorbable ion is negatively charged, or non-ionic, the activity of which may be due to polar and non-polar groups in the molecule.

The wetting agents are salts of organic amines, alkali soaps, sulphated oils, glycol ethers, etc. There are numerous products available with various claims of effectiveness. Some of them, or combinations of them, which may be critical, are standard constituents of waterborne (in particular emulsion) paints to provide ionic stabilization. Suitably designed carboxylated acrylics and hydrolytically stable alkyds, when neutralized with an amine or inorganic base, may also be used to provide a combination of steric/ionic form of stabilization, in particular for the manufacture of tinters where flow is of some importance.

In general, excessive use of surfactants is detrimental to many properties of a paint, hence the determination of the exact amount to give the right degree of dispersion and stabilization is of considerable importance. Prior to this, however, the wetting efficiencies of the various resins and their effect on each of the pigment/extenders present in any paint formulation have to be determined when formulating the dispersion stage to obtain the maximum use of raw materials, machine time, and quality of the end product. Wetting agents should only be used when the paint medium has little or no potential to accomplish the necessary degree of wetting.

8.3   Methods of optimizing millbases for dispersion

The formulating techniques, applied in the paint industry for the dispersion of pigments, have been based mainly on empirical values, long experience, and time- and material-consuming laboratory evaluations, in order to establish working — but not necessarily optimum — conditions for the process of dispersion.

It is fairly safe to say that at least six laboratory scale ballmill trials have to be carried out on a single pigment single vehicle and single solvent system with variations in pigment volume percent (PV%) and vehicle solids percent (VS%) to obtain some idea of reasonable processability. The same applies to beadmills and high speed type dispersions (HSD).

The time spent, including testing of the experimental paint, may be as long as three days, and if the paint is a multipigment and resin system, weeks of experimentation is common. The subsequent semi-technical and plant scale proving trials often require further modifications.

Prior to the pigmentation of a millbase for dispersion, the paint chemist may mix known weights of pigment and resin solution in a can, and when the mix flows, scales up the ingredients to fill the dispersion volume (D_v) of the appropriate milling machine. This technique is called the 'can test' and may or may not result in a satisfactory dispersion. Researchers have long recognized the importance of the process of dispersion and its profound effect on the economics and quality of the subsequent product. No satisfactory dispersion of the pigments/extenders can be achieved with arbitrary millbase formulations. F K Daniel [1] developed the 'flow point' technique by which improved millbase formulations were obtained.

The technique consists of the titration of known weights of pigment with resin solutions of varying concentration while the mixture is being agitated with a palette knife. The end point is reached when the mixture begins to flow from the palette knife with a break, then starts flowing again within a second. The results, volume of resin solution used for the known weight of pigment, are then worked out. The concentration of resin solution which gave the lowest value is considered the best wetting medium and is proportionally scaled up to give a millbase for ballmill or beadmill type dispersion as there is little difference in the millbase consistency required for these machines. The complication is that not all premixed millbase will flow satisfactorily to be certain about the end point, e.g. organic pigments and fine particle extenders, in which case a series of ballmill trials are recommended.

Work by Guggenheim [2] was based on a number of production scale experiments with high-speed disc impeller dispersers. It resulted in an empirical expression to determine optimum millbases for this type of dispersion process. Non-volatile (NV) vehicle solids, vehicle viscosity (η), and Gardner–Coleman oil absorption (OA) are related to obtain pigment/vehicle ratios by weight:

$$\frac{W_v}{W_p} = (0.9 + 0.69 \text{ NV} + 0.025\eta)\frac{\text{OA}}{(100)}$$

where W_v = weight of vehicle
W_p = weight of pigment
NV = non-volatile fraction of vehicle
η = viscosity of vehicle in poises
OA = Gardner–Coleman oil absorption (lb/100 lb)

The above equation is based on a base value 0.9 OA/100 with upward adjusting factors for the *NV* resin content 0.69 NV·OA/100 and for the viscosity 0.025 η·OA/100.

In practice a millbase formulation for high-speed dispersion technique using Guggenheim's empirical method could be worked out. For example:

NV of vehicle = 40% by weight, and its viscosity = 5 poises
The OA = 25 for the pigmentation

$$\frac{W_v}{W_p} = (0.9 + 0.69 \times 0.40 + 0.025 \times 5.0)\frac{(25)}{(100)}$$

$$= (0.9 + 0.28 + 0.125)(0.25) = 1.30 \times 0.25 = 0.325$$

The ingredients of the millbase are therefore:

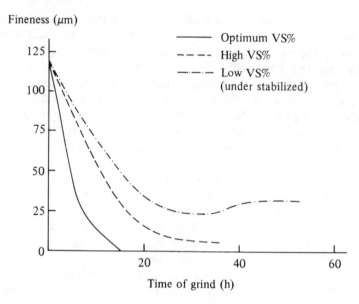

Fig. 8.2 — The effect of VS% on the rate and fineness of dispersion.

$$\text{NV vehicle } (0.40 \times 0.325) = 0.130 = 9.81\% \text{ by weight}$$

$$\text{Solven } (0.60 \times 0.325) = 0.195 = 14.72\% \text{ by weight}$$

$$\text{Pigment } = \frac{1.000}{1.325} = \frac{75.47\%}{100.00\%} \text{ by weight}$$

There is no doubt that these formulating techniques are useful and help the chemists to do a better job. However, they lack the accuracy essential for the measurement of the order of wetting efficiencies of the vehicle system, an equally important factor required to produce a stable paint. Furthermore, the results obtained are very dependent upon the operator variable which can be as much as ±25% depending on the amount of work put into the mix by hand.

The ratios of pigment/binder obtained in the undispersed state of the pigment do not therefore provide a firm and reproducible base for subsequent formulating, and millbases can become deficient in dispersant during the process of dispersion.

Further separation and stabilization of primary particles from the aggregates becomes increasingly difficult. The dispersion process comes to a halt and the pigment remains in a partly reaggregated state, irrespective of prolonged machine time.

Higher vehicle solids (VS%) than the optimum, in order to induce some flow, will only prolong the time of the dispersion process (see Fig. 8.2).

8.4 The instrumental formulating technique

Wirsching et al. investigated the oil adsorption characteristics of pigments using a Brabender Plastograph (Torque Rheometer) [3], and found remarkably reproducible values.

This prompted the present author to examine the possibility of an instrumental formulating technique. For over 20 years of research at ICI Paints Division, this technique has been developed and used to formulate optimum millbases for the various dispersion techniques [4].

From the data of well over 11 000 tests carried out on a vast range of pigments, extenders, film-forming media, solvents, and additives, it was found that the relationship and quantitative interactions of these ingredients were specific and accurately measurable.

The criticality of the relationships greatly influences the viscosity, rate of dispersion, and utilization of pigments during the process of dispersion, and determines the order of addition of subsequent make-up ingredients to yield stable products with the highest quality attainable within the limits of that system. Each pigment behaves preferentially in an environment which contains more than one surface-active agent, hence reaches the optimum and proportional amount of the preferred surfactant in the dispersion stage if stabilization is to be maximized.

The degree of dispersibility of the pigment, or alternatively the wetting efficiency of the polymer, surface-active agent solution, is determined instrumentally in the dispersed state of the solid particle (pigment) with a torque/rheometer e.g. Brabender Plastograph. Such an instrument is capable of dispersing pigmented millbases under the standardized conditions and rate of shear to a sufficient degree.

The process of dispersion and the end result cannot be influenced by the operator using a specially developed technique which also simulates the temperature of millbases likely to be encountered in plant scale dispersion machinery.

The end result of a set of tests (optimization) may be converted directly into a practical millbase formulation for the dispersion of the said particle(s) for the different types of wet dispersion techniques well known in the art. Alternatively it may be calculated from the individual specific data of PV% (pigment volume) and VS% (vehicle solids) previously generated and stored.

Data storage is most conveniently carried out by a computerized system which is capable of producing optimized formulations from the accrued data for any number of pigments, extenders, resins, solvents, and additives which may be required for a product.

It is possible, therefore, to analyse the pigmentation of a colour-matched test panel using a spectrophotometer, and with the ratios obtained to compute the best possible formulation and method of manufacture for the selected resin (film-forming) system. This obviously saves a lot of time and money which would otherwise be spent on the trial and error method still being used in the industry, to obtain a workable but not necessarily optimized formulation. In addition the technique provides the basis of accurate quality control of pigments, extenders, resins, solvents, wetting agents, and additives, provided that standards have been established for comparison.

The final PV% at the end of the test and the torque curve obtained under specified and standardized conditions will quickly show differences, if any, between batches which are otherwise unmeasurable and can cause manufacturing and/or product problems. The technique also provides the means to monitor the manufacture of resins and dispersants by measuring the effect of rate of stir, feed, and temperature on the wetting efficiency of the product.

As a research tool it is invaluable to develop new dispersants, resins, and pigmentation of the same leading to new products.

Plate 8.1 — Brabender Plastograph.

The Brabender Plastograph (Plasticorder) (Plate 8.1) is a torque rheometer. The measuring principle is based on the display of the resistance of test material sheared by rotors in the measuring head. The corresponding torque moves the dynamometer from its zero position, and a curve is recorded (torque vs time). The unit of torque is Newton-metre (N m). The force required to accelerate 1 kg weight to $1\,\mathrm{m\,s^{-1}}$.

Parameters that influence the viscosity of material under test such as temperature and rate of shear can be varied over a wide range, hence test conditions similar to the milling conditions can be simulated. The first Brabender Plastograph was built over 40 years ago for the rubber and food industry and has found wide application since, in particular in the field of thermoplastics, raw material control, and general research. It is a sensitive and robustly built instrument which requires small amounts of materials for the test. The formulating technique consists of the dispersion of a specified volume of pigment in a specified volume of resin solution to provide a stiff paste at around 70% loading by volume of the measuring head. After the dispersion, the millbase is titrated with the same resin solution till the torque drops to a predetermined and standardized value at which the total amount of resin solution used is recorded.

The initial loading data are then used as standard for any subsequent tests employing the same pigment or worked out proportionally if mixtures of pigments are being tested.

8.4.1 Determination of the best-wetting resin solution and optimum millbase formulation for a system

Following the operating technique a range of vehicle solids solutions of the appropriate vehicle(s) or dispersant(s) and solvents should be tested on the pigment or pigment mixture of the system to obtain the wetting efficiency curve(s). This is obtained by plotting the pigment volumes obtained at the end point against the range of vehicle solids tested. This curve is characteristic of the system and shows the best-wetting resin or dispersant solution to be used for the dispersion of the pigment. It also shows the order of wetting efficiencies of the various vehicles in a system, which is the basis of the optimized let-down procedure.

It is advisable to omit the various wetting agents and additives from the standard millbase formulation when evaluating the vehicle system. However, they should be included, and their effect, if any, measured on the optimized millbase formulation. In a very large number of cases they give no further improvement in an optimized system.

The PV% of the millbases which were obtained at the end point on the Plastograph are calculated with the different vehicle solids solutions used for the tests.

Each of the millbases is composed of a fixed amount of pigment and the total volume of vehicle solutions:

$$\frac{w_p}{\rho_p} = V_p \quad V_v \rho_v = w_v \quad PV\% = \frac{V_p}{V_p + V_v} \times 100$$

where:

    w_p = weight of pigment
    V_p = volume of pigment
    w_v = weight of vehicle solution
    V_v = volume of vehicle solution
    ρ = specific gravity (subscript p for pigment and v for vehicle solution).

Example: A carbon black pigment was tested with an alkyd resin (a) and with a nitrogen resin (n) with the following results:

Total volume of resin solution (ml)			The corresponding PV%		
VS%	Resin (a)	Resin (n)	VS%	Resin (a)	Resin (n)
10	50	43	10	14.25	16.23
15	45	38	15	15.47	17.98
20	43.6	42	20	16.04	16.55
30	57	48	30	12.75	15.00

Resin (n) is a better resin for wetting and therefore should be used for the dispersion of carbon black pigment at 15% VS concentration which gives the highest volume of pigment incorporable/unit volume of millbase and dispersible in the shortest time possible.

The shape of the wetting efficiency curve is different and specific for the different pigments and resin solutions. Curves for several pigment types are shown in Fig. 8.3a and for a single pigment in a range of dispersant solutions in Fig. 8.3b.

The reason why the PV% is worked out instead of its weight becomes apparent when practical millbase formulations are worked out from the test data.

Using the conversion figures it is possible to obtain optimum millbase formulations for all major types of dispersion techniques used in the paint industry by a single calculation. Scale-up is direct from laboratory to plant, and only the optimum formulation is checked on a laboratory scale for the rate and fineness of dispersion prior to implementation.

The wetting efficiency curves of Fig. 8.4 also show that resin (a) is inferior to resin (n) in general wetting properties on the carbon black pigment tested, and although essential for the paint formulation it will not interfere with the stability of the millbase provided that it is added after further amounts of resin (n) are incorporated in the dispersed millbase as a second stage.

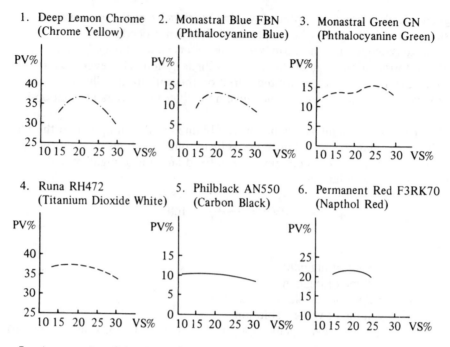

1. Deep Lemon Chrome
   (Chrome Yellow)

2. Monastral Blue FBN
   (Phthalocyanine Blue)

3. Monastral Green GN
   (Phthalocyanine Green)

4. Runa RH472
   (Titanium Dioxide White)

5. Philblack AN550
   (Carbon Black)

6. Permanent Red F3RK70
   (Napthol Red)

7. A composite of the six graphs, illustrated above, to give a direct comparison.

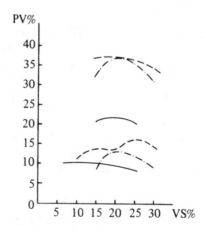

Fig. 8.3a — Wetting efficiency curves for several pigments.

In general (with reference to Fig. 8.5a and 8.5b), if lower than the optimum VS% (understabilized region) is used for the dispersion of pigment the millbase may run short of the stabilizing resin and may partly reaggregate during the process. Higher than the optimum VS% (overstabilized region) will, on the other hand, disproportionally extend the time of dispersion to reach the required fineness.

Where compromise is required (e.g. sensitive pigments; formulation is short of solvents) higher than the optimum VS% may be used with the corresponding PV% as determined by the wetting efficiency curve.

As an additional and important factor the amount of work done on the mix during the dispersion stage (the area covered from the coherency point) can be used

1. A medium oil alkyd in an aliphatic hydrocarbon.

2. MMa polymer in a blend of an ester and an aromatic hydrocarbon.

3. An acrylic copolymer in an aromatic hydrocarbon.

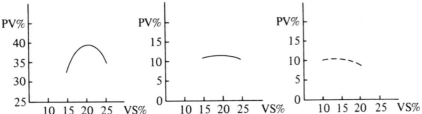

4. A long oil alkyd in a blend of an aliphatic and aromatic hydrocarbon.

5. A 2nd long oil alkyd in a blend of an aliphatic and aromatic hydrocarbon.

6. A 2nd acrylic copolymer in a blend of an aliphatic and aromatic hydrocarbon.

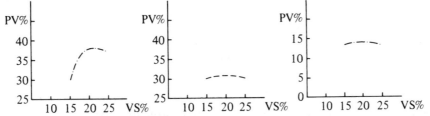

7. A composite of the six graphs, illustrated above, in direct comparison to each other.

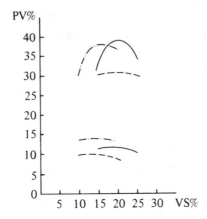

Fig. 8.3b — Wetting efficiency curves for Monolite Red Y in different dispersant solutions.

to differentiate between dispersants and resin solutions regarding their wetting and dispersing efficiencies. The less work that is done in Plastograph Units, the better the wetting efficiency of the dispersant or resin solution. A typical Plastograph recording as shown in Fig. 8.6, is obtained after each test, and from it the wetting efficiency value is worked out.

Since the conditions are standardized for the system under test the use of this factor in doubtful cases is of considerable value.

The efficiency of optimized millbase formulations is also apparent when the colour strengths of subsequent finishes are compared with non-optimized standards.

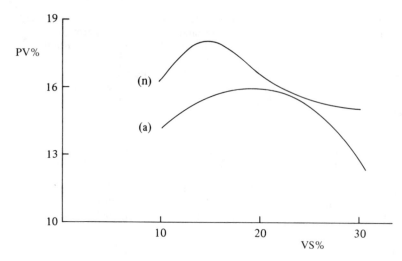

Fig. 8.4 — Wetting efficiency curves of resin (a) and (n) measured using a carbon black pigment.

Fig. 8.7(a)–(f) shows that economies in both raw material and processing time may be obtained. The graphs show the results of colour strength versus time of grind for three pigments dispersed in solutions of:

- a medium oil alkyd in an aromatic hydrocarbon
- a long oil alkyd in an aliphatic hydrocarbon before and after optimization.

Principles and rules for the dispersion of pigmented millbases and their subsequent make-up to paint are:

1 The wetting-out and dispersibility characteristics of pigments and extenders are different and specific to the various dispersants, film-forming materials, and additives.

2 Each pigment will therefore behave preferentially in an environment which contains more than one surface-active agent. Hence it is necessary to use the optimum and proportional amount of the preferred surfactant for each pigment in the dispersion stage if stabilization is to be maximized.

3 These characteristics can be measured instrumentally using the special technique employing the Brabender Plastograph. The order of dispersibility or the wetting efficiency of the film-forming materials, etc. within the limits of a product formulation, defines the dispersion, let-down and make-up stages, and yields the best possible utilization of raw materials and process time.

4 Even small differences in wetting efficiency between component resins if used in the wrong order in the dispersion and let-down stages can cause long-term stability problems (e.g. colour drift).

   Interaction of 'dispersing aids' and additives have a similar effect, and their use is often unnecessary in an optimized system.

5 Higher than the optimum VS% may be used in the following cases:
   (a) To protect certain pigment surfaces (coatings) from damage brought about by the highly efficient dispersion conditions of the optimum VS%.
   (b) To conform to the tight solvent (viscosity) limits of a particular product.
   (c) To give more flexibility to plant operations where required.

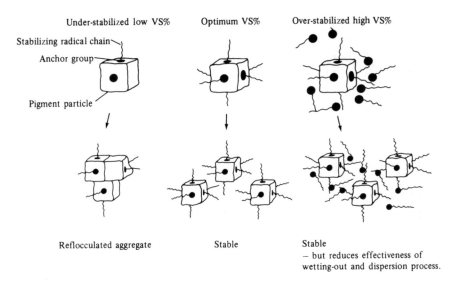

Fig. 8.5a — Particle arrangements.

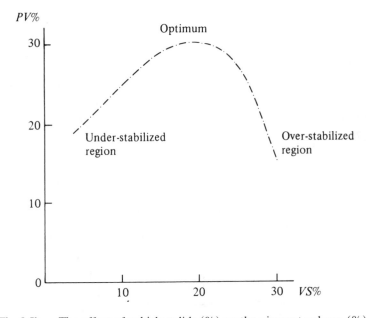

Fig. 8.5b — The effect of vehicle solids (%) on the pigment volume (%).

6   Never disperse any pigment by any method at more than 5% below its optimum
    VS% or level of dispersant, e.g. if optimum VS% is 20 then 15% is the bottom
    limit. Otherwise the millbase will run short of stabilizer during dispersion, will
    reaggregate, and may not reach the fineness of dispersion required irrespective
    of extended time of grind. In general, 10% higher VS% than the optimum will
    extend the time of grind disproportionally.

7   Never 'second stage' a dispersed millbase with solvents or very high solids
    resins. This causes partial or total shock and may result in a flocculated base

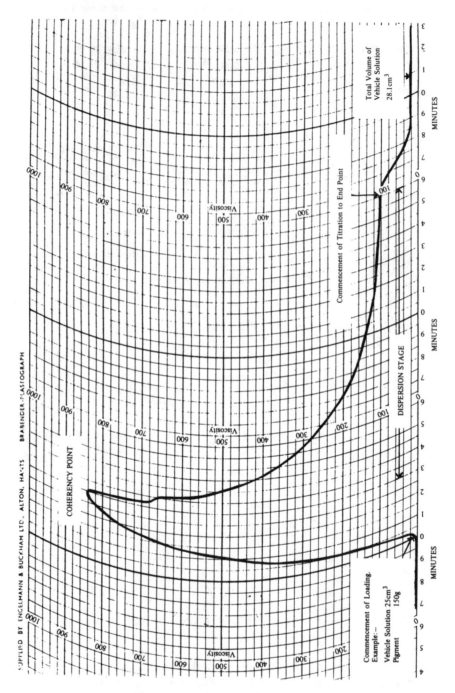

Fig. 8.6 — Plastograph recording.

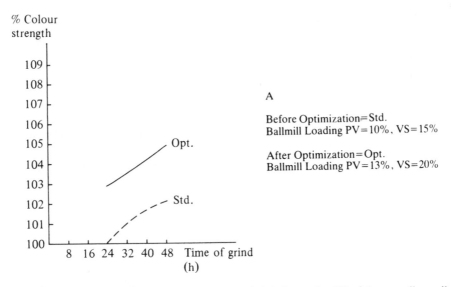

Fig. 8.7(a) — Dispersion of Monastral Blue FBN (Phthalocyanine Blue) in a medium oil alkyd–aromatic hydrocarbon solution.

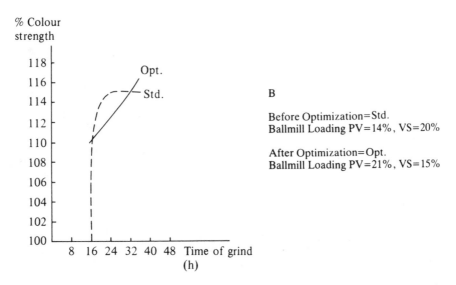

Fig. 8.7(b) — Dispersion of Monastral Blue FBN (Phthalocyanine Blue) in a long oil alkyd–aliphatic hydrocarbon solution.

difficult to recover. Instead, the vehicle solids should be increased 8–10% above that of the optimum dispersion VS% during the process of second stage, preferably with solutions of the appropriate resin added incrementally. This ensures the necessary stabilization of the dispersed millbase for short-term storage and freedom from shock on subsequent make-up.

8   If a tinter contains a more efficient dispersant than the finish to be tinted, care must be taken. Some pigments in the finish could be destabilized and/or floccu-

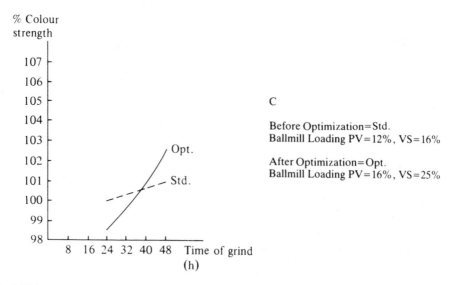

Fig. 8.7(c) — Dispersion of Monastral Green GN (Phthalocyanine Green) in a medium oil alkyd–aromatic hydrocarbon solution.

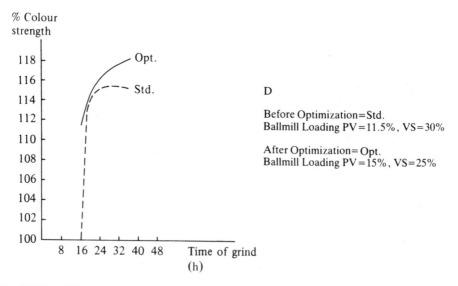

Fig. 8.7(d) — Dispersion of Monastral Green GN (Phthalocyanine Green) in a long oil alkyd–aliphatic hydrocarbon solution.

lated if more than half the amount of stabilizer required by them is introduced via the tinter. If, however, the tinter is in a less efficient dispersant than the resin system of the finish, partial or total flocculation will occur. The extent of this will depend on the degree of wetting efficiency differences between the tinter resin and the most efficient resin or resin mixtures in the finish. It is assumed that in both cases the media of the tinters are compatible with the film-forming materials of the finish.

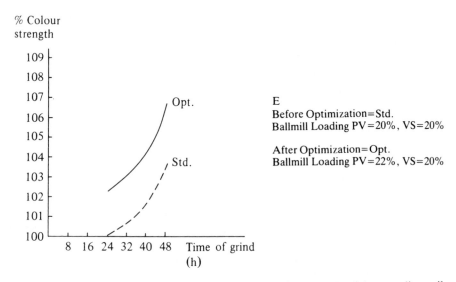

Fig. 8.7(e) — Dispersion of Permanent Red F3RK70 (Napthol Red) in a medium oil alkyd–aromatic hydrocarbon solution.

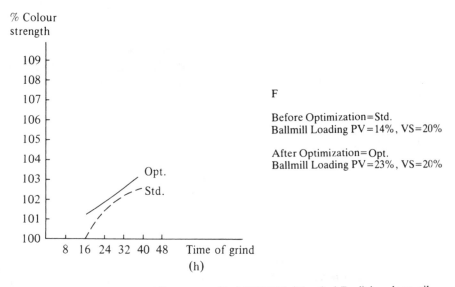

Fig. 8.7(f) — Dispersion of Permanent Red F3RK70 (Napthol Red) in a long oil alkyd–aliphatic hydrocarbon solution.

9   It is important to maintain the solvent balance at all stages of manufacture to prevent film-forming materials coming out of solution. Even partial insolubility or incompatibility of components can cause irreversible shock and flocculation. Transparent dispersions are particularly sensitive in this respect.

10   To obtain maximum benefit from optimized millbase formulations it is to be ensured that the dispersing machinery required for the job is operated at its optimum conditions.

Finishes, and therefore corresponding millbases, can be classified as follows:

1   Single pigment–single vehicle.
2   Two or more pigments–single vehicle.
3   Single pigment–two or more vehicles.
4   Two or more pigments–two or more vehicles.

In cases (1) and (2) the pigment(s) has little 'choice' and will disperse to its limitations at the optimum vehicle solids (VS%).

In cases (3) and (4) the pigment(s) has a choice and therefore must be dispersed in the best resin or resin blend at the vehicle solids which is determined or calculated from the individual optima.

Practical example (1) below shows how to calculate a millbase formulation from PV% and VS%; it therefore applies to case (1). If the VS% is the same for two pigments in a single resin system, case (2), the same calculation is to be used. It is assumed that the PV% is also the same. If, however, the VS% and PV% are different for two or more pigments in the same resin, proceed as practical example (2), and in different resins as practical example (3).

8.4.2   How to use the test data for formulating millbases

Ideally, the data obtained by the systematic optimization of millbases for dispersion should be filed, and may be used to formulate different combinations of pigment(s) and resin(s) from the individual values without further tests.

The basic statement of PV% and VS% defines a millbase formulation from which a unit of 100 by volume should be calculated first before scale-up. This is to be multiplied by the appropriate factor to conform to the different loading requirements of dispersion machinery.

Calculations

To find weight: $\rho\, V$

To find volume: $\dfrac{w}{\rho}$

To find ρ: $\dfrac{w}{v}$

To find N/V: $\dfrac{w_r \times R_s}{100}$

To find w_r and w_s at the VS% wanted for the weight of dispersion solution required:

$$w_r = \frac{w_{ds} \times \text{VS\%}}{R_s}$$

$w_s = w_{ds} - w_r$ therefore $w_r + w_s = w_{ds}$

where

ρ	= density
V	= volume
w	= weight
w_r	= weight of resin
w_s	= weight of solvent
w_{ds}	= weight of dispersion solution

R_s = resin solids
N/V = non-volatile resin weight
VS% = vehicle solids %.

To find composite pigment volume (CPV%) at different weight ratios (w_{pr}):

$$CPV = \frac{T_{vp}}{T_{vp} + V_{ds}} \times 100$$

$$\frac{W_{pr1}}{\rho_1} + \frac{W_{pr2}}{\rho_2} + \frac{W_{pr3}}{\rho_3} + \frac{W_{prx}}{\rho_x} = V_{pr1} + V_{pr2} + V_{pr3} + V_{prx} = T_{vp}$$

where:

CPV = composite pigment volume %
T_{vp} = total volume of pigment
W_{pr} = weight ratio of pigment
V_{pr} = volume ratio of pigment.

To find the proportional amount of dispersant solution (w_{dsr}) and their corresponding volume (v_{dsr}) for each pigment from the individual optima.

$$w_p : w_{ds} = w_{pr1} : w_{dsr1} \text{ therefore } w_{dsr1} = \frac{w_{ds}(w_{pr1})}{\rho_1} \text{ and } \frac{w_{dsr1}}{\rho_1} = v_{dsr1}$$

$$w_p : w_{ds} = w_{pr2} : w_{dsr2} \text{ therefore } w_{dsr2} = \frac{w_{ds}(w_{pr2})}{\rho_2} \text{ and } \frac{w_{dsr2}}{\rho_2} = v_{dsr2}$$

$$w_p : w_{ds} = w_{pr3} : w_{dsr3} \text{ therefore } w_{dsr3} = \frac{w_{ds}(w_{pr3})}{\rho_3} \text{ and } \frac{w_{dsr3}}{\rho_3} = v_{dsr3}$$

$$w_p : w_{ds} = w_{prx} : w_{dsrx} \text{ therefore } w_{dsrx} = \frac{w_{ds}(w_{prx})}{\rho_x} \text{ and } \frac{w_{dsrx}}{\rho_x} = v_{dsrx}$$

therefore

$$w_{ds} = w_{dsr1} + w_{dsr2} + w_{dsr3} + w_{dsrx} \text{ and}$$
$$v_{ds} = v_{dsr1} + v_{dsr2} + v_{dsr3} + v_{dsrx}$$

where:

w_p = weight of pigment and/or extender in individual optimum millbase;
w_{ds} = weight of dispersion solution in individual optimum millbase;
w_{pr} = weight ratio of pigment;
w_{dsr} = weight ratio of optimum VS% solution;
v_{dsr} = volume ratio of optimum VS% solution;
v_{ds} = volume of dispersion solution required.

To calculate composite vehicle solids (CVS%):

$$CVS = \frac{NV}{w_{ds}} 100; \quad \text{for example}$$

$$\frac{w_{dsr1}(VS)}{100} = NV_1; \quad \frac{w_{dsr2}(VS)}{100} = NV_2$$

$$\frac{w_{dsr3}}{100} = NV_3; \quad \frac{w_{dsrx}(VS)}{100} = NV_x$$

$$NV = NV_1 + NV_2 + NV_3 + NV_x$$

where:

CVS = composite vehicle solids %
NV = non-volatile resin solids
NV_x = non-volatile resin solids ratio
VS   = optimum individual vehicle solids %.

Calculate weight of resin(s), (w_r) and weight of solvent(s) (w_s) to complete the formulation.

$$w_r = \frac{NV}{R_s} 100 \text{ and } w_s = w_{ds} - w_r, \text{ for example}$$

$$\frac{NV_1}{R_s} 100 = w_{r1} \text{ therefore } w_{dsr1} - w_{r1} = w_{s1}$$

$$\frac{NV_2}{R_s} 100 = w_{r2} \text{ therefore } w_{dsr2} - w_{r2} = w_{s2}$$

$$\frac{NV_3}{R_s} 100 = w_{r3} \text{ therefore } w_{dsr3} - w_{r3} = w_{s3}$$

$$\frac{NV_x}{R_s} 100 = w_{rx} \text{ therefore } w_{dsrx} - w_{rx} = w_{sx}$$

$$w_{s1} + w_{s2} + w_{s3} + w_{sx} = w_s$$

To obtain 100 by volume of millbase, multiply the weight of each ingredient by the factor of

$$\frac{100}{CPV + v_{ds}}$$

Practical example 1
TiO_2 Kronos RN45. This pigment is to be dispersed in a 15% VS solution of an aliphatic alkyd resin. The corresponding PV% is 43.5%

	Volume units		Density (g/cc)		Weight units
Pigment	43.5	×	4.1	=	178.35
15% VS solution of alkyd	56.5	×	0.788	=	44.52
	100.0				222.87

Now calculate the components of 44.52 weight units of 15% VS solution of the alkyd resin.
The solids of alkyd = 40%
To find the amount of NV resin in 44.52 of 15% VS solution:

$$N/V = \frac{44.52 \times 15}{100} = 6.678$$

To find the amount of 40% solids alkyd required to give 44.52 15% VS solution:

$$\frac{6.678}{40} \times 100 = 16.695 \text{ weight units of 40% alkyd}$$

To find the amount of solvent required:

$$\frac{\begin{array}{r} 44.520 \\ -16.695 \end{array}}{= 27.825}.$$

The millbase therefore consists of:

Pigment	178.35 weight units
40% solids alkyd	16.695 weight units
Solvent	27.825 weight units

Now multiply all weight units with the 'factor' to give the requisite volume of mill-base; for example, if a ballmill with a total volume of 1000 is to be used then at 60% loading and charge to voids ratio of 1/1.5 the volume of millbase required will be 240 and the factor is 2.4.

The factor will vary according to conditions and dispersing machinery.

Practical example 2
A co-grind formulation to be calculated from individual data of TiO_2 pigment and blanc fixe extender at a weight ratio of 20/80. The dispersant is an aliphatic alkyd resin (40% NV) in white spirit.

Step 1. Write individual optimum PV% and VS% for the pigments required to formulate the millbase.
The following data was obtained using the Brabender Plastograph:

For TiO_2	PV = 43.5% at VS = 15% dispersant in white spirit
and Blanc Fixe	PV = 38.35% at VS = 20% dispersant in white spirit.

Step 2. Convert above to actual volume and weight units.

	V	ρ	wt
TiO_2	43.5	4.1	178.35
15% VS alkyd solution	56.5	0.788	44.52
	100.0		
Blanc Fixe	38.35	4.1	157.24
20% VS alkyd solution	61.65	0.798	49.20
	100.00		

Step 3. Write determined ratios of the different pigments in millbase and work out the volume of that pigment mixture.

	wt.%	ρ	V
TiO_2	20	4.1	4.88
Blanc Fixe	80	4.1	19.51
	100		24.39

Step 4. Work out the proportional amount of dispersant solution for each pigment and their corresponding volume for each pigment.

TiO_2

$$178.35 : 44.52 = 20 : x \quad x = \frac{44.52 \times 20}{178.35} = 4.99 \quad 0.77.88 \text{ therefore } \frac{4.99}{0.788} = 6.34$$

wt ρ V

Blanc Fixe

$$157.24 : 49.20 = 80 : x \quad x = \frac{49.20 \times 80}{157.24} = 25.05 \quad 0.798 \quad \text{therefore } \frac{25.05}{0.798} = 31.37$$

Step 5. Determine composite pigment volume CPV%:

Volume of pigments   24.39 +

Volume of disp.solutions $\dfrac{37.71}{62.10}$ therefore CPV $= \dfrac{24.39}{62.10} 100 = 39.28\%$

Step 6. Calculate composite vehicle solids CVS%

required wt of 15% VS alkyd = 4.99 NV $= \dfrac{4.99 \times 15}{100} = 0.75$

required wt of 20% VS alkyd = 25.03 NV $= \dfrac{25.03 \times 20}{100} = 5.06$

CVS $= \dfrac{5.76}{30.02} \times 100 = 19.15\%$ therefore total NV = 5.76

Step 7. Calculate weight of resin (w_r) and weight of solvent (w_s) required to complete the formulation:

$$w_r = \frac{NV}{R_s} \times 100 \text{ and } w_s = w_{ds} - w_r$$

w_r $w_r - w_s$ $= w_{s(required)}$

for 15% alkyd $\dfrac{0.75}{40} \times 100 = 1.86$ therefore $4.99 - 1.86 = 3.13$

for 20% alkyd $\dfrac{5.01}{40} \times 100 = 12.52$ therefore $25.03 - 12.52 = 12.51$

therefore total $= 15.64$

To obtain 100 by volume of millbase multiply by the factor of

$$\frac{100}{62.10} = 1.61$$

Step 8. Write up the completed millbase formulation:

(PV = 39.28% VS = 19.15%)

	V	ρ	wt
TiO_2	7.86	4.10	32.2
Blanc Fixe	31.42	4.10	128.8
alkyd (40% NV)	28.44	0.84	23.83
white spirit	32.28	0.78	25.18
	100.00		

Practical example 3
To find composite vehicle solids (CVS%) and composite pigment volume (CPV%) from individual data of VS% and PV% to give the proportional amounts of stabilizer for the pigments present in 100 volume units of millbase (co-grind).

Step 1. Write individual optimum PV% and VS% for the pigments required to formulate a millbase for an automotive topcoat.

Pigment
Monolite Yellow PV = 8% VS = 20% dispersant melamine resin 1 in xylol
TiO$_2$ PV = 42% VS = 15% dispersant acrylic resin in xylol
Titan Yellow   PV = 36% VS = 25% dispersant melamine resin 2 in xylol
Red oxide   PV = 35% VS = 20% dispersant alkyd resin in xylol

Step 2. Convert above to actual volume and weight units:

	V	ρ	wt
Monolite Yellow	8	1.6	12.8
20% VS Melamine R1	92	0.91	83.72
	100		
TiO$_2$	42	4.1	172.2
15% VS acrylic resin	58	0.89	51.62
	100		
Titan Yellow	36	4.3	154.8
25% VS Melamine R2	64	0.92	58.88
	100		
Red oxide	35	4.3	150.5
alkyd resin	65	0.90	58.5
	100		

Step 3. Determine weight ratios of the four different pigments in the millbase to give the requisite colour of the subsequent finish and work out the overall volume of the pigment mixture. It is assumed that the weight ratios required are as follows:

	$w\%$	ρ	$V\%$
Monolite Yellow	2.85	1.6	1.78
TiO$_2$	18.36	4.1	4.48
Titan Yellow	46.72	4.3	10.87
Red oxide	32.07	4.3	7.46
	100.00		24.59

Step 4. Work out the proportional amount of dispersant solution for each pigment and their corresponding volume for each pigment:

Monolite Yellow

$$12.8 : 83.72 = 2.85 : x \quad x = \frac{83.72 \times 2.85}{12.8} = 18.64 \quad 0.91 \quad \frac{18.64}{0.91} = 20.48$$

TiO_2

$$168 : 51.62 = 18.36 : x \quad x = \frac{51.62 \times 18.36}{168} = 5.64 \qquad 0.89 \quad \frac{5.64}{0.89} = 6.34$$

Titan Yellow

$$154.8 : 58.88 = 46.72 : x \quad x = \frac{58.88 \times 46.72}{154.8} = 17.77 \quad 0.92 \quad \frac{17.77}{0.92} = 19.32$$

Red oxide

$$150.5 : 58.5 = 32.07 : x \quad x = \frac{58.5 \times 32.07}{150.5} = 12.47 \qquad 0.90 \quad \frac{12.47}{0.90} = 13.86$$

$$\text{Total wt} = 54.22 \qquad \text{Total vol} = 60.00$$

Step 5. Determine composite pigment volume CPV = volume of pigments 24.70 + volume of disp.soln. $\dfrac{60.00}{84.70}$

$$CPV = \frac{24.70}{84.70} \times 100 = 29.16\%.$$

To obtain 100 by volume of millbase multiply by the factor of $\dfrac{100}{84.70} = 1.181.$

Step 6. Calculate composite vehicle solids CVS%

required wt of 20% VS Melamine resin R_1 = 18.64 NV = $\dfrac{18.64 \times 20}{100}$ = 3.728

required wt of 15% VS Acrylic resin = 5.64 NV = $\dfrac{5.64 \times 15}{100}$ = 0.846

required wt of 25% VS Melamine resin R_2 = 17.77 NV = $\dfrac{17.77 \times 25}{100}$ = 4.442

required wt of 20% VS Alkyd resin = $\dfrac{12.47}{54.52}$ NV = $\dfrac{12.47 \times 20}{100}$ = 2.494

Total NV = 11.510

$$CVS = \frac{11.510}{54.22} \times 100 = 21.11\%.$$

Step 7. Calculate weight of resin (w_r) and weight of solvent (w_s) to complete the formulation.

$$w_r = \frac{NV}{R_s} 100; \quad w_s = w_{ds} - w_r, \text{ therefore}$$

$$\qquad\qquad\qquad\qquad w_r \qquad\qquad w_{ds} - w_s \qquad = w_s$$

for Melamine resin R_1 $\dfrac{3.728}{67} \times 100 = 5.56$ and $18.64 - 5.56 = 13.08$

for acrylic resin $\dfrac{0.846}{65.5} \times 100 = 1.29$ and $5.64 - 1.29 = 4.35$

for Melamine resin R_2 $\dfrac{4.442}{60} \times 100 = 7.40$ and $17.77 - 7.40 = 10.37$

for alkyd resin $\dfrac{2.494}{50} \times 100 = 4.99$ and $12.47 - 4.99 = 7.48$

$$\text{Total} = 35.28$$

To obtain 100 by volume of millbase multiply by the factor of $\dfrac{100}{84.70} = 1.181$.

Step 8. Write up the completed millbase formulation.
(PV = 29.82% VS = 21.15%)

	V	ρ	w
Monolite Yellow	2.10	1.6	3.37
TiO$_2$	5.41	4.1	21.18
Titan Yellow	12.84	4.3	55.21
Red oxide	8.81	4.3	37.87
Melamine R_1	6.32	1.04	6.57
Acrylic resin	1.52	1.00	1.52
Alkyd resin	6.20	0.95	5.89
Melamine R_2	8.74	1.00	8.74
Xylol	48.06	0.868	41.72
	100.00		

8.5   Methods of dispersion and machinery

Millbases are dispersed in various equipment depending upon the nature of the pigmented millbase, and the quality and volume required. It is not the intention here to deal with the design, characteristics, and performance of the large variety of dispersing equipment available for the paint-maker, but to give sufficient guidance for the most commonly used machinery.

8.5.1   Ballmills

Ballmilling is a batch process and probably the oldest form of manufacture. It still has a useful role to play in the paint industry. It is eminently suitable for the manufacture of high quality paints, and particularly with pigments that are difficult to disperse. Owing to their enclosed mode of operation no change in the constants of the millbase can take place. Ballmilling is a reliable system which requires little supervision and maintenance. However, the processing time can be long, 8–24 hous, and sometimes in excess of 36 hours.

There are restrictions regarding the solvents allowed in the mill, because of vapour pressure. Cleaning requires time and a lot of cleaning solvent between colour changes.

A ballmill is a cylindrical unit with roughly equivalent length/diameter ratios, and it is rotated about its horizontal axis. The inner wall of the cylinder is lined with non-porous porcelain, alumina, or hard silica blocks, to prevent contamination of charge due to abrasion. The mill is filled with grinding media (balls) to specified volume loading. Some mills are still filled with pebbles. Other than spherical-shaped grinding media (cylinders) may be used for specific purposes.

The balls may be different in diameter and made of porcelain, steatite, alumina, and different grades of steel, all of which have different densities. The process of dispersion is, in fact, the simplest in a correctly loaded ballmill rotating at the correct speed. The grinding medium takes up a concentric and approximately parabolic line of motion, and the individual rotation of the balls results in, through shear and abrasion, the reduction of agglomerates and aggregates to primary particles. Impact of balls is less important in wet grinding. From this it follows that the smaller the size, the greater will be the number of points of contact and higher the active surface area of the balls and greater the efficiency of work by shear and abrasion. Relevant data are given in Table 8.1 and Fig. 8.8a and 8.8b).

In practice, the lower limit of ball diameter is determined by the average viscosity/density of millbases to be dispersed, and the ease of separation thereafter. In general, high density grinding media will disperse high viscosity and/or high density millbases faster. In practice it is difficult to allocate ballmills according to the densities of millbases, therefore the general use of steatite balls ($\rho = 2.9$) or alumina balls ($\rho = 3.6$) will give very satisfactory results. The various types of grinding media, listed below, are most efficient when used within the density limits of the corresponding millbase.

Grinding medium	ρ	Millbase ρ
Glass	2.0	up to 1.7
Porcelain	2.4	up to 2.0
Pebbles	2.6	up to 2.2
Steatite	2.9	up to 2.5
Alumina	3.6	2 to 3.0
Zirconium	5.0	3 to 4.0
Steel	7.9	4 to 6.5

Table 8.1

Ball diameter in mm	Surface area of 1 litre of balls m^2	No. of contacts between spheres
0.5	6.90	
1.0	3.44	
2.0	1.72	576000
3.0	1.17	
4.0	0.87	
5.0	0.68	37000
6.0	0.57	
7.0	0.50	
8.0	0.43	
9.0	0.38	
10.0	0.35	4620
12.7	0.27	
25.4	0.13	
30.0	0.12	171
40.0	0.08	
50.0	0.07	37

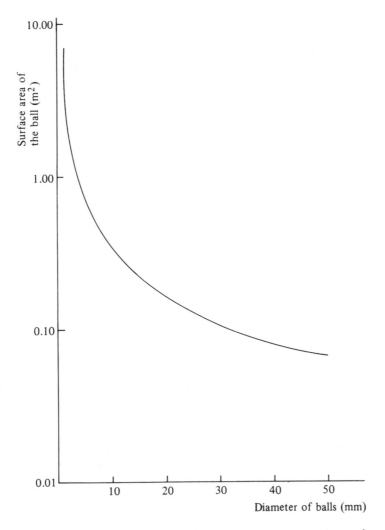

Fig. 8.8a — Characteristic effect of ball diameters on surface area of ball. 1 m^3 T_v mill at 60% loading (D_v).

A thin millbase or large difference between the density of the millbase and that of the grinding media will cause reduced efficiency, overheating, and excessive wear. This should be kept in mind when using steel ballmills with steel grinding media, despite their detuned mode of operation. The rheological characteristics of a pigmented millbase change during the process of dispersion owing to a number of factors, and they are influenced considerably by the nature of the pigment, its particle size, shape, and surface characteristics. Therefore it is difficult to define an ideal viscosity in absolute units for this stage of the process.

Water-cooling plays an important part in plant-scale milling, as this helps to keep the consistency of the millbase within the right shear range.

Laboratory ballmills are essential tools to carry out meaningful processability tests on millbase formulations, provided that they are operated according to the con-

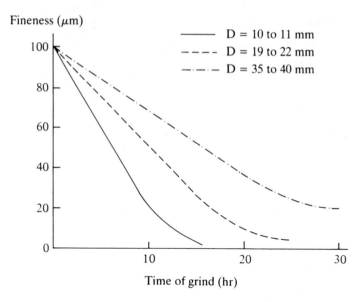

Fineness (μm)

D = 10 to 11 mm
D = 19 to 22 mm
D = 35 to 40 mm

Time of grind (hr)

Fig. 8.8b — The effect of ball diameter on the rate and fineness of dispersion. (Example: Carbon Black/Alkyd Resin millbase. 400 Ltd. T_v porcelain ballmill).

ditions described previously. It is important to remember that owing to geometrical considerations, the time needed to achieve a specified fineness will be shorter in a larger mill than in a small mill.

Critical speed
N_c = critical speed is the speed at which the balls are just held against the wall of the mill by centrifugal force. It is expressed by the mathematic formula

$$\frac{54.14}{r}$$

where r = the internal radius of the mill measured in feet.
    The critical speeds of different size laboratory ballmills are:

Nominal capacity	Volume (ml)	r	Critical speed (rev/min)
1 pint	500	0.385	140.6
2 pints	1000	0.428	126.5
½ gallon	2500	0.460	117.7
1 gallon	5000	0.559	100.5
2 gallons	9300	0.614	88.2

(*Editor's Note*: It is still quite common to find a mixture of imperial and metric units in use in the paint industry. SI units have not gained general acceptance.)

The optimum ballmill speed of rotation, N rev/min is a fraction F of N, and

$$\frac{N}{N_c} = F_c,$$

where F_c in rev/min is given by the formula

$$F_c = 43.3\sqrt{\frac{1}{D-d}}$$

where D = internal diameter of mill in feet;
$\quad\quad d$ = diameter of grinding medium in feet.

The optimum speed is about 65% of N_c, but acceptable conditions are obtained between 60–70% of N_c. Above 75% there will be overheating in larger ballmills, therefore this figure should not be exceeded. It is essential to check and adjust, if necessary, the rotation of plant-scale ballmills.

Calculation of millbase volume c and weight of grinding medium w at different charge-to-voids, C/V, ratio and dispersion volume D_v
The dispersion volume is a percentage of the total volume of the mill, and 60% is recommended as the optimum. However, there are circumstances where changes may be made in D_v and/or in C/V, owing to mechanical conditions to ease the load, or in non-critical millbase formulations to increase the yield. In these cases the following will help to obtain the conditions required:

$$D_v = \frac{T_v \% D_v \text{ required}}{100}$$

The voids (v) is the air space by volume % on the total apparent volume of the grinding medium. For convenience it is usually taken as 40% for close-packed spheres.
    The weight of the grinding medium is obtained by

$$\text{mass volume} \times \rho$$

where the mass volume, M_v = the apparent volume – voids, for example $100 - 40 = 60\ M_v$. The weight of the grinding medium essential to give correct mill conditions is obtained by the following equation:

$$w = \frac{3D_v\rho}{3+2C/V}$$

C = the volume of millbase charge to be dispersed, and it is obtained by

$$C = \frac{D_v(2C/V)}{3+2C/V}.$$

A ballmill is shown diagrammatically in Fig. 8.9 and 8.10 and a typical mill in Plate 8.2.

8.5.2   Attritors
The attritor is a vertical and cylindrical dispersion vessel which is stationary. The centralized rotating shaft agitator is fitted at right angles with bars (spokes) evenly spread, which agitate the charge in the vessel. An attritor is shown diagrammatically in Fig. 8.11, and a typical attritor and circulating system in Plate 8.3.
    The units are available in sizes up to about 100 gallons total capacity. They can be used as batch or as a continuous process. The original Szegvari attritor, in fact, was a continuous unit. Loading is by hand or by pipeline of liquid ingredients if used as a batch process. The following empirical conditions apply:

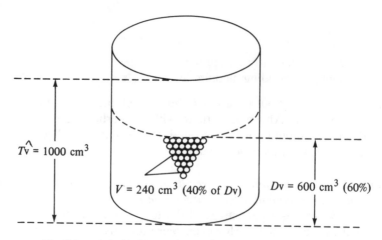

Fig. 8.9 — A ballmill at 60% loading and *C/V* ratio of 1/1.

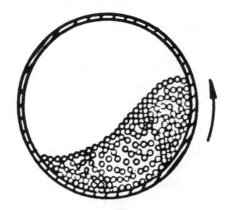

Fig 8.10 — Operating principle of a ballmill.

Plate 8.2 — Ballmill (Sussmeyer).

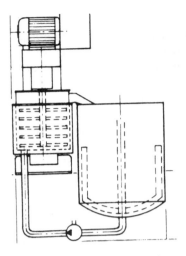

Fig. 8.11 — Diagram of an attritor.

Plate 8.3 — Attritor and circulating system (Torrence & Sons Ltd.).

Loading	= 70% by volume
Charge-to-voids ratio	= 1:1
Tip speed of agitator spokes	= 165 – 190 m min^{-1} (an empirical figure only)
Vessel diameter-to-agitator ratio	= 1:0.75
Size of grinding medium	= 1/4in, 3/8in to 1/2in (6.4, 9.5 to 12.5 mm) steatite balls.

Sufficient recirculation of millbase during dispersion and adequate water cooling are necessary. The charge-to-voids ratio is rather critical; for example, 0.9:1 may lead to excessive wear on working surfaces and may cause ball breakage. The system requires inspection at frequent and regular intervals, in particular with respect to ball weight and general wear of the agitator. The process is on average about three times faster than a ballmill, but requires constant supervision. It can handle higher viscosity millbases than ballmills. A definite disadvantage of the process is that if no top cover is used, solvent evaporation due to the heat generated during dispersion changes millbase conditions, to the detriment of productivity. This should be checked several times by measuring the viscosity of the millbase during dispersion, especially when longer runs are necessary, the lost solvent being replaced as required. If this is not done the constants of the subsequent finish may be affected. It is not ideal for millbase systems containing low-flash solvents.

8.5.3   Sand/bead mills

Sandmills have been used for the dispersion of pigmented millbases since the early 1950s, using on average 30-mesh Ottawa sand; but during the following decade improved results were obtained with glass beads and other synthetic grinding media, hence the process is now called beadmilling.

Typical mills are shown in Fig. 8.12 and Plates 8.4 and 8.5.

The operating principle is to pump the premixed, preferably predispersed, millbase through a cylinder containing a specified volume of sand or other suitable grinding medium which is agitated by a single or multidisc rotor. The disc may be flat or perforated, and in some units eccentric rings are used.

The millbase passing through the shear zones is then separated from the grinding medium by a suitable screen located at the opposite end of the feedport.

There are several makes, for example Torrance, Sussmeyer, Drais, Vollrath, Netzsch, and Master, with vertical dispersion chambers. These can be divided into two types, open-top and closed-top machines, all being multi-disc agitated in a fluidized bed-type situation. The latest machines use horizontal chambers and are available in various sizes from 0.5 to 500 litres. All beadmills, with the exception of the latest Vollrath twin, have cylindrical dispersion chambers and employ glass, ceramic, alumina, and, in certain cases, steel balls as grinding media. The factors affecting the general dispersion efficiency of these units are known reasonably well.

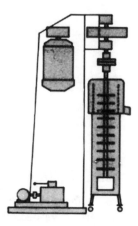

Fig. 8.12 — Original Du Pont de Nemours open unit.

Plate 8.4 — Sandmill (Sussmeyer).

Plate 8.5 — Horizontal beadmill (Netasch LME series).

The selection of the right type and diameter of grinding medium and the product range, in particular the right millbase formulation for a given type, is important for maximum utilization. In theory, the smaller the grinding medium and higher its specific gravity, the more efficient it becomes, because of the much increased surface area, as shown in Table 8.1, and because the centrifugal force transmitted to the grinding medium at the tip of the rotating disc increases considerably its weight, and applies greater shear to the millbase. The speed transmitted to the individual members to grinding media at the tip of the disc assumes that speed, and therefore the force, can be calculated. Let tip speed of the disc be 670 m min⁻ = (2200 ft/min) and the radius of the disc 0.10 m (3.94 in). Then

$$F = \frac{V_2}{rg} = \frac{(670/60)^2}{0.10 \times 9.81} = \frac{124.69}{0.981} = 122 \times \text{weight increase}$$

where F = centrifugal force
    r = radius of disc
    v = velocity
    g = acceleration due to gravity (981 cm s⁻¹).

In most paint media and in practice, however, 1 to 1.2 mm diameter balls is about the limit, because of separation problems, and 2 to 3 mm is the best general-purpose grinding medium. In aqueous low-viscosity media, however, 0.7 mm or smaller diameter balls will further increase the efficiency of the mill, subject of course to relatively Newtonian flow characteristics of the millbase. However, the upper viscosity limitations of vertical units, irrespective of whether or not they are pressurized systems, seems to be 5 to 6 poises of millbase viscosity. In practice they cannot handle difficult-to-disperse systems, for example iron oxides, extenders, and Prussian Blue satisfactorily, and in many instances multi-passing of the millbase is necessary. The efficiency of these units could be improved by the seletion of the right type of grinding medium for the manufacture of more difficult systems, but these media have probably reached the end of development. The difference between these and horizontal units seems to be that in the horizontal beadmill the kinetic energy of the same bead charge is greater, therefore they are more efficient and can handle higher viscosity and more difficult-to-disperse millbases. These units have not yet reached the end of the development stage, and seemingly have more potential with regard to the range of pigmented systems they can handle, and are easier to clean. All units must be water-cooled to keep operating temperature below 50 °C, as even the best-formulated millbase will lose its dispersibility potential above that temperature, owing to a drop in viscosity which can lead to considerable wear of the working surface and bead breakage, and to badly discoloured dispersion. Similar problems arise if the charge to voids ratio is, or drops below 1:1. The normal operating charge to voids ratio is 1:1 or 1:1.2. The accepted shell diameter-to-disc diameter ratio is 1.4:1. The speed of the rotor is about 67 m min⁻¹ (220 ft/min). All units need premixes, but more efficient use of the machines can be made if HSD type predispersed millbases are used. Auxiliary equipment consists of pumps, pipes, storage vessels, and mixers. Maintenance can be costly. Cleaning is effected by passing through the additional resin or solution of resin, essential for the final or part make-up of the paint. Final cleaning is effected by solvent, later to be recovered or used to adjust viscosity. A spare shell per unit is advisable to maintain continuity of production and for ease of colour change. Horizontal units seem to be more flexible in operation; they can handle larger sections of paint dispersions, and probably have the potential to replace ballmills in the long run.

8.5.3.1   Correlation between laboratory and plant-scale units

Single disc batch or multi-disc continuous beadmills from 1 pint to ½-gallon (0.6–2.3 l) vessel volume are ideal for laboratory investigation of the processability of millbases, but at least a 5-gallon (3l) machine is required (with a 20 gallons/hour (11 l h^{-1}) flow rate) to establish scale-up factors with some confidence.

8.5.4   Centrimill
The centrimill is a batch type beadmill with centrifugal discharge. It can be difficult to load the large units. It is a good system if smaller batch size production is required. Owing to the centrifugal discharge facility it gives very good yield, with the exception of very thixotropic bases. It has been found sensitive to millbase formulations and charge to voids ratio. Wrong conditions lead to excessive wear on glass beads and contamination of colour. A diagram of a centrimill is shown in Fig. 8.13, and a Sussmeyer machine in Plate 8.6.

8.5.5   High-speed kinetic energy types of disperser
There are several makes of high-speed disperser (HSD): Torrance, Cowles, Voll-rath, Mastermix, and Silverson. The operating principle is a free rotating disc of 'limited' design (saw type blades) in an open vessel. The exception here is the Silverson type machine where the rotor is enclosed in a stator device which is a somewhat pressurized dispersion chamber. Since there is no grinding medium present, the pigment disperses on itself if the millbase is formulated properly by being accelerated on both the surfaces of the rotor, and the high speed small-volume streams are discharged into a relatively slow-moving vortex, ideally of laminar flow pattern, also created by the rotor. The units come in different sizes and can handle products with viscosities up to 32–35 poise. However, as a true pigment dispersion machine it has its limitations, therefore a refining process, for example, beadmilling, should be added when high quality products are made. HSD is, however, an essential process

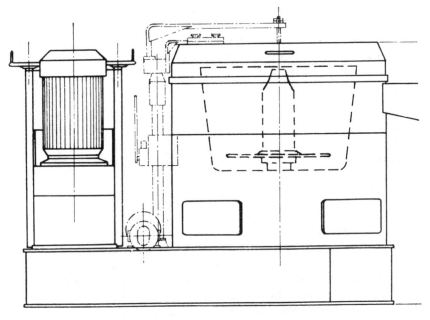

Fig. 8.13 — Diagram of a centrimill.

Plate 8.6 — A centrimill (Sussmeyer).

when properly used, and will do an effective predispersion on most of the pigmented products; and some of these, for example undercoats and primers, will not need further refinements by beadmill, owing to their low standard of fineness. The predispersion on the high quality bases will cut down the number of passes in subsequent beadmills, which means improved productivity. HSD is a batch process, and therefore it is loaded manually in most cases. This creates dust nuisance, in particular when organic pigments and silica extenders are used, therefore an effective dust-extraction system is required. For a more effective batch turnover two dispersion vessels per unit are required unless the unit is arranged in a vertical process system feeding a mixer, storage vessel, or collecting drums for packaging. Alternatively, units may be mounted onto a suitable platform. Simple products can be made up to finishes because there is adequate free vessel volume left, or they can be second-staged to feed a beadmill of suitable size. This second stage in these cases is a formulation modification and ideally should provide the optimum millbase formulation for the subsequent beadmilling process.

For efficient operation, the following specifications are ideal:

Horsepower	15	20	30	45	60
Rotor diameter in inches	12	13	14	16	20
Vessel diameter in inches	30	33	38	40	43
Height of vessel in inches	30	37	40	46	47
Batch size in gallons	30	46	77	100	140
Maximum rotor speed in feet/min.	5000 in all cases				
Horsepower per gallon of millbase	0.5	0.44	0.43	0.45	0.43

The vessel should be water-cooled to give an operating temperature <45 °C. The temperature rises rapidly during the dispersion, and contact temperatures at the tip of the rotor can exceed 70 °C. This causes rapid loss of solvents, changes the viscosity of the millbase, and reduces dispersion efficiency. The speed of the shaft should be infintely variable to ease loading and second-staging. The rotor position should also be vertically adjustable. In practice the distance between the bottom of the vessel and the position of the rotor in the vessel in centimetres equals roughly the viscosity of the millbase in poises; for example, if a 25 poise millbase is to be dispersed, then the distance should be 25 cm (10 in) to obtain a vortex with laminar flow at around 4000–5000 ft/min 1220–1520 m min⁻¹ rotor tip speed. When operating conditions are right the vortex is rolling steadily without surge and splashing. Rise in millbase temperature will, however, lower its viscosity, which is indicated by the onset of splashing, and under these conditions no further dispersion of the pigment will take place. To correct this: reduce rotor speed, lower the position of the rotor, and step up water cooling.

The process needs constant supervision. The average time of dispersion is about 30 min, and batch turnover about 90 min. Dispersion efficiency does not significantly increase with the increased rotor speed above 5000 ft/min (1520 m min⁻¹), but increases only to a limited extent with reduced vessel diameter. This, however, reduces the batch size and considerably increases the power requirements. Scraping the inside wall of the vessel should be carried out several times during the process of dispersion, as the millbase tends to cake at the top edge fo the vortex, which is much worse with non-water-cooled vessels owing to the rapid loss of solvent. For this operation, of course, the unit must be switched off.

Practical operation
Cleaning is relatively straightforward by using solvents. This process is rather wasteful, although the solvent is recovered later. It creates, however, the potential hazard of exceeding the TLV values in the atmosphere in the immediate vicinity of the unit. Adequate ventilation is therefore required.

In conclusion, high speed dispersers have an important function in production. They are relatively simple, and they have improved little in efficiency during the last decade. Nevertheless they should be exploited to the full and not used as mixers only. They are too expensive for that purpose, and there is room for improvement.

Correlation between laboratory 1-horsepower (745 w) machines and up to 60-horsepower (44, 700 w) production units is quite good, but one has to bear in mind that the amount of power available on the laboratory scale is about twice as much as can be obtained from production units per volume unit of millbase.

Two HSD machines are shown in Plates 8.7 and 8.8.

Plate 8.7 — High speed disperser (Sussmeyer).

Plate 8.8 — High speed disperser with hydraulic transmission (Torrance).

8.5.6   Rollmills

The operating principle of rollmills is to feed a premixed millbase between a roller and a bar or between two or three rollers rotating at different speeds, and to apply high pressure onto the thin film in the nip.

The pigmented film which is spread onto the rollers is subjected to very high shear. The dispersed material is removed with a scraper tray (apron), the blade of which lightly touches the front roller.

Rollmills are not suitable for large-volume production; they are labour-intensive, requiring high skill and are therefore costly to operate. Their use is confined to the manufacture of very high quality paints, printing inks, and high viscosity pigmented products where the fineness of dispersion, cleanliness of colour, and colour strength are of great importance. In this sphere of activity it is hard to beat the performance of a well set up triple roll mill.

Efficient extraction is essential over each unit, to eliminate hazards associated with solvent vapours in the atmosphere.

Two rollmills are shown in Plate 8.9.

8.5.7   Heavy-duty pugs

Heavy-duty pugs are used mainly for the manufacture of very high viscosity putties, filled rubber solutions, and fillers where the fineness of dispersion is not critical.

Plate 8.9 — Two Torrance rollmills.

Pugging is a batch process and may be carried out under thermostatically controlled conditions, circulating water or oil through the jacket. The most commonly used rotors are Z-shaped (sigma blade), and two per unit rotate in an intermeshing mode inwards at different speeds in relation to each other to maximize shear in the dispersion chamber. The bottom of the chamber is curved to form two half-cylinders with a dividing ridge in the middle. The shear forces developed by the rotors can be very high, and they depend on the viscosity of the mix. The discharge of the processed batch is carried out by tilting the dispersion chamber or by a screw transporter which is built into the chamber. The machines are available from 0.5 kW to several hundred kW units.

8.5.8 Heavy-duty mixers

Paste mixers are designed to handle intermediate and smaller batches of high viscosity pastes, and are used mainly for the dispersion and manufacture of sealers, PVC plastisols, certain printing inks, and pigmented intermediates for paints.

The advantages of the latest hydraulic drive units are that high torque and infinitely variable speeds can be obtained from 0 rev/min, which ensure that optimum mixing conditions can be established for the different systems. The vessel is cylindrical and must be clamped securely in position owing to the high torque characteristic of the unit. The rotor is of variable design and mounted onto a central shaft which is adjustable for height.

In general, scale-up from laboratory to plant production is not linear, owing to the greater power/unit volume of mix available on laboratory-scale units. However, the correlation curve from the Plastograph Pug mill can quickly be established and matched to plant machinery.

A Torrance paste mixer is shown in Plate 8.10.

Plate 8.10 — A Torrance paste mixer.

8.6 Mixing

The process of paint-making requires extensive use of various types of mixers for the blending and mixing of resins, solvents, additives, and intermediates in order to complete the make-up of paint formulations and obtain a homogeneous product after the dispersion stage. The mixer is a cylindrical vessel which may be fitted with paddle, propellor, turbine, or disc-type agitators to cover a variety of applications and systems. It can be portable or fixed and controlled by automatic timing devices to give periodic mixing if the vessel is used for storage.

Mixing is still often considered as an art, but it has become more sophisticated during the last decade. Many articles have been published on the design and efficiency of mixers; references [5–7] are expecially useful.

The selection of mixers to do a specific job, and the corresponding data regarding power, tank turnover, and performance, are best known by those engaged in the design and sale of such equipment, with the exception of qualified engineers and scientists working for large industrial organizations. We offer here only general guidance. The main factors affecting the efficiency of the mix are:

- viscosity of material(s),
- specific gravity of material(s), and
- solid content.

These factors determine the geometrical dimensions of the mixer, the shear rate, and the power required. The mixing velocity should be adjustable in order to overcome the settling-out of solid particles. In practice this speed is at least twice that of the settling-out rate, and is predetermined for a particular product to be processed.

The intensity of mixing using the tip speed of the impellor as the unit can be classified as

Slow up to	500 ft/min (150 m min^{-1})
Medium	500–900 ft/min (150–270 m min^{-1})
Fast	1000–1200 ft/min (300–365 m min^{-1})

It is important to use the right speed to avoid the introduction of excessive air into the mix. The viscosity of the finished paint is seldom Newtonian; the mix may possess this property at different stages of the process, but it ultimately becomes pseudo-plastic or thixotropic.

Since mixing is not necessarily a thinning process, the viscosity may increase during certain stages, which means increase in the inertia of the material leading to the flattening of the flow pattern and a reduction of the height up to which the impellor will pump.

The viscosity handling characteristics of the various types of impellers used in the paint industry are such that at a cut-off point of 100 or even 200 poise, the paddle, turbine, and propellers can easily deal with the mixing problems.

However, on the basis of equal vessel volume circulation, with at least two vessel turnovers/minute, the trubine type impeller is more economical as it needs 30–50% less power than the propeller at the same peripheral speed. Furthermore, on an equal power basis the pumping/circulating capacity of the turbine is 30% higher than that of the propeller, again at the same tip speed. One of the least efficient impellers in this respect is the paddle which perhaps requires less power than the turbine for the same size of vessel, but produces liquid movement tangential to the device itself with

Plate 8.11 — Portable mixer (Sussmeyer).

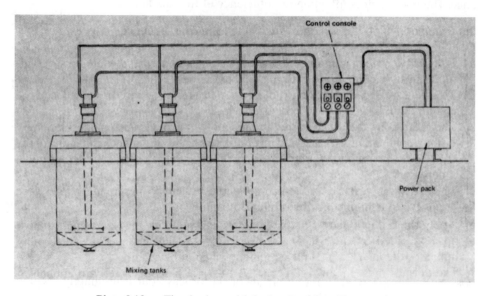

Plate 8.12 — Fixed mixer with hydraulic drive (Torrance).

reduced tank turnover. The paddle is probably the oldest mixing implement and the simplest one. It consists of a shaft with one or more arms either horizontal or pitched to 45°, operating in a vessel which may be baffled to improve efficiency.

The paddle diameter is usually up to 0.9 of the vessel diameter, and it operates in the tip speed range of 250–450 ft/min (76–140 m min^{-1}) with a blade width-to-paddle diameter ratio of 1:8 to 1:12.

The anchor-type agitator is a variation on the paddle, giving better sweep of the bottom of the vessel.

Baffles, where required should be placed not higher than the level of the mix, and an oversweep blade on top may be advisable to stop the build-up of dry bits which develop because of evaporation.

Disc-type impellers are not very efficient in an open vessel, owing to their low pumping efficiency. The flow is generated by the surface friction and by the centrifugal force created by the rotation.

Stator/rotor type impellers can subject the mix to very high mechanical and hydraulic shear.

Portable and fixed mixers are shown in Plates 8.11 and 8.12 respectively.

8.7   Control techniques

8.7.1   Quality control of dispersions

8.7.1.1   Fineness

To assess the degree of dispersion of the pigments and extenders in pigmented products the use of a fineness gauge is common practice in the industry. It is no more than a control test, and the figures specified for a particular product must be established with a batch known to be satisfactory in practice. The gauge does not measure particle size distribution but displays only the largest undispersed aggregates. It is not beyond criticism regarding the size of the aggregate (particle), which may not correspond to the depth of the groove, but proper use by trained operators produces good agreement in the results obtained.

The gauge consists of a case-hardened steel block 7in $\times$ 2$\frac{1}{2}$in $\times$ 15/32in (178 $\times$ 64 $\times$ 12mm^3), all faces being machined flat. A tapered groove is machined down the middle of one of the broad faces of the block. It starts from zero depth at one end and increases to either 50 or 100μm at the other end. One side of the groove is graduated in 10μm intervals, with a dot for each 5μm division. A set of two fineness gauges, with groove depths 0–50μm and 0–100μm, is advisable.

A liquid sample is placed at the deep end of the groove, and is drawn to the shallow end with a doctor blade. The distance from the deep end at which continuous pepperiness in the film becomes apparent is a numerical indication of the state of dispersion of the pigment particles in the material.

Variations in ambient illumination can influence the operator's decision; strip lighting, rather than daylight, is therefore advisable. The groove and blade edge must not be unnecessarily handled; and the gauge and knife must be stored in the box which is supplied with them.

8.7.1.2   Colour strength

Determination of colour strength entails the following sequence of operations.

1   Weigh into a clean container 100g $\pm$ 0.1g of a standardized white finish †. This finish must be compatible with the millbase under test.
2   To give a suitable reduction ratio, add, by careful weighing, 10g $\pm$ 0.1g of the coloured millbase to the standardized white.
3   The reductions should be stirred immediately to avoid inducing flocculation.

---

† Standardized white finish:
The TiO$_2$ pigment is dispersed and made up to a standard recipe. The finish is then adjusted with vehicle to give the same tinting strength as the preceding batches.

4   Using a convenient stirrer ‡, and mix until homogeneous. Check that none of the coloured millbase has adhered to the stirrer or to the can sides, and scrape down if necessary.

5   The reductions can be applied by spraying, brushing, or spreading, therefore viscosity adjustment may be required.

6   Apply the reductions to suitable panels § and build up the film weight to give opacity, and by the use of chequered tape ¶. The degree of opacity may influence the instrumental measurements of colour strength.

7   The finishes must be air-dried or stoved to give films that are hard enough to handle without damaging.

    (a)   Most air-drying paints can either be dried at room temperature or 'force dried' e.g. 30 min at 65 °C.

    (b)   Stoving systems are cured at higher temperatures, e.g. 30 min at 127 °C for a typical alkyd/MF finish.

    Before stoving, time is allowed for solvent evaporation from the paint film; for example, 30 min 'flash-off time' is to avoid bubbling etc. while curing.

    Before examination the panels must be completely cold, as hue changes occur during the cooling process; thus, allow for example, 30 min.

8   The panels can now be assessed for strength either visually or instrumentally ‖. For colours checked repeatedly, a panel can be selected as a reference standard. In this case the reduction must be made in a non-yellowing system, e.g. nitrocellulose.

Notes

When comparing millbases containing different % weights of the same pigment, corrections can either be made by:

- altering the weight of millbases added, or
- adjusting the strength figures using the simple proportions method.

To test white millbases, make the millbase up to a finish and tint using a standardized coloured millbase.

8.7.1.3   Strength and hue determination

Using the graphs for strength and hue determination (Fig. 8.14), the standard with which the sample is to be compared is plotted on the cross wires of both the Y against X and Y against Z graphs. The differences in the X, Y, and Z values between

---

‡ Stirrer type:
Multispeed Stirrer from Anderman & Co. Ltd, Battlebridge House, Tooley Street, London SE1.
§ Suitable panels:
6 in × 4 in (152 × 100 mm²) tinplate panels.
¶ Chequered tape:
¾ in (19 mm) black and white chequered tape from Sellotape.
‖ Instruments used for analysis:
(a) Absorption or reflectance spectrophotometer.
(b) Spectrum/profile instruments.
(c) Photoelectric filter colorimeter.
The values of colour co-ordinates X, Y, Z of standard and test panels may then be calculated.
(a) by a suitable computer program for hue and strength;
(b) by plotting the values as illustrated, obtaining the same;
(c) by a less satisfactory experimental reduction technique adding small amounts of the standard white to the appropriate proportions, and from the measured weight calculate the difference.

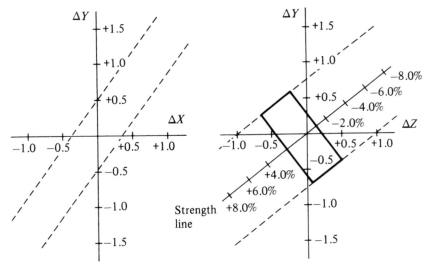

Fig. 8.14 — Strength and hue determination graphs.

the sample and the standard are calculated and plotted on two graphs. For the sample to be within specification the plot of Y against X must fall between the parallel lines, and the plot of Y against Z must fall in the solid line box. Dependent upon the 'tightness' of the specification required, the size of the box can be altered.

Acknowledgements

Thanks are due to Torrance Ltd, Bitton, Bristol; Netzsch GMBH and Sussmeyer GMBH for permission to reproduce photographs and diagrams of their respective dispersion machinery. I would also like to thank Ms Eunice Burridge for her help in the compilation of data included in the chapter.

References

[1] DANIEL F K, Natl Paint, Varn, Lacquer Assoc Sci Sect Circ No 744 (1950).
[2] GUGGENHEIM S, Off Dig 30 (402) 729 (1958).
[3] WIRSCHING F, HAUG R, & HAMANN K, Deutsche Farben Zeits Oct (1957).
[4] FARKAS F K, unpublished work.
[5] PARKER N H, Chem Eng June (1964).
[6] JEANNY G H, Amer Paint J. Aug (1971).
[7] OLDSHUE J Y and SPRAGUE J, Paint Varnish Prod May (1974).

9

Coatings for buildings

J A Graystone

9.1   Introduction

Coatings for buildings represent the largest sector of the total paint market which estimates place at over 50% of the total volume [1]. The market is diverse and at the heavy duty end overlaps with industrial paints. In this chapter the emphasis is on the products that would be typically used on and around houses, schools, hospitals etc., and applied by a hand process rather than in a factory. In common with other coating types, building paints must protect the surfaces to which they are applied and indeed the use of materials such as ferrous metals would be considerably restricted without a protective coating. However, it is also a requirement for most building paints that they should enrich the surfaces to which they are applied with colour, texture, or other appearance characteristics even where, strictly speaking, protection is not required. Paints also have a role in increasing the availability of both natural and artificial light, and the actual colours chosen can influence mood and feeling. Hence this market sector is often referred to as the 'decorative market', other terms used include 'architectural coatings' or simply 'house paints'.

In this chapter some of the choices and factors that influence the selection or development of decorative coatings are considered. Specific detailed formulations are not covered as much information has been compiled in the literature (e.g. [2]) and is also available from raw material suppliers. Emphasis is placed on broad principles and constraints that will govern selection from the building blocks (pigments, binders, solvents, modifiers, etc.) which are described elsewhere in this book, and the ratios in which they are combined. An overarching theme is the influence of the substrate particularly for exterior use. Coatings on an inert substrate will normally last longer than they do on many common building materials such as wood, masonry, or metal. There is a strong interaction between substrate and coating which in extreme cases is antagonistic. Movement in wood for example can cause flaking of paint; conversely an impermeable paint might contribute to decay if moisture becomes trapped. Where standards are quoted they will often be BS standards, in

Table 9.1 — Some typical product descriptions, or classification terms, for building paints

Type	Examples
Generic	Pain, stain, varnish
Appearance	Sheen (matt, silk, eggshell, gloss)
	Opacity (opaque, transparent)
	Colour (hue), broken colour (stipple, marble, graining)
	Build (high, medium, low)
	Texture (fine, coarse)
System function	Primer, undercoat, finish
Market sector	Trade, retail, contract, ready mixed, in-store, woodcare
End use	Trim, broadwall, floor, ceiling, kitchen & bathroom
Composition	Solvent-borne, water-borne, low VOC, high solids
	Alkyd, vinyl, acrylic, polyurethane
Property	Thixotropic, permeable, anti-corrosive
Substrate	Wood, metal, masonry

many cases there will be equivalents or near equivalents from other sources such as ISO or CEN.

9.1.1   Types of decorative coating
Decorative coatings can be classified in a variety of ways which reflect the differing perspectives of marketing, R&D and operations as well as the proprietary and descriptive terms used in the market-place. The terms are not mutually exclusive and give an indication of the attributes that are considered significant in this sector. A selection of those in common use is given in Table 9.1.

9.2   Formulating considerations and constraints

9.2.1   The role of the formulating chemist
Formulation may be usefully defined as:

The science and technology of producing a *physical mixture* of two or more components, (ingredients) with more than one *conflicting* measure of product quality.

Although this is but one of several possible definitions [3] it highlights two essential elements of the formulator's problem. The *first* is that formulators are dealing with *physical mixtures*. A distinction can be made between chemical and physical interactions, though inevitably there is a degree of overlap. Just about all the things found useful in real life are mixtures, yet this point is not always given enough emphasis in explaining structure/property relationships. A great deal of science concentrates on 'pure' materials, but combining things creates enormous degrees of freedom in achieving a desired property mix — often with multiple solutions. More recently science has rediscovered the power of mixtures and renamed them as 'composites'. Since the 1960s there has been a growth in the literature dealing with composite materials and the role played by parameters such as phase volume, interfacial stress transfer, aspect ratio, strength, modulus, toughness, and many others. In their *final physical form* paint coatings are very much 'composites' and can benefit from the new insights gained.

The extent to which a formula is dependent on the proportions of components used, i.e. the 'mixture properties', is a key element in paint formulation: this is illustrated by the importance of PVC (see Section 9.3) as a formulating parameter. Within a formulation (expressed in either weight or volume terms) the proportions of each component *cannot be varied independently* of each other because the proportion of each component must lie between 0 and 1.0, and the sum of the components must also equal 1.0. A consequence of this is that factorial designs are not always the best way to study the response of multicomponent systems. On the other hand mixture theory combined with contour plotting provides a powerful method for solving formulation problems [4], to approach formulation without some understanding of mixture theory, is to accept a severe handicap!

Ultimately the task of the formulator is to meet business needs (Fig. 9.1) by designing a product which has the right appearance, and a balance of properties suited to the envisaged application. Most coatings must satisfy aesthetic and functional considerations, which range from storage through application to film durability. These requirements are met by making an *informed* selection of components and then combining them in *specific ratios* and in the *correct sequence* at a specified state of *subdivision*. But the formulator must also contend with the *second* of the key elements implicit in the formulation definition, namely the certainty that there will be 'conflicting measures of product quality'. This is true of any artefact. In a motor car for example one cannot maximize both acceleration and fuel consumption and a trade-off must be made which would be different between a sports car and a family saloon. In coatings there are analogous choices, for example finding a rheology to maximize brushmark levelling without sagging or settlement, and the ever-present need to maximize key properties at minimum cost. The number of combinations and permutations that can be made from even a small subset of raw

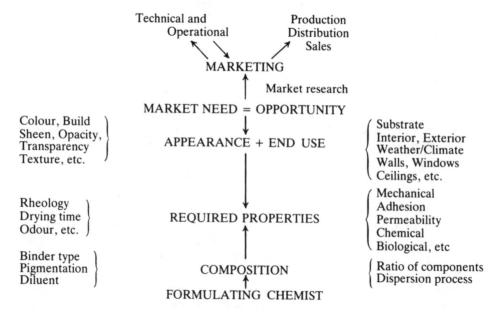

Fig. 9.1 — The role of the formulating chemist.

materials is quite staggering, and formulators need to take full advantage of computers in data analysis, optimization, and information retrieval [3].

Although from time to time it will be necessary to start formulation from a 'green field' situation, by far the most common practical approach to formulating is to modify an existing formulation through substitution of one or more components. Such an approach recognizes that an established formula encapsulates a considerable amount of know-how. The process is most effective if there is *a priori* knowledge to guide the selection of components and a structured approach to experimental design. Analysis of existing formula in terms of volume relationships (PVC), surface areas, morphology, etc. is a means to make the underlying rules more explicit and maximizes the learning from past work. Formulation at its best is a professional discipline which requires:

- access to all available knowledge including historical formulation and relevant technical data;
- experience with experimental design, mixture theory and optimization routines;
- conceptual models relating coating structure and morphology to target properties (e.g. pigment packing, composite material science, colloid science);
- in-depth knowledge of potential formula components at functional, physical and chemical levels;
- understanding of physical phenomena such as
  — mechanical properties, adhesion
  — durability (internal, external, weathering)
  — film formation
  — optical properties (colour, opacity, sheen, etc.)

9.2.2 Constraints operating in the decorative paint market

9.2.2.1 Safety, health, and environment (SHE)
The past decade has shown the continuation of a trend already apparent in the 1970s and continued through the 1980s for SHE issues to have a major impact on the course of paint formulation and even the most cursory glance through current publications will show this to be a dominant theme in research programmes (e.g. [6, 7]). SHE legislation is aimed at both global issues, such as the impact of solvent emissions on the ozone layer, and at the level of personal protection. The latter is important as interior paints will often be used in confined spaces where the build up of solvent fumes can exceed the recommended short-term exposure limits [8].

Decorative paints can be divided into two broad groups according to whether the carrying solvent is predominantly water, or an organic solvent such as a hydrocarbon of the 'white spirit' type. Solvent-borne paints typically contain 50% by volume of volatile material thus their VOC (volatile organic component) will easily reach $400\,\mathrm{g\,l^{-1}}$. Legislation will require this figure to be lowered [9], hence there has been a growing interest in high solids and water-borne polymers including alkyd emulsion binders. Despite their name 'water-borne' paints have traditionally contained cosolvents to impart specific properties such as freeze–thaw resistance. Ethylene glycol is an example where other toxicological problems have been identified [10] and in general there has been a movement to reduce the organic solvent content of water-borne coatings. At the limit this can be reduced to zero with the additional benefit that paint odour is significantly reduced.

Other raw materials that raise environmental issues include heavy metals, formaldehyde, specific biocides, nonyl phenol ethoxylate surfactants, crystalline silica, and specific monomers.

9.2.2.2   Film formation

Coatings must undergo a process where the liquid film is converted to dry. It is an obvious, but nonetheless significant, factor that the means by which this can be achieved are considerably constrained for building paints in general. Unlike motors and industrial markets, the use of high temperature as a route to conversion is ruled out; also, many chemical reactions of the sort used in 'two-pack' products are impractical, or too expensive, or have unacceptable toxic hazards. In consequence, the conversion and curing mechanisms used in general decorative paints are almost entirely confined to some combination of evaporation, coalescence, and oxidation, the implications of which are discussed in Section 9.6. The drive to zero VOC water-borne paints requires that latexes are developed that do not depend on the use of coalescing solvents to aid film formation.

9.2.2.3   Storage stability

A characteristic feature of the decorative market is that the products must have a long shelf-life. Stock turn may be slow for specific items and users generally expect products to remain useable for several years after purchase. Particular attention must be paid to property retention on storage, including application and drying characteristics. Products must resist heavy settlement and irreversible changes such as skinning or coagulation. Where low temperatures are expected, it is desirable to build in freeze–thaw stability. In aqueous paints, hydrolysis of ester linkages is pos-sible and will mitigate against the use of some polymers as binders. Many coating compositions show rheological changes with the passage of time, with pronounced effects on application. Aqueous and, to a lesser extent, non-aqueous paints will often show a steady increase in viscosity after manufacture. Allowance for this can be made in formulation, but it is vital to confirm that a stable viscosity plateau will be reached. A consequence of these constraints is that confirmation of good storage stability becomes an important part of the formulating process and may even become the rate-determining step for a new product launch. It is possible to speed up the acquisition of some data, for example, by incubator storage ($60\,°C$) and 'trav-elometers' (a vibrating device to simulate the effect of transport) but in general it would be most unwise to launch a new product into the mass market with less than 12 months' storage data. Even so, accurate records must be kept, so that trends can be extrapolated from an early stage.

9.2.2.4   Application

The method by which a coating is applied is a constraint that influences all coating formulation. Although it is sometimes feasible to design an applicator to suit specific paints, in general decorative coatings are applied by established tools in particular by brush or roller. This situation is deceptively simple: avoiding drips, brushmarks, roller spatter, etc., while maintaining good application characteristics requires subtle rheological properties. Many of these are difficult to measure and equate to sub-jective opinions. It is not unusual for the professional tradesman and the DIY user to have different preferences. Likewise national preferences can be discerned. In the UK, for example, paints are usually more structured and thixotropic than in the USA.

9.3   Pigment–binder–solvent relationships

In Section 9.2.1 it was stressed that paint is a mixture; consequently properties will be very dependent on the proportions in which components are combined. Each industry involved with formulation problems will have an expression which reflects the ratio of the binder phase to the dispersed 'hard' phase. In rubber for example the index PHR (parts pigment per hundred parts weight of rubber) is used. Paint formulators tend to favour P:B (pigment binder weight ratio) or P:V (pigment binder volume ratio); the latter is often expressed as a PVC (pigment volume concentration) as defined below. Of these indices PVC is widely used in the decorative sector.

If a paint property is plotted as a function of PVC it will show a characteristic shape though clearly the exact values at a given PVC will be very dependent on the specific components used. A good illustration of the forms taken by typical mechanical, optical, and barrier properties as a function of PVC will be found in [11] and [12]. These relationships can be expressed mathematically in phenomenological terms or in some cases as a theoretical model reflecting the underlying physical processes. For example the relationship between opacity and PVC can be explained in terms of a dependent or multiple scattering theory [13, 14].

Despite the usefulness of PVC as a relative reference point it conveys only limited absolute information. If compositions based on, say, titanium dioxide and carbon black are compared at a common PVC in a common binder their properties will be very different. Apart from the obvious differences in optical properties, there will be extreme differences in mechanical and barrier properties which reflect the huge difference in surface area and surface energy even though the pigment volume content is the same. Normally less extreme comparisons are required, such as the difference between two grades of the same pigment type, but clearly it would be useful to have some normalizing constant to facilitate comparison. A clue to what this constant might be can be found by inspection of the PVC/property graphs which almost invariably show a more or less sharp point of inflexion at a characteristic PVC which is known as the 'critical PVC' (CPVC). Interpreting the significance of the CPVC and its determination has engendered considerable debate within the coatings fraternity and is returned to below.

The significance of PVC and other ratios is conveniently illustrated with reference to a trilinear diagram as illustrated in Fig. 9.2. Such 'mixture' diagrams are useful conceptual devices for illustrating relationships between coatings. For specific compositions they may be used as a framework onto which properties may be superimposed and mapped [15].

In Fig. 9.2 composition 'X' comprises 20% pigment, 40% binder, and 40% solvent. A line from (S) through X indicates compositions that have a constant pigment/binder ratio (which can be read off the PB axis) and increasing solids content, while the line AA′ indicates compositions at constant solids but varying pigment/binder ratio. Because the solvent axis is horizontal, total solids content is indicated by distance from the base line which equals 100%. A line such as (P) through X is at a constant binder solids (vehicle solids, v/s) which may be read off

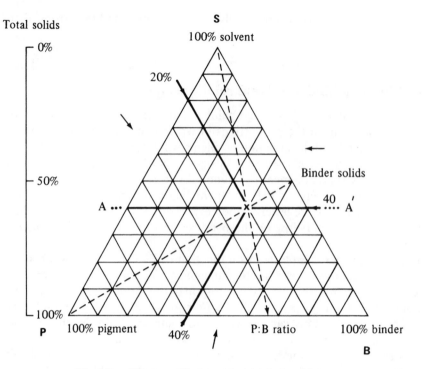

Fig. 9.2 — Pigment–binder–solvent relationships.

the SB axis. A useful property of the diagram is that it allows the geometric solution of certain formulating problems and lends itself to simplex lattice design experiments [4, 5]. Diagrams can be in units of weight or volume; clearly the latter is used for PVC relationships.

9.3.1   Pigment volume concentration

Pigment volume concentration (PVC) is simply defined as the fractional (or percentage) volume of pigment in the total volume solids content of the dry paint film:

$$PVC = \frac{V_p}{V_p + V_b}$$

where V_p = pigment volume and V_b = binder volume.

This deceptively simple relationship is now around 50 years old; its historical development has been usefully reviewed by Stieg [16, 17], also active in developing concepts derived from the PVC and other pigment–binder relationships [18]. Even today there is still debate over the interpretation of PVC and CPVC, some of which has been engendered by the introduction of 'plastic pigment' and other non-pigmentary opacifying aids (see below) as well as developments in clustering and percolation theory.

Understanding the implications of pigment–binder geometry is made easier by starting with a simple model system from which useful relationships can be derived. Such relationships can be used as a qualitative framework in interpreting paint properties, but it is important when doing so not to forget the simplifying assumptions that have been made. Real systems may show significant deviations from the model, especially when interpreted in quantitative terms. For example, calculated PVC may differ from actual PVC if the binder undergoes shrinkage on drying or curing.

The simple PVC definition given above presupposes that the composition of the dried paint film can be expressed only in terms of pigment and binder, whereas in actuality some additives do not fit unequivocally into these categories. The difficulty can be avoided by defining $V_b + V_p$ as total dry film volume (excluding air). V_p (the pigment volume) is usually easier to define, but paint chemists are at liberty to include that part of the total pigmentation in which they are interested provided it is clearly defined. Plastic pigments that may be quasi film forming are a case in point. Clearly, there are a number of variants of the basic PVC that could be useful for a specific purpose. Unless these are defined, it can only be assumed that the simplest definition (i.e. percentage or fractional volume of pigment in the *total* dried film) applies.

9.3.2   The critical PVC

This is one of the most useful concepts in decorative paint formulation. It is commonly associated with paints that are formulated 'above critical' to gain dry hiding. However it should be stressed that the estimated or measured value of the CPVC for *any* given set of components, regardless of whether a derived coating is formulated above or below critical, contains information relevant to important properties including dispersion, rheology, and mechanical properties. The CPVC value can also be an indicator of the degree of flocculation.

In conceptual terms it is instructive to consider the changes which take place when binder is progressively added to a bed of dry pigment until the binder is in excess. The process can then be continued by removing pigment until only binder is present. The initial and final conditions represent the extremes of 100% and 0% on the PVC scale.

During the process of adding binder to a bed of dry pigment, the point will be reached when added binder just fills the voids between the pigment particles. The PVC at this point is known as the critical PVC or CPVC, and represents a significant transition point in paint film morphology which more or less coincides with the discontinuity in properties referred to earlier. Most liquid coatings contain solvents which are lost during the drying process. It is thus possible to formulate them to have PVCs that are above, below, or equal to the CPVC. In the dry film, paints formulated above their CPVC will contain insufficient binder to fill the space available, and the film will become porous on drying. In a solution binder, deflocculated, well-dispersed pigments below their CPVC are not in contact with each other.

The process described above, though a useful introduction, is an oversimplification. Real systems will be subject to many modifying variables including adsorption of binder onto the pigment surface. There are important differences between solution and dispersion binders in their penetration of voids and interaction with particle surfaces. Real pigments are far from spherical. These differences affect the quantitative value of the CPVC, but they do not undermine its qualitative value in explaining key facets of coating behaviour.

9.3.2.1 PVC and CPVC as reference points

It is common practice to use the PVC of a coating as one of the parameters in its qualitative description. Typically, gloss paints have a low PVC (15–25%), while primers, mid-sheen wall finishes, undercoats, and flat paints are progressively higher. Knowledge of a coating's PVC may indicate an expectation of other properties such as opacity and sheen level, and it is common to use PVC as abscissa in many graphical property representations. But in general it would be more useful to relate PVC to the potential CPVC of a coating even when the composition is formulated below the CPVC. The latter acts as an internal reference point automatically compensating for some of the differences between alternative pigmentations. In recognition of this, the term 'reduced PVC' (Λ) has been defined where

$$\Lambda = \mathrm{PVC/CPVC}$$

The reduced PVC will often prove a more appropriate parameter than PVC when comparing formulations or displaying them in graphical form. Λ will be greater than 1 for coatings that are formulated above critical [19].

That expressions containing a PVC:CPVC ratio should be useful is to be intuitively expected. In qualitative terms such expressions indicate whether there is a surplus or deficit of binder relative to the volume occupied by the pigment. If pigments were perfect spheres with no interaction with the binder then a CPVC approaching that of random close packing (0.64) might be expected. In real systems which may be flocculated the packing will be less. Thus the CPVC indicates the packing consequences of pigment–binder interactions and will be *system specific*. There is a parallel here with the expressions used to describe the rheology of suspensions which will normally include an expression to denote the phase volume of hard particles in relation to the maximum packing fraction $\emptyset_{max}$. For example the Krieger–Dougherty equation [20] contains the expression $(1 - \emptyset/\emptyset_{max})$. Toussaint has reviewed the choice of rheological models for application to steady flow [21] which discusses in detail how a shear dependent effective maximum packing fraction has been introduced into phenomenological viscosity equations.

The expression $\emptyset_{max}$ is usually used in relation to the liquid composition before solvent is lost, whereas the term CPVC is more likely to be used in relation to dry film properties. There has always been considerable debate in the literature about the measurement and definition of CPVC which is partly explained by the difference between a conceptual definition, and the 'real' world of anisotropic particles with specific absorption sites. Another perspective is to consider the CPVC as an *effect* rather than a *cause*. A composition such as 'X' in Fig. 9.2 will be in a dynamic state subject to colloidal, Brownian, and viscous forces. As solvent is lost, the composition moves towards the pigment binder axis and assumes a morphological state dictated by its recent history. This state might be determined by for example direct microscopic examination, or inferred from property measurements (optical, mechanical, etc.), each of which will have a different sensitivity to the morphology. Thus it is possible to see the CPVC as an apparent transition point with underlying causal factors, rather than a predestined end point. In this sense there are underlying continuities between $\emptyset_{max}$ and the CPVC.

For the reader interested in more detailed background on this topic attention is also drawn to the March 1992 edition of *Journal or Coatings Technology* [22] which devoted a whole issue to the CPVC debate with contributions from Asbeck and other acknowledged experts. The paper by Bierwagen explains why local fluctuations in pigment concentration, and percolation theory, will cause both variability in CPVC determination and a different sensitivity according to the method of investigation.

It has already been indicated that a CPVC value is system-specific rather than pigment-specific. This does not necessarily mean that there will be a single CPVC for a given combination of components since the CPVC will also depend on the concentration of individual components, particularly those that participate in stabilizing a pigment dispersion. This fact is often overlooked but is very clear from published work. For example Asbeck's data [23] and reproduced in the article cited above, shows that the CPVC of TiO_2 in white mineral oil varies from 0.28 to 0.43 according to the level of dispersant used. Asbeck was able to derive an expression relating the CPVC to the number of primary particles in each agglomerate. This leads to the important conclusion that 'the CPVC represents the densest degree of packing which the pigments can occupy, *commensurate with its degree of dispersion*' [23].

9.3.2.2   Effect of pigment packing geometry

Although CPVC is a system-specific interaction parameter, as discussed above, the actual value will be strongly influenced by geometric packing factors. This is particularly apparent if a mixture of large and small particles is used, since the smaller particles will to varying degrees fit inside the spaces between the larger. A plot of CPVC against the ratio of large to small particles will not be linear and will often show a pronounced peak. Curves of this type have been widely published, and further information is available from suppliers of pigments and extenders.

9.3.2.3   CPVC and oil absorption

A simple way of establishing the approximate CPVC for a given pigment is to add linseed oil to the dry pigment until a coherent mass is formed. The test may be carried out as a spatula rub or by using mechanical mixing against a torque gauge. By tradition, the result (expressed as weight of linseed oil absorbed by 100 g pigment) is known as the 'oil absorption'. If expressed in volume rather than weight terms, then this is equivalent to a CPVC. A high CPVC implies low oil absorption and vice versa. Although there are mathematical methods [24] to predict the CPVC of mixed pigments it is often simpler to carry out the spatula rub on the actual mixture.

Simple linseed oil absorption determinations are widely available and provide a useful qualitative yardstick for selection of pigments and extenders. However, bearing in mind the system-specific aspects described previously, formulating chemists must make informed judgements as to the value of OA-type measurements and with what liquids the test is to be carried out. Table 9.2 illustrates the range of results that was obtained with three proprietary pigments and a selection of test liquids. Organic pigments do not show a very precise end point and the repeatability of the test between operators is not good. The results were significantly different when repeated using mechanical means (in a Brabender Plastograph), reflecting a different degree of pigment deagglomeration.

Solution binder associated with a given pigment at its CPVC can be considered as comprising two components: that absorbed onto the pigment surface V_{ab} and that necessary to fill voids between the particles V_f. This has been illustrated as [15, p. 209]:

Table 9.2 — Example of 'Oil absorption' data with different pigments and liquids (figures in brackets are the derived CPVCs)

	Bayferrox® 130M	Irgazin Red® BO	Irgazin Yellow® 5GLT
Acid refined linseed oil	17 (52)	39 (60)	49 (54)
Solsperse RX50 in aromatic hydrocarbon	28 (41)	71 (47)	62 (48)
Dibutyl phthallate	25 (45)	56 (55)	51 (55)
Solsperse 20,000 in water	37 (35)	78 (46)	68 (48)

$$\text{CPVC} = \frac{V_p}{V_p + (V_{ab} + V_f)}$$

In principle, V_f can be calculated from particle packing models. When this is done, V_{ab} is found to vary more than can be accounted for by absorption theories. Huisman [25] emphasizes that the above equation can only be used to calculate V_{ab} if the pigment paste at the end point of an oil absorption determination consists of *single dispersed particles*. In practice, aggregates of unknown surface area are generally present. The OA end point will be controlled by the effective density of particle aggregates, their packing, and the wettability of aggregates by binder. Huisman proposes a model based on effective particle density which allows for the presence of aggregates. Asbeck used the expression 'ultimate critical pigment volume concentration' to denote the monodispersed state, and proposes an expression which relates CPVC, UPVC, and the diameters of primary pigment particle, d, and pigment agglomerate, D, respectively [23].

If the PVC of a coating is progressively increased above the CPVC, a second transition point may be reached where there is only sufficient binder to coat the particles (i.e. the term V_{ab}), and all the 'free' binder is replaced by air. Castells *et al.* [26] have published experimental data which corroborate the existence of this point, which may be of value in establishing interrelationships between pigment and binder and hence factors which influence optical efficiency. Cremer has used a rheological method to investigate the thickness of absorption layers on pigment and extender particles [27].

9.3.2.4   Critical PVC for dispersion binders
Up to this point the discussion of CPVC has assumed that binder can readily penetrate and totally fill spaces between pigment particles; by implication the binder is a solution. But for many coatings and especially for the majority of water-borne paints the binder is, in fact, a polymeric dispersion comprising particles that must undergo deformation in forming a film. As there will be situations in which this deformation process is not complete, there are likely to be differences in CPVC, however measured, between solution and dispersion binders. Intuitively, it may be anticipated that packing of pigment and dispersion binder will be influenced by par-

ticle size of binder (as well as pigment) and the glass transition temperature (T_g) of the dispersion polymer. T_g is a useful, though not absolute, guide to the ability of a dispersion binder to coalesce and form a film. This ability is further modified by fugitive plasticizers — usually known as coalescing agents — which temporarily increase flow and deformation of polymer particles. However, even with the use of coalescing aids the penetration of particles into intimate contact with pigment is likely to be less than that of a solution, with the result that more dispersion polymer than solution polymer is needed to bind a fixed quantity of pigment, i.e. the CPVC of a dispersion paint is less than that of a solution equivalent.

The ratio of solution to dispersion polymer needed to bind a given mixture has been referred to as the 'binder efficiency' or 'binding power index' e [28]. That is,

$$e = \frac{\text{volumetric oil absorption}}{\text{volumetric (dispersion) absorption}}$$

Inasmuch as it has already noted that CPVC is system-specific it might be argued that there is little point in using oil as a reference point for a latex CPVC (or LCPVC), although it does underline the differences in morphology arising from dispersion and solution binders. Differences in CPVC for pigments in different systems are better estimated from other properties. It is also worth stressing that the value of the 'LCPVC' reflects more than either the difficulty of latex coalescence, or some degree of flocculation. The value reflects all factors that may affect film formation including packing effects and random clustering which would be statistically *unavoidable* even if the system showed no tendency to flocculation. Percolation, clustering, and their consequences have been the subject of much discussion some of which is covered in the papers in [22]. The 'apparent CPVC' (see below) of a pigment in a dispersion binder may not be independent of PVC owing to excluded volume effects. In a dispersion binder the CPVC will also depend on the particle size and distribution of the polymer as well as the pigment particles [29].

9.3.2.5   CPVC and porosity

If a coating is formulated such that PVC > CPVC then the dry film will contain voids and pores, the void structure will have a complex morphology and as the PVC is raised will become increasingly interconnected through percolation of the air and other effects. Floyd and Holsworth have described the CPVC as the point of phase inversion where air becomes the continuous primary phase [22]. Although the consequences of porosity are often undesirable its presence has a major effect on opacity. Typically around 50% of the opacity of a commercial matt emulsion paint can be attributed directly or indirectly to the presence of pores small enough to scatter light. Control of porosity, and pore morphology, is thus an important formulating parameter.

If a coating such as composition 'X' in Fig. 9.2 is formulated to be above the CPVC, then as it loses solvent during drying the composition will progress down the constant P:B line until it crosses (or approaches) the line of pigment concentration equivalent to the CPVC. As there is now insufficient binder to fill the total volume, air enters, and the composition becomes a four-component mixture which could be represented as tetrahedron with air as the fourth axis. As drying continues the representation of the composition will pass through the tetrahedron until when fully dried it 'reappears' on the face representing pigment/binder/air compositions, as illustrated in Fig. 9.3. Such diagrams provide an instructive conceptual aid to visu-

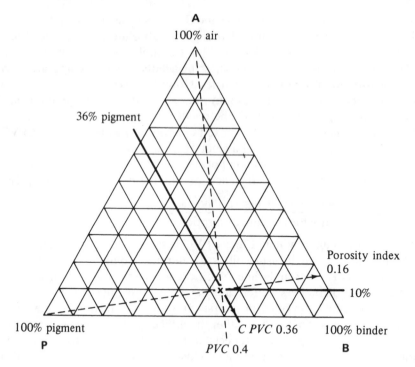

Fig. 9.3 — Relationship between CPVC and air content.

alizing porosity and as Stieg has pointed out can in principle be used to estimate the CPVC without the need for extensive PVC ladders [30]. Such estimates work reasonably well for solution binders. For dispersion binders packing effects interfere and the CPVC will not be independent of the PVC from which it is estimated.

In Fig. 9.3 the composition of any dry film formulated below critical will lie along the baseline P–B. Any composition formulated above critical will contain air (as indicated by the scale A–B) and will lie within the triangle. Any line drawn from the apex A to intercept P–B represents compositions of constant PVC and decreasing air content.

If a coating is made at a given PVC above critical and the air content is determined, then the composition of the dry film is at the intersection of the PVC and % air axes. This is illustrated in Fig. 9.3 for a film containing 0% air and a PVC of 0.4. If it is assumed that voids between pigment particles do not change in volume as binder is removed, then the % pigment can be extrapolated to the baseline of zero air content, indicating a CPVC of 0.36. A PVC ladder would result in a steadily increasing air content whose intercepts with the appropriate PVC line should lie on the same straight line also indicating a CPVC of 0.36. However, for the reasons that were discussed earlier it cannot be assumed that the CPVC is independent of the binder concentration. For real, rather than hypothetical, compositions, the CPVC line is often curved and to some extent reflects the same shaped contour that would be obtained by translating Daniel flow point measurements onto the same diagram. In simple cases a good approximation to the CPVC may be estimated from a single measurement but sensitivity to concentration effects should be determined before assuming this.

Inspection of Fig. 9.3 shows that the proportion of air present may be expressed as a total porosity (as indicated by the horizontal scale) or alternatively as a % of the volume, excluding the pigment. The second way of expressing porosity is usually known as the porosity index PI, and may be read off the graph by extrapolating the line from point P through the composition to the air–binder axis. In the example shown in Fig. 9.3 the composition has a total porosity of 0.1 and porosity index of 0.16. If it is intended to change a given formulation, it may be possible to change one or both of these parameters. As they have a different effect on paint properties, this is an additional degree of formulating freedom.

Porosity and porosity index are conveniently expressed by numerical expression (see [30] for derivation) such that:

$$\text{Overall porosity} = 1 - \frac{\text{CPVC}}{\text{PVC}} \tag{9.1}$$

$$\text{Porosity index} = 1 - \frac{\text{CPVC}(1 - \text{PVC})}{\text{PVC}(1 - \text{CPVC})} \tag{9.2}$$

If the porosity of a film is known at a given PVC, then the CPVC can be estimated from (9.2) providing a relatively quick method which relates directly to the composition under investigation. One method of estimating porosity is from the difference between the measured D_a and theoretical density D_t using expression (9.3)

$$\% \text{ air} = \frac{D_t - D_a \times 100}{D_t} \tag{9.3}$$

Some care may be needed in calculating the theoretical density if the composition contains opacifying beads. Density can also be measured by other techniques including hydrostatic weighing [31] and mercury porosimetry [32].

9.3.2.6 Summary

In view of the sensitivity of many important paint properties to PVC/CPVC it is clear that there are numerous ways in which this transition point could be detected. The method chosen should take into account the reason for which the information is needed. For coatings that are formulated *below* the critical PVC and that show a sensitivity to concentration as discussed above there is clearly a question mark over using values that can only be obtained from a different composition (i.e. by inferring the CPVC from another compositional point). Simple methods such as oil absorption can provide useful qualitative data but may not relate to specific systems. There is a case for arguing that for coatings that are formulated *above* critical, then methods that detect the onset of porosity are particularly relevant. A review of experimental methods is given in [33]. The Cleveland Society for Coatings Technology has published a series of detailed studies on changes in hiding during latex film formation [34] in which the reflectance of the film is measured during drying. The pattern of reflectance change will be different for films above and below the CPVC. The technique has enabled the role of specific components (extender, opaque polymer, coalescing agent) on film formation and the position of the CPVC to be studied.

9.3.3   PVC property relationships

Most paint formulation takes the properties of the film former (binder) as a starting point and the chemistry of a wide range of organic film formers has been described in Chapter 1. Selection of binders suited to the decorative market will depend on many factors (see Section 9.6) including the dry film properties. These will be systematically modified by the inclusion of hard particles such as pigment and extender. Film properties which exhibit a sensitivity to the phase volume of the inclusion (i.e. PVC) have been conveniently grouped under three main headings [11, 12] in Table 9.3.

9.3.3.1   Mechanical properties

The mechanical properties of polymers depend on molecular weight, crosslink density, free volume, length and number of rotational bonds in the repeat unit, and T_g [35]. Quantitative structure–property relationships (QSPR) and group contribution theory can be used to estimate mechanical properties from material constants providing the polymer chemist a starting point for design. However for paint formulators, especially when using proprietary materials, it will usually be more convenient to measure the mechanical properties of the unmodified polymer directly.

A general account of mechanical properties and their determination is given in Chapter 16. Properties of particular interest to the decorative sector include modulus, extension to break, and elastic recovery. Toughness and ultimate tensile strength must reach minimum levels adequate for the service conditions, but in general this is not such a stringent requirement as for, say, car paints where impact damage from stones is a problem. High strength and modulus can be a positive disadvantage in decorative coatings which will often be subjected to a strain rather than stress damage regime. Thus if a decorative coating is subjected to a fixed strain, a high modulus (the slope of the line in Fig. 16.3) results in a high interfacial stress which may detach or weaken adhesion to the substrate. Some elastic recovery is necessary to mitigate the effects of strain damage. Strain cycling of a coating between fixed limits will provide a pattern of hysteresis curves which provide a characteristic 'fingerprint' providing a useful basis for a quick comparison between compositions of known and unknown performance. Mechanical property testing is a specialized field and requires care in sample conditioning particularly for thermoplastic glassy polymers which may be slow to reach equilibrium. For polymers of a more thermosetting nature the overall age will have a dominant effect. It should also be noted that much laboratory testing employs uniaxial tests whereas coatings are normally subject to biaxial or multiaxial tensions. In addition to direct measurement, qualitative information on deficiencies in mechanical properties can often be inferred from service performance, exterior durability provides a good example.

Table 9.3

Mechanical	Transport	Optical
Density	Permeability	Light-scattering
Strength	Blistering	Contrast ratio
Modulus	Staining/rusting	Tinting strength
Adhesion	Enamel hold-out	Gloss

The effect of filler content on the mechanical properties of polymer has been widely studied. In general, the modulus of elasticity (stress/strain) will show a steady increase; at the same time there is usually a decrease in the extension-to-break. The net result may be an overall rise in the strength at break, but this will depend on a number of features including the efficiency of stress transfer between pigment and binder. One convention classifies pigments and extenders as 'reinforcing' if the strength rises and as 'fillers' if only density and hardness are affected. With surface treatment (e.g. silane coupling agents) fillers can sometimes be changed to reinforcers. Where higher strength is achieved it will be at a lower elongation and the maximum extension to break is usually reduced. Morphology of pigments (spherical, rod-like, or laminar) has a pronounced effect on the form of the stress–strain curve. Reinforcing of pigments and extenders interact with polymers through their surface energy and will alter the mobility and flexibility of adjacent polymer chains, the net result is a change in T_g of the polymer. Generally the T_g will increase, though with selective adsorption of specific functional groups has been reported to give a reduction in T_g by increasing thermodynamic degrees of freedom [36].

Quantitative interpretation of the effect of hard inclusions on modulus and strength properties show a parallel with rheological equations. Expressions for modulus can be related to the Einstein equation which contains the term $(1 + 2.5\emptyset)$, later expressions include the term $(1-\emptyset/\emptyset_{max})$, and where $\emptyset_{max}$ allows for an adsorbed polymer layer it is clearly analogous to a CPVC. The derivation of these equations and other influences of pigmentation on the mechanical properties of polymers is reviewed in an excellent paper by Toussaint [36]. Asbeck [37] has related changes to the ultimate properties of pigmented coatings to strain amplification of the binder and derived expressions for tensile strength which include PVC/CPVC ratios. A maximum tensile strength at the CPVC is predicted and generally found in practice. Strain amplification leads to increasing film tension as the PVC increases and will reach a maximum at the CPVC. This is demonstrated by the warping of thin coated strips and is another way used to estimate the position of the PVC in ladder experiments [38]. The author has noted characteristic changes in acoustic emission noise at the CPVC.

It has also been reported that adhesion will be maximized at the CPVC. Some caution is needed in interpreting such results and it is important to draw a clear distinction between 'adhesion' as an intermolecular concept and 'adhesive performance' (or practical adhesion as measured by mechanical testing). Adhesion is a complex combination of mechanical adsorption, diffusion, and electrostatic factors where consequences will be greatly modified by the way the test is carried out (including strain rate), and by the condition of the substrate (cleanliness, porosity, etc.). Bikerman has stressed the role of weak boundary layers. Specific examples of using adhesion to detect CPVC have been described [39]. Because adhesive performance depends on the viscoelastic response of the film tested it will be sensitive to pigment volume effects but is more readily controlled through specific interactions arising from functional groups in the binder. With paints based on aqueous dispersions residual surfactants will have complex effects on practical adhesion.

9.3.3.2 Transport properties

The permeability of coatings to liquids, gases, and vapours — and in particular water vapour — is an important property. Moisture vapour permeability is of especial importance in the corrosion of metals and in controlling the rate of moisture change in wood.

Since vapour passes through the binder and not the pigment, it would be predicted that an increase in PVC would at first decrease permeability — this is usually found to be the case unless the pigment is not wetted by the binder. Anomalous permeability can be used as a sensitive measure of pigment wetting [40]. With or without good pigment wetting a fairly abrupt increase in permeability will occur at the CPVC as definite micropores appear, and this will be reflected by changes in rust prevention and blistering behaviour. The sharpness of the transition in behaviour will be modified by percolation of entrained air, or air intimately associated with the pigment, and flocculation effects. Enamel holdout (i.e. resistance to penetration by glossy finishing coats with consequent gloss loss) will also change.

9.3.3.3 Optical properties

White paints achieve opacity and whiteness from their light-scattering ability (in coloured paints absorption becomes more important). Functionally, titanium dioxide is the major source of such scattering. It is convenient to break down the scattering potential of particles into three groups of phenomena: surface reflection, refraction, and diffraction as discussed in Chapter 17.

Diffraction

Wave theories of light account for bending phenomena which give rise to scattering. Light scattering by this mechanism increases with decreasing particle size until an optimum is reached at approximately half the wavelength of light. Particles of the optimum size can bend approximately four times as much light as actually hits them, because the scattering cross-section is around four times the geometric cross-section. The crowding of TiO_2 particles will therefore detract from TiO_2 particles achieving their full scattering potential. This makes it difficult to increase the opacity of a given paint in a way that makes the most efficient use of extra titanium. Optical particle size for scattering can be calculated, and also depends on wavelength. Practical results will also be influenced by particle size distribution and shape. In paints which contain extender the titanium will be packed into the voids between the larger particles — thereby acting as though it were at an even higher TiO_2 PVC. However, the TiO_2 cannot be packed into the voids of extender system any more tightly than at its own CPVC. For this reason prime pigments designed to be used in high PVC formulations are sometimes given a relatively thick coating, typically silica. The 'fluffy' layer provides spacing to retain efficiency under conditions of close packing. However it should be noted that silica coatings will significantly alter the isoelectric properties of the pigment with other formulating consequences. It has also been argued that such highly coated pigments are an expensive way to purchase TiO_2 and the alternative is to use fine particle size extenders. Such extenders are sometimes described as 'spacers'. Fine particles extenders do not improve efficiency in absolute terms but avoid the excluded volume crowding associated with coarser extenders. Stieg uses the term 'dilution efficiency' to rank extenders on this basis [41]. Temperley et al. [42] have developed a mathematical model to predict the effects of extenders on pigment dispersion and distribution.

The excluded volume effect of extenders has its counterpart in the effect that dispersion polymers can have on pigment distribution and this has engendered interest in pigment encapsulation as a means to reduce percolation and flocculation crowding [43].

Refraction
Light may be refracted by particles and hence scattered; the change in direction is dependent upon the refractive index. The smaller the radius of the pigment particle, the greater the total area available to interact with light. Scattering by refraction increases until a critical point where light rays no longer 'see' the particle and go past it as though it were in solution. Microfine TiO_2 [44] is thus transparent.

Reflection
Surface reflection obeys Fresnel's law (equation 17.3); since air has a lower refractive index than any binder, its introduction into a film will increase light-scattering by the introduction of pigment/air interfaces. Thus in paints formulated above their CPVC, TiO_2 will show enhanced opacity known as 'dry hiding'. Hence the importance of the porosity index described earlier. Dry hiding can only be taken advantage of if the other consequences of porosity are not a major problem. Dry hiding paints are frequently used on ceilings and internal walls.

9.3.4   Formulating above-critical decorative coatings
Matt emulsion paints may be formulated above or below the critical PVC; the former will show better scrub and stain resistance. For many applications, such as ceilings and walls in low traffic areas, more porous coatings are acceptable and a majority of commercial paints are formulated in this area. The presence of air voids contributes around 50% of the total opacity, a fact that can be demonstrated by wetting the dried film with a penetrating solvent such as white spirit. The enhanced opacity can be thought of as arising directly by scattering from the air voids, and by the effect their lower refractive index has on the scattering of TiO_2 itself. Furthermore since the films from 'above critical' coatings occupy a greater volume they will reduce crowding effects between the pigment particles.

A typical strategy in formulating is to raise the PVC as high as possible until deterioration of other properties becomes unacceptable. From the preceding discussion it follows that the choice of components (e.g. extenders, polymer type, dispersants) offers the formulator compositions covering a range of CPVC values. A given value of porosity can be achieved with different combinations of PVC and CPVC. Initial comparisons are best compared on a basis of equal porosity, rather than equal PVC. The size of extenders used will control the void size. Finer voids are more effective at scattering than coarse ones. The formulation of paints to a constant total film porosity has been discussed by Stieg [45] and Casarini [46]. A problem with above-critical paints, particularly if the CPVC is low, is a tendency towards mudcracking.

Mudcracking is caused by isotropic shrinkage on a fixed substrate which causes a biaxial tension [47], it is very sensitive to film thickness and a wedge film test can be used to compare different formulations. In latex paint coalescing solvents reduce mudcracking. The move towards solvent-free compositions has thus exacerbated this tendency. The problem can be reduced by introducing an extender which reinforces the drying film. Laminar extenders such as wet milled mica have proved effective [48].

9.3.5   Air and polymer-extended paints

Whiteness, when it occurs in nature as in snow, flower petals, or sea foam, does not require the presence of TiO_2! Such whiteness arises from the interaction of light with interfaces and voids between air and a transparent medium as described above. Clearly this is a potential source of cheap opacity for the coatings industry and one that has attracted a great deal of attention in the decorative market sector. The introduction of microvoids gains opacity in a manner analogous to the dry hiding described previously; but in principle, depending on the mechanism used, it is possible to increase opacity without sacrificing other paint properties (such as dirt retention) which are coincident with an increase in porosity.

When a microvoid containing air is totally encapsulated, it is conceivable to have coatings of low TiO_2, high opacity, and high gloss; but the greater part of work is more appropriate to be used in conjunction with conventional pigments and extenders and is especially relevant to matt and silk paints.

The use of plastic pigment, encapsulated air, opaque polymer, etc. poses interesting questions in interpreting CPVC and similar calculations. It is not always clear whether a plastic particle should be considered as part of the pigment or as part of the binder, and to some extent it is because these materials fall between the two that they offer the chemist new formulating options. To get the best from these options requires an understanding not only of the new pigments but also how best to exploit the old! It will frequently be found that claims made for a new organic pigment (microvoid or solid) are based on a formulation box which is disadvantageous to conventional pigmentation, or vice versa.

Weighing up the pros and cons of two dissimilar formulations each 'optimized' within its own preferred formulation box is by no means straightforward. Formulae may be optimized around a number of properties, each with a different market appeal. However, the relatively high cost of titanium dioxide has ensured continued interest and the use of these materials in interior products is widespread. The use of opaque polymer in external coatings is less common but evidence shows that substitution into a durable conventional product is possible and replacement of non-durable extenders a positive advantage. The pattern of colour change on external weathering is likely to be different due to chalking differences and loss of voids in the surface layers.

9.3.5.1   Air/particle configurations

The use of microvoids in man-made materials was extensively reviewed by Seiner in 1978 [49]. Seiner has also published an interpretation of theoretical analysis by Kerker et al. as applied to coatings [50]. Since then there have been several different practical approaches to the problem resulting in much patent, and some commercial, activity. Chalmers and Woodbridge reviewed the better known commercial routes in 1983 [51]. Four methods of pre-forming polymeric beads have been extensively: 'Spindrift', 'Microbloc', 'Plastic pigment', and 'Ropaque'.

'Spindrift' [52, 53]

Dulux Australia Ltd disclosed the use of both solid and vesiculated polyester/styrene beads. The solid beads are comparatively large (up to about 40m) and may be clear or pigmented. A major use has been to formulate matt paints with exceptionally good polish resistance. Unpigmented beads act as windows in the film and are used only for highly saturated colours. A requirement of very flat finish combined with good polish resistance has created a market for beads of this type in Australia and South Africa, even though they are more expensive than conventional extenders. However, for cost savings in terms of opacity, the vesiculated bead is used.

Vesiculated beads are prepared by a double emulsion technique in water, starting with an unsaturated polyester/styrene precursor and using a redox-free radical system for curing. The manufacturing route requires careful process control. Beads are produced in a size range up to around 25 mm, though 11–14 mm is more typical, the vesicles are around 0.7 mm and contain TiO_2 within the vesicle. In use, Spindrift beads enhance TiO_2 efficiency while maintaining film integrity, and with a suitable latex very high PVCs can be achieved (i.e. counting beads as pigment). More recent developments have led to both smaller beads and a higher degree of vesiculation, thereby extending the range of application. As with most specific formulations the usefulness of this technique depends on relative costs and local market requirements. Beads which have been successfully exploited commercially in one market may not have the same perceived advantages when translated to another. The difficulty in choosing between an optimized bead formulation and a suitably formulated conventional paint is highlighted by the correspondence reported in [54]. Guidelines on the effective use of vesiculated beads have been published by Goldsbrough *et al.* [55].

'Microbloc' (Berger Jenson & Nicholson)

'Microbloc' particles are actually aggregates of finer particles formed by shearing an addition polymerization reaction in aqueous medium; they are used in a similar way to Spindrift. It has been claimed that their irregular shape gives higher film strength then spherical beads. Unlike Spindrift vesiculated beads, it is not claimed that internal pigmentation with TiO_2 is efficient and 'Microbloc' is usually combined with external TiO_2 and film extender to produce high PVC paints, which are very flat.

'Plastic pigment' (Glidden)

Ramig and co-workers [56, 57] propose fine particle size polystyrene 'beads', i.e. a polystyrene latex in the range 0.1–0.6 μm to cause microvoids in paint films and therefore enhance opacity. The beads are of the same order of magnitude as the latex in emulsion paints, which they are typically blended in a 1:1 ratio. They are used in silk and matt paints effectively to raise the PVC above critical and gain opacity by dry hiding, but with less loss of film integrity.

The use of polystyrene pigment in this way is a good example of the difficulties encountered in applying conventional PVC calculation to these organic pigments. On the one hand the particles from a high T_g hard latex could be seen as fine particle size extender (i.e. above the line in the PVC equation), and on the other hand as a latex with zero or low binding power. The truth probably lies between these extremes. By appropriate choice of coalescing aids it should be possible to ensure better integration of the hard polymer particles than would be possible with an inorganic extender, but without causing the particles to coalesce totally. Careful matching of coalescing, and other paint constituents, would be necessary to optimize performance, but the route to an alternative formulating box is indicated [58].

'Ropaque'

Rohm & Haas [59] have developed opacifying aids, comprising hollow acrylic styrene beads suspended in water (typically 37% by weight, 52% by volume). As paints containing this 'opaque polymer' dry, water is lost from the cores of the particles to be replaced by air. Resulting air voids act as scattering sources and make a direct contribution to hiding. It is claimed that four parts by volume of opaque polymer is approximately equal in hiding power to one part of TiO_2 [60]. Commercially available material ('Ropaque' OP-62) has a uniform fine particle size around 0.4–0.5 μm and may also contribute to hiding power by uniformly spacing the titanium dioxide particles and helping to prevent crowding. Because the particles have less surface area than corresponding volumes of titanium dioxide, it has been argued that there is a reduction in binder demand. In other words, the CPVC has been increased, allowing formulation at a higher PVC. Clearly this argument defines opaque polymer as pigment rather than quasi-binder.

As with all new opacifying aids, reformulation of existing paints must be carried out sensibly and on a volume rather than a weight basis. Detailed formulating protocols have been published by the manufacturers and are discussed by Woodbridge [61].

9.4   The nature of the paint binder

9.4.1   General

Formulation of most coatings usually starts with the selection of the main film former or 'binder' even though, as described elsewhere, these properties are extensively modified by pigmentation. Typically the specification for a decorative coating will contain elements relating to appearance, application, and durability. The latter may refer to exterior durability (including resistance to weathering) or interior durability (covering toughness, stain, and scrub resistance). Choice of binder to meet the specification for a given subsector is governed by three major 'elements':

- the nature of the carrying solvent;
- the physical form of the binder;
- the chemistry and physics of the dry film.

In Section 9.2.2, SHE issues were noted as constraints governing the choice of binder. While this has been made more explicit by legislative pressures, there have always been implicit constraints in this market sector where coatings are applied under a wide range of conditions often by non-specialists. Hence many of the film formers described in Chapter 2 are unsuited to all but the most specialized subsection of the building sector. This is particularly true of technologies requiring stoving or two-pack reactions. For many years the decorative market was dominated by two broad types of technology, namely 'emulsion polymers' and 'alkyds'. The former are polymeric *dispersions* carried in water, while the later are *solutions* of oil-modified polyesters carried in hydrocarbon solvents. It can be seen that these two groups differ in all three of the elements described above, and *all* must be invoked to explain the balance of properties in derived coatings. However, for simplicity these broad technologies are often identified on the basis of the carrying solvent alone as 'water-borne' and 'solvent-borne'

In general terms there is a continued swing from solvent- to water-borne which proceeds at different rates around the world [62, 63]. It is dominated by SHE considerations but strongly influenced by practical considerations (e.g. climate), economic aspects and sometimes historical preferences. For example in Germany it is estimated that over 90% of coatings in the decorative sector are water-borne. The figures for the UK, France, and Norway are 68%, 51%, and 47% respectively [62]. The percentage of water-borne paints is expected to increase further. Solvent-borne paints have been virtually eliminated from interior broad wall applications but retain an important share of trim paints, external paints, and specific application where the property balance cannot be matched by water-borne technologies.

9.4.2   Possible binder combinations

Taking the simplest extremities of the elements identified above could be interpreted as giving a choice of water or solvent, solution or dispersion and crosslinking or non-crosslinking chemistries. This defines eight broad categories of binder. Polymers and polymer combinations within each category will have certain features in common, but there will of course be significant differences arising from their individual chemistry. The net result is a pattern where some properties are in clusters while others will overlap across the categories. As noted above, the traditional decorative market has only made use of two out of the eight possibilities defined here. Consideration will now be given to why this has been the case and what changes can be made to this simple model in order to explain emerging technologies.

9.4.2.1   Solvent vs water

Solvent-borne (sometimes called non-aqueous) paints based on hydrocarbons might be expected to be flammable, have strong primary odours and exposure limits that can be exceeded in confined spaces. Water-borne or aqueous paints, on the other hand, will be non-flammable, non-toxic but susceptible to freezing and may cause rusting of ferrous substrates. A wider range of solvents is available giving solvent-borne paints more control of drying time. Water has a high latent heat of evaporation compared with solvents which can cause problems in those market sectors (e.g. industrial, factory applied) where force drying is necessary. However in building paints under normal drying conditions the problem is often to slow the rate of drying and this is difficult without the use of cosolvents or humectants which have undesirable side effects. On the other hand under very humid conditions water-borne paints will not dry at all, unlike their solvent-borne counterparts. Also water-borne coatings used outside are more vulnerable to rain damage when freshly applied. Another problem with an aqueous continuous phase is a high surface tension which must be lowered especially where wetting of low energy substrates is required. This can be achieved with suitable surfactants which may in turn cause foaming.

9.4.2.2   Solution vs dispersion

A significant practical difference between solution and dispersion systems arises from their rheological behaviour. As discussed in Chapter 14 and illustrated in Fig. 14.10 and 14.12, the viscosity of solution polymers is very dependent on molecular weight and rises progressively with concentration, the rate of increase is steeper in the concentrated region. Dispersion systems in contrast show rheological behaviour virtually independent of molecular weight, viscosity will initially be relatively low but rises steeply as the phase volume approaches $\emptyset_{max}$. This difference in behaviour has a major influence on the way rheology is controlled during the liquid phase including application properties. Furthermore, the rate at which viscosity rises as liquid is lost will be different. With dispersion systems irregularities in the film such as brushmark disturbance are more likely to be trapped [64 and Chapter 15], impairing the appearance of the more glossy systems. In contrast the better flow of solutions can lead to sagging or runs on vertical surfaces, and a failure to cover sharp edges. It is generally true to say that in order to minimize flow problems, the solids content of aqueous dispersion paints will be lower than that of solution paints.

Control of rheology in alkyd-based paints is usually with polyamide-modified alkyds (see Chapter 2 Section 2.5.7). Organically modified bentones are also used, these have a high 'low-shear' viscosity and are more common in undercoats and primers. For water-borne dispersions the most common rheological modifiers (or thickeners) were sodium carboxymethyl cellulose and hydroxyethyl cellulose. These are still used but increasingly displaced by new types of synthetic polymers collectively known as 'associative thickeners'. The two most common types are carboxylated acrylic alkali soluble (CAAS) and hydrophobically modified ethoxylated urethanes (HEUR). This is an active area of development. Associative thickeners offer new degrees of freedom in controlling the rheological profile of water-borne dispersion paints. However the interaction with other components including the pigment [65], pigment dispersant, coalescing aid, and latex particle [66] is complex. A practical approach to optimizing latex paint rheology with associative thickeners through blending has been described by Anwari and Schwab [67]. Associative thickeners have the advantage over cellulosics of being resistant to enzyme degradation. Their main disadvantages are cost and the formulating complexity alluded to above.

A feature of aqueous solution polymers is to show a very non-linear rheological response to water dilution. The viscosity may show a pronounced rise and fall which has been described as 'humpy' or as 'the water mountain'. An explanation for this behaviour can be found by consideration of the quasi-solubility which will be concentration and pH dependent. Trilinear phase diagrams are a useful way of mapping the formulation options and the pathways that the composition will take on drying [68]. Coupling solvents are sometimes used to influence rheology through their effect on resin solubility, and to avoid any phase separation during drying. However avoidance of coupling solvents is preferable for low VOC applications and the alternative is careful control of the other components and the design of the resin itself.

9.4.2.3 Crosslinking vs non-crosslinking

This element focuses more on the chemical nature of the film former; alternative headings would include 'low and high molecular weight' or 'thermoplastic and thermosetting'. To be useful as a paint binder in the building sector a polymer must be strong and tough and also available in a liquid form during application. Everyday experience and theoretical considerations (including the viscosity relationships referred to above) show that high molecular weight polymers will not be soluble in the solvents acceptable to the market. For solution paints it is therefore necessary to use a low molecular weight material that is soluble, and then increase molecular weight during or shortly after drying. Although there are many types of crosslinking, relatively few are safe or convenient for this market. In view of the nature of our atmosphere, oxidation reactions are the most convenient source of crosslinks.

Thus, for many years the autoxidation of solutions of oil or oil-modified polymers was a mainstay of the decorative coatings industry.

The exact mechanisms of oxidation have been the subject of considerable work but the broad characteristics are explained by the 'hydroperoxide theory' postulated by Farmer in 1942. During the primary stages of oxidation, viscosity changes are small, during which time hydroperoxides are formed on active methylene groups. This is accompanied by an increase in conjugation due to a free radical mechanism where detachment of hydrogen from active methylene groups is followed by rearrangement. Secondary oxidation reactions, which are catalysed by metal 'driers', involve peroxide decomposition followed by recombination forming C—C, C—O—C and C—O—O—C bonds [69]. Unfortunately, the autoxidation of oils leads to unwanted side effects causing secondary odours and yellowing. Furthermore the autoxidation process is difficult to stop and will eventually lead to embrittlement, with an obvious detrimental effect on durability.

The need for crosslinking can be avoided by using a pre-formed high molecular weight polymer, but the aforementioned solubility considerations dictate that this should be in dispersion, rather than solution form, in which case the film formation process becomes one of coalescence. By a suitable choice of T_g and other properties film integration is achievable for a useful range of polymers. Such films do not need to crosslink (though some subsequent crosslinking may offer specific advantages in special applications), and will remain thermoplastic and tough. According to Dillon *et al.* [70] the driving force for coalescence is surface tension followed by viscous flow of the polymer. Brown [71] stresses the role of capillary pressure. Subsequent work combined features of both theories [72], more recently a theoretical treatment by Padget and Kendal distinguishes between the elimination of 'triangular' and 'cubic' pores. Coalescence is favoured by a decrease in polymer modulus, a decrease in particle size and an increase in contact surface energy [73]. In principle dispersion (latex) films can fully integrate and experimental evidence has shown similar barrier properties between identical films cast from solution or dispersion. In practice this is not always the case and microscopic examination may show the original dispersed state. Such films may be heterogeneous with water-soluble material in pockets or as a network throughout. More significantly the ability to integrate fully with pigment is different from solution polymer and will be influenced by particle size as noted in the earlier discussion on 'binder power'.

From the above discussion it is clear that water *vs* solvent, and dispersion *vs* solution are useful distinct categories for classifying binder types. Crosslinking and film formation, however, are likely to become conflated with solution and dispersion. On this basis four major formulation 'boxes' can be defined (Table 9.4). In practice the majority of decorative paints can be categorized into box I (emulsion or latex paints) and box III (oil or alkyd paints), and in many ways their properties are diametrically opposed, arising from the three main underlying causes described above. Some of the more general consequences are summarized as 'positive' or 'negative' traits in Table 9.5. Clearly there is no overwhelming advantage to either type, selection will depend on the intended end use and appearance, weighted by consideration of market, economic, and legislative aspects. As the perceived weighting changes (e.g. by growing environmental pressure), and as technical innovation minimizes the negative traits, there will be a change in the preferred technology. Consider for example technology used for interior matt, mid-sheen (silk, satin, eggshell, etc.), and gloss paints 15 years ago and today.

Table 9.4 — Some formulation 'boxes' available to the decorative coatings market

	Water-borne (or aqueous with minor co-solvents)	Solvent-borne (non-aqueous; especially white spirit)
Polymer **dispersions**: film formation through coalescence only	**I** 'Emulsion' or 'latex' paints	**II** Non-aqueous dispersions (NAD)
Polymer **solutions**: film formation though autoxidation	**IV** Solubilized alkyds	**III** 'Oil' or alkyd paints

Matt paints are used on large areas such as walls and ceilings, where the easy application, quick dry, and lower odour of water-borne dispersion paints are major advantages. The consequences of poorer flow and lapping are not readily visible and were never seen as an insurmountable disadvantage. Matt paints have thus been firmly within box I for many years. With mid-sheen paints the choice was less clear cut. The consequences of flow deficiencies were more visible, and durability requirements more demanding, e.g. for cleaning in corridors of schools, hospitals, factories, and in areas of high condensation such as kitchens and bathrooms. As a consequence this sector was split between water- and solvent-based technologies. However, technical advances such as associative thickeners, and latexes with improved wet adhesion, combined with legislative pressures, have swung the balance strongly towards water-borne. Solvent-borne interior mid-sheen paints have a considerably reduced market share. In the case of full gloss paints this process is less advanced. Some 15 years ago the market was dominated by solvent-borne alkyd chemistry. Latex glosses were available but made relatively little market penetration. Today all major manufacturers have acrylic latex gloss paints available and the sector is growing, although penetration is still only around 10% and variable among countries [74]. Growth can however be expected as a result of continuing legislative issues and continued technical advances which will also widen the choice of technology as discussed in the next section.

9.4.3   Current and future developments

The pattern of binder development in the coatings industry continues to be one of technical innovation to design polymer architectures with enhanced properties but overlaid with the constraints of SHE legislation. This is also reflected in the building sector with the additional constraints noted earlier. Within the traditional solvent-borne sector efforts have concentrated on higher solids alkyd systems [75] with the possibility of reactive diluents, typically allylic or methacrylate, replacing some of the solvent [76]. This approach can meet specific intermediate VOC targets but is not a viable route to eliminating all solvent for decorative paints as the high amount of crosslinking required is difficult to control. Alternative ways of using alkyd type chemistry are to solubilize or emulsify the polymer in water. At the other end of the spectrum, research into aqueous dispersion polymers has produced a plethora of morphological variants, new comonomers, alternative stabilizing systems and many others. Each of these developments raises specific formulation problems and creates new property combinations, but there still remain a number of charac-

Table 9.5 — Generalized differences between water-borne and solvent-borne decorative coatings

Main cause of Property difference	Water-borne coatings ('emulsion' paints)		Solvent-borne coatings (oil or alkyd paints)	
	Positive traits	Negative traits	Positive traits	Negative traits
Nature of thinner: aqueous or non-aqueous?	Non-flammable	Poor early shower resistance	Films resist water at an early stage	Flammable
	Easy clean-up Quick-drying	Freeze–thaw stability problems Poor dry in cold, damp conditions	No low-temperature storage or can corrosion problems	Needs special thinner
	Low primary odour Low toxicity Cheap, readily available Thinner Same-day recoat	Rusting of ferrous fittings Can corrosion Grain-raising Poor wetting of linseed oil putty Prone to biodegradation	Less grain-raising Longer open time Compatible with linseed oil putty	High primary odour Relatively low irritancy threshold Usually overnight recoat
Physical state: dispersion or solution?	Relatively high permeability		Relatively low permeability	
	Easy application Good edge cover	Poor lapping Poorer flow Lower solids (build) Difficult to achieve full gloss Forms crust on container rim	Good flow Good lapping Higher solids (build) High gloss possible Good penetration	Stickier application
Chemical state: thermoplastic or crosslinking?	Non-yellowing Little embrittlement on ageing, can be very flexible Retains initial sheen No secondary odour	Dirt pick-up Blocking Transparent to UV	Harder film Easier to clean Less prone to blocking	Yellowing tendency Will embrittle with age Secondary odours May skin in part-full can

teristic properties which are indicated qualitatively by consideration of the three elements identified in Section 9.6.1.

Taking Table 9.4 as a reference point, then box I has a growing choice of new latexes with enhanced properties greatly increasing the formulating scope. Box III is moving to higher solids which lowers the VOC but raises problems of tinting and pigmentation. In box IV the use of solubilized alkyds reduces the disadvantages of solvents but retains such disadvantages as the yellowing, odour, and embrittlement characteristic of alkyds. It is possible to blend binders from I and IV to produce 'hybrids' which have intermediate properties which have been exploited in wood coatings. If the alkyd is in emulsified form then there will be a mix of some of the characteristic properties from I and III. Thus the compositions will become non-flammable and lower VOC, but will retain certain alkyd and rheological disadvantages. Alternatives to alkyd include silicone resins which can be emulsified to produce very hydrophobic coatings suitable for masonry.

The area represented by box II is usually referred to as non-aqueous dispersion technology (NAD) [77]. To a first approximation, the polymeric dispersions in this box may be considered as solvent-borne counterparts to the water-borne latexes of box I, though there are numerous differences of detail. In comparing two broadly similar polymers in either solvent or water it would seem obvious that the latter would be preferred. However NAD paints have one unique advantage in that they allow the exploitation of thermoplastic polymeric dispersions under adverse weather conditions and at least one company has exploited this opportunity [78]. However the inexorable VOC pressures are likely to constrain growth in this sector. Like their water-borne counterparts, NADs can be combined with alkyds to produce hybrid binders [79].

9.4.3.1 *Classifying water-borne technologies*
While the classification of binders implied in Table 9.4 distinguishes certain broad characteristics of binders, it does not go far enough in explaining the finer detail of current developments in water-borne technology. To do this requires a system that considers both the synthetic procedures that are used to produce the polymers and their likely physical state in water over a wide pH range which recognizes more intermediate possibilities between boxes I and IV. Padget [80] has produced a comprehensive overview from this perspective. Features of this classification from a decorative paint perspective are included in the discussion below.

Insoluble aqueous polymer dispersions
Emulsion polymerization is very well established as the route to dispersion binders with diverse properties. Characteristically free radical initiators are used to polymerize monomers of low water solubility to produce an insoluble polymer (see Section 2.8). Paint properties will be influenced by the choice of monomers, particle size, and stabilizer system. The effect of monomer type on mechanically related properties can be interpreted through the T_g; in somewhat simpler terms it is possible to talk of soft and hard polymers. Hardening monomers included methyl methacrylate, vinyl acetate, styrene, and vinyl chloride, while the softer monomers are exemplified by n-alkyl acrylates, alkyl maleates, and vinyl esters of 'Versatic' 10 acid (a highly branched neodecanoic acid) [81]. Most latexes used in the decorative market are copolymers or terpolymers with the ratios chosen to suit specific sectors. In very general terms softer polymers will favour extensibility, exterior durability,

and film integration, while hard polymers give lower dirt retention, gloss retention, and scrub resistance [82]. Properties will be influenced by the choice of coalescing solvent and occasionally external plasticizers may be used.

Aqueous polymer dispersions also show very significant differences according to the nature of the stabilizer. This has important consequences for formulators and is also an area of change as suppliers develop alternative approaches to specific market needs. Particle stabilization is often through the addition of anionic and or non-ionic surfactants. Cationically stabilized latexes are much less common and would not be compatible with many established paint components. Non-ionic and anionic combinations show a generally greater robustness resulting from the combination of charge and steric stabilization. As described in Chapter 2, further robustness is conferred by the presence of a protective colloid (e.g. polyvinyl alcohol, hydroxyethyl cellulose). The colloid aids in forming structure in derived coatings in combination with additional cellulosic or other 'thickeners'. Colloid-free latexes do not show this interaction with conventional thickeners and must be structured with the more expensive acrylic or urethane associative thickeners. Interactions between surfactant, dispersant, thickener, and cosolvent are complex and fine tuning of the rheology in the presence of the many interactions possible is a major formulation task, usually system-specific and greatly aided by good experimental design.

Carboxylated comonomers such as acrylic acid can confer anionic charge stabilization if the carboxyl group is ionized. Since the presence of surfactant in a polymer dispersions can have undesirable side effects on water resistance and foaming, anionic charge stabilization through acid groups is one route to surfactant free dispersions. High T_g styrene acrylic latexes have been described [83], which offer specific advantages on metal substrates since the high pH avoids flash rusting and the acid groups confer good adhesive performance. The permeability of such coatings is low because the films are below the T_g during service. On the negative side a relatively high amount of coalescent is needed to ensure film integration and external plasticizers to improve flexibility. Other formulation constraints arise from the alkaline conditions in the wet which means avoiding acid pigments or components prone to hydrolysis. This would include, for example, ester-based coalescents and plasticizers.

The need for high levels of coalescent is against the general trend which is to lower the solvent component. Water-borne dispersion polymers are at the low end of the VOC spectrum but even so there has been interest in reducing the VOC component to zero. Apart from meeting the most stringent legislative requirements this has the added advantage of reducing the odour of derived paints. If odour is the prime target then attention must also be paid to reducing free monomer and neutralizing amines. The simplest way to eliminate the need for coalescent is to lower the T_g by increasing the proportion of softer monomer(s). Taken too far this can give rise to films that remain tacky unless some post-crosslinking is introduced. The relationship between T_g and MFFT (minimum film-forming temperature) is not direct and some monomers are more effective than others in achieving coalescence without excessive softening. Vinyl acetate/ethylene copolymers, for example, have been shown to have advantages in this respect which is attributed to the ability of water to act as a plasticizer for the polymer backbone and thus take on the role of the coalescent in lowering the MFFT relative to the T_g [84].

Other approaches to the problem of lowering the MFFT without sacrificing film properties have included polymer blending and the use of sequential emulsion polymerization to obtain a 'gradient morphology'. The blending of a high and low T_g latex can certainly give intermediate properties but percolation effects mitigate against getting the ideal balance this way. Latex blending will also introduce interactions between the two latexes and other ingredients which complicate formulating and also increase operational complexity. Alteration of latex particle morphology can be achieved through control of both kinetic and thermodynamic aspects of emulsion polymerization. In principle a 'core–shell' morphology offers an additional degree of control between the process of film formation and subsequent dry film properties. A number of characteristic morphologies have been described (e.g. multi-lobe, current bun, raspberry, and half moon); ref. [80] provides a good overview and sources of further information.

Water and partially soluble polymer binders

Fully water-soluble polymers over the entire pH range are of limited use as binders but find an important role as additives. More useful as binders are materials which show an intermediate degree of solubility which may be pH-specific. These 'hybrids' [85] have characteristics intermediate between boxes I, III, and IV. Unlike the dispersion polymers the majority are not synthesized directly in water but preformed to varying degrees and introduced at a separate stage. The range of preformed polymers includes epoxies, polyurethane, polyesters, and alkyds. In the case of the latter hydrolytic stability is a problem which restricts the range of materials that can be used if a long shelf life is required. However, considerable development of alkyd emulsions has taken place in the past decade and some of the stability problems overcome.

One way to take an alkyd into water is to use an external emulsifier. Such an approach places some restrictions on the viscosity of materials that can be handled and by implication the choice of raw materials. A process of phase inversion is often used. Externally emulsified alkyds are not highly shear stable and it is recommended that predispersed pigment pastes are used. Alkyd emulsions require the same driers as their solvent-borne counterparts but have the additional complication that partitioning between the phases may reduce drier availability and this is influenced by the choice of emulsifier. Greater shear stability can be achieved with self-emulsifying resins. This adopts the mechanism of anionic stabilization via ionizable carboxylic groups noted above for surfactant-free latexes. The degree of ionization will depend on the pH relative to the pK_a of the acid. With sufficient acid groups the initially dispersed polymer can be taken to complete solubility. Alkali-soluble carboxylated polymers are sometimes known as hydrosols and used in temporary coatings and strippable floor polishes. In the case of self-emulsifying alkyd emulsions only sufficient hydrophillic groups to confer stability are introduced. In principle sulphonate and phosphate groups could be used but these are reported to have adverse effects on drying [86] and amine-neutralized carboxylic groups are the most widely used. Some formulating details and comparisons with conventional alkyd paints are given in [86, 87]. Needless to say the properties of such emulsions are very pH dependent and this must be carefully controlled during resin and paint manufacture.

Polyurethane dispersions are another group of preformed aqueous dispersions which have shown considerable development and are finding some applications in the building sector, particularly for wood coating. Typically a hydrophilically modified NCO terminated prepolymer is chain extended with polyamine compounds during or after dispersion. Like the alkyd emulsions the majority of these dispersions are non-ionic, or anionic with tertiary amines used to neutralise carboxylic functionality.

Mixed polymer hybrid systems

Although polymers of intermediate solubility have been described as 'hybrids' the term is also used where two different polymer types are combined either as a cold blend or by a pre-condensate route. Urethane acrylates, epoxy acrylates, and urethane acrylics all find application in coatings. Alkyd acrylic blends have found particular application in exterior wood coatings [88, 89]. The addition of an acrylic dispersion to an emulsified alkyd improves drying and initial film hardness and dilutes the yellowing tendency of the alkyd. The combination of acrylic, alkyd, and pigment dispersion leads to a complex rheology and some impairment of brushmark flowout can be expected.

9.5   Colour delivery

9.5.1   General

In most countries a significant proportion of decorative paint is white, followed in order of popularity by pastel colours and the stronger darker colours, in order of rapidly diminishing volume. Naturally, there is national variation and there will also be differences reflecting the intended application. All manufacturers face the problem of balancing the limited demand for specific colours against the level of service required without tying up capital in large stock holdings. The operational problem is made worse if, as is normal, there are a number of product lines to support.

A solution to this problem is to strike a balance between ready-mixed coloured paint (i.e. a tint, co-grind, or blend made in the factory) and that which is tinted at or near the point of sale. The latter are often referred to as 'in-store' tinting systems and comprise a device for dispensing tinters in quantized units to prefilled paint bases which are then mixed by shaking.

9.5.2   Factory-made colours

Traditionally solvent-borne decorative paints were made by dispersing the main pigments in part of the main film former, the transient millbases were subsequently let down with the remaining resin and other components. Final adjustment of colour was by addition of single pigments dispersions also dispersed in the main film former. The proliferation of such 'tinters' in a multiproduct plant led to a growing interest in 'multipurpose tinters' which could be used to tint a broader range of products [90, 91]. Universal or multipurpose tinters required a new generation of polymeric dispersants with an amphipathic character to ensure strong adsorption to the pigment surface and compatibility with the film former and these are now widely available commercially. Both 'universal' and conventional single pigment dispersions also lend themselves to an alternative way of assembling a coloured paint, namely the blending of pre-dispersed pigments with the other ingredients. This approach is sometimes described as a 'load' recipe. It offers operational flexibility to meet right-first-time (RFT) and just-in-time (JIT) criteria and is increasingly used. Such an approach requires good strength and hue control of the single pigment concentrates which must also show good storage stability.

Water-borne decorative coatings tended to be at the pastel end of the colour spectrum and were usually made by tinting rather than a cogrind. Typical pigment pastes were stabilized with a non-ionic–anionic surfactant blend. With the swing from solvent to water there is an increasing demand for deeper colours, new technologies such as those discussed in Section 9.5.3 also place a higher demand on the pigment dispersant. There is likely therefore to be increased use of polymeric dispersant in water-borne as well as solvent-borne pigment concentrates [92, 93].

9.5.3   In-store tinting

In-store tinting systems are now widely distributed but their development and exploitation has occurred differently around the world and within specific market sectors. Currently, for example, the bulk of coloured paint in North America, Australia, and Scandinavian countries is delivered via tinting systems, whereas in the UK retail sector the established practice has been to have a limited range of popular colours in ready-mix form supplying the bulk of the volume, which is supplemented by a tinting system which provides less than 10% of the total volume. There is no single reason for the differences found between countries, which reflect geographical, economic, and social differences. For example, the UK has a high population density which enables a daily delivery service from centralized warehouses to be viable in a way that is less practical in a situation where the consumers are widely dispersed. Basic manual tinting systems require the assistance of a trained operative, and have an operating cycle which will take 5–10 min. This will not be sufficient to meet peak demand in a busy 'superstore' though at other times the machine will be lying idle. Operating cycles can be shortened by various degrees of automation, though at an increasing cost premium, but even so the degree of automation necessary for user-friendly self-selection has yet to be approached, and may never be economically justified. Outlets operating some sort of tinting scheme will face an increasingly difficult choice of competing options with a need to balance savings in shelf space and reduced stockholding, against capital running costs, and the intangible 'halo' effect of a visibly modern system. In the trade (professional and specifier) sector, however, there is a need for a much wider colour choice and tinting is much more common. This is also true in the USA where it is accepted that orders can be collected the following day [94] thus allowing for managing peaks and troughs in demand.

Decorative paint tinting systems differ from those used in the automotive refinishing market. The latter deal with a much wider colour range which must include very deep colours, metallics, and pearls, and also cover technologies ranging from nitro cellulose to 2-pack isocyanates. Consequently, refinish systems have developed as intermixing schemes; that is the operative starts with an empty container into which is combined by *weighing* pigmented bases and blending clears. By contrast, decorative systems start with a container that already contains the bulk of the paint which is coloured by the *volumetric* addition of tinters. A problem facing decorative systems is to control the final volume of paint. Tinter additions are usually limited to a maximum of around 7% and one solution is to fill base containers to 95% of their nominal size which must be labelled accordingly. Consequently, the paint as purchased may contain between 95% and 102% of the nominal size, presenting some costing problems and an inconsistency in comparison with ready-mixed paint. An alternative solution is to aid for 'full fill' by adjusting the volume of tinter to, say, 5% with 95% base. Such an approach further constrains the number of colour hues that can be achieved and if it is not to be too restrictive, usually requires some increase in the number of bases and tinters. During the past decade there has been considerable progress in the exploitation of cheaper microprocessor and computing technology both in dispensing, and in the generation of recipes and databases. More recently the composition of tinters and bases has had to respond to the environmental issues affecting the market in general.

9.5.3.1 The dispensing machine

With few exceptions the heart of any tinting scheme is the dispensing machine, which must accurately dispense known volumes of tinter. Such machines can be divided into manual, semi-automatic, and automatic groupings. Paradoxically the development of automatic machines tended to precede that of semi-automatics and in any case, the boundaries between the two have become somewhat blurred [95].

All machines must contain a reservoir capable of holding several litres of tinter and usually a mechanism — manual or automatic — to provide periodic stirring. Typically tinter is dispensed through a piston pump, with volume being controlled by the length of stroke, but gear pumps are also used. Wear of the pumps can be a problem with abrasive pigments and additives may be used in the formulation to aid lubrication. In manual machines a common arrangement effectively connects the piston rod to a notched scale which is graduated in fractions of a fluid ounce (a measure which varies from country to country). A system common in the UK was based on 1/48 fluid oz. increments, known as a 'shot' with most machines able to dispense 1/2 or 1/4 shots. In SI units a shot is close to 0.6 ml.

The persistence of archaic or hybrid units reflects the difficulty and cost of changing either machine or formula once a system has become established. This method of dispensing imposes a scale of quantized colour differences which will be modified by tinter and base strength. With automatic machines electronic control of stepper motors can give a continuous range of colour additions, but there will be practical constraints imposed by the minimum drop volume that can be controlled. The accuracy of dispense depends on the volume and is unlikely to better ±10% for 0.15 ml, ±5% for 0.6 ml, and ±2% for 6 ml.

The arrangement of tinter reservoirs in a machine varies between manufacturers. One of the more common is to use a 'carousel' allowing each reservoir to be rotated to a common dispense point. Alternative arrangements in banks may require moving the paint container. Carousel layouts are usually combined with a sequential tinting sequence; with other layouts a simultaneous dispense may be optional. This speeds up part of the overall cycle but long pipe runs can be troublesome if a blockage occurs.

In a typical manual tinting sequence, the operative must look up the recipe in a file which might be in printed card, microfiche, or Eprom format. Having selected the correct paint base, this must be opened and placed at the tinting station where the various operations to set and activate pumps are carried out. With many colours requiring addition of at least three separate tinters, there is considerable scope for operator error! After dispensing, the container must be resealed and taken to a suitable mixing machine. The so-called 'fully' automatic machines, of which there are several variants on the market will normally have an electronic recipe store, and through electric motors the ability to carry out the dispense cycle automatically. Operatives, usually prompted by the machine, must select the correct base type, but many machines have the ability to confirm that the correct size in relation to the selected recipe is in position.

Frequently tinting is carried out through a hole punched in the lid at the beginning of the tint cycle, which is later sealed by a plastic plug. Using either internal or externally linked computers it is possible to print labels and other product information at the same time. Despite their advantages automatic machines are more than an order of magnitude more expensive than their manual counterparts, and therefore not cost-effective for small or medium size outlets. Semi-automatic machines have evolved which bridge the gap between the two extremes. Such machines differ considerably in detail but often include features such as an electrically driven pump and tint sequence after the receipt has been entered manually. This is an advantage for small multiple orders.

In terms of future developments it is clear that the continued price fall in real terms of microchip technology will provide opportunities for more sophisticated control systems which, as well as providing unlimited recipes, will provide facilities for other activities. Possibilities include stock control and a databank to provide product information. There are already commercially available packages [96] in which a spectrophotometer is linked with an automatic tinting system, providing a colour-matching service. But many customers are looking for complementary colour schemes rather than an exact match, and here too there are possibilities as the field of colour graphics advances. To utilize these advances fully will require strong tinters and dispensing pumps of very high accuracy. Incorporating all the features mentioned above will increase capital costs, probably putting them out of reach of smaller outlets.

Advances in dispensing technology will also increase the pressure for a corresponding improvement in paint mixing which, for automatic and semi-automatic machines, is already the rate-determining step. The difficulties of cleanly mixing a full can in the confines of a store require a non-intrusive mode of mixing. Machines based on either a shaker or some biaxial rotation mode are widely used, and to date experiments with a magnetic, ultrasonic, or vibrational mixing have proved impractical. Improved engineering has led, and will continue to lead, to quicker and quieter mixing. Some machines increase throughput by a capability for multiple loading, which may be combined with automatic hydraulic clamping. The energetic nature of the mixing process creates formidable design problems in combining the mixer with the dispenser.

For completeness, it should be noted that although dispensing machines are integral to nearly all tinting schemes, there are some non-machine alternatives. A possibility used in some stores is to use graduated syringes containing tinter which customers use to create their own colours. Clearly this gives only a limited choice and reproducibility. A commercially successful alternative used in the USA is based on small plastic cartons containing a powder tinter which is tipped into paint bases at the point of sale [97]. Less successful commercially was a system introduced in Germany in which preweighed quantities of tinter were sealed in water-soluble films protected in transit by an outer blister pack [98]. Another system incorporates a collapsible syringe within the lid of a paint can which dispenses colorant into the base as the lid is pressed home [99]. Although such systems can, in principle, produce a reasonable colour range, there is a practical limit to the number of sizes and colours that can be stocked and such systems cannot offer the wide colour choice offered by the machine routes.

9.5.3.2   Tinters and bases

Colour and compatibility
The demands placed upon tinters used 'in-store' are similar in many ways to those upon tinters in general, but there are some additional constraints. Of these, one of the most difficult concerns the question of compatibility. Factory tinters are often dedicated to a single product line and even where multipurpose tinters are used they will be used in product groups of broadly similar solvency. For in-store tinting, cost and space consideration have dictated common tinters for both water and hydrocarbon thinned products. This difficult demand has led to compromises and product quality may be affected, especially at high levels of use, leading to differences between 'ready mixed' and 'tinted' versions of the same colour. The balance between solvent and water paints is continually changing in favour of the latter and it should eventually be possible to service the market with water-based systems alone. Even so, the requirement of different product groups will not be easily met particularly as new high performance technologies are introduced.

The most common formulating basis for broad spectrum tinters is ethylene glycol with various surfactants, often known as 'glycol pastes'. Such compositions can retard dry (especially in alkyds) and increase the water sensitivity of water-borne paints. In countries such as the UK where a structured paint is preferred, the effect of tinters on the viscosity profile is more marked, a problem further exacerbated by mechanical shaking. Minimizing these effects is a further reason for limiting the maximum tinter addition. Some tinter manufacturers have reduced the problems of adverse tinter reaction by producing 'low glycol' tinters. While generally successful, they are more prone to caking and may require dispense machine modification to prevent the nozzles becoming blocked. More recently toxicological concerns have led to the requirement for eliminating ethylene glycol altogether. Propylene glycol has been used as a substitute, but with a growing interest in solvent-free coatings there is also a move to eliminate volatile organic components. Non-volatile components may still be needed to prevent nozzle blocking and this has been the subject of patent activity. Another toxicity issue relating to tinter formulation arises from concerns that have been expressed over the aquatic toxicity of alkylphenol ethoxylates. New classification and labelling regulations are likely. Nonyl phenol ethoxylate is widely used as the non-ionic surfactant in tinters. A number of alternatives have been evaluated, including for example tridecylalcohol ethoxylate [100]. Substitution will generally require fine tuning of the formulation to maintain rheology, stability, and compatibility [101].

An important consideration when designing a tinting system concerns the number of tinters and bases, and their hue tone and strength. In an ideal world it would be possible to have one colourless base per product line and five very strong tinters in white, black, and the three primary colours. In reality there are practical and economic reasons why this cannot be so. The effect of pigments on hue is a complex combination of subtractive and additive mixing and there are no pure 'primary' pigments. Even so, a surprising number of colours can be produced from, say, phthalocyanine blue, azo red, arylamide yellow (plus white and black), but there remain large areas of colour space that are inaccessible. Adding further tinters increases the range but a point of diminishing returns is reached where the addition of each new tinter introduces so few new colours that the increased cost of maintaining another reservoir is not justified.

Pigments in tinters must also be chosen on the basis of economy, opacity, and durability, and there are likely to be common features among the ranges used by different manufacturers. A typical core of tinters would include black, white, organic red, organic yellow, inorganic red, inorganic yellow, and an organic blue. The choice of further colours begins to diverge and will depend on the selection of bases available, fashion trends, and constraints inherited from the recent past (bearing in mind the cost of changing an established range). Additional tinters will be chosen for a group which includes violet, orange, green, brown, and other hues of organic yellow and red. (See Chapter 3 for details of pigment types used in the decorative market.)

Another important consideration is colour (or tinting) strength, which will be closely related to pigment concentration modified by the degree of deagglomeration. For many purposes it would be true to say the higher the pigment concentration, the better. Within the constraints of a maximum addition of 5–10%, it is necessary to have high pigment concentration tinters in order to achieve the maximum number of deep colours. Increasingly, manufacturers are meeting this range with high strength tinters; even so, it is not possible to reach all the deeper colours. A disadvantage of high and, in some cases, even normal strength tinters, is that they must be dispensed with a high accuracy if pale shades are to be reproducibly achieved. Such accuracy is outside the reach of most machines and inherently strong tinters — such as phthalocyanine blue — are usually held in reduced as well as full strength form. For a given level of accuracy the addition of stronger tinter might have to be matched with further reduced tinters, thus placing further demands on the limited amount of reservoir space. High colour strength does not necessarily equate to hiding power since the former is a measure of the ability of a colouring matter to impart colour, through its absorbing power to other substances (BS 3483), and the latter the ability to obliterate colour differences in the substrate. Maximizing the former will ultimately result in a transparent pigment concentrate [102], thus pigments with a coarser particle size than the optimum for colour strength may be required to meet opacity needs economically. For less saturated shades blending of high chroma organic pigments with cheaper but more opaque oxide pigments is the preferred option. Modern colour matching systems have the ability to produce alternative recipes to meet cost/opacity criteria.

To exploit a given set of tinters fully requires a corresponding set of bases, but as with tinters a point of diminishing returns is reached. All systems will include around three to five bases which are in effect a TiO_2 PVC ladder covering a range from zero to an upper TiO_2 level typical of ready mixed white paint, thus enabling deep, medium, and light tints to be produced. To widen the availability of colour space further requires the addition of coloured bases with red and yellow being the most common, although others are used including some mixed pigment bases. Since each new base must be held in a variety of paint sizes and product types there is a disincentive in introducing too many new ones and generally there is a trend away from coloured bases. A major benefit of strong tinters is a reduction in the need for coloured bases.

An important difference between tint bases and their ready mixed counterparts is the need to facilitate tinter acceptance. Both water-borne, and, more importantly, solvent-borne bases will contain additional stabilizers to reduce pigment–pigment interactions and surfactants and compatibilizing agents to aid incorporation of the tinter. This is an area of considerable proprietary knowledge though some additive suppliers will provide formulating protocols.

Consistency and reproducibility
Once a paint has been tinted, no further adjustment is practical and the system is required to give a close and reproducible match to colour cards at point of sale displays. This, in turn, requires a higher level of standardization compared to ready mix in both tinter and base, which must be accurately strength controlled. Tests must be capable of controlling to the required specification through control limits which reflect the test capability. Controlling the strength of the bases is usually carried out with a coloured tinter, often black, and it is vital that this has a robust stability and is representative of the whole range in its interactions. Tinters themselves are controlled using a standard white base but will often not show the same relative colour strength in different types of base. This can be allowed for in the colour matching software. Vigilance is needed to avoid a drift in strength or colouristic properties due to circular standardization. Because tinters are usually dispensed by volume, it is important that their specific gravity is known and stable; de-aeration equipment is often used during manufacture.

As well as showing reproducible strength, it is highly desirable that tinters produce the same hue in different product bases, thus enabling the same recipe to be used for different products. Where this is not so, the size of the recipe bank is increased, with additional possibilities for error in manual systems. In practice, it is difficult to achieve identical hues in different media and it is not unusual for at least the water-borne and solvent-borne paints to require separate recipes. Once the composition of a tinter has become established it becomes very operationally complex to substitute different pigments since alternatives will often show a different pattern of colour development across a product range.

Finally, it is noted that tinters must also have satisfactory rheological properties with sufficient fluidity to pass easily through nozzles and tubes. At the same time they must neither cake nor settle — though most dispensing machines include a stirring cycle. The effect of the tinting medium on any materials of construction should also be taken into account. Aluminium, for example, is corroded by glycol.

9.6   Meeting the needs of the substrate

Decorative coatings can sometimes be chosen with little regard for the underlying substrate, but often it will be the substrate which has a major influence in deciding their suitability. Choice of a coating may be dictated by the need for a specific protective function or it may arise from potential interaction between substrate and coating. In some cases these needs will require a totally dedicated system while in others the needs of the substrate may be met by specific primers which are subsequently overcoated with coatings of a more general-purpose nature.

It is convenient to divide the substrates for building paints into three broad categories, namely *wood*, *metals*, and *masonry*. Organic plastics represents a small but growing sector of the building market which may require more consideration in the future, as plastic components such as UPVC windows become weathered and discoloured. In redecoration it is seldom possible to ignore the substrate totally but it will also be necessary to take the nature and condition of previous coatings into consideration. For example, bitumen coatings and some preservation pre-treatments on wood can present problems of bleeding and discoloration when overcoated, while strong solvents may lift previous coatings. A build-up of film thickness will change some properties, including permeability and mechanical properties.

In addition to meeting the needs of the substrate a coating must also resist climatic and environmental influences and these may interact with the substrate. Thus it is often desirable to formulate interior and exterior paints differently; this partly arises from the direct effect of weather on the coating also because the need to protect the substrate is different internally and externally. Each of these has, or creates, characteristic problems in service which require specific properties from the coating. For example, wood must contend with biodegradation and moisture movement; concrete is a highly alkaline surface, and metals are subject to corrosion. In each of these processes water plays a major role and it would be true to say that one of the prime purposes of the coating is to control some aspect of a process in which water plays an essential part — hence the importance of liquid and vapour permeability. However, the permeability required for each substrate type will be different.

In the remainder of this chapter particular attention is paid to the nature of the substrate, the influences that will cause deterioration and some of the consequences this has for the design and development of surface coatings in the decorative market sector.

9.6.1 Wood-based substrates

Wood is widely used in buildings for roof trusses, timber frames and joists, and non-structurally in doors, window frames, cladding, and fencing. This widespread use reflects wood's numerous attractive properties which include ease of working and high strength and stiffness-to-weight ratios. It is also an attractive material and when properly husbanded is naturally renewable, but wood is vulnerable to extraneous degrading influences which include light, moisture, and biological attack. Surface coatings have a major role to play in preventing or reducing the consequences of these influences but must be considered in conjunction with design and preservation.

9.6.2 Characteristics of wood

Wood is a naturally occurring composite material with a complex structure which occurs over a wide range of magnification. It is necessary to understand something of this complexity in order to select or design the most appropriate surface coating [103]. The properties of wood will differ greatly from one species to another, but it is often useful to divide wood into two broad groupings: hard-wood and softwood. This is a botanical distinction; by convention the wood from coniferous trees is known as softwood and that from broadleafed trees as hard-wood. A majority of hardwoods are actually harder than softwoods but this is not always so: balsawood, for example, is defined as hard wood despite its soft physical character. Hardwoods and softwoods show significant differences in their cell structure.

9.6.2.1 Macroscopic features

A slice taken across the trunk of a tree shows a number of characteristics which are apparent even to the naked eye. Beneath the outer and inner bark is a thin layer of active cells (cambium) which appear wet if exposed during the growing season. It is here that growth takes place. In spring, or when growth begins, the cells are thin-walled and designed for conduction, whereas later in the year they become thicker, with emphasis on support. The combined growth forms an annual growth ring which is usually divided into two distinct areas known as *springwood* (or earlywood) and *summerwood* (or latewood). Each type has different properties and this can lead, for example, to differential movement and is a potential problem to coatings.

When viewing a transverse wood section it is often noticeable that the central area is darker than the outer circumference. This is a characteristic difference between *heartwood* and *sapwood*. Sapwood is the part of the wood, the outer growth rings, which actively carries or stores nutrient. Young trees have only sapwood, but as the tree matures the cells in the centre die and become a receptacle for waste matter including tannin and gum. Once it starts to form, the growth of heartwood keeps pace with the growth of sapwood and the ratio of heartwood to sapwood gradually increases. Some heartwoods are distinctly coloured and they all show important differences from sapwood. Heartwood is less porous, shows greater dimensional stability to moisture movement and generally shows greater resistance to fungal and insect attack. However, the resistance of sapwood to attack can be upgraded with preservative treatments.

The weathering and durability of coatings is greatly influenced by the consequence of the differences mentioned above. When comparing the performance of different coatings, it is not unusual to find that the difference among coatings is masked by the variability of the substrate and this must be taken into account in the experimental design, in that the wood itself must be treated as a variable.

Because wood is anisotropic, there is the further complication of the way in which the wood is cut. Transverse tangential and radial sections each present a different alignment of cells, resulting in the familiar grain patterns. The two main types of cut are known as plain sawn and quarter sawn but a commercial through and through cut will produce both plain and quarter sawn timber. Quarter sawn boards are less inclined to shrink in their width and give more even wear. The weathering of many paints is better on quarter, rather than plain sawn timber. The new European weathering test method (prEN 927-3) gives a detailed specification for a simple panel test.

Another macroscopic feature of timber which may interact with the coating, is the presence of knots which arise from junctions between trunk and branch. Knots may be considered a decorative feature but they can be an unwanted source of resins and stains when coated.

9.6.2.2 Microscopic features

At low magnification the cellular structure of wood is readily discernible. A majority of the cells align vertically with a small percentage horizontal in bands or rays giving a characteristic 'figure' to quartersawn wood. Functions of cells include storage, support, and conduction. In softwoods these are carried out by two types of cell (trachieds and parenchyma); hardwoods have a more complex structure, with four main types. A characteristic feature of hard wood is the presence of vessels (e.g. the pores in oak). Vessels can absorb large quantities of paint and can transport water behind the coating. Individual cells walls are made up from cellulose microfibrils which form a series of layers as the cell grows. Fluid moves between cells via tiny pits which occur in matched pairs. Softwood pits may close after seasoning, making it more difficult to force preservatives into the wood.

9.6.2.3 Molecular features

Wood contains 40–50% cellulose which is in the form of filaments or chains built up in the cell walls from glucose units. Amorphous regions of cellulose are hygroscopic. Cellulosic components are embedded in matrix of hemicellulose and lignin, the latter a three-dimensional polymer built up from phenyl propanol blocks. Lignocellulose is largely inert to coatings but wood may contain up to 10% of extractives, some of which are highly coloured and chemically active. They can cause staining or loss of dry in coatings. Examples include stilbenes, tannin, and lignans.

9.6.3 Causes of wood degradation

Wood is affected by moisture, light and living organisms, often acting simultaneously, with harmful consequences for both substrate and coating.

9.6.3.1 Moisture

The moisture content of wood is conventionally expressed as a percentage based on the weight of dry wood. When newly felled, timber can contain up to 200% moisture contained in cell cavities and walls. As wood is dried, seasoned water is lost from the cell cavities, with no corresponding change in volume. Eventually the point is reached where cavities contain no liquid water and the remaining water is held in the cell walls; this is known as the fibre saturation point which occurs around 30% moisture content; further loss causes dimensional change. Wood will establish an equilibrium moisture content dependent on the temperature and humidity of the surrounding air.

It is customary to describe the changes in dimension which occur in drying timber from the green state as *shrinkage*, while changes which occur in seasoned wood in response to seasonal or daily fluctuations are known as *movement*. Typical shrinkage values are 0.1% longitudinally and up to 10% tangentially; radial movement is around half that of the tangential. By convention, movement is reported in the UK as a percentage change occurring between 90% and 60% RH at 25 °C. Tangential movement can be up to 5% but differs considerably among species.

In service, timber is subjected to a constantly fluctuating environment; this includes changes in humidity internally and externally, and the effects of both rainwater and condensation. This will lead to movement and when considering the applicability of specific coating types it is useful to divide wooden components into those such as joinery where dimensional stability requires control, and those such as fencing, where movement is less critical. The rate at which movement occurs depends on the water permeability of the coating and this is an important factor to be considered when designing a wood coating system.

9.6.3.2 Effect of water on coating performance

Although coatings can influence the rate of movement, they cannot physically restrain it. Clearly a wood coating must have sufficient extensibility to expand and contract with the wood and adequate adhesion to resist interfacial stress between substrate and coating. Coatings with a low modulus of elasticity will generate less interfacial stress for a given movement than those with a high modulus.

Water acts as a plasticizer when absorbed by a coating and although this increases extensibility it will also aid removal by peeling. Furthermore, it is possible for low molecular weight soluble material to diffuse to the interface where it concentrates, giving a weak boundary layer. While adhesion is low the coating will be vulnerable to blistering, especially if vapour pressure is high and permeability low. Blisters are caused by a loss of adhesion combined with sufficient flexibility but there must also be a driving force which might be vapour pressure, swelling, osmotic pressure, or resin exudation. Some polymers, for example those containing ester linkages, are prone to hydrolysis and are best avoided in permanently damp areas.

9.6.3.3 Sunlight

Solar radiation causes wood to undergo physical and chemical reactions which are largely confined to the surface. Many of these reactions also require the presence of water and oxygen. In dry conditions unfinished wood tends to turn brown but becomes progressively grey on normal wet weathering. Such changes reflect radiation-catalysed reactions which involve oxidation depolymerization and general breakdown of lignocellulose. Free radicals have been detected in irradiated wood. Breakdown products are soluble and are leached out by water leaving a grey denatured surface. Although the high energy of UV light is responsible for much of this damage, work by Derbyshire and Miller [104] underlined a significant contribution from ordinary visible light. The implications of this are serious, for it means that the interface between any transparent coating and wood may be prone to photodegradation, undoubtedly a factor contributing to the tendency of conventional varnishes towards flaking. Problems may also arise with opaque coatings if they are applied to an already degraded surface. The reduction in durability of coatings applied over wood exposed to light for only a matter of weeks is quite striking and may halve the period before maintenance is required.

In considering the effect of sunlight on wood coating interactions the role of infrared, i.e. heat, should not be neglected. Dark coatings will reach temperatures up to 40 °C higher than white ones and this may exacerbate problems such as resin exudation. On the positive side the higher temperatures will dry the wood and reduce mould growth.

9.6.3.4 Biodegradation

Wood is vulnerable to attack by bacteria, fungi, insects, and marine borers. Bacteria play little obvious part in the decomposition of wood though their presence is sometimes a precondition for more serious attack. Wood which is stored in water, or ponded, can become very permeable as a result of bacterial attack and this may affect the uptake of water and preservatives with consequent influence on paint properties. Fungal attack is more serious and can cause both disfigurement or structural damage, according to the species. Softwoods are particularly prone to a blue staining which is caused by several types of fungi, including *Aureobasidium pullulans*. 'Blue stain in service' is caused by further infection arising from colonization of surfaces during or after manufacture, and is not the final stage of growth of blue stain fungi originally present in the timber.

In addition to the disfiguring surface moulds and blue stain fungi, there are a substantial number of fungal species which cause serious structural damage by a process which, in its advanced stages, is known as rot or decay.

Soft rots including fungi from Ascomycotina and Deutermycotina groups are limited in their growth by the availability of nitrogenous nutrient, but degraded surface layers can have an effect on coating performance. More serious in their effect on structural properties are the fungi that destroy wood. Most are from the Basidiomycetes group. They attack both lignin and cellulose and are less restricted by nitrogen availability. This group includes 'wet' rot (*Coniophora puteana*) and 'dry' rot (*Serpula lacrymans*).

During the past decade there has been an increase in decay attributed to *Dacrymyces stillatus* ('Orange Jelly') in Scandinavian countries, most particularly Norway [105]. Many of the affected houses had been coated with opaque water-borne stains of the alkyd emulsion hybrid type and as a result the paint manufacturers were sued for damages. All cases were acquitted but during the intensive investigations that followed a number of potentially causal factors were identified including climatic change, forestry and wood processing issues, and the practice of leaving uncoated wood exposed for lengthy periods. A number of changes were recommended including the use of hydrophobic solvent-borne oil-based primers as the first coat [105]. The incident underlines the care that must be taken in formulating new products in areas where long-term effects will be slow to show and even though the manufacturers were exonerated there remains a concern about the role of coating moisture permeability and the nature and composition of the interface between coating and substrate.

Influence of coating permeability

Decay in wood is normally prevented by keeping the moisture content below about 22%, and by suitable preservatives. Coatings have an important role to play in controlling moisture uptake but could be counter productive if moisture is trapped. The latter concern furthered an interest in coatings with a higher vapour permeability than traditional oil-based systems, and terms such as 'microporosity' were introduced, albeit with some controversy [106]. The *Dacrymyces* incident has raised awareness that relating moisture permeability to the distribution of water in structures is a complex issue. It is currently the subject of further investigation in an EU funded 'AIR' Programme [107], and elsewhere including the Building Research Establishment [108] who are also participants in the AIR programme. Despite all this work a clear link with moisture transport and decay prevention has yet to be established. Recent work in the Netherlands did not show a correlation between window frame moisture content and coating type [109]. In contrast the influence of wood species was clear cut. (For further discussion of permeability see Section 9.9.6.1.)

9.6.4 Defensive measures

Although this discussion is concerned primarily with coatings, it is important to stress that in isolation coatings cannot ensure maximum life for timber components. This can only be achieved by an alliance between good design, appropriate preservative, and the most suitable protective coating.

9.6.4.1 Design aspects

Good design includes selection of the appropriate timber, attention to detailing [110] and the adoption of good site practices. It is outside the scope of this book to consider design aspects in detail but a few pointers to some of the more important aspects are appropriate.

Joinery

Design considerations with serious implications, from the point of view of decay, include the nature of joints, profiles, and the way the wood is handled and fixed on site. Water is most likely to enter at joints and exposed end-grain. Moisture uptake through end-grain, that is through transverse sections, is very much greater than other faces. Work carried out at the Building Research Establishment and elsewhere has shown the moisture content of uncoated wood with sealed end-grain is, on average, lower than that of coated wood with unsealed end-grain. This arises because joint movement can break the protective coating seal, allowing access of water which cannot readily escape. It underlines the need for joints and any exposed end-grain to be sealed using boil-resistant glues and an impermeable end-grain sealer. Two pack polyurethane fillers are particularly effective but a number of other candidate materials have also been screened [111].

Flat surfaces allow water to collect against joints and window profiles and must be designed with a run-off angle of 10–20°. Sharp corners are a frequent source of coating failure and should be rounded. The water resistance of windows is greatly improved by weathersealing using, for example, Neoprene or PVC weatherstrips. Where timber is in contact with brick or blockwork, a damp-proof course should be provided [112].

Cladding

Properly designed timber cladding should make provision for both movement and ventilation. Failure to allow for movement leads to warping, while inadequate ventilation can lead to dangerously high moisture levels being reached. Trapped water will support decay and has a generally disruptive effect on film-forming coatings [113].

Glazing

Paint and other coating failures are often localized at the joint between timber and glass, underlining the care that must be taken to ensure coating and glazing methods are compatible. A wide variety of glazing combinations is available, but for wood windows the glass normally sits in a rebate, with or without a bead. The simplest and longest established method of glazing low-rise buildings which are to be painted, is to use linseed oil putty which, when properly painted, gives remarkably good service, but putty glazing is not suitable for frames that are to be stained, varnished or coated with water-borne paints. The appearance will be marred and the putty will not be protected, leading to early failure. Bead glazing is generally recommended instead, but where sealants are used their compatibility with coating should be confirmed (BS 6262).

9.6.5 Preservation of timber

Timber species vary in their resistance to decay, with nearly all sapwood vulnerable. Heartwood of some species, such as oak and teak, is very resistant but ash and beech have little resistance. A widely accepted classification divides timber into five grades in increments of five years [114]. Hazard classes are defined in European Standard EN 335-1. Preservation should be considered for all sapwoods and non-durable heartwoods, where the equilibrium moisture content is likely to rise above 20%. Such a situation is likely where ventilation is poor, where the timber is in ground contact and where design features allow contact with water. Preservation is essential where insect or fungal attack is endemic. EN 599-1: 1996 specifies for each of the hazard classes defined in EN 335-1, the minimum performance requirements for wood preservatives for the preventative treatment of solid timber against biological deterioration. Since it is the role of preservatives to be toxic they present significant environmental problems and are the subject of new legislation both in composition and application. Within Europe relevant legislation includes the Biocidal Products Directive and the Environmental Protection Acts.

The three main established classes of preservative are tar oils, water-borne preservatives, and organic solvent preservatives.

Tar oils are typified by creosote which may be derived from coal or wood distillation. Timber treated with creosote is not suitable for painting but can be stained, preferably after a period of weathering.

Organic solvent preservatives include pentachlorophenol, tributyl tin oxide, and copper and zinc naphthenates as active ingredients, carried in hydrocarbon. The two former are under increasing environmental pressure, and as a group they are flammable with a high VOC. Additives include waxes, oils and resins. The labelling of wood preservatives in Europe is described in EN 599-2. Because hydrocarbons do not interact strongly with wood they penetrate deeply, and such preservatives established a dominant role for industrial pretreatment.

Water-soluble preservatives have long been available based on copper, chrome, and arsenic compounds, with sodium dichromate as a fixative. Ammoniacal/amine copper systems provide an alternative to chromium for fixation. With quaternary ammonium compounds fixation may be through the acidic and phenolic groups of lignin [115]. An alternative water-borne type employs disodium octaborate which is not fixed and must be protected during and after installation. Fluoride salts may be employed for *in situ* remedial treatment. More recently and in response to the environmental pressures there has been a renewed interest in water-carried preservatives. The challenge has been to overcome the poorer penetration of active ingredients, and the dimensional changes caused by water. This has been partly met by emulsified systems. Coarse macroemulsions are generally less effective than solvent-borne equivalents but show some efficacy against beetle emergence. Finer microemulsions are more effective, and can show similar penetration depth to solvent-borne albeit at lower active ingredient concentration. Although used more for remedial purposes they are being developed as replacements for solvent-based products in industrial applications [115].

Preservatives are applied by a variety of methods which include brushing, spraying, dipping and more effective, double vacuum or vacuum pressure processes (see, for example, BS 5589 and BS 1282).

Although most preservative manufacturers attend to compatibility problems between treated wood and paints, glues or glazing compounds, it would clearly be unwise to assume compatibility in every case. Formulating chemists should be alert to possible problems of intercoat adhesion especially of water-borne coatings over water-repellent preservatives. Other compounds, including some of the copper-based ones, can inhibit autoxidation.

In the UK specification of preservatives has focused on the treatment process and the type of preservative in relation to the durability class of the timber and service environment. This is being displaced by new European Standards, in response to the Construction Products Directive. The European Standards are performance rather than process-based and relate to defined hazard classes. The relevant standards are EN 350-2, EN 335-1 and EN 335-2.

9.6.6 Coatings for wood

Coatings for wood are conveniently described by three archetypal terms: 'paint', 'varnish', and 'stain'. These terms cover some expectation of appearance and performance but are not exact. They offer the user a choice of appearance and while individual products vary, each has certain inherent advantages and disadvantages. Until comparatively recent times the terms unless otherwise qualified, would probably be interpreted as implying:

- *Paint* — an oil or alkyd-based (solvent-borne) opaque paint system comprising primer, undercoat, and a glossy topcoat.
- *Varnish* — a solvent-borne transparent, clear glossy coat.
- *Stain* — a low solids penetrating composition semi-transparent containing an 'in-film' fungicide.

Of these three types the third group was the last to find widespread use in the UK influenced by growth in North America, Scandinavia, and Europe. Clearly each of the three product types offer a different appearance and might be chosen for purely aesthetic reasons. However, a major reason for the growth in exterior woodstains (known as 'lasurs' in some European countries) was an expectation of easier maintenance. To many users, though, it became increasingly clear that each group had characteristic disadvantages, thus:

- *Paint* — embrittlement on ageing led to cracking, flaking, and subsequent expensive maintenance. Low moisture permeability believed to trap moisture and promote decay.
- *Varnish* — even more prone to flaking and disfiguration. Difficult to maintain and restore appearance.
- *Stain* — easy to maintain but thin film and high permeability resulted in considerable movement in joinery. Prone to allow wood splitting and warping. Did not maintain dimensional stability (e.g. of windows).

To a large extent the generic differences between the three classes arise from the proportions in which the three main components are combined as illustrated in Fig. 9.4. Thus in developing solutions to the problems associated with established product there are three principal avenues to explore:

- change the proportion of the components;
- change the nature of the components;
- change the coating system (e.g. pretreatment or primer).

All these options have been extensively explored leading to a plethora of new products. In some cases these are described by derivative terms which can be related to

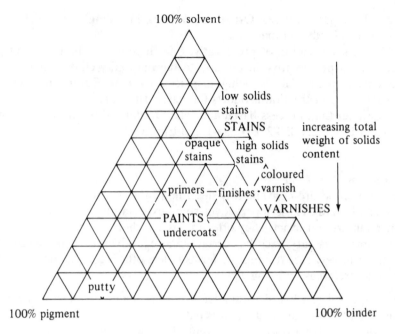

Fig. 9.4 — Relationship between wood coating types according to relative weight of main components.

the compositional diagram. Examples are 'opaque stains', 'high build stains', and 'varnish stains'. More commonly materials have been described by proprietary names causing some confusion to users and specifiers. In the UK this led to pressure for the drafting of a British Standard giving guidance on the classification of exterior wood coatings [116]. The published standard BS 6952 defines 48 categories of appearance. More recently this has been adapted to a European Standard (EN 927-1) in which the number of categories has risen to 60! These are derived from four levels of build (minimal, low, medium, and high), three levels of hiding power (opaque, semi-transparent, transparent), and five levels of gloss (matt, semi-matt, semi-gloss, gloss, and high gloss). This classification does not cover the multiplicity of film-forming binders now available and increasingly the water-borne binders described earlier are displacing the traditional oil- and alkyd-based compositions. In the European Standard performance is not inferred from appearance or composition, but has to be proven by meeting performance criteria including natural weathering [117].

Designing a wood coating starts with a specification in terms of appearance and the projected end use including the severity of exposure and climatic conditions. End uses are conveniently categorized according to the degree of dimensional stability required. 'Stable' categories cover joinery, windows, and doors, 'semi-stable' would cover many types of cladding, while 'non-stable' includes most types of fencing. The important property related to substrate stability is water permeability. Other important film properties are extensibility, modulus, and adhesive performance [118]. There are no single solutions to the formulating problem raised but some general remarks may be addressed to the three main generic classes.

Stains
Archetypal stains are designed to penetrate and will generally have a low (<20%) volume of solids. Penetration can be studied by a combination of microscopy with staining or labelling with a fluorescent agent. Radioactive labelling and radiography have also been used. Such studies show that there can be differential penetration among solvent, fungicide, and polymer. Different molecular weight species will also show different depths of penetration. Penetration is usually deeper with solvent-borne compositions, whereas water causes swelling. However as with preservatives penetration by alkyd emulsions has been shown to be closer to solvent-borne alkyd solutions than to water-borne acrylic dispersions [89]. The physical form of the binder has a major effect and polymeric dispersions will not penetrate unless very fine. Penetration by pigment is also constrained by size and only very finely ground pigments such as transparent iron oxide will penetrate the wood surface.

It is important to distinguish between the functions of a stain and a wood preservative. The latter are designed to prevent decay and require special application methods to ensure deep penetration. Stains should contain fungicide to inhibit surface colonization but normal brush application will *not* give sufficient penetration to ensure protection against decay. Stains, and indeed all wood coatings on non-durable wood, should be applied to preservative treated wood.

Penetrating stains can be formulated with little binder, often no more than is needed to stabilize the pigment. In solvent-borne products oils and alkyds are often used but care must be exercised when formulating at low vehicle solids as there can be problems of phase separation or oxidative gelling. Water-borne stains can be based on low volumes of aqueous dispersion polymer. Recent years have seen a proliferation of low-cost water-borne stains, especially for fencing and garden applications. These are often positioned against creosote and have handling benefits with little or no effect on adjacent plant life. Such products are essentially decorative and have little protective or preservative effect.

Low solids stains may be modified with silicone oils, waxes, etc. to give some, albeit temporary, water repellency but they exert virtually no control over water vapour. Their use is thus best confined to situations where free movement of the substrate is acceptable. Typical applications might include fencing and some types of cladding, though the latter must be of a design that allows for movement. Good performance on cladding requires a sawn, rather than smooth, planed surface. As a generalization it would be true to say that virtually all wood coatings, even gloss paint, will perform better on sawn surface, though naturally the appearance of gloss products will be affected.

In the case of joinery, especially softwood, there *is* a need for moisture control, otherwise constantly fluctuating movement can cause splitting, loosening of glazing, and the seizure of opening lights. There is also a body of opinion which holds that too low a permeability will allow moisture to become trapped and increase the chance of decay. As noted earlier this line of thinking led to the promotion of some coatings as 'microporous' and other terms implying a higher level of permeability than a traditional paint system. As discussed earlier evidence for a critical lower level of permeability has not been forthcoming and more weighting is currently given to design and wood quality factors. However, the *Dacrymyces* incident (Section 9.9.3.4) has caused concern and the subject is still one of on-going research. Attention is also now focused on the distribution of water just below the coating surface and the role that the coating composition might have on this. Such considerations suggest that until more evidence is available a prudent compromise will be to control permeability for joinery between an upper and lower limit defined by the wood species and end use. As a guide line, a film permeability between 30 and $60 \mathrm{g\,m^2/24}$ hours (Payne cup method, ASTM E96B) should be adequate to control movement in joinery but actual values are very dependent on film thickness and the method of measurement [119]. Standard test methods for liquid and vapour permeability have been developed as further parts of the European Standard and are circulating as drafts for public comment [prEN 927-4 and -5:1997]. Recent work suggests that more sophisticated methods involving absorption and desorption of water should be used [120] but correlation with field studies is very time consuming. This latter study also indicated some possible advantages in combining water- and solvent-borne products as a combined system.

The rate of moisture transmission in woodstains can be reduced by raising the volume solids, i.e. moving to a high solids stain, thus effectively giving a thicker film. However, care is needed for a given binder; raising the film thickness could change the failure mode from one of erosion to flaking [118] which negates one of the benefits of using a stain. Binders must be selected with good extensibility and adhesion. Because penetration of higher build stains is reduced it is usually beneficial to pre-treat the wood with a lower solids product.

Solvent-borne high build stains can be successfully formulated using alkyds, bearing in mind the remarks made above. A number of proprietary alkyds have been developed specifically for this purpose [121, 122]. Water-borne *transparent* stains proved more difficult to formulate. There are three major problems to overcome. Firstly, the much higher permeability creates moisture control problems; secondly, fungal attack is more common; and finally transparency of acrylic and vinyl polymers to UV light makes it more difficult to prevent photodegradation of the surface. It has been found that many of the vinyl and acrylic latexes which perform well in paints (i.e. including opaque stains) are not suitable for transparent coatings. Attempts have been made to overcome this problem by incorporating components capable of absorbing UV light and by the use of hybrid systems described earlier. This is an area of continued development.

Pigmentation of woodstains is relatively straightforward; the demand for 'natural' wood shades makes red and yellow iron oxides particularly useful. The best colour, brightness, and transparency is achieved with synthetic iron oxides which achieve full transparency when ground to colloidal dimensions. However, these are expensive materials and the standard grades require long dispersion times. Their high surface area is quite active and some care in selection of dispersant must be exercised in order to achieve long-term stability. Not all alkyds are suitable. When fully dispersed these grades confer exceptionally good UV screening.

The choice of fungicides for woodstains is vast but their compatibility and long-term storage must always be checked very carefully. For fencing and ground contact applications copper and zinc soaps perform well, and have relatively low mammalian toxicity. Stains and other transparent coatings for wood should include fungicides which are specific against blue stain.

Varnish

Exterior varnish usage has declined steadily in favour of woodstains, owing to a high incidence of flaking and discoloration with consequent increased maintenance costs. However, there remains a demand for good quality varnish to set off the aesthetic qualities of wood. Varnishes also have better wear characteristics in high traffic areas such as doors and door frames.

Problems with conventional varnish arise from the fact that they are high build films of low permeability, becoming brittle with age. As mentioned earlier the problem is made worse by their transparency which allows photodegradation at the wood–varnish interface. Traditionally good quality varnishes were based on tung and linseed oils modified with phenol formaldehyde resins. Straight alkyds are generally less effective. It is probable that the good performance associated with tung phenolics is associated with their UV absorbing characteristics and improving this aspect of performance has received considerable attention, notwithstanding the fact that visible light is also detrimental to the substrate. Improvements in performance have been claimed with several types of UV absorber, including the benzophenone groups and hindered amine light stabilizers. Benefits have also been noted with inorganic fillers, including pulverized fuel ash [123]. Yellow transparent iron oxide, as used in woodstain, also confers a marked improvement, though clearly the colour will darken. In practice, this may be no worse than the yellowing that occurs with many varnishes on weathering.

In the past it has not been customary to include fungicide in varnishes but logically they should also contain agents to inhibit blue stain and other fungi. Many fungicides are themselves degraded by UV radiation so the benefits are seldom realized unless the varnish does contain some protection.

Adhesive performance, and hence resistance to flaking, can be improved by thinning the coat first applied to the wooden surface or by substituting a low solids woodstain. Such an approach further blurs the distinctions that can be drawn between high solids stains and coloured varnish. Both two-pack and one-pack polyurethanes have been investigated as varnishes. Externally the results are disappointing but urethane alkyds are useful as the basis for quick drying interior varnishes. Demand for exterior varnishes is usually in the full gloss form, but satin and matt are also popular for interior use. Settlement of matting agents requires careful formulation, for some can destroy the structuring effect of polyamide-based resins.

By far the majority of external varnishes remain solvent-based, reflecting the difficulty of good flow and high gloss with aqueous dispersion binders. Externally there is the problem of durability already mentioned in the context of stains. Aqueous latices with in-built UV protection can provide the basis of an exterior varnish [124]. For interior use there has been a steady growth in the adoption of new water-borne technologies such as polyurethane dispersions [125]; this technology is also showing promise for external coatings.

Paint and paint systems for wood

As noted earlier, the term 'paint' may be used as a generic umbrella term to cover opaque pigmented coatings, regardless of sheen, build, or system. In many countries there has been a tradition of using high-build full gloss systems on woodwork, especially outside. During the past decade the supremacy of gloss systems has been challenged first by woodstains, and more recently by paints with a sheen level lower than full gloss [126]. Despite these changes, the demand for a full gloss system remains high, especially for redecoration, and this will remain an important market sector in France, Germany, and the UK for the immediate future. In other parts of the world, including North America and Australia, the move away from high gloss exterior systems has gone further. This situation can be explained in a number of ways, but it is likely that different construction methods and the greater use of wooden siding would show up potential defects of the traditional system (such as flaking) on a larger scale, thus paving the way for greater use of more permeable off-gloss water-borne systems.

Traditionally high gloss paints were part of a three-product system comprising primer, undercoat, and gloss. Many of the new paints are two-product, or one-product multicoat systems. Before embarking on the design of a new coating, some consideration must be given to the merits of a *system* as opposed to a *single-product* approach.

An obvious advantage of single products is simplicity, both to the user and the stockist. Against this it must be recognized that a coating has many different functions to perform which might include sealing, adhesion promotion, and filling, as well as protection and appearance attributes. Combining these in one product is likely to entail compromise which may or may not be acceptable to the final customer.

To illustrate this point further, consider a traditional liquid gloss paint. If two or three coats are applied directly to bare wood the durability can be remarkably good, better in many cases than a traditional system. What then has been lost? Answers to this question include build (affecting appearance), speed of re-coat, and flexibility in redecoration. What weighting is given to these factors varies among individuals and user groups, but it is a mistake to underestimate them. Many of the newer products will inevitably be used for redecoration of existing coatings rather than new wood. They may have good durability but often fail to measure up to the practical problems encountered in both trade and retail use. Whatever the detail of the final chosen system, formulators must consider how the needs of priming, filling, and finishing are to be met.

Wood primers
An important function of wood primers is to provide adequate bonding between the substrate and subsequent finishing coats, especially under damp conditions. Primers should be able to seal end-grain while resisting hydrolytic breakdown over long periods. Some building practices expose building components for variable periods without the protection of a full system, hence primers should have sufficient intrinsic weather resistance to provide a sound base for subsequent recoating.

The archetypal wood primer was based on linseed oil and white lead. Lead carbonate has the useful property of forming with linseed oil, fatty acid soaps which have good wetting properties and yield tough, flexible films. White lead was often admixed with red lead (lead tetroxide) giving the well-known, and still imitated, pink wood primer. Recognition of the cumulative toxicity of lead has led to the withdrawal of BS 2521 (the UK standard for lead-based primers) and the development of products without lead pigment. These are usually based on oil, oleoresinous, or alkyd binders with conventional pigmentation. To maintain flexibility the PVC should be relatively low (typically 30–40%). Early low lead primers were often over-filled to reduce blocking and this, combined with fast-drying inflexible alkyds, resulted in poor performance, a factor contributing to dissatisfaction with solvent-borne systems in general. In the UK, standards for low lead solvent-borne primers are set by BS 5358 including a six months' exposure period, which used a white lead primer as control. The corresponding standard for water-borne primers, BS 5082, includes the same durability criteria. To reduce variability introduced by the wood itself, the test paint was exposed alongside the control on the same piece of wood. (In the future the separate standards will be redrafted into a composite standard with no reference to composition and the lead paint control omitted.) Differences between good and poor paints are quickly shown up. Solvent-borne paints tend to fail by cracking of the primer, while water-borne paints (due to higher permeability) may allow cracking of the wood itself. Such cracks, when they do occur, usually stop at the junction between the test paint and the control. With careful formulation, both types of primer will pass the test which correlates reasonably well with subsequent system performance. For conventional solvent-borne products a key requirement is to use an extensible binder. Weathering trials have shown advantages in using a combination of free drying oil, as well as that which is chemically bound to the polyester backbone of an alkyd resin. This may reflect the ability of free oil to penetrate the wood while the alkyd is retained on the surface to maintain flexibility. It used to be said that the oil was 'feeding the wood'; however overpenetration by the binder will leave the primer underbound.

An important feature of the old and revised primer standards is a blisterbox test. This has shown good ability to predict adhesive performance under wet conditions. Generally systems that pass the blisterbox test show good exterior durability and will show up for example the difference between an interior and exterior grade latex when used in a primer. Improving the adhesive performance of aqueous dispersions has been the subject of much patent activity [121] with the chemistry of the ureido group being especially productive. The term 'adhesive performance' rather than 'adhesion' is used here quite deliberately. Not only is the concept of adhesion rather elusive, but products that perform well on the blisterbox test do not necessarily give the highest apparent bond strength when measured on peel and other tests. The role of so-called adhesion promoters is more subtle than meets the eye and may involve a dynamic interaction with water. Blisterbox tests correlate less well with the durability of solvent-borne solution paints, but there is a good correlation with solvent-borne dispersion binders (NAD). This group fitting box II in Table 9.4 has a useful role to play for a quick-drying, flexible exterior primer with good early shower resistance [128]. Exterior wood primers will benefit from the inclusion of fungicide.

Aluminium sealers and wood primers

Primers containing aluminium flake offer good barrier properties both to water and certain types of staining. They are useful for sealing end-grain and against resin exudation, creosote, bitumen, and coloured preservatives. Barrier properties are dependent on the amount of flake pigment used. Established vehicles for this type of product include phenolic-modified resins; typical formulations are specified in BS 4756.

Aluminium primers are not advisable on large areas where high movement may be expected. A high incidence of flaking has been found on softwoods, reflecting poor intercoat adhesion. Performance is usually better on hard woods. The darker colour of aluminium primers can be a disadvantage when it comes to overcoating with white or light colours. However, some specifiers have seen this as advantageous as a means of ensuring a full finishing system over a primer that is readily detected!

Aluminium primers suffer from two specific defects on storage: binder exudation and gassing. The former results from the low viscosity, low surface tension binder separating from the pigment and passing through the most minute gap, especially welded seams, rivet holes, and the lid seal. Prevention is usually achieved by a good quality lacquered container. Gassing is the result of contamination by water or alkali residues from mixing equipment. Contamination of this sort must be rigorously excluded; the consequences of exploding lids can be very damaging. The gassing propensity of aluminium has inhibited commercialization of water-borne aluminium primers though advances have been made as water-borne metallic paints become established in car painting and refinishing.

Preservative primers

This term is sometimes used to describe an unpigmented fungicide containing binder. Such products are designed for maximum penetration at the expense of any contribution to film build. It is thus easier to achieve penetration of fungicidal components, especially into the vulnerable end-grain.

Inclusion of a suitable low viscosity binder enables the product to seal end-grain to a greater depth than can be achieved with a pigmented product, though more than one application may be necessary. The penetrating nature of these products can help stabilize a denatured surface and improve subsequent coating performance. For completeness, it should be noted that the term 'preservative' has also been used as an adjective to describe any products including stains and primers which contain fungicide.

Dual-purpose stain primers (stain basecoats)

Because there is a possibility that new joinery may be finished with either stain or paint, it is an advantage to joinery manufacturers to prime with a primer that can be overcoated with either type of product. This has created an apparent niche for dual-purpose primers [129]. Their appearance is inevitably dictated by the needs of the semi-transparent stains and such products tend to be closer to stain primers than traditional wood primers. In consequence they offer much less temporary protection than a wood primer, and this has caused problems. There is also a danger that this approach obscures from users the fact that stains and paints are not always fully interchangeable for a given construction. Highly permeable products may require a better quality of joinery wood if excess movement is to be avoided and will require alternative glazing and non-ferrous fixings. A performance specification for this category has been issued in the UK as BS 7779: 1994.

Undercoats

Undercoats can play an important role in contributing build and opacity to the traditional paint system. They also improve adhesive performance when old gloss is repainted and help provide cover on sharp edges. Achieving a high gloss appearance when renovating weathered gloss paint is more difficult without an undercoat to fill damaged areas and provide a contrast between coats. In order to fulfil these functions, undercoats are normally heavily filled and habitually have the highest PVC of any exterior coating. Moreover to aid sanding, they are usually formulated on a brittle binder. As a consequence they lack extensibility and, from the point of view of durability, are the weakest link in a paint system. Replacing undercoat by an extra coat of gloss significantly improves the durability of many paint systems, though this would be regarded as less practical by many decorators.

The fact that some undercoats are inextensible does not invalidate their usefulness, but exterior undercoats for wood should be formulated differently from those for interior or general purpose. In particular, they must have greater extensibility which can be achieved by reducing the PVC or by using a more flexible binder. This will, of course, increase sheen and makes sanding more difficult. Requirements with respect to adhesion are less stringent if a good primer is used and there are many potential formulating routes via water-borne or solvent-borne technology.

Finishes

Full gloss solvent-borne gloss finishes have remained an important feature of UK and other European markets, especially in maintenance painting. As noted above, when used with primers and undercoats (optional) which have been correctly formulated for use on wood, they can give performance with a typical maintenance period of 5–7 years. There are sufficient differences between interior and exterior conditions to make separate formulations worth while. In particular, exterior gloss finishes for wood will benefit from increased flexibility and the presence of fungicide to inhibit blue stain and other surface moulds. Gaining the optimum balance between longer term durability and initial drying properties is a difficult and long process. Alkyds, that is, oil or fatty acid modified polyesters, lend themselves to almost limitless modifications, both alter mechanical properties, permeability, adhesion, etc. and can have a marked effect on durability. Over the four or five decades that alkyd resins have been in commercial use there have been many attempts to relate their structure and chemistry to properties such as durability, but owing to the complexity of the situation, clear and unequivocal relationships have not been established. A diligent search of published literature shows many combinations in respect of oil, polyol, and fatty acid type. Not infrequently important data, such as the molecular weight distribution, are omitted. Alkyds with long oil lengths often show greater initial flexibility but are more prone than shorter oil alkyds to rapid change on weathering, though exceptions to this 'rule' have been reported [130, 131]. There are now an increasing number of alkyds on the market which have been formulated specifically for wood (e.g. [121]).

Urethane alkyds have specific advantages in terms of quicker surface and through dry, but this is accompanied by a decrease in extensibility which invariably leads to a greater incidence of cracking and flaking on exterior wood. The same problems beset silicone alkyds which show excellent gloss retention but poor extensibility. Silicone alkyds have outstandingly good chalk resistance but although this may seem an advantage it can result in very high dirt pick-up as there is no self-cleansing action. This highlights the careful balancing of properties that must be made in developing an exterior coating. The choice of TiO_2 grade can have far-reaching effects and considerable data have been published by major TiO_2 manufacturers.

In selecting one of the many fungicides now offered for wood coatings [117], it is extremely important to assess the longer-term storage stability; some have shown seeding or discoloration effects, which may take several months to appear.

Off-gloss finishes for exterior wood are usually formulated to be applied direct to bare wood or over an appropriate primer. To some extent they combine the function of both undercoat and finish. They do not form a very clearly defined product group and are sometimes described as 'opaque stains' or by proprietary names (see Section 9.9.6.1). Ideally they should be based on a fairly permeable binder and, like gloss finishes, be flexible and fungicidally protected. Choice of matting agent is critical as it is necessary to raise the PVC but without sacrificing mechanical properties. Pigments with a high oil absorption will, as discussed in Section 9.3.2.3, lower the CPVC, in effect reducing the availability of 'free' binder. Partly to compensate for any loss of extensibility in this way, extenders with a reinforcing effect such as talc or mica, may be employed. Many other new binders (alkyd emulsions, hybrids, etc.) described earlier have found application in this area and also for water-borne gloss systems.

Water-borne acrylic gloss finishes suitable for exterior woodwork are well established in many parts of the world, especially where large areas of wooden cladding are found. A major advantage, compared with alkyd-based paints, is the good extensibility which is maintained for long periods of weathering. Advances in the development of new thickeners, such as the associative types, have greatly improved flow properties, though considerable optimization is required to obtain a balanced formulation for a specific group of raw materials. Although the highest initial gloss levels and distinctness of image achieved by alkyd systems is not fully matched by water-borne systems, they normally show a much slower rate of gloss loss and are generally superior after a period of weathering. The thermoplastic nature of acrylic dispersion polymers means that dirt pick-up is high with soft polymers. Dirt becomes ingrained into the film itself and becomes very difficult to remove. This problem was more apparent when gloss finishes were based on dispersions designed for semi-gloss finishes where the higher level of pigment reduces thermoplasticity. More recently, latexes based on harder but still flexible resins have become available, effectively eliminating the problem. A related problem is that of blocking, which can be caused by a soft polymer and will be exacerbated by water-soluble material being resolubilized. A major problem area is the space between rebate and opening window lights. This problem is also reduced by the newer generation latexes available, but care in the selection of other ingredients must be exercised. Blocking resistance is sometimes improved by the incorporation of a proportion of non-film-forming hard latex which, in formulation terms, is considered as part of the extender PVC [132]. Another characteristic problem with dispersion gloss finishes is a short open time, especially in windy conditions; clearly the effect of this on appearance will be more noticeable with a gloss, as opposed to mid-sheen finish. It is this fact combined with practical considerations, such as application in damp or cool weather, that have held back the market penetration of acrylic gloss paints and alkyd gloss is still the preferred choice in many countries. However new developments will continue to change the balance. For example improvements in the open time can be gained with a sterically stabilized polymeric dispersion [133]. Each of the technologies described in the binder section will continue to erode the alkyd gloss sector as legislative pressure increases.

Off-gloss water-borne finishes are readily developed from either similar or softer polymer dispersions than used for gloss finishes. In comparison with solvent-borne mid-sheen finishes it is easier to compensate for the consequences of a higher PVC on mechanical properties.

The higher permeability of water-borne dispersion paints is usually seen as an advantage for wood coatings, but for some situations it can prove too high (leading to excess movement) and pigmentation or choice of polymer should be adjusted accordingly. Surface moulds may grow readily on water-soluble components in the coating and it is essential that they are fungicidally protected.

9.7   Masonry and cementitious substrates

Among the most widely encountered surfaces in buildings are plaster, concrete, external rendering, and brick. Although these are individual materials with their own characteristics they also have a number of general similarities which influence coating formulation. Of particular significance are alkalinity, the porous friable nature of the surface and the general consequences of moisture and its interactions with the substrate.

9.7.1   Implications of moisture

Water is often present in large quantities in the materials referred to here, especially in new buildings. This is particularly true with hydraulic cements and plaster, but will also result from the storage of materials in the open during construction. Surfaces created from 'wet' materials of construction may require an initial coating of very high permeability to allow drying out, though subsequent redecoration with less permeable coatings is possible.

Moisture content is conveniently quantified by quoting the relative humidity in contact with the surface. BS 6150 divides surfaces into four groups according to their equilibrium moisture content:

- Dry — <75% RH
- Drying — 75–90%
- Damp — 90–100%
- Wet — 100% (with visible surface moisture)

Wet surfaces are very difficult to paint, but damp and drying surfaces can be coated with emulsion paints which are usually formulated above the critical PVC to increase permeability. As with wood, the presence of water which has not been allowed to dry out, will reduce adhesive performance, may cause blistering and creates the possibility of mould growth. Moisture is also a key factor in relation to alkaline attack, efflorescence and staining.

9.7.1.1   Alkaline attack

Portland cement is highly alkaline, as are some plasters. Such alkalinity in the presence of water will saponify many oil-based paints and may discolour pigments.

9.7.1.2   Efflorescence

Unsightly deposits of salt on the surface of plaster or brickwork, etc. are known as efflorescence and appear in bulky and dense form. Bulky efflorescence is usually a form of sodium sulphate and may disrupt coatings of low permeability, although oil paints have a limited ability to hold back efflorescence. Permeable coatings, such as above-critical emulsion paints, allow efflorescence to pass through, where it can be wiped from the surface but heavy eruptions may cause adhesion failures and under these circumstances painting should be deferred until efflorescence has largely ceased. The denser type of efflorescence is usually calcium carbonate (known also as lime bloom) and is more difficult to remove but easier to overpaint after light abrasion.

9.7.1.3   Staining
Brown stain may appear on emulsion paints on some types of brick, clinker, and
hollow clay blocks. The colour derives from soluble salts or organic matter capable
of reacting with alkali. Alkali-resistant primers will normally prevent this type of
staining. A useful summary of typical defects in walls will be found in [134].

9.7.2   Cement and concrete

Cement, as a component of concrete, finds very widespread application in all types
of building and, indeed, its per-capita consumption has been suggested as an indicator of economic development. Such widespread use reflects the attraction of a
material which is relatively cheap, can be moulded or cast, and will withstand high
compressive stress. Iron or steel reinforcement of concrete also allows the carrying
of tensile loads, although cracking of the concrete is not necessarily prevented. Load
carrying is much improved by putting the concrete into compression which is
achieved in pre-stressed concrete through tensioning the steel reinforcement. In
reinforced concrete the function of the concrete is to resist compression and buck-
ling and also to protect the reinforcement from corrosion. Insufficient thickness or
inadequately prepared concrete has shown failure in this area and in turn, created
markets for both remedial and preventative coatings [135]. A potentially large
market, though outside the scope of this chapter, is in reinforced concrete brick
decks. These are especially vulnerable to the corrosion exacerbated by de-icing salts.
Epoxy powder coated reinforcements for bridge decks are mandatory in several US
states [136].

   An alternative to the massive reinforcement described above is to use a fibre
reinforcement. Asbestos cement is a well-established example, though under
increasing pressure for health reasons. Cement panels and components can also be
made using glass, metal, or plastic reinforcing fibres, and such materials are becoming more common. Although described as a reinforcement, the function of the fibres
is to prevent crack propagation rather than to carry a major tensile load. Coating
requirements of such products are largely dictated by the nature of the cement
matrix, though surface appearance will be modified by the presence of fibre.

9.7.2.1   Characteristics of cement and concrete [137]
Concrete is made by binding an aggregate of sand, broken stone, gravel, etc. in a
matrix of cement paste and water. The most widely used binder is Portland cement,
though 'high alumina' and other more specialized cements find a number of important specific uses. Hardened cement paste may be regarded as a cement gel matrix
which will contain unhydrated cement particles, air and water; mortar is a combination of sand and cement paste.

   Cement is a well-established building material; both Romans and Greeks used
cements based on hydrated lime, and this type of cement still persists. Strength
development depends partly on an initial slaking reaction:

$$CaO + H_2O \rightarrow Ca(OH)_2$$

but long-term durability derives from the conversion of colloidal hydroxide to carbonate by reaction with atmospheric carbon dioxide:

$$Ca(OH)_2 + CO_2 \rightarrow CaCO_3 + H_2O$$

Portland cement is produced by heating a lime-bearing material such as lime-stone with a material containing silica, alumina, and some iron oxide — typically clay. After compounding, the material is fused to a clinker at around 1400°C and ground with gypsum and other materials (which act as retarders) to produce cement powder.

The main compounds in anhydrous Portland cement correspond to tricalcium silicate, dicalcium silicate, tricalcium aluminate, and tetra calcium alumino ferrite (C_3A, C_2S, C_3A, and C_4AF in cement notation). When hydrated separately these compounds gain strength at different rates. C_3A is the most rapid hardening, pro-ducing considerable heat evolution; its percentage is reduced in cements which are to be used in thick sections. C_3A is attacked by sulphate and must be omitted from sulphate resistant cements.

9.7.2.2   Hydration mechanisms of Portland cement

Hydration of cement is a reaction in which a solid of low solubility reacts with water to form products of even lower solubility. There has been much debate as to the mechanisms of cementing action. Competing theories from the past include Le Chatelier's Crystallization Hypothesis (1887) and Michaelis's Gel Hypothesis (1893). Today there has been a tendency for some synthesis of crystallization and gel theories but there is still no general agreement for all the mechanisms involved.

Initial mixing of cement and water produces a dispersion, the particles of which become quickly coated with hydration products. Hydration products are largely col-loidal (10–100 Å) though some larger crystals of calcium hydroxide may be formed. The solution becomes saturated with Ca^{2+}, OH—, SO_4^{2-}, and other ionic species. On further reaction the coating of hydration products extends partly at the expense of the grains and partly at that of the liquid. A $1 cm^3$ of cement produces $2.3 cm^3$ of cement gel; 45% of the gel must fall inside the boundary of the cement grain and 55% in the surrounding capillary space. Rupture of the coating may take place (according to Powers [142]) as a result of high osmotic pressure. When the coatings begin to meet, the cement is at the setting stage. With further reaction the particles become increasingly densely packed; this further hydration and crystallization from super-saturated solution involves a complex diffusion process, in which water from capillary space diffuses 'inwards' through the gel pores, while hydration products move in the opposite direction. No further cement gel forms in the gel pores and it is assumed that these are too small to allow nucleation of a new solid phase. Crys-talline particles such as calcium hydroxide are disseminated through the gel and may form in the pores by crystallization. Calcium hydroxide crystals are usually thought to be detrimental to the strength of cement, but it has been suggested that it could be a major factor in bonding inert particles and fibres.

The latter stages of hydration are slow and can take 25 years to complete. During the early stages of hydration free water is in the form of capillary channels, but as these block, once cavities become filled hydration must cease because there is no room for gel formation (although in certain cases, formation of hydrate by disrupt-ing the existing gel may occur — leading to very weak cements). If all the capillary cements are blocked it should be virtually impermeable.

The space structure of cement gel as first formed is expanded and unstable with a tendency to change, which is accompanied by a diminution in surface area and shrinkage. Properties of hardened cement paste are considerably modified by aggre-gate. Among the many parameters which modify the properties of hardened cement paste are the volume fraction size and shape of the aggregate, water/cement ratio, and degree of hydration.

Water requirement

Stoichiometrically the requirement of cement powder for complete hydration is met by a water/cement ratio of 0.23. However, if all the cement is physically to have room to hydrate (remembering the volume nearly doubles) then a water/cement ratio of 0.38 is required. Another constraint is that combined water undergoes 'compression', with a specific volume reduced from 1.0 to 0.72; even gel pore water only has a specific volume of 0.9. This reduces the relative vapour pressure within a sealed sample, and if it falls below 80% hydration virtually stops. To avoid this a water/cement ratio of 0.5 is required though this will not necessarily produce other strongest concretes. Even in the presence of excess water it is common to find unhydrated cores of the original cement particles.

Dried cement gel characteristically has the chemical composition $3CaOSiO_2 \cdot 2H_2O$ but in saturated gel an extra mole is present as intercrystalline water, and can only be removed at water vapour pressures less than 0.1 (mm Hg). Gel pore water is strongly held and is present at relative vapour pressures of 0.1 to 0.5. Smaller capillaries are full at vapour pressures of 0.5–0.8, but larger capillaries require a vapour pressure above 0.8 to fill [137, p. 129].

The chemistry of the hydration reactions taking place in Portland cement is extremely complex. Of special relevance to the coatings technologist is the reactive and alkaline nature of the surface, combined with its variable porosity and permeability.

Retarders

Calcium sulphate, usually as gypsum, is universally added to ground cement to control the otherwise rapid 'flash set'. Many other compounds have a retarding effect and these have been put on a systematic basis by Forsen, according to their effect on the solubility of alumina. Following his categorization, retarders may be divided into four sets depending on their actions as a function of concentration. Typical examples from each group as (i) $CaSO_4 \cdot 2H_2O$, (ii) $CaCl_2$, (iii) Na_2CO_3, (iv) Na_3PO_4. Type (iv) retarders may hold up setting and hardening indefinitely if used in sufficient quantity, but they are not all harmful and some, such as the calcium lignosulphonates, are used as water-reducing agents.

9.7.2.3   Structure and morphology

At the macroscopic level the structure of concrete, etc. is dominated by the dimensions of aggregate and the extent to which they have been compacted. Poor compaction and mixing leads to obvious heterogeneity. Within the cement paste air entrainment pores are typically around 50 μm. Typically capillary and gel pore 'diameters' are 500 Å and 15 Å respectively. Under the optical microscope hardened cement paste appears amorphous and semi-transparent but the electron microscope reveals a wealth of micro structure [138]. Care is needed in interpreting some micrographs if the hydrates are produced at untypically high water/cement ratios. Various cement hydrates have been identified and their structure related to that of known minerals such as 'Toberemorite' and they have been observed in crumpled sheets or rolled-up tubes. Midgley described splines and plates as important morphologies [139]. A model described by Double and Hellawell explains the formation of hollow tubes by a similar mechanism to that seen in the silica 'gardens' that form when metallic salt crystals are placed in solutions of sodium silicate [140].

9.7.2.4   Volume changes in concrete

Volume changes during hydration and subsequent sensitivity of cement gel to moisture content ensure that the overall shrinkage or expansion of cement paste is complex. Re-immersion of dried cement paste causes swelling as water penetrates interstices within the gel, but not all the drying shrinkage is recovered. In the presence of carbon dioxide any calcium hydroxide present may be converted to calcium carbonate with a subsequent irreversible carbonation shrinkage — this can be as much as 50% of the initial drying shrinkage but is normally limited to external surfaces. The inclusion of aggregate changes the magnitude of volumetric strain largely in proportion to its volume fraction. Concrete is a multiphase material and some of the shrinkage of the paste may show up as cracks at the aggregate paste interface. The range of length change in hydrated cement paste specimens subjected to wetting and drying cycles is characteristically spread around 0.5%.

9.7.2.5   Porosity and permeability of concrete and cement

Moisture movement in concrete (as in timber) may be usefully considered in terms of both permeability and diffusion. Although these are derived from the same physical processes, the mathematical forms differ. Permeability is associated with a pressure difference and is associated with saturated materials, while diffusion is more useful in considering partially dry materials with the fluid driven by chemical or moisture potential [141].

Permeability is influenced by porosity and hydration of material within pores and capillaries will greatly reduce flow. Normally water movement will occur within capillaries, rather than the pores. Powers has published data showing the relationship between permeability and capillary porosity [142]. Drying cracks or flaws at the aggreagate interface are likely to increase permeability. Porosity for hardened cement is characteristically 25% and once capillaries are blocked the permeability falls to about 10–$12 \, cm \, s^{-1}$, which is less than many natural rocks.

The most important driving force for diffusion is the gradient between internal moisture and surface or capillary forces. Solutions to the diffusion equation are discussed in Chapter 8 of [141].

The permeability of concrete is a major indicator of its potential durability, both in the sense of mechanical strength and resistance to chemical attack. Permeability may be tested by measuring the flow through a saturated specimen subjected to pressure; a penetration test is more appropriate where moisture is drawn in by capillary action. A test known as the Initial Surface Absorption Test (ISAT), which can be applied to concrete in situ is described in BS 1881, and can be used to assess the rate of water absorption — clearly this is relevant to some coating properties.

9.7.2.6   Resistance of concrete to destructive processes

In service, concrete may be subject to a physical process such as freezing or heat, which can cause damage. Other potential disruptive reactions include expansive alkali–aggregate reactions which are an increasing cause for concern. However, these are outside the preventative scope of coatings. In principle, concrete is a highly durable material which should not require coating to achieve good weathering, but there are circumstances where the protection offered by a coating becomes necessary. These include protection against acidic environments, and to prevent attacks of reinforcement if the thickness of concrete is insufficient or otherwise unable to provide protection. It may also be necessary to provide protection against specific agents to which it is sensitive.

The resistance of concrete to chemical attack is very dependent on its permeability, the nature of the aggregate and any additives, and the care with which it is made and placed. The life of good quality concrete can be 20 times that of indifferent material in identical conditions.

Portland cement (but not high alumina) is markedly attacked by sulphates, with magnesium and ammonium sulphates being particularly severe. Strong aluminium sulphate will also attack high alumina cement. Attack by sea-water stems from the presence of magnesium sulphate which is modified by the presence of chloride and can also attack metal reinforcement. Pure water can dissolve the lime from set cement but the action is slow unless water is able to pass continuously through the mass. Attack increases with pH as caused by the presence of CO_2 above the amount needed to maintain the equilibrium $CaCO_3 + H_2O + 2CO_3 = Ca(HCO_3)_2$. Displacement of this equilibrium means CO_2 is more aggressive in saline solution.

The resistance of all hydraulic cements to inorganic acids is low, but they are also attacked by many organic acids though there are anomalous concentration effects: acetic (vinegar), lactic (dairies), butyric (silage), and tartaric (fruit) acids all have adverse effects as do higher molecular weight acids, including oleic, stearic, palmitic, and most aliphatic acids.

Portland cement is resistant to strong alkalis including sodium and potassium hydroxide. High alumina cement, in contrast, undergoes severe strength deterioration. Even alkaline detergents should be avoided when washing high alumina cement floors. Other agents known to attack concrete include sugar, formaldehyde and the free fatty acids in vegetable oils and animal fats. Glycerol reacts with lime to form calcium glycerolate [143].

9.7.2.7   *Organic growths on cementitious substrates* [144]

Although not biodegradable in the sense that wood is, the rough surface of most cementitious and masonry substrates leads directly to conditions which will support organic growth; that is collection of, and holding of, moisture and nutrient. Organic growths found on building surfaces include those requiring light such as algae, lichens, mosses, as opposed to fungi, which do not. Sulphate-reducing bacteria in gypsum plasters can cause staining of lead-containing paints and have been known to promote corrosion of steel. It is vital that, prior to coating, any organic growth should be physically removed by scraping and brushing to prevent rapid re-infection. Surfaces should also be treated with a toxic wash which has been approved for safety in use.

9.7.3   Other cementitious substrates

9.7.3.1   *External renderings* [145]

Cement rendering
Many renderings are based on Portland cement, possibly with incorporation of lime; they are therefore highly alkaline. Sand cement renders become extremely friable

with time. Lime/sand renders, as found in older buildings, are usually known by the term 'stucco'. BS 5262 contains recommendations for repairs to rendering, including stucco. Characteristic properties of other renderings including 'rough cast', 'pebble-dash', and 'Tyrolean' are described in [146].

Lightweight concrete blocks
Not normally used outside, these materials present problems in consequence of their open pore structure.

Asbestos cement sheets
When new, these sheets are strongly alkaline and vary in porosity, even within the area of a single sheet. Good resistant sealing will be necessary under many coatings. It is important not to seal one face of a sheet with an impervious coating as there is a risk of warping caused by differential carbonation.

9.7.4 Plaster and related substrates

Plasters are used extensively for finishing internal walls; characteristic problems include variable porosity and alkalinity. The latter will be especially high with lime and cement plasters and in practice lime is often added to gypsum derived plasters to facilitate spreading.

Most commonly 'plaster' is based on calcium sulphate hemi-hydrate (Plaster of Paris) formed by partial dehydration of gypsum (the dihydrate) which will be reformed on addition of water. The reaction of water with Plaster of Paris is rapid, and to aid application on large areas it is retarded with alum or borax. In those plasters containing lime the action of potassium sulphate (alum) and calcium hydroxide (lime) in producing potassium hydroxide may affect paint properties. BS 1191 broadly distinguishes between the above possibilities with:

- Grade A — Plaster of Paris;
- Grade B — retarded hemi-hydrate plasters;
- Grade C — anhydrous.

The latter are likely to contain added lime. On occasions plasters fail to hydrate fully; not only will this lead to a powdery surface layer, but subsequent wetting results in expansion with adverse consequences for plaster and coating. All plasters are likely to soften if wetted and this too leads to coating failures.

Keene's cement (Grade D in BS 1191) is atypical in being slightly acid, though the acidity will often be neutralized by more alkaline backing materials.

9.7.5 Brick and stone

Brick and stone may generally be regarded as durable materials which with a few exceptions do not require coating for protection. Coatings are more likely to be required for aesthetic reasons; this is especially the case internally where painting will also facilitate cleaning and improve lighting.

Efflorescence is always a possibility on brick and stonework. Paint is best avoided in any situation where major moisture penetration is possible, such as below the damp-proof level.

Types of brick include clay bricks, classified by BS 3921 as common, facing, and engineering.

Common bricks include 'Fletton' which are notoriously difficult to paint unless sand-faced or rustic [147]. One potential difficulty is coating those areas of bricks which were in contact during baking. So-called 'kissmarks' appear almost glazed and have a coarse pore structure. Not only are they more difficult to wet, but their greater porosity is more likely to allow efflorescence.

Facing bricks are made to give a specific surface texture or colour and do not present the same adhesion problems as Flettons. *Engineering bricks* are much denser, virtually non-porous, and more likely to give adhesion problems. They may have similarities with glazed surfaces, as does non-porous stone, especially if polished, and may require specific primers to promote adhesion. This area is not well documented but silane coupling agents are worth consideration in difficult situations.

Calcium silicate bricks are classified by compressive strength, reflecting porosity (BS 187). Normally more uniform than clay bricks, they are not regarded as problematical in painting.

Stone varies considerably in different locations with extremes ranging from limestone to granite. BRE digest 177 discusses some of the factors affecting decay of stone masonry and the conservation measures used in protection [148]. Normally stone masonry is not painted but where a need exists coatings must be formulated to suit the characteristics of the material in question.

9.7.6   Coatings for masonry and cementitious substrates

Masonry and cementitious substrate coatings present a similar difficulty to wood in defining up a neat, logical framework. This reflects the complexity of the many substrates and products with an inevitable degree of overlap. Jotischky [136] has reviewed aspects of the market structure which can be divided, for example, by substrate (masonry, concrete, etc.), by application (civil engineering, residential, local authority, etc.), by purpose (protective, remedial, etc.) or by function (e.g. chemical resistance, anti-graffiti). Probably the most visible market sector is for masonry, essentially decorative products for stucco, rendering, pebble-dash, and brick. The largest product offering in this sector is offered by exterior emulsion paints based on vinyl and acrylic latexes.

The extension of this market sector to cover architectural coatings on concrete has been very slow, though in Germany the percentage is higher. Many architects would maintain that concrete does not require painting, but apart from aesthetic aspects there are situations when a coating can prevent water penetration and reduce attack by carbon and sulphur dioxides [135]. Leading on from this is the repair market for reinforced concrete mentioned in Section 9.10.2. This has called for a variety of remedial and preventative products, with liquid epoxy resins finding a useful niche. The market is largely in the hands of specialists but the value of buildings at risk is high [136]. In the UK the market for coating new reinforcement coatings is small and is more likely to be met by galvanizing, but in North America protecting new reinforcements accounts for an appreciable component of the epoxy powder coatings market.

9.7.6.1   European Standard classification

In response to the difficulty of classifying masonry and related coatings a CEN committee has been working to draw up a classification and performance specification for exterior systems. This exactly parallels the activity referred to in Section 9.9.4.3 with respect to wood coatings and there are a number of similarities between the two approaches. There are also some differences which reflect the approach of the two committees, but also reflect real differences between certain essential features of wood, as opposed to masonry coatings. The classification terms adopted, after a great deal of debate, provide a useful insight into some of the issues of importance in this sector, and have been published as European Standard EN 1062-1 (May 1994). In common with the wood coatings standard, part of the classification deals with appearance with three levels of specular gloss and five levels of film thickness (build) with a much greater range than wood coatings. There is no equivalent terminology relating to transparency, but the availability of textured masonry coatings is reflected in a 'Largest Grain Size' term with four levels: fine, medium, coarse, and very coarse. Thus just as for wood coatings there is the opportunity to classify 60 appearance categories.

The masonry standard also includes reference to the chemical types of binder and to its state of dispersion as (a) water-dilutable, (b) solvent-dilutable, and (c) solvent free. It is also required that the chemical types of binder should be given e.g:

- *inorganic* — hydraulic lime, cement, silicate;
- *organic* — acrylic resin, vinyl resin, oil, alkyd resin, polyester, chlorinated rubber, organosilicone, epoxy, polyurethane, bitumen.

Coatings are also classified according to 'end use', here the emphasis is on the intended function of the coating, rather than the function of the coated substrate (as is the case with the wood coating classification). The masonry coating end-use categories are (a) preservation, (b) decoration, and (c) protection. Clearly the term 'preservation' has a different meaning from that for wood coatings and means 'to maintain the original state and appearance of the substrate' (e.g. a clear water-repellent treatment). The terms overlap to a degree, but it is possible to stress the principal function.

The masonry standards also place emphasis on controlling moisture, both vapour and liquid permeability are used as prime classification terms. This is a somewhat different emphasis from the wood coatings standard where moisture requirements are built into the performance criteria but the net effect is much the same. The test method for moisture movement (prEN 1062-2) uses calcium silicate bricks as a substrate, the test face is immersed in water and weight gain plotted as a function of time. Permeability calculations use a coefficient derived from the linear portion of the weight gain curve and are expressed in the units $kg\,m^{-2}$ per SQRt (h). Uncoated bricks show a typical value of 0.38, a typical exterior emulsion 0.15 and a moisture curing polyurethane 0.04 [149]. Work is still in hand to interpret these in terms of a performance specification. The following properties are identified as important where there is a special protective function, e.g. for steel reinforced concrete:

- crack bridging;
- mould, fungi, and algae resistance;

- carbon dioxide permeability;
- alkali resistance.

For the remainder of this section a selection has been made which illustrates some of the products in the market; in principle these can be related to the above classification.

9.7.6.2   Sealers and colourless treatments

The porous and sometimes friable nature of masonry surfaces has created a market for water repellents and in some cases for sealers to act as primers prior to painting.

Water repellents for masonry

These materials are intended to improve resistance to rain penetration with minimal effect on appearance (i.e. 'preservatives' in CEN terminology). They function by inhibiting direct capillary absorption but do not normally provide a continuous surface film. Properties of interest include resistance to water penetration, water vapour transmission rate (permeability), resistance to efflorescence, and longevity of the effect. Such treatments will not necessarily decrease water uptake through cracks which may, in fact, increase if the treatment causes more water to run across the surface.

Waxes, oils, and metallic soaps have been used as the basis of water repellents but these have tended to be supplanted by silicone resins in various forms. A survey carried out by the GLC [150] showed silicones to be the largest group of proprietary agents available in the UK but also noted were siliconates, silanes, epoxy resins, and acrylates. User guidance and performance standards for silicone-based repellents only are given in BS 3826. When using proprietary silicone resins as the basis of a formulation, attention should be paid to manufacturer's literature as the different grades may have restrictions in use; some, for example, are not suitable over limestone. Silicone resins are available in both solvent- and water-borne form; the latter can be highly alkaline.

It must be emphasized that although water repellents can be effective in reducing rain penetration, there is a danger that under some circumstances they can cause serious spalling. This is caused by trapped crystallization salts (normally showing as efflorescence) forcing the surface off. It has also been suggested that differential thermal and moisture movement between bulk and surface is another cause [151]. The vulnerability of repellents to efflorescence pressure reflects their high vapour permeability. Although this has the benefit of allowing trapped moisture to escape, it allows rapid absorption of water when humidity is high. There are parallels with the problems of high permeability in woodstains discussed earlier. This problem underlines the difficulty in striking a correct balance between permeability and efflorescence resistance. Water repellents are best avoided if efflorescence is a known problem [152]. In the GLC survey [150] with a single exception, all the products that passed the water penetration test caused spalling of bricks in a simulated efflorescence test.

Sealers

Masonry sealers, also known as *stabilizers*, are intended to consolidate friable surfaces. Typically they are based on alkyd solutions carried in white spirit; a tung-modified alkyd is traditional in order to improve alkali resistance; tung phenolic resins are also suitable. Paradoxically some masonry paints, including the emulsion type described below, do not adhere well to a continuous stabilizer film. It is, therefore, important to ensure (a) that the viscosity is sufficiently low to aid penetration, and (b) that stabilizers are only applied to truly friable surfaces. If the surface is sound, then a stabilizer should not be necessary.

An alternative to alkyd or other resin solution is to use a very fine particle size latex at relatively low concentration; styrene, acrylic, and polyurethane latexes have proved suitable.

9.7.6.3   Alkali-resisting primers

Another product associated specifically with the masonry market is that of the alkali-resistant primer designed to hold back alkali attack on essentially dry alkaline substrates. Although normally used below oil finishes, they can sometimes be used to improve the adhesive performance of emulsion paints on plaster surfaces. Variants specifically for preparing plaster are also marketed.

Alkali-resistant primers have been successfully formulated for many years on tung phenolic and tung coumarone resins. Resistance may be further upgraded with isomerized rubber. PVC and volume solids content are typically around 30% and 50% respectively. Water-borne primers can be formulated on acrylic resin dispersions. To aid penetration the latex should be of fine particle size and the primer pigmented to a low PVC and solids content.

9.7.6.4   Inorganic paints

Hydraulic cement paints

Once widely used, the market for 'cement paints' is declining but still filling a useful niche. Normally used outside, they have a specific advantage in being applicable to wet surfaces. Cement paints are based on white Portland cement with further additions of titanium dioxide and coloured pigment as appropriate (BS 4764: 1991); they also require agents to control flow and structure.

The rough surface of cement paints encourages dirt pick-up and algal growth; they will be eroded rapidly in polluted acidic environments. Interaction between cement and gypsum precludes their use over the latter substrate. They are supplied in dry form, to which water is added prior to application. Stored product must be tightly sealed to prevent hydration in the container.

Although often regarded as old-fashioned, cement paints are unusual in being essentially inorganic and clearly have a very different balance of properties from organically bound paints. In principle, their properties could be modified with additives such as spray-dried polymer particles, to give coatings outside the current formulating box and with modified properties.

Alkali metal silicate-based paints

Alkali metal silicates of a wide range of composition can be made by the fusion of sand and metal hydroxide, followed by water dilution. They fall within a large phase diagram which also includes non-soluble glasses. Sodium silicate solutions were also known as water glass and used to preserve eggs. As a binder sodium silicate is brittle, highly alkaline, and remains soluble. The film can be rendered insoluble by using a second metal oxide anion. This can be provided in some cases by an extender, by iron or tin chelates, and various aluminium phosphates. Potassium silicate can be rendered self-curing by partial condensation of the silicate to give a colloidal form which must be stabilized. Lithium silicate gives insoluble films directly. The very brittle nature of the dry films reduces their usefulness unless substrate movement is relatively low. Use of silicate binders has found a useful application on inorganic substrates and are well known in Europe under their proprietary name 'Keim', after the inventor [153, 154]. Their use appears to be growing. Silicate paints also react with zinc dust and in this form are used as a base for anticorrosive paints on metal.

9.7.6.5   Masonry paint — 'normal emulsion'

Masonry (but not reinforced concrete) surfaces, represented by stucco, rendering, brick, pebble-dash, are a potentially large market for decorative paints where the dominant reason for painting is one of aesthetics. In the UK, about a quarter of the available surface is said to be painted [136]; the volume of 20 million litres representing 6–7% of the total UK decorative paint market. Within the sector, emulsion paints account for the major volume.

A distinction is sometimes drawn between 'general-purpose', 'contract', and 'exterior emulsion' paints; the former represent good quality paints as used indoors. Many of these give adequate performance on exterior masonry without further modification and some manufacturers do not restrict their use to interior only. 'Contract' grades are, in the main, cheaper products taken well above critical PVC with extra extender; they are characterized by high opacity and low scrub resistance, and perform poorly outdoors. In general exterior emulsion masonry paints should be formulated specifically for this purpose. Important formulating parameters include the choice of binder, PVC, and coalescing solvent, which must be selected or designed to meet specific needs during and after the drying process. It is also advisable to include fungicide and/or algicide to inhibit organic growth on the film, though it should be noted that the prevalence of such growth depends also on other factors, including the availability of water-soluble colloid or other material in the film.

Two major groups of binders used to formulate exterior masonry paints are vinyl acetate and acrylic copolymers, common comonomers being 2-EHA and VeoVa. Maximum alkali resistance is usually associated with acrylic resin, though much depends on individual formulating practice. Acrylic resin has also shown improved chalk resistance in comparison with other types. The ratio of hard to soft monomer in the copolymer is adjusted to minimize dirt pick-up while maintaining sufficient flexibility to cope with substrate movement which may include the opening of fine cracks. Chalk resistance is also strongly influenced by the type of binder, with best resistance given by small particle size binders of low MFFT [155].

Another type of acrylic latex binder [156] is modified with vinyl and vinylidene chloride. These latexes are anionic in character and have a pH below 2 influencing their compatibility with other polymer emulsions. High alkali resistance makes these polymers suitable for masonry surfaces, and detailed formulating principles are available from the manufacturers. An outstanding feature of these chlorine-modified copolymers is their low water vapour permeability which is typically between one and two orders of magnitude below that of other water-borne latex films.

The PVC of masonry paints should be chosen to achieve desired physical properties and can be used to modify the properties of the binder. Typically, the PVC of an exterior masonry paint will lie between that of an interior matt or silk — probably in the range 30–45 — but as discussed previously in Section 9.3.2, PVC in isolation is a fairly meaningless parameter which should be adjusted in relation to the CPVC of the composition. CPVC, will, in turn, be influenced by the binding power of the emulsion and the water demand of pigment and extender. These should be chosen to give a high CPVC, i.e. extender of low water demand. Choice of extender is also important for other reasons; laminar/fibrous extenders such as talc can either reinforce the film or provide stress-relieving mechanism [157]. Although most exterior emulsion paints are formulated below the CPVC it has been suggested that above critical formulations are viable if a hydrophobic polysiloxane resin is included in the formulation [158].

Application of emulsion paints over a 'hill and dale' topography such as pebbledash, accentuates mudcracking (simulated pebble-dash wallpaper is a useful test substrate for simulating mudcracking in general) which tends to be a problem at low temperatures. Mudcracking is caused by three-dimensional shrinkage in a coating which has at least one dimension confined, usually by adherence, to the substrate. Internal stress may exceed the rather weak tensile strength characteristic of a dispersion coating during coalescence, resulting in the familiar mudcrack pattern. The role of coalescing solvent in controlling mudcracking is critical and with new formulations it is advisable to investigate the effect of coalescent level and type. Solvent-free emulsions obviously present a greater challenge and especial care is needed in the pigment/extender balance. Talc and mica extenders are useful to inhibit mudcracking.

9.7.7 Textured masonry paints

Emulsion paints for masonry are readily modified with fine sand-like aggregates creating market subsectors such as 'rough textured' and 'smooth' masonry coatings. As noted above the CEN classification defines four levels of granularity which range from <100 to >1500 µm in grain size.

The most immediate effect of adding sand to emulsion paint is a change in appearance which is obvious at short range, but less so at a distance. Sand-filled paints tend to achieve a lower spreading rate and in certain specific systems have led to improved chalk resistance [159]. Against this they may show worse dirt pick-up which is more difficult to remove from the rougher surface. Sand is normally stirred into the composition at the end of the other production process at levels in the range 25–45% by weight. In effect these paints comprise sand embedded in a matrix of pigmented emulsion paint. PVC calculations are usually more easily related to other film properties if the volume of the aggregate is omitted from the calculation. Although sand is the most common additive for the fine textured paints, other materials including polymer powder or fibre have also been used. Fibres have a dramatic effect on rheology and are likely to present application problems.

9.7.7.1 'High build' textured coatings ('organic' renderings)

Some aspects of durability are related to film thickness and when properly formulated thick emulsion coatings can give protection for up to ten years, though dirt pick-up remains a problem. A large range of thick textured coatings is available in Europe, though the market is less developed in the UK where their use is largely remedial on repair. Significant formulating features are a generally higher PVC

(50–60%) than the group described above; the presence of much coarser aggregate (up to 2 mm diameter) and the use of rheological modifier to give a very high low-shear viscosity. Applied film thickness can be up to 3 mm and is achieved by trowel or spray, often followed by roller texturing; colour is normally achieved by over-painting with conventional masonry paint. Variants of these coatings include some coloured chips dispersed in an essentially clear latex binder which can give multi-coloured texture effects.

Guidance to the formulation of these products is available from the manufacturers of the specialist emulsions used. The vapour permeability of these coatings lies in the middle range, i.e. between the high conventional cost and the low values of bitumen and solvent-borne two-pack coatings. Water resistance is good — generally better than a sand/cement render.

9.7.7.2 High performance latex-based systems
Related to high build textured coatings is a group of latex-based products designed to give higher durability. Often described as 'high performance' these are designed as a system comprising primer, topcoat, and, in many cases, a separate undercoat. Practices differ between countries, but when used to give maximum protection, up to five coats will be applied. It is usual for these products to be supplied on a contract 'supply and fix basis' — some companies offer up to 15 year guarantees.

Primers in these systems are based on very fine particle size latexes with both acrylic and PVA finding use. Bitumen emulsion can be used to upgrade water and vapour resistance either alone or blended with acrylic. In either case a sealer which can be acrylic-based, will be necessary to stop the bitumen bleeding through into the topcoat. Several of the high preform systems employ undercoats, some fibre-reinforced, which are used on a remedial basis. Topcoats are generally structured and textured, often requiring specialized application equipment. Acrylic, styrene acrylic, and vinyl acetate variants are all represented in the market-place. A feature of this market is the need for ancillary products including fillers and mastics. Epoxy-based mortars are used to repair badly damaged surfaces prior to renovation.

9.7.8 Solvent-borne masonry paints
The solvent-borne sector of the masonry market, like the emulsion section, has a few clearly defined general-purpose areas, while the higher performance end is more complex and diversified. General-purpose one-paint products include alkyd styrene acrylic, chlorinated rubber, and PVC 'solution' products.

9.7.8.1 Alkyd masonry paints
Although their alkali resistance is not good, alkyd coatings perform adequately over masonry, provided an alkali-resistant primer is used. Alkali resistance is improved by the use of urethane alkyds, though this approach will limit flexibility. In the UK both mid-sheen ('eggshell') and gloss paints are used externally on masonry. The market is a significant niche as evidenced for example by the number of buildings in London which are still coated with alkyd gloss. Alkyd gloss paints will be vulnerable to replacement by latex-based gloss paints and other water-borne technologies, though early shower resistance remains a practical problem.

9.7.8.2 Solvent-borne thermoplastic coatings

Solvent-borne thermoplastic paints are not widely used in the UK, though they are well established on the Continent, especially in France [160]. Normally they are used as single product two-coat systems though special primers may be needed on poor surfaces. Essentially, they compete for the same market sector as the general-purpose 'emulsion' paints described above. Their performance is generally good in terms of film integrity, but with a greater tendency towards chalking than emulsion types. Being closer to a solution, rather than dispersion-based binder system, leads to more restrictive application properties.

A characteristic of matt or mid-sheen solution paints is to show a patchy mottled sheen when applied over substrates of uneven porosity and on overlaps. This is a familiar problem in respect of wall paints and has been extensively discussed in the literature [161]. The problem is largely caused by loss of binder into the substrate, causing a variable PVC in the coatings; it can be reduced by formulating at the CPVC. However, if a paint is at its CPVC and still loses binder it will go above critical, thus reintroducing sheen variation and increasing the severity of chalking. Other formulation parameters must be adjusted to prevent this, namely high capillary forces within the film and a high solution viscosity. Both are factors that will resist suction of the binder into the substrate.

For reasons of both economy and sheen control, paints are conventionally formulated on a blend of coarse extender finer prime pigment such as TiO_2. Maximizing the capillary forces will require a high CPVC achieved usually through the judicious blending of coarse extender, fine extender, and TiO_2. Changes in the particle size distribution or level of any of these ingredients will require re-balancing of the formulation. It is a point worth stressing in these specific formulations, which is also of general significance in the decorative market, namely that transferring a formula from one country to another may require adjustments, to allow for local raw material differences. Particle size distribution of extender pigments is an example of a parameter likely to change.

Commercially available thermoplastic resin normally requires exterior plasticization to achieve the best balance of mechanical properties, and also to reduce the cost of the binder component. A substantial amount of data listing the solubility, parameter, and compatibility of other plasticizers is available [162]. Plasticizer type and level must be adjusted to suit climatic exposure conditions and is one of the variables which may be adjusted to compensate for other raw material variations such as extender type.

In common with other exterior masonry coatings, these paints will normally contain a fungicide. Some fungicides contribute to chalking and this is a point which should be ascertained during screening trials. As noted earlier, these paints require an aromatic content to maintain solubility; however, excess aromatic solvent can cause lifting on re-coat. More recently, available resin can be dissolved in aliphatic hydrocarbon [163]. High styrene/acrylate copolymers are available and function as solution thickeners.

An entirely different type of solvent-borne thermoplastic masonry paint can be formulated using NAD technology. Such paints have many of the properties associated with water-borne acrylics but are very resistant to early shower damage. This is a particular advantage when painting in inclement weather [164].

9.7.8.3   *Chlorinated rubber masonry paints*
Solution carried chlorinated rubber and poly (vinyl chloride) have good resistance to attack and can be used as the basis of masonry coatings, with a typical lifetime of between 10 and 15 years.

9.7.8.4   *Two-pack polyurethanes*
Although relatively expensive, two-pack polyurethanes have found applications around the world which call for very high performance. They have been successfully used on all types of masonry and mineral surfaces, including reinforced cement. Coating life is claimed to be 20–25 years. Formulation guidelines are available from resin manufacturers (e.g. [165, 166]). Typically, systems comprise a penetrating primer, basecoat, and topcoat. The first two coats in particular must be alkali-resistant. Normally the first coat is unpigmented and based on polyether resin which may also be blended with vinyl copolymers. The basecoat is formulated to a high PVC and both it, and the primer, crosslinked with an aromatic isocyanate adduct. The topcoat can be formulated to a lower PVC and, to avoid yellowing, should be based on aliphatic isocyanate. Increasing use is being made of hydroxy acrylic as the polyol component.

The re-painting of old polyurethane coatings may lead to a problem of intercoat adhesion, which may be overcome with adhesion promoting primer.

An obvious disadvantage of two-pack polyurethanes is the limited pot life. More recently moisture curing products have been developed based on blocked isocyanates.

9.7.8.5   *Two-pack epoxy coatings*
Another expensive two-pack group of products is based on epoxy resins. It finds application at the very high performance end of the market where protection against aggressive environments and high adhesion is required [167].

9.8   Metallic substrates

The third major substrate encountered in buildings is metal, with ferrous metals being of greatest significance. Whereas wood and masonry are most conveniently grouped under the general heading of 'building paints', ferrous substrates are encountered in virtually all market sectors, including industrial, motors, marine, and heavy duty sectors. Preparing metal surfaces for coating is a specialized subject in itself and is dealt with separately in Chapter 13. Information concerning the main characteristics of metals will be found in other places in this book. For completeness, and to align with the type of information given in the previous sections on wood and masonry, the main characteristic of metals are also summarized here.

9.8.1   Characteristics of iron and steel
Pure metals have an underlying crystal structure which is substantially modified by impurities or deliberate inclusions used to form an alloy. Alloys will normally show a mixture of grains comprising different crystal structures or phases which have a marked effect on mechanical properties. Controlling the movement of crystal

dislocations is a powerful method of altering strength and toughness. The most common way to achieve this is by inclusion of carbon which is extremely efficient in controlling dislocation movement within the iron crystal, but other elements such as silica and manganese have specific advantages. Manufacturing processes will also alter both physical and chemical properties; established techniques include quenching, tempering, and work hardening. Among the most commonly encountered ferrous substrates in general building are mild steel, cast iron, and wrought iron. Mild steel normally contains between 0.2% and 0.8% carbon, whereas cast iron contains 4% (i.e. 20% by volume). Wrought iron has been worked so that the morphology is changed and the carbon is present as glossy inclusions.

Mild steel will normally be covered with millscale which, over a period of time, loosens and may fall away. It can contribute to corrosion and presents an unsound basis for coating. Techniques for removal and preparation are well documented [168]. Cast iron has a more adherent scale with some protective value; it corrodes at a similar rate to mild steel, though the residue is less obviously coloured. Wrought iron is generally similar to mild steel, though the corrosion rate may differ. Iron and steel will frequently be coated with less corrodible non-ferrous metals such as zinc or aluminium, which have other characteristics as described below.

9.8.2   Corrosion of ferrous metals

By far the major concern in protecting ferrous metals is the problem of preventing them from returning to their naturally occurring state, that is to prevent corrosion. Iron is far more vulnerable than wood or masonry to exterior exposure damage without some form of protection. There are several categories of corrosion which include dry (oxidative) corrosion and stress corrosion. Here only atmospheric corrosion is considered.

Free energy considerations show that the reaction of iron with oxygen and water is energetically favoured. Ferrous hydroxide is produced which, in its hydrated oxide form, is known simply as 'rust'. Many theories have been invoked to explain corrosion but today electrochemical theories are standard [169].

In a neutral saline environment the principal reactions involved are anodic oxidation of iron and cathodic reduction of oxygen. Iron goes into solution as ferrous chloride and is randomly distributed where areas rich in oxygen — such as the edge of a drop — become alkaline:

$$Fe \rightarrow Fe^+ + 2e^-$$

$$O_2 + 2H_2O + 4e^- \rightarrow 4OH^-$$

Typically, a flow of electrons, i.e. current, will occur between the centre and edge of a droplet which act as anode and cathode respectively. Hydrogen will form a thin layer on the surface, more rapidly in acid solution, than alkaline. As the reacting proceeds, ferrous chloride at the droplet centre reacts with sodium hydroxide to produce ferrous hydroxide, and as hydrogen is removed, the hydroxide is hydrated.

$$FeCl_2 + 2NaOH \rightarrow Fe(OH)_2 + 2NaCl$$

$$2Fe(OH)_2 + \tfrac{1}{2}O_2 \rightarrow Fe_2O_3H_2O + H_2O$$

Note the regeneration of chloride and the fact that the reaction products are not precipitated directly at anode and cathode, where they might stifle the reaction.

However, this principle is behind the use of some inhibitive pigments. Differences among metals in joints or within alloys have a simple, more direct, and usually faster mode of corrosion. Crevices in metal favour differential aeration, which is a prerequisite of the mechanism described above.

Paint defects related to corrosion are complex and not easily related to the above simplified mechanism. Such defects include blistering, delamination, undercutting, under-rusting, and filiform corrosion. Blistering will involve a lack of adhesion which is normal under moist conditions and may be driven by osmosis or alkaline attack. Alkyds are very prone to swelling under alkaline conditions and this provides and drives the energy for blistering. Undercutting, which occurs laterally under a film, is aided by the cathode areas which result from a local oxygen concentration, as oxygen diffuses through a film adjacent to existing rust. Filiform corrosion also involves differential aeration, though the mechanism of growth is still unclear. The relationships between these mechanisms have been reviewed by Funke [170].

9.8.3   Corrosion and coatings
Corrosion requires:

- an aqueous phase;
- the presence of ions;
- a conductive path between anodic and cathodic areas [171].

Ideally a coating should prevent one or more of these conditions being met, but once a corrosion cell is established the function of the coating is to contain and retard any further spread. Factors which influence the extent to which a coating will protect ferrous substrates from corrosion depend largely on:

- permeability to water and oxygen;
- ionic migration through the film;
- electrical resistance of the coating;
- adhesive performance under wet conditions;
- alkali resistance;
- the presence of substances which act in some way to inhibit corrosion;
- the absence of substances which promote corrosion [172].

Many of the mechanisms through which the factors operate are still the subject of debate as to the exact mechanisms involved. References [172] and [173] both include useful bibliographies for further reading. Dickie has stressed the need to maintain adhesion to suppress corrosion [174] and identifies a hierarchy of failure modes which depend on the composition of the coating and chemistry of the substrate. Steel failure modes identified are water disruption of the interface, followed by cathodic disruption and degradation of interface, coating, or substrate.

9.8.3.1   The role of water and oxygen
Understanding and quantifying the role of water at the interface is a crucial step in developing improved systems and in common with other aspects of corrosion and adhesion is amenable to study by modern analytical techniques and mathematical modelling [174, 175].

Water plays a major role in the corrosion processes of painted metal, by creating the electrolyte which completes the corrosion cell and through its influence on adhesive performance. However, consideration of the amount of moisture necessary to allow corrosion shows that an excess will be available through the water permeability of typical coatings [176]. Since painted steel normally corrodes much more slowly than unpainted, water permeability alone is not the rate-determining step. Oxygen is perhaps a more likely candidate and it has been reported that permeation rates of oxygen are more comparable to oxygen consumption in corrosion. Oxygen diffusion decreases markedly as the PVC increases (up to the CPVC) but is strongly affected by temperature. Mayne [177, 178] concluded that oxygen availability was no more the common rate-determining step than water, but the question is still an open one. It has been pointed out [172] that there are examples of films with high permeabilities and good corrosion resistance, also examples of low permeabilities and poor corrosion. It is probable that different mechanisms can become rate determining according to circumstance. The performance of both barrier and inhibitive coatings is strongly influenced by the PVC/CPVC relationship [177, 178]. Studies carried out with AC impedance and electrochemical noise techniques on alkyd paints showed poor corrosion resistance above the CPVC but considerable differences between zinc chromate and barium metaborate at different PVC/CPVC ratios below critical. Barium metaborate showed a much more rapid deterioration in performance as the CPVC was approached [179]. A very similar study with zinc phosphate [180] confirmed system-specific interactions.

9.8.3.2   Ionic migration through films

Coatings can contain the process of corrosion by resisting migration of ions, though opinion differs as to the rate of permeability. Results have been published which show chloride permeability can, in some coatings, be comparable to oxygen mobility. High electrical resistance will inhibit the flow of mobile ions. Chloride ions are particularly aggressive in promoting corrosion and are known to break down protective oxide films and reduce the effectiveness of oxidizing inhibitors by competing for adsorption sites. Ammonium ions accelerate rusting and British Rail reports heavy corrosion in steel wagons which carry ammonium sulphate, but there is some controversy over this point [181]. Coatings with high ion exchange capacity tend to have low corrosion resistance.

9.8.3.3   Alkaline attack

As noted during the discussion on masonry paint, some coating binders including alkyds and epoxy esters, are vulnerable to saponification and therefore damaged by the alkali formed during corrosion reactions. This may ultimately lead to a cohesion failure with the coating.

9.8.4   Pigmentation for corrosion resistance

Although unpigmented film can prevent corrosion, this will require very thick films or an effectively impermeable binder, high electrical resistance, and good resistance to alkali. Corrosion protection is substantially improved by a suitable pigmentation.

One of the most widely used anticorrosive pigments is zinc dust which provides a measure of cathodic protection. Pigments may also act by reducing permeability or interact chemically with the metal surface to inhibit the corrosion process. The term 'passivator' is sometimes used to denote inhibition which specifically promotes the formation of a protective film. Pigments may also increase the corrosion resistance of a coating by reducing permeability through the film.

9.8.4.1　Barrier pigments

As has already been noted, most pigments if properly wetted and dispersed, will decrease permeability until the CPVC is approached. Part of this effect is a simple lengthening of the diffusion paths available. In principle, such diffusion paths are greater for lamella pigments. It is usually argued that to be effective lamella pigments should 'leaf', that is orientate themselves parallel to the coating surface. However, the barrier effect might be even greater if the lamella were randomly oriented and in contact.

Any reduction of water and/or oxygen permeability may check the progress of corrosion and should improve other properties such as adhesion performance which is degraded in the presence of moisture. These mechanisms of protection have been reviewed by Funke [170] who has demonstrated a number of situations where undercutting and under-rusting tendencies could be correlated with water permeability.

Currently the most important barrier pigments are micaceous iron oxide and aluminium. Glass flake has also shown promise and could be of use where pale colours are important. Micaceous iron oxide has similarities to mica; grades vary in particle size and aspect ratio. It is important that the lamella shape is not appreciably damaged during processing. (For a review of formulating procedures see Carter [182].)

Aluminium will also reflect UV light to a high degree and this is of practical benefit in protecting the binder which gives aluminium a practical advantage over other barrier pigments which may not be apparent in some types of accelerated corrosion testing [183].

9.8.4.2　Inhibitive pigments

Inhibition is a complex subject, the details of which are well outside the scope of this chapter. Reference [171] contains a useful review of inhibitive mechanisms.

For many years lead and hexavalent chromium compounds have been the most effective inhibitive pigments and still provide the yardstick by which others are judged. However, they are increasingly under environmental pressure (with initial legislation generally directed against building paints).

Lead chromate pigments employ at least two inhibitive mechanisms. Chromate acts as an oxidizing inhibitor and seems to form a thin oxide layer on the surface of the metal to be protected, which also provides the oxide cation. The lead cation may inhibit corrosion by forming moities such as soaps which are adsorbed on the substrate surface, but note that Leidheiser records 11 mechanisms through which metallic cations may inhibit corrosion [171].

Calcium plumbate has found widespread use as an inhibitor pigment over galvanized metal and as a tie-coat over inorganic zinc rich primers. It is believed that its mode of action is related to the alkalinity of the pigment and the formation of calcium, rather than lead soaps. Calcium plumbate is able to inhibit the formation of zinc soaps which act as a brittle interlayer [177, 178].

Among the pigments which have been introduced, an alternative to the lead and chrome are phosphates, phosphate borate, and molybdate, and a number of novel composite inhibitors [184]. Hare and Fernald draw attention to good results with magnesium tetroxide used in combination with calcium borosilicate [185]. It is also noted that alkaline extenders, including wollastonite (naturally occurring calcium metasilicate), have a synergistic effect with some inhibitors, perhaps by acting as pH buffer [186]. Goldie [187] describes inhibitive pigments which operate on an ion-exchange principle. Leblanc has published a useful protocol for substituting and formulating primers on non-toxic anticorrosive pigments [188]. The paper also provides a useful illustration of the use of triangular mixture diagrams in formulation and the role of the CPVC as a normalizing constant.

9.8.4.3   Zinc phosphate
For many general purpose building applications zinc phosphate has become a useful 'standby'. The mechanism of zinc phosphate as an inhibitive pigment is not fully established. Proposed mechanisms include the absorption of ammonium ions, complex formation and passivation through a phosphating process. Turner and Edwards found little evidence of phosphating [189] but noted anodic and cathodic polarization.

9.8.4.4   Cathodic protection with zinc-rich primers
Paints containing high levels of zinc can protect steel by a sacrificial cathodic mechanism. Zinc dust still remains the most widely used corrosion-inhibiting pigment. Protection by zinc-rich primers often lasts longer than can be explained by the cathodic protection ability, and it is though that this extra protection arises from precipitated corrosion products which fill pores in the paint and reduce further attack [190]. The use of zinc pigment is associated with both inorganic and organic binders; ethyl silicate and two-pack epoxy respectively, are typical. In designing zinc-rich primers attention should be paid to CPVC and percolation effects, as there will be a significant difference in performance according to whether or not the zinc particles are in contact and the electrical properties or the binder that separates them. The geometrics of zinc-rich primers and associated effects on pigment loading have been reviewed by Hare et al. [191].

9.8.5   Design of anti-corrosive paints
Requirements of an anti-corrosive paint include:

- a binder resistant to alkaline hydrolysis (i.e. resistant to cathodically formed alkali);
- an adhesion mechanism which is not destroyed by the presence of alkali (thus carboxy groups are likely to be vulnerable);
- the presence of a mechanism through which the corrosion process is actively inhibited;
- low moisture and water vapour permeability;
- some means of neutralizing or buffering alkali.

9.8.5.1   Paint systems for ferrous metals
When designing coatings to suit a specific application, it is usually advantageous to design the complete system including primers and topcoats, and this will certainly be true at the 'heavy' end of the building paints market. At the 'lighter' end of the market it may be sufficient to combine an anticorrosive primer with general-purpose

undercoats and topcoats. Primers for metal are thus an important subsector of the general buildings paint market. For reasons which have been outlined above, low permeability is likely to be advantageous in protecting metals without any of the corresponding problems which beset wood or masonry. Binders for metal coating parts are thus chosen for impermeability and used at high film thickness — up to 250 µm — in a severe environment.

Water-borne systems

The need for low permeability has constrained the use of water-borne dispersion polymers, many of which are very permeable, though the chlorine-modified types referred to earlier [156] are more suitable. Other problems include flash-rusting and the problem of coalescence on cold metal in a damp environment. The problem of flash-rusting is a different phenomena from that of early rusting. Early rusting can be expected from any water-borne coating which is applied thinly and/or under unsuitable conditions where the film does not properly integrate. Flash-rusting happens almost immediately on application while the film is still damp and is more noticeable on fresh metallic surfaces (it may also be a problem over nail heads for water-borne wood primers). Flash-rusting is reduced by designing a resin with barrier properties and incorporating corrosion-inhibiting pigments. The presence of sodium nitrite (also used to reduce can corrosion) is also beneficial.

For general maintenance of metal, in domestic and light industrial buildings, increasing use of water-borne binders can be expected where good drying conditions can be expected, i.e. warm and dry. Both styrene butadiene, acrylic, acrylate modified vinyl chloride — vinylidene chloride copolymer, and alkyd emulsion binders have become established [192]. Starting formulations are widely available from latex manufacturers. The problems which have restricted the use of water-borne coatings on metal as opposed to other substrates are being overcome by technical progress and their use is increasing [193]. Nonetheless film formation under adverse weather conditions will remain a problem.

In discussing the needs of different substrates, the importance of preparation has been stressed. This is especially true of ferrous metals and in constructional work; steelwork should be prepared and primed prior to delivery on site. BS 6150 and BS 5493 review major methods of surface preparation which include blast cleaning, acid pickling, and flame cleaning. The manual methods likely to be used on site are not really satisfactory for best performance, and it will be difficult to remove all rust and scale deposits, underlining the need for anticorrosive as well as barrier film properties to provide maximum protection.

9.8.5.2   Primers for ferrous metal

Etch primers

'Etch primers' (wash primers) are more appropriately considered as part of the pre-treatment process. They are used to promote adhesion (especially with non-ferrous metals) and give a measure of temporary protection to ferrous metals. They must be overcoated with a fully pigmented primer.

Etch primers have been traditionally based on polyvinyl butyral resins which show good adhesion to metal. This may be improved with phosphoric acid (as a separate component in a two-pack product) or phenolic resins in one-pack products.

Traditionally UK products are tinted yellow and blue respectively for identification purposes. Two-pack products are often pigmented with zinc tetroxychromate but at a low (10%) total volume solids. The one-pack products frequently contain zinc phosphate with a total volume solids around 20%.

Drying oil primers
A substantial number of general-purpose metal primers have always been based on raw or processed drying oils.

Improved alkali resistance and, therefore, performance is gained by modification with phenolic resins. Problems of drying and re-coat have led to greater use of alkyd resins which may themselves be further modified with phenolic resins, or with styrene or vinyl toluene, to speed dry further. Alkyds can be modified with chlorinated rubber to improve water resistance. Hardness is gained by forming a urethane alkyd. Other typical binders containing drying oils are the epoxy esters which are also fast drying but prone to yellowing.

Chlor rubber-based primers
Chlorinated rubber, with its good chemical resistance, is a very suitable base for anticorrosive paints and has become widely used in all aspects of steel protection, including bridge paints, chemical plant, and pipelines. Chlor rubber paints form a film through solvent evaporation and dry quickly relative to oil and alkyds, especially in cold weather where application as low as −20 °C is possible. They have inherently low vapour permeability and can be applied at high film thickness, and when applied by airless spray will give dry film thicknesses of up to 125 μm per coat.

Primers for metal covered by British Standards include red lead (BS 2523, types A,B,C), calcium plumbate (BS 3698, types A&B) and zinc-rich (BS 4652, types 1,2,3).

9.8.5.3   Non-ferrous metals [192]
Non-ferrous metals likely to be encountered as coating substrates in domestic and light industrial buildings include galvanized steel, aluminium, and relatively minor amounts of copper, brass, and lead. These metals will often have been left unpainted for many years; thus cleaning and surface preparation will usually be an important part of the overall painting process. With the softer metals it is inadvisable to use wire wools for cleaning, as these may become embedded in the surface and even cause electrochemical reactions; nylon pads are an acceptable alternative. Long exposure of coated steels may allow rusting of the base metal which will require careful preparation. Etch primers are often found to improve adhesion. A phosphoric acid treatment known as British Rail 'T-wash' is sometimes effective (for formula see BS 6150: 1962, 28.3.1.5).

When priming non-ferrous metals, many of the ferrous primers are suitable but some pigments such as red lead and graphite may accelerate corrosion of aluminium.

Zinc-coated metal (galvanized, sheradized, or electroplated)
Metallic zinc coatings present a specific problem, caused by a reaction between the metal and acidic oil fragments. The dominant corrosion mode is anionic dissolution of metal and the resulting voluminous soluble salts cause intercoat failures which are often apparent on garage doors and window frames. Primers based on calcium plumbate have a proven track record in overcoming this problem, but there is

increasing concern over lead in the environment, and this type of primer is largely phased out. Calcium borate has proved a viable alternative in alkyd paints [194]. Suitable alternatives to alkyd include acrylic latex-based paints, though these may not prevent adverse reaction when overcoated with oil and alkyd-based paints. The low permeability of vinyl and vinylidene chloride-modified acrylics is a specific advantage here. Coating performance is almost invariably better over-weathered rather than unweathered zinc.

Aluminium and other metals
Aluminium and its alloys will not necessarily require painting but after long weather period may show unsightly salt deposits which should be removed. Zinc chromate primers are well established but zinc phosphate is more suited to domestic situations. Aluminium is increasingly used in the building sector, for example in glazing units, coating where carried out is an industrial process with powder coatings emerging as a preferred technology [195]. A curious form of corrosion with aluminium, comprising hair-like tracks, is known as 'filiform' [196]. A mechanism developed by Ruggeri and Beck [197] relates this to the existence of differential aeration cells below the coating. It is largely independent of the nature of the coating though some dependence on Tg has been observed. Aluminium composition and chromatization are also influential factors.

Lead and copper can usually be coated directly with alkyd-based gloss paint.

9.9   Plastic as a substrate [198]

The common plastics used in buildings do not present any special coating difficulties but it should be noted that abrasion can initiate cracks, and preparation is often best limited to washing. Expanded polystyrene will be attacked by some solvents and the overcoating of polystyrene ceiling tiles with gloss paints creates a fire hazard; only 'emulsion' paints should be recommended. BRE information paper 11/79 gives more detailed information.

Since around 1970 unplasticized poly (vinyl chloride) (uPVC or PVC-U) has taken an increasing share of both new and replacement windows [199]. Some of these materials are beginning to show sufficient surface degradation, chalking and dirt build up such that recoating is of increasing interest. Conventional alkyds have been used with some success but attention to problems of adhesion is required. In coating any plastic it is necessary to check that the mechanical properties, particularly impact, have not been seriously degraded.

The need for coating plastics is much more developed in the automotive market where a much wider range of plastics is used. Successful coating usually requires a pretreatment such as corona discharge (in air), glow discharge (vacuum) or 'CASING' (inert gas), flame, and chemical treatments are also used. An FSCT monograph on this topic gives a good overview [200].

References

[1] FAZEY I A, *Paints Environment* **April 8** II (1994).
[2] FLICK E W, *Waterborne Paint Formulations* (1975); *Solvent Based Paint Formulations* (1977); *Industrial Water Borne Trade Paint Formulations* (1980); *Handbook of Paint Raw Materials* (1982); *Waterbased Trade Paint Formulations* (1988), Noyes Data Corporation, NJ.

[3] BOHL A H, *Computer Aided Formulation*, VCH Publishers (1990).
[4] HESLER K K & LOFSTROM J R, *J Coatings Technol*, **53** (674) 33 (1981).
[5] GUTHRIE J T et al., *J Oil Colour Chem Assoc* **1992** (2) 66–73.
[6] PRA International Conference Proceedings, 'Paint and the Environment', Copenhagen, November (1990).
[7] STOREY R F & SHELBY F T, Proc 21st Waterborne, High Solids and Powder Coatings Symposium, New Orleans (University of South Mississippi), Feb (1994).
[8] '*Solvent vapour hazards during paint application*'. BRE Information Paper IP3/92, BRE Advisory Services, Garston, Watford WD2 7JR (1992).
[9] SYKES D, *European Coatings* J **1995** (5) 372–375.
[10] DAWOODI Z, *Polym Paint Colour J* **176** (4159) 41–45 (1986).
[11] PATTON T C, *Paint Flow and Pigment Dispersion*, Fig. 7.1, John Wiley (1979).
[12] ASBECK W K and VAN LOO M, *Ind Eng Chem* **41** 1470 (1949).
[13] FITZWATER S & HOOK J W, *J Coatings Technol* **57** (721) (1985).
[14] BOHREN C F et al, *J Coatings Technol* **61** (779) (1989).
[15] CUTRONE L, in *Principles of Paint Formulation*, chapter 6, ed. R Woodbridge, Blackie & Son (1991).
[16] STEIG F B, *Prog Organic Coatings* **June 1** (4) 351–73 (1973).
[17] STEIG F B, *Pigment/Binder Geometry Pigment Handbook*, Vol.III, pp. 203–217 ed. T Patton, Wiley-Interscience (1973).
[18] STEIG F B, in *Principles of Paint Formulation*, chapter 4, ed R Woodbridge, Blackie & Son (1991).
[19] BIERWAGEN G P & HAY T K, *Prog Org Coatings* **3** 281–303 (1975).
[20] KRIEGER I M & DOUGHERTY T J, *Trans Soc Rheol* **3** 137–152 (1959).
[21] TOUSSAINT A, *Prog Org Coatings* **21** 255–267 (1992).
[22] *J Coatings Technol*, Special Issue, **64** (806) 1–104 (1992).
[23] ASBECK W K, *J Coatings Technol* **49** (635) (1977).
[24] MEADOWS W D, *Amer Paint J* **70** (5), 42–48 (1985).
[25] HUISMAN H F, *J Coatings Technol* **56** (712) 65–79 (1984).
[26] CASTELLS R et al, *J Coatings Technol J C T* **55** (707) 53–59 (1983).
[27] CREMER M, Euro Supplement to *Polym Paint Colours J* **Oct 5** 86–93 (1983).
[28] BERNARDI P, *Paint Technol* **27**, 24 July (1963).
[29] DEL RIO G and RUDIN A, *Prog Organic Coatings* **28** 259–270 (1996).
[30] STEIG F, *J Coatings Technol* **55** (696) 111–114 (1983).
[31] COLE R J, *J Oil Colour Chem Assoc* 776–780 (1962).
[32] RASENBERG C J F M & HUISMAN H F, *Progr Org Coatings*, **Sept 13** (3/4) 223–235 (1985).
[33] BRAUNSHAUSEN R W et al, *J Coatings Technol* **64** (810) (1992).
[34] Cleveland Society For Coatings Technology. *J Coatings Technol* **65** (821) (1993).
[35] SEITZ J T, *J App Polym Sci* **49** 1331–1351 (1993).
[36] TOUSSAINT A, *Prog Org Coatings* **2** 237–267 (1973/74).
[37] ASBECK W K, *Amer Chem Soc Div Org Coatings Plastics Chem*, **A2** (7) 175 (1966).
[38] HOLZINGER F et al, *Optimium Formulation of Emulsion Paints etc.*, Fatipec Congress Book XVIII (Vol 1/B), p. 327 (1986).
[39] REDDY J N et al, *J Paint Technol* **44** (566), 70–75 (1972).
[40] FUNKE W, *J Oil Col Chem Assoc* **50** 942–957 (1967).
[41] STIEG F B, *J Coatings Technol* **53** (680) (1981).
[42] TEMPERLEY J et al, *J Coatings Technol* **84** (809) (1992).
[43] TEMPLETON-KNIGHT R L, *J Oil Colour Chem Assoc* **1990/11** 459–464 (1990).
[44] BALFOUR J G, *J Oil Colour Chem Assoc* **1992/1** 21–23 (1992).
[45] STIEG F B. *J Coatings Technol* **42** (545) (1970).
[46] CASARINI A, *Paint Manufacturer* **April** 10–12 (1973).
[47] BIERWAGEN G P, *J Coatings Technol* **51** (658) 117–126 (1979).
[48] Cyprus Industrial Minerals, *Efficient use of TiO2 with platy extenders in emulsion paints*, August (1989).
[49] SEINER J A, *Ind Eng Chem Prod Res Dev* **17** (4) 302–317 (1978).
[50] KERKER M et al, *J Paint Technol* **47** (603) 33–42 (1975).
[51] CHALMERS J R & WOODBRIDGE R J, *Euro Suppl Polym Paint Colour J* **Oct 5** 94–101 (1983).
[52] Dulux Australia Ltd, PO Box 60, Clayton, Victoria 3168 Australia.
[53] HISLOP R W & MCGINLEY P L, *J Coatings Technol* **50** (642) 69–77 (1978).
[54] Letters to the Editor, *J Coatings Technol* **50** (645) 112–113 (1978).
[55] GOLDSBROUGH K et al, 'Formulating paints with vesiculated beads', *Prog Org Coat* **10** (1) 35–53 (1982).
[56] RAMIG A & FLOYD L *J Coatings Technol* **51** (658) 63 (1979).
[57] RAMIG A & RAMIG P F, *J Oil Colour Chem Assoc* **64** 439–447 (1981).
[58] Letters to the Editor, *J Coatings Technol* **52** (663) (1980).
[59] HILL W H et al, *Resin Rev* **XXXV** (3) 3–10 (1985).
[60] Rohm & Haas Co, Philadelphia, PA 19105, Ropaque OP-62, Trade Sales Data Sheets.

[61] WOODBRIDGE R, *Principles of Paint Formulation*, pp. 106–107, Blackie & Son (1991).
[62] JOTISCHKY H, *Towards Waterborne Technology: An Upsurge or Trickle?* COMET, Paint Research Association, Teddington (1997).
[63] *Polym Paint Colour J* **175** (4156) 825 (1985).
[64] SMITH N D P, ORCHARD S E & RHIND TUTT A J, *J Oil Colour Chem Assoc* **44** (9) 618–633 (1961).
[65] MELVILLE I et al, *Polym Paint Colour J* **177** (4187) 174–184 (1987).
[66] ALAHAPPERUMA K & GLASS J E, *J Coatings Technol* **63** (799) 69–78 (1991).
[67] ANWARI F M & SCWAB F G, *Optimising Latex Paint Rheology with Associative Thickeners* ed, J E Glass ACS Monograph 223 (1989).
[68] HAZEL N J, *Surface Coatings Internat* **1994** (5) 213–219.
[69] SOLOMON D H, *The Chemistry of Organic Film Formers*, 2nd edn R E Krieger (1977).
[70] DILLON W E et al, *J Colloid Sic* **6** 108 (1951).
[71] BROWN G L, *J Polym Sci* **22** 423 (1956).
[72] SHEETZ D P, *J Appl Polym Sci* **9** (11) 3759–3773 (1965).
[73] KENDALL K & PADGET J C, *J Adhesion Adhesives* **July** 142 (1982).
[74] SYKES D, *Europ Coat J* **4/95** 270–271 (1995).
[75] HAYON A F, *Paint Resin*, **Oct/Nov** 5–9 (1993).
[76] CHANG-PEI KUO, *Appl Polym Coatings J* **Aug 15** (1994).
[77] BARRETT K E J (Ed), *Dispersion Polymerisation in Organic Media*, John Wiley (1975).
[78] BROMLEY C B & GRAYSTONE J A, *Coating Compositions*, GB 2,164,050 A.
[79] JOHNSON R, *Paint Coatings Ind* **October** 80–82 (1994).
[80] PADGET J C, *J Coatings Technol* **66** (839) 89–105 (1994).
[81] WATKINS M J, *Modern Paint Coatings* **Oct.** 120–126 (1993).
[82] BONDY C, *J Oil Colour Chem Assoc* **51** 409–427 (1968).
[83] FREAM A & MAGNET S, 'Water based metal protection coatings using surfactant free latices', Paper 23, 3rd Nürnberg Congress, March (1995).
[84] BOODAGHIANS R, 'Waterborne polymer dispersions for coalescent-free matt and semi-gloss paints', Paper 9, 3rd Nürnberg Congress, March (1995).
[85] PADGET J C, *Additives for Waterborne Coatings*, Royal Society for Chemistry Special Publication No 76, 1–20 (ed. Karsa) (1990).
[86] GOBEC M & MERTEN G, 'Waterborne Resins for Architectural Paints', Paper 40, 3rd Nürnberg Congress, March (1995).
[87] RODSRUD G & SUTCLIFFE J E, 'Alkyd emulsions — solvent free binders for thye future', Paper 14, 3rd Nürnberg Congress, March (1995).
[88] DERBY R et al, *Amer Paint Coatings J* **Jan 30** 57–64 (1995).
[89] RODSRUD G & SUTCLIFFE J E, *Surface Coatings Internat* **1** 7–16 (1994).
[90] WALBRIDGE D J, 'Multipurpose pigment dispersions for solvent based paints', PRA 5th International Conference, paper 1 (1983).
[91] SCHOLZ W, 'Pigment concentrates — selection and use of the right additives', Paper 33, 3rd Nürnberg Congress, March (1995).
[92] HEILEN W, 'A new dispersant for waterborne pigment concentrates', Paper 31, 3rd Nürnberg Congress, March (1995).
[93] CARTRIDGE D J et al, *Polym Paint Col J* **184** (4342) (1984).
[94] TYREMAN H, *Painter Decorator*, **April** 24 (1994).
[95] WIDE E G, *Polym Paint Colour J* **175** (4146) 454–461 (1985).
[96] ATKINSON D, *Paint Resin* **Dec** 29–30 (1985).
[97] Colony Powder Colorants, Colour Corporation of America, 200 Sayre Street Rockford, Illinois 61101.
[98] Herbol Colour Pill, Hoechst.
[99] TOENSMEIER P A, *Mod Plast Int* Vol **24** (4) 13–14 (1991).
[100] BIELEMAN J H, *Polym Paint Col J* **March** 16–17 (1995).
[101] BIELEMAN J H, *Polym Paint Col J* **184** (4349) 212–218 (1994).
[102] WÜRTH H, 'New possibilities with isoindolin pigments', XIX Fatipec Congress, Aachen, IV/49–65 (1988).
[103] GRAYSTONE J A, *The Care and Protection of Wood*, ICI Paints, Slough SL2 5DS (1985).
[104] DERBYSHIRE H & MILLER E R, *The Photodegradation of Wood during Solar Radiation*, pp. 341–350 Holz als. Roh-und Werkstof (1981).
[105] AAGAARD P K et al, 'Wood rot in coated exterior cladding: occurrence of *Dacrymyces stillatus*', XXII Fatipec Congress, Budapest, IV/22–38 (1994).
[106] GRAYSTONE J A, British Decorator **Jan. Feb.** 8–9 (1984).
[107] AHOLA P, AIR Project AIR3-CT94-2463. Performance and durability of wooden window joinery Technical Research Centre of Finland (VTT).
[108] MILLER R E, EU AIR Project Reports (1995) 'Performance and Durability of wooden joinery painted with new types of paints with low VOC', BRE Technical Consultancy, Garston, Watford WD2 7JR.

[109] ABLAS B P, and VAN LONDEN A M, *Surface Coatings Internat* **12** 555–572 (1996).
[110] HANDISYDE C C, *Everyday Details*, Architectural Press (1976).
[111] BOXALL J, *Surface Coatings Internat* **5** 189–196 (1994).
[112] *Principle of Weathersealed Timber Window Design*, The Swedish–Finnish Timber Council, 21/25 Carolgate, Retford DN22 6BZ.
[113] *Exterior Cladding, Redwood and Whitewood*, Swedish–Finnish Timber Council, 21/25 Carolgate, Retford DN22 6BZ.
[114] BRE Technical Note No. 40, BRE Technical Consultancy, Garston, Watford WD2 7JR.
[115] WORRINGHAM J H M, 'Water-based preservatives and allied low build coatings for wood', Paper 3, OCCA Symposium 'Environmentally Friendly Wood Preservatives and Coatings', 18 April (1996).
[116] GRAYSTONE J A, 'Wood coatings and the problem of classification', Paper 1 PRA/BRE Symposium The Care and Maintenance of Exterior Timber, June (1986).
[117] MILLER E R & GRAYSTONE J A, 'Progress in Standardisation for Exterior Wood Coatings', BRE Inform paper, IP 5/96, May (1996).
[118] GRAYSTONE J A, *Formulating for Durability*, PRA Progress Report No. 5, pp. 27–33 (1984).
[119] WOODBRIDGE R J, *The Role of Aqueous Coatings in Wood Finishing*, PRA Progress Report No. 5, pp. 33–38 (1984).
[120] VAN LINDEN J A & OPT P, 'Water permeability; comparison of methods and paint systems', XXII Fatipec Congress, Budapest, IV/135–148 (1994).
[121] Synolac 6005W, Cray Valley Products Ltd, Farnborough BR6 7EA.
[122] Jagalyd Antihydro, Ernst Jager, Fabrik chem. Rohstoffe GmbH, 4000 Dusseldorf 13, PO Box 130 380.
[123] BOXALL J, HAYES G F, LAIDLOW R A & MILLER E R, *J Oil Colour Chem Assoc* **9** 227–233 (1984).
[124] Experimental Emulsion E-1656. Rohm and Haas Co. European Operations, Lennig House, 2 Masons Avenue, Croydon CR9 3NB.
[125] SATGURU R et al, *Surface Coatings Internat* **1** 424–434 (1994).
[126] PALMER R, *Paints Resin* **2** 17 (1994).
[127] See also, for example, BP 1,209,108; USP 3,719,646, BP 1,088,105.
[128] TUCKERMAN R, *Polym Paint Colloid J* **179** (4250) 832–838 (1989).
[129] BOXALL J, *Factory-applied Stain Basecoats for Exterior Joinery*, BRE Information Paper IP 2/92 (1992).
[130] VITTAL RAO R et al, *J Oil Col Chem Assoc* **51** 324–343 (1968).
[131] DANNEMAN J H & SMITH A C, *Amer Paint Col J* **Jan 17** (1994).
[132] RHOPLEX R, AC-73, Rohm and Haas Co. Philadelphia, PA 19105, Trade Sales Coatings (1979).
[133] HIRSEKORN F, 'Improved application properties of high gloss latex paints', Paper 20, 3rd Nürnberg Congress, March (1995).
[134] Building Research Digest 198 *Painting Walls: Part 2*, Building Research Station, Garston, Watford WD2 7JR (1982).
[135] ROTHWELL G W, *J Oil Colour Chem Assoc* **74** (12) 395–439 (1988).
[136] JOTISCHKY H, *Polym Paint Colour J* **175** (4153) 740–742 (1985).
[137] LEA F M A, *The Chemistry of Cement and Concrete* 3rd edn, E Arnold (1970).
[138] WILLIAMSON R B, *Prog Mater Sci* **15** (3) (1972).
[139] MIDGLEY H G, *Proc Brit Ceram Soc* **13** 89–102 (1969).
[140] DOUBLE D D and Hellawell *Proc R Soc Lond A* **359** 435–451 (1978).
[141] ILSTON J M, DINWOODIE J M & SMITH A A, *Concrete, Timber and Metals*, Chapter 8, Van Nostrand (1979).
[142] POWERS T C, *Proc Ameri Concrete Inst* **51** 285 (1954).
[143] ROBSON T D, *High Alumina Cements & Concretes*, Wiley (1962).
[144] Building Research Establishment Digest 139 *Control of Lichens, Moulds and Similar Growths* (1982).
[145] Building Research Establishment Digest 197 *Painting Walls Part 1* (1982).
[146] SIMPSON L A, 'Influence of certain latex paint formulation variables on exterior durability', XIVth Fatipec Congress, Budapest, 623–9 (1978).
[147] Building Research Establishment Information Paper IP22/79 *Difficulties in Painting Fletton Bricks* (1979).
[148] Building Research Establishment Digest 177 *Decay and Conservation of Stone Masonry* (1975).
[149] DAVIES H & BASSI R S, *Surf Coat Int* **77** (9) 386–393 (1994).
[150] GLC Bulletin No. 129, *Brickwork Waterproofing Solutions* (1980).
[151] BS 6150 Code of Practice for painting of buildings (1982).
[152] Building Research Station Digest 125, *Colourless Treatments for Masonry* (1971).
[153] DAVIES G W, 'New long life coatings'. Proc PRA Symp. 'Protection and Maintenance of Building Materials', Paper 9, pp. 59–67 (1989).
[154] KIERCHIE B, *Coatings Agenda* 75–76 (1995).
[155] CLARK J, FULLER H & SIMPSON L A, *Some Aspects of the Exterior Durability of Emulsion Paints*, Tioxide Group PLC Tech Service Report D8672GC.

[156] Haloflex vinyl acrylic copolymer lattices, Zeneca Resins, PO Box 14, The Heath, Runcorn WA7 4QG.
[157] ARONSON P D, *J Oil Colour Chem Assoc* **57** (2) 59 (1974).
[158] HERZ S L, *Paint Coating Ind* **Oct**. 54–78 (1994).
[159] SIMPSON L A, 'Influence of Certain Latex Paint Variables on Exterior Durability', XIV Fatipec Congress Budapest, p 623 (1978).
[160] *Pliolite Resins*, The Goodyear Tyre and Rubber Co., Akron, Ohio.
[161] MYERS R R & LONG J S (eds) *Treatise on Coatings*, Vol. 4, Chapter 6, Marcel Dekker (1975).
[162] *Solubility Fractional Parameters for Pliolite Resins, Etc.*, Compagnie Francais Goodyear, Avenue des Tropiques, BP 31-91402 Orsay.
[163] SANDFORD R & GINDRE A, *Polym Paint Colour J* **176** (4166) 346–347 (1986).
[164] *All Seasons Masonry Paint*, ICI Paints, Wexham Road, Slough SL2 5DS.
[165] O'HARA K, *J Oil Colour Chem Assoc* **12** 413–414 (1998).
[166] MOSS N S, *J Oil Colour Chem Assoc* **12** 415–416 (1998).
[167] JONES P B, *J Oil Colour Chem Assoc* **12** 407–409 (1998).
[168] BS 5493: 1977. Code of Practice for protective coating of iron and steel structures against corrosion.
[169] WHITNEY W R, *J Amer Chem* **25**, 395 (1903).
[170] FUNKE W, *Ind Eng Chem Prod Res Dev* **24** 343–347 (1985).
[171] LEIDHEISER H, *J Coatings Technol* **53** (678) 29–39 (1981).
[172] New England Society for Coatings Technology, *J Coatings Technol* **53** (683) 27–32 (1981).
[173] GURUVIAH S, *J Oil Colour Chem Assoc* **53** 669 (1970).
[174] DICKIE R A, *Prog Org Coat* **25** 3–22 (1994).
[175] POMMERSHEIM J M et al, *Prog Org Coat* **25** 23–41 (1994).
[176] HARE C H, *Federation Series on Coating Technology — Unit 27* Federation of Societies for Coatings Technology, USA (1979).
[177] MAYNE J E O, *The Mechanism of the Protective Action of Paints Corrosion*, Vol. 2, pp. 15:24–37 2nd edn, Shreir L L, ed., Newnes-Butterworths (1976).
[178] MAYNE J E O & MILLS D J, *Surf Coat Internat* **4** 154–155 (1994).
[179] SKERRY B S et al, *J Coatings Technol* **64** (806) (1992).
[180] GOWRI S & BALAKRISHNAN K, *Prog Org Coat* **23** 363–377 (1994).
[181] Symposium on the Protection and Painting of Structural Steelwork, 11–12 April 1984, PRA Progress Report No. 6, p. 12 (1984).
[182] CARTER E, *Pigment Resin Technol* **March** 3–12 (1984).
[183] HARE C H, *Modern Polym Coatings* **Dec**. 37–44 (1985).
[184] KALEWICZ Z, *J Oil Colour Chem Assoc* **Dec**. 299–305 (1985).
[185] HARE C H & FERNALD M G, *Paint Resin*, **Dec**. 22–25 (1985).
[186] HARE C H, *Modern Paint Coatings*, **Nov**. 32–40 (1993).
[187] GOLDIE B P F, *Paint Resin*, **Feb**. 16–20 (1985).
[188] LEBLANC O, *J Oil Colour Chem Assoc* **8** 288–301 (1991).
[189] TURNER W H D & EDWARDS J B, *Zinc Phosphate as an Inhibitive Pigment*, PRA Technical Report TR/7/78 (1978).
[190] DICKIE R A & SMITH A G, *Chemtech*. **Jan**. 31–34 (1980).
[191] HARE C H et al, *Modern Paint Coatings* **June** 30–36 (1983).
[192] BRE Digest 71, *Painting in Buildings: 2 Non-ferrous Metals and Coatings* June (1966).
[193] CHAPMAN M, *Surf Coat Internat* **7** 297–300 (1994).
[194] KOSKINIEMI M S, *Amer Paint Coat J* **Feb. 27** 49–53 (1995).
[195] SHEASBY A, *Surf Coat Internat* **9** 394–395 (1994).
[196] STEELE G, *Polym Paint Col J* **184** (4385) 90–96 (1994).
[197] RUGGERI R T & BECK T R, *Corrosion* **139** (11) 452 (1983).
[198] RISBERG M, *J Oil Colour Chem Assoc* 197–200 (1985).
[199] BRE Digest 404 *PVC-U Windows*, April (1995).
[200] RYNTZ R A, *Painting of Plastics*, FSCT Series on Coating Technology (1994).

10

Automotive paints

D A Ansdell

10.1 Introduction

The demands and performance required for automotive coatings are considerable. There is a need for 'body' protection, such as anti-corrosion and stone-chip resistance, and for a durable and appealing finish. Products also have to be appropriate to mass-production conditions, and in this respect, must be robust, flexible, and economic to use.

Vehicle construction has gone through significant changes since its inception at the turn of the century. At the present time the substrate is generally of mild steel but may also contain other alloys and include plastic components; the shape is inevitably complex and certain parts of the vehicle are almost inaccessible and difficult to paint. Production rates are high, e.g. 45 units/hour, and this requires process and material technology to meet the limitations this imposes.

Vehicle production (passenger cars and commercial vehicles) throughout the world in 1995 was 47.6 million utilizing of the order of 900 million litres of paint. In relation to these figures, vehicle production in Western Europe was 12.6 million, requiring 254 million litres of paint.

These figures are broken down and represented in Table 10.1 and diagrammatically (refer Figs. 10.1 and 10.2).

The paint products used are principally primers and surfacers (fillers), designated the undercoating system, and the finish or topcoat. In a modern painting system the relative use is broadly in percentage terms, primer:surfacer:topcoat, 30:20:50. In different parts of the world there is often variation in product technology, particularly in topcoats, and this can have a significant influence on paint performance, specifications and details of the process.

The basic objectives of the painting process are to protect and decorate. In order to achieve this the process is broken down into a number of different component parts. These parts, or 'layers', are applied in a specific order and although the function of each 'layer' is specific it relates very closely to the others to provide the desired balance of properties.

Table 10.1 — Western European vehicle production 1995 (millions)

	Passenger cars	Commercial vehicles	Total
United Kingdom	1.376	0.193	1.569
France	2.836	0.319	3.155
Germany	3.753	0.237	3.990
Italy	1.117	0.150	1.267
Spain	1.506	0.250	1.756
Other	0.748	0.138	0.886
	11.336	1.287	12.623

The component parts of the painting system are:

Metal pretreatment
Primer } Protection

Surfacer (filler)
Finish } Decoration

This fundamental process may be divided into three basic systems which may be classified as follows:

- spray priming system;
- dip priming system;
- electropriming system (currently most widely used).

The reason for this classification is that the method of priming is the major difference between them; the subsequent process is virtually identical. All these three systems are used over suitable pretreatments.

Spray application is used for surfacers and topcoats. There are a variety of different types of spraying systems, e.g. air atomized (hand and/or automatic) and electrostatic methods. These are described in more detail later.

10.1.1 Spray priming system
This comprises:

- Zinc-rich primer: internal sections only.
- Primer surfacer: 40–50 µm.
- Finish: 45–55 µm.

The zinc-rich primer is applied to the internal surfaces of body components before welding of the body is undertaken. This is to give corrosion protection to such surfaces which are otherwise inaccessible for painting in following operations.

The primer surfacer provides a measure of corrosion protection to the outer skins of the phosphated body.

This system was used before the advent of dipping primers but is now only found where low volume production does not justify the cost or plant to dip or electroprime bodies.

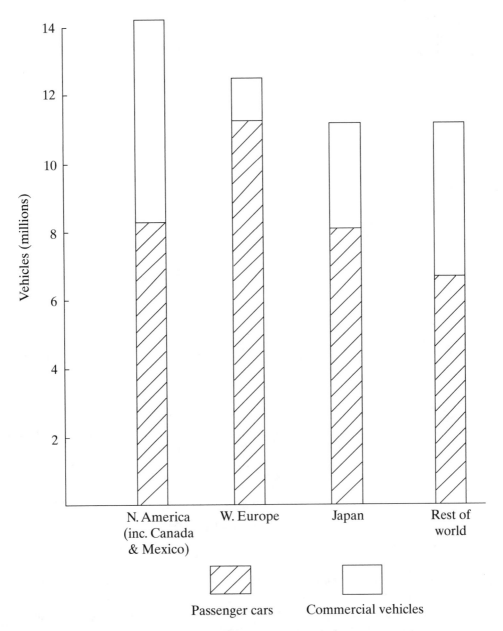

Fig. 10.1 — Vehicle production 1995.

10.1.2   Dip priming system

- Anti-corrosive dipping primer: 12–18 μm.
- Primer surfacer: 40–50 μm.
- Finish: 45–55 μm.

This system was widely used on mass-production lines before the introduction of electropriming in the early 1960s. In addition to its normal function the primer

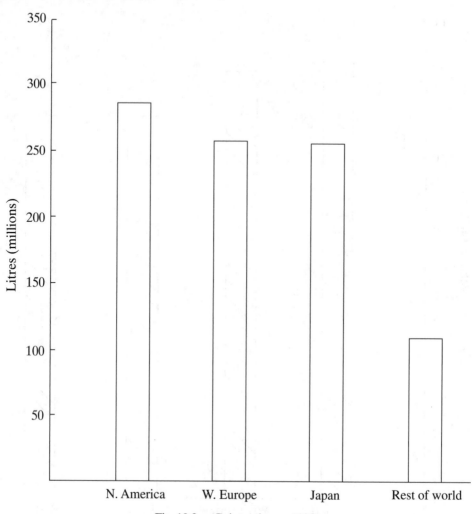

Fig. 10.2 — Paint volumes 1995.

surfacer needed to provide a measure of corrosion protection since the underbody dip primer only covers the lower sections of the unpainted car body.

Now more or less obsolete, this process still finds limited use where low volume production or other factors such as cost prevail.

10.1.3   Electropriming system

- Electrocoat (anodic or cathodic): 18–25 µm.
- Surfacer: 35–40 µm.
- Anti-chip coating: 50–100 µm.
- Finish: 45–55 µm.

Electropriming, predominantly cathodic, is currently the accepted standard in mass-production plants. It is an efficient, relatively simple operation with a high degree of automation. Introduced in the early 1960s with anodic technology, subsequently replaced by cathodic technology, it has set new standards in processing and corrosion protection.

Inclusion of an anti-chip coating to protect vulnerable areas has grown considerably over the past ten years. It simply reinforces the surfacer's resistance to stone-chipping and forms a highly important part of the overall painting system.

In the processes outlined above all coatings need stoving; this is required to achieve very high levels of performance and to facilitate processing in a conveyorized production line. Stovings vary from as high as 180 °C for cathodic primers to as low as 80 °C for repair finishes.

A typical modern paint process/plant is shown in Fig. 10.3 in schematic form.

10.1.4   Performance

Individual manufacturers have their own performance specifications and, for example, while they may vary on the type of finish and local plant/processing conditions, there are basic aspects which are common to all:

- Appearance: gloss and distinction of image.
- Durability: film integrity, colour and gloss retention, and free from blistering and corrosion.
- Mechanical properties and stone-chip resistance.
- Substrate: acceptable performance over a variety of substrates including plastics.
- Adhesion.
- Workability: amenable to modern methods of application.
- Ability to meet environmental requirements/regulations: product and process.
- Corrosion and humidity performance.
- Petrol and solvent resistance.
- Etch resistance.
- General chemical resistance, e.g. acid and alkali resistance.
- Hardness and mar resistance.
- Repair properties.

The products required to achieve these properties are described later. They require extremely sophisticated technology and a lengthy process involving various methods of application and stoving. Costs in energy, labour, space, and capital are considerable and future processes will have to take other factors into account, such as anti-pollution, while still continuing to improve performance standards. For example, new developments have to back up improved warranties from car producers such as a six year anti-corrosion guarantee and, with certain manufacturers, up to three year's overall coating performance.

10.2   Pretreatment

The pretreatment process has three purposes:

- To remove the mill and pressing oils ingrained in the steel and any other temporary protective coatings.

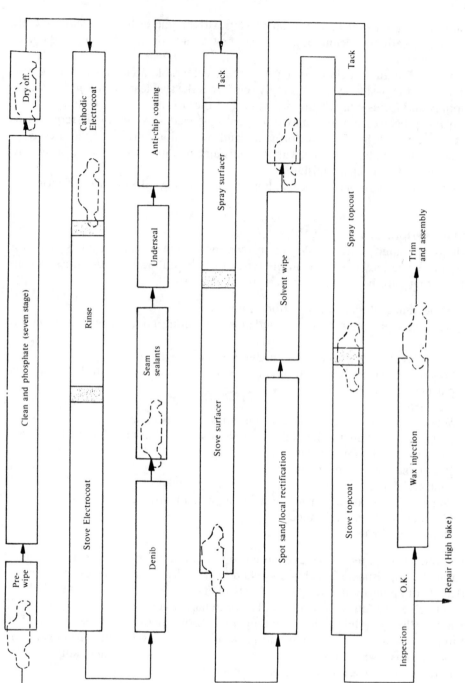

Fig. 10.3 — Paint plant process.

- To improve paint adhesion by providing an inert surface of metal phosphate which will give a better key for the subsequent primer layer.
- To provide a resistance barrier to the spread of corrosion under the paint film.

A normal process for pretreatment may be summarized as follows:

1   rust removal;
2   alkali degrease;
3   water rinse;
4   metal phosphating;
5   demineralized water rinse.

Depending on individual requirements and throughputs, processes may be spray, spray-dip, or dip; the latter two are preferred for modern high volume installations.

10.2.1   Rust removal
The best method is by the use of mineral acid particularly where rust deposits are heavy or mill scale is present; phosphoric acid-based materials are normally preferred.

Although phosphoric acid is slower acting than hydrochloric or sulphuric, any salt contamination is less due to the low solubility of most metal phosphates.

10.2.2   Alkali degrease
Alkali cleaners are widely utilized in dip and spray installations where the oil and grease are partly saponified and emulsified into the alkali solution. The physical force generated by the jetting of the cleaner significantly assists in the removal of any solid matter present on the surface of the unit.

Formulations vary and are dependent on a number of factors such as the type of oil or grease to be removed, metals to be processed, etc. However, there is a need to restrict the strength of the alkali to prevent deactivation of the metal surface. Typical alkali chemicals are caustic soda, trisodium phosphate, and sodium carbonate: these are used in conjunction with various types of detergents for emulsification of oils and lubricants.

10.2.3   Metal phosphate (conversion coating)
There are two basic types of phosphates:

- Iron phosphate — coating weight $0.2–0.8\,\mathrm{g\,m^{-2}}$.
- Zinc phosphate — coating weight $0.5–4.5\,\mathrm{g\,m^{-2}}$.

As a generalization, increasing the phosphate coating weight increases the corrosion resistance and decreases the mechanical strength or adhesion of the subsequent paint coating. Clearly coating weight is an important parameter, the value being a compromise designed to achieve an acceptable balance of properties.

Iron phosphate produces an amorphous, non-crystalline coating of low weight used mainly on components such as refrigerators, washing machines, and metal furniture which do not have to stand up to rigorous corrosion conditions. The prime requirement for such components is mechanical strength/adhesion which makes the low coating weight or iron phosphate ideal for such purposes.

Zinc phosphate is used almost universally in automotive paint processes and is an integral part of the total paint system. Zinc phosphate coatings are unique in that they can be made to crystallize from solution onto the metal surface because of the existence of metal phosphates of different chemical form and the equilibrium that exists in aqueous acid solution.

A chemical equilibrium is set up as follows:

$$3Zn(H_2PO_4)_2 \rightleftharpoons Zn_3(PO_4)_2 + 4H_3PO_4$$

Zinc dihydrogen phosphate	Tertiary zinc phosphate	Phosphoric acid
(soluble)	(insoluble)	(free acid)

This equilibrium will remain unchanged until an influence is brought to bear which will affect the concentration of the chemicals present. The initial reaction, therefore, in the phosphating process is the attack on the metal by the free phosphoric acid.

$$Fe + 2H_3PO_4 \rightarrow Fe(H_2PO_4)_2 + H_2$$

Substrate	Free acid	Soluble iron phosphate	Hydrogen

The equilibrium in the first equation is disturbed and the reaction moves to the right, the soluble zinc phosphate disproportionates to form insoluble zinc phosphate and free acid. Phosphate crystals start to grow and continue until the surface of the substrate is completely covered, at which time there is no more iron left with which the free acid can react, and the process stops. A build-up of soluble iron phosphate, which would ultimately slow down the reaction and poison the phosphate solution, is controlled by the addition of an oxidizing agent (toner) which precipitates the soluble iron phosphate as ferric phosphate (sludge).

10.2.4 Pretreatment as a corrosion inhibitor: mechanism

The steel surface, or substrate, is made up of a lattice of iron, carbon dissolved in iron, iron carbides, and undissolved carbon particles.

Rusting, or corrosion, is essentially an oxidation process aided by electrolytic action. On a moist steel surface electrolytic cells can be set up by the impurities in the metal (carbides, carbon, etc.) which give rise to cathodic (negative) and anodic (positive) areas. In the presence of water and salts (an electrolyte), these areas become the electrodes in a corrosion cell. Ferric ions form at the 'cathode', where the metal becomes pitted, combine with oxygen in the air and hydroxyl ions from the electrolyte to form iron oxide and hydroxide. Corrosion consequently starts to take place and to prevent it the corrosion currents must be prevented from flowing by insulating the metal surface.

Paint alone is a poor insulator since it is not impervious to an aqueous environment. Consequently, any damage to bare metal, by for example stone-chipping, results in corrosion taking place which spreads under the film owing to the inherent lack of adhesion of virtually all paint films to untreated metal surfaces.

Phosphate, however, acts as the perfect insulator. The phosphate crystals grow from active sites on the surface of the metal until the coating is completely impregnated and covers the whole surface. This crystalline structure acts as a type of 'ceramic' insulator unaffected by water, particularly when sealed with the paint coating. Even damage due to stone-chipping, or other forms of mechanical abuse, is restricted.

10.3   Priming

The function of the primer is to provide corrosion protection. This is achieved by formulating a product which incorporates anti-corrosive pigments carried in a resin system providing unnecessary mechanical, as well as anti-corrosive, properties.

Vehicle construction has changed considerably over the past 50 years to the extent that different techniques in applying the primer layer are necessary in order to ensure all parts of the vehicle are coated. Table 10.2 identifies the evolution of the vehicle priming up to the present day.

Historically there have been two major advances in the priming of motor vehicles. The first was a direct result of the automated techniques developed during the Second World War, i.e. the advent of the spot welder produced the welded unit or monocoque body shell. In this type of construction many areas of the unit are virtually inaccessible, for example the various strengthening cross-members and the front end near door posts. This meant that spray application of primer was no longer feasible. As a consequence dipping procedures were introduced to make it possible to prime all parts of the vehicle, particularly the underbody.

Initially, dipping primers were solvent-borne, but in the 1960s water-borne primers were introduced to minimize fire and health hazards. However, although the water-borne systems failed to improve efficiency or economics they did lead on to the second major advance in autobody priming, i.e. applying the primer by electrodeposition.

10.3.1   Spray priming

This system utilizes a zinc-rich primer, and sometimes other anti-corrosive coatings, which is applied to the internal surfaces of body components before 'welding-up' of the body. As a result, corrosion protection is given to such surfaces that are otherwise inaccessible for painting in subsequent operations.

Before the advent of dipping primers this system was quite widely used. It is now used only where low-volume production does not justify the cost of plant to dip or electrocoat bodies, or where capital is not available to install such plant.

10.3.2   Dip priming

10.3.2.1   Products
These materials are used to provide protection to internal parts of the vehicle body. Since no further coats of paint are applied in such areas, dipping primers are formulated to have good corrosion resistance and good adhesion to various types of phosphating pretreatment, as well as untreated metal.

Because of the size and complexity in shape of autobodies, dip primers must have good flow characteristics. Also since the size of dip tanks is large, to accommodate car bodies, there is also a need for good stability.

Table 10.2 — Evolution in carbody priming

Period	Substrate	Body type	Primer type	Features
Pre-1945	Steel with iron phosphate conversion coatings.	Body/chassis.	Spray primer (solvent-based).	All areas accessible. Uneconomic/polluting.
Post-1945–1960	Steel with iron/zinc conversion coatings.	Welded unit construction (monocoque).	Solvent-based dip primer.	Laborious. Uneconomical. Poor internal protection. Fire/health hazard. Low production output.
1960–1966	Steel with iron/zinc conversion coatings.	Welded unit construction (monocoque).	Water-based dip primer.	Reduced fire hazard/pollution. Otherwise similar to solvent/dip process.
1966–1977	Steel with zinc phosphate coatings of controlled 'film' weight, composition, and morphology.	Welded unit construction (monocoque).	Anodic electropaints.	Automated. High production rate. Reduced hazard. Efficient. Uniform coating. Good external and internal protection.
1977–1982	Steel with zinc phosphate conversion coatings of controlled 'film' weight, composition, and morphology.	Welded unit construction (monocoque).	Cathodic electropaints.	Superior corrosion protection, especially in thin films. Good stability. Greater penetration. More sustainable technology.
1982–1988	Steel. Growing use of zinc and zinc alloy coated steels (single or double sided). Specialized conversion phosphate coatings. Tricationic technology.	Welded unit construction (monocoque).	Cathodic electropaints.	Higher builds attainable (~25 μm). Enhanced protection. Improved throwing power. Improved mechanical properties. Highly compatible with 'mixed metals'.
1988–1992	As above.	Welded unit construction (monocoque).	Cathodic electropaints.	Lower stoving (165 °C). Reduce VOC. Improved coating flexibility. Reduced coating weight loss. Improved efficiency. Reduced topcoat yellowing.
1992–	As above.	Welded unit construction (monocoque).	Cathodic electropaints.	Improved edge protection. Reduced lead content. Improved throwing power. Improved corrosion protection. Reduced 'combined oxygen' demand. Improved mechanical properties.

There are four basic types of dip primers still in use:

- alkyd;
- alkyd/epoxy;
- epoxy ester;
- full epoxy.

The majority of dip primers are solvent-based but water-borne variants are available. The most widely used in the industry were alkyd-based since they gave an acceptable balance between performance and cost. Epoxy types were introduced later to give improved corrosion resistance but are more costly and less easy to control in tip tanks.

10.3.2.2   Pigmentation
Traditionally, these were red oxide type products but the more common colour now is either black or grey.

Prime pigments:

- Red oxide and carbon black.
- Zinc chromate is often included for anticorrosive properties.
  Extenders:
- Barytes (barium sulphate): hard with good blister resistance.
- Blanc fixe (precipitated barium sulphate): helps maintain dip tank paint stability.

Pigment volume concentration (PVC) is normally 20–30%.

10.3.2.3   Process
This process is now only used on a very limited scale where low volume production or other factors such as economic considerations are prevalent.

Dipping of the car body can be shallow ('slipper' dipping) up to windows or full immersion. In the 'slipper-dip' process the short dipping distance involved gives comparative freedom from runs and sags and the next coating of surfacer can be applied 'wet-on-wet' after appropriate draining and air drying. The two materials are then stoved as a 'composite'.

In the case of deeper dips the complex shape of the car body gives more runs and sags which have to be removed by solvent wiping, etc., before surfacer application. Sometimes the primer is stoved and rectification carried out prior to surfacer application.

An interesting form of dip priming used in the mid-1960s was roto dipping. In this process the body was mounted transversely on a spit and, by means of a suitable conveyor, carried through the pretreatment (dip zinc phosphate), primer, and then stoving operation.

The advantages of this method were the thorough pretreatment and even distribution of primer to all surfaces. However, the plant required was expensive and high in maintenance costs and could only be operated at relatively low line speeds. These and other factors led to it being superseded by electropriming in common with most other dipping processes.

In summary, dipping methods are dirty, laborious, and wasteful. The advent of electropriming showed great advantages (refer to Section 10.3.3), giving greater internal protection and eliminated solvent 'reflux' in box sections, a feature inherent in both solvent- and water-based dip primers.

10.3.3 Electropainting

Electropainting of motor vehicles began in 1963, some time after the principle had been established in other areas, although the technology was conceived in the 1930s. The electrocoat system consists of the deposition of a water-borne resin electrolytically, the resins employed being polyelectrolytes.

At the present time virtually all mass-produced vehicles are primed in this manner; the body shell is immersed in a suitable water-borne paint which is formulated as an anodic or cathodic resin system at very low viscosity and solids.

10.3.3.1 Anodic electrocoat

Until 1977 all electropaints used in the automotive industry were of the anodic type, primarily because the resin chemistry was relatively simple, readily available, and adaptable to the needs of the motor industry.

Resin systems
There are four main resin systems:

- maleinized oil;
- phenolic alkyd;
- esters of polyhydric alcohols (epoxy esters);
- maleinized poly butadiene.

Early anodic electropaints were based on maleinized oils and phenolic alkyds but these were superseded by epoxy esters and maleinized polybutadiene because of their superior corrosion performance.

Pigmentation
The pigmentation used in anodic electrocoat (and cathodic) should be:

- simple;
- non-reactive, i.e. stable (e.g. to stoving) and pure, containing no water-soluble contaminants (e.g. chloride and sulphate);
- in one form, or another, reinforce or improve corrosion performance.

Pigment volume concentration is normally in the region of 6%.

Prime pigments are high grade rutile titanium dioxide, together with a minimal amount of carbon black, in greys, and a very good quality red (iron) oxide for reds.

Extender pigments are high grade, free from soluble ionic species, such as chloride and sulphate.

A number of anti-corrosive pigments are available; a typical example is lead silico chromate.

Mechanism of deposition
During electrodeposition three basic reactions occur: electrolysis, electrophoresis, and electroendosmosis.

Note: solubilization of the resin:

$$RCOOH + KOH \rightarrow RCOOK + H_2O$$
$$\text{(insoluble)} \qquad\qquad \text{(soluble)}$$

e.g. R = epoxy/ester.

(The base used here is potassium hydroxide, but others such as amines are also utilized).

1 Electrolysis

$$RCOOK \rightleftharpoons RCOO^-K^+$$

$$4H_2O \rightleftharpoons 3H^+ + 3OH^- + H_2O$$

Reduction at cathode:

$$H_2O + 3H^+ + 4e \rightarrow 2H_2 + OH^-$$

Alkaline reaction from OH^- present; can be removed by electrodialysis.
   Oxidation at anode:

$$3OH^- - 4e \rightarrow H_2O + O_2 + H^+$$

acidic reaction from H^+ present, i.e.

$$RCOO^- + H^+ \rightarrow \qquad RCOOH$$
insoluble (electrocoagulation)
deposited coating

Metal dissolution (minor reaction)

$$M - ne \rightarrow M^{n+1}$$

(favoured by presence of contaminating ions)

2 Electrophoresis. When a voltage is applied between two electrodes immersed in an aqueous dispersion the charged particles tend to move towards their respective electrodes. This effect is called electrophoresis.
   In the electrocoat process, when a current is applied, negatively charged paint ions move towards the anode (positive). Electrophoresis is the mechanism of movement, and does not contribute to the actual deposition of the paint on to the anode.

3 Electroendosmosis. The layer of electrodeposited paint on the anode will contain occluded water. This is driven out through the porous membrane formed by the paint by the process called electroendosmosis.

Practical considerations

In practice the process simply involves the passing of a direct current between the workpiece (anode) and the counter-electrode (cathode). The resultant coating deposited on the workpiece is insoluble in water and will not redissolve.
   The film is compact, almost dry, and has very high solids which adhere strongly to the pretreated (zinc phosphate) metal substrate. It is covered by a very thin dip layer which is easily removed by rinsing. After stoving (165 °C) the resin cures to form a tough, durable polymeric film.
   The amount of paint deposited is largely dependent on the quantity of electricity passed. As the film is deposited its electrical resistance increases and the film becomes polarized when the resistance is so great that it stops the flow of current and halts the deposition. Thus once an area has been coated the local increase in resistance forces the electrical current to pass to more remote areas. This effect ensures progressive deposition, or 'throw', into box sections and internal or recessed areas.

The result is exceptional uniformity of film thickness, free from runs, sags, and other conventional film defects. The process is very suitable for mass-production conditions, facilitating a primer coating over the whole surface of a welded mono-coque carbody. It can be fully automated and is highly efficient in terms of paint utilization.

Basic plant requirements

- A powerful paint circulating system to preclude pigment sedimentation at the low solids (~12%) of operation and maximizing paint mixing.
- Sophisticated refrigeration and filtration to maintain stability and remove contamination.
- A smooth and stable direct current supply with suitable 'pick-up' gear to the car body.
- A paint rinsing system to remove the adhering dip layer. Incorporation of ultra-filtration (see Section 10.3.3.2) maximizes water and paint utilization as well as serving as a method of decontamination.

Control methods

The alkali bases generated at the cathode (see Mechanism of deposition above) by electrolysis during electrodeposition will increase the pH of the paint and must be removed to maintain optimum properties. This can be done in two ways:

1   Electrodialysis. This uses a selective membrane (ion exchange) around the cathode. Base ions are allowed to pass through the membrane, but the anions (colloidal paint particles) are prevented from doing so.
2   Acid feed/base-deficient feed. This is self-explanatory. The feed, or top-up, material is partly neutralized to compensate for the base generated during the process of deposition.

The difference in method of pH control gives rise to differences in other aspects such as resin system, means of neutralizing, and type of plant. Electrodialysis has the advantage of being fully automated with a consistent feed material to the tank; the ion exchange membrane having a considerable life unless subjected to physical damage.

Deficiencies of anodic electrocoat primers

The two main deficiencies are phosphate disruption and poor saponification resistance.

During anodic electrodeposition very high electrical field forces occur which rupture many of the metal–phosphate bonds (phosphate disruption). This leads to a weakening of the adhesion of the phosphate coating to the steel substrate.

When subsequent paint coats are applied, stresses are set up in the total paint film which when damaged to bare metal, tend to cause the total paint film to curl away from the damage point. In early anodic systems, as soon as corrosion started at any point of damage, the weakened phosphate layer allowed the paint to peel back, exposing more metal already chemically clean and very prone to corrosion. This failure was designated 'scab corrosion'.

This property was improved by reducing the incidence of rupture of bonds by simply increasing their number, i.e. phosphate coatings of densely packed fine crystals.

Such phosphate coatings are of low coating weight $(1.8\,\mathrm{g\,m^{-2}})$, and their introduction led to a marked reduction in the incidence of 'scab corrosion'.

Poor saponification resistance is due to the fact that the chemistry of anodic electropaints is such that they are formulated on acid resin systems. As a consequence the deposited stoved film, when exposed to an alkaline environment, will tend to form metal soaps soluble in water.

When damage occurs to bare metal, salt (as caustic) will simultaneously attack the steel substrate and the electrocoat film, producing rust and the sodium salt (soap) of the anodic resin. This dissolves the primer coating, leading to a loss of adhesion of the remaining paint film and general corrosion problems.

10.3.3.2   Cathodic electrocoat

Cathodic electroprimers had always been acknowledged to be theoretically desirable because of:

- their anticipated freedom from substrate disruption;
- the fact that cationic resins being 'alkaline' in nature would tend to be inherent corrosion inhibitors free from saponification.

However, the complexity of the required resin systems precluded their early introduction.

In the early 1970s successive outbreaks of serious motor vehicle corrosion in North America, combined with legislation for minimum corrosion standards, prompted the US motor industry to develop a process of guaranteed corrosion protection. The situation gave the impetus for the development of cathodic electrocoat in the USA, and it was introduced into the North American motor industry in 1978.

Resin system

All current cathodic products are based on epoxy/amine resin systems which are stabilized in water by neutralization with various acids. These products require a crosslinking agent to be present with optimum film properties being developed at high stoving temperatures (165–180 °C).

Pigmentation

The basic properties are similar to those required for anodic electrocoat — simplicity, high purity, and the ability, in some form, to supplement corrosion performance and catalyse the curing process.

Pigment volume concentration is normally in the region of 10%.

Colour

The most popular colours vary from medium grey through to black, and the prime pigments and extenders align very closely to those used in anodic technology.

Mechanism of deposition

The same principles of electrochemistry apply as in anodic deposition. Key differences are as follows.

Solubilization of the resin:

$$\begin{array}{c} R \\ | \\ R\!-\!N\!: + R'COOH \to R\!-\!N\!-\!H^+ + R'COO^- \\ | \\ R \end{array}$$

(R = epoxy) (R′ = solubilising acid)

Electrolysis of water:

$$4H_2O \rightleftharpoons 3H^+ + 3OH^- + H_2O$$

Reduction at the cathode:

$$H_2O + 3H^+ + 4e \to 2H_2 + OH^-$$

$$\begin{array}{c} R \\ | \\ R\!-\!N\!-\!H^+ + OH^- \to R\!-\!N\!: + H_2O \\ | \\ R \end{array}$$

Insoluble deposited coating

Oxidation at the anode:

$$3OH^+ - 4e \to H_2O + O_2 + H^+$$
$$\text{acidic reaction}$$

H^+ *removed by electrodialysis*

$$R'COO^- + H^+ \to R'COOH$$

Performance characteristics
Cathodic technology rapidly replaced anodic products in North America since results from test track evaluations, and the field, confirmed the superior corrosion protection of cathodic systems. Where the cathodic system is particularly good is in its throwing power, being better than most anodics, and in its thin film (~12 µm) performance particularly on unpretreated steel. This is found to be extremely important in box sections, etc., where the durability of the product is most vulnerable.

The Japanese automotive industry was also quick to change, with Western Europe following, even though European anodic technology was adjudged as being quite satisfactory at the time. Cathodic plants are now to be found in all parts of the world, e.g. Taiwan, Malaysia, Indonesia, Brazil, and the Philippines, with the overall objective to maximize standards.

Figure 10.4 clearly demonstrates the improvements made in corrosion protection, as measured by the ASTM salt spray test.

Plant requirements (refer Fig. 10.5)
Cathodic installations are basically similar to anodic plants in terms of paint circulation, refrigeration, filtration and the requirement for a smooth and stable direct current power supply.

However, there is a fundamental need to introduce either materials which are fully acid resistant or to use suitable acid-resistant coatings on mild steel. Stainless steel is the most common material, and pumps, valves, and piping are often made

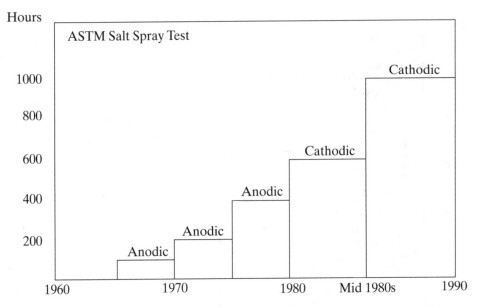

Fig. 10.4 — Improvements in corrosion protection. Note that anodic and cathodic electropaints were compared at equivalent film thickness (18–22 μm) over a suitable zinc phosphate pretreatment.

of this. The high cost of stainless steel can often preclude its use and it is replaced with PVC and other plastics where feasible.

The electrocoat tank and rinse tunnels also need to be lined with acid-resistant materials although there are instances where stainless steel is used for rinse tunnel construction.

Dip rinsing
Cathodic electropaints operate at relatively high solids (20–22%) and the most effective method of rinsing has been found to be dipping. In this way 'drag-out' materials can be removed more effectively than the normal spray rinse procedure. Clearly, the more effective the rinsing, the better the economics.

The principal feature in design is to ensure turbulence in the dip rinse tank to effectively remove the dip layer (or 'cream coat').

Ultrafiltration
Ultrafiltration is pressure-induced filtration through a membrane material of very fine porosity. In the case of electropaints it can be regarded as a process which separates the continuous phase (water, solvent, and water-soluble salts) from the dispersed phase (resin and pigments). The velocity of the paint required to drive the process is about 4–4.5 m s^{-1} (12–15 ft/s) or 150 litres min^{-1}. To avoid pigment precipitation in the tubes, and subsequent flux decay, this flow must be maintained.

The purposes of ultrafiltration are as follows:

- *Saving of paint.* This is the most important consideration. By utilizing the ultrafiltrate to rinse or wash the 'drag-out' paint (dip layer or cream coat) back into the tank, the efficiency of the paint usage can be improved by 15–20%.

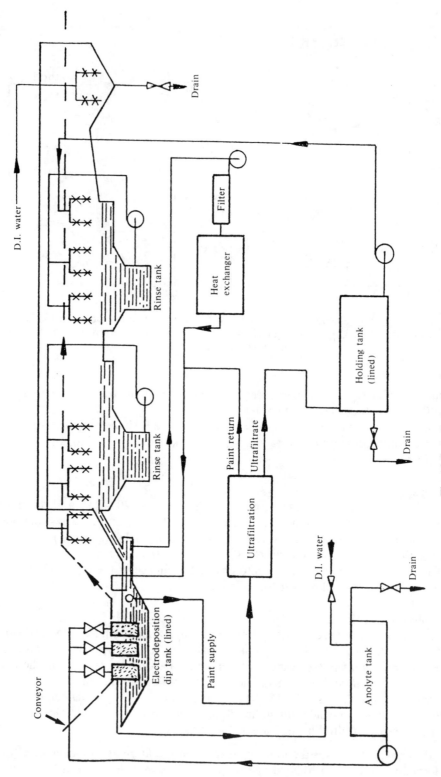

Fig. 10.5 — Cathodic electrocoat plant.

- *Decontamination.* Soluble salts, which are derived from carry-over materials and which can significantly degrade performance and processing, can be purged or removed from the dip tank by simply discarding ultrafiltrate.
- *Control of paint performance.* In general terms, with cathodic electrocoat, purging is essential to maintain the paint within certain specified limits (normally 800–1200 μS). This is needed to avoid any reduction in performance and for ease of processing.
- *De-watering.* When the tank solids are too low and tank value is high, solids may simply be increased by discarding ultrafiltrate and replacing with fresh paint.

Control method

The pH is controlled by the system of electrodialysis. A selective, ion exchange membrane is used around the anode and acid ions, generated during the electrolytic process, are allowed to pass through the membrane and are simply flushed away. Control is automated, using a recycling anolyte system, and the conductivity of the recycling liquid is maintained by integrating a conductivity meter with a de-ionized water supply.

Pretreatment

In relation to cathodic electropaints there have been quite significant advances in pretreatment development.

Spray phosphate systems suitable for anodic systems were found to give loss of adhesion under 'wet' conditions with cathodic products. The cause of this adhesion loss has been identified as cleavage of a secondary phosphate layer (chemically known as hopeite, $Zn_3(PO_4)_2 4H_2O$). This secondary phosphate has been termed 'grass' because of its appearance and because it 'grows' on top of the primary phosphate layer (chemically known as phosphophyllite, $Zn_2Fe(PO_4)_2 4H_2O$). In the anodic electrodeposition process this 'grass' is largely detached because of substrate disruption, but remains basically intact during cathodic deposition.

The high cohesive strength of the cathodic epoxy film exerts a pull on the weak secondary layer when swollen with the absorption of water. The net result is that cleavage takes place at this interface, which manifests itself in loss of adhesion.

As a result of this problem, with its associated mechanism, iron-rich homogeneous zinc phosphate coatings of fine tight crystal structure were developed which had a minimal secondary phosphate layer. In fact, it is generally accepted that the greater the phosphophyllite content (i.e. the higher the iron content of the coating) and the lower the hopeite content, the better the likelihood of performance.

The phosphophyllite content of the coating can be derived by measuring the iron content of the coating, or better still by crystal structure analysis using X-ray diffraction. The result can be expressed as the so-called P/P + H ratio where P = phosphophyllite and H = hopeite contents of the coating.

Added insurance for optimum performance is often provided by a post-phosphate rinse with a weak chrome solution. This acts as a 'knitting process' which simply 'knits' the crystals together, aiding homogeneity.

European and North American manufacturers saw this post-phosphate route as an essential feature to maintain consistency of quality and performance. However, because of environmental restrictions, it was unacceptable in Japan. This led to dip phosphating being adopted in Japan since it had already been established that, by utilizing the more controlled diffusion conditions afforded by dip processes, a higher degree of homogeneity in the phosphate coating could be achieved, i.e. the process favours the formation of the primary phase.

More recently there have been further developments in pretreatment technology. It has been clearly established that the achievement of optimum crystal structure is more important for good performance than simply obtaining a coating of low hopeite content.

Objectives have included developing products having a capability of being amenable to a truly multi-metal process (car-bodies now being constructed with a variety of metals apart from mild steel, e.g. aluminium, galvanized steel, etc.). Also there is a need for a reduction in energy demand (operating at <45 °C).

Significant improvements have been achieved by incorporating other heavy metals rather than zinc, i.e. tricationic philosophy — zinc, manganese, and nickel in combination with crystal-modifying agents.

Benefits from this newer pretreatment technology are numerous:

- excellent performance under cathodic electrocoat;
- ease of operation of high zinc processes with none of the disadvantages;
- coating weight less critical;
- good crystal morphology;
- lower energy usage;
- multi-metal processing;
- optimum chemical efficiency;
- requirement of chrome rinse minimized and in most circumstances eliminated.

General appraisal and current developments

Cathodic electroprimers have led to major advances in corrosion protection. The levels of improvement have been significant and, in association with wax injection which is often applied in box sections, six years of anti-corrosion and freedom from perforation can be confidently expected. The product/process has shown itself to be stable and consistent in performance, a most important feature in mass-production terms.

However, the international motor industry continues to demand improvements to produce and process performance. The numerous influences on material formulation and the electrocoat process in association with market needs, and resultant development activities, may be briefly summarized as follows:

1 *Engineering*. Compatibility/paintability of a variety of substrates, i.e.:
   (a) plastics in body components,
   (b) bake hardened metals,
   (c) alloy coated steels,
   (d) pre-primed coil including 'Bonazinc', 'Durasteel', etc.,
   (e) smooth steels, e.g. 'Laser Etch', etc.,
   (f) body shape and styling.
2 *Environmental*:
   (a) lower VOC and oven losses,
   (b) reduced effluent disposal/treatment (combined oxygen demand),
   (c) eliminate heavy metals and lower lead content.
3 *Performance*:
   (a) improved mechanical properties,
   (b) improved film smoothness and crater resistance,

(c)   upgrade edge corrosion resistance,

(d)   lower/non-yellowing.

4   *Process*:

(a)   ease of usage, e.g. tank and ultrafiltration control,

(b)   accommodation of line speed changes, e.g. effects on deposition time, stoving schedule, etc.

5   *Economics*:

(a)   lower stoving temperatures (target 150 °C),

(b)   improved material utilization — uniformity of film thickness,

(c)   reduction in 'bake off' losses.

Significant progress has already been made with many of the above properties but there is no doubt that developments will be on-going as performance standards continue to rise. In addition other pressures/needs such as environmental regulations and economic considerations will undoubtedly become more demanding.

10.4   Surfacers

10.4.1   Background

In spite of the development of more sophisticated priming and topcoat technologies, surfacers, sometimes referred to as fillers or middle coats, continue to play an important role in automotive coating systems. Originally designated primer surfacers, because of the basic undercoating system, their role as a primer diminished with the introduction of anodic electrocoat systems in the 1960s.

The role as a primer declined even further in the late 1970s with the advent of cathodic electrocoat. As a result of these changes in priming systems so newer surfacer technologies were developed not only to satisfy these changes but also to cope with more novel topcoats, new application techniques, higher performance standards, environmental requirements, and consumers' expectations.

10.4.2   Introduction

When the assembled untreated metal car-body (often a mixture of steel and galvanized steel) arrives for painting the process, apart from protection and decoration, is expected to provide a degree of 'filling'. This is necessary to hide any minor imperfections arising from the pressing and assembly operation. However, improvements in metal quality have been very significant and this has consequently affected the way surfacers are formulated.

The phosphating process has no filling potential and conventional dip and electroprimers have, because of their extremely low solids, minimal filling properties. Also, in the case of electroprimers, the inherent nature of the deposition process tends to emphasize any metal defects. As a result of this one of the primary functions of the surfacer was simply to 'fill' but with better standards of metal preparation this requirement is now much less important.

Furthermore, there are other highly important properties to be embraced, namely:

• mechanical properties such as impact and flexibility;

• stone chip resistance;

- resistance to water/moisture;
- UV resistance — particularly significant when used in the basecoat clear process;
- good sandability/rectification properties;
- easy, efficient, and economical to use;
- amenable to a range of methods of application (i.e. good workability);
- tolerant to a variety of topcoats;
- provide a good even surface (with minimal or no sanding) to maximize the appearance and performance of the appropriate finish;
- good adhesion to both the primer (most probably cathodic electrocoat), anti-chip coatings, seam sealants, and the subsequent topcoat;
- adequate performance over phosphated or galvanized steel;
- tolerance to under- and over-baking at a recommended stoving schedule.

In addition to the highly important role of embracing the above properties, surfacers offer additional benefits such as flexibility and safeguards to the production process.

10.4.3 Product types and formulation

The development of suitable surfacers for mass production conditions has been strongly influenced by the following factors:

- new priming techniques/methods;
- improved quality of metal pressings;
- new topcoat technologies;
- overall improvement in performance standards;
- newer methods of application leading to more efficient and economical painting processes;
- requirements for lower levels of pollution.

Early primer surfacers were often formulated to give high build (40–50 μm), were highly porous in nature with very high pigment volume concentrations (PVC) of 35–50% requiring heavy sanding, often 'wet' and by hand, to obtain an acceptable surface for finish application. Furthermore, because of the porous nature of the primer surfacer, a lowly pigmented sealer was sometimes applied to improve the appearance of the final topcoat.

These early products were also required to perform over pretreated steel and provide a level of corrosion protection. Applied in two coats, in a wet on wet procedure, the first coat was generally formulated with red oxide, supported by extenders and anti-corrosive pigments. This was then followed by a second coat of a grey or off-white primer surfacer. The difference in colour also minimized 'rub throughs' to the substrate, the 'red oxide' coat acting as a guide coat during the sanding operation. With the advent of electrocoat in the late 1960s surfacer film thicknesses were reduced (~35 μm) with lower pigment volumes (PVC ~ 30%) and no need for different colour coats. Improvements in metal quality also had a significant role in reducing the level of 'filling' and current surfacers have PVCs down to ~15% giving the product topcoat-like properties.

In addition, there is now considerable use of 'colour keyed' surfacers formulated to align to specific coloured topcoats. There are two reasons for this. One is the drive for cleaner and brighter colours where opacity levels are often borderline and the second is to economize in the use of highly expensive and sophisticated finishes.

In terms of gloss and appearance these types of surfacer have certain similarities to topcoats and are used quite satisfactorily inside engine and boot compartments. Also, because they are 'colour keyed' to individual topcoats they help bolster the opacity of the finish and minimize the use of expensive pigments.

In some Japanese plants 'inner' and 'outer' surfacers are often used. The 'inner' type of product has a high gloss (often using organic pigments in the formulation) and is very akin to a topcoat. They are applied directly to the electrocoat primer on all interior surfaces with no topcoat application. The exterior surfacer is of a more conventional nature and topcoat is applied in the normal way.

As far as resin systems are concerned a variety of alkyds were used in earlier products. These were superseded by epoxy-modified alkyds, epoxy esters, and more recently by polyester and polyester/polyurethane resins. Although solvent-based surfacers still tend to predominate, water-borne materials, because of their pollution advantages and good levelling, are attracting more and more attention and are now widely used in Europe. Water-borne surfacers are described in more detail later (see Section 10.13).

Epoxy-based products were commonly utilized over anodic electrocoat paints because of their excellent adhesion and water resistance. They were also used in the early days of cathodic electroprimers and minimal or non-sanding versions were developed to reduce the costly and labour-intensive sanding operation. However, during the late 1970s the advent of basecoat/clear metallic topcoats had a marked effect on the use of epoxy-based surfacers. Their poor resistance to erosion from UV radiation (chalking at the interface between topcoat and surfacer) led to delamination. As a result of this, polyester surfacers were developed which had far superior UV resistance.

Polyester surfacers, suitably modified with polyurethane resins, are now universally recognized as the best for all-round properties in modern undercoating systems. These products, with obviously accepted standards for chemical and physical properties, are designed for minimal or non-sanding to maximize their suitability for production purposes.

10.4.3.1   Resins systems
There are three distinct resin systems (Table 10.3) used in the formulation of surfacers:

Table 10.3 — Main resin systems: comparison of properties

	Alkyd	Epoxy ester	Polyester
Mechanicals	1	3	2 (3)*
Holdout	1	2	3
Durability	2	3	3
Chalking	2	1	3
Levelling	3	1	2
Sandability	1	3	3
Filling	1	3	3

1 = poor, 2 = good, 3 = very good.
* PU modification.

- alkyds;
- epoxy esters;
- synthetic polyesters.

All are reacted (crosslinked) with a suitable nitrogen resin and, in more modern systems, blocked isocyanates are incorporated. The crosslinking reacting is completed by stoving at a suitable stoving temperature, normally 20 min at 165 °C (effective). Modification with epoxy resins was, and still is, included to improve certain properties and this will be described separately.

Alkyds

Alkyds surfacers, although now being replaced by other products, are based on a type of resin produced by the reaction of an alcohol (glycerol, glycol, etc.) and a dibasic acid (phthalic anhydride) and modified with a natural oil or fatty acid to give the desired balance of properties within an appropriate stoving schedule. For cost and other reasons typical oils are linseed, tall or dehydrated castor oil. Oil lengths are normally short, i.e. ~35%.

The crosslinking reaction needed to form a suitable film takes place between the alkyd and usually a melamine formaldehyde condensate involving N-methylol groups or their ethers.

Epoxy esters

As epoxy esters were developed they tended to replace alkyds in surfacer products because of their better all-round performance.

The terminal epoxide groups and the secondary hydroxyl groups of epoxy resins can be reacted with fatty acids to produce epoxy esters. Oil lengths of 30–50% (i.e. short) are normally utilized in surfacers and esters are usually based on linseed or tall oil. After stoving, in combination with a nitrogen resin, a harder film of superior adhesion, flexibility, and chemical resistance to similar alkyd/melamine formulation is formed.

The major weakness of epoxy esters is their instability to UV and this has led to problems of delamination of metallic finishes, particularly basecoat/clear materials.

Polyesters

Polyesters form the principal resin system of modern surfacers used currently in the motor industry. They are usually hydroxy functional saturated polyesters and are made from a range of synthetic fatty acids suitably balanced to generate an acceptable balance of properties. Polyesters have particularly good UV resistance and this was one of the reasons for their introduction. A typical example is a blend of the following monomers: isophthalic acid/teraphthalic acid/adipic acid/tri-methylol propane/neo-pentyl glycol.

Now 'modified' with polyurethane additions to improve mechanical properties they represent an excellently balanced product.

Epoxies: film modifiers

Epoxy resins have been used in surfacers, in one form or another, for a number of years because of their inherent strengths of adhesion, toughness, and corrosion resistance. They are still present in current polyester products to improve adhesion to bare steel and to provide a degree of corrosion protection. In some cases stone chip performance is upgraded.

When formulating, epoxies should be considered as part replacement for the main resin at levels of about 10%. This amount is not normally exceeded because of UV instability. The low molecular weight epoxies give good adhesion but not particularly good flow. Increasing the molecular weight leads to slightly poorer adhesion but better flow.

The current trend is to move towards higher molecular weight due to carcinogenic concerns associated with lower molecular weight species. In fact some users are now specifying epoxy-free systems because of this concern. Often a blend of high and low molecular weight epoxies are used to achieve an optimum balance of properties.

Crosslinking resins
Traditionally melamine formaldehyde resin and, to a lesser extent, urea formaldehyde were used as the main crosslinking agents in surfacer formulations. Although cheaper, urea formaldehyde was dropped in favour of melamine formaldehyde over the years because of inferior performance, particularly over electrocoat primers.

More recently, isocyanates and benzoguanamine resins have been included (to improve certain properties).

Blocked isocyanates, normally hexamethylene diisocyanates, unblock at around 120 °C to form a strong urethane linkage. These resins have been introduced to improve stone-chip resistance, film toughness, and impact resistance without leading to a loss in hardness. Surfacers formulated in this way are designated PU-modified.

Benzoguanamine resins are added to improve adhesion, flexibility, and overbake tolerance, i.e. there is no degradation in mechanical properties at elevated temperatures (up to 200 °C) and/or extended stoving time. Conventional melamine formaldehyde resins would begin to break down and become brittle at such temperatures.

10.4.3.2   Pigmentation
Surfacers are highly pigmented products although more recent developments, in line with improved metal quality, have quite low levels of pigment and an appearance which almost lends itself to a topcoat. Traditionally, pigment volume concentrations (PVC) vary from 30 to 55% depending on physical and chemical requirements and the topcoat to be applied. The very high levels (55%) are limited to use with acrylic lacquers which have very poor adhesion properties. Modern surfacers, because of

Table 10.4 — Specific pigments: colour keyed surfacers

Description	Comments
Carbon black	Standard carbon black pigment
Inorganic yellow — iron oxide	Two types — both 'dirty yellow' but bluer alternative available
Organic blue — phthalocyanine blue	Standard prime pigment
Inorganic red — iron oxide	Two types — standard and 'bluer' alternative
Organic reds (range) ⎫	Normally restricted to 'inner surfacers' for
Organic yellows (range) ⎭	use in interiors only

the reasons above, have PVCs as low as 15–20% with gloss levels of 80% measured on a 60° head.

The basic pigmentation is normally a blend of prime pigment (for opacity and colour) and extender pigment(s) (for economic reasons and where performance considerations allow).

In the past surfacers were spray applied in two coats wet-on-wet over dip primers or phosphated metal (film thicknesses were ~45 μm). The first coat was a red (iron) oxide followed by a second, normally grey. As mentioned earlier the first (red oxide) coat acted as a 'guide coat' to control sanding and to provide a measure of corrosion protection.

The advent of electrocoat has meant that the 'red oxide' coating has been dropped and now grey, off-white, and sometimes specific coloured surfacers are used, at a film thickness of 30–40 μm.

For example, a typical off-white surfacer would have a volume ratio of high grade rutile titanium dioxide (prime pigment) to barytes (extender) of 3:2 to 2:3 depending on required properties. The barytes contributes to reducing cost and resistivity, and to improving filling capacity: titanium dioxide to gloss and opacity.

Prime pigments

A good grade of rutule TiO_2, designed for high performance coatings, provides gloss and opacity. In the past, for reasons of economy, its content was kept to a minimum provided overall performance could be maintained. Nowadays, where high gloss levels are generally specified, TiO_2 predominates in the formulation with minimal and, sometimes, no extender present.

Anatase has poor UV resistance which is the reason it is not used.

Synthetic iron oxide used in past formulations provided colour and opacity. Although difficult to disperse it does have the important feature of good chemical resistance. This is important in surfacers.

Extenders

Extender pigments are much cheaper than prime pigments and, apart from reducing cost, can by careful selection improve certain properties of the finished product or dry coating. Proper choice can upgrade properties such as consistency, levelling, and pigment settling of the paint. They can also reinforce the dry coating's mechanical properties and improve sanding and resistance to moisture and blistering.

The principal extenders are as follows:

- *Barytes* (barium sulphate). Originally the 'natural' form of barytes was used. It is extremely insoluble in water which gives good blister resistance but is hard with poor sanding properties and is difficult to disperse.

  It has now been replaced by blanc fixe, a precipitated form of barium sulphate. Precipitated forms of barytes are softer and easier to disperse than natural ones and are available in a range of particle sizes. All have very good blister resistance but they can lead to a level of thixotropy in the liquid paint. Fine grades may be used to give relatively high gloss levels in the final product.
- *China clay* (aluminium silicate). China clay, being inert and non-reactive, can play a useful role in the formulation of surfacers. It has a good colour and rubbing/sanding properties but its high oil absorption tends to reduce gloss.

In a typical surfacer pigmentation, to improve sanding properties and resistivity, 5–10% of barium sulphate may be replaced by china clay without significantly affecting overall properties.

- *Winnofil* (stearate-coated calcium carbonate). Provides 'structure' and anti-settling properties.
- *Talc* (magnesium silicate). The inert and hydrophobic nature of modern treated talcs, together with their availability in plate-like forms, can improve such properties as: water and humidity resistance; sanding properties (better than china clay); and film toughness.
- *Calcium carbonate* (chalk/Paris white). A widely used extender in the past, having properties of stability and cheapness. However, because of its reactivity with acids it has been removed from surfacer compositions. Evidence has suggested it contributed, in part, to the delamination of basecoat/clear finishes.

10.4.4   Polyurethane-modified polyester surfacer (including 'colour keyed' products)

10.4.4.1   Summary of basic parameters

Solids content
Unthinned material (45″/DIN 4/25 °C)

> Normal solids: ≮63%
> High solids: ≮68%
> Colour keyed: dependent on colour (see below)

Thinned material (at spray viscosity)

> Normal solids: (~25″/DIN 4/25 °C) ≮60%
> High solids: (~36″/DIN 4/25 °C) ≮64%
> Colour keyed: (~36″/DIN 4/25 °C)
> Light Mid Grey Std: ≮64–66%
> Inorganic pigments, e.g. off whites, pastels: ≮62–64%
> Dark colours, e.g. blues, black grey, reds: ≮56–58%
> Bright/transparent colours, e.g. light reds, yellows: ≮56–58%

Specific gravity (ASTM D1475)
1.2–1.4 (could be higher at high solids).

Flash point
≯21 °C.

10.4.4.2   Film properties (stoved film)

Appearance
After stoving the film must have a good appearance with no surface defects such as craters or 'solvent popping'. Also it should have good flow (or levelling) and be free of mottle (orange peel) without tendency to 'sag' at high film thicknesses.

- *Film Thickness* (dry): ~35 ± 5 μm.
- *Colour*: off white/light grey: carried out visually. Needs to be within acceptable limits to original 'approved' sample.
  Colour keyed: A match to original accepted standard.
- *Gloss* (60° head): 75–85%.
- *Opacity* (dry hiding): 25 μm (black and white chart).
- *Sandability*: the stoved surfacer must be amenable to wet sanding (P400 paper) or be dry sandable (P800 paper) by hand. Clogging of the paper, excessive sanding marks, and dusting are considered unacceptable.

10.5   Anti-chip coatings

10.5.1   Background and resin types

Introduced over the past decade the prime function of anti-chip coatings is to upgrade stone chip resistance on vulnerable areas. Particular areas are the door sills, front and rear ends below the bumpers, the underbody, and leading edges such as the front end of the bonnet. Sometimes, depending on the design of the car, these coatings are used up to 'waistline level' and, in some instances, as a complete coat but this is rather exceptional and uneconomic.

They are applied as high build products (50–100 μm), wet-on-wet with the surfacer. Compatibility between the two coatings is essential to avoid problems such as cratering or poor 'wetting'. The anti-chip coating is subsequently stoved (normally 20′ @ 165 °C) as a composite with the prior coat of surfacer.

There are three resin types:

- *Polyester/PU.* These are formulated with similar resin compositions to current polyester surfacers. Generally softer, they are crosslinked with melamine formaldehyde resins, applied at high film thickness of 50 μm and have the advantage of superior appearance to the blocked isocyanate products described below.
- *Isocyanate/cycloaliphatic diamine blends.* One-pack products composed of a blocked aromatic isocyanate blended with a cycloaliphatic diamine. The final stoved film has a rubber-like character giving it exceptional mechanical properties but poor appearance.
- *Powder coatings.* Powder coating stone chip primers have found some limited use in the USA. These are described in more detail later.

10.5.2   Pigmentation

All types are pigmented in a conventional way with prime and extender pigments being included, although the level of extender has to be carefully controlled to avoid degradation of the mechanical properties in any way. Thixotropic aids such as Aerosil or Bentone are often used since they are extremely beneficial in reducing sagging tendencies at the very high film thicknesses of 50–100 μm.

10.6   Inverted or reverse process

The established undercoating system, following the introduction of anodic electropainting was pretreatment/electroprimer/surfacer or sealer. Later the concept of inverting, or reversing part of this process, was developed; the objective being to overcome some of the fundamental weaknesses of anodic electrocoat and maximize the performance of the paint components of the system.

A typical inverted process may be designated as follows:

Pretreatment (spray zinc phosphate)
↓
Spray surfacer to outer skin panels
(stove or partially stove)
↓
Electroprime (normal procedure)
↓
Final stove

The main advantages are seen as:

- Higher film build and thus better corrosion protection on internal surfaces (due to insulating effect of coating on external and carbody interior).
- Improved filiform, scab corrosion, and blister resistance (no phosphate disruption on external surfaces).
- Better adhesion and stone-chip resistance of full finishing system — inverted surfacer can be formulated with suitable properties and greater tolerance to phosphate variations than electroprimers.
- Easier painting of composite metal bodies, e.g. steel, zinc, and aluminium: the different electrical properties of metals can give rise to problems during electrodeposition.

A number of manufacturers did experiment with this process utilizing solvent-based surfacers or powder coatings. Its adoption was inhibited by a number of factors:

- Poor appearance/performance at the surfacer/electroprimer interface. Although this problem may be resolved by the use of a water-borne surfacer having suitable electrical and compatibility properties.
- Capital cost required to modify existing plants.
- The advent of cathodic electrocoat and the simultaneous development of improved pretreatments and process control.

Electro powder coating (EPC)

This is another example of a reverse process but incorporating cathodic electro-primer. Originating in Japan, EPC is essentially a process in which a powder surfacer carried as a slurry in water is electrodeposited in a matter of seconds on to the phosphated carbody. Since the EPC has little or no throwing power it is confined to the outer skin panels. After rinsing and a set-up bake the car body is then cathodic electrocoated.

The cathodic electrocoat does not deposit on the electrically insulated EPC surfacer but rather is used to maximum effect in protecting the floor area, box sections, etc., and formulated accordingly.

The process has much to commend it from the standpoints of automation, quality consistency, elimination of a spraybooth, and the virtual absence of organic solvents. On the debit side, however, the present generation of EPC surfacers deposit films of variable thickness and exhibit pronounced mottle or surface texture which requires considerable sanding before topcoat application.

10.7   Automotive topcoats

The differing and stringent demands made by the user, and the ultimate customer, have led to a number of different topcoat technologies. All have a different balance of properties closely aligned to particular test specifications and process requirements.

Automotive topcoats, being the final coat in the painting process, have need to both decorate and protect and this has to be reconciled with use in mass production. These requirements may be summarized as follows.

For decoration

* appearance/high gloss and smoothness;
* aesthetic/customer appeal.

For protection

* UV protection/colour stability/durability;
* humidity resistance;
* water resistance;
* chemical resistance;
* resistance to insect/bird effects;
* physical properties, e.g. mar resistance;
* mechanical properties;
* distortion resistance.

Materials to satisfy these needs are extremely sophisticated with can be subdivided into two basic forms — solid (or straight) colours and metallics:

* Alkyd or polyester finishes: solid colours.
* Thermosetting acrylic finishes (including NAD technology): solid colours and metallics.
* Thermoplastic acrylic finishes: solid colours and metallics.
* Basecoat/clear (solvent- and water-borne metallic systems): now include solid colours mainly in water-borne basecoat technology.

Alkyd/polyester finishes remain the most widely used throughout the world for solid colours, principally because of their low cost and ease of processing.

Thermosetting acrylic systems (or NAD) were originally introduced because of their good durability in metallic finishes, alkyds having poor performance. This type of finish was also adopted by many producers in solid colour form. However, at the present time, the use of this technology has greatly decreased. It can still be found in some truck plants (mainly in the USA), and in isolated plants in Japan where it is preferred to alkyds because of better acid resistance. Otherwise, its use is confined to various plants in the 'rest of the world'.

Thermoplastic acrylic lacquers were widely used by General Motors for both solid colours and metallics, However, because of their very low solids content, which led to high levels of organic solvent emission, their use has diminished in recent years. They have been replaced by alternative solvent- or water-borne technologies to satisfy economic and environmental considerations.

Basecoat/clear metallics, both solvent-borne and water-borne, are now considered the 'norm' in the automotive sector for both appearance and performance. They have enhanced gloss, stylistic appeal, and outstanding durability — the clearcoat being based on a thermosetting acrylic resin 'reinforced' by a suitable UV absorber. This technology is also showing significant growth in solid colour technology, mainly because of performance and for reasons of technology rationalization where water-borne systems are used.

Water-borne basecoats, in metallic form, were originally introduced in the early 1980s to reduce the excessive solvent emissions from conventional basecoats. In addition, the proportion of metallic finishes versus solid colours was rising sharply, representing about 60% of production in the major manufacturing areas (Europe, USA, and Japan) further aggravating the emission problem. This technology now embraces solid colour basecoats as environmental demands have become more stringent and, as already mentioned, to rationalize the production process.

Their level of usage continues to grow rapidly throughout the world as stricter controls on solvent emissions have been introduced. Nevertheless growth did slow in the early nineties, principally because of the shortage of available capital for investment in new plant. Virtually all new plants which are being built have the capability of using water, e.g., stainless steel circulating systems, even if they are not doing so. This includes areas such as Eastern Europe.

Water-borne basecoats are now very widely used in North America, Japan, Germany, Scandinavia, the United Kingdom and other European countries mainly in line with environmental legislation. Current estimates suggest somewhere between 55% and 65% of European production is water-borne and that the worldwide figure is broadly very similar.

As far as solid colour basecoats are concerned a number of European manufacturers, e.g. Volkswagen, Volvo, BMW, and Mercedes, have all gone this way and are totally water-borne. Others are considering options and are likely to continue this way but they will be under increasing pressure to follow the water-borne route as environmental pressure increases.

10.7.1   Alkyd or polyester finishes

10.7.1.1   Basic chemistry
Alkyd finishes are based on a class of resins produced by the reaction of alcohols (glycerol, glycol, etc.) and dibasic acids (phthalic anhydride), and modifying with a natural or synthetic oil to give the designed balance of durability, flexibility, hardness, etc. The 'oil' or fatty acid is selected for its good colour and non-yellowing (non-oxidizing) characteristics, e.g. coconut oil fatty acid or 3,5,5-trimethyl hexanoic acid and simply acts as a reactive plasticizer. Typical alkyds are short/medium oil length (~35%).

Polyesters are typically isophthalic or adipic acid with pentaerythritol, neopentyl glycol, 1,6-hexanediol, or similar polyfunctional alcohols.

However, because of the ever-increasing demands of performance specifications and emphasis on higher solids there is much more use of alkyds produced from synthetic fatty acids rather than oils (e.g. iso-nonanoic) which can give lower viscosity alkyds. Similarly, both alkyds and polyesters are frequently modified with Cardura E (see Section 2.5.1) to improve durability, chemical resistance, colour retention, and higher solids.

The crosslinking reaction, need to form an insoluble film, takes place between the alkyd (or polyester) and a melamine formaldehyde condensate involving N-methylol groups or their ethers. The melamine formaldehyde is normally alkylated but, as performance requirements and demands for higher solids have increased, isobutylated or 'mixed' are used rather than the traditional n-butylated type.

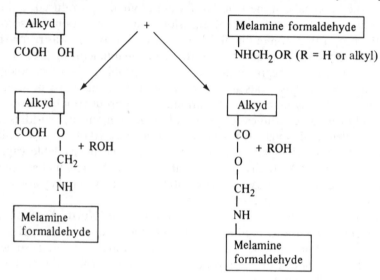

A stoving temperature of 20 minutes at 130 °C is suitable for effecting suitable cure.

10.7.1.2   General properties

These may be summarized as follows. Widely used for solid colours in Europe, Asia, Pacific Rim, and the rest of the world. In North America acrylics predominate.

- Relatively high solids (~50% w/w).
- Low cost.
- Ease of processing.
- Good durability.
- Tolerant to a wide range of undercoats.
- Proneness to dirt pick-up (film wetness).
- Relatively poor polishing properties.

Their high solids, low cost, and ease of processing (40–50 μm in two coats wet-on-wet application) has helped to maintain their popularity. These properties enable good build (and filling) properties and very good appearance after stoving. Also they have good durability, being equivalent to thermosetting acrylics after two year tests in Florida (5° south).

However, on the debit side they do have relatively poor solvent release, compared with acrylics, and the wet film is prone to pick up dirt. Clean operating conditions are particularly important as a consequence. Also their somewhat limited polishability makes minor rectification on a production line difficult.

More recently there has been a move in some plants to use basecoat/clear technology (refer Section 10.7.4) for dark colours to improve appearance, mar, and etch resistance. Also for 'lead-free' red topcoats when used in combination with a red 'coloured surfacer', to overcome appearance, durability, and coverage problems.

10.7.2 Thermosetting acrylic/NAD finishes

The use of thermosetting acrylic finishes has decreased significantly in recent years although the technology forms the basis of clearcoat formulations. In addition, NAD technology, i.e. the stabilization of discrete polymer particles in organic solvent, originated in the automotive sector and has since been extended into water-borne systems.

10.7.2.1 Basic chemistry

Thermosetting acrylics are based on complex acrylic copolymer resins produced by the reaction of a number of acrylic monomers selected to give the desired balance of properties. The required crosslinking reaction, with a suitable melamine for-maldehyde resin, is facilitated by the presence of hydroxyl groups in the polymer backbone. The hydroxy-containing polymers are readily prepared by the use of hydroxy-acrylic monomers.

A stoving of 20 minutes at 130 °C produces an insoluble crosslinked film:

$$\boxed{\text{Melamine Formaldehyde}} - \text{NHCH} - \text{OR} + \text{HO} - \boxed{\text{Acrylic}}$$

$$\downarrow$$

$$\boxed{\text{Melamine}} - \text{NHCH} - \text{O} - \boxed{\text{Acrylic}}$$

$$+ \text{ROH}$$

Early thermosetting acrylic finishes were made from these resin systems, and they allowed the formulation of solid colours and metallic finishes. Durable metallic paints were made by incorporating non-leafing aluminium flake (via an aluminium paste) into the film. The non-leafing flake (15–45 μm in length) distributes randomly through the coating and produces specular reflection from almost any angle of view. Suitably pigmented with coloured pigments, extremely attractive colour ranges were produced.

These types of finish, introduced in the late 1950s and early 1960s, found wide usage through the world and in modified form still find use today. Their deficiencies are the requirement for three coats, because of solids content, and the difficulty in metallic application. The introduction of NAD versions of these finishes in the late 1960s enabled problems to be overcome.

Whereas with conventional thermosetting acrylics the resin is carried in solution, NAD finishes are based on the dispersion of similar polymers in a solvent mixture such that a significant proportion of the polymer is insoluble, even at application viscosity.

These dispersion are stabilized by the presence of aliphatic soluble chains (i.e. by the use of an amphipathic graft copolymer) which are chemically linked to the polymer particles during the polymerisation process. Such dispersions are referred to as 'super-stabilized' to distinguish them from systems where stabilizer is held on the surface of the particle by physical or polar forces.

As long as a substantial proportion of aliphatic hydrocarbon is present in the continuous phase of the dispersion these chains are extended and provide a stabilizing barrier around each particle. In order to promote eventual coalescence of these particles into a continuous film the volatile portion of an NAD finish also contains some solvents for the polymer itself. The amount of such solvents is controlled so that the stabilizing chains are not collapsed while the finish is in the package or the circulating system at the automotive plant. These solvents can be shown to partition themselves between the polymer (leading to softening and swelling of the particles) and the continuous phase.

On application of the finish, evaporation of the aliphatic non-solvents leads to fusion and coalescence of the solvent-swollen polymer particles to give a continuous film; the process being completed by the stoving process. The structure of the final film is almost identical to that laid down from a conventional solution polymer.

Although the basic technology allows the preparation of dispersions where practically all the polymer is in the disperse phase, the rheology of 'all disperse' systems is not necessarily the optimum for achieving latitude necessary under production line conditions. However, definite benefits can be gained by having some polymer in solution and an important part of the technology consists of establishing the optimum ratio of disperse to solution resin for each polymer system. The rheological characteristics of such thermosetting acrylic formulations are unique and lead to greatly improved control of metallic finishes, better utilization of paint solids, and greater resistance to sagging and running.

10.7.2.2   General properties
These may be summarized as follows:

- Durable in solid colours and metallics (single coat).
- Solids lower than alkyds (~30%).
- Applied in two or three coats.
- Good polishing properties.
- Good solvent release.
- High performance undercoat required, e.g. polyester/PU.

Thermosetting/NAD acrylic finishes can be formulated in both solid colours and metallic and it is possible to produce a wide and attractive colour range. Their rapid solvent release minimizes dirt pick-up and their good polishing properties make them more amenable to local rectification in production lines. They require high grade undercoats, such as a polyester/PU type, to maximize their performance. One other important feature is that thermosetting acrylics can be used directly over electroprimers because of their inherent adhesion properties. Alkyd finishes cannot, and poor adhesion can result. It is for this reason that acrylic finishes are always used in the two-coat electroprimer/finish system adopted for commercial vehicle production.

10.7.2.3   Metallic appearance
A good metallic finish is designed to achieve a pronounced 'flip' tone, i.e. the polychromatic effect seen when viewed from different angles. It is simply an optical effect and depends on the orientation of the metallic flake parallel to the surface so

that the amount of light reflected varies with the angle of viewing. Thus at glancing angles the surface appears deep in colour. In fact, the aluminium flake can be regarded as a small plane mirror.

In order to achieve this optimum orientation it is necessary to ensure:

- spray application is uniform;
- maximum shrinkage of the film after application so that the aluminium flake is physically 'pulled down' parallel to the surface;
- minimum tendency for the flake to reorientate randomly after application.

Typical solution thermosetting acrylic finishes go some way to meeting these requirements by changes in solvent composition, resin flow and design and differing types of aluminium. High levels of operator skill during application are also necessary since uneven application leads to areas of different colour and overwetness allows the aluminium flake to move around producing lighter and darker patches (sheariness or mottle). Dark lines (black edging) can also form around the edges of holes in the body shell or along styling lines. To overcome these problems solution enamels are sprayed as dry as possible consistent with gloss and flow with low film thicknesses often resulting.

Introduction of dispersion systems, often in combination with solution technology, eased the situation considerably by introducing improved rheology to the paint film. As previously stated a dispersion does not exert such a strong viscosity influence in a film as a solution version of the same polymer. However, in passing from the dispersed state to the solution state as the non-solvent part of the liquid evaporates, a sharp increase in viscosity is introduced. As this change occurs in the wet paint film after spraying there is a faster increase in viscosity at this stage than would be produced purely by evaporation of the liquids. This effectively reduces the amount of movement available to aluminium flakes, restricts any tendency to reorientation and allows an even appearance to be produced more easily.

10.7.2.4 'Sagging'

The same rheological control described above, exerted by the change from disperse to solution phase, has resulted in a marked reduction in 'sags' and 'runs' with dispersion coatings. A higher build can also be applied to fill minor defects. These improvements are particularly marked in highly pigmented colours such as whites and oranges.

10.7.2.5 'Solvent-popping' resistance

'Solvent-popping' (sometimes designated as 'boil') is caused by the retention of excessive solvent/occluded air in the film which, on stoving, escapes by erupting through the surface. It invariably occurs on areas where there is above-normal wet paint thickness. Solution acrylic thermosetting finishes are very prone to this because of their fast 'set-up' rate immediately after application.

It would be imagined that dispersion systems would suffer the same problem since the polymer compositions are similar. However, dispersion systems do show advantages. The reasons are probably two-fold:

- Better atomization during spraying leading to a finer droplet size.
- The utilization of non-solvents, not associated with the polymer; this leads to a faster and more effective 'solvent' release.

10.7.3   Thermoplastic acrylic lacquers

Acrylic lacquers were widely used throughout the world for many years by General Motors and prestige car producers such as Jaguar. Their enhanced appearance, particularly metallics, proven durability, and in-process flexibility (self-repair and polishability), made for an attractive technology.

Nevertheless, their high level of solvent emission during processing and high cost of application have led to a significant decline in their use. Recently, they have been replaced by other technologies more environmentally acceptable, particularly water-borne basecoats which are described in more detail later.

Although this technology is considered outdated it is still worth including in this chapter. One reason is for completeness, since the author would like to cover the whole range of topcoat technologies, particularly one that found such considerable use over the years. The other reason is that this technology forms the basis of repair lacquers because of its air dry and low bake capability.

10.7.3.1   Basic chemistry

Acrylic lacquers, in common with all lacquers, dry simply by the evaporation of solvent, and are based on a hard poly (methyl methacrylate) polymer that is suitably plasticized. Plasticizers are normally external and include butyl benzyl phthalate and linear polymeric phthalates derived from coconut oil fatty acid. The external plasticizer ensures a good balance of properties, i.e. improved crack resistance, adhesion to undercoats, solvent release properties, and flexibility.

Most common acrylic polymers for this type of finish have average molecular weights of approximately 90000: these give outstanding gloss retention on external exposure. Polymers with average molecular weights greater than 105000 tend to cobweb or form long filaments when applied by spray at commercially acceptable solids contents. Low molecular weight polymers result in poor film properties and poor durability. Furthermore, the improvement in gloss retention when the molecular weight is increased above 105000 is proportionally small, and is more than offset by the reduced solids at application viscosity.

Solvent blends used for lacquers are balanced compositions, albeit expensive, chosen to given acceptable viscosity, evaporation, and flow characteristics. To avoid excessive solvent retention in the film it is necessary to use solvents free from high boiling 'tail' fractions, and to balance carefully the evaporation rates of the remainder. The external plasticizer assists in solvent release by maintaining a fluid film for as long as possible; this allows shrinkage stresses (considerable in acrylic lacquers) caused by the drying/stoving process, to be relieved.

Although acrylic lacquers will ultimately air-dry, in practice drying is accelerated either by a short stoving, 30 minutes at 90 °C, where polishing is required to achieve acceptable gloss, or by the bake–sand–bake process. In this latter process surface imperfections of the film are removed by sanding after a short set-up bake (15 minutes at 82 °C), and the film is reflowed (20 minutes at 154 °C) to give a glossy flaw-free film.

10.7.3.2   General properties

These may be summarized as follows:

- Very good durability in solid colours and metallics.
- Robustness and adaptability in production.

- Good polishing and self-repair properties.
- Reflow (bake–sand–bake) process
    set-up 15 minutes at 82 °C
    reflow 20 minutes at 154 °C (effective metal temperature).
- Excellent metallic appearance.
- Low application solids (15–20%)/multi-coat process.
- Requirement for large quantities of expensive thinners.
- High raw material costs.
- Special undercoats needed.

It is worth noting that the outstanding appearance of acrylic lacquer metallics is due to its inherent properties:

- Low solids products.
- Multi-coat process.
- Carried in high viscosity/high molecular weight acrylic polymer.

These properties meet the necessary criteria for optimum 'flip' tone, i.e. high shrinkage, rapid solvent release, and thin coats restricting by shear geometry reorientation and movement of aluminium flake.

The disadvantages of acrylic lacquers also stem from their fundamental characteristics:

- Low application solids means up to four coats are required to achieve the film thickness (55–60 µm) necessary for reflow. Multi-coats also mean long spray booths with the inherent cost.
- Appearance on force drying at low temperature (80–90 °C) is poor, requiring excessive polishing to achieve acceptable gloss.
- High raw material costs, especially for solvents.
- Poor intrinsic adhesion entailing the use of special undercoats (high PVC, 55% epoxy-ester). In fact the choice of undercoat has a greater influence on general performance than other types of finish. Special adhesion-promoting sealers are also used, which add even more to the cost of the process.

These weaknesses, particularly the high level of solvent emissions and processing costs, have been the main cause of the demise of this technology and it has been replaced by more economic and less polluting products, i.e. alkyds, basecoat/clear and water-borne topcoats.

10.7.4   Basecoat/clear technology

10.7.4.1   Solvent-borne
As has been described earlier, metallic car finishes have been used by the majority of large car producers for a considerable time. Metallics, because of their stylistic appeal, have been considered a very desirable feature of finishing by both colour stylists and designers alike. In the form of what are designated 'single coat' metallics they have been supplied either in thermosetting or thermoplastic acrylic technologies.

However, they do have a number of inherent disadvantages compared with solid colours:

- Lower gloss levels — particularly in light metallics.
- Limitation in certain pigment areas such as organic pigments.
- Poorer resistance to acidic environments.
- Application difficulties: eased by the introduction of NAD thermosetting acrylics.

The concept of putting the aluminium flake in a separate foundation or basecoat and then overcoating with a clear resin was first thought of and applied several decades ago. 'Flamboyant' enamel technology used on bicycle frames is a case in point.

Certain European car manufacturers saw this type of technology as overcoming the weaknesses described above and introduced basecoat/clear technology into production in the late 1960s. The basecoat provided the opacity and metallic appearance while the clear imparted gloss, clarity, and overall durability. The use of this technology has subsequently grown considerably and is now considered the 'norm' for metallic finishes in Europe, Japan and North America.

In some plants there has been a move to solid colour basecoats for use in dark colours to improve appearance, mar, and etch resistance and also in red topcoats when used in combination with a coloured red surfacer.

Basic chemistry
The function of the two components has been described above, and this is achieved in the following manner.

The properties of the *basecoat* are:

- high opacity (~10 μm) to facilitate application in thin films;
- rapid solvent release — short drying time (2–3 min) before application of clearcoat;
- low solids (<20%) to achieve maximum metallic effect;
- 'compatibility' with the clearcoat, i.e. good adhesion with no sinkage of clear.

Basecoats are thermosetting products modified with resins such as cellulose acetate butyrate to promote 'lacquer dry' and to accelerate solvent release. The basic resin component is either an oil-free polyester or a thermosetting acrylic polymer suitably reacted with nitrogen (melamine) resin. The features of these two types may be summarized as follows:

- Polyester type: initially a low solids product (10–12% w/w) recognized for outstanding metallic appearance and ease of application. In recent years there has been a tendency to move to 'medium solids' polyester basecoats (15–18% w/w) to minimize solvent emissions. This change has had a minimal effect on metallic appearance. (Main basecoat used in Western Europe.)
- Acrylic type: higher solids than polyester type (15–20%), less pronounced metallic effect, but better filling properties and shorter processing time. (Main basecoat used in Japan and USA.)

The properties or *clearcoats* are:

- good clarity of image and 'compatibility' with basecoat;
- offers a high level of protection to ultraviolet, i.e. >3 years' Florida exposure.

There are two types of technology in use at the moment: thermosetting acrylic (solution and NAD), formulated on crosslinking thermosetting acrylics reacted with a melamine resin and modified with UV absorbers and light stabilizers. These materials are applied at a film thickness of 35–50 μm to achieve maximum gloss and UV protection.

The *features* are:

- Solution acrylic type: a high gloss product giving outstanding clarity. Applied in one or two coats.
- NAD acrylic type: A single coat product with lower clarity than the equivalent solution type. (Has a low usage in the motor industry.)

Two-pack 2K clearcoats are two component acrylic compositions using an isocyanate crosslinking mechanism. Their features are:

- High solids (60–70%), applied in one coat.
- Low stoving temperature (80 °C).
- Very good etch and mar resistance.
- Health risk/toxicity due to free isocyanate. Precautions needed.
- Expensive.

Application/process
The basecoat is applied as a two-coat wet-on-wet process with a short air drying time between coats. This is necessary to give acceptable opacity and evenness of appearance. After application of the second coat of basecoat has been completed, a short air drying time (2–3 min) is allowed, sometimes supplemented by a warm air blow, before the clear is applied in one or two coats.

Typical film thicknesses for the system are:

- Basecoat: 15 μm,
- Clearcoat: 35–50 μm.

Stove for 30 min at 130–150 °C to effect crosslinking.

Colour/pigmentation
The pigments used in metallics in general are chosen for their potential transparency, realized when they are correctly dispersed and stabilized. Full transparent coloured pigments leave the metallic flakes free to contribute the maximum of brightness, sparkle, and flip tone.

Provided that satisfactory transparency exists, metallic appearance will depend upon the orientation of the flakes. As described earlier, if each flake is parallel to the substrate then this will give the optimum metallic effect or 'flip tone'. The light-reflecting quality of metallic coatings can be measured by a goniophotometer. This instrument is particularly useful for measuring the reflectance of unpigmented silvers, since the performance of different silver paints can be compared without any additional reflection and absorption by coloured pigments.

In Fig. 10.6, instrumental comparisons are made of various metallic technologies demonstrating the excellence of low solids polyester basecoats. At a standard angle of incidence (45°) reflectance is measured at various viewing angles, and a curve is plotted; the higher the peak the better the reflectance, indicating the extent of parallel metallic orientation.

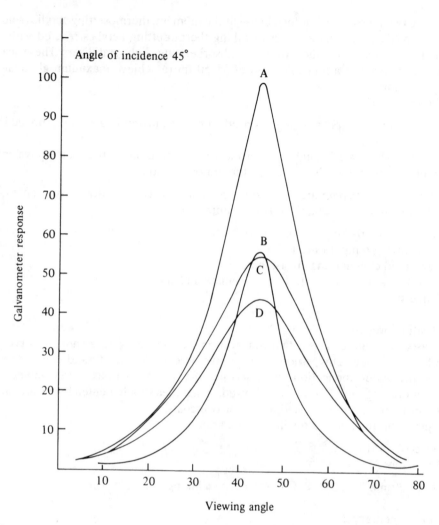

Fig. 10.6 — Goniophotometric curves.

Colour silver metallic
Key:

Curve		Application solids
Curve A: Polyester basecoat	}	12% }
Water-borne basecoat		15% }
Curve B: Thermoplastic acrylic lacquer		15%
Curve C: Acrylic basecoat		18%
Curve D: Thermosetting acrylic		25%

Aluminium flake orientation
It has already been stated that the principal factor regulating aluminium flake orientation is film shrinkage during the drying process. Loss of solvent from the applied film during the flash-off and baking periods presents the flakes (typically of length from 10 to 25 μm) with an ever-decreasing freedom of movement. Surface tension, together with the large size of flakes, ensures that the flakes will align more or less parallel to the substrate.

However, the state of 'almost parallel' alignment can still cover great differences in visual appearance and metallic systems where all the flakes are aligned nearly perfectly parallel to the substrate will exhibit a far brighter appearance than systems containing many flakes at angles of, say, up to 20° to the surface. This is particularly true of very bright low solids basecoats. In addition dry film thickness is of the same order as, or even frequently less than, the flake length. This constrains flake mobility very effectively, particularly as the dry film thickness in the final stage of the drying process involves a very high degree of shrinkage.

All these factors combine to give low solids basecoats such an attractive, stylistic appeal.

Undercoats

In general terms basecoat/clear technology requires polyester/PU surfacers to maximize performance, particularly in the resistance to delamination. As described in Section 10.4 epoxy products tend to 'chalk' at the interface between undercoat and basecoat due to UV radiation. In the past this has led to breakdown in the field of early basecoat/clear systems.

Nowadays the use of polyester surfacers, combined with UV absorbers and light stabilizers in the clearcoat, has resolved this problem. In fact the performance of basecoat systems both on test at Florida and in service is exceptional (see below).

Performance/durability

Durability testing at Florida (5° south) is a universally accepted measure of exterior durability in the automotive industry. Florida is very suitable for such testing because it is high in ultraviolet and humidity.

Early clearcoats were based on alkyds but failed due to cracking (UV degradation) within 12 months. Modern thermosetting acrylic clearcoats have quite outstanding durability — a minimum of three to five years can be confidently expected. Such high levels of durability are unique to basecoat/clear technology, since the normal accepted standard is two years at Florida free from defects.

It is also possible to use a wider range of pigments in basecoat technology than thermosetting or thermoplastic acrylics; not only because of the pronounced face/flip contract but also because it is feasible to use a much wider range of organic pigments than hitherto without sacrificing colour stability on exposure.

Organic pigments have high transparency but poor coverage and often poor durability. However, since there are no gloss constraints with basecoats, high levels of such pigments can be used at low film thicknesses (15 μm). The clearcoat provides gloss and offers the necessary ultraviolet protection.

10.7.4.2   Water-borne

During the past 20 years car manufacturers have been under increasing pressure to reduce, and possibly eliminate, volatile organic solvents present in such coatings.

This pressure has come in two forms:

- From environmental and health and safety agencies, anxious to preserve the environment.
- As an economic pressure to maximize the efficiency of production processes to meet competitive market conditions.

The environmental and ecological pressures were, and still are, particularly strong in North America. Europe likewise has introduced stricter controls on solvent emissions over the past 20 years.

In terms of the stoving operation the use of mechanical, thermal, and chemical techniques, including after-burners, scrubbers, and carbon absorption units, has improved the situation. However, there still remains the vast amount of solvent-laden air from spraybooths to cope with. Figure 10.7 shows the level of solvent emission for a typical low solids solvent-borne basecoat system. Clearly, the basecoat contributes the highest level of solvent emission and if you consider the introduction of more and more stricter controls over such emissions then a technology needed to be developed to resolve this problem.

Water-borne systems were considered to be the best solution to the problem particularly when compared with the alternatives (see Section 10.13). Development began in the late 1970s with the following objectives:

- To achieve the highest standards of appearance.
- Current ease of application.
- Use of existing spraybooth/processing conditions.
- Greatly reduced solvent emission levels.
- Use of current and future clearcoat technologies.

Once the objectives were clearly defined the following formulating principles were used in the development programme:

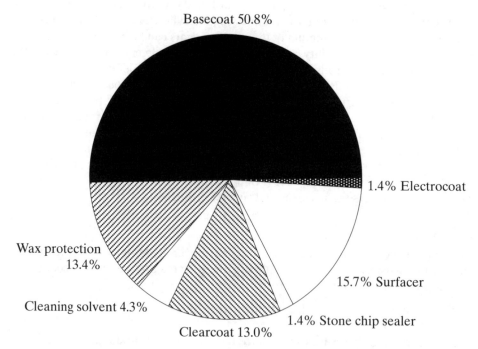

Fig. 10.7 — The basecoat clear process (total emissions).

- Use of aqueous polymeric dispersions.
- Water release aided by low basecoat film thickness (12 μm) and a high pigment content.
- Novel rheological control introduced via the polymer system.

The development and introduction of this technology has been quite rapid and its use continues to grow throughout the world (detailed in the introduction to this section).

The main features of water-borne basecoats are summarized below and its effect on reducing emissions is shown in Fig. 10.8:

- Film properties at least comparable to existing standards.
- Maintains the highest appearance standards (Fig. 10.6).
- Excellent durability under a wide range of clearcoats.
- Similar processing requirements to existing products.
- Adaptable to automatic (robotic) and electrostatic application.
- Low solvent emission.
- Improved usability and user environment.

Processing
The general application and processing of water-borne basecoats has been less of a problem than would have been anticipated from the experience with early water-borne technologies. It is also being broadened into solid colour basecoat technology because of the need for process rationalization and the technology's outstanding performance.

Characteristics
Detailed plant and processing requirements may be summarized as follows:

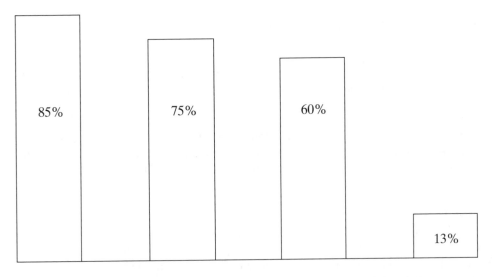

Fig. 10.8 — Emissions of metallic basecoats, — % organic solvents (100% = wet paint).

- Force dry (blow off, 2 min at 40–70 °C) required before clearcoat application.
- Operating humidity needs to be controlled: 40–80%.
- Stainless steel paint circulation equipment.
- Isolated material feed needed for electrostatic application.

In summary, water-borne basecoat technology is not only highly innovative but it is unique. This uniqueness is twofold in that it has maintained and improved topcoat appearance and performance and also, with the aid of rotory atomizers, made a significant impact on reducing solvent emissions.

Its future is soundly based. It is expanding into solid colour basecoats and, with newer clearcoat technologies in the pipeline, it is likely to become a permanent feature of the car painting process for many years to come.

10.7.5   Pigmentation of automotive topcoats

The choice of pigmentation for any particular colour must be considered in the context of the requirements of the market in which the product is to be used. In this respect the motor assembly market is probably the most demanding of all, requiring class 'A' matches at all times, excellent durability under severe conditions, and good opacity at minimum cost. Thus, for most types of topcoat many of the lower-cost pigments with inferior lightfastness cannot be considered, and the range of colours which can be used is limited in comparison to other markets, e.g. decorative paints.

10.7.5.1   Solid colours

Pigmentation practice for all types of automotive technology is very similar. As a general rule (apart from some relatively minor differences in performance) alkyds/polyesters and thermosetting acrylics can be treated as identical. However, in acrylic lacquers whites are more prone to chalking, reds and violets tend to fade more in pale shades, and colour retention of some phthalocyanine blues is not as good as in the other technologies. The following general constraints apply.

Durability
Many organic pigments are completely excluded or can be used only at certain concentrations or when combined with other pigments of excellent durability.

Opacity/gloss
Very clean, bright, pure colours are often non-feasible because the pigment loading required to achieve opacity reduces gloss to an unacceptably low figure. Inorganic pigments should always be used where possible because of their opaque nature but stronger organics such as blues, greens, and violet are also helpful.

Cost
Most organic pigments are very expensive and in many cases of relatively poor tinting strength. The most notable exceptions are phthalocyanine blues and greens which are less costly than most and, because of good tinting strength, can be used in lower concentrations.

Bleed
Many organic reds and yellows are excluded because of a tendency to bleed.

Metamerism
This describes the phenomenon observed when two samples are similar in colour under one set of lighting conditions but different under another (commonly daylight and a tungsten filament lamp). This usually occurs where pigments of a different type to those in the original 'master pattern' have been used. Pigments especially prone to this are iron oxides and phthalocyanines.

Use of lead chromate pigments
Automotive topcoats are heavy users of lead chromate/sulphate/molybdate pigments because of their brightness of colour, good tinting strength, and low cost. Their limitations are, however, considerable:

- On exposure to sunlight lead chromate pigments darken. This is particularly obvious in the bright yellow shade and varies with seasonal changes at the exposure site and the degree of pollution in industrial atmospheres.
- Lead chromates are also susceptible to attack by dilute acids, the colour being bleached out. This can show as white spots or as an overall effect.
- Acid-catalysed repair films will also darken more readily because of the influence of the acid on the surface coating of the pigment.

Environmental pollution problems are leading to more and more demands for automotive paints to be lead/chromate-free. Unfortunately, many of the alternatives available for production of clean, bright colours are very costly, difficult to achieve opacity/gloss to meet specifications and, particularly with yellows, of poor durability. However, a reasonable range of lead-free reds can now be produced although there are still many problems to be overcome before the totally lead-free situation is reached to the satisfaction of all concerned, i.e. colour stylist, paint manufacturer, end user and environmentalist.

10.7.5.2 'Single coat' metallics
General pigmentation practice is the same for both thermoplastic and thermosetting acrylics, although some organic pigments, particularly certain yellows and phthalocyanines, perform slightly better in the former technology. The following constraints apply:

Durability
There are even more constraints on organic pigments than in solid colours, particularly with pale shades. These are often borderline for feasibility unless performance is boosted by the use of a UV absorber or the inclusion of transparent iron oxide pigment. This is normally the only inorganic pigment used in metallics. Both red and yellow shades are available, can be prepared in highly transparent form, are of excellent durability (apparently having some UV-absorbing property), and are cheap. Their only disadvantage is weakness/dullness of colour.

Opacity/gloss
In any metallic finish aluminium flake contributes most of the opacity, since transparency of the tinter is essential for tone metallic appearance (flip tone). Deep, clean metallic colours (low aluminium level) are therefore usually more difficult to formulate than pale shades unless some black can be included. The most difficult areas of all are the clean, bright red shades where the majority of durable pigments avail-

able have poor tinting strength and the aluminium level must be kept low to avoid producing a greyish pink tone.

Cost
The cheapest durable pigments available for metallics are the transparent iron oxides which account for the popularity of metallic gold, beige, brown, and bronze colours. Phthalocyanine blues/greens also feature prominently because of their high tinting strength and relatively low cost.

Colour matching
Metallics must match the master pattern at all angles of viewing. Colour achieved from a metallic is very considerably influenced by conditions during film formation. The paint must be formulated and applied such that during drying the aluminium flakes align parallel to the substrate surface to ensure maximum brightness and degree of flip. Obviously any variation from one sprayout to another will result in a different colour and for this reason automatic application is used wherever possible, particularly when colour matching. Spraybooth conditions, i.e. temperature, humidity, and air-movement, also need to be closely controlled.

Choice of aluminium flake
Various grades are available differing principally in particle size. As a general rule, as size increases sparkle increases, colour becomes brighter (less grey), and flip tone increases. On the debit side, gloss, image clarity, opacity, and tinting strength diminish. Therefore the usual compromise is between adequate opacity/gloss and degree of brightness. The medium/fine grades are most popular and least likely to give problems. Very coarse flakes should be avoided whenever possible.

10.7.5.3   Basecoat/clear metallics
General pigmentation practice is the same for both types. Constraints are as follows:

Opacity
This is normally expected to be in the range of 10–20 μm. As with 'single coat', most of the opacity is obtained from aluminium, therefore the same basic constraints apply, particularly with the low solids polyester type where a very high degree of flip is expected. Again, bright reds are the most difficult. The big advantage over 'single coat' is that no consideration need be given to gloss, therefore much lighter colours can be achieved because of the higher aluminium levels permitted.

Cost
The same constraints as for single coat apply.

Colour matching/durability
A much wider range of pigments is available for use in basecoat systems because of the protection afforded by the use of UV absorbers and light stabilizers in the clearcoat (which serve a dual purpose in preventing breakdown of clearcoat due to UV and also protect pigments in the basecoat). Many of the organic pigments prone to fade on exposure can therefore be used in base/clear systems, and the range of colours available is much greater. This is particularly true of the bright pastel shades where a combination of wider pigment range and better aluminium laydown produces some very attractive colours.

The same constraints on application parameters apply to basecoats as to single coat.

Choice of aluminium flake
As with 'single coat', medium-fine flake is preferred. Coarse flakes cannot be used. In addition to poor opacity, with the very low (10–20 μm) film thickness of basecoat films applied, the flakes size (up to 30 μm) is such that if application is not perfect, flakes may protrude through the surface of the basecoat, giving a seedy appearance and the danger of film breakdown.

In addition to this, coarse flakes can cause safety problems on electrostatic application (discharge to earth through the pipework of the circulating system). Table 10.5 provides a comparison of the basic processing properties associated with automotive topcoats.

10.8   In-factory repairs

After the initial paint process the car body is trimmed and final assembly completed. During this latter part of the process minor paint damage can often occur which will require repairing; the quality of such repairs must of need align to the quality of the original paint finish.

Final repairs can be effected in two ways:

- Panel repairs, e.g. doors, bonnets, etc.
- Spot repairs.

The type of repair possible is directly related to the type of technology used for the original finish.

10.8.1   Thermosetting finishes (panel repairs)

Technologies include alkyds, thermosetting acrylics and basecoat/clear (solvent-borne) systems. The original enamel, or the clearcoat in the case of basecoat/clear, is catalysed by a small addition of an acid catalyst and stoved 30 minutes at 90 °C/10 minutes at 100 °C. Higher temperatures are not feasible because of possible damage to trim, plastics, etc.

The use of two-pack '2K' (polyurethane and acrylic) materials for in-factory repairs has grown rapidly. Such products have good durability and appearance needing only minimal temperatures, 15 minutes at 80 °C, to cure. However, toxicity problems due to isocyanate catalysts require particular respiratory precautions. '2K' clearcoats are mandatory for use with water-borne basecoats since performance (humidity resistance) is unacceptable with acid-catalysed thermosetting acrylic clears.

Another recent trend in basecoat technology is that there is much more use of spot repair utilizing techniques developed in the refinish market, i.e. 'fade out' thinners, etc., even for in-line repairs.

10.8.2   Thermoplastic acrylic lacquers (spot repair)

The very nature of these products facilitates self-repair. The technique of spot repair is mainly employed, although if the damage is extensive a panel repair can be done

Table 10.5 — Automotive topcoats: comparison of basic processing properties

Type / property	Alkyd/melamine or polyester/melamine	Thermosetting acrylic/NAD	Thermoplastic acrylic	Basecoat/clear (solvent-borne)	Basecoat/clear (water-borne)
Solid colours/ metallics	Solid colours	Solid colours and metallics	Solid colours and metallics	Solid colours and metallics	Solid colours and metallics
Appearance	High gloss 85% at 20°	Slightly lower gloss. Microtexture 80% at 20°	High gloss 85% at 20°	High gloss 90% at 20°	High gloss 90% at 20°
Solids at Spray (weight)	Alkyd: 50–60% Polyester: 50–60%	30–35%	12–18%	Basecoats (polyester) Metallics 12–18% Solid colours 25–40% Clearcoat TSA: 46–50% 2K: 60–70%	Basecoats Metallics: 15–20% Solid colours: 25–40% Clearcost TSA: 46–50% 2K: 60–70%
No. of coats (air spray)	Two	Two/three	Three/four	Two basecoat One/two clearcoat	Two basecoat One/two clearcoat
Sensitivity to undercoat	Tolerant (poor adhesion to electrocoat)	Reasonably tolerant (good adhesion to electrocoat). Polyester/PU surfacer recommended	Specific-high PVC (~55%) surfacer required. Sealer optional (poor adhesion to Electrocoat)	Polyester/PU Surfacer recommended	Polyester/PU Surfacer recommended
Stoving temperature	20 min at 130°C	20 min at 130°C	30 min at 155°C (reflow)	20 min at 130–150°C	20 min at 130–150°C
Polishability	Poor	Good	Excellent	Good	Good
Repair	Panel repair (acid-catalysed)	Panel repair (acid-catalysed)	Spot (self) repair	Limited spot repair TSA clearcoat — acid-catalysed or 2K clearcoat	Limited spot repair 2K clearcoat

just as easily. Their excellent polishability also allows minor imperfections, such as dirt or dry spray, to be conveniently removed by polishing.

This use of the same product, unmodified, makes for considerable process flexibility and minimizes any colour matching problems. A minimum stoving of only 15 minutes at 80 °C is required, followed by polishing to maximize gloss.

10.9   Painting of plastic body components

Since the 1970s there has been considerable growth in the utilization of plastic materials in motor vehicle construction. These products are now finding more and more use as exterior components such as bumpers, 'spoilers', 'wrap-arounds', and ventilation grills. Apart from advantage in weight-saving there are benefits in styling, resistance to minor damage, and corrosion resistance. The use of plastics for major body parts, such as door panels or bonnets, is seen as the next logical step forward.

The use of plastic does, however, bring particular painting problems such as adhesion, stoving limitations, solvent sensitivity, and matching for appearance. Currently, the autobody painting process is designed to coat a mild steel monocoque construction, and such processes involve high temperatures such as 165 °C for primers and 130 °C for finishes. As more plastics are used, either they will have to align with current practice or new systems will have to be developed.

There are many different types of plastics with a variety of different properties which affect their painting. The principal ones are as follows.

10.9.1   Sheet moulded compound (SMC) and dough moulded compound (DMC)
The main advantages over other plastics are its high flexural modulus (stiff enough for horizontals), its high distortion temperature which will withstand the 180 °C electrocoating schedule, and its high solvent resistance.

Disadvantages are a surface profile prone to waviness and 'outgassing' on stoving, causing topcoats to bubble. Normally, it is usual to seal the surface either by an in-mould coating or with a spray-applied polyurethane.

10.9.2   Polyurethane: PU RIM and PU RRIM
Reaction injection moulded (RIM) polyurethane and reinforced reaction injection moulded (RRIM) polyurethane have the following main advantages: a very wide range of moduli (from rigid to rubber-like) and toughness. Disadvantages are variable porosity which is sometimes at, or very near, the surface, dimensional instability (the material 'grows' when overstoved particularly when glass reinforced), and solvent sensitivity, which is aggravated by the presence of glass fibre.

10.9.3   Injection moulded plastics
These materials include polycarbonate, ABS (acrylonitrile–butadiene–styrene), polyamide (mainly glass reinforced), and polypropylene (modified with EPDM, ethylene propylene diene methylene). Their general advantages are toughness and strength, good surface quality, and a wide range of flexibility from rigid to ductile. Their general disadvantages are that some are excessively brittle at lower temper-

atures, heat distortion temperature (HDT) is fairly low in many cases, and the amorphous types are notch-sensitive.

10.9.4   Painting problems

10.9.4.1   Adhesion
It is difficult to get consistently adequate adhesion to some modified polypropylenes and special adhesion promoting primers are required. With some other plastics adhesion is no problem.

10.9.4.2   Heat distortion
Plastic parts will warp or sag at elevated temperatures. The temperature to which a part can be taken without distorting depends not only on the polymer but also on fillers, reinforcement, shape, size, and degree of mechanical support.

Thus for a given polymer there is an absolute upper limit at which it begins to melt or decompose, but well below this temperature there will be a practical heat distortion temperature for a particular moulding and mounting.

Depending on the heat distortion temperature the plastic may have to be painted entirely off-line (e.g. most PU RIM) or fitted after the electropainting oven (e.g. some PBT, glass reinforced polypropylene, most polyamides) or can be fitted in the 'body in white' (e.g. some polyamides, SMC, and related materials).

Where painting is off-line, colour matching to the body is difficult.

10.9.4.3   Surface texture
For true appearance matching of different materials meeting in the same plane, it is virtually essential to use a common undercoat.

10.9.4.4   Solvent sensitivity
Some plastics are affected excessively by common paint solvents, causing the surface to craze and degrading the mechanical properties of the component. On the credit side a mild degree of solvent attack can be beneficial to adhesion.

10.9.4.5   Degradation of mechanical properties
If the paint film fails by cracking when the painted part is impacted or flexed the effect in some cases is to induce failure by cracking of the plastic substrate. Thus an unsuitable paint system will weaken the part.

10.9.5   Paint processes and products

10.9.5.1   On-line
The conventional painting process, using a monocoque construction, has already been well described. Of the possible plastics only SMC-related materials and some grades of polyamide can withstand electropaint stoving schedules (165–180 °C) without distorting, and also accept standard body finishing systems.

A typical process for an SMC component is to use either an in-mould coating or a sprayed polyurethane coating to 'bridge' or seal the surface. Subsequently the component is fitted to the 'body in white', passed through cleaning, phosphating, and electropainting (where it remains relatively unaffected) before receiving the standard spray surfacer and topcoats.

10.9.5.2 Off-line

In this process the plastic part is painted off-line, maybe at a moulders or painting subcontractor and fixed to the car-body after the final paint oven. This practice enables the paint system and process to be designed to suit the plastic, but if the part is to be in body colour it makes colour matching, particularly in metallics, difficult.

PU RIM is always painted off-line. If the parts are fully jigged they can be painted with a one-pack flexible polyurethane/melamine formaldehyde finish and stoved at 120 °C. This is standard practice in the USA but in Europe two-pack flexible clears and non-metallics are used for RIM and ductile thermoplastics. The two-pack method gives lower stoving temperatures (100 °C), more formulation scope, and a shorter painting process: problems of toxicity are diminishing as automatic processes are introduced.

Colour matching in off-line painting may be more or less critical according to the shape and position of the component. A styling break or a change of plane where the off-line and on-line painted parts meet will conceal minor differences. With basecoat/clear metallics use of the same basecoat on plastics and metal is a major help, alternative clears (normal and two-component) complete the finishing process. In the longer term a common clear such as 2K is a good objective.

10.9.6 'Part-way' down paint line

A final method is to fit the particular plastic component between the electropaint oven and the colour spraying station, provided the paint system is suitable for the part. This procedure is commonly used for small parts such as ventilation grills in polyamide. Some other reinforced plastics will withstand normal topcoat schedules and also retain satisfactory impact performance when painted in a standard system.

10.10 Spray application

The introduction of the automobile mass production line, in combination with the development of new synthetic paint technology, was the major factor in the adoption of spray application. The basic principle of spray painting is to atomize the liquid paint into a fine spray and subsequently direct this spray on to the car body. Originally, compressed air was the atomizing medium but other techniques such as electrostatic spray are being more and more employed because of improved transfer efficiency.

This method of application, however, generates overspray and provision must be made to carry off this overspray and exhaust it to the atmosphere or, in some instances, retain it through a recovery system. Nowadays water washed spraybooths are used, being efficient and free from fire hazard. In this type of booth the exhaust is drawn through water which 'carries' the overspray into a tank for effluent disposal.

The main basic properties required for spray application are as follows:

- To offer a fast, flexible, reliable and robust method of application which will achieve the necessary film thickness in the limited time available.
- To provide a smooth film with good flow free from defects such as mottle (orange peel), 'popping' (air entrapment), sags, and craters. Surface-active agents are often used to good effect to ease problems such as cratering.
- In the spraying process to produce a wet film capable of absorbing its own overspray.
- With metallics, to produce shear-free films.
- To maximize material utilization/transfer efficiency without impairing film appearance and processing.

For many years conventional air spray was the most favoured application. Nevertheless, although offering an acceptable level of film appearance and meeting most of the criteria above, it is inefficient and wasteful in terms of paint usage.

To improve transfer efficiency newer methods such as air-assisted electrostatic and full electrostatic have been developed. These systems, particularly the latter, are now mostly fully automated and constitute the main methods of application in modern high volume car plants.

Sometimes a mixture of air assisted (for interiors) and automatic airspray (for exteriors) is used as a compromise. Alternatively, a mixture of air assisted and full electrostatic is incorporated. Such compromises are often necessary because of facility limitations and cost implications.

10.10.1   Air spray
The principal method used in mass-production is the 'pressure feed' system although suction feed techniques are used for small-scale operations such as minor repairs and laboratory work.

In pressure feed, thinned paint (suitably stored and circulated) is fed by pressure through paint lines to the spray gun and leaves the gun through a needle valve — the amount of paint being controlled by a trigger and the pressure applied. The fine stream of paint leaving the gun is atomized by jets of compressed air flowing out of openings in a removable air cap at the head of the gun. The jets can be directed to produce an even spray pattern. This is diagrammatically represented in Fig. 10.9.

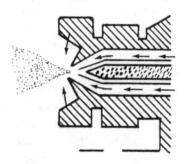

Fig. 10.9 — Spray air-cap.

Because of its versatility and speed the spray gun has dominated since its inception. In spite of significant paint losses caused by the fact that not all the atomized paint is deposited it remains widely used for surfacer, sealer, and topcoat, including metallic application.

10.10.1.1   Spray losses

As already stated the spray gun is essentially a device for atomizing liquid with compressed air and projecting this cloud of fine droplets so formed onto the surface of the car body. When the fan of paint droplets sprays from the gun, a part of the fan passes to the side of the unit as it hits the surface being coated. This loss is acceptable as edges need to be well covered, but a portion of the atomized paint is pulled away by the compressed air current deflected by the surface, and the paint appears to rebound or 'bounce-back'.

At that particular moment the paint droplets are subjected to two actions. One is caused by their forward velocity which tends to push them along their course towards the object; the other is the air current which will deflect them from their trajectory and drag them in a different direction. Obviously this latter action will be stronger if the droplets are very fine and the air current very fast. A coarse spray needs less air pressure and the rebound can be reduced, but the quality of the finish will be poor as a result.

A compromise has to be reached between the air pressure necessary to produce the desired quality of the coating and an acceptable degree of 'bounce-back' and overspray. Other factors that affect the air pressure used are the surface tension and the viscosity of the paint. The higher these are, the greater the energy required to ensure atomization and the more powerful the air jet.

The actual efficiency of an air spray gun under normal production conditions is low. Only 40% of the paint reaches, and remains, on the car body. Of the 60% lost, 20% is due to 'bounce-back' and 40% to overspray.

The need to improve efficiency is obvious. Automatic machines are beneficial but there have been other developments in recent years which have offered very significant paint savings. The means of reducing losses and associated developments are:

- Air spray gun
  - lower pressure (already described)
  - automatic spray
  - low pressure hot spray
  - air-assisted electrostatic
- Electrostatic spray
  - air-assisted
  - bells
  - discs

The newer techniques, which are described later, are often fully automated and offer a range of benefits such as personnel savings and consistency of performance apart from marked improvements in transfer efficiency.

10.10.2   Automatic spray

A common method to improve spray efficiency is to use automatic machines with some 'hand reinforcing' in difficult recessed areas and interiors. It has the added advantage of reducing labour.

Automatic machines are either attached to a fixed base or to a moveable carriage and controlled by electronics to prevent the equipment spraying when the car body is not in the correct proximity. The gun carriage normally reciprocates transversely for painting vertical and horizontal surfaces of the carbody. A typical installation is shown in (Plate 10.1).

This method is ideal for the exterior painting of vehicles with interiors being sprayed by hand. In the past systems did exist for spraying interiors automatically. This was achieved by means of extension nozzles which were actuated by pistons and thereby inserted into interiors and then withdrawn. It is suitable for, say, the interior of vans but it is not ideal and problematical compared to modern procedures.

10.10.3   Low-pressure hot spray

This technique has found some use in the application of primer surfacers, but it is rarely used in modern installations and is probably now more or less obsolete in the automotive sector.

Plate 10.1 — Automatic spray application (Courtesy De Vilbiss Ransberg).

Heating of the paint to 60–80 °C reduces viscosity and surface tension, making atomization easier. Consequently, lower pressures can be used and higher solids at application reduces 'bounce' or rebound losses. However, overspray losses can increase because of the wastage of high solids and can limit solvent selection. Evaporation losses can also be a problem.

10.10.4   Airless spray
This procedure is often used for applying anti-chip coatings to sills, lower sections, etc. Atomization is produced by a combination of pressure and heat. The high pressure forces the paint through the gun nozzle at a greater velocity than the critical velocity of the liquid coating breaking it into small droplets (i.e. atomizes the stone-chip primer).

The anti-chip coating can be heated to reduce viscosity to improve atomization, care being taken to control temperature thermostatically. It is most likely applied automatically with the guns located in fixed positions.

10.10.5   Electrostatic spray
The principle of electrostatic spray is simple. If paint particles are atomized in an electric field they will become charged and drawn towards the article to be painted, which is usually at earth potential. There are many different types of electrostatic spray systems but the most widely used in the motor industry are rotating bells or discs, covering the whole range of undercoats and topcoats.

Figure 10.10 represents a typical system. The paint is pumped to the atomizers from pressure feed containers via a hollow drive shaft. The rotation of the atomizers spins the paint to the periphery where it is partially atomized by centrifugal force but mainly by the electrostatic field.

As the paint particles leave the atomizer, under the attraction of the electrostatic field, they are drawn to the unit being painted. Any particles which pass are attracted to all sides giving electrostatic its very high efficiency of paint utilization. Generally adopted parameters are rotation speeds of 30,000–40,000 rpm and high voltages in the region of 90–110 KV for maximum efficiency. The reduced overspray also means less efficient and expensive spraybooths with reduced effluent. This system also lends itself to full automation with the attendant benefits.

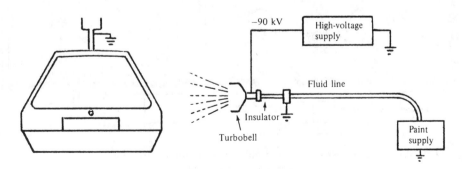

Fig. 10.10 — Electrostatic spray: basic system.

There are basically three main bell sizes currently in use, 75, 68 and a 50 mm mini-bell. Rotation speeds vary between 25,000 rpm for the larger 75 mm bell to 48,000 rpm for the small 50 mm type.

The small bell (designated mini-bell) does have other benefits:

- More easily mountable for robotic use.
- Capable of reciprocating at different speeds.
- Amplitude is programmable.

A modern electrostatic application system is shown in Figs. 10.11 and 10.12. However, in some instances certain recessed and other areas are difficult to paint with this method since paint is attracted to the nearest part of the complex shaped car body and does not have sufficient velocity to penetrate further. Air-assisted electrostatic guns can overcome this weakness. These use compressed air and electrostatic techniques for atomization.

Air-assisted equipment has an insulating barrel having an earthed handle with a high voltage electrode in the air cap which is usually charged from 30 to 60 KV. The discharge current flowing from the high voltage electrode creates a region, adjacent to the atomization zone, rich in unipolar ions that attach themselves to, and charge, the paint spray particles.

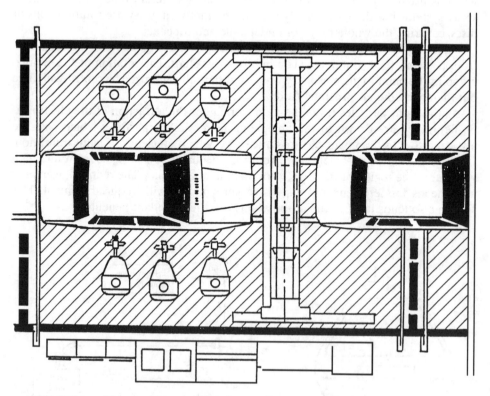

Fig. 10.11 — Electrostatic spray installation (automatic turbo bells) (Courtesy of De Vilbiss Ransberg).

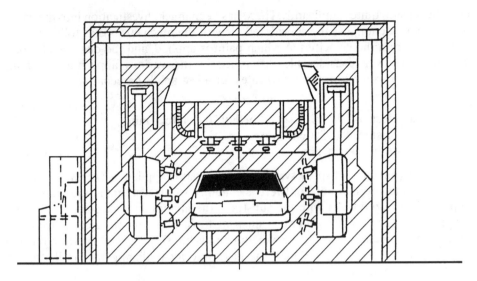

Fig. 10.12 — Electrostatic spray installation (automatic turbo bells) — Cross-section
(Courtesy of De Vilbiss Ransberg).

In the vicinity of the spray gun the inertia of the aerodynamic forces usually dominate, but as the paint particles get nearer to the earthed car body the forces on the
charged particles (owing to the electrostatic field between the charging electrode of
the spray gun and the unit) become more significant. These forces drive the spray
particles towards the workpiece.

Maximum paint savings are generally obtained by maintaining the charging
voltage as high as possible, giving a high depositing field strength between the gun
and the workpiece and by keeping the spray velocity in the vicinity of the article
being sprayed as low as possible to give good atomization and flow of the 'deposited'
film.

Air-assisted guns can be used either manually or automatically. They offer a measurable compromise between the less efficient air spray and the highly efficient electrostatic system, and compared to the latter process give better film uniformity and
smoothness.

In general terms the advantages of electrostatic spray may be summarized as
follows:

*   Speed of production.
*   Uniformity of production — free from the 'human element'.
*   Economizes on labour — highly automated.
*   Versatile and flexible.
*   High transfer efficiency — leading to large paint savings/less effluent.

Electrostatic spray — metallic appearance
The controlling factors in metallic appearance have been clearly stated in previous
sections. In practice, electrostatic spray diminishes the high flip (tonal contrast) of

metallic finishes quite significantly. There has been much investigation into the cause of this phenomenon and the considered view of the mechanism is as follows.

In comparing the velocities at which paint is applied, and the degrees of atomization, there is a considerable difference in the various application systems employed. For example, typical flow rates for a low solids basecoat are:

	Atomization
Full electrostatic (high speed bells): 200–300 cm^3 min^{-1}	Fair
Air-assisted: 450–500 cm^3 min^{-1}	Good
Air spray: 800 cm^3 min^{-1}	Very good

In an air (pneumatic) spray system the basecoat is applied at very high velocity and is extremely well atomized. There is evidence to suggest that at the moment of impact of a spray droplet with the substrate, the aluminium flakes are thrown flat against the surface by a considerable shearing force. Subsequently a certain amount of reorientation takes place as the wet film builds but shrinkage and surface tension effects quickly take over and 'pull' the flake into ideal alignment (parallel to the substrate) for optimum effect.

The lower velocities and relatively poorer atomization of electrostatic systems means that, on impact, shearing forces are lower and the flakes are not forced totally into parallel alignment and tend to be more randomly orientated. The shrinkage and surface tension effects improve the orientation but, because of the initial starting point, the final film does not align in the perfect manner of hand spray. It is for this reason that metallic basecoats are applied by a 'mix' of application methods to optimize appearance and efficiency and to minimize repair colour matching, i.e. two coats — first coat by electrostatic spray followed by a second coat using air spray (either hand or automatic).

There is, however, a new development in electrostatic systems to improve aluminium orientation. This uses a hybrid bell with very high 'shroud air'. This gives a compromise between the alignment of the aluminium flake and transfer efficiency.

Resistivity
An extremely important feature in the spraying of a product electrostatically is its electrical resistance. This is easily measured by one of the various commercially available meters. If the resistance is too high, the paint will not accept the charge and is not sufficiently attracted to the car body, i.e. 'wrap around' is poor. On the other hand, if the resistance is too low, as is the case with water-borne systems, the high voltage at the head of the gun is likely to be conducted back to earth via the paint-feed pipeline and the main paint mixer which is more frequently earthed.

The correct resistance (i.e. specification) for a solvent-borne coating is regularly established by a system of trial and error for an individual plant making addition of polar solvents or special additives, e.g. Ethoquad C25® to lower resistance and increase conductivity.

(Refer to Tables 10.6, 10.7, and 10.9 for typical resistivity ranges.)

10.10.6 'Interior' application (electrostatic spray)
Robotics using electrostatic application for 'interior' application are increasingly to be found in Europe and North America. However, they are very expensive to install

and to maintain and are 'out of date' very quickly. They also require meticulous maintenance to overcome the inherent dirty application caused by overspray from the use of high fluid rates and air pressures.

10.10.7 Electrostatic application of water-borne automotive coatings

At the present time, the most favoured method of application is automatic electro-static spray using high speed rotary atomizers. This combines the low polluting nature of the water-borne products with a high transfer efficiency, thereby maxi-mizing emission reduction.

However, by their very nature water-borne coatings are highly conductive. So as to ensure that high tension does not leak through the paint circulating lines special systems have been developed to isolate the 'charge':

- Circulating systems designed specifically to insulate the paint from the internally charged bell or atomizer.
- Externally charged or 'crown of thorns' bells.

10.10.7.1 General plant design features
All metal in contact with water-borne technology must be stainless steel, including spraybooth panelling. Nevertheless, certain plastics such as polyamide '11' and '12' are suitable for low diameter pipework.

10.10.7.2 Paint circulating system for electrical insulation
A typical system designed to isolate the main paint circulation equipment from the electrostatic atomizers is shown in Fig. 10.13. It is principally used for water-

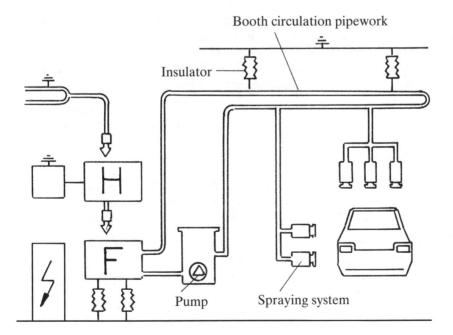

Fig. 10.13 — System for electrical insulation.

borne surfacers where internally charged guns are adopted. In the case of other water-borne products, e.g. basecoats, its use is restricted to only a few individual producers.

The main paint circulating system is isolated by using separate header and feed thanks 'H' and 'F'. The feed tank 'F', the circulation pipework and atomizers are insulated from earth and connected to the high tension equipment. Tank 'H' is switchable by a high tension potential or to earth. It is filled with paint from the main mixer through earthed circulating pipework. During this filling procedure the header tank is at earth potential.

The lower tank is replenished automatically from the header tank during gaps between carbodies at designated intervals or, sometimes, at the end of a shift.

10.10.7.3   Externally charged atomizers

In the application of water-borne (and conventional) basecoats the externally charged or 'crown of thorns' electrostatic bell is preferred by most car manufacturers. Although this system has marginally lower transfer efficiency (approximately 3–5%) there is no need to isolate the paint.

The original so-called 'crown of thorns' bell had an array of 32 electrodes at the gun head, recently modified to 6 electrodes.

Most production lines are now equipped with externally charged bells (first coat of basecoat) followed by a pneumatically applied second coat, most often by automatic reciprocator. Solid colour basecoats are applied in one coat. Hand reinforcing after bell application is very seldom necessary.

10.10.8   Application efficiency — practical considerations and processes

In terms of absolute efficiency for any particular type of equipment it is difficult to be specific. There is considerable variation from plant to plant (whether the system is manual or automatic, available spraybooth space, and the differing shapes of the unit being sprayed) and practical figures can differ significantly from the ideal.

However, some mean figures are quoted below based on data available from various installations. They are intended only to give the reader a realistic indication of the transfer efficiencies of systems currently in use.

	Variance ± 5%
High voltage electrostatic (high speed bell)	75%
Air-assisted electrostatic	60%
Air spray	40%

In practice the principle of 'mixing' or 'blending' the various methods is often used in mass production. Such systems balance the various strengths and weaknesses of the individual spray methods and can realize the following benefits:

- Maximizes appearance standards at required film thicknesses.
- Maintains optimum level of transfer efficiency.
- Copes with the complex nature of a monocoque unit which comprises internal and external surfaces.
- Provides an optimum application procedure within the time and space available.
- Minimizes labour.

10.10.9 Modern spraybooth design — ventilation modes

The air flow through spraybooths is extremely high and 'once through' ventilation means high energy consumption for heating and humidification. In fact, spraybooths consume approximately 50% of the energy consumption in the painting process. A feasible method of reducing energy demand is to recirculate the air, with or without intermediate solvent separation.

Recirculation of 70–90% of the booth air, and treatment of the bled-off air, can be used in booth zones where increased solvent concentration up to $2–3\,\mathrm{g\,m^{-2}}$ can be expected, i.e. zones with automatic operation, electrostatic bells, etc.

For zones with manual operation the air has to be cleaned from solvents before it is reintroduced into the booth. The cleaning is preferably done by rotor concentrators.

There are, however, limiting factors with direct recirculation:

- Solvent concentration must not exceed 25% of the lower explosion limit (LEL).
- Finishing quality can be influenced by solvent concentration, e.g. evaporation, flow, sagging, and 'solvent popping'.
- Extreme accuracy is required in booth construction, particularly the ducting since 'outwards' leaking must be avoided and 'inwards' leaking must be controlled.

10.10.9.1 Preconditioning the air

When recirculating air, or when taking it through a concentrator with absorbent, it is necessary to recondition the air necessitating the following:

- To clean the air further of paint overspray.
- To dehumidify and reheat the air to be recirculated to the booth.
- To adjust the relative humidity of the air — generally by heating it 6–8 °C before it enters an activated carbon or zeolite absorbent filter.
- To 'smooth' solvent concentration variations in the air.

10.10.9.2 Concentrators

Rotors with absorbents are used as concentrators for solvent-laden air. There are two types, disc or cylinder, with activated carbon or zeolite material on a carrying structure often with a honeycomb geometry. The rotor turns slowly, 0.5–4.0 revolutions per hour.

The air to be treated passes through typically 75% of the face area where the solvents in the air are absorbed. In the remaining part of the rotor a small amount of heated air passes in the opposite direction and the entrapped solvents are desorbed.

The dissolved air, with a high solvent concentration (often ten times the concentration of the process air), is then fed either to an incinerator for combustion, to a fixed bed absorber, or to a condenser for solvent recovery.

10.10.10 Process details: typical application parameters — turbo bells

Tables 10.6–10.9 summarize some general application conditions for a number of different products, solvent-borne and water-borne, using high speed rotary atomizers under mass-production conditions.

Table 10.6 — Typical application parameters (turbo bells): solvent-borne polyester/PU surfacer

Application viscosity (at 25 °C)	19–40 s DIN4
Resistivity	650–1000 kΩ
Bell rotation speed	20–25 000 revs/min
Shrouded air 0.6–2.0 bar	Spraybooth temperature 23 °C
Voltage 70–80 kV	Spraybooth humidity 60–65%
Nozzle 1.2 mm	Spraybooth air velocity 0.2 m s^{-1}
Typical track speed	7.5 m min^{-1}
Fluid delivery	Varies across body but normally 150–300 cc min^{-1}
Bell body distance	250–300 mm
Number of coats	Two
Flash time (between interior and exterior)	5 min
Spraying sequence	
1st Station	Interior
2nd Station	Exterior
(3rd Station	Back-up)
Flash time (before stoving)	5–7 min
Film thickness	~35–40 μm

Table 10.7 — Typical application parameters (turbo bells): polyester solid colour

Application viscosity (at 25 °C)	15–17 s DIN4
Application solids	~50%
Resistivity	700–1000 kΩ
Bell rotation speed	23 000 revs/min serrated edge bells
Shrouded air 0.6–2.0 bar	Spraybooth temperature 21–24 °C
Voltage 70–90 kV	Spraybooth humidity: normally 60–65%
Nozzle 1.2 mm	Spraybooth air velocity: 0.2 m s^{-1}
Fluid delivery	66–250 cc min^{-1} (depends on location of bells and track speed)
Bell body distance	220–250 mm
Number of coats	One
Flash time (between interior and exterior)	1–5 min
Flash time (before stoving)	4–7 min
Film thickness	45–50 μm

10.11   Stoving procedures

The various types of materials used in autobody painting have already been described. They vary from water-based epoxy technology for electropainting to sophisticated solvent-borne finishes such as alkyds/polyesters and acrylics. The organic solvents involved are complex blends of aliphatic and aromatic hydrocarbons, alcohols, esters, and ketones. These paint materials are either applied by some form of dipping process or spray-applied by one form or another. Subsequent to application all of these products require stoving, for the following reasons:

• To achieve the high level of performance demanded by the motor industry, i.e. to form durable and protective films.

Table 10.8 — Typical application parameters (turbo bells): water-borne basecoat

Application viscosity (typical)	500–3000 mPas @ 0.2 s^{-1} shear
Application solids (range)	Solid colours: 20–35%
	Metallics: 14–20%
Resistivity	Not normally measured
Bell rotation speed	50 mm bell: 45 000 revs/min
	68 mm bell: 28 000 revs/min (flat edged bell)
Shrouded air 1.2–2.5 bar	Spraybooth temperature 21–23 °C
Voltage 70–90 kV	Spraybooth humidity 65% ± 5%
Nozzle 1.2 mm	Spraybooth air velocity 0.2 m s^{-1}
Fluid delivery	70–300 cc min^{-1} depending on zone track speed
Bell body distance	250–350 mm
Number of coats	Solid colours: one
	Metallics: two
Flash time (between coats)	1–2 min typically
Flash time (between coats interior/exterior)	Depending on line construction 1–6 min
Flash time (before stoving)	2–5 min before infrared preheat zone
Film thickness	Solid colours: 15–25 μm
	Metallics: 10–14 μm
Hand reinforcing	Only when 'robots' are used

Table 10.9 — Typical application parameters (turbo bells): clearcoat (thermosetting acrylic)

Application viscosity (@ 25 °C)	26–30 s DIN4
Application solids (range)	40–45%
Resistivity	700–1000 kΩ
Bell rotation speed	23 000 revs/min
Shrouded air 0.5–2.0 bar	Spraybooth temperature 21–23 °C
Voltage 60–80 kV	Spraybooth humidity normally 60–65%
Nozzle 1.2 mm	Spraybooth air velocity 0.2 m s^{-1}
Fluid delivery	80–200 cc min^{-1} per bell
Bell body distance	250–300 mm
Number of coats	Normally two (gives better appearance and 'hold-up' on verticals)
Flash time (between coats)	1 min typically
Flash time (between coats interior/exterior)	1–5 min
Flash time (before stoving)	4–8 min
Film thickness	45–55 μm

- To facilitate processing in the limited time available on a conveyorized production line.
- To control or accelerate solvent release and minimize dirt pick-up.

The stoving operation simply cures the film by effecting certain chemical or crosslinking reactions which form a protective and durable coating. During such film formation solvent and by-products are released which can, and do, provide pollution problems.

Stoving temperatures vary. Modern cathodic electropaints require up to 20 minutes at 180 °C to cure, while at the other end of the scale repair requirements can be as low as 15 minutes at 80 °C. In the initial stages of stoving the temperature must be increased at a controlled rate to prevent possible film defects such as 'solvent popping' or 'boil' (owing to entrapped solvent or occluded air). Once the required temperature is reached it is maintained for a period sufficient for effective cure to occur. These times are typically 8–10 min to heat up of the paint film, and 20 min held at a specified temperature.

10.11.1   Oven technology

There are two types of oven that can be used to stove automotive coatings, although the former are by far most commonly used:

- Convection heating ovens.
- Radiant heating ovens, e.g. infrared.

Convection ovens rely upon air movement and its even distribution over the unit, ideal for the complex shape of a car body. Radiant heating is better suited to regular shaped items but it is often used in conjunction with convection ovens, e.g. to offer preheating in the initial zone to minimize dirt pick-up or to supplement the heating of heavier sill areas of the car body.

10.11.2   Design considerations of convection ovens

Optimum design requires that a number of parameters be considered to effect the best compromise between process requirements, plant layout, energy consumption, anti-pollution requirements, maintenance, and capital cost.

10.11.2.1   Oven configuration

There are two stages in the stoving of the painted body, i.e. the controlled heat-up followed by a 'hold' period during which the resin system either reacts or, in the instance of acrylic lacquer, reflows. This requires specific oven zones, each with their own heating and control systems. The convected air is introduced at high velocity into the oven enclosure, normally via distribution ducting mounted at roof level. Individual heat-up and 'hold' zones may be further divided into two or more zones.

10.11.2.2   Oven ventilation

The stoving of automotive paints releases into the oven atmosphere combustible compounds which can cause an explosion hazard. For this reason fresh air is introduced into the enclosure to ensure a safe installation. This requires that a 'balancing' volume of air be exhausted from the enclosure; obviously this contains the diluted combustible compounds released from the painted surface, which are also a source of fume and odour and generally undesirable.

10.11.2.3   Oven heating

Each zone of the oven is fitted with a recirculation/supply fan which extracts a volume of air from the oven, mixes this with any fresh air required for ventilation purposes, supplies heat to that mixture to satisfy the zone heat load, and then it returns it via either distribution nozzles or slots to the oven zone.

10.11.2.4   Fresh air requirements

The fresh air quantity supplied to an oven is calculated from the anticipated solvent quantity entering the body, and to ensure that a maximum of 25% of the LEL of that solvent is not exceeded. This air quantity may be introduced into the oven as filtered fresh air into the zone heating system, as air infiltrating at the oven airseal, or as combustion air to direct heating zone burners.

10.11.2.5   Fuel available/heating method

In the United Kingdom, and most of Europe, it is normal to use natural gas as the heating fuel. As this is a sulphur-free fuel it has resulted in the adoption of direct fired heating systems on installations for all parts of the painting process since testing, and experience, has shown this to have no detrimental effect on the quality of the coating where natural gas is not available. Butane or propane are utilized, both having negligible sulphur contents.

Other systems involve burning a distillate of oil; this results in the use of a less efficient indirect fired system (an indirect fired heater being about 70% efficient).

In the past, ovens tended to be direct fired using oil or 'synthetic' gas as the heating fuel. This often lead to problems, particularly with surfacers, of what was designated 'gas fouling'. It showed itself as a surface defect, i.e. low gloss and a 'frost-like' appearance which, in the case of surfacers, lead to major intercoat adhesion problems.

The cause was premature oxidation/cure at the surface of the coating owing to impurities (sulphur) in the heating fuel. The main types of materials to suffer were oil-modified products such as alkyd topcoats and alkyd or epoxy ester surfacers. It often could be alleviated by the addition of small amounts of acid catalyst or increasing the 'acid value' of the resin to effect a consistent 'through cure'. Alternatively, retard the crosslinking reaction by the addition of an amine, e.g. triethylamine.

10.11.3   Fume and odour emission

As has been mentioned earlier, the stoving of automotive coatings releases certain materials during the curing process. These are either mixtures of organic solvents, products of the chemical reaction and some decomposition products. This can lead to visible fume and odour problems from the exhaust stack and to condensation in the stack.

While the problem of visible fume, odour, and exhaust stack condensation are not over-serious, the high level of solvent emission can be a major concern. Legislation exists in the USA to control such emissions and there are significant safeguards in Europe.

The control of the exhaust emissions from industrial plant, and in this instance stoving ovens, originated in the USA in the 1960s. Since the majority of coatings are stoved and based on organic solvents, such solvents became recognized as potential sources of pollution because of the toxic nature of the products formed by their photochemical reaction in the atmosphere. In the late 1960s the Los Angeles County and San Francisco Bay area in the USA introduced regulations which limited the amount of certain organic solvents to be used in organic compositions. These regulations were called Rule 66 or Regulation 3 by their respective bodies; and where pollution control is required under this code it is stipulated that 90% or more of the hydrocarbons from the process be oxidized to carbon dioxide before exhausting to the atmosphere.

More recently in the USA, federal regulations have provided guidelines on the permitted hydrocarbon emission. The actual limits depend upon the location of the plant. These can vary from 1.4 to 2.3 kg per hour and 100–300 tonnes per annum. This level of emission can be achieved by adopting pollution control of the oven exhaust or by a process modification such as a change of paint formulation/technology, or by a combination of both.

In France and the UK the Rule 66 legislation is still widely adopted but in other parts of Europe the German TA-Luft is becoming more widely used. In this latter instance precise limits are set for permissible emissions for all solvent types and for various flow quantities.

The 1986 TA-Luft stated that for automotive paint shops these limits should be $60\,\mathrm{g\,m^{-2}}$ for solid colour topcoats and $120\,\mathrm{g\,m^{-2}}$ for metallic finishes. These figures are maximum levels — local authorities can stipulate even lower limits.

Proposed European Union Regulations are on a similar level, i.e. $45\,\mathrm{g\,m^{-2}}$ E D coated surface area (or 3 kg/unit); $50\,\mathrm{mgC/m^2}$ carbon in the exhaust of the oven.

Besides new paint technology a number of methods of achieving pollution control are now available. These include thermal incineration, catalytic combustion, carbon absorption, liquid scrubbing, and odour masking. Of these, thermal incineration is the most widely accepted and reliable method of achieving control but, more recently, catalytic combustion has proved successful in a number of installations.

10.11.3.1   Thermal incineration
Thermal incineration consists simply of passing the fume-laden exhaust air through a highly efficient combustion system where a primary fuel is burned in order to raise the temperature of the effluent to a critical reaction point and holding this for a specific period. In this way the complex hydrocarbon compounds are oxidized to carbon dioxide and water vapour. The primary fuel may be gas or distillate oil.

10.11.3.2   Catalytic combustion
The application of catalytic combustion techniques has been gaining acceptance for the control of exhaust gases emitted from various processes and show a substantial reduction in fuel consumption when compared with thermal incineration systems, with limited or no heat recovery equipment included.

The exhaust gases supplied to the catalyst cell contain organic compounds which, when passed over the catalyst surface, react with the oxygen present in the airstream.

The application of catalyst combustion to stoving in the automotive industry is a fairly recent development and was, in fact, introduced in the mid-1970s. Catalyst cells do not have indefinite life but if contamination or poisoning can be avoided then a service life of five years can be expected.

10.11.4   Future stoving developments
The development of alternative curing systems will obviously be linked to developments in paint technology. The two overriding considerations from the point of view of the curing system will be the most economical use of energy and the elimination of the effluent problem.

Any changes in the curing process will involve the adopting of new radiant heating technology, as the present convection type oven is very close to its optimum efficiency. There are a number of radiant curing techniques at present used in the curing of flat stock such as boards, sheet metal, fabrics, paper, and plastics. Perhaps these technologies will be adopted for use in a more general way in the metal fin-

Table 10.10 — Typical stoving schedules

Dipping primers	30 min at 150 °C/20 min at 165 °C
Electroprimers	Anodic type 20 min at 165 °C/20 min at 175 °C
	Cathodic type 20 min at 165 °C/20 min at 180 °C
Primer surfacers	20 min at 140 °C/20 min at 165 °C
Topcoats	
Alkyds, thermosetting acrylics/NAD	10–20 min at 130 °C/140 °C
Thermoplastic acrylic lacquers	Reflow 30 min at 155 °C
Repair schedules	
Thermoset products	
Catalysed enamels/clearcoats	30 min at 90 °C/10 min at 100 °C
Two-pack enamels	15 min at 80 °C
Thermoplastic products	
Self-repair	Minimum stoving 15 min at 80 °C

All temperatures are metal (effective) temperatures not air temperatures.

ishing industry. Among these are electron beam and ultraviolet curing and induction heating.

At this stage it is not certain which of these processes will be adopted. Electron beam and UV curing have the advantage of being suitable for plastic parts which are being used increasingly to reduce body weight. Infrared stoving has the advantage of rapid curing time, reduced length of line, minimal dirt, and flexibility (such as its use in conjunction with convection ovens).

Typical store schedules are given in Table 10.10.

10.12   Performance/testing

The performance of automotive coatings has been outlined. The testing of the paint system, and the component 'layers', is designed to simulate conditions likely to occur in practice so as to give some measure of performance in the field. Obviously the testing of undercoats has a different emphasis from that of topcoats but the interaction of these products in a total system is of equal, or more, importance.

Performance standards have improved dramatically in recent years as warranty periods have been extended. For example, the 240 hours Salt Spray test used as an acceptable standard for anodic electropaints has been extended to 1000 hours for current cathodic primers and further improvements are being aimed for. The acceptable standard of exterior durability was considered to be satisfactory after 12 months Florida testing; for modern basecoat/clear systems it is three to five years.

There is also now a greater emphasis on chemical (environmental) resistance properties such as resistance to acid and insect/bird effects. For example, rather than very dilute acids/alkalis being used in testing for long periods at room temperature, more modern procedures involve shorter times, stronger acids/alkalis at elevated temperatures. In fact a variety of tests exist and a number are described later in this section.

The subject of testing and durability is a complex and detailed subject and is dealt with in Chapter 19 of this book. However, key properties, performance, and testing of automotive paints/systems are summarized below and comparisons are made. Also included is a brief description of important and suitable test methods.

A comparison of various topcoat processing properties was made in Section 10.7.

In terms of performance, apart from utilizing standard stoving schedules, under-bake and overbake of both surfacer and topcoat are assessed, sometimes at vari-able thicknesses. This is to simulate every possibility that can occur in the production process.

Testing may be divided into two categories: appearance and performance. The qualities to be assessed are given in note form in the next two sections.

10.12.1   Appearance

Colour; opacity (coverage); smoothness, i.e. freedom from defects such as cratering, solvent 'popping', mottle, or orange peel; gloss and distinction of image (DOI) — important for consumer appeal.

Note: apart from the formulation and quality of the finish the quality of the undercoating system has a significant effect on the final appearance.

10.12.2   Performance

10.12.2.1   Physical properties
Hardness; flexibility; impact resistance; mar resistance; adhesion; stone-chip resis-tance; cold crack resistance, i.e. stability to extremes of temperature and humidity; curing efficiency.

10.12.2.2   Chemical resistance
To petrol, acid, insect/bird effects, alkali, water, humidity, corrosion, scab corrosion. External durability, i.e. resistance to ultraviolet irradiation and humidity.

10.12.3   Test procedures

Typical test procedures and performance properties are listed in Table 10.11.

The following descriptions are mainly in summary form and are fairly typical. Different car manufacturers have their own variants and different emphasis. These procedures are used for full painting systems and for primer only, as appropriate.

10.12.3.1   Cure (test for crosslinking products)
Twenty double rubs with a clean white cotton cloth soaked in MIBK (methyl isobutyl ketone)

Pass: no removal or marking of paint film.

10.12.3.2   Sandability (surfacers)
The stoved surfacer must be amenable to wet sanding (400 paper) or be dry sand-able (P800 paper) by hand. Clogging of the paper, excessive sanding marks and dusting are considered unacceptable.

10.12.3.3   Adhesion: crosshatch test (1.5 mm or 2.0 mm template)
Test panels are evaluated by crosshatch before and after water immersion, normally 120, 240, and 480 hours. (Refer below for details of water immersion test.)

Pass: <5% removal (after taping with Scotch tape).

Table 10.11 — Typical performance properties: various finishing systems

Type of topcoat	Alkyd melamine	Thermosetting acrylic/NAD	Thermoplastic acrylic lacquer	Basecoat/clear (solid colours metallics)
Performance test				
Gloss (20°)	85%	80%	85%	90%
Hardness (Tukon-Knoop)	5–7	8–14	14–18	8–12
Adhesion (cross-hatch)	<5% removal	<5% removal	<5% removal	<5% removal
Petrol resistance (some Super Shell — slow drip)	Excellent	Very good	Fair	Very good
Acid resistance (non-staining 1N H_2SO_4 × 48h)	Pass	Pass	Pass	Pass
Alkali resistance (non-staining 1N NAOH × 48h)	Pass	Pass	Pass	Pass
Surface distortion	—	—	55–65°C (depending on colour)	—
Impact resistance	Pass	Pass	Pass	Pass
Stone-chip Resistance	Excellent	Very good	Good	Excellent
Water-soak (40°C)	No blistering >500h	No blistering >500h	No blistering >500h	No blistering >500h
Humidity resistance (100% RH at 40°C)	No blistering >240h	No blistering >240h	No blistering >240h	No blistering > 240h
Corrosion (salt spray) resistance	1000h	1000h	1000h	1000h
Scab corrosion resistance	Pass	Pass	Pass	Pass
Florida exposure/ 20° gloss (washed) after 2 years	70%	65%	75%	80%
Primer	Cathodic electrocoat	Cathodic electrocoat	Cathodic electrocoat	Cathodic electrocoat
Surfacer	Polyester or polyester/PU	Polyester or polyester/PU	High PVC (~55%) epoxy ester	Polyester or polyester/PU
Pretreatment	Zinc phosphate	Zinc phosphate	Zinc phosphate	Zinc phosphate

10.12.3.4   Hardness ('Tukon indentation'–Knoop hardness)

The Tukon hardness is the generally accepted test in the automotive industry, particularly for topcoats. Other tests include Pencil Hardness, Sward Rocker, Pendulum Hardness (Köning, Perzoz, etc.).

10.12.3.5   Stone-chip resistance

This is a key test for the full paint system and, in particular, for surfacers and anti-chip coatings. Poor chip resistance can be a major source of paint warranty claims and, as a result, test procedures have become more and more rigorous as newer products have developed.

There are a variety of pieces of apparatus and procedures for checking this property. Originally the equipment and test were relatively simple but now they are much more sophisticated. Typical examples are as follows:

1   In the past, although the method was relatively basic it was actually quite efficient. Carried out on the full paint system as follows:
   (a) $100 \times \frac{1}{4}''$ (6 mm) Whitworth nuts are dropped down a 15 feet (457 cm) pipe (2.5 inches or 6.0 cm in diameter) onto a coated panel held at 45°.
   (b)   Panels are graded by degree of removal of paint. The test was normally carried out at ambient, room, temperature.
   (c)   Pass: <5% removal of paint.
2   A more up to date test is to 'shot blast' the test panel for a prescribed time using a constant pressure at constant temperature. The fully painted panel is 'shot blasted' for $\not> 10$ seconds with 0.75 kg of angular iron grit (4–5 mm in size) at $0.1\,MN\,m^{-2}$ @ 20–26 °C.

   After 'shot blasting' the panel is subjected to 72 h salt spray (ASTM B117) then rinsed with deionized water, dried and retested as above.

   Performance is measured against a prescribed standard usually in the form of photographs.

   Pass: <5% paint removal to substrate. (Sometimes the performance required is even more stringent.) Note: all stone-chip failure to be noted, i.e. intercoat and down to metal substrate.
3   Stone-chip resistance — anti-chip coatings. Testing in this case is even more severe. In the procedure just described the weight of 'shot blasted' iron is increased to 2 kg and the pressure to 2 bar at 20–26 °C. As before performance is related to prescribed standards.

   Pass: <5% removal (all chipping — see 'Note' above).

10.12.3.6   Impact test

A 1 kg weight dropped from 24 cm and 50 cm onto a painted test panel (impacted from reverse side).

   Pass: no cracking of paint film.

10.12.3.7   Flexibility

This is quite common in topcoat specifications. Typical are the Erichsen Indentation and/or Bend Test. Erichsen figures of 5–6 mm minimum are typical.

10.12.3.8   Acid resistance

Test panels are immersed for 96 h at room temperature in 0.1 N sulphuric acid.
   Pass: film is not affected.

10.12.3.9   Alkali resistance
Test panels are immersed for 240 h at room temperature in 0.1 N sodium hydroxide.
    Pass: film is not affected.

10.12.3.10   Acid and alkali resistance (alternative procedure)
A more modern, and demanding, test to the above is carried out by subjecting the
test panel to consistent sized droplets of various test solutions which are covered
by watch glasses (the surface of the panel having previously been 'activated' by
exposure to UV light).
    Duration of test: 24 h ± 1 h @ 23 °C.
    (Test solutions removed by rinsing with deionized water and drying in a stream
of air.)
    Test solutions:

0.5%	by weight sodium hydroxide
2.0%	by weight sodium hydroxide
0.5%	by weight sodium carbonate
2.0%	by weight sodium carbonate
0.5%	by weight hydrochloric acid 36% by weight
2.0%	by weight hydrochloric acid 36% by weight
38%	by weight sulphuric acid 96% min by weight
10.0%	by weight citric acid

    Pass: film is not affected.
    Note: tests above are alternatives to those shown in Table 10.11.

10.12.3.11   Water immersion (continuous)
Panels are immersed in a stirred/agitated demineralized water bath at 40 °C.
    Pass: no loss of adhesion, blistering, colour change, loss of gloss, or 'sinkage' (of
the topcoat) after a minimum of 240 h. In excess of 500 h is expected.
    Note: before testing panels would, by necessity, be aged at least 16 h after immer-
sion.

10.12.3.12   Humidity resistance (continuous)
Panels are placed in a humidity cabinet — 100% RH at 40 °C; some auto-
producers use 50 °C.
    Pass: no loss of adhesion, blistering, colour change, loss of gloss, or sinkage (of
the topcoat) after a minimum of 240 h. (Also refer 'Note' above.)

10.12.3.13   ASTM B117 Salt Spray (5% salt solution at 35 °C)
Test panels in a cabinet at the conditions stated above. Panels are 'X' scribed before
testing.
    Results are evaluated at 240, 480, and 840 h (often over 1000 h and sometimes to
'destruction') by inspection and tape strip (over 240 h is normally only applicable
to cathodic electroprimers.
    Pass: not more than 2 mm corrosion at scribe (assessed by taping).
    This is a key test for primers and is carried out in primer only and a full system
situation.

10.12.3.14   Scab corrosion test
This procedure is designed to simulate scab corrosion of a car body after water
contact and stone-chipping.

Initially, the painted panel (full system) is immersed in a stirred water bath at 40 °C. It is then subjected to a stone-chip test and subsequently to ASTM B117 Salt Spray. In addition it can be further exposed.

Pass: no scab corrosion (interaction of the pretreatment and primer is critical in respect of this phenomena).

10.12.3.15   Florida exposure (5° south)

Florida is the site preferred by the majority of motor manufacturers for the exterior durability testing of automotive finishes. It has a high humidity combined with a high level of ultraviolet; this provides an extremely rigorous environment for topcoat technology. Ideal for testing colour stability of pigments, film (polymer) integrity, gloss retention, and erosion.

Although it can be considered a good absolute test of a coating, certain factors should always be taken into account. Performance can often be affected by the time of year when exposure begins and the particular weather conditions in that year. Another hazard, which can often be misconstrued as microblistering, is fungal growth — characterized by small 'pits' in the film and threads or filaments leading from this defect. Consequently testing is normally 'relative' with known standards included in any test programme, to avoid misinterpretation or results.

Pass: typical results are shown in Table 10.11. The accepted standard is good gloss and colour retention, free from defects after up to two years (e.g. alkyds, thermosetting/NAD acrylics and acrylic lacquer). Basecoat clear systems have expected Florida performance up to 5 years.

10.12.3.16   Peel resistance: Florida 5° south

This test procedure is designed to simulate the UV resistance of the surfacer under basecoat/clear topcoat systems.

The stoved surfacer is coated with an approved thermosetting acrylic clearcoat (±UV absorber) at a film thickness of 30–35 μm (i.e. low and critical).

Exposure times are:

+   UV absorber (i.e. normal in practice): 12 months
−   UV absorber: 3 months

After exposure the surface of the clearcoat is assessed mainly for blistering and whether there is any indication of peeling at the interface between the clearcoat and surfacer.

10.12.3.17   Accelerated weathering

The development of a new paint formulation requires an early appreciation of the exterior durability characteristics and it is not possible to wait for a two to three year exposure at Florida to obtain this information. Recourse is therefore made to an accelerated weathering device which can indicate probable durability performance.

A variety of machines are available which subject the paint film to UV light in combination with humidity and temperature. No machine can accurately predict the Florida performance because of the difficulty of producing a UV spectrum identical to natural sunlight by artificial means. Nevertheless, a composition which has withstood 2000 h on one of the more severe accelerated cycles can confidently be predicted to show acceptable Florida performance.

Typical machines are marketed by Atlas (carbon arc or xenon lamp UV source), Q-Panel Company (QUV apparatus using xenon lamp), Xenotest (xenon lamp). Natural sunlight concentrated by mirrors is the basis of an accelerated weathering process devised by Desert Sunshine Exposure Tests of Arizona (EMMA and EMMAQUA cycles).

10.13   Future developments

The future development of automotive coatings technology and/or processes will depend on a number of factors, namely, economics, energy saving, environmental considerations, and consumerism (the demand for improvements in durability and appearance). There are a number of options open, each having their respective advantages and disadvantages and a compromise, to satisfy the various conflicting demands, will undoubtedly evolve.

As far as environmental controls are concerned, they have become more and more stringent, for example, the European Union proposal for 1996 introduction brings emission levels down from the current $60\,g\,m^{-2}$ to $45\,g\,m^{-2}$ for all new plants producing more than five thousand cars per year. Existing plants will have to come down to $90\,g\,m^{-2}$ within five years and $60\,g\,m^{-2}$ within ten years.

Nevertheless progress has been made in reducing the level of solvent and stoving effluent released to the atmosphere. The introduction of water-borne basecoats and surfacers, higher solids products, e.g. polyester topcoats, and the limited use of powder coatings have all played a significant part in reducing emissions. Modern application systems have also contributed by significantly improving transfer efficiency.

In addition, mechanical, thermal, and chemical methods are already available and in use for reducing stoving effluent. These include 'after burners', scrubbers, and carbon absorption paints.

However, there still remains a vast amount of solvent-laden air to be dealt with from spraybooths, and the high amounts of energy consumed during the painting process (>50% energy is, in fact, consumed by the spraybooths simply moving and heating the air and water). High curing temperatures are also an area requiring consideration, particularly if the concept of plastic components is to be fully realized.

In economic terms the cost of organic solvents will continue to rise in line with oil prices. This increases the pressure to move away from solvent-borne to water-borne, not only from an environmental point of view but also from an economic standpoint.

Three basic routes are being followed as a means of reducing or, ultimately, eliminating organic solvents in spray coatings: high solids technology, powder coatings, and water-borne products. These will now be discussed.

10.13.1   High solids technology

The benefits and problems associated with this type of technology may be summarized as follows, in relation to existing products:

Benefits	Problems/Disadvantages
Use current manufacturing techniques	More difficult to achieve equivalent film properties
Use current application equipment	More difficult to achieve equivalent appearance in metallic finishes
Use similar process parameters	Rheological control difficult
	Automatic spray equipment with high transfer efficiency essential for good economics

Solids range:
Pigmented products — up to 70%, metallic (basecoats) 20–40%, clearcoats up to 55%.

Resin types:
Alkyd, polyester, acrylic.

Types of product involved:
Primers, surfacers, sealers, anti-chip coatings, solid colour and metallic finishes.

The obvious advantages are that high solids products normally have similar manufacturing, application and process methods to present low(er) solids technology. Problems and disadvantages arise basically from the fact that lower molecular weight/viscosity resins must be used. These, however, tend to give excessive flow on application, leading to lower run/sag resistance and, therefore, some form of rheological control is essential to facilitate suitable application properties. Furthermore, lower molecular weight resins require higher degrees of reactivity to form coatings with acceptable chemical and physical properties.

10.13.1.1   Higher solids surfacer technology
At the present time higher, or 'medium solids' as they are sometimes designated, are used fairly extensively particularly in Europe. In comparison with normal solids products the solids difference is small but significant.

	Application solids
Polyester/PU surfacer (normal solids):	≮60%
Polyester/PU surfacer (higher solids):	≮64%

Rheological control is normally achieved by the addition of small amounts of aerosil or bentone which are thixotropic aids and help to provide the necessary balance between flow and sagging.

10.13.1.2   High solids polyester topcoats
Solids are already fairly high and not likely to go much above current levels (up to 60% at the gun for white). However, because of the lower molecular/low viscosity resins needed, application is difficult particularly now that most application is 'one-coat' by electrostatic high speed bells where 35–60 μm dry film thickness is expected. Sag resistance and 'solvent popping' problems are major issues and rheological control is often effected by the use of microgels. Microgels are essentially organic extenders and are used in small amounts; they also have crosslinking potential and some form an integral part of the film after curing.

The requirement for improved appearance, mar, and acid etch resistance is pushing more plants towards basecoat/clear technology particularly for dark colours. However, this causes problems with higher levels of solvent emission and cost which is why it has not been more widely adopted.

10.13.1.3   *Higher solids basecoats*

There has been a trend with solvent-borne basecoats to move to what are designated 'medium' solids basecoats where a change to water-borne systems is difficult for one reason or another. The increase in solids is only marginal (~12% low solids/~18% medium solids) but it is significant enough to reduce the level of solvent emissions. A slight decrease in 'flip tone' is not seen as a particular disadvantage.

10.13.2   Ultra high solids coatings

This relates to products with solids contents in excess of 70%. There are, however, severe constraints in their development. For example, metallic finishes cannot be considered in this context since the loss of cosmetic appearance would be so dramatic as to make them unacceptable.

Pigmented (solid colour) products would have to be formulated as two-pack products using a reactive monomer or high solids activator because of their limited pot life. In any mass-production plant such products would undoubtedly call for dual-feed application equipment with a complex metering system. While equipment manufacturers have made certain progress, the cost and complications involved have severely limited the exploitation of such products, and currently they are of purely local interest.

10.13.3   Water-borne products

Benefits	Problems/disadvantages
Environmentally acceptable; cheap and available diluent	Water release on stoving
Use normal manufacturing methods	Need for tighter control of temperature/ humidity for application
Use normal application equipment	Excessive flow: need for rheological control
Use similar process parameters	Resin limitations (stability); prone to surface defects through contamination

Solids range:
No environmental restriction.

Resin types:
Alkyd, polyester, epoxy, acrylic.

Product types:
Primers, surfacers, sealers, anti-chip coatings, solid colour, and metallic basecoats and clearcoats.

10.13.3.1   Surfacers

In Europe, over the past few years, there has been a significant growth in the use of water-borne surfacers and, to a lesser extent, in Japan. This trend is likely to continue as environmental demands become more severe and car producers aim for a total water-based coating system.

Water-borne surfacers have actually been around for a number of years. Early products were based on water-soluble resins, mainly alkyds. However, their film performance was marginal and application properties led to difficulties.

The present range of resin systems are in emulsion form and polyesters with PU modification (where applicable) predominate. These developments have been in line with the improvements in the stability of polymer systems in water in general. As already mentioned, emulsion systems tend to release water more rapidly than solution types. Add to this the relative high levels of pigments (and extenders) then you have a system which is very usable under mass-production conditions minimizing the need for tight control of humidity.

The need for rheological control is important to assist in the balance between flow and sagging on application. This is normally provided by the use of a thixotropic aid mainly in the pigmentation, e.g. small amounts of either bentone or aerosil are the most common.

At the moment, the overall performance of water-borne surfacers is slightly inferior to the best solvent-borne products. Yet they are considered, for the reasons detailed previously, the most favoured future option.

10.13.3.2   Basecoats

While the aqueous dispersion route has greater scope for formulation it is not been possible to produce aqueous finishes which can be used in the 50 μm range (e.g. automotive solid colour topcoats) without strict control of temperature/humidity. However, such polymer systems can be used to good effect in the formulation of aqueous basecoats where basecoat film thicknesses of 10–15 μm need to be applied. In these, the bulk of the film thickness is contributed by the clearcoat.

The aqueous dispersion controlled rheology approach permits good atomization of the basecoat through a conventional spray gun, while providing suitable control of flow in the applied film. Aqueous metallic basecoats can be formulated to give similar brightness and varied hue to the best low solids (12%) solvent-borne but at an 'equivalent' solids when related to organic solvent level of 55–60%. Such basecoats are much less dependent on temperature/humidity levels than would be ordinarily expected and they can be used in similar facilities to the solvent-borne type, with the assistance of a warm airblow before application of the solvent-borne clear.

Water-borne basecoats can now be considered an established technology. Its use has been well described in Section 10.7 and formulating, processing, and performance details appear under a separate heading (Section 10.7.4.2).

As emission legislation becomes more widespread and tighter (1996 in the European Union) the future direction towards water-borne basecoat/clear for both metallics and solid colours seems more and more likely. This will require major capital investment which is possibly the main reason of its non-adoption by certain plants/manufacturers.

10.13.3.3   Water-borne clearcoats

This technology is described in more detail later but it should be noted its development is still at a very early stage in what is considered something that could complete the 'total water' painting process.

10.13.4   Powder coatings and aqueous slurries

Benefits	Problems/Disadvantages
100% solids (solvent-free) High application efficiency Good film properties (at high film thicknesses)	Need for specific manufacturing equipment Need for specific application equipment Limited colour range for good economics; colour ranges difficult; metallics non-viable; difficult to repair; need for high film thickness to achieve acceptable appearance; surfacers require heavy sanding to achieve acceptable topcoat appearance; difficult to coat complex shapes economically; higher temperature needed for flow/cure; dust explosion risk

Resin types: epoxy, polyester, acrylic

Internationally the use of powder coatings in the automotive sector is patchy. There is no real activity in the European motor industry but in the USA some powder stone-chip primers are being run and there is some minimal use of powder surfacers in truck/commercial vehicle plants.

Their single major advantage is the absence of solvent pollution and its recycling capability. This ability to be recycled is now becoming a key issue in some companies.

However, there have been inherent problems concerned with appearance (flow), high stoving temperatures, and cost, although there has been progress in the problem of reducing the baking schedule (to 165 °C) and general appearance.

Nevertheless, this technology is generating a lot of interest as a clearcoat with a number of manufacturers who see it as the only alternative to water. This is described in more detail later (Section 10.13.6).

10.13.4.1   Aqueous powder slurries

Such materials would overcome some of the inherent problems associated with powder coatings. However, there would still remain the question of high energy and cost, repair difficulty, and lower standard of appearance because of the coarse powder particles.

10.13.5   Solid colour basecoats

This type of basecoat can be formulated either in solvent-borne or water-borne technology. They have a high pigment content with excellent coverage in a thin film and are independent of gloss level. The main performance criteria are solvent release, smoothness, and film properties.

The principal benefits are the alignment with metallic analogues, more assured durability, and scope to use pigments which are difficult to employ in, say, polyester type solid colours (the performance being mainly dependent on the clearcoat). This latter property is particularly beneficial in the use of lead-free pigmentations.

The requirement for improved appearance, mar, and acid etch resistance is pushing more plants towards solid colour basecoat/clear technology, particularly for dark colours. However, in solvent-borne systems this causes problems with higher levels of emissions and cost which is why solvent-borne solid colour basecoats have not been widely adopted.

10.13.6   Clearcoats

At the present time normal thermosetting acrylic clears have an application solids of 45–48% by weight and are generally referred to in Europe as 'medium solids' clearcoats. A move to higher solids products is not presently considered to offer enough benefits compared to more novel types under development, i.e. water-borne and powder.

The main concerns at the present time are performance, i.e. (a) acid etch resistance and (b) scratch/mar resistance. In a typical hydroxyacrylic/MF system any steps to improve (a), would be to cause deterioration in (b). As a consequence of this significant developments in resin technology have been necessary.

Improvements in 'acid etch', which is a prime target in North America, have been achieved by a number of routes. These include blocked isocyanate resins, siloxane functional acrylics and epoxy/acid crosslinking chemistry. Scratch/mar improvement has a higher priority in Europe and this has been achieved by blending in more flexible polyester or acrylic resins, acrylourethanes or by changes in the MF resin.

Whereas these two areas are where most clearcoat activity is currently concentrated, i.e. Europe and North America, some Japanese producers have gone to the so called 'low maintenance' clearcoats incorporating fluoropolymer resins. However, due to cost implications, they are generally confined to the luxury car market.

'2K' clears, because of their greater ability to 'reflow' after scratch damage and inherently better acid resistance, would be a possible solution to these problems. However, they still suffer because of health and safety concerns and have not been very widely adopted. A few European plants use them but in all cases they are processed in combination with water-borne basecoats. In North America about 15% of plants use '2K' and a number of these are truck installations.

The one area where usage of '2K' is high is on plastic parts but in overall volume terms this is very small.

Water-borne clear is still in its infancy. There is only one plant in the world running this technology and this is in Europe in a total water-borne installation.

Most European manufacturers are involved in test programmes but it is generally accepted that the development of water-borne clearcoats has not proceeded at the pace many people expected. This is for a variety of reasons:

- Application is more difficult than for water-borne basecoats.
- Appearance standards are not so easily achieved.
- The advantages of water- over solvent-borne clears are much less than for basecoats.

Powder clearcoats are still generating a lot of interest with a number of producers seeing it as the only real alternative to water. Recycling capability is one of its plus points and this property is considered highly important by some companies.

No one is yet running powder clears on a regular basis and use is presently limited to coating small parts in North America. Problems of inadequate appearance, compared with other technologies, and contamination during the recycling process have delayed introduction on a large scale.

10.13.7   Pigmentation

All the general principles described in Section 10.7.5 still apply but there is a major requirement for cleaner, brighter tones. This has led to a high usage of lead-free (organic) reds and yellows.

These types of pigment have become the norm because of the toxicity problems associated with lead chromates/molybdates which has led to their near demise.

It is now much easier to formulate with organic pigments. This is because of the high usage of basecoats, where opacity (and gloss) considerations become less difficult, and the increasing use of colour keyed surfacers.

The biggest change in pigmentation technology in recent years has been the use of 'effect' pigments. Micas, particularly coloured ones, are widely used either in combination with aluminium or alone with pigments. This has led to a better range of cleaner, brighter metallic/mica shades to be produced, particularly reds, greens, and blues.

Currently, mica containing colours probably constitute 60% or more of total non-solid colours. Opacity, in many cases, is not as good as its metallic counterparts but again this is less of a problem because of the use of coloured surfacers.

10.13.8   Painting of plastics

The problems of painting a composite car body built of different metals and plastic has been described in Section 10.9. The particular requirements of such an integrated unit will have to be taken into account in the design of new paint systems as the growth in the use of plastics continues.

A major factor is the temperature constraint, and the development of novel low-curing resin systems is the principal solution to the painting of the wide range of plastics available.

It is worth noting that there has not been the growth in plastic in body construction as previously envisaged. However, this has not minimized the problems associated with the numerous plastic parts that require painting, quality standards having to align closely to the overall coating system.

10.13.9   Electrodeposition and spray application

The process of electrodeposition (EDP), using cathodic technology, is established world-wide and will continue to be improved. The EDP process is an extremely elegant method of coating complex units, being virtually solvent-free, and is the most efficient and economic of all methods of application. Its virtues are such that its future is assured for some time to come.

In the field of spray application, which applies to surfacers and topcoats, the use of robotics and automatic high speed rotory atomizers (electrostatic) is now very much the norm and growing in use. Such equipment provides a high level of transfer efficiency, reducing solvent emissions, with additional benefits such as consistency of application/quality with a low labour content.

Finally, improvements in oven and spraybooth design have also played, and will continue to play, an important part in reducing emissions, effluent, and energy consumption. Their role cannot be underestimated as newer technologies have been introduced so designs have been continuously upgraded to maximize performance and satisfy environmental demands.

10.13.10   Summary

It is evident that the high solids approach to reduce solvent emissions has limited potential compared to the more novel water-borne and powder technology. As a consequence the likelihood is that less and less effort will be put into high solids coatings in the longer term.

The use of water-borne systems in surfacer and basecoat technology have been widely adopted world-wide and will no doubt increase as environmental regulations are more strongly enforced. Usage of solid colour (water-borne) basecoats will also undoubtedly broaden so as to rationalize production processes and upgrade overall performance.

A major issue still to be settled is the future of clearcoat technology. The contenders are water-borne, powder, and '2K' products each having its particular strengths and weaknesses. At the present time there are major developments in each area, progress is being made and some use can be found in various locations. Europe appears to favour the water-borne option, North America powder technology and 2K is common to both continents. In the longer term much will depend on the resolution of certain basic problems and it is most likely that it will take until the end of the century before one of these clearcoat technologies predominates throughout the international motor industry.

The future of the EDP process looks assured, its benefits having been well documented. Developments leading to continued improvements in performance and process refinements should see this system being retained as a permanent feature of the car painting process for the foreseeable future.

In conclusion, a combination of low polluting coatings with new equipment, i.e. electrodeposition and automatic/robotic high speed electrostatic spray, is clearly the way ahead. Such a package should satisfy the growing demand for energy constraints, ecological considerations, high performance standards, and the need for economical and shorter processes.

Acknowledgements

The author would like to acknowledge the assistance given by many past colleagues at ICI Paints, Slough, and Dupont Automotive Coatings. Particular recognition should be given to the late Michael Waghorn for his advice and consultation with the original publication. Also to Barrie Wood, David Griffiths, and Steve Wrighton for their respective contributions in this revision on topcoats, electrocoat technology, and modern application methods.

To my wife Kathleen and son David, much gratitude is due for the original and additional high quality illustrations.

Finally, special thanks to Margaret Powers, my former secretary, for a magnificent effort in completing an excellent typescript at very short notice. She showed exceptional diligence and a high level of competence and the final document owed much to her ability to produce work of the highest standard.

Bibliography

ANSDELL D A, *Ind Production Eng* **3** 30–35 (1980).

BAYLIS R L, *Trans Inst Metal Finishing* **50** 80–86 (1972).

CLANCY F, *Finishing* 25–30 (1990).

EATON M J, *Product Finishing*, July (1983).

FETTIS G C, *Automotive Paintings and Coadings*, VCH Publishing (1995).

INSHAW J L, *J Oil Colour Chem Assoc*, **July** 183–185 (1983).

PAYNE H F, *Organic Coating Technology*, Volume I (Oils, resins, varnishes and polymers), John Wiley (1954).

PAYNE H F, *Organic Coating Technology*, Volume II (Pigments and pigmented coatings), John Wiley (1961).

SOLOMON D H, *The Chemistry of Organic Film Formers*, Robert E Krieger (1977).

STRONG E R, *FIRA Technical Report 21* November (1965).

WAGHORN M J, 'Non-aqueous dispersion finishes', *IMF/OCCA Symposium Warwick University*, pp. 228–233 (1973).

WARING D J, Automobile paint systems', Institute of Production Engineers, February (1984).

Finishing Handbook and Directory (1991).

11

Automotive refinish paints

A H Mawby

11.1 Introduction

11.1.1 Definition

The term 'automotive refinish' refers to paint products applied to a motor car at any time subsequent to the initial manufacturing process. In the factory, a car body is finished as a piece of metal only, permitting the use of products which are cured at any temperature, commonly up to 150–160 °C.

Once the vehicle has been fitted with plastics, rubber tyres, and fabrics, and indeed may be on the road with petrol in the tank, it is no longer feasible to cure finishes at these temperatures. Hence different types of products must be used for *re-finishing* cars, and the refinish paint manufacturer must strive to achieve finishes of equivalent performance to the original finish within this constraint.

The size of the market for refinish paints, in volume terms, is comparable with that for original automotive products. Approximately 250 million litres of topcoat colour is sold worldwide (1984), with a roughly equal additional volume of primers, solvents, and ancillary products.

Commercial vehicles — trucks and buses — are usually painted in materials which are more closely related to automotive refinish products than to either original automotive or general industrial products. The volume market for paint products on commercial vehicles is approximately 40% of automotive refinish. The performance requirements for commercial vehicle paints are generally less critical than for car refinishing, although in some applications resistance to corrosive chemicals or environments is needed.

Automotive refinish products are applied to motor cars for three different purposes:

- Repair or respray following accidental damage in use.
- Rectification of damage or errors during manufacture or during transit before sale.
- Voluntary respray to improve the appearance or change the colour of a vehicle.

In developed countries the bulk of the market is covered by the first two. Voluntary respray is largely limited to the secondhand car market, where a new coat of paint can make the difference between selling or not selling the vehicle. A change of vehicle colour is very rare. A large proportion of car refinish work is paid for by accident insurance.

In developing countries cars tend to be older, the accident insurance business is less well developed, and voluntary resprays with colour change are much more common.

Even within developed countries cultural differences in attitudes to motor cars are quite marked. These show up in the average standard of appearances of vehicles, and can be traced through to effects within the refinish market. The difference in the average standard of bodywork appearance between, say, Lisbon and Zurich is very marked, and it is reflected in the level of sophistication of bodyshop equipment and techniques.

The purpose of most car refinish work is to return a motor vehicle to a standard of appearance equal to that of a new car direct from the production line.

In most Western countries, a damaged motor car represents a severe blow to the ego, morale, and lifestyle of the owner, and its repair to 'original' conditions is correspondingly significant. When the repaired car is returned, the only standard by which the owner can assess the quality of work done is the exterior finish. Every owner will inspect the paint job, and above all will judge the work by the standard of the *colour match*. Colour capability and accuracy is the most important feature of car refinish systems.

Naturally the colour to be matched stems from the car manufacturer, and eventually from the supplier of automotive paints. However, car colours are chosen for their customer appeal, not for their repairability, and the car refinish suppliers, although often the same as the automotive paint suppliers, have historically had to react to changes initiated by them.

11.1.2 Air-drying and low stoving

Automotive refinish products are applied under a very wide variety of conditions. In many developing countries it is common to find cars being resprayed in the open air, even at the side of a street. At the other end of the scale many large body repair shops are highly sophisticated industrial installations. This diversity of conditions, together with the frequently low level of operative skill and training, represents a very different challenge to the paint formulator in comparison with the automotive original market, where a production line has highly trained operatives and a strictly controlled environment. The major diversity in refinish workshops as it affects automotive refinish products, lies in the division between air-drying (drying at ambient temperatures) and stoving. In the automotive refinish context, stoving temperatures range up to 80° metal temperature, although modern refinish materials normally have optimum stoving schedules which reach 60 °C.

The absence of stoving facility in a refinish workshop usually also implies a relatively poor environment for painting, often dusty and variable in temperature and humidity. It is essential for an air-drying product used in such an environment to be 'dust-free' in an extremely short time, normally a few minutes. This implies 'lacquer' products drying by physical processes. Products curing through reaction with atmospheric oxygen or two-pack materials cure relatively slowly at ambient temperatures (since otherwise they would inevitably be unstable in storage or have short pot-lives), but may be used in refinish if a dust-free environment is available. In proper conditions such products have great advantages over lacquers, in both application and performance.

Hence, increasingly in all markets, and almost exclusively in the most highly developed, cars are refinished in full spray booths which are either linked to a 'low-stoving' oven or can themselves be converted into an oven (a combination spray booth/oven).

11.2   Topcoat systems

1.2.1   Introduction

The restriction on stoving temperatures eliminates all the common automotive orig-inal topcoat technologies from use in automotive refinish, i.e.

- thermosetting acrylic;
- reflow thermoplastic acrylic;
- high bake alkyd melamine.

In addition, it is a consequence of the structure of the refinish industry, with a very large number of small body repair establishments, that the pace of technology change is slow, so that all the main basic technologies used in the industry through-out its history are still in use today.

The driving force for changes in technology has been two-fold:

- The need to improve productivity in body repair shops.
- The requirement by many car makers to use materials capable of sustaining extended anti-corrosion warranties.

The result has been a long trend to higher solids, higher build materials needing fewer coats and giving 'gloss from the gun', i.e. requiring little or no polishing, and towards more durable resin systems, particularly acrylics.

11.2.2   Nitrocellulose

In the early days of the motor vehicle, essentially all automotive original and refinish paints were nitrocellulose-based. On the production line nitrocellulose was soon replaced by acrylic and alkyd systems. In refinish, nitrocellulose has persisted until the present day, though by now it has been superseded in most developed markets. The major advantage of physically drying products like nitrocellulose (and thermoplastic acrylic) in automotive refinish is the very rapid air-drying which, combined with easy polishing, permits good results to be obtained in ill-equipped and dusty workshops. The major disadvantage is poor durability. In summary, the strengths and weaknesses of nitrocellulose refinish products are as follows:

Strengths: Very fast drying
             Easy application
             Polishability
             Recoatable at any time

Weaknesses: Poor durability, poor UV resistance
              Brittleness
              Low gloss — polishing required
              Poor gloss retention
              Low solids/low build
              Poor solvent (petrol) resistance

Nitrocellulose lacquers are usually plasticized with a combination of solvent plasticizers, polymeric plasticizers, and non-drying alkyds.

11.2.3   Thermoplastic acrylic (TPA)

TPA automotive refinishes have many of the same characteristics as nitrocellulose products, and, except where durability and resistance to UV is essential for climatic reasons, are seen as largely interchangeable with nitrocellulose by the refinisher.

Like nitrocellulose, TPA lacquers harden by solvent evaporation only. Unlike the products used for original finishes, refinish TPA products are not subjected to stoving to achieve 'reflow'. Polymers used in TPA lacquers always contain a high proportion of methyl methacrylate, copolymerized with several other acrylic monomers to achieve the appropriate blend of hardness, flexibility, and adhesion, and blended with solvent plasticizers.

The strengths and weaknesses of TPA for refinish are as follows:

Strengths: Very fast air-drying
            Good durability (UV resistance)
            Easy rectification, polishable
            Excellent metallic effects
Weaknesses: Poor solvent (petrol resistance)
              Low solids/low build
              Brittleness
              Low gloss — polishing required

11.2.4   Alkyd

Alkyd finishes used in refinishing are typically short oil, fast-drying by nature. They may be air-dried with conventional cobalt and lead driers or applied as two component products by the addition of a melamine or polyisocyanate resin. When used with melamine the alkyd must be stoved at 70–80 °C metal temperature: addition of isocyanate accelerates hardening at any temperature, but is commonly associated with 'force drying' at 40–60 °C.

Alkyds were the first 'enamels' (as opposed to physically drying 'lacquers') to be used for automotive refinish, bringing with them the great advantages of higher solids on application and hence higher build and 'gloss from the gun', eliminating the necessity for labour-intensive polishing.

Alkyds are, however, slower to become 'dust-free' than lacquers, whether nitrocellulose or TPA, and hence must be applied in a dust-free environment. Their introduction was accordingly accompanied by the development of suitable spray booths for the car repair industry.

A major disadvantage of alkyds is the difficulty in rectification, arising from the tendency of the partly cured film to soften if overcoated. As a result, any error in painting or damage caused subsequently, cannot be rectified until the film is fully cured. For alkyds dried conventionally by air oxidation this process can take several days, but films crosslinked by melamine or polyisocyanate addition can be recoated much more quickly.

In addition, the characteristics of the alkyd resin system are not well suited to the inclusion of aluminium flake for metallic finishes.

Metallic pigmented alkyd finishes give a poor 'flip' effect because the orientation of the aluminium particles is too random.

11.2.5   Acrylic enamel

Acrylic enamels are alkyd/TPA copolymers which represent a hybrid product type, developed to improve the build and gloss of conventional TPA without losing the very rapid dust-free performance of the lacquer. The properties of these products fall accordingly between the two types. Like conventional alkyds they may be used as two-pack materials with a polyisocyanate second component. This confers benefits of through-drying speed, durability, and hardness, but loses all the benefits of the 'lacquer' drying, thus requiring dust-free application conditions.

11.2.6   Acrylic urethane

The most recently introduced major product type in the refinish industry is based on a hydroxy functional acrylic resin, used exclusively in association with a polyisocyanate second component. Although slow to be dust-free compared with lacquers, in all other respects acrylic urethanes have the best range of those properties which are most relevant to automotive refinish, and are hence rendering other products obsolescent in most developed markets for other than specialized applications.

The acrylic urethane combination is particularly well suited to the refinish market. Acrylic resins confer benefits of:

- durability, gloss retention;
- hardness;
- flexibility;
- easy pigmentation;
- high gloss.

By using a relatively low molecular weight acrylic resin, solids can be high by refinish standards, while the crosslinking reaction with polyisocyanate takes place at rates which allow acrylic urethanes to be used across the range of temperatures required by the market, curing at an acceptable rate even below 5 °C.

Application of heat accelerates through-drying, and for optimum bodyshop throughput, acrylic urethanes are typically cured for 30–40 minutes at 80–100 °C air temperature, leading to a metal temperature of about 60 °C maximum.

11.2.7   Basecoat/clear systems

The development of basecoat/clear metallics for automotive finishing (Section 10.7.4) had necessarily to be matched by refinish systems capable of repairing and respraying them to achieve an identical appearance.

The main requirements for basecoat formulation are high opacity and good inter-coat adhesion to primer and to clearcoat. Gloss and durability are conferred by the clearcoat, and are not needed in the basecoat. Conventional refinish systems are designed to have gloss and durability and cannot satisfactorily be used to repair basecoat/clear original finishes. On the other hand, the main resin constituent of automotive basecoat technologies, CAB, hardens by solvent evaporation only, and is therefore equally suited to refinish use.

The introduction of basecoat/clear metallics therefore brought about the development of refinish basecoat repair technologies based on CAB, modified with polyesters, acrylics, or nitrocellulose. Many refinish basecoat metallics also contain polyethylene wax or other aids to aluminium flake control.

Indeed, automotive basecoat colour is commonly supplied for refinish use with little if any modification. Although some automotive basecoats contain melamine, which almost certainly contributes little to the curing process in a refinish environment, the hardness, durability, and intercoat adhesion performance of automotive basecoats under acrylic urethane refinish clearcoats is more that adequate.

Where refinish basecoats are not identical to the original, the technology is nonetheless similar, based on CAB blended with polyester or acrylic resins and with similar additives to automotive products.

11.3 Colour

11.3.1 Introduction
Colour is the central feature of refinish paint technology. The number and diversity of car colours on the roads rose very rapidly from about 1970, when colour became a significant part of the marketing package used by the car makers in an increasingly competitive market. This was particularly true of European markets where at the same time the expectations of colour-matching quality in bodywork repairs also increased.

It is estimated that car manufacturers' production colours on the roads in the main European markets in 1985 exceeded 10000, a five-fold increase from 1965.

The colour range required to be matched in the process of repair is further extended by the variation in colours coming from the car factories. These variations may be caused by:

- Drift over time in the standard or technology of paint supplied, conditions, or equipment on the production line.
- Differences between the paint, equipment, or techniques used at two or more units producing cars in nominally the same colour.

Refinish paint makers monitor the colours of cars actually coming from the factories on to the roads, rather than simply matching the original 'styling pattern' which is the car maker's master for the colour.

11.3.2 Factory-matched colour and mixing schemes
The supply of colour matched at the factory to specific car colours is the most obvious and simplest approach to the refinish market. This approach is viable so

long as a manageable number of colours represents the major part of the total market demand. Once it became impractical to supply the whole of the colour demand in this way, paint manufacturers introduced the concept of the refinish mixing scheme to permit colour mixing initially at distributor level, subsequently at end-user level. At first the mixing scheme supplemented the factory-matched range by providing a means to supply the less common, less frequently needed, colours. However, in many markets the mixing scheme has almost wholly supplemented factory-matched colour, because the enormous diversity of colours on the road renders the potential sale of an individual factory-matched product too small to be worth manufacturing or stocking.

Factory-matched colour is still a significant feature of markets which are either less well developed, with a majority of small repair shops, and/or a low expectation of colour accuracy, or else are dominated by a relatively small number of car makers, so that each individual colour can still command a reasonable sale. The biggest market in the world, the United States, remains in this position, though the rapid increase in imports of cars has initiated the trend towards mixing schemes which is by now nearly complete in Europe. The sheer size of the USA market naturally allows the paint maker to achieve an economic batch size for an individual colour in the range of one of the giant car makers: the eventual limitation lies in the size of stock inventory which the distribution outlet is prepared to hold. Experience in developed markets suggests that once a range of 700–1000 colours is able to service less than 60–75% of the demand, then the trend to mixing schemes becomes steady and irreversible.

11.3.3   Refinish mixing schemes

A typical refinish mixing scheme consists of a series of single-pigment paints, very carefully controlled for batch-to-batch consistency of both hue and tinting strength. These are mixed together, according to formulae supplied by the paint manufacturer, to match car colours. Formulations are normally supplied on microfiche in the form of parts by weight or by volume required to achieve a specific volume of accurately matched colour.

The hardware of a refinish mixing scheme consists of a stirrer bank, capable of simultaneously stirring all the tinters to maintain homogeneity within the tins, a microfiche reader for formula presentation, and a weigh scale or volumetric measuring device.

The number of pigments which can be included in a mixing scheme is naturally finite, generally amounting to 30–40 pigments, including a selection of grades of aluminium flake for metallic colours. Often a number of pigments are duplicated at 'full' and 'reduced' strength, the latter allowing small colour adjustments, for example in tinter white car colours. A typical range of pigments included in a refinish mixing scheme, and designed to give maximum coverage of typical car colours, is shown in Table 11.1.

Naturally, the range of pigments available to automotive paint formulators for new colour styling is much wider than that available within a refinish mixing scheme. As a result, refinish colour matches are frequently to some extent metameric to the original car colour. A satisfactory repair may then require a large area of the car to be painted; in extreme cases the entire car may have to be resprayed, because no acceptably close colour match can be achieved from the available pigments.

Table 11.1 — Typical mixing scheme pigmentation

Colour area	Pigment type	Example
White	Titanium dioxide	SCM Tiona RH472
Blue	Copper phthalocyanine	BASF Heliogen Blue L6975F
	Copper phthalocyanine	Hoechst Hostaperm Blue BFL
	Indanthrone	ICI Monolite Blue 3R
	Potassium ferro/ferricyanide	Manox Blue 10SF
Green	Chloro-brominated copper phthalocyanine	ICI Monastral Green 6Y
	Chlorinated copper phthalocyanine	ICI Monastral Green GN
Yellow	Lead chromate	
	Lead chromate/sulphate	
	Flavanthrone	
	Tetrachloroisoindoline	Ciba Geigy Irgazin Yellow 2 GTLN
	Azomethine copper complex	
	Hydrated iron (III) oxide	Bayer Bayferrox 3910
Red/orange	Iron (III) oxide	Bayer Bayferrox 110–160
	Molybdated lead chromate	BASF Sicomin L3135S
	Monazo naphtol AS	Hoechst Novoperm Red F3 RK-70
	Quinacridone	Ciba Geigy Cinquasia Red Y
	Dibromoanthrone	ICI Monolite Red 2Y
	Naphtol AS type	Hoechst Hostaperm Brown HFR
	Perylene	Bayer R6418/R6436
	Dimethylquinacridone	Hoechst Hostaperm Pink E
	Thioindigo derivative	Sandoz Sandorin Bordeaux 2RL
	Linear, *trans*, quinacridone	Ciba Geigy Cinquasia Violet RT-887-D
	Dioxazine	Hoechst Hostaperm Violet RL Special
Black Blue	Carbon Black/Furnace Black	Degussa Special Black 10
Jet	Carbon Black/Channel Black	Degussa Carbon Black FW200
Aluminium	Fine	Eckert 601
	Medium	Toyo 8160
	Coarse	Silberline 3166

Refinish paint suppliers have to invest heavily in providing formulae to match all the car colours which might require to be mixed, in dissemination this information, and in giving the refinisher the means to identify the car colour. Production of mixing formulae is greatly assisted by the use of computer-linked reflectance spectrophotometers. By comparing the reflectance spectrum of the car colour with the spectra of the available mixing scheme pigments, the computer can suggest a range of possible mixing formulae from which the colour matcher selects according to the criteria he or she chooses to impose, e.g. level of acceptable metamerism, pigment cost, or total number of pigments used. Computer-matching is not yet capable of handling metallic colours fully satisfactorily, because of the angular dependence of the reflectance in such colours; but it is nevertheless useful in shortening the process of mixing formula production even for metallics.

11.4   Future developments

11.4.1   Relationship to automotive technology

The basic difference between automotive and refinish technology which arises from the constraint on stoving temperature in refinishing a complete vehicle, is likely to continue to divide the two technologies, in spite of pressure to reduce stoving temperatures in car manufacture. This pressure comes from, first, a need to reduce energy costs, and second, the trend to the use of plastics in vehicle construction. Few plastics are capable of withstanding conventional automotive stoving schedules without distortion. Hence, unless the plastics are to be painted separately in a different, lower stoving process (with obvious dangers of poor colour matching), the new low stoving processes are required for automotive finishing. Nevertheless, stoving temperatures of 100–130 °C can be used without distortion of most structural plastics, and these temperatures are still well in excess of those available to refinish.

However, we have seen that the introduction of basecoat/clear processes has reduced the difference between automotive and refinish technologies in these finishes to the clearcoat alone. It is apparent that in the next few years a much larger proportion of new car colours, whether conventional solid colours, metallic, or pearlescent (mica) in pigmentation, will be formulated on basecoat/clear technology. Possibly only pastel colours and whites will be conventional one-coat technology, so that the colour technology in automotive and refinish will be equivalent if not identical.

The trend towards more variety in automotive pigmentation, rendered possible by the separation of the decorative and protection functions of the paint finish implied by basecoat/clear, could start to reverse the trend towards on-site mixing of colour, and refinishes may find themselves increasingly using colour identical to that applied to the car on the production line. There is a limit to the number of pigments which may be added to a mixing system to permit an acceptable match to be achieved, and the addition of coated mica pigments, in particular, may cause that limit to be exceeded. However, the convenience of the mixing scheme principle to the bodyshop should not be underestimated, and much research effort is likely to be put into further extending the usefulness of refinish mixing schemes.

11.4.2   New resin systems

The development of acrylic urethanes has resulted in a resin system which meets most of the requirement of refinish. But for one major drawback it would be difficult to foresee a fundamentally new technology superseding these products in a generation, especially since substantial further improvement in already excellent performance is almost certainly possible.

The major drawback is the inherent toxicity of the isocyanate. There is a constant danger to operatives of becoming sensitized through inhalation of spray mist (aerosol), arising from bad practice or equipment malfunction. Subsequent exposure to isocyanate even at extremely low concentration can cause asthmatic symptoms, and sensitization is therefore likely to oblige an operative to leave the refinish industry.

Research into refinish resin systems is therefore concentrated on achieving the performance of acrylic urethanes without the need for isocyanates. The great flexibility and established durability of acrylic systems make it likely that any new generation of refinish products will still be acrylic-based.

The patent literature contains a number of alternative approaches to the chemistry of crosslinking in hydroxy-acrylics, but the urethane reaction is singularly difficult to match in its flexibility and capacity to proceed even at very low temperatures. Such patented crosslinking resins include those based upon:

- melamine;
- polyepoxides;
- polyanhydrides.

All of these should be suitably externally or internally plasticized.

It remains to be seen which, if any, of these technologies has the capacity to make obsolete the present generation of acrylic urethane automotive refinish products.

Bibliography

The Refinishers Handbook: published by Imperial Chemical Industries in association with the Vehicle Builders and Repairers Association (1983).

12

General industrial paints

G P A Turner

12.1   Introduction

To the paint industry 'industrial paints' are those coatings used by industry at large, as opposed to painters and decorators, painting contractors, and do-it-yourselfers. Some parts of the industry prefer, as in this book, to treat automotive, automotive refinish, marine, and heavy-duty coatings separately, leaving the remainder as 'general industrial paints', the subject of this chapter.

General industrial paints are therefore all paints, except those excluded above, which are used by industry in factory finishing processes. They include coatings and end uses as diverse as wire enamels, clear and pigmented furniture finishes, can lacquers, tractor finishes, paints for toys, paper coatings, aircraft finishes, domestic appliance finishes, protection for under-body automotive parts, coatings for plastics, and so on. Industrial articles may be as large as road-grading machines or as small as dice. They are often made of metal, but may frequently be made of wood, wood composites, paper, card, cement products, glass, or plastic. Metals may be steel in any of its forms, with or without protective surfacing, such as galvanizing or tin, or they may be aluminium, zinc, copper, or any of the numerous alloys. Each substrate and end use is a different painting problem, which must be solved within the commercial and other constraints of factory processes.

There is therefore no such thing as a typical general industrial paint or painting system. Most sub-classifications of general industrial paints are based upon the industries served with those paints, e.g. drum paints for the steel and plastic drum industry. These classifications are often used in statistical data on paint usage. Some typical data are shown in Tables 12.1–12.4 illustrating the volumes of paint used and the industrial markets served in the UK and elsewhere.

Let us now consider the factors which shape an industrial painting process, leading to such a diversity of paints and systems.

Table 12.1 — The UK market for industrial finishes (1994)

Industrial market	Volume, litres ($\times 10^6$)	%
Agricultural, construction and earth-moving equipment	6	4.7
Auto components	2	1.6
Aviation	4	3.1
Can	19	14.9
Coil	12	9.4
Domestic appliance	2	1.6
Drum	3	2.3
Electrophoretic	4	3.1
Furniture	11	8.6
General machinery	10	7.8
Joinery	2	1.6
Metal fabrication	14	10.9
Paper	4	3.1
Plastics	5	3.9
Powder (allowing 1.5-1kg^{-1})	30	23.4
Total	128	100.0

Table 12.2 — West and Central European market for industrial finishes (1992)

Industrial market	Volume, tonnes ($\times 10^3$)	%
Agricultural, construction and earth-moving equipment	90	4.7
Aviation	25	1.3
Can	155	8.1
Coil	100	5.2
Domestic appliance	180	9.4
General industrial uses	650	33.9
Plastics	45	2.3
Powder	160	8.3
Wood finishes	320	16.7
All others	195	10.1
Total	1920	100.0

12.2 Factors governing the selection of industrial painting processes

12.2.1 The substrate

Industries make products for sale, and these articles need finishing. Often the finishing process does not involve painting, e.g. the metal-plating of plastic car trim or costume jewellery. More often it does, or it involves a number of finishing processes, including painting or lacquering. Thus the specification of an industrial painting process begins with an article to be finished. The surface of this article is the substrate to be painted. This substrate has a major influence on the finishing system that is finally chosen for the article.

Table 12.3 — The US market for industrial finishes (1994)

Industrial market	Volume, gallons (×10⁶)	%
Aviation, trains	5.0	1.7
Can and drum	51.2	16.9
Coil and sheet	29.1	9.6
Domestic appliance, heating and air conditioning	6.6	2.2
Electrical insulation	5.0	1.7
Flatboards	11.5	3.8
Machinery, equipment for building and farms	17.1	5.7
Non-wood furniture and business equipment	18.3	6.1
Paper, film and foil	16.1	5.3
Powder (allowing 1 gal/3 lb)	49.0	16.2
Wood furniture	36.0	11.9
All others	57.3	18.9
Total	302.2	100.0

Table 12.4 — The Japanese market for industrial finishes (1992)

Industrial market	Volume, tonnes (×10³)	%
Building materials, mainly coil and flatboard	93	10.2
Domestic appliances	47	5.2
Electrical industry	104	11.5
General machinery	82	9.0
Metal products	143	15.7
Structures	128	14.1
Wood products	99	10.9
All others	212	23.4
Total	908	100.0

12.2.1.1   Physical and chemical characteristics
Some questions need to be asked about the substrate:

1   Is it susceptible to attack by features of the normal everyday environment (heat, light, oxygen, water, microorganisms)?
   (a)   If so, will it need protecting from these things by the paint system? For example, the sensitivity of iron and steel to water and oxygen may lead not only to the design of a quality protective coating system, but also to the use of a passivating chemical pretreatment prior to coating. Sensitivity to light may compel the use of a pigmented coating for UV protection, though a clear coating would otherwise give sufficient protection and an attractive appearance.
   (b)   Will the substrate have been altered by environmental conditions immediately before painting? As a result, will it need conditioning, preparing, cleaning, or treating first? For example, wood absorbs moisture from,

or surrenders it to, its environment. If, after painting, it is to be stored or used in an environment of markedly different humidity, it can warp or distort. It should be conditioned to the right moisture content first by storage in air of the appropriate relative humidity.

2   Is the substrate susceptible to attack by the paint or is the paint likely to be adversely affected by the substrate?

For example, a plastic substrate may be susceptible to attack by the solvents in some types of paint, or a paper substrate may be susceptible to attack by water in water-borne coatings. This may lead to avoiding certain coatings, or to the use of sealer coats to insulate the substrate.

Alternatively, the substrate may have such a low critical surface tension (e.g. untreated polyethylene or polypropylene) that most — possibly all — coatings will not wet it or adhere to it. In this case, the chemical nature of the surface must be changed, e.g. by oxidation by corona discharge, so that paint will wet and adhere.

3   Will the substrate be adversely affected by the paint application or curing conditions, or will it have an adverse affect on paint application or cure? For example plastic or wood substrates may be distorted by a hot oven, but accept short bursts of radiant energy directed at the coating. The melting point of tin (232 °C) determines the maximum metal temperature acceptable in the stoving of coated tinplate in can manufacture. Conversely, the high heat capacity of heavy metal castings can make the stoving of coatings on them impracticable or very inefficient, and some woods contain addition polymerization inhibitors which can be extracted by the solvents into the coating, where they inhibit its crosslinking drying mechanism.

4   Is the physical nature of the surface as supplied suitable for wet painting? If it is too porous the pores must be sealed first. If it is too rough it may require abrasion or sanding to make it smoother. If it is too dirty it will require cleaning or degreasing. For example, chipboard is too porous and must first be sealed, either with a layer of paper impregnated with melamine resin, or with a putty-like chipboard filler applied by reverse roller coater.

12.2.1.2   Shape and size

Finally, there are the shape and size of the article to be painted to consider. These have a major influence on the methods of application and cure to be used.

For example, metal screws present problems of efficient paint application and subsequent handling, and the answer may be to use the old-fashioned technique of tumbling the screws in a drum with just sufficient paint to coat them. A jumbo jet, on the other hand, can only be housed in a practical oven with great difficulty, and will normally require a paint that dries at room temperature. Flat articles, such as sheets of paper, hardboard, or coils of strip steel, are very amenable to a wide variety of painting and curing techniques which are readily automated.

12.2.2   The finished result

The industrial paint user has an article to be painted, and the nature of this article, as we have seen, has a major impact on the paint system and painting process chosen. The paint user also has in mind an end result that he or she wishes to obtain from the finishing process. This finished result is the next major factor determining the finishing system and process finally chosen.

12.2.2.1   Appearance

Undoubtedly the manufacturer of the article is after a certain finished appearance. Colour will be important, as will uniformity of colour and consistency of colour from article to article. Gloss also has a major effect upon appearance, and the paint user may require anything from the high gloss and crisp image reflection that gives showroom appeal to a motor car, to the satin-smooth look and feel of a furniture finish, to the dead-matt appearance used to hide minor imperfections in the substrate.

12.2.2.2   Cost

The process of selection of a new industrial finishing system is a matter of give and take between paint user, paint suppliers, and suppliers of application and curing equipment. Initially, the user will have a general concept of the desired appearance, which he or she will examine further in discussion with paint suppliers. Overhanging this and all subsequent discussions will be the factor of cost. The finished appearance must be achieved within a given cost, which may be tightly or loosely defined. If, for example, the finisher has envisaged a high gloss, high quality appearance and presents the paint supplier with a porous surface of rough texture, the latter will be obliged to suggest a thorough surface preparation followed by a multistage finishing system; possibly thorough sanding, a filler or sealer, followed by two further coats. The total film weight may also have to be relatively high. The finisher may be forced to conclude that this cannot be afforded, and will have to think again.

12.2.2.3   Protection

This re-think will also have to take account of the other main requirement in the finished result: the protection afforded by the finishing system to the final product. This may range from protection from damage in the shop or showroom, to a complex specification including flexibility; impact and scratch resistance; wear resistance; water, humidity, and corrosion resistance; resistance to specified chemicals; and exterior weathering resistance. To combine, for example, good corrosion resistance with good weathering properties may require the use of a system, e.g. an anti-corrosive primer followed by a durable topcoat, as in aircraft finishing. Extremes of performance, such as 15 years or more outdoors without repainting, are also likely to narrow dramatically the choice of film formers available to the paint formulator, as in the coating of strip metal for cladding of buildings. The options will, as before, be subjected to the constraint of cost.

12.2.3   The required output rate

The manufacturer has an article which he or she wishes to paint to get a certain appearance and a certain amount of protection. We have seen that these factors alone almost define the options open. However, there is another major factor: there must be a certain output rate of painted articles from the factory, and again it must be done within a certain cost.

Given unlimited capital resources, this output could probably be achieved in a number of different ways. In practice, though, unless manufacture is commencing on a 'green field' site, there will be the constraints of an existing factory and site, as well as limitations on finance. Perhaps the existing factory already has a paint line with existing equipment installed. The question arises as to how much money the manufacturer is prepared to spend to improve the paint line to finish the new article, and what constraints of space and services there will be. Looking ahead to running costs, personnel will be an important factor, and the manufacturer is likely to define at an early stage a constraint on the number of people per shift acceptable for the new painting process.

Once the required production rate has been coupled with the acceptable capital and labour costs for the painting plant, and the nature of the article has been considered as discussed above, a limited number of options for applying and curing the paint will have been identified.

For example, the problem may be that there is an existing paint line for flat boards doing an excellent job, but it has reached the limit of its capacity. The new article can be painted in the same way, but an upsurge in sales is expected. There is insufficient capital to move the painting line to a new and larger site. More production must be obtained from the same space, but there is some money for new equipment. This creates an opportunity for a complete review of the painting process since, as the line cannot be duplicated, it must run faster. This is almost certain to throw strains upon the existing paint system, which it may not be able to endure. A likely outcome is a modernized line with updated application and curing equipment, e.g. infrared lamps in place of hot air, and new paint products or even a new system.

In another example, the complexity of shape of a metal article means that the only suitable methods of application are some sort of spraying or electrodeposition. The manufacturer wants to keep paint and labour costs as low as possible, and is prepared to spend capital to achieve this. He or she is therefore attracted to electrodeposition, but a high-gloss finish is needed. No electrodeposition coating meets this requirement. He or she therefore reconsiders spraying, and may well end up with some variant of electrostatic spraying. The amount of automation that can be built in will depend upon the complexity of the shape and whether other articles are to be finished on the same line. Curing will almost certainly have to be by hot air, but ovens may be shortened by high-velocity hot air.

At this stage of the planning *safety, toxicology, and pollution* will also have to be considered. The shape and cost of the paint line are emerging. The paint options are, at this stage, limited. Cost criteria seem to be attainable. But is the finishing process emerging safe to use? Will it produce unacceptable gaseous effluent or require toxic liquid or solid waste disposal? Sometimes these questions are enough to rule out certain paint options. In other cases, they may be resolved, for example, by installing superior fume extraction facilities to protect the working environment, or afterburners to reduce pollution and recover energy from the paint solvents.

12.2.4 Summary
The interactions of the above factors in the selection of an industrial painting process are summarized in the two complementary Tables 12.5 and 12.6. In subsequent sections more will be said about the application and curing processes used in industrial finishing, and examples will be given of how these factors have operated in specific cases, when selected industrial finishing processes are discussed.

Table 12.5 — Constraints imposed by or on the paint user, and the paint and process options they affect

Constraint	Paint and process options affected
A Substrate	Preparation, pretreatment Need for a paint system Film former type Solvent-borne or water-borne (solvent types) Application method Curing method
B Required dry film properties (finished result)	Preparation, pretreatment Need for a paint system Film thickness Film former type Curing method
C Required production rate	Nature of paint system Film former type Application method Curing method
D Capital expenditure and existing plant	Preparation, pretreatment Nature of paint system Application method Curing method

E Running costs	Preparation, pretreatment
	Nature of paint system
	Film thickness
	Film former type
	Paint type: High, medium or low solids
	           solvent types
	               one-pack or two (film-former)
	Application method
	Curing method

F Health, safety, and the environment	Film former type Solventborne or waterborne Paint solids

Table 12.6 — Summary of the paint and process options and the constraints which affect them

Paint and process option	Constraints affecting choice (see Table 12.5)
Need for a paint system	A, B
Preparation, pretreatment	A, B, D, E
Nature of paint system	C, D, E
Film thickness	B, E
Film former type	A, B, C, E, F
Paint solids	E, F
Solvent type	A, E, F
Application method	A, C, D, E
Curing method	A, B, C, D, E

12.3   Industrial application and curing methods

In Chapters 9–11 the techniques of applying and curing architectural, automotive, and refinish paints have been described. Application in those markets is by brush, roller, dipping, electrodeposition, and a variety of spraying techniques, manual and automated, including airless and electrostatic spraying. Cure is under ambient atmospheric conditions or in convected hot-air ovens, or sometimes by radiant heating (infrared). In general industrial painting, because of the wide variety of articles to be painted and other requirements (as discussed above), a number of additional methods have been introduced. As we have seen, these have a significant effect on painting systems and processes.

12.3.1   Application methods

In any method of application, either an excess of paint is applied and the surplus is removed, or the desired thickness of paint is put on directly during application. Of the methods described in earlier chapters, dipping is in the first category, and electrodeposition, spraying, brushing, and roller application are in the second. The additional methods used in industrial painting can be similarly subdivided.

12.3.1.1   Application of excess paint and removal of surplus

Flow-coating
Paint is applied under pressure through a number of carefully positioned jets, so that all parts of the article are coated. Excess paint drains from the article and coating chamber walls back into the sump. Articles are usually hung from a monorail; accurate jigging is important. After the coating chamber comes a draining zone, in which the solvent content of the atmosphere can be controlled to delay evaporation and to permit paint flow and the escape of aeration. A sloping floor takes drips back to the sump. Next comes a drying zone, which in most cases includes a stoving oven.

Flow-coating can be used for large articles with relatively complex shapes. Penetration into recesses is better than in dipping, and the paint sump is much smaller ($\leqslant$10% of dip tank size). However, solvent losses due to evaporation are much higher, thanks to the large paint/air interface continually being recreated in the application chamber. This increases painting costs. However, with the increased use of water-borne primers, the cost of replenishing the tank after evaporation losses has become lower, and this method of application has been used for painting agricultural machinery.

Centrifuging
Small articles, such as jewellery or furniture fittings, are placed in a basket, which is immersed in and then withdrawn from a paint tank. The basket is then centrifuged at several hundred rev/min to remove excess paint. The articles may be practically dry when tipped out onto net drying trays. The film thickness is dependent on centrifuge speed and paint viscosity.

Vacuum impregnation
Vacuum impregnation is used for the application of relatively high solids, viscous coatings to complex articles, such as electrical windings. Penetration into the

extremely small spaces would be difficult without the aid of vacuum. The winding is preheated, and vacuum is applied while it is still hot. A varnish cock is then opened, causing varnish to rise under atmospheric pressure from a lower tank into the coating chamber. When the windings are submerged, the varnish cock is closed, the vacuum is released, and additional air pressure is applied to force varnish into the windings. The varnish cock is then opened, and the surplus varnish drains back into the tank. Solvent may be partly removed from the windings by further application of vacuum.

Knife-coating

Knife-coating can be used for the coating of flat and usually continuous sheet material, e.g. paper, plastic film. Excess coating is applied by any suitable technique, e.g. roller coating, and the coating thickness is then reduced by passing the web under a doctor knife (an angled metal blade) or an air knife (a curtain of high velocity air directed onto the web). This technique is particularly suitable where very thin coatings indeed are required. Viscosity can be relatively high, since the pressure of the knife determines film thickness, and the appearance requirement is not exacting.

Extrusion

Extrusion is a form of knife coating suitable for rods, tubes, and wire. The article to be coated is passed through a reservoir of coating and is then extruded through a die or gasket. The die acts like a doctor blade, permitting only a controlled thickness of coating to pass through with the article. Wire enamels are applied in this way.

12.3.1.2   Direct application of the required paint thickness

Tumbling or barrelling

Tumbling or barrelling is a method suitable for coating small articles such as screws, buttons, knobs, or golf tees. The articles are loaded into a barrel which can be rotated about its axis, the axis being horizontal or inclined at about 45°. By experiment the amount of paint required to coat the interior of the barrel and the surfaces of the articles to the required thickness is determined. This is then added, so that the barrel is one-third to one-half full. It is then rotated at 20–40 rev/min [1] for a predetermined time. Film thickness and appearance are controlled by paint viscosity, solids, and quantity, and barrelling time, speed, and temperature. The coated articles are tipped out onto screens or trays, where the coating dries by solvent evaporation or by thermal cure in ovens. Tumbling and centrifuging are variations on the same theme.

Autophoretic deposition [2]

This process was developed by AmChem and can be described as electrodeposition without electrodes. It is a specialized method suitable only for ferrous metal parts, which permits the controlled application of 7–25 μm of coating from a dip tank. The bath is mildly acidic, free from solvent, and contains latex-based paint at 3–5% volume solids. The steel articles must be scrupulously cleaned before they enter the bath, where they are attacked by acid, causing iron ions to be released. These destabilize the latex in the immediate vicinity of the article, causing coagulation of the particles and deposition on the article. As the coating thickness grows, deposition slows, but thinner areas continue to be coated faster, resulting eventually in overall uniformity of thickness, even on cut edges and in recesses. Corrosion resistance is excellent, even though no conversion coating pretreatment has been applied. Film thickness is controlled by time of immersion. Application must be followed by thermal cure. The colour range of coatings is restricted.

Forward roller-coating

Forward roller-coating is a widely used method of applying low to medium thicknesses of coating to flat surfaces, such as board, sheet, or a continuous web of metal coil, paper, or plastic, one or both sides being coated in a single coating station. Coating is transferred from a reservoir via two or more rollers (see Fig. 12.1) to the substrate. Coating thickness is controlled via a polished metal doctoring roller, and application is from a rubber or gelatine-coated roller rotating in a 'forward' direction, i.e. with, rather than against, the forward movement of the substrate. Other factors controlling film thickness are relative roller speeds, roller pressures, and web or sheet speed. Coating solids and viscosity also affect thickness, and viscosity is critical to flow. The roller process imparts a wavy surface to the coating so that, in the worst cases, 'tramlines' of paint are laid down on the article. Paint viscosity must be low and Newtonian, so that these striations may flow out and level to a uniform film. Except when the machinery is emptied and cleaned, no paint is wasted, since what is not applied stays on the roller. Coating speeds in excess of $100\,\mathrm{m\,min^{-1}}$ are possible.

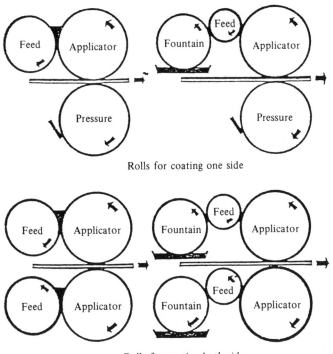

Rolls for coating one side

Rolls for coating both sides

Fig. 12.1

A variant on this process, borrowed from printing, is the use of a gravure rather than a plane or grooved roller to transfer paint to the applicator roller. The thickness of coating is then directly related to the depth of the gravure cells etched in the roller surface; excess is returned to the reservoir by a doctor blade. Coaters of this type are used to apply very accurate coats of a few micrometers, and are sometimes called 'precision coaters'. The coating is laid down as a series of closely spaced 'dots', which must flow out to produce a continuous film.

Reverse roller-coating (coil)

Reverse roller-coating is essentially the same as forward roller coating, but with the web of strip metal running in the opposite direction, i.e. counter to the direction of the application roller (see Fig. 12.2). Thus the paint is almost scraped off the application roller, and a high shear is applied, leading to better flowout of the coating. Since web speeds in coil coating are extremely fast (30–200 m min^{-1}) and the coating enters the oven only seconds after application, every effort is made to avoid tramlines and to encourage flow. Consequently, reverse roller coating is the preferred method for topcoats, and is also used for some primers.

The comments made on viscosity and film thickness control in the previous section also apply to reverse roller coating on a coil line.

Reverse roller-coating (wood)

In the coating of flat sheets of chipboard and other porous surfaces, a level impervious surface can be obtained by applying a viscous paste filler to the substrate, curing it, and then sanding. A reverse roller coater is used to apply the filler, but the applicator roller rotates in a *forward* direction. It is, however, followed immediately by a large doctor roller rotating in a reverse direction. The reverse roller applies pressure to the filler, forcing it into the board and smoothing out the surface.

Coat weights of filler are high (50–120 g m^{-2}), and the material must be high in solids content and relatively viscous. The usual controls on film thickness are supplemented by varying the pressure and speed of the doctor roller.

Curtain-coating

Curtain-coating is an ideal method of applying thicker coatings (60 μm and upwards) to flat boards or sheets. A curtain of paint is allowed to fall from a head (which may be pressurized) through a V-shaped slot into a trough below (see

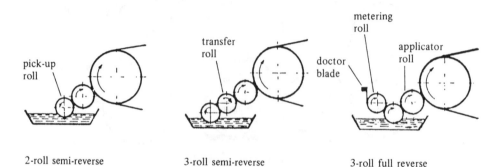

2-roll semi-reverse   3-roll semi-reverse   3-roll full reverse

Fig. 12.2

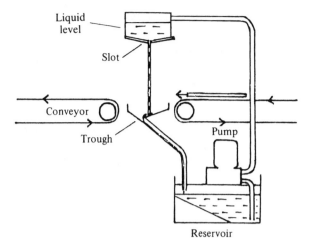

Fig. 12.3

Fig. 12.3). From the trough it is returned to a reservoir tank, from which it is pumped again to the head. The curtain width depends upon the size of machine, and widths up to 2 m are common. The curtain thickness is controlled by adjusting the width of the slot between 0.1 and 5 mm. Excessively thin curtains are unstable.

Running at right angles to the curtain on either side are two conveyor belts. The leading belt picks up the panel from the main conveyor line and accelerates it at up to 100 m min^{-1} so that it passes through the curtain and is coated. It is then picked up by the conveyor on the other side and carried away.

Some machines have twin parallel heads, and these can be used to apply similar or different coatings.

Coating film thickness is controlled by the width of the slot, the pressure in the head, the viscosity of the coating, and the speed of the conveyor belt.

12.3.2   Curing methods

There are two stages in the drying of paint films:

1   removal of solvent or diluent;
2   (a)   fusion of particulate polymer into a film, or
    (b)   crosslinking of polymer molecules (curing), or
    (c)   a combination of (a) and (b).

In lacquers, (1) is the only stage in the drying process. Fast air movement is even more important than heat for this stage. In 100% polymerizable coatings, only stage (2) is required, and air movement is only important in so far as it aids heat transfer.

Of the methods of transferring energy to the coating, conduction from a hot substrate can be used to supplement other forms of heating, and is the mode of heat transfer in induction heating. Forced convection is widely used in conventional stoving ovens. Radiation and oscillating electric fields are used in various special curing techniques.

Drying methods specific to general industrial painting will now be discussed.

12.3.2.1   Energy transfer by conduction: induction heating

If a metal object of simple shape, e.g. a rod or sheet, can be placed in close proximity to an induction heating coil, then the eddy currents induced in the object will heat it. Heating can be localized within the skin or surface of the object. Heating is rapid and readily controlled. If the object is a painted object, then heat is rapidly induced in the substrate and transferred to the coating by conduction from the substrate.

In spite of the attractive features inherent in the method, induction heating has had very limited use in the drying of industrial paints, the main reason being, probably, the need to match very closely the contours of the painted article and the induction heater. The method is suitable, therefore, only for processing similar articles of simple shape, e.g. metal coil or sheet, day-in day-out on a given painting line.

12.3.2.2   Energy transfer by convection

High velocity hot-air jet heating

In this variant of the convected hot air oven, drying times are considerably reduced by directing jets of extremely hot air (180–550 °C) onto the moving surface of the painted article. This method has been used successfully for the curing of painted flat boards (e.g. for 'knock-down' furniture) or coil metal strip (in air-flotation ovens).

In the former use, solvent is partly removed in a more conventional hot-air zone, so that the coating is viscous enough to resist the impingement of jets of air at about 180 °C, moving at speeds of 25–$36 \, \mathrm{m \, s^{-1}}$ and striking the coating at right angles to the film plane, without being disturbed by them. Since the boards are moving at speeds of up to $30 \, \mathrm{m \, min^{-1}}$, the time of contact with the hot air can be short, and boards do not reach temperatures which cause distortion.

In the latter use, the whole weight of the strip metal is supported along its length by hot air from jets located in the base of the oven. Line speeds of $150 \, \mathrm{m \, min^{-1}}$ can be used with oven stay times of about 15 seconds, so air temperatures of up to 550 °C are necessary for cure. These ovens are claimed to have advantages over the so-called catenary ovens traditionally used for coil. The latter do not support the strip at all within the oven, weight being borne by infeed and outfeed drive bridles. This leads to strip sagging (the name catenary deriving from the coil's shape), distortion, and inefficient heat transfer. Air flotation ovens reduce distortion and the oven with cooling zone can be 30% shorter [3].

Flame drying

In this method, hot air impinging on the film is provided in the form of an air curtain surrounding a smooth high velocity flame. This air barrier prevents solvent ignition. Contact times are as short as 0.02–0.04 seconds. This method has been used successfully to dry inks and coatings on plastic and metal containers. Equipment is also available for sheet-fed and moving-web processes.

12.3.2.3   Energy transfer via radiation or electric fields

Most types of electromagnetic radiation have been employed to cure or dry industrial coatings. The high frequency oscillations of an electric field are employed in radiofrequency drying, and radiations varying in wavelength from long wave infrared to the ultrashort wavelength β-radiation (better known as electron beams) have been used.

Radiofrequency drying

In radiofrequency drying, otherwise known as 'dielectric heating' [4], pairs of electrodes (rods or platens) of opposite polarity are arranged parallel to one another and to the conveyor, but at right angles to the flow of work pieces. The workpieces should be flat boards or sheets, e.g. hardboard sheets, since the coating should pass within about 5 cm of the electrodes. The electrodes are connected to a highfrequency generator so that the polarity of each electrode oscillates at about 20 MHz.

In this arrangement the fringe field of the stray field between adjacent electrodes selectively passes through and heats the material or dielectric in the vicinity which is of highest 'loss factor' (the product of dielectric constant and power factor of the material). Water-based coatings have higher loss factors than hardboard, thus the field passes through the coating and does not heat the substrate. The coating is heated because the water molecules are small dipoles, and, every time that the polarity of the electrodes changes, they attempt to realign to the new field. With millions of field changes per second, this causes great friction between the water molecules, and the temperature rises rapidly, causing water to evaporate. Once the water has gone, heating ceases, so the method is not very suitable for curing thermosetting coatings. However, water can be removed from 50 μm films in about 20 seconds (faster removal may cause blistering).

The method is used for board and paper coatings based on water.

Infrared heating

In this method IR heat is emitted by the radiation source and directed at the coating, which uses the energy efficiently if it absorbs most of it.

IR sources are hot bodies which emit radiation over a broad spectrum of wavelengths, the peak wavelengths varying with the temperature of the hot body. White-hot sources at 1200–2200 °C emit with peak wavelengths in the short wavelength region (1.0–2.0 μm), red-hot sources at 500–1200 °C in the medium wavelength region (2.0–3.6 μm), and dull emitters (90–500 °C) in the long wavelength region (3.6–8 μm). The peak wavelength is the highest energy point radiated from the source, with 25% of the waves of a shorter length than the peak and 75% of a longer wavelength [5]. Emitters may be heated by gas (flames impinging on the backs of curved or flat panels) or electricity (heated wires or filaments in ceramic sheathing, lamps with built-in reflectors, or quartz tubes with external reflectors).

Since the high temperature sources emit more energy, they may seem the natural choice for all uses. However, absorption by the coating is an equally important factor. Most polymers contain groups which absorb in the IR. These groups absorb strongly around 2.9–3.7 μm and above 5.5 μm [5, 6]. Thus medium- and long-wave IR is better for heating clear coatings; short-wave heats the substrate. If the coating contains pigment, pigment colour and scattering are important. Black is highly absorptive, other colours less so. Scatter due to reflection is most pronounced at the shortest wavelengths and declines to zero between 2 and 7.5 μm [7, 8]. Thus a good general compromise for efficiency of absorption lies in the region above 5.5 μm, but faster (though not necessarily more cost-effective) heating can be achieved with short–medium wavelength emitters because more energy is emitted.

Radiant heating has to be directed, and heaters and workpieces have to be arranged to avoid shadowing and cool spots. For the reasons given above, for a given arrangement of emitters and a given exposure time, paint film temperature can vary from paint to paint and colour to colour. It will be higher over thin-gauge metal rather than thick. These problems lead to difficulties of control with more complex articles or mixed products on one line. On the other hand, IR is good for substrates such as chipboard, hardboard, and plastic, where substrate heating is to be avoided and where articles may be in sheet form or of simple shape.

Ovens combining zones of IR and convected hot air avoid many of the difficulties and capitalize on the ability of IR to bring paint to temperature very rapidly.

Ultraviolet curing

UV curing involves only the crosslinking of molecules. Ideally, no solvent is present to require evaporation, the coating being kept fluid by the use of low molecular weight polymer (oligomer) and monomers (see Chapter 2, Section 2.19). If solvent must be used, it is removed in a flash-off section, possibly with the aid of IR. Curing is triggered by the absorption of UV by photo-initiator molecules, which decompose, normally to free radicals. These then initiate polymerization.

UV sources are tubular quartz lamps in a suitable housing which contains a reflector. The lamps contain mercury which is vaporized and ionized between electrodes or via excitation by microwaves. The light radiation emitted has the line spectrum of mercury, with principal UV lines at 365 and 313 nm, and several smaller lines at or below 302 nm, merging into a continuum between 200 and 250 nm. Strong lines in the visible occur at 405, 436, 546, and 578 nm. The most widely used lamps are of the medium pressure type with a power rating of 80 W per cm of tube length.

Lamps are set over the conveyor at right angles to the direction of travel and usually at a height such that the radiation is focused in the plane of the coating. The number of lamps is sufficient to give cure at the conveyor speed required, and the curing rate is expressed in $m\,min^{-1}$ of conveyor speed per lamp, e.g. $15\,m\,min^{-1}$/lamp. Coatings are exposed to radiation for times well below one second; but, since most lamps emit about 15% of their energy as IR, coating temperatures rise at slower line speeds.

UV curing is widely used for clear coatings on flat surfaces such as wood-based boards, paper, card, and floor tiles. Pigments absorb UV strongly, preventing decomposition of photo-initiator and hence cure. However, with inks and with thinner, low-opacity paint layers, cure is possible, and UV has found its widest use for the cure of printing inks. Lamps can be introduced in the very limited space available on printing lines, and can be used to produce cure between printing stations of different colours. UV energy costs are lower than for thermal methods of cure.

Electron beam curing (EBC)

In an electron accelerator, electrons are generated at a heated wire or rod within a vacuum chamber, and are directed through a thin titanium window as a narrow beam which scans rapidly backwards and forwards across a conveyor, or as a 'curtain' beam covering the full width of the conveyor. The electrons are highly energetic (voltages of 150–600 kV are used), penetrating the coating and creating free radicals on impact with molecules therein. Photo-initiators are not needed, otherwise coating compositions are similar to those used for UV curing, and free radical addition polymerization ensues. Line speeds of several hundred $m\,min^{-1}$ are theoretically possible.

Since pigments have only a limited effect on beam penetration, thick coatings of any colour can be cured, though the depth of penetration is directly related to voltage. At 150 keV penetration to a depth of 120 μm is possible in a coating of density 1.0. Electron accelerators emit X-rays, so they must be screened to protect the operators. At 150 keV, 5 mm lead screening is sufficient, but accelerators of 300 keV and above need to be housed in concrete bunkers. This increases the cost of an already expensive machine: the cheapest type (150 keV) costs about ten times as much as a UV installation. Running costs are lower than for UV in terms of energy, but are boosted by the need for a continuous supply of inert gas over the coating. As with UV, surfaces must be flat or nearly so. Thermoplastic substrates can be coated, since temperature rises are low.

The very high capital cost of this process limits its use to high volume production outlets. EBC has been used on wood panels and doors, car fascia panels, car wheels, and in certain reel-to-reel processes, e.g. silicone release papers and magnetic recording tapes.

12.3.2.4   *Vapour curing*
This method of curing is not a process for energy transfer; cure normally occurs at room temperature and if higher temperatures are desired they must be arranged in a conventional manner. Instead, this process introduces a catalyst for the curing reaction *via the vapour phase*. The catalyst is either injected into the air of the curing chamber or atomized into the spray fan during the application process. Curing times are considerably shortened without shortening the pot-life of the paint. Commercially this process has been offered with polyurethane paints and varnishes and amine catalysts. It offers advantages where faster cure is needed without the use of heat and at relatively low capital cost.

12.3.3   Summing up
A wide variety of application and curing methods is available to the industrial finisher. In Tables 12.7 and 12.8 the main features of these alternatives are summarized.

12.4   Finishing materials and processes in selected industrial painting operations

12.4.1   Mechanical farm implements
Mechanical farm implements are large and complex machines with many component parts, some very large, some small, some sheet metal, and some cast metal, but mainly steel. Equipment is made by large international companies and painted in their distinctive house colours.

These manufacturers require a uniform and attractive colour on all parts, with a high gloss, giving an excellent showroom appearance. Less emphasis is put on the quality of protection, though the coating must be durable outdoors and must withstand wear and tear, oils, and fuel.

The complexity and variety of size and shape rules out many application and curing options at once. The uniformity of colour makes dipping and flow-coating acceptable and economic options for primers and for one-coat finishing. Airless spraying is used for other topcoats. Room temperature and convected hot air are the only practical drying options, but drying temperatures vary from 15 °C to 150 °C according to the size and heat capacity of the painted part.

Table 12.7 — Characteristics of industrial methods of application

Application method	Applies excess (E) or required amount (R)	Suitable for flat (F), simple (S), or complex (C) surfaces	Suitable (S), fairly suitable (F), or unsuitable (U) for rapid colour change	Paint transfer efficiency: E = excellent G = good F = fair	Speed of application F = fast S = slower	Continuous (C) or batch (B) process
Dipping	E	S/C	U	G-E	S	C
Flow-coating	E	S/C	U	G	F	C
Centrifuging	E	S	U	E	S	B
Vacuum impregnation	E	C	U	E	F	B
Knife-coating	E	F	S	E	F	C
Extrusion	E	S	F	E	F	C
Electrodeposition	R	C	U	E	S	C
Autophoretic deposition	R	C	U	E	S	C
Forward roller coater	R	F	S	E	F	C
Reverse roller coater (coil)	R	F	S	E	F	C
Reverse roller coater (wood)	R	F	S	E	F	C
Curtain-coating	R	F	F	E	F	C
Spraying, various	R	C	S	F-G	F-S	C

Table 12.8 — Characteristics of industrial methods of drying and curing

Drying or curing method	Energy transfer: conduction (Cd), convection (Cv), radiation (R), electric field (F)	Suitable for flat (F), simple (S), or complex (C) surfaces	Speed of dry/cure: VF = very fast, F = fast, M = moderate, S = slow	On/off control	Start-up time: L = long, M = moderate, S = short	Capital costs: energy costs: H = high, M = medium, L = low, VL = very low		Notes
Induction heating	Cd	S	F	Yes	S	M	M	For metal substrate
Room-temperature cure	Cv	C	S	No	S	VL	VL†	
Vapour cure	Cv	C	M	No	S	L	VL*	
Convected hot air	Cv	C	M	No	L	M	M/H	
Jet-drying	Cv	F/S	F	Yes	M	M/H	H	
Flame-drying	Cv	F	VF	Yes	S	M	M	
Radiofrequency	F	F	F	Yes	S	L/M	M	For water removal
Infrared	R	S	M/F	Yes	S	L/M	M/H	
Ultraviolet§	R	F	VF	Yes	S	L/M	M	For curing by addition
Electron beam	R	F	VF	Yes	S	H	L‡	polymerization

† Air movement needed; § not for thicker pigmented films; ‡ but nitrogen running costs; * but catalyst running costs.

Paint costs are extremely important in this market and must be minimized. Production is on mechanized lines, but output rates need be only moderate.

What is required, therefore, is an inexpensive coating capable of room-temperature drying, force-drying, or stoving to give exterior durability and moderate protection on steel. Solids must be relatively high in the paint, and the overall appearance full and glossy. It would be difficult to think of polymers better suited to meet all these requirements than the short–medium oil length alkyds. With driers, these give the necessary room temperature or force-dried curing. With melamine resin as crosslinker, they provide an even higher standard of protection on parts that can be fully stoved. The coatings remain stable in dip-tanks and flow-coaters, provided that these are monitored and adjusted at regular intervals. The same materials can be thinned to spraying viscosity for spray application to completed machines, assemblies, or parts.

Farm machinery parts are normally degreased, but not pretreated. Trouble is taken to prepare a relatively smooth metal surface for painting, but the filling and natural levelling properties of the medium-low molecular weight, higher solids alkyd resins are important for giving the final high quality appearance.

Recent trends are to increase safety, reduce pollution, and avoid loss of expensive solvents by moving to water-borne alkyds, for primers at least, and, in some cases, to cathodic electrodeposition. However, water-borne alkyds must be made from more expensive ingredients than their solvent-borne counterparts (for hydrolytic stability), and the water-miscible solvents still necessary in minor amounts cost more than hydrocarbons. Paint costs do not fall, therefore, but savings can be made on thinners. There is also a move to higher standards of exterior durability, towards better alkyds, methacrylated alkyds or two-pack polyurethane topcoats [9].

12.4.2 Panel radiators

Panel radiators are the heating panels used in the circulating hot-water heating systems of European houses. In their simplest forms, they consist of moulded, hollow steel panels. More complex are the extended surface radiators, in which corrugated sheet metal has been spot-welded to the backs of the panels to produce a large, hot, 'extended' surface from which convection may take place. Then there are double radiators, comprising parallel pairs of panels, with or without extended surfaces in between. Even with these complications, panel radiators remain essentially laminar and suitable for dip application.

Appearance standards are good, but not necessarily of the highest quality. In some countries, house purchasers paint the radiators even before using them, to match their preferred decorating schemes. Thus the coating must have the appearance of a topcoat, but act as a primer and accept overcoating with decorative house paints. Coatings must resist knocks and scratches, but other protective requirements are modest. Good coverage of all parts of the sometimes complex shape is, however, essential. Economy is important in the paint process. Production rates are moderate.

In view of the panel shape, radiators are ideal for dipping. For some years the most popular method was dipping in trichloroethylene paints. This solvent has a high vapour density and, although evaporation from the paint still readily occurs (boiling point 87 °C, evaporation rate six times as fast as butyl acetate), the vapour does not escape from a properly constructed tank, but hangs about above the liquid layer. As the radiator is withdrawn from the paint, it passes through this zone saturated with trichloroethylene vapour, and evaporation of solvent from the paint is delayed and flowout of paint on the radiator takes place readily. As the radiator leaves the vapour layer, evaporation proceeds rapidly and flow ceases. The overall result of this is that a much more uniform film thickness is obtained than with conventional dipping. Resin systems are alkyd–amino resin or styrene- or vinyl toluene-modified alkyds.

In recent times manufacturers have been concerned about both the toxicology of chlorinated solvents and also the ability of the coating to cover all recesses of the more complex modern designs of radiators. As a result, electrodeposition has become the most frequently used method of coating, giving complete coverage of all surfaces and good control of film thickness, coupled with full automation. Anodic acrylic or polyester coatings are generally preferred, since pale colours (near white) are usually applied, and cathodic resin systems are prone to discolour on baking. These coatings require high bake temperatures (170–180 °C) and stoving times of about 20 minutes for 20–25 μm films. Curing mechanisms usually entail crosslinking of hydroxyl-functional polymers with amino resins.

Good-quality metal is degreased and then pretreated with iron phosphate to give enhanced adhesion and a sufficient degree of corrosion protection at minimum cost.

There is an increasing trend to fully finished radiators, on which an electrocoat or acrylic water-borne dip primer is topcoated with a spray polyester or a powder coating [10, 11].

12.4.3 Refrigerators

The refrigerator market saw fierce international price competition in the 1980's. There was heavy pressure on painting costs, with a requirement to achieve more with less paint and to apply and cure it most economically.

Coating colours in Europe are mainly white, with a larger colour range in the USA. Colours must be clean and bright, and the finished appearance good. The coatings must be hard and scratch-resistant and must withstand household hazards, such as fats, fruit juices, ketchups, and various polishes. They must perform well in a humid environment. A steady production rate is required.

Good-quality metal is therefore cleaned and pretreated with iron phosphate, and 25–35 μm of finish is applied directly to the treated metal. Good paint utilization is sought by electrostatic spraying from disc or bell. The hardness requirement, resistance properties, and clean white colours call for thermosetting acrylic or polyester/melamine resin finishes. Acrylics are usually of the self-condensing type and contain crosslinking monomer, such as N-butoxymethyl acrylamide. Sometimes epoxy resin may also be included in either type of finish to improve resistance properties in corrosion tests. Solids contents have risen steadily. Although the shape is essentially box-like, it is complex enough to require convected hot-air curing. Stoving schedules for these resin systems (with the aid of acid catalyst such as p-toluene sulphonic acid) are 20 minutes at 150–175 °C.

Under pressure to upgrade quality while reducing solvent emissions, some manufacturers have turned to powder coatings. Acrylic, epoxy, polyester/TGIC, polyester–epoxy or polyurethane types are used [12]. Very tough, high quality finishes are produced, but there are difficulties with colour-matching, colour changes and with controlling coat weights to the economic levels required by the industry. Cure times are around 15 minutes at metal temperatures of 170–190 °C. Electrostatic powder spray guns are used, and economy is improved by recovering unused powder via cyclones. All this calls for special plant and capital investment, but coatings are virtually non-polluting, and further economy comes from low reject rates, owing to the tough finish resisting damage during assembly.

An alternative approach to the needs of the industry has been to remove the paint shop altogether and fabricate the refrigerator from prepainted electrozinc-plated steel coil finished by a coil coater. This has called for a redesign of the doors and casing to deal with the problem of exposed unpainted metal edges where the sheet is cut. Spot-welding has been replaced by adhesive bonding. Special flexible, but resistant, polyester/melamine coil coatings have had to be developed to withstand the forming operations in refrigerator construction. Some sacrifice in hardness can occur, but hardness can be maintained by forming the metal at temperatures of about 60 °C using coatings less soft and less flexible at room temperature. Most refrigerators in the USA are now made this way [13].

12.4.4   Coil-coated steel for exterior cladding of buildings

Many large industrial buildings are constructed with a steel framework, which supports an outer skin of cladding and a roof. The cladding and roof are frequently made from prepainted hot-dipped galvanized steel, which has been formed and corrugated after painting to give extra strength and stiffness. The product from the coil-coater is a coil of painted metal, which may vary in width and uncoiled length. Coils may be bought in a fairly wide range of colours. The coil-coater is expected to give a prediction (and in some cases a guarantee) of the life expectancy of the painted metal.

A coil line is a large-scale reel-to-reel operation, with metal at one end and pre-treated, painted metal at the other (see Fig. 12.4). The capital investment is very high. The process is economic if (a) the final article is a high quality product, (b) paint thicknesses can be minimized and carefully controlled without waste, and (c) painting can be done quickly (minimizing floor space needed and hence capital costs). Coil lines run, therefore, at speeds between 30 and 200 m min^{-1}.

Application by roller-coater (mainly reverse roller-coater) is ideal for these speeds (see Section 12.3.1.2) and permits a rapid change of colour, since a 205 litre drum is the paint reservoir, and only the limited amount of paint on rollers and in the pump, troughs, and piping needs cleaning out. In view of the flat nature of the painted surface, a wider variety of drying processes are theoretically suitable; but in practice coil coaters have found UV and electron beam too limiting on the nature of the coating used, and have preferred high velocity hot air to IR.

The coatings must therefore be suitable for rapid application and must cure in 15–60 seconds, though a lot of heat is available. The metal temperature peaks at 180–250 °C just before the paint leaves the oven, and is 'quenched' by water spray. When it has cured, it must be capable of forming around sharp bends, must have excellent exterior durability, and must withstand corrosion in marine or industrial environments. Specific chemical resistance may be needed, and resistance to building-site damage is obviously required.

Although the zinc layer will protect the steel cathodically in the event of damage that exposes the metal, a pretreatment is required to provide a suitable surface

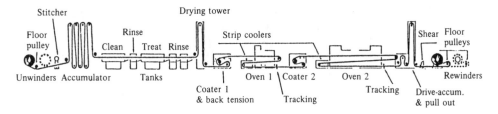

Fig. 12.4

for the coating to adhere to, and to reduce 'white rusting' of the zinc. This is usually a spray or immersion coating of a complex metal oxide pretreatment, followed by a passivating chromate rinse. These are applied in the first stages of the coil line.

A paint system is normally required in order to provide the very high standard of durability and resistance properties. The first coat is 5μm of primer containing chromate anti-corrosive pigments. The topcoat varies with the durability required. The most expensive coatings are applied at 20μm, and vary in life expectation from thermosetting acrylic (7 years), polyester/melamine (10 years), silicone polyester/melamine (12–15 years) to polyvinylidene fluoride (20 years). Flexible, abrasion-resistant, durable polyurethane finishes have also been introduced [14]. Alternatively, the cheaper PVC plastisol can be applied at 100–250μm to give a very damage-resistant coating with 10–15 years durability in temperate climates. The primer for plastisol is usually acrylic-based, while the thinner finishes are applied over epoxy–urea or polyester–melamine primers.

The back of the coil is coated in the same sequence of operations as the front (see Fig. 12.4) with 10μm of polyester–melamine backer or 3–5μm of primer and 8–10μm of backer.

A considerable amount of work has been done on water-borne coatings for coil, and thermosetting acrylic emulsion-based materials have met with limited success on cladding. However, the use of after-burners to incinerate solvents from the coatings, with recycling of hot gases into the oven, has provided an economical alternative to water-based coatings with fewer problems in use. Nevertheless, driven by CO_2-abatement pressures, development work on water-borne and powder coatings continues. Problems with powder include the slow rate of application, the need for infrared or induction heating to sinter particles before the oven and the difficulty of holding film thickness down while maintaining quality of film. Work on chromate-free pretreatment continues [15]. Chromate-free primers are being introduced successfully [14], as are dried-in-place pretreatments, which reduce sludge formation with its disposal problems [15].

2.4.5 Interiors of two-piece beer and beverage cans

The beer and beverage can is a form of food packaging, and must not add excessively to the cost of its contents. Can-makers are constantly seeking ways of making the package cheaper. Once the can was made in three pieces: the body (from a flat sheet) and two ends. Now most beer and beverage cans are two-piece cans. The body is produced from one piece of metal by a process known as drawing and wall ironing.

This method of construction allows much thinner metal to be used and the can has maximum strength only when filled with a carbonated beverage and sealed. Spin-necking saves metal by reducing the diameter of the neck. Between 1970 and 1990, beer and beverage containers became 25% lighter [16]. In the USA, where aluminium is cheaper, most beer and beverage cans are made from that metal. In Europe, tinplate is often cheaper, and many cans are made of this. Modern beer and beverage tinplate has a low tin content at the surface, the main functions of the tin being cosmetic and lubricating (in the drawing process). So a lacquer with excellent protective properties is required, to be used at minimum coat weight (6–12 μm, dependent on metal type).

Can-making is economical only if the cans can be made very quickly. Some 800–1000 cans a minute will be produced from one coating line, with bodies and ends coated separately. Bodies for beer and beverage cans are lacquered after being made and degreased. The rapid application is achieved by short bursts of airless spray from a lance positioned opposite the centre of the open end of the horizontal can. The lance may be static or may be inserted into the can and then removed. The can is held in a chuck and rotated rapidly during spraying to obtain the most uniform coating possible. Coating viscosities must be very low, and solids about 25–30%. The shape is relatively simple, but interiors are cured by convected hot air, in schedules around 3 min at 200 °C.

Carbonated soft drinks are acidic. Resistance to corrosion by such products is provided by coatings such as epoxy–amino resin or epoxy–phenolic resin systems. Beer is a less aggressive filling for the can, but its flavour may be spoilt so easily by iron pick-up from the can or by trace materials extracted from the lacquer, that it also requires similar high-quality interior lacquers.

The majority of these coatings have been successfully converted to water-borne colloidally dispersed or emulsion polymer systems, especially on the easier substrate to protect, aluminium. Water-based coatings have reduced overall costs and lowered the amount of solvent that has to be disposed of by after-burners to avoid pollution. Most successful systems are based on epoxy–acrylic copolymers with amino or phenolic crosslinkers.

There continues to be commercial interest in the electrodeposition of water-based lacquers in beer and beverage cans. Such a procedure avoids the need to apply in two coats, and is potentially capable of giving defect-free coatings resistant to the contents of the can at lower dry film weights. In water-borne spray coatings, solvent contents lower than 10–15% are being sought [17].

12.4.6 Aircraft

Jet aircraft are made principally from a range of aluminium alloys. Most parts are coated before assembly, but eventually the completed aircraft must be finished, providing a huge surface and bulk, difficult to cure with any of the stoving or curing techniques described above. Coatings that dry at room temperature are required. Nevertheless, a high quality of appearance and protection is necessary. Resistance to corrosion is essential, and also to the operating fluids of the aircraft, including aggressive phosphate ester hydraulic fluids. Exterior durability is, of course, necessary, with the coating being exposed to shorter-wavelength high energy UV at high altitudes and to cycles of temperature varying from −50 °C in flight to over 70 °C on a tropical airstrip.

Thus a high quality room-temperature curing system is essential. Most widely used primers are of the epoxy–polyamide (or polyamine) two-pack type, with leachable chromate anti-corrosive pigmentation. The preferred topcoat formulations are two-pack polyurethanes, incorporating hydroxyl functional polyesters and aliphatic polyisocyanates, also two-pack. Cleaning and pretreatment of the metal is important, with chromate conversion coatings and chromic acid anodizing being widely used.

Because of the size and complexity of shape of the aircraft, these coatings have been applied manually by airless or conventional spray, with operators protected by wearing air-fed hoods, but work is being carried out widely to introduce various forms of electrostatic spray [18] and also application by robots [19, 20]. There have been trials with infrared drying [20] and with warmed air in application hangars [21]. In the latter, fumes and overspray are extracted and trapped in water and solvent is subsequently recovered.

The recycling of solvent is one of a number of measures being taken by the industry to meet stringent new regulations limiting solvent emissions. Others include the introduction of water-borne primers and high solids topcoats based on variants of proven polymer types [22–24].

The search continues for effective chromate-free pretreatments and primers [22, 24].

12.4.7 Flush doors for interior frames

The modern flush door consists of a rectangular wooden frame between two 'skins' of hardboard sheet. These are kept apart by a lattice pattern cardboard spacer, the whole structure being given rigidity by glueing. The skins may be plain hardboard or plywood, or the hardboard may have a wood veneer coating. A thin wooden lipping protects the long edges of the door.

The two skins of the door are large flat surfaces and as such are ideal for coating and curing by the modern techniques described above. Equipment is situated in a conveyorized coating line. The final coating is required to look good, to be hard and abrasion-resistant, and to have resistance to humidity changes and a limited number of domestic liquids. These properties can be provided by materials as long established as acid-catalysed alkyd–amino resin or nitrocellulose–amino resin types or as modern as 100% solids polyester–styrene or acrylic oligomer-plus-monomer types.

In a 'green field' situation the choice between these types will depend on the required production rate, the available floor space and capital, acceptable running costs, etc. The more traditional finishes will involve sealer coat application by forward roller and finish application by curtain–coater. Cure will be by convected hot air, jet-drying, or a combination of convection and infrared. The 100% solids finishes will be applied by roller (possibly even by precision coater) and cured by UV (clears) or electron beam (clears and pigmented).

Like the modern furniture industry, this industry has long accepted two-pack products for use in sophisticated application machinery, provided that they can be formulated to give at least eight hours pot-life. For acid-catalysed products this is easily exceeded (1–4 days). Radiation-cured coatings are one-pack materials.

Some of the earliest commercial uses of electron beam curing were for the finishing of doors by major European door manufacturers.

12.4.8   Paperback book covers

For many years the covers of paperback books were finished after printing by laminating polyester film onto the card. This gave a glossy finish which withstood continual use.

The same effect can now be produced by coating the print with $4 g m^{-2}$ of UV-curing acrylic overprint varnish. Although the varnish is expensive by conventional paint standards, the low coat weight is economical against plastic film and laminating adhesive, and it can be roller coated and cured with two UV lamps at speeds of 60m/min. This high output with low capital cost leads to overall finishing economies without loss of quality.

The UV coating must crease without cracking and peeling, must be scuff-resistant, have good adhesion over inks, be solvent-resistant, and withstand over-wrapping at 160°C. The high-gloss finish should have good 'slip' (low coefficient of friction), yet it must accept foil blocking with metal foils.

Very high standards of worker safety are required in the printing industry, and modern UV coatings are based on materials of such low toxicity that skin irritation indices (Draize ratings) of the monomers used must be no higher than those of long-established coatings solvents.

Other end uses for clear film laminates are also amenable to UV-curing finishing processes.

12.5   Developments and trends in general industrial finishing

The past decade has been one of considerable change in industrial finishing, change that has been brought about as the result of the following pressures:

- Legislation and pressure against pollution.
- Legislation and pressure for increased safety in the work-place.
- Rising costs and increasing competition.
- Economic recession.

12.5.1   Anti-pollution and safety pressures

Economic pressures are always the most immediate for end users of paint, but the major pressure for change has undoubtedly been the ever-intensifying drive against pollution. It has been claimed that 70% of all expenditure on research into new coatings has been directed towards this objective [17].

Most publicity has been given to the need to reduce solvent emissions into the atmosphere, but pressure has also intensified against pollution of water and land by dumping solid and liquid waste. This has added impetus to the search, previously begun for reasons of safety in the work-place, for less toxic ingredients in paints and their pretreatments, including pigments, film-forming ingredients, and additives. As toxicology research expanded, so ingredients previously thought to be harmless came under suspicion. This led to their replacement if possible or, if no acceptable alternatives were available, to improved handling procedures and protective measures.

It has also directed a spotlight onto those methods of paint application with low transfer efficiencies, since paint not arriving on the article to be coated is initially an air-pollutant and then, if it is removed from the air, presents a disposal problem as, for example, spray booth sludge or dirty cleaning liquid. Thus spray equipment manufacturers have sought to improve transfer efficiency by developing high volume low pressure (HVLP) equipment [22]. There has been a move towards all types of electrostatic spraying or, where possible, to even more efficient methods of application.

Some end users have got rid of the pollution problems associated with paint altogether by closing their paint shops and moving to prepainted coil [13]. Others have minimized disposal problems by changing product and process, e.g. avoiding sludge from coil pretreatment tanks by changing to rollercoated dried-in-place pretreatments. There has been more interest in using solvents that evaporate in ovens by burning them in after-burners and recovering the energy to heat work places or pretreatment tanks. Recovery of useful materials from sludge is also being investigated.

Solvent levels of conventional coatings have continued to drop and sales of solvent-borne paints have declined in favour of water-borne paints particularly, which — although they are not solvent-free — have the additional benefit of being generally non-flammable. Growth continues in other coatings that contain little or no air pollutants, such as powder coatings and unsaturated coatings cured by UV or electron beams.

12.5.2 Economic pressures

In the face of rising costs and static prices all conceivable economy measures have been taken. The most obvious of these is to reduce overall paint thickness, either by reducing the number of coats in a system, or by reducing the thickness of one coat. Few, if any, manufacturers were prepared to accept a lowering of finished appearance or performance. Often this led to a change to new materials, which might have been more expensive in themselves, but gave superior results at lower coat weights.

Reductions in energy costs have been achieved by reduction in stoving time or temperature (with attendant paint modification or change), or change to less energy-consuming methods of cure, such as EBC, UV, and vapour cure, or by moving to 'cold' pretreatments.

But for those able to afford the capital investment, the most impressive long-term economies have been achieved by the introduction of automated painting processes with application techniques which controlled coat weights closely and minimized paint losses or waste.

In the recession, companies that did not or could not take measures like these have gone out of business, and other companies have ensured their survival by amalgamating with their rivals.

In the next decade the above trends can be expected to continue and the anti-pollution pressures to intensify. Some of the problems holding back the growth of powder coatings are likely to be resolved. Water-based coatings with very low solvent levels will become available. The use of radiation-curing will be expanded by coatings which have built into them a second curing method. Some new curing chemistry for low temperature processes is likely. Towards the end of the decade attention may return to petroleum as a finite resource and to the manufacture of coatings from renewable resources.

References

[1] TATTON W H & DREW E W, *Industrial Paint Application*, p. 47, Newnes (1971).
[2] MARINO F P, *Proc SME Conf. 'Finishing '87'*, Cincinnati, Paper FC87-611, 10–11 (1987).
[3] GARSIDE J S, *Product Finishing*, **46** (9) 12–15 (1993).
[4] Intertherm Ltd, *A First Guide to Dielectric Heating*, London (1973).
[5] PRAY R W, *Radiation Curing* **5** (3), 19–25 (1978).
[6] WHITE R G, *Handbook of Industrial Infra-red Analysis*, p. 178, Plenum Press (1964).
[7] *Kronos Guide*, Kronos Titanium Companies (1968).
[8] PATTON J C, *Pigment Handbook*, Vol. III, John Wiley (1973).
[9] SCHWALM H, *Ind-Lack-Betrieb*, **56** (7) 239–240 (1988).
[10] *Ind.-Lack.-Betrieb* **56** (11) 372–375 (1988).
[11] *Finishing* **15** (9) 58–59 (1991).
[12] DONATI O, *Pitture Vernici* **64** (9) 95–104 (1988).
[13] MIRANDA T J, in *Organic coatings: Their Origin and Development* eds Seymour R B & Mark, H F, pp. 69–85, Elsevier (1990).
[14] European Coil Coating Association, *Pitture Vernici* **68** (3) 7–10 (1992).
[15] SIMMONS G C, *Polym Paint Colour J* **183** (4333) 372–373 (1993).
[16] WINNER P P, *Proc Tappi Polymers, Laminations and Coatings Conf, Boston*, **2**, 501–505 (1990).
[17] POHL E E, *Volatile Organic Compounds (VOC): Policy of the European Paint Industry*, pp. 32–39, Comité Europeen Peintures Encres (1991).
[18] *Fahrz. Metall Lack* **32** (3) 31–32 (1988).
[19] DOBSON D, *Proc SME Conf 'Finishing '89'*, Cincinnati, Paper FC89-608 (1989).
[20] DANIELS C, *Products Fin.* **57** (3) 75–78 (1992).
[21] *Natural Gas*, pp. 14–15, March/April (1990).
[22] CHATTOPADHYAY A K & ZENTNER R, *Aerospace and Aircraft Coatings*, Federation of Societies for Coatings Technology, Philadelphia (1990).
[23] FUJIHARA G, *Mod Paint Coatings* **78** (9) 28–30 (1988).
[24] *Mod Paint Coatings* **78** (9) 35 (1988).

13

The painting of ships

R Lambourne

13.1 Introduction

The painting of ships and structural steel work, such as oil and gas rigs, poses some of the most formidable problems in respect of corrosion protection that the paint technologist has to face. Salt-water immersion or partial immersion, spray followed by drying winds, are extemely aggressive environments. They call for the development of paints which protect the steel from which the ship's hull is constructed, by means of a number of different mechanisms which we shall discuss briefly. In addition to problems of corrosion, the fouling of ships' bottoms with marine organisms leads to increased drag on the vessel and hence to increased fuel consumption. Gitlitz [1] cites an example of the economic effects of fouling, in relation to the operation of a VLCC (very large cargo carrier) with an operational speed of 15 knots. Fuel is consumed at the rate of about 170 tonnes per day. Assuming that the ship is at sea for 300 days in the year, the fuel cost (at the 1981 price of $80 per ton) exceeds $4m. He suggests that even moderate fouling can increase the fuel required by as much as 30% to maintain the optimum operating speed, i.e. by over $1m in a year. The availability of effective anti-fouling compositions is clearly of prime interest to the ship owner. The conclusions drawn by Gitlitz are just as relevant today — perhaps even more so in the case of the faster container-ships — although the cost of fuel oil fluctuates and at the time of writing the cost of heavy fuel oil is about £65/tonne. It is not uncommon for 10lb per square foot of fouling ($50\,\mathrm{kg\,m^{-2}}$) to accumulate in some marine environments within six months in the absence of an effective anti-fouling. Fouling contributes to the breakdown of protective coatings, and hence to the earlier onset of corrosion. It is the aim of the ship owner to avoid frequent and expensive dry-docking. He or she has therefore a vested interest in improved anti-corrosive and anti-fouling paint systems. It is interesting to note, however, that the break up of the Eastern Bloc countries has led to a significant deterioration in the maintenance of their merchant navy fleets, many of the independent nations being unable to fund the purchase of marine paints in the free markets of the world. Examples have been described of Russian factory ships and fishing vessels rusting, through lack of necessary maintenance, to the stage at which they are barely seaworthy. No doubt these economic problems will be overcome when these countries have developed basic market economies.

Of course only the part of the hull that is immersed is subject to fouling, and it is important to identify and classify other types of surface which call for painting, each requiring a different type of paint. We have already mentioned the anti-corrosive/anti-fouling paints that are used for ships' bottoms. The area between the light load line (the water line when the ship is in ballast) and the deep load line (the water line when the ship is carrying a full cargo) is known as the boottopping. This is sometimes below the water level and sometimes exposed to the marine atmosphere, alternating between the two in service. Also, this area is subject to damage, e.g. abrasion by contact with jetties or wharves and with other vessels, such as tugs, in spite of the use of fenders. Above the boottopping are the topside and superstructure, terms that are self-explanatory. In addition to these areas there are speciality paints for decks, interiors, engines, etc. In naval vessels intumescent paints, designed to prevent the spread of fire, have been developed, but doubt exists as to the efficacy of such systems after ageing or subsequent repainting. In the light of the experience gained during the Falklands war, it is doubtful if these paints would have had more than a marginal effect on the spread of fires following a direct hit from a missile. However, they may be useful in reducing the spread of minor fires.

The painting of ships can be considered in two contexts: the painting during construction and the repainting during service. The difficulties of painting ships in service arise from the limited opportunities available to apply the paint under anything like ideal conditions. Paints as supplied must therefore be sufficiently 'robust' to enable them to be applied under adverse conditions, for example, with the most rudimentary surface preparation. Certain areas such as the topside and superstructure may be repainted at quite frequent intervals, the latter often while the ship is at sea.

The paint formulations given in Section 13.5 represent in the main the conventional technology of the early 1980s. As such, they exemplify the diversity of the products in relation to the variety of applications of marine paints. Many of these types of formulation or similar formulations derived from them remain in use today, except in certain areas such as in anti-fouling paints, where research has been necessary to develop novel, even more effective paints that are 'environmentally friendly'.

In this chapter we shall consider briefly the requirements for the main types of marine paints. However, before doing so we shall examine in simple terms the mechanism of corrosion of steel and the methods by which it can be eliminated or at least retarded.

13.2   Corrosion

Corrosion is an electrolytic process involving the reaction between iron metal, oxygen, and water resulting in the formation of hydrated ferric oxide or rust. The overall reaction is:

$$4Fe + 3O_2 + 2H_2O \rightarrow 2Fe_2O_3 \cdot H_2O \tag{13.1}$$

The generally accepted reaction mechanism is one which involves anodic and cathodic processes. The surface of the iron or steel in contact with water develops localized anodes and cathodes at which these processes take place. Electron flow (constituting a 'corrosion current') occurs through the metal, and this is complemented by an equivalent transport of charge through the water or electrolyte by hydroxyl ions. The process is shown diagrammatically in Fig. 13.1. The individual electrode processes are as follows.

At the anode: the formation of ferrous ions by the loss of electrons,

$$4Fe \rightarrow 4Fe^{++} + 8e \tag{13.2}$$

and at the cathode: the formation of hydroxyl ions,

$$4H_2O + 2O_2 + 8e \rightarrow 8OH^- \tag{13.3}$$

The initial product of oxidation is thus ferrous hydroxide,

$$4Fe^{++} + 8OH^- \rightarrow 4Fe(OH)_2 \tag{13.4}$$

In the presence of excess oxygen the ferrous hydroxide is oxidized to hydrated ferric oxide, the all too familiar red oxide which is rust.

The anodic and cathode regions at the surface of the metal arise from compositional heterogeneity of the surface. This may be due to a number of factors including grain boundaries, stresses, and microscopic faults that cause local concentration gradients of electrolyte or oxygen in solution. Any of these giving rise to a potential difference between adjacent areas in the surface is sufficient to cause galvanic action.

In dealing with ships under construction one is faced with painting steel that has at its surface a layer of oxides up to 60 μm in thickness that is formed when the steel is manufactured. The steel is subjected to hot rolling to the required thickness, and this is carried out in the temperature range 800–900 °C. Oxidation of the steel takes place during cooling. This oxide layer is known as 'millscale'. Millscale, unless removed, can cause corrosion because there is a significant potential difference between the millscale and the bare steel (about 300 mV) when the steel is immersed in electrolytes such as sea-water. The millscale is cathodic, and the bare steel anodic. Thus where there is a crack or fissure extending through the millscale layer to the metal a galvanic cell is formed and the corrosion process begins. The rust is formed in the vicinity of the anode which is in contact with the salt-water, but not directly on the anodic site. Thus the anode is slowly dissolved, and pitting occurs. The cathode site undergoes no such dissolution. All millscale is now removed by blast cleaning,

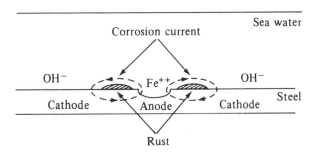

Fig. 13.1 — The corrosion process.

and the steel is 'shop primed' before fabrication. However, this does not mean that corrosion is entirely eliminated.

How, then, can corrosion be reduced or eliminated? Two approaches are possible:

- The elimination or inhibition of the electrolytic processes.
- The exclusion of water or oxygen from the potential corrosion site.

The cathodic process requires the presence of water and oxygen at the metal surface, so that it can be seen that the two factors indicated above are interrelated. In practice it is very difficult to prevent oxygen and water reaching the metal surface because most paint binders have quite high permeability to these agents. Materials that are low in permeability to water and oxygen are, for example, waxes, because of their crystalline nature. Similarly, crystalline polymers exhibit low permeability, but these are difficult to formulate into air-drying compositions. In general, conventional marine paints do not confer cathodic protection. At the anodic surface, for corrosion to take place ferrous ions are required to pass into the electrolyte. This may be suppressed by the use of paint systems of two kinds. The first uses metallic zinc as pigment at a very high pigment concentration, such that contact between the zinc particles and the iron surface takes place. In this case, because zinc is more electronegative than iron, a galvanic couple exists between the two metals, the iron being cathodic and the zinc anodic. The net result is the suppression of the dissolution of ferrous ions and the consequent corrosion of the zinc. The second kind of cathodic suppression can be achieved by using paints containing inhibiting pigments such as red lead, white lead, or zinc oxide. These pigments are used in conjunction with binders based upon drying oils that undergo autoxication. In addition to the formation of crosslinked films some scission of the drying oil triglycerides occurs with the formation of oxidation products such as azelatic acid. These acids react with the pigment to form lead or zinc soaps as the case might be. The soaps react with the surface oxide film on the metal and prevent the ingress of water which is necessary for the formation of rust. A second group of pigments that act in a similar way are those that exhibit limited water solubility. These include the chromates of zinc, barium, strontium, and lead. In these cases the oxide film is protected by the formation of an iron/chromium oxide complex.

Recently much concern has been expressed about the toxicity of these pigments, and work is in progress to eliminate lead and chromate pigments from all paints.

Two other ways of suppressing the electrolytic processes that are responsible for corrosion have been used on ships:

- The use of 'sacrificial anodes', in which anodes are built onto the hull, using alloys that are electropositive with respect to the steel of the hull. These anodes are unpainted and their purpose is to corrode away, being replaced when necessary.
- The application of a controlled external current to the hull to suppress the dissolution of ferrous ions. This is called the 'impressed current method'.

Neither method eliminates the need for painting the hull for other obvious reasons.

The foregoing is undoubtedly an oversimplification of the processes involved in the corrosion of steel when immersed in natural sea-water. For example, bacteria and other microorganisms can cause corrosion to take place even under anaerobic conditions. The reason that a film of microorganisms can influence an electrochemical process such as corrosion is because of its ability to change the chemistry of the environment at the locus of corrosion. Thus, a biofilm can act as physical barrier to the diffusion of oxygen through the film and also act as an active oxygen scavenger during respiration; it can also have a significant effect on the local pH. The influence of biofilms formed on metal surfaces in natural sea-water is thought to account for differences in rates of corrosion observed between metal panels tested in natural sea-water and those tested in artificial sea-water used in laboratory testing. It is not possible within the scope of this book to pursue this topic further. Readers wishing to delve more deeply are directed to Dexter's authoritative paper on the role of microfouling organisms in marine corrosion [2].

13.3   Surface preparation

To obtain satisfactory performance from a paint system it is essential for millscale to be removed from the surface of the steel. Not only does it contribute directly to corrosion, but it will eventually become detached from the steel surface and is therefore an unsatisfactory substrate for painting. Surface preparation is vitally important when one is seeking to obtain the best performance from a paint system.

It is only since the early 1960s that cost-effective surface treatments for millscale-coated steel have been developed. Before World War 2, when ships were built from a keel and a frame and were riveted, it was common for millscale to be removed by weathering over periods often in excess of a year. The steel plate was simply exposed to the elements and allowed to rust. The rusting enabled the millscale to be detached by scraping and wire brushing. After World War 2, with the development of welded ships, the shorter time to build did not allow for millscale to be weathered off. More recently a number of alternative methods have been used, including acid pickling, flame cleaning, and grit blasting. The last-named is the most satisfactory and commonly used method of surface preparation today.

Two main methods are in general use, mechanical blast cleaning and 'open' or 'nozzle' air blasting. In the former, steel shot or grit are commonly used with the abrasive being recovered after being separated from the millscale particles. The abrasive is normally coarser than the dust that is produced in the process, and it is capable of being separated by a cyclone separator. Some attrition of the abrasive occurs, so that in spite of recycling there is a gradual consumption of abrasive. The machines using this method employ a rotary impeller which projects the abrasive particles at high velocity at the surface. The open or nozzle blasting method uses compressed air to project the abrasive at the surface. This method is more versatile, since it can be used on site on preformed structures, over welds, etc., but suffers from the disadvantage of dust generation. A modified method allows the abrasive and the dust to be removed by applying a vacuum to an annulus around the cleaning nozzle, e.g. the Vacublast method. However, this method is much slower and is used only on small areas where dust cannot be controlled or permitted. After blast cleaning it is important to apply a protective coating as soon as possible to prevent rusting of the clean surface. Blast or prefabrication primers are used for this purpose.

13.4   Blast primers

Blast primers are applied by spray immediately after blast cleaning. They must dry quickly (within a few minutes of application) to enable the steel to be handled without significant damage to the coating. The primer must not adversely affect the subsequent fabrication process. For example, it should not prevent successful welding. It must also provide a suitable surface for subsequent painting.

Four types of blast primer are in common use:

- phenolic/polyvinyl butyral (PVB), acid cured, one- and two-pack compositions;
- two pack cold curing epoxy compositions, pigmented with red iron oxide (and optionally an inhibitive pigment such as chromate);
- two-pack cold curing epoxy compositions pigmented with zinc dust; and
- zinc silicate one- and two-pack compositions.

The pot life of the paint has to be about 8 hours to be of practical use. The epoxies are usually crosslinked with a polyamide such as Versamid 140 or 115. In the two-pack zinc silicate, pigmentary zinc is stirred into the paint at the time of use.

Typical formulations of these types of primer, as described by Banfield [3], are shown in Table 13.1.

The phenolic/PVB types are the most widely used, and are capable of being over-coated with most systems except in situations where they are likely to encounter strong solvents. The two-pack epoxy is the second most widely used. Both the red oxide and zinc-rich types can be used under practically any paint. They do, however, suffer the disadvantage of being two-pack, and can be more expensive than the phenolic/PVB types. The zinc silicate primers are mainly used when they are to be over-coated with zinc silicate systems.

Zinc-metal-containing paints give the best overall protection against corrosion, but they do have several disadvantages in use that will sometimes dictate the use of one of the other types of primer. The corrosion products of zinc are water-sensitive, and they must be removed before being overcoated with another paint. This can be done, for example, by high-pressure hosing with fresh water. Another problem associated with zinc-containing blast primers is the formation of zinc oxide fume when coated plate is welded. Welders often object to this, although it is not established that it presents a long-term health hazard. The inhalation of zinc vapours may give rise to a condition known as 'zinc fume fever'. This has been known in the galvanizing industry for over a century.

13.5   Paint systems for ships

The protection of steel plate after manufacture has been described briefly above. The subsequent painting of ships during and after construction is a highly complex technology. As has been indicated, different surfaces require different treatment. Each surface requires a *system*, since it is impossible to combine all the requirements for any one surface within one paint. Thus multi-coat systems are essential, and for any one type of paint it has become the convention to apply several coats to build up film thickness to minimize corrosion occurring, for example, through pinholing. Typical systems for topside and superstructure paints, boottopping, and

Table 13.1 — Formulations for blast primers

One-pack phenolic/PVB blast primer

Paint base	% by weight
PVB (e.g. Mowital B6OH)	4.7
Phenodur PR263 (phenolic resin)	4.7
Red oxide (synthetic)	8.6
Asbestine (magnesium silicate)	1.4
Phosphoric acid, SG 1.7	1.0
Methyl ethyl ketone	30.0
n-Butanol	19.6
iso-Propanol	30.0
	100.0

Two-pack epoxy blast primer

Paint base	% by weight
Epikote 1004	12.0
Zinc phosphate (to BS 5193:1975)	8.0
Red oxide (synthetic)	10.0
Asbestine	1.5
Talc (5 µm)	15.3
Bentone 27 (anti-settling agent)	0.4
Toluene	23.0
iso-Propanol	11.5
Crosslinking agent	
Versamid 140 (polyamide)	1.8
Toluene	11.0
iso-Propanol	5.5
	100.0

Two-pack zinc-rich epoxy blast primer

Paint base	% by weight
Zinc dust (Zincoli 620 or Durham Chemicals 'Ultrafine')	80.0
Bentone 27	1.0
Calcium oxide	0.3
Epikote 1001	4.0
Toluene	6.6
iso-Propanol	3.3
Crosslinking agent	
Versamid 115	1.8
Toluene	2.0
iso-Propanol	1.0
	100.0

Note: The calcium oxide acts as a water scavenger, reacting with moisture in the solvents. The moisture would otherwise react with the zinc to form hydrogen in the can.

Zinc/ethyl silicate blast primer

Paint base	% by weight
Ethyl silicate (e.g. Dynasil H500)	20.0
Bentone 38, anti-settling agent	1.4
Talc (5 µm)	4.0
Toluene	5.3
iso-Propanol	5.3
Cellosolve	4.0
Zinc dust (e.g. Zincoli 615)	60.0
	100.0

The zinc dust is stirred into the base at the time of use. The product then has a useful life of about 8 hours.

bottoms are summarized for what are regarded as traditional or conventional marine paints in Table 13.2.

In recent years major changes have taken place in painting practices and paint formulation to meet the needs imposed by the development of the supertanker. The development of the supertanker came about with the closing of the Suez Canal in 1967 as a result of the Middle East war. Supertankers were required to make the transport of oil from the Middle East an economic proposition. This could not be achieved with the typical tanker of the early 1950s, having a tonnage of about 30 000 gross. The average supertanker of the 1980s has a gross tonnage of 300 000, and it operates with a smaller crew than its predecessor. This means that fewer crew are available to do maintenance painting on a much larger ship. The overall effect of these changes has required the development of paints that require the ship to undergo less frequent and shorter periods in dry dock. To achieve this, paints giving higher build with fewer coats yet providing improved performance have had to be produced.

Thus a new generation of paints has been developed that enable the periods between dry docking to be increased from about 9–12 months to 24–30 months.

Table 13.2 — Summary of 'conventional' marine paint types

Surface	Paint system	Binder type	Coats	Dry film thickness (μm)	Total average film thickness (μm)
Bottom	Primer	Oil-modified phenolic; bitumen	3–4	40–60	
	Barrier coat No.1 anti-corrosive	Bitumen/limed rosin	1	50–60	~265
	Anti-fouling	'Soluble matrix' based on boiled oil/limed rosin; Cu or tributyltin oxide (TBTO) toxicant[†]	1	50–75	
Boottopping	Primer 655	Tung/linseed-modified phenolic	2–3	40–60	~200
	Finish	Tung-modified phenolic	2	40–60	
Topside and superstructure	Primer (quick-drying)	Medium oil length (a) linseed alkyd (55–60% OL)	2	50	
	Undercoat	Medium oil length linseed alkyd (50–55% OL)	1	50	~220
	Enamel	Long oil length soya alkyd (65% OL)	1–2	35–50	

[†] Three grades are available, 'Atlantic', 'Tropical', and 'Supertropical' in which the level and choice of toxicant vary.

Even this extended period is not considered to be the limit of modern technology, and some paint manufacturers are aiming to extend the period between dry docking up to five years.

These requirements have been met by developing systems which are capable of being applied as thick coats, each dry film being at least 100 μm thick. These modern products are most commonly applied by airless spray. One airless spray gun is capable of spraying between 50 and 80 litres of paint per hour, i.e. covering 150–400 square metres per hour at the required film thickness. To avoid sagging on vertical surfaces the paints must exhibit non-Newtonian rheology, that is, they must be shear-thinning to facilitate flow through the gun and atomization, but must rapidly develop structure in the liquid film. Thus they may have a viscosity of about 5–10 poise at $10000 \, s^{-1}$, and will probably reach a viscosity of ~1000 poise at the low shear rate applicable to flow under gravity on a vertical surface, e.g. $10^{-2} \, s^{-1}$. This type of behaviour is achieved by using thixotropic or gelling agents such as the montmorillonite clays or polyamides. Many of the high performance systems are two-pack. The anti-corrosion performance obtained with these systems relies more on the suppression of electrolytic action by acting as thick barrier coats than the specific action of inhibitive pigments, although the primers used do often contain such pigments. The toxicity of lead and chromates has called for a concerted effort to find replacements for them. In this context, zinc phosphate has increased in use.

13.5.1   Topside and superstructure paints

Conventional paints meeting the requirements of topsides and superstructures come closest to those used in the trade paint market, requiring primers, undercoats, and gloss coats that dry at ambient temperature. These have for many years been based upon oleoresinous and alkyd binders. Thus, a primer would use an oleoresinous vehicle (e.g. an oil-modified phenolic) with an inhibitive pigment, usually red lead. An undercoat would be formulated on a medium oil length linseed alkyd. The topcoat might be based on a long oil alkyd or a vinyl-toluenated alkyd. Pigmentation in the latter case would be to achieve opacity and other aesthetic purposes. The vinylated alkyd confers an improved drying rate, but suffers from the disadvantage that special care has to be exercised in recoating, because the initial dry occurs as a result of solvent evaporation (lacquer-drying), and lifting can arise if the first coat is recoated after 12 hours but before 16 hours have elapsed, particularly if the ambient temperature is low. This effect is due to the relatively slow rate of crosslinking of this type of binder.

Modern high-performance systems are based on a wide range of binders including epoxies, polyurethanes, and chlorinated rubber. The last-named is often used in blends with acrylic or alkyd resins.

A typical epoxy system for topsides would, for example, involve applying two coats of a high-build primer/undercoat followed by one coat of epoxy enamel, giving a total dry film thickness of about 300 μm. An alternative for the superstructure, where gloss retention is more important, would be to use one coat of primer, followed by an epoxy primer/undercoat and a polyurethane topcoat. The dry film thickness would be approximately 200 μm in this case. Examples of a red oxide epoxy thick coating and a white epoxy enamel are given in Table 13.3.

Epoxies can be crosslinked with amines, diethylene triamine and triethylene tetramine commonly being used. Amines are often pre-reacted with epoxides to

Table 13.3 — Formulation of a red oxide thick coating and a white epoxy enamel

Paint base	% by weight
Synthetic red oxide	5.0
Asbestine (magnesium silicate)	22.0
β Crystoballite (silica) to pass through a 325 mesh sieve	18.0
Epikote 1001	16.0
Bentone 27	2.0
Xylene	14.0
n-Butanol	5.0
Curing agent	
Versamid 115 (polyamide)	8.0
Cellosolve	10.0
	100.0

A typical white epoxy enamel formulation:

Paint base	% by weight
Rutile titanium dioxide (e.g. Tioxide R-CR2)	30.0
Aerosil 380 (anti-settling agent)	0.5
Epikote 1001	23.0
Solvesso 100 (naphtha)	16.5
Methyl isobutyl ketone	8.0
n-Butanol	3.0
Curing agent	
Versamid (polyamide)	11.0
Accelerator (e.g. Anchor K54)	1.0
Xylene	7.0
	100.0

form adducts which have certain advantages over amines *per se*. The adduct will usually be formed with part of the epoxide that will be used in the paint base. By using this approach, the odour of the free amine can be reduced or eliminated, and the mixing ratio is more convenient, e.g. 2 or 3:1 instead of about 10:1, paint:activator.

Two-pack polyurethanes are also used in topcoat enamels. The paint base contains a hydroxyl-containing saturated polyester resin (e.g. Desmophen 650); the curing agent will be an isocyanate-containing moiety (such as Desmodur N) as a 75% solution in a 1:1 mixture of xylene and cellosolve acetate. Polyurethanes are used when good gloss and gloss retention are required.

High build, high performance paints can also be formulated in one-pack compositions based, notably, on chlorinated rubber and vinyl resins. Chlorinated rubber paints have gained wider acceptance than vinyls although the latter have been developed for use by the US and Canadian navies. Typical formulae for a primer, undercoat, and topcoat all based upon chlorinated rubber are given in Tables 13.4–13.6.

The composition given in Table 13.4 calls for careful formulation and manufacture. The plasticizer is a preformed blend of a solid (waxy) chlorinated paraffin with a liquid chlorinated paraffin. The thixotropic agent is based on hydrogenated castor oil. This develops structure only when it is incorporated during the milling stage, when the temperature rises, depending on the scale of manufacture, to over 40°C. The temperature should not exceed 55°C, otherwise the thixotropic

Table 13.4 — Chlorinated rubber primer

	% by weight
Aluminium paste, non-leafing (65% in naphtha)	20.0
Zinc oxide	1.0
Blanc fixe	10.0
Thixatrol ST (thixotropic agent)	0.5
Alloprene R10 (chlorinated rubber)	15.0
Cerechlor 70 (chlorinated wax plasticizer)	7.5
Cerechlor 42 (chlorinated wax plasticizer)	7.5
Propylene oxide (stabilizer)	0.1
Xylene	28.4
Solvesso 100 (Naphtha)	10.0
	100.0

Table 13.5 — Chlorinated rubber undercoat: white

	% by weight
Titanium dioxide (e.g. Tioxode R-CR2)	16.0
Blanc fixe	14.0
Zinc oxide	1.0
Thixatrol ST (thixotropic agent)	1.0
Alloprene R10 (chlorinated rubber)	14.0
Cerechlor 70 (chlorinated plasticizer)	7.0
Cerechlor 42 (chlorinated plasticizer)	7.0
Propylene oxide (stabilizer)	0.1
Xylene	29.9
Solvesso 100	10.0
	100.0

Table 13.6 — Chlorinated rubber topcoat: white

	% by weight
Titanium dioxide (e.g. Tioxode R-CR2)	18.0
Blanc fixe	2.0
Zinc oxide	1.0
Bentone 38 (thixotropic agent)	1.0
n-Butanol (swelling agent)	0.5
Alloprene R10	16.0
Cerechlor 70	12.0
Cerechlor 42	4.0
Propylene oxide	0.1
Xylene	30.4
Solvesso 100	15.0
	100.0

effect is lost. The aluminium paste is not milled in with the other pigments, but is stirred in when the batch has cooled. The propylene oxide is an 'in-can' stabilizer which 'mops up' Cl^- ions that are generated slowly from the chlorinated rubber on storage.

In the formulation given in Table 13.6, the reduced amounts of filler (cf. undercoat), the use of Bentone 38 instead of Thixatrol ST, and the higher proportion of Cerechlor 70, are all measures designed to improve the gloss of the topcoat. Even so, the gloss is low compared to epoxide or polyurethane paints. These paints are all applied by airless spray. A typical system would be:

	Dry film thickness (μm)
Primer, one coat	50
Undercoat, two coats	200
Topcoat, one coat	50
Total	300

13.5.2 Boottoppings

As distinct from the traditional boottoppings mentioned previously (Table 13.1), high performance compositions for this area are based essentially on the same binders as the topside paints. In this category the chlorinated rubbers perform very well because of their excellent intercoat adhesion. This is particularly important because of the damage to the paint system from abrasion and impact with fenders and quays, and the consequent need for frequent repainting. With VLCCs, fouling can also be a major problem in the boottopping area, and it is common practice to use an anti-fouling composition as the finishing paint. Thus it is usual to employ the same anti-fouling as is used on the bottom.

13.5.3 Bottom paints

The high performance/high build systems described for topsides and boottoppings are also used for bottoms. However, since asethetic reasons are not predominant it is possible to formulate such paints at lower cost, without sacrificing performance, by the incorporation of coal tar. The paints are thus usually chocolate brown or black, but this is of no real significance. The proportions of coal tar in an epoxy coal tar thick coating can be varied over a wide range. The higher the coal tar content, the poorer the oil or chemical resistance, but this is less important in underwater compositions. The coal tar pitch in a formula complying with the UK Ministry of Defence (Ship Department) Specification DGS 5051 would represent about 60–65% of the total binder. It would be selected according to its compatibility with the resin; and, although regarded as essentially unreactive, it can react with the epoxy groups because it contains some phenolic hydroxyl groups. For this reason it is often incorporated into the curing agent part of the two-pack composition, otherwise the storage life of the base would be inadequate. A typical formulation, as given by Banfield [3], is given in Table 13.7.

After mixing, the paint is applied by airless spray to give a dry coat thickness of 125 μm per coat. The pot life is about 4 hours at 15 °C. The reaction rate doubles, approximately, with a 10 °C rise in temperature, or is halved with a 10 °C drop in temperature. Thus the paint can have a pot life varying from about 8 hours at 5 °C to only 2 hours at 25 °C. The rate of cure also affects recoatability. To achieve satisfactory intercoat adhesion and to avoid lifting, maximum and minimum recoating times are specified, according to the temperature of application.

Table 13.7 — Typical formulation for two-pack bottom paint

	% by weight
Base	
Barytes	20.0
Asbestine	20.0
Epikote 1001	11.0
Bentone 27	1.0
Xylene	10.0
Solvesso 100	5.0
n-Butanol	3.0
Curing agent	
BSC Norsip 5 (75% aromatic coal tar in xylene)	25.0
Versamid 140	2.5
Synolide 968	0.3
Xylene	2.2
	———
	100.0

Table 13.8 — Aluminium tie coat

	% by weight
Aluminium powder, non-leafing	14.0
Basic lead sulphate	14.0
Gilsonite (bitumin)	22.4
Linseed stand oil, 50 poise	5.6
Coal tar naphtha, 90/160	44.0
	———
	100.0

This type of composition may be used both on ships under construction and those undergoing 'in service' repainting. The former poses a problem in that the application of an anti-fouling will not take place until shortly before the ship goes into service, and this may well be long after the optimum time for overcoating the coal tar epoxy paint. In this case a 'tie coat' is used. Tie coats are also specified by the UK Ministry of Defence (Specification DGS 5954). A composition that meets this requirement is given in Table 13.8.

The tie coat is applied by airless spray to give a dry film thickness of about 50 µm. The anti-fouling coating can then be applied shortly before launching.

Alternative bottom paints may also be based on chlorinated rubber and vinyls. A fuller treatment of the formulations commonly used is beyond the scope of this chapter. The reader seeking more information is referred to Banfield's excellent monographs on this subject [3].

13.5.4 Anti-fouling coatings

13.5.4.1 The problem of fouling
Marine organisms, both plant and animal, are able to attach themselves to the hulls of ships. Plant growth requires some daylight to sustain it, so attachment occurs to the sides of the hull in the upper regions of the underwater area. The most common

of these are of the genus *Enteromorpha*, seaweeds that have long green tubular filaments that grow very rapidly under favourable conditions. The brown and red seaweeds require less light, and these can grow at lower levels or even on the bottom of the ship. Firm attachment of these plants can take place in a few hours unless they are prevented from doing so.

Aquatic animals that are capable of attaching themselves to ships include barnacles, mussels, polyzoa, anthozoa, hydroids, ascidians, and sponges. The best known (and most troublesome) are the barnacles. These animals do not require light to sustain growth, so they are to be found on all immersed parts of the hull. The larvae require about 48 hours to become firmly attached. The consequences of fouling have already been mentioned, but it is worth repeating that the most important consequence of fouling with large modern ships is the increased cost of propelling the vessel through the water as a result of the increased drag that fouling imposes.

The first 'anti-fouling' paints based upon the use of toxicants were introduced in the middle of the 19th century. These were oleoresinous compositions containing mercuric and arsenious oxides. Subsequently, because of the high human toxicity and the increasing cost of mercuric oxide, alternative paints using cuprous oxide were developed. Cuprous oxide remains the most commonly used toxicant in the 'conventional' types of composition today.

13.5.4.2 Conventional anti-foulings
There are two types of conventional anti-fouling paints, classified according to their mode of action. These are generally known as 'soluble matrix' and 'contact' types.

Soluble matrix type
In the soluble matrix type the toxicant is dispersed in a binder that is slightly soluble in sea-water. The toxicant is slowly released as the binder dissolves. The rate of dissolution has to be controlled very carefully since an inadequate concentration of the poison at the surface will not prevent the attachment of the marine organisms, whereas too rapid dissolution will mean that the effectiveness of the anti-fouling will be too short-lived. The binder used in this type of anti-fouling might be a 3:1 limed rosin/boiled linseed oil mixture.

A better 'supertropical' grade might have the following composition given in Table 13.9. 'Atlantic' or 'tropical' grades would not necessarily call for the use of tributyltin oxide.

Table 13.9 — Formulation for supertropical anti-fouling composition % by weight

Cuprous oxide	24.0
Zinc oxide	5.0
Tributyltin oxide	2.0
Red iron oxide	8.0
Paris white	9.0
Talc (10 μm)	6.0
Bentone 38	0.3
n-Butanol	0.1
Limed rosin/boiled linseed oil, 3:1, 60% in naphtha	41.0
White spirit	4.6
	100.0

Contact type

The contact type of anti-fouling composition is so named because it is formulated at a very high pigment volume, such that the pigment (toxicant) particles in the disperse phase are in contact with each other in the dry film. The binder is largely insoluble in sea-water, so that when dissolution of the toxicant takes place a porous film of the paint binder remains on the surface. The binder is not entirely insoluble, and the balance between binder solubility/insolubility plays a part in controlling the rate of leaching of the toxicant. Although these paints are formulated to give a close-packed pigmentary structure in the dry film, it is doubtful if this is achieved or is desirable in practice. Any significant flocculation of the particles of toxicant will lead to a more open structure, and the paint may in fact be underbound. This does not appear to be a major problem in practice. A typical formulation for a contact type anti-fouling is given in Table 13.10.

One of the most important features of anti-fouling compositions is the rate at which they release the toxicant. Various organisms exhibit different degrees of sensitivity to the poison. In the case of barnacles about $10\,\mu g$ of copper per cm^2 per day is sufficient to prevent their attachment. There is some evidence of synergism between poisons. This is one of the reasons that copper and tributyltin oxide are used together to meet more demanding situations, for example, in 'supertropical' grade anti-foulants.

13.5.4.3   Modern anti-foulings

Just as the development of larger vessels (e.g. supertankers) has required paint systems that call for less frequent maintenance occasioned by paint breakdown and corrosion, the prevention of fouling over a longer period is similarly required.

In 1954, Van der Kerk & Luijten [4] reported on an investigation into the biocidal properties of tributyltin compounds. They concluded that these materials had a broad spectrum effect; that is, they were toxic to a wide range of marine organisms. Subsequently, tributyltin oxide was used in combination with copper oxide. However, with the expectation that the effective life of an anti-fouling coating should exceed 18 months, it became clear that this could not be achieved with cuprous oxide at practical film thicknesses. Thus, the use of organo-tin derivatives has increased, and a 'family' of organo-tin derivatives is available to the paint formulator. In laboratory experiments tributyltin derivatives were shown to be one hundred times as effective as copper oxide, weight for weight [5]. Under more practical test conditions later work showed that $\frac{1}{10}$ to $\frac{1}{20}$ the amount of organo-tin compound, compared to copper, would provide equivalent algae and barnacle control [6]. It is important that the organo-tin compound should be *available*, otherwise it

Table 13.10 — Typical formulation for contact type anti-fouling % by weight

Cuprous oxide	57.4
Asbestine	2.4
Rosin	16.1
Modified phenolic/linseed stand oil, 1:2, 60% in 90/190 solvent naphtha.	9.0
Chlorinated plasticizer	5.4
Solvent naphtha, 90/190	9.7
	100.0

will not be any more effective than copper. Thus the paint requires very careful formulation in order to make the maximum effective use of the toxicant.

Several organotin derivatives have been used successfully as toxicants in the marine paint market, the principal ones being tributyltin oxide, tributyltin fluoride, and triphenyltin fluoride.

Tributyltin oxide is a liquid, miscible with common paint solvents. It has a relatively high salt-water solubility, 25 ppm, that enables it to be used when a high leaching rate is required. It has a plasticizing action on the film in which it is incorporated, and this limits the amount that can be used to about 13% by weight in a typical vinyl system. In current commercial use it is usually a cotoxicant with cuprous oxide at about 2–5% by weight of the dry film. It is apparently biodegradable. Photolysis and biodegradation processes convert it ultimately to tin oxide which is quite low in toxicity. Moreover, although tributyltin oxide is soluble in sea-water to a limited extent, it is readily adsorbed onto suspended matter and tends to become associated with the silt after release where it undergoes degradation [7].

Tributyltin fluoride is a white, high melting waxy solid that is insoluble in common paint solvents. It is therefore incorporated as a pigment up to about 30% by weight of the dry film. It is less soluble in sea-water than the oxide, 10 ppm, and is often used as the sole toxicant in vinyl/rosin or chlorinated rubber/rosin-based paints. Tributyltin fluoride is available either in powder form or as a paste. The paste offers more convenient handling and can be readily incorporated by high-speed mixing.

Triphenyltin derivatives were also found to be useful toxicants. Triphenyltin fluoride (TPTF) has been used as an agricultural fungicide. It is a white powder with a sea-water solubility of less than 1 ppm. Paints containing it can give effective protection against marine growth for up to two years. The hydrolysis products of triphenyltin fluoride in sea-water are triphenyltin chloride and triphenyltin hydroxide, both of which are solids that will not diffuse through the film. TPTF is therefore most effective when used in compositions that dissolve or erode.

Conventional anti-fouling systems, relying on the leaching of a toxicant, exhibit a logarithmic decay in the concentration of available toxicant at the paint/water interface. Thus to achieve a long service life for the coating it must deliver a much higher concentration of toxicant initially than is required for the effective control of fouling. When the concentration of available toxicant falls below that required to prevent fouling there is still some residual toxicant in the film. The process is thus inefficient in two ways, in the use of what is an expensive component of the paint. Looked at another way, doubling the film thickness increases the effective life of the coating by only about 13%, if the only release mechanism is that of diffusion. Clearly, if it is possible to achieve a constant and optimum rate of release of the toxicant throughout the service life of the coating, this would be a great step forward. A breakthrough of this kind was made in the early 1970s through the pioneering work in the UK by International Paint in the field of organo-tin polymer compositions.

Polymers of trialkyltin acrylates were first described by Montermoso et al. [8] in 1958. These polymers were prepared in a search for thermally stable polymers, and they were not tested for their biocidal activity. Leebrick [9] in 1965 was the first to claim the biocidal use of organotin polymers, but it was not until 1978 that they received approval by the US Environmental Protection Agency. By this time they were well established in Europe and Japan.

Polymers that incorporate organo-tin moieties are simple random co(or ter)polymers having, typically, the following structure derived from tributyltin acrylate and methyl methacrylate:

$$
\left[
\begin{array}{c}
Bu_3Sn-O \\
| \\
C=O \\
| \\
-CH_2-CH-
\end{array}
\right]_x
\left[
\begin{array}{c}
CH_3 \\
| \\
C-CH_2- \\
| \\
C=O \\
| \\
O \\
| \\
CH_3
\end{array}
\right]_y
$$

Binding the toxicant into a polymer offers many advantages over the use of non-polymeric derivatives that generally have to be incorporated into the paint in a pigmentary form. The toxicant content of the polymer can be varied at will between wide limits, and the T_g of the polymer can be varied to meet the requirements of use. Perhaps the biggest single advantage is the opportunity to provide controlled release of toxicant through the chemical process of hydrolysis rather than by physical means alone.

The advantages claimed for the newer organo-tin polymer anti-foulings include:

- constant rate of release of toxicant;
- controllable rate of erosion and toxicant release;
- self-cleaning (and at high erosion rates, self-polishing);
- high utilization of toxicant is achieved, and as a result of the erosion there are no significant problems due to residues when repainting is necessary.

The formulation of organo-tin polymer-based anti-fouling paints did, however, present very special problems. The use of organo-tin biocides in rapidly eroding systems increases the initial cost to the ship owner, but this may be more than offset by the reduced requirement for dry docking. The main protagonists of organo-tin polymer anti-fouling paints, International Paint (in the UK) and the Nippon Oils and Fats Company in Japan, developed their products in different directions. International Paint claimed the combined benefits of anti-fouling coupled with self-polishing, and they offer products that underwent more rapid erosion than those offered by the Japanese company. Thus, for the polishing/anti-fouling type the rate of erosion was typically 10–12 µm per month. Up to four 100 µm coats are required to give 2–3 years of effective service life. The Japanese system eroded at about 3–6 µm per month, so that two 100 µm coats would however give adequate protection. However, as we have observed earlier, surface roughness can affect the energy requirements for propelling large vessels at their service speed. Additional savings claimed for the polishing type of composition with respect to full costs were not readily realizable with the slower-eroding type. Harpur and Milne [10] point out that conventionally painted ships increase in hull roughness by about 25 µm per year from an initial mean amplitude of 110–125 µm, if well painted. Hydrodynamicists have shown a $\frac{1}{3}$ power law relationship between propulsive power and hull roughness, such that 10 µm of increased roughness can contribute up to an additional 1% in increased fuel consumption which was hardly significant. However, the benefits of self-polishing were less likely to be realizable on old vessels which had already undergone considerable corrosion and exhibited gross pitting.

Over the past two decades, antifouling coatings based on TBT have been developed that have set new standards with respect to antifouling performance. Systems of the self-polishing type can provide protection against fouling for up to five years, and are well established in the industry. In the period 1977–93, International Paints has supplied TBT-containing self-polishing paints for over 13 000 applications, representing 600 million tonnes dead-weight of shipping [11]. However, during the 1980s, a question concerning the environmental acceptability of tin-based antifoulings was raised. Data on the toxicity of these systems are conflicting. Studies at the Plymouth Marine Laboratory suggest that TBT is not as toxic as has been suggested. For example, one type of dog-whelk has been found to overcome the initial impact of TBT on its ability to reproduce. This observation alone would not be sufficient for it to be concluded that there are no adverse effects on the environment from the continuing use of TBT. It has been considered necessary to introduce a ban on the use of TBT-containing coatings on small boats; this has resulted in reduced amounts of tin residues being found according to studies of in-shore sea-waters on the coasts of France, the UK, and the USA. Further legislation has been enacted which restricts the use of TBT-containing compositions notably in Japan, where all such compositions are now banned.

Doubts about the continued use of TBT-based paints have prompted the development by all the major marine paint companies of tin-free alternatives. The concept of self-polishing compositions is in general retained and other toxicants (not generally disclosed by the manufacturers) have been employed; these toxicants presumably meet acceptable environmental standards. Thus, the Norwegian Company, Jotun Protective Coatings, introduced a TBT-free antifouling ('Seaguardian') in 1990, when a trial was initiated on BP's VLCC *British Resolution*. When dry docked in Singapore in 1993, the ship was found to be free of fouling [12]. A 'third generation' tin-free antifouling has also been introduced by the Danish Company, Hempel's Marine Paints, with the tradename 'Combic 7199'. This material is described as environmentally friendly and cost effective; it has a low cost per unit area, low polishing (erosion) rate, good roughness control, and low dry film thickness requirement. It is claimed to be effective for 'up to five years'. Since being introduced to the market early in 1993 it has been approved for use on 35 ships world-wide [13].

Alternative approaches to the problem of fouling continue to be explored. Some are based on the identification of alternatives to heavy metal toxicants that do not present a threat to the environment; others start from a very different point of view.

Thus, the use of naturally occurring toxicants has been suggested. One such study has been the evaluation of extracts from certain species of sponges obtained from the coastal waters of Curaçao [14]. In this study extracts from 51 species of sponge were evaluated in laboratory tests using the barnacle, *Balanus crenatus*. It was found that the sponge extracts inhibited at least the first stage of fouling by affecting the barnacle's metabolism and respiration. However, it is doubtful if studies of this kind will lead to commercially viable products in the short term. Nevertheless, naturally occurring bioactive materials may provide effective antifoulings that are environmentally acceptable in the long term.

The use of low surface energy ('non-stick') coatings is another way of controlling fouling. One system, based on crosslinkable silicone resins, has already been used on some 55 vessels, including nuclear submarines. This is International's 'Intersleek' system [15], which provides a smooth, rubbery, flexible film to which fouling organisms do not readily adhere. Even organisms which form some attachment to the hull when it is stationery are easily removed at the normal service speed as the ship moves through the water. It has been employed successfully on fast ferries and naval vessels that have aluminium hulls. Newer environmentally friendly copper-based antifoulings cannot be used on these vessels because of electrolytic action between the two metals. The 'Intersleek' system does not seem to have been employed with VLCCs, possibly because of its lack of resistance to damage, and for economic reasons.

Yet another approach using the low surface energy concept involves coatings based on fluorinated polyols crosslinked with conventional polyisocyanates, and pigmented with 35–40% (by vol.) PTFE particles [7]. These systems, although initially expensive, may become less so with further development and when carried from the experimental stage to potential commercial exploitation. It is claimed that fouling loosely attached to the coating can be readily removed using low pressure water washing and/or using sponges and nylon brushes. The latter methods may prove to be unattractive when one considers the size of the task involved with a VLCC.

A novel antifouling system based on the use of a conducting paint has been pioneered in Japan by Mitsubishi Heavy Industries, and is marketed as MGPET-200, ('Marine Growth Prevention System by Electrolysis Technology'). It is effective as an antifouling paint, but is free from heavy metal pollutants. The electroconductive coating is insulated from the ship's hull by a non-conductive paint system. A small potential difference is maintained between the hull and the conducting film, such that anodic and cationic processes involving the electrolysis of salt-water occur according to the polarity of the system, which is changed at regular intervals. Thus, the conducting film is alternately the anode or the cathode depending on the polarity of the applied voltage. The active biocidal materials are chlorine and the hypochlorous ions that are generated at the anode. It has been shown that marine growth is prevented when the hypochlorous ion concentration in sea-water is greater than 0.05–0.1 ppm. At present the system is offered for use on ships up to 500 tonnes. Further work to extend its use to larger ships is in hand [16, 17].

13.6   The painting of off-shore structures

Drilling rigs, off-shore platforms, and underwater pipelines all require protective coatings similar to those used in the painting of ships. They are commonly referred to as 'heavy duty coatings'. Drilling rigs are of three types: semi-submergible, 'jack up', and vessel (i.e. floating). All can be considered to be static during drilling operations. Platforms at the junction of the pipeline from a gas or oil field and the main connecting pipeline to the mainland are static structures which resemble drilling rigs, but do not have drilling equipment; nevertheless because they are located permanently in hostile environments they also require effective protective coating systems that call for the minimum maintenance. Corrosion protection is to

Table 13.11

1.	**Main exterior surfaces — (hull, tanks and ladders, etc.)**	
	Primer:	Zinc silicate primer (one coat, 70 µm dry film thickness)
	1st build coat:	Micaceous iron oxide-pigmented polyamide-cured-epoxy paint. (1 coat, 120 µm dry film thickness (dft)
	2nd build coat:	High build vinyl paint (one coat, 100 µm dft)
	Finish:	Vinyl acrylic paint (one coat, 50 µm dft)
2.	**Main exterior decks — (including heliport deck)**	
	Primer:	Zinc silicate paint (one coat, 70 µm dft)
	Build coat:	Two-pack, high build polyamide-cured epoxy paint (one coat, 120 µm dft)
	Finish:	Two-pack, high build polyamide-cured epoxy paint (one coat, 120 µm dft) (incorporating flame dried silica sand 300–500 µm).
3.1	**Submerged zone — Legs, topsides of spud tanks, jack-up rig**	
	Primer:	Two-pack zinc phosphate pigmented polyamide-cured epoxy (one coat, 50 µm dft)
	Build coat:	Two-pack polyamide-cured epoxy paint (three coats, 360 µm dft)
3.2	**Below drilling load line (pontoons, columns, bracings, etc.)**	
	Primer:	Vinyl wash primer (one coat, 5 µm dft)
	Build coat:	Two-pack high build polyamide-cured epoxy paint (three coats, 360 µm dft)
	Intermediate coat:	Aluminium pigmented chlorinated rubber paint (one coat, 40 µm dft)
	Finish:	High build, self-polishing, antifouling paint (two coats, 200 µm dft)
4.	**Mud zone**	
4.1	**Mud pits room and mud pump room; mud pits, mud tanks and mud pumps (exterior)**	
	Primer:	Two-pack, zinc phosphate-pigmented polyamide-cured epoxy primer (one coat, 50 µm dft)
	Build coat:	Two-pack, high build, polyamide-cured epoxy paint (two coats, 240 µm dft)
4.2	**Mud pits and mud tanks (interior)**	
	Primer:	Vinyl wash primer (one coat, 5 µm dft)
	Build coat:	Two-pack, high build polyamide-cured coal tar epoxy paint (two coats, 240 µm dft)
4.3	**Drilling floor**	
	Primer:	Zinc silicate paint (one coat, 70 µm dft)
	Build coat:	Two-pack, high build polyamide-cured epoxy paint (three coats, 360 µm dft)

some extent more important than antifouling, except where marine organisms contribute to the onset of corrosion; thus, unlike marine paints the energy savings obtainable by the reduction of drag (applicable to fast moving ships) do not arise.

Like ships, off-shore structures require multicoat systems, which together have a passivating effect on the steel substrate, and a barrier effect. The initial painting of the structure is carried out on shore in a more desirable microclimate. In service, painting is confined largely to the retouching or repair of the original coatings as a result of damage to the coating or its eventual breakdown. Three main zones, which are exposed to different environmental conditions, are recognized. These are the 'atmospheric zone' (akin to the topside and superstructure of a ship), the 'waterline zone' (akin to the boottopping of a ship), and the 'submerged' zone (akin to the bottom of a ship). Giudice [18] identifies some 12 types of surface, both interior and exterior, requiring different paint finishing systems. The main exterior zones employ systems which are exemplified in Table 13.11. Other surfaces include those of ballast and freshwater tanks, machinery, etc.

References

[1] GITLITZ M H, *J Coatings Technol* **53** 46 (1981).
[2] DEXTER S C, *Biofouling* **7** 97–127 (1993).
[3] BANFIELD T A, *Protective Painting of Ships & Structural Steel*, SITA Ltd, Manchester (1978); *J Oil Colour Chem Assoc* **63** 53 (1980); **63** 93 (1980).
[4] VAN DER KERK G J M & LUIJTEN J G A, *J Appl Chem* **4** 314 (1954).
[5] EVANS C J & SMITH P J, *J Oil Colour Chem Assoc* **58** 160 (1975).
[6] CHROMY L & UHACZ K, *J Oil Colour Chem Assoc* **61** 39 (1978).
[7] HARE C H, *J Protective Coat Linings* **10** (2) 83–89 (1993).
[8] MONTERMOSO J C et al, *J Polym Sci* **32** 523 (1958).
[9] LEEBRICK J, US Patent 3,167,473 (1965).
[10] HARPUR W & MILNE A, *Proc 5th PRA Internat Conf*, pp. 45–50 May (1983).
[11] ANDERSON C D, *Drydock*, **Dec.** 5 (1993).
[12] ANON, *Pigment Resin Technol* **23** (2) 16–17 (1994).
[13] ANON, *Europ Paint Resin News* **31** (10) 11 (1993).
[14] WILLEMSEN P R & FERRARI G M, *Surface Coatings Internat* **10** 423–427 (1993).
[15] WAKE M, *Seatrade Rev* **October** 33, 35 (1992).
[16] ANON, *Europ Paint Resin News* **31** (3) 11 (1993).
[17] NISHI A et al, *MHI Technical Rev* **29** (1) 30–35 (1992).
[18] GIUDICE C A, *EuroCoat J* **5** 344–354 (1993).

14

An introduction to rheology

T A Strivens

14.1 Introduction

The science of rheology is concerned with the deformation and flow of matter, and with the response of materials to the application of mechanical force (stress) or to deformation. Such responses include irreversible (viscous) flow, reversible (elastic) deformation, or a combination of both. In the former process, energy is dissipated (mainly in the form of thermal energy (heat)); in the latter, energy is stored and released, when the mechanical force is removed. The balance of such responses is dependent on the speed (time scale) at which the mechanical force is applied as well as the material temperature. Thus, at normal room temperature substances such as glass and pitch will shatter when hit with a hammer, but will slowly stretch and deform irreversibly when weights are hung on sheets or rods of the material. Familiar materials, such as 'bouncing putty', can be kneaded and stretched between the fingers (long time scale), but, when formed into a ball and dropped onto a surface they bounce like a rubber ball (short time scale for impact). Plastic sheet such as Perspex is reasonably pliable at room temperature, but becomes hard and brittle like glass when immersed in liquid nitrogen and soft and permanently deformable when immersed in boiling water. Even simple liquids like water or substituted paraffin hydrocarbons, such as 6,6–11,11 tetramethyl hexadecane, can show elastic responses if the time scale is short enough: 10^{-9} to 10^{-12} seconds [1]. Many commercially important materials, such as paint, are dispersions of one or more liquid or solid phases in a liquid or solid matrix, e.g. emulsions, dispersions, composites. Such materials often exhibit very complex responses to the application of quite small mechanical forces. For example, thixotropic paints look like solids or very viscous (thick) liquids, when at rest in the can; but, when stirred gently or stressed by the insertion of the paint brush, they become thin, mobile liquids. When left at rest, they recover their original appearance. The science of rheology covers all the complex and varied responses of this whole range of different materials.

Whilst, academically, rheology is often seen as a branch of applied mathematics or physics, proper understanding of the results of rheological measurements (as well as, sometimes, a sensible choice of rheological measurement technique) must involve other scientific disciplines, in particular, physical chemistry. Thus a knowledge of colloid science is essential to a proper understanding of the rheology of emulsions, dispersions, and suspensions, as much as polymer science is to the rheology of polymer melts and solutions. It is (or should be) a truly multidisciplinary science involving the skills of mathematicians, physicists, chemists, and engineers, as well as others, such as biologists, on occasion. Rheology can involve considerable mathematics, but, in this account, it will be reduced to the minimum necessary to clarify relationships and concepts.

The control of rheology is essential to the manufacture and usage of large numbers of products in a modern industrial society, e.g. food, plastics, cosmetics, petroleum derivatives, and paints. Few manufacturing industries are devoid of material forming or coating processes, mixing operations, transport of liquids or slurries through pipelines, liquid–solid separation processes, such as sedimentation or filtration, etc. All of these processes require, to a greater or lesser extent, control of material rheology. The use of such products often involves rheology in their application; for example, the smoothing of cosmetic creams on the face, the taste and texture of foodstuffs such as sauces and mayonnaise in the mouth, and the application of paints to a surface by spraying, brushing, etc. as well as the flow-out after application to give a smooth, uniform film.

14.2 History of viscosity measurements

E C Bingham, one of the founders of the modern science of rheology, in 1929 defined rheology as the science of deformation and flow of matter. However, historically, the measurement and definition of viscosity were established well before the foundation of the science of rheology. Some centuries before the Greek philosopher Heraclitus of Samos (c. 540–475 BC) categorically stated 'παντα ρει', i.e. everything flows; the Jewish prophetess Deborah talked about 'The mountains flowed before the Lord' (Judges 5,5; the King James authorized version is wrong to say 'melted'; the Hebrew word is quite definitely 'flowed'†). Some understanding of the flow of simple liquid through capillaries was shown by the designers of water-clocks in ancient Egypt (c. 1540 BC). However, it was not until our own era that Newton, in his *Principia* (2nd ed. 1731) gave the modern definition of viscosity, and Hooke, in 1678, discovered experimentally the proportionality between stress and strain for elastic materials (*ut tensio, sic vis*). During the eighteenth century came the development of the branch of physics known as hydrodynamics by, among others, the Swiss brothers Bernouilli, as well as the measurement of gas liquid viscosity by Coulomb in 1798, using the decay of oscillation of a disk suspended in the test sample. This method is still used by a torque measurement on a rotating disk in the Brookfield viscometer, much favoured by paint technologists. The reference method for measuring the viscosity of Newtonian liquids, namely capillary viscometry, arose from the work of Hagen (1839) and Poiseuille (1846), the latter of whom was trying to understand the flow of blood in the human body. The use of concentric cylinders as an experimental method of measuring viscosity was established by Couette in 1890, and Bingham's own studies on the viscosity of paint were published in 1919 [2].

† The Revised Version corrects this error.

14.3   Definitions

Before describing rheological measurements and their interpretation, a number of definitions must be given, which form the conceptual basis of the science of rheology.

14.3.1   Modes of deformation

The commonest mode of deformation for liquid or deformable solid materials is *shear* deformation. This is achieved by confining the material between the walls of the measuring instrument and by setting up a velocity gradient across the thickness of the material, i.e. causing it to flow. This may be by forcing the fluid to flow through a pipe or capillary tube or by moving one wall relative to another. This latter can be achieved in a number of flow *geometries* which will be described under measurement techniques (Section 14.4, below).

In simple shear, the change in shape brought about by the application of mechanical force is not accompanied by a change in volume. In other modes of deformation, a change in volume rather than shape may occur (*bulk compression* or *dilatation*), or both may change together as in *extensional deformation*. This latter mode can be of practical importance in the paint industry; for example, it has been established that it affects spatter and filament formation in roller application of paint [3].

If the macroscopic deformation of the material sample is the same as that of any of its constituent elements, then the deformation is referred to as being *homogeneous*. However, in many experimental deformation geometries, such as torsion between concentric cylinders, flow through a tube, etc., the deformation and deformation rates vary from point to point. If the deformations are small, the mathematical relationships between the rheological parameters and the external forces and displacements can be simplified, and the relationships are not dependent on the deformation geometry used. However, if the deformations are sufficiently large, this is no longer the case, and the relationship becomes geometry-dependent, and this can cause difficulties in both the design and interpretation of experiments. This is an important point when considering experimental methods for materials with complex rheological properties, such as the thixotropic paints mentioned above.

14.3.2   Shear stress

Consider the experimental set-up in Fig. 14.1 (simple shear deformation mode). Two parallel plates, each of area A, are separated by a thickness, h, of the material under test. Suppose a force, F, is applied to the upper plate at its left-hand edge, and the bottom plate is held rigid; then the stress applied to the material is given by F/A (cgs units, dyne cm^{-2}; SI units, Pascal, or Newton m^{-2}).

14.3.3   Shear strain

When the force is applied as shown in Fig. 14.1, the upper plate responds immediately by moving a distance, y, from left to right. The shear strain (y) is then defined as y/h and is dimensionless.

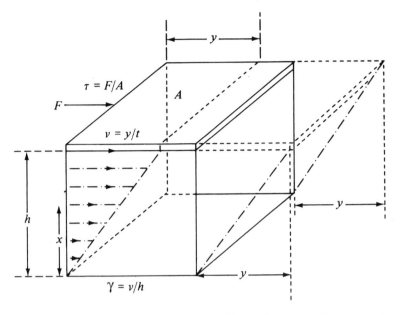

Fig. 14.1 — Simple shear experiment: definitions of stress, strain, strain rate.

14.3.4 Shear strain rate (shear rate)
If the upper plate moves through the distance y in time t, then the upper plate moves with a velocity v, given by y/t.

If the material filling the gap is an ideal liquid, then the plate will continue to move at a steady velocity v for as long as the force F is applied. Furthermore, the velocity, v, will be proportional to the force F applied. As soon as the force is removed, movement will stop (assuming both plate and fluid have negligible inertia). Because the liquid velocity at the surface of the bottom plate is zero, and, assuming that the liquid sticks to both plates, there will be a velocity gradient through the thickness of the liquid. If it is further assumed that this velocity gradient is linear as shown in Fig. 14.1 (i.e. the liquid velocity at any point distance x from the fixed plate is given by xv/h) then a shear strain rate or shear rate ($\dot{\gamma}$) can be defined by the ratio v/h (sec^{-1}).

14.3.5 Shear viscosity
This quantity is simply defined by the shear stress divided by the shear rate. In Fig. 14.1, the shear viscosity (η) is given by

$$\eta = \frac{\tau}{\dot{\gamma}} = \frac{Fh}{Av} \tag{14.1}$$

where τ is shear stress and $\dot{\gamma}$ shear rate. Note that $\dot{\gamma}$ is equivalent to $d\gamma/dt$.
The units are poise or dyne sec cm^{-2} (cgs) and Pascal seconds or N s m^{-2} (SI).

14.3.6 Shear elasticity modulus
Suppose the material between the plates is a perfectly elastic body. Then when the force, F, is removed, the top plate will return immediately to its initial position,

assuming again perfect adhesion between material and plate. Moreover, once the material has moved the distance y under the influence of the force, movement will cease; in fact, y will be proportional to the magnitude of F.

The modulus of elasticity (G) is then defined as the shear stress divided by the shear strain, i.e.

$$G = \frac{\tau}{\gamma} = \frac{Fh}{A\gamma} \tag{14.2}$$

14.3.7   Normal force

This is a difficult process to describe without using rather sophisticated mathematics, such as tensor analysis. Basically, in the experiment shown in Fig. 14.1, assume the gap between the plates is filled with a liquid, possessing elasticity as well as viscosity. Suppose the upper plate is moved at a constant velocity, v (or strain rate $\dot{\gamma}$). As a result of the resistance to continuous deformation exerted by the material, owing to the presence of elasticity in the material, the total force exerted on the moving plate is at an angle to the direction of motion. This total force can then be resolved into its components (Fig. 14.2), which include a force parallel to and in the same plane as the plate, as well as a component in a plane vertical to the plane of the motion and at right angles to the direction of motion. This latter is the normal force. In practical terms, this force will tend to try to push the plates apart while there is motion. In practical instruments, the moving plate must either be held rigidly in the vertical plane or it can be allowed to move and kept in position by applying an equal and opposite restoring force to counteract the normal force. Clearly, this last approach provides a means of measuring normal force. Familiar examples of this normal force effect are the rod climbing effect (Plate 14.1) and the tendency for some flour mixtures or doughs to climb up the stirrer rod when kitchen mixers are used.

14.3.8   Types of rheological behaviour

14.3.8.1   Newtonian viscous liquid and Hookean elastic bodies
If the viscosity, as defined in equation (14.1), is independent of the shear rate (or shear stress) applied to cause the liquid to flow, then the liquid is referred to as a Newtonian liquid, after Isaac Newton, who first enunciated the law of liquid flow in his *Principia* (1718).

Equally, if the elasticity is independent of the value of shear strain (or shear stress) applied, the material is referred to as a Hookean elastic body (Hooke's Law, 1678).

As has been indicated above, both these materials represent physical ideals to which most real materials will approximate over limited ranges of stresses, strains, or strain rates. However, if the materials do show variation in their viscosity or elasticity moduli, then it is still often useful to define so-called *apparent* viscosity or elasticity moduli (more usual with the former) using the definitions of equations (14.1) and (14.2).

For the purposes of considering paint rheology, we will examine a number of types of behaviour in which the apparent viscosity is a function of shear rate and time.

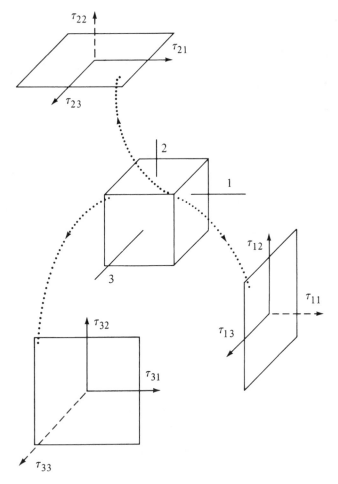

Fig. 14.2 — Components of stress in a viscoelastic liquid under shear. Normal stresses, i.e. stresses at right angles to plane concerned, are shown dashed. For simple shear, $\tau_{13} = \tau_{23} = \tau_{31} = \tau_{32} = 0$.

14.3.8.2   Pseudoplastic materials

The important features of pseudoplastic materials are that (1) the apparent viscosity decreases with increasing shear rate (or shear stress) values, and (2) the apparent viscosity value at a given shear rate value is independent of the shear history of the sample.

The last statement implies two things: (1) during a given measurement, the apparent viscosity value is *independent of the time* for which the shear rate has been applied to the sample, and (2) the apparent viscosity value is *not dependent on the previous measurements made* (whether these are at higher or lower shear rates). These statements need some qualification in practice. Firstly, there is nearly always some rapid change in shear stresses with time when a measurement is started, owing to *inertia* in the moving parts of the viscometer and perhaps also a less rapid change due to elasticity in the sample (the so-called stress overshoot or undershoot effects). Secondly, either high sample viscosity or high shear rate applied or both can lead

Plate 14.1 — Rod climbing (Weissenberg) effect.

to substantial energy dissipation in the form of heat. Unless sample temperature control is very good, this leads to a rise in sample temperature and an apparent drift down in viscosity value with time.

Also implicit in the definition is that the shear stress increases less than proportionally with shear rate. The chief features of pseudoplastic behaviour are illustrated in Fig. 14.3 for an experiment in which shear rate is increased uniformly with time until a maximum value is reached and then reduced at the same rate (shear rate cycle experiment).

14.3.8.3 Dilatant materials
The important features of dilatant materials are that (1) the apparent viscosity increases with increasing shear rate, and (2) the apparent viscosity value at a given

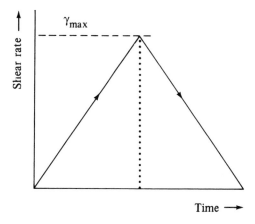

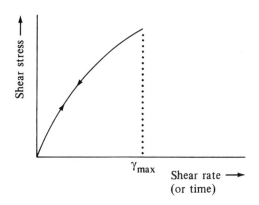

Fig. 14.3 — An example of pseudoplastic behaviour.

shear rate is independent of the shear history of the sample. Implicit in the first statement is that the shear stress increases more steeply with shear rate than proportionality (see Fig. 14.4).

In addition to the qualifications discussed under pseudoplastic materials, there is an important restriction to be added to the second half of the definition. Many dilatant materials, such as concentrated suspensions, behave like hard solids and break or fracture in the viscometer, if the shear rate is too high or it is applied too quickly [4, 5].

To this extent, they become history- or time-dependent, but these effects arise from different mechanisms and are more limited in extent than the time effects characteristic of thixotropic or rheopectic materials (see below). Also true dilatancy must be carefully distinguished from the artefacts introduced by the viscometric measurement technique applied to these materials [5].

14.3.8.4   Power law materials
The three types of viscous materials just defined (Newtonian, pseudoplastic, and dilatant) may all be classified as *power law fluids*, because the shear stress — shear

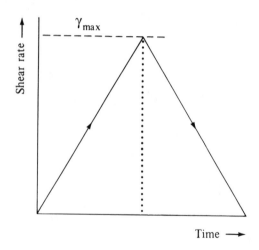

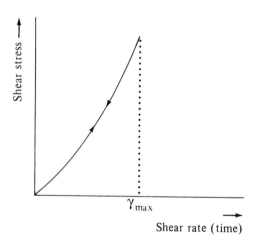

Fig. 14.4 — Dilatant behaviour.

rate relationships can all be fitted more or less accurately and usually over limited ranges of shear rate by a power law relationship.

$$\tau = K\dot{\gamma}^n \tag{14.3}$$

or

$$\eta_a = L\dot{\gamma}^{n-1} \tag{14.4}$$

If n equals 1, then the shear stress is proportional to the shear rate, the material is a Newtonian liquid, whose viscosity is equal to the values of the constants K and L in equations (14.3, 14.4). If n is less than 1, then the material is pseudoplastic, and it is dilatant if n is greater than 1.

Such a relationship is often used by engineers for interpolation and (dangerously!) for extrapolation of viscosity — shear rate data.

14.3.8.5 Thixotropy and rheopexy

If the definitions given above for pseudoplastic and dilatant materials are changed in the second half, so that the apparent viscosity value at a given shear rate is now *dependent* on sample shear history, then the materials may be described as *thixotropic and rheopectic*, respectively.

The resulting effect of shear history can be seen for the shear rate cycle experiment by comparing the behaviour shown in Fig. 14.5 for a thixotropic material with that shown in Fig. 14.3, for a pseudoplastic material. It will be seen that a plot of shear stress against shear rate (or, equivalently, time) is a loop, where the shear stress value for a given shear rate value is higher if the shear rate is increasing than if it is decreasing. Moreover, the area of the loop, A, is dependent not only on the rate of shear rate change ($d\dot{\gamma}/dt$), but also decreases with the cycle number if successive shear rate cycles are applied to the material.

Rheopectic materials are rather rarer in occurrence than thixotropic materials [6, 7].

14.3.8.6   *Yield stress, Bingham bodies, etc.*
If on a shear stress — shear rate plot for a material, the data are extrapolated to zero shear rate and the plot appears to cut the shear stress axis of the graph at a positive stress value, then the material is said to possess a *yield stress*.

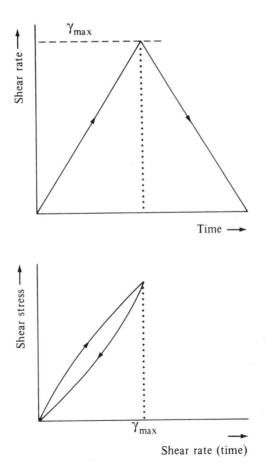

Fig. 14.5 — An example of thixotropic behaviour.

In practical terms, this means that it appears that a certain minimum force must be applied to the material before it will flow.

If, above the shear stress value, the shear stress is proportional to the shear rate, then the material is referred to as a *Bingham body*. If not, then it is referred to as a *pseudoplastic, thixotropic, etc. material with yield stress*, whichever description is appropriate.

14.4   Methods of measurement

14.4.1   Geometries

In practice, the simple shear geometry illustrated in Fig. 14.1 is difficult to realize. Most shear deformation experiments are more easily achieved by causing shear by rotation around a central axis (torsional shear mode). Viscometers using this mode are referred to as *rotary viscometers*. There are three main geometries of this kind:

- cone and plate;
- concentric cylinder; and
- parallel plate.

In addition, the capillary flow geometry is frequently used in steady-state measurements (see below) as a reference standard for Newtonian liquids, as well as in other applications. Other geometries have been, or are being, developed, usually for rather specialized applications, and the books by Walters [8] and Whorlow [9] should be consulted for details of these.

In the torsional shear mode, the sample is confined between the two components (components A and B) of the measurement geometry. The measurements can be made in two ways: (a) controlled stress and (b) controlled shear rate. In either case, one of the components (component A) will be mounted on some form of spring or torsion bar, by means of which torque can be applied to, or measured on, component A. In the controlled stress experiment, a known torque is applied to component A by means of the spring, or in more recent instruments by an induction motor system, while component B is held fixed. The resulting movement of A is measured as a function of the time and the torque applied. In the controlled shear rate experiment two methods of operation are possible. In the first, component B is rotated at a controlled speed and the torque transmitted to component A by the sample is measured by the deflection of its mounting spring. Alternatively, component A and its spring are rotated at controlled speed with component B fixed and the torque on component A again measured as a function of rotation speed.

In the capillary flow mode, the liquid sample is caused to flow through a capillary of known length and diameter. Either the pressure driving the fluid through the capillary is controlled and the consequent flow rate measured (usually, volumetric flow rate), or the reverse procedure with flow rate control, is applied.

14.4.1.1   Cone and plate geometry

In cone and plate geometry (Fig. 14.6) the sample is held by surface tension in the gap between the cone and the plate, either of which may be rotated in controlled shear rate measurements. Obviously, in a parallel plate system, the velocity at any point on the rotating plate will vary with the distance from the centre, and,

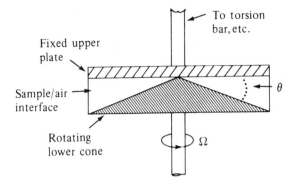

Fig. 14.6 — Cone and plate geometry.

consequently, so will the shear rate across the gap. By using a cone, whose surfaces are at a small angle to the horizontal (usually, less than 4°), it is possible (at least, in theory) to ensure uniform shear rate across the gap and across the whole diameter of the cone and plate system.

With this system, the shear stress (τ) on the sample is given by:

$$\tau = \frac{3T}{2\pi R^3} \qquad (14.5)$$

and the shear rate (γ) by

$$\dot{\gamma} = \frac{\Omega}{\theta} \qquad (14.6)$$

where T is the measured torque,
R the radius of the cone and plate system,
θ the (internal) angle of the cone (radians), and
Ω the rotational speed of the system (radian s^{-1}).

In practice, the cone is often slightly truncated to avoid frictional contact between the two components, and then gap setting becomes critical.

Apart from the uniform shear rate, the major advantages of the cone and plate system is the small volume of sample required and the ease of cleaning the system.

Major disadvantages of the geometry are the free liquid edge and difficulties associated with measuring concentrated dispersion systems. Because of the large air–liquid boundary area, the formation of a film or thick viscous layer, owing to evaporation of volatile solvent, will exert an extra torque, in addition to the torque exerted by the bulk of the sample, leading to serious errors. Also, at high rotary speeds (shear rates), sample may be thrown out of the gap by centrifugal force.

In concentrated dispersions, apart from the danger of mechanical abrasion or particles jamming in the gap if they are large, there is a subtler effect, due to hydrodynamic forces, and a purely geometric factor. Because the particle centres cannot approach closer to the wall than a distance equal to the particle radius, there is a thin 'particle-depleted' layer separating the component wall from the bulk of the sample. This acts like a layer of lubricant. In addition, it is possible that hydrodynamic forces may prevent the close approach of particles to the walls. These effects are difficult to estimate, but it appears that they can be of the order of the particle diameter for the depletion layer thickness [10]. Unless the gap width average is very much greater than the particle diameter, this will lead to measurement errors. Obviously, the influence of such effects should be assessed by measuring a dispersion sample with several cone and plate geometries of different cone angle and diameter; or, better still, using the parallel plates geometry, in which the gap width may be freely varied [5].

14.4.1.2 Concentric cylinder geometry (Couette flow)

In this geometry, the sample is used to fill the gap between a cylindrical cup and a cylindrical bob suspended within it. By using a small gap in relation to the cup and bob radii, it is possible to ensure that the shear rate is nearly constant across the gap and only dependent on the rotation rate. However, this makes mechanical design difficult, if high shear rate values are required, and increases the likelihood of viscous heating if prolonged measurement periods are necessary.

The shear stress (τ) at the surface of the inner bob is given by:

$$\tau = \frac{M}{2\pi R_b^2 h} \tag{14.7}$$

If the sample is known to be non-Newtonian, then the shear rate at the bob surface can best be determined by the formula of Moore and Davies [11], using the first and second derivative of the experimental stress — rotation speed plot, i.e.

$$\dot{\gamma} = \frac{\Omega}{K}\left[1 + \frac{K}{a} + \frac{K^2}{a^2}\left(1 - \frac{b}{a}\right) + \ldots\right] \tag{14.8}$$

where $K = \ln(R_c/R_b)$,
$\quad a = d\ln\tau/d\ln\Omega$,
$\quad b = da/d\ln\Omega \ (\equiv d^2\ln\tau/d\,(\ln\Omega)^2)$,
$\quad \Omega$ = rotation speed (radian s^{-1}),
$\quad M$ = measured torque,
$\quad R_b$ = the inner (bob) radius,
$\quad R_c$ = the outer (cup) radius, and
$\quad h$ = depth of immersion of the bob in the sample.

Corrections still have to be made for end effects; the review by Oka [12] should be consulted for details.

Taylor [13, 14] showed that for Newtonian liquids, the shear rate at a point, distance r from the centre of the system, is

$$\dot{\gamma}_r = \left[\frac{\Omega R_b^2}{R_c^2 - R_b^2}\right]\left[1 + \left(\frac{R_c}{r}\right)^2\right] \quad \text{(inner cylinder rotated)} \tag{14.9a}$$

and

$$\dot{\gamma}_r = \frac{2\Omega R_b^2 R_c^2}{r^2(R_c^2 - R_b^2)} \quad \text{(outer cylinder rotated)} \tag{14.9b}$$

Both of these equations reduce for small gaps in relation to cup and bob radii to

$$\dot{\gamma} \simeq \frac{\Omega R_{\rm I}}{R_{\rm)} - R_{\rm I}} \simeq \frac{\Omega R_{\rm)}}{R_{\rm)} - R_{\rm I}}. \tag{14.9c}$$

While the data treatment and theory are more complicated for concentric cylinder geometry, this geometry does have a number of advantages over cone and plate geometry: (1) there is no loss of sample due to centrifugal forces at high rotation speeds, (2) there is less likelihood of dependence on particle size with suspension or dispersion samples, and (3) because the liquid level is below the rim of the cup and the liquid–air boundary area is very small, compared to the measuring surface area on the bob, there is less likelihood of loss of volatile solvent, and any surface skin or thickening will have less effect on the measured torque.

14.4.1.3   Parallel plates

Parallel plate geometry has the advantage over the cone and plate geometry that the gap width can be varied freely. As discussed in Section 14.1.1, this is an advantage when measuring suspension or dispersion systems. Against this advantage, the shear rate on the sample varies with the distance from the plate centre, and thus data are more difficult to evaluate.

When oscillatory (dynamic) measurements are being made, parallel plates possess considerable advantages over either cone and plate or concentric cylinder geometries, as will be discussed in Section 14.4.2.3.

The maximum shear rate (at the plate rim) is given by

$$\dot{\gamma}_{\rm m} = \frac{\Omega R}{h} \tag{14.10}$$

and the apparent viscosity ($\eta_{\rm a}$) corresponding to this maximum shear rate ($\dot{\gamma}_{\rm m}$) may be evaluated, using the following equation [8, p. 52]:

$$\eta_{\rm a} = \frac{3M}{2\pi R^3 \dot{\gamma}_{\rm m}} \left[1 + 3 \frac{{\rm d} \ln M}{{\rm d} \ln \dot{\gamma}} \right] \tag{14.11}$$

where M is the measured torque,
  R is the plate radius,
  h is the plate separation, and
  Ω is the speed of rotation.

Like the cone and plate geometry, the parallel plate geometry has the advantage that in theory (if not always in practice) the liquid velocity distribution in the gap is determined by the geometry and not by the liquid properties.

14.4.1.4   Capillary (Poiseuille) flow

If a liquid is caused to flow through a circular section tube or capillary of radius R, then it can be shown [12, pp. 21–29] that (for a Newtonian liquid and assuming no slip at the capillary wall, i.e. $v_{\rm w} = 0$),

- the velocity distribution across the radius is parabolic (liquid moving fastest at the tube centre); and
- the stress varies linearly across the tube radius, being maximum at the wall.

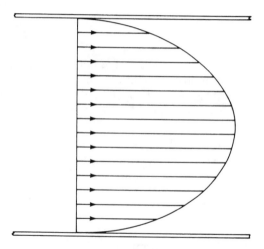

Fig. 14.7 — Liquid velocity distribution for a Newtonian fluid in capillary (Poiseuille) flow.

Measurements can be made of the driving pressure (P) and the volume flow rate (Q) of the liquid. For a Newtonian liquid, Q and P are related by the Hagen–Poiseuille equation

$$Q = (\pi R^4 / 8l\eta)P \tag{14.12}$$

where l is the tube length and

η the liquid viscosity.

The maximum stress (at the wall) is given by

$$\tau_w = PR/2l \tag{14.13}$$

Corrections must be made for tube end and kinetic energy effects. For non-Newtonian liquids, the velocity distribution deviates from the parabolic form (see Fig. 14.7), and some assumption must be made about the apparent viscosity shear rate relationship, before the dependence of apparent viscosity on the measured pressures and flow rates can be evaluated [12, pp. 21–29]. Alternatively, the shear rate at the wall can be evaluated from the volume flow rate — wall shear stress data [8, p. 100].

$$\dot\gamma_y(\tau_w) = \bar q \left[\frac{3}{4} + \frac{1}{4} \frac{d \ln \bar q}{d \ln \tau_w} \right] \tag{14.14}$$

when $\bar q = -4Q/\pi R^3$.

14.4.2   Measurement techniques

14.4.2.1   Steady state
In the definitions given in Section 14.3, and the discussion of measurement geometry (Section 14.4.1), it has been tacitly assumed that steady-state measurements were being made. This means in practice, for example, that a known shear rate is

applied to the sample and the resulting torque is observed. After a few seconds, during which instrument inertia effects die away, the measured torque usually becomes steady in value within the next half to one minute. This steady value is then noted before changing the shear rate value. In the case of thixotropic materials, a steady torque value may not be obtained even after prolonged shearing.

Other measurement techniques are available, and these may give more useful information in some circumstances. They will now be described.

14.4.2.2 Transient techniques

If, in the experiment described in the previous section, the torque is recorded as a function of time, and perhaps also the decay with time of the torque when rotation is stopped, then a transient type of experiment is being performed.

For more solid materials, the more conventional techniques of stress relaxation and creep may be used. In the former, stress (equivalently, torque) is monitored as a function of time when a known deformation is applied to the sample, and vice versa for the latter (creep) experiment. Typical results are shown in Fig. 14.8.

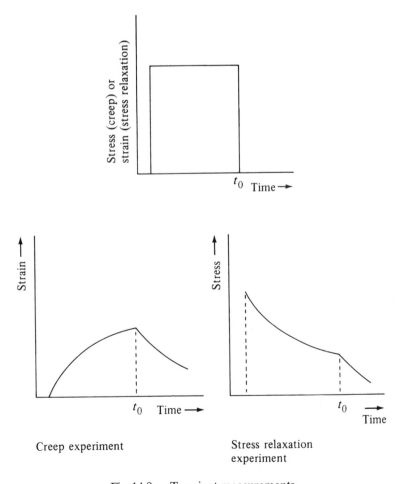

Creep experiment   Stress relaxation experiment

Fig. 14.8 — Transient measurements.

Analysis of the resulting stress or deformation–time plots can give useful information on the elasticity as well as the viscosity of the sample and the dependence of these parameters on the experimental time scale [15, pp. 9–12, 38–43]. Such results can supplement the results of dynamic measurements (see next section). However, care must be taken in interpreting the results, owing to artefacts introduced, particularly at short times, by the inertia of the measuring instrument's moving parts.

Such techniques are potentially of great utility in probing rheological structure, such as occurs in thixotropic paints. Because of the greater control exercised on the sample's deformation history, the techniques are much more useful than the thixotropic loop type of experiment described in Section 14.3.8.5.

14.4.2.3   Dynamic (oscillatory) techniques

So far, all the experiments described have involved liquid flow in one direction only. If, after torque or rotation has been applied for a short time, the direction of application is reversed for the same time, and this cycle is repeated until a steady response (strain or stress) cycle is observed, then a dynamic experiment is being performed. The applied oscillation is usually sinusoidal with time, although there is no reason (except more complicated theory and mathematics) why sawtooth or square wave may not also be used. Such experiments are usually made with small input amplitudes, and the corresponding (steady) output amplitude and phase shift are measured. The frequency of the cycle is also controlled. From these measurements, values of the dynamic viscosity and dynamic shear modulus of the sample may be calculated, and the variation of these values with frequency studied.

Because of the limited frequency range available with many instruments, it is customary when dealing with polymer solids or melts to vary the temperature, and then to use the time–temperature superposition principle to obtain data at one reference temperature over a larger range of frequency [15, pp. 641–643]. This technique is of limited use in paint measurements, as, apart from practical problems of solvent loss, subtler irreversible changes in the paint may result owing to changes in solubility of polymers, stability of dispersions, etc., brought about by large changes in temperature.

Because the theory of viscoelasticity tacitly assumes linear response, it is essential, particularly with structured or thixotropic paints or suspensions, to check for linearity by varying the input amplitude and monitoring the output waveform. While non-linearity may show as distortion of the output waveform, it may also appear as a change in phase angle and/or change in input/output amplitude ratio as the input amplitude is varied. The theory of non-linear viscoelasticity is less well developed, and experimental results are more difficult to analyse [15, pp. 140–151].

While capillary flow can be used for dynamic measurements, it is more usual and more convenient experimentally to use torsional flow geometries. Of these, the parallel plate geometry is the most attractive. Not only is it experimentally convenient (for example, the gap may be easily varied without changing any of the components of the geometry), but also the theory is simplest for this geometry, particularly if fluid inertia effects cannot be neglected. (These are severest for low viscosity liquids and high frequency measurements) [15, pp. 125–140].

If the lower plate is driven at a forced (sinusoidal) oscillation of frequency, ω (rad sec^{-1}) and angular amplitude, θ_2 (rad), and, because of the sample between the plates, the upper plate oscillates at the same frequency with angular amplitude, θ_1

(rad) (lagging behind the lower plate by an angle, ϕ (rad)), against the constraint of its suspension (torsion wire or bar of stiffness, E dyne cm rad^{-1} (Nm rad^{-1})), then the dynamic viscosity (η', poise) and elasticity modulus (G', dyne cm^{-2} (Nm^{-2})) are given by the following equations [8, pp. 125–127]:

$$\eta' = \frac{-AS\sin\phi}{A^2 - 2A\cos\phi + 1} \tag{14.15}$$

and

$$G' = \frac{\omega AS[\cos\phi - A]}{A^2 - 2A\cos\phi + 1} \tag{14.16}$$

where A is the amplitude ratio (θ_1/θ_2)

$$S = \frac{2h[E - I\omega^2]}{\pi a^4 \omega}$$

where h is the gap between the plates,
$\quad$ a the radius of the plates, and
$\quad$ I the moment of inertia of the upper plate about its axis.

E may be determined by direct measurement and I from the natural frequency of free oscillation of the upper plate in the absence of sample.

In the previous type of dynamic measurement, the strain amplitude is controlled. It is often advantageous, particularly when studying structured paints or dispersions, to control the stress amplitude applied. This may readily be done by fixing the lower plate and applying an oscillatory (sinusoidal) torque to the suspension of the upper plate. Again, amplitude and phase lag of the upper plate are measured in relation to the applied torque, but this time the upper plate motion is constrained by the sample between the plates. In this case, the equations for η' and G' are simpler than equations (14.15, 14.16), [16], but there is no advantage when fluid inertia effects have to be included.

14.5 Interpretation of results

Before discussing in detail the rheology of paint (see Chapter 15), a brief review of the factors controlling the rheology of liquid systems will be given to aid the interpretation of results.

14.5.1 Simple liquids

In a low density gas, the molecules are widely separated in space and the viscosity is determined by momentum transfer between the molecules when they collide. As the density increases, for example, owing to cooling, the momentum of the molecules, because of their thermal energy, is reduced and they come closer together, so that finally they are close enough for intermolecular forces to play a significant role in determining bulk properties like viscosity. When the gas has become a liquid, momentum transfer during molecular collisions contributes an insignificant amount to the liquid viscosity. The liquid viscosity is now determined entirely by the balance between intermolecular forces and the much reduced movements of the liquid molecules, owing to their thermal energy. These movements will be in all directions between existing positions and neighbouring spaces and at random, and against opposition provided by the intermolecular forces. They will be very frequent (in a simple liquid at room temperature, typically, 10^9 to 10^{12} movements or 'jumps' per second).

If a shear stress is applied to the liquid, then the molecular 'jumps' in the direction of the shear stress will be favoured, and those in other directions opposed. The macroscopic consequence of this is that the liquid flows in the direction of shear. Clearly, the stronger the intermolecular forces, the larger the molecules, and the closer the molecules are together (for example, owing to application of pressure), the higher the viscosity of the liquid. Equally, reducing the temperature reduces the thermal energy of the molecules and the frequency of their 'jumps'. So if the shear stress is applied fast enough — in particular, if the time scale of application is equal to, or less, than the time for a molecular jump — then the molecules do not have time for their jumps, and the energy supplied is stored rather than dissipated by molecular movement. In other words, the liquid shows an elastic response. The time scale required for a simple liquid will be very short (10^{-9}–10^{-12} s) [1], but will be much longer for a liquid with large molecules such as a polymer melt. Examples of this have already been discussed in the introduction to this chapter.

Finally, when the intermolecular forces predominate over the thermal movements of the molecules, the molecules can only oscillate about a fixed position by means of their thermal energy, and the material has become a solid. Conventionally, the liquid/solid boundary is set at a viscosity of 10^{15} poise.

14.5.2 Solution and dispersion viscosity

If two simple liquids are mixed together and form a homogeneous mixture, then the molecules are of similar size and the viscosity of the mixture may generally be calculated by some sort of mixture law, using mole or volume fractions of the components of the mixture.

However, if the solute molecule or the suspension or dispersion particle size is much greater than the size of the solvent or medium (continuous phase), then the solvent or suspension medium can be treated as a featureless, uniform medium (a continuum) of viscosity, η_s. In this case, it is usual to define a relative viscosity,

$$\eta_r = \eta/\eta_s \qquad (14.17)$$

or a specific viscosity,

$$\eta_{SP} = \frac{\eta - \eta_s}{\eta_s} = \eta_r - 1 \qquad (14.18)$$

where η is the viscosity of the mixture (solution, dispersion, or suspension). This can be done over a range of sizes from ions of strong electrolytes in aqueous solution through micellar or colloidal solution 'particles' to dispersion particles or very large suspension particles. Microscopically, the viscosity increase over that of the solvent alone is determined by the total 'particle' volume, as well as the interparticle forces and Brownian motion, owing to the impact (thermal energy) of solvent molecules on the 'particle' surfaces.

Einstein [17] formulated a quite general viscosity concentration relationship for such systems. Assuming the particles are (1) spherical, (2) rigid, (3) uncharged, (4) small compared to dimensions of measuring apparatus but large compared to the medium molecular size (continuum hypothesis), (5) with no hydrodynamic interactions between the particles (large interparticle distance or very low concentration), (6) with slow flow (negligible inertia effects) and (7) with no slip at the particle–liquid interface), his equation reads

$$\eta_r = 1 + 2.5\phi \tag{14.19}$$

However, as Goodwin has pointed out, Einstein's working out of this equation also yields [18]

$$\eta_r = \frac{1+\phi/2}{(1-\phi)^2} \tag{14.20}$$

which expanded as a polynomial yields

$$\eta_r = 1 + \frac{5}{2}\phi + 4\phi^2 + \frac{11}{2}\phi^3 + 7\phi^4 + \tag{14.21}$$

In equations (14.19–14.21) ϕ denotes the dispersed phase volume fraction.

Many workers have attempted since to produce theories which build on Einstein's theory and relax some of the assumptions made (1–7, above). Particular effort has been made to extend the theory to higher concentrations, anistropic particles, and the role colloidal forces play in determining dispersion rheology. These topics have been reviewed many times, but most recently by Goodwin [18], Russel [19, 20], and Mewis [21] amongst others.

It is beyond the scope of this chapter to review this work in detail, but brief consideration will be given to concentration effects, as they have important practical consequences for paint rheology, particularly during the drying or flow-out period. The large amount of experimental work on concentration effects has shown that the coefficients of the polynomial terms in the series expansion form of Einstein's equation represent minimum values, obtaining only for very dilute dispersions or suspensions. Departures from spherical shape lead to increases in the coefficient of the linear term. As the concentration increases, so binary and n-tuplet collisions of particles lead to temporary associations between particles, owing to hydrodynamic forces (in the absence of any significant colloidal or other interparticle attractive forces), as has been demonstrated by the elegant experiment work of Mason and his co-workers [22] through the past 30 years. These associations lead to some occlusion of medium, and thus the associated group of particles occupies more phase volume than the total volume of the individual particles comprising the group. Vand [10] appears to have been the first to grasp the significance of this effect for dispersion rheology. Equally, the phase volume of individual particles must include a contribution due to the volume occupied by solvating molecules from the medium, adsorbed molecules of surfactants, and polymeric dispersants, as well as the effective volume resulting from strong electrostatic fields generated by ionization at the particle surface (e.g. aqueous colloidal dispersions).

As the concentration increases, it becomes experimentally more difficult to fit polynomials accurately to the data. For this reason, many workers have sought to establish theoretical and empirical relationships based on fractional, exponential, or

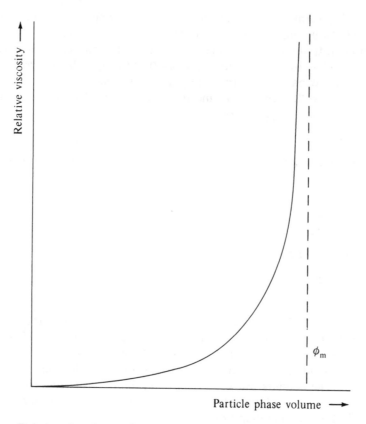

Fig. 14.9 — Relative viscosity — phase volume relationship for hard particle suspension and dispersion systems.

power law forms. Much of the variability in experimental results used to test these relationships can be attributed to limited understanding and characterization of the systems studied, e.g. imperfect stabilization leading to an unknown and uncontrolled degree of flocculation or aggregation, wide particle size distribution, etc., as well as experimental difficulties due to the onset of non-Newtonian effects at high concentrations. As common factor to these alternative forms is the introduction of a factor which appears in most cases to be related to the critical packing phase volume fraction, ϕ_m. This is because, as the concentration increases, the particles come closer and closer together, until finally they are too close together to be able to move. At this point, the concentration has reached ϕ_m and the dispersion has become solid and will exhibit fracture if stressed too much. The precise value of ϕ_m will depend on the packing geometry, but will be in the range $\phi_m = 0.52$ to 0.74. This implies that for hard particles, the viscosity of the suspension increases very sharply in a limited region of the concentration range around $\phi = 0.60$, as shown in Fig. 14.9, and ultimately reaches very high values. As the particles become closer, so ordering processes become essential for flow, and these require a longer and longer time scale to accomplish. Such processes in the presence of a shear stress are now assisted by Brownian motion (at least for particles below about $2\,\mu m$ in diameter). The longer time scale implies the appearance of non-Newtonian and viscoelastic effects in exactly analogous fashion to the case of molecular liquids discussed above, except that now the thermal or kinetic energy of the medium molecules impacting the particles provides the energy for the particle movement (Brownian motion) in the absence of a shear field. If a significant number of particle aggregates or floccules are present when the system is at rest, then the sharp increase in viscosity, as particle concentration is increased, will occur at a lower phase volume, owing to the volume contribution of occluded medium. Also non-Newtonian effects may occur at lower concentrations than usual, as the shear field may be strong enough to break down some of the aggregates or floccules, releasing occluded medium and thus reducing the effective volume and the relative viscosity. This breakdown process will be balanced by the rebuilding process when particles collide in the flow, owing to the net attractive interparticle forces resulting from imperfect stabilization, for example.

If the particles are liquid droplets (emulsion systems), then when the droplets are close-packed, they will compress and deform. Even if the stabilizer layer on a hard particle forms a significant part of the effective particle volume, then the strong dilatancy and shear fracture effects observed at or close to ϕ_m will be reduced or disappear [23]. However, unlike emulsion droplets, they cannot phase invert, which is what happens finally to emulsions when the concentration is pushed beyond ϕ_m. As a result of this, the disperse phase volume of the emulsion decreases sharply with a corresponding drop in relative viscosity (Fig. 14.10) [24]. Just as with hard particles, increases in the time scale lead to the appearance of non-Newtonian and viscoelastic effects in the vicinity of the phase inversion concentration. Equally, these effects can be seen with mixtures of partly miscible simple liquids [25].

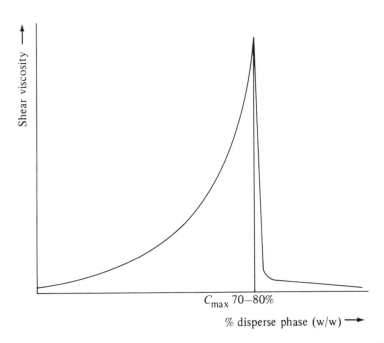

C_{max} 70–80%

% disperse phase (w/w) →

Fig. 14.10 — Relative viscosity — phase volume relationship for an emulsion (liquid dispersion) system.

Summarizing this discussion, the relative viscosity–disperse phase volume relationship requires the use of the effective phase volume (ϕ_e) instead of the nominal or physical phase volume fraction (ϕ), and the introduction of a critical packing volume fraction (ϕ_m) beyond which the dispersion changes its nature (becoming a solid for hard particles and phase-inverting with liquid droplets). It is the author's judgement that of all the theoretical treatments of this relationship so far published only those which try to take account of hydrodynamic interactions between particles have any predictive utility, for example, those based on cell models, such as the early theory of Simha [26] and the later theories of Yaron and Gal-Or [27] and Sather and Lee [28].

14.5.2.1 Polymer solution viscosity

The relation between the relative viscosity of a polymer solution and the concentration of polymer in solution shows a more gradual increase with increasing concentration, than is observed for dispersion systems. In general, three concentration regions can be distinguished; namely, dilute, semi-dilute, and concentrated regions (see Fig. 14.11 and cf. Fig. 14.9). The viscosity behaviour derives from considering the polymer molecules as loose, randomly arranged coils, and from considering the balance of three types of intermolecular forces, i.e. those between the polymer molecules, between the solvent molecules, and between the polymer and the solvent molecules.

In the dilute region, the coils are widely separated and not able to interact hydrodynamically. If the solvent is a thermodynamically 'good' solvent for the

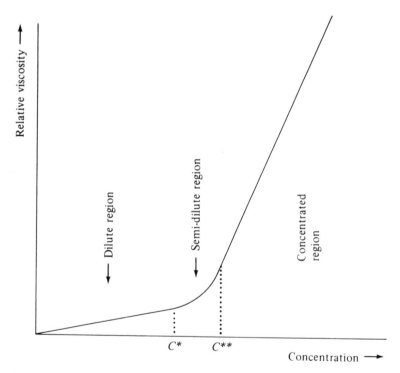

Fig. 14.11 — Relative viscosity — concentration (phase volume) relationship for a polymer in solution.

polymer (polymer–solvent intermolecular forces are strongest), then solvent molecules penetrate the coils, causing them to approach their maximum possible size. Consequently, the effective volume occupied by each coil and the relative viscosity of the solution both approach a maximum. As the thermodynamic quality of the solvent is reduced (for example, by adding non-solvent or changing the temperature), intramolecular forces between neighbouring parts of the polymer coil become stronger in relation to the other intermolecular forces, and the coil contracts with a consequent reduction in effective volume and relative viscosity. Finally, the 'theta state' is reached where the coil volume is reduced to its minimum fully saturated size. Thereafter, further reduction of solvent quality leads to coil–coil association and finally to precipitation (bulk phase separation) of the polymer from solution.

In the semi-dilute region, the relative viscosity starts to change more rapidly with concentration. In this region, the coils are closer together and interact with each other. Put crudely, they compete for space. In 'good' solvents, this results in the coils reducing in effective volume as the concentration increases until at the upper boundary of this region the coil's effective volume is again at a minimum (unperturbed dimensions). In a 'good' solvent, this region tends to cover a wider range of concentration for a given polymer, than for the same polymer in a 'poor' solvent, because of the greater volume of coil contraction available in the good solvent.

In the concentrated region, further contraction of effective volume is obtained by overlap of the coil volumes. The polymer chains are now forced close together, and intermolecular forces between them are increased, particularly if the solvent is a 'poor' one for the polymer. This results in a much more rapid increase in viscosity with concentration, and, unlike the dilute concentration region, 'poor' solvents give higher viscosity solutions than 'good' solvents.

It should be emphasized that the coils cannot be thought of as discrete entities with definite boundaries, unlike dispersion or emulsion particles. The coil volume has dimensions determined by the statistical average positions of the individual polymer molecular segments in space, averaged over a sufficient length of time. This explains why polymer coil volumes can apparently overlap or interpenetrate in the concentrated region, although they have apparently reached their minimum dimensions at a lower concentration. As before, the interaction of the polymer coils leads to the appearance of non-Newtonian and viscoelastic effects in the semi-dilute region, whilst, even in the dilute region, strong shear or extensional flows can lead to such effects because of polymer coil deformation and orientation effects.

More detail will be found in textbooks on the subject [29, 30]. The range of each concentration range depends on a number of factors, but, principally, the polymer molecular weight and weight distribution (determining individual coil dimensions).

14.5.3   Rheological structure

Liquids possessing rheological structure have this feature in common: they appear to be solid-like or to be thick (high-viscosity) liquids at rest, but, when relatively small forces or deformations are applied to them (for example, by gentle stirring or shaking) they become relatively mobile or low-viscosity liquids. This apparent breakdown process may reach a steady state quite quickly while the force is applied (pseudoplastic materials), otherwise it may continue for as long as the force is applied and, if the force is removed, the structure may recover relatively quickly (thixotropic materials) or only very slowly or not at all (irreversible materials). Some solid-like materials may also appear to require the application of a minimum force value before they will flow: they are said to possess a yield stress.

The origins of thixotropy, methods of measuring materials possessing thixotropy, and modelling of such behaviour have been reviewed by Mewis [31].

References

[1] LAMB J, *Rheologica Acta*, **12** 438 (and refs. cited in this paper) (1973).
[2] BINGHAM E C & GREEN H, *Proc ASTM* **19** 640 (1919).
[3] GLASS J E, *J Coatings Technol* **50** (641) 56 (1978).
[4] HUTTON J, *Rheologica Acta* **8** (17) 54 (1969).
[5] STRIVENS T A, *J Colloid Interface Sci* **57** (3) 476 (1976).
[6] BAUER W H & COLLINS E A, in *Rheology: Theory and Applications*. ed. F R Eirich, Academic Press (1967).
[7] ELIASSAF J, SILBERBERG A & KATCHALSKY A, *Nature* **176** 1119 (1955).
[8] WALTERS K, *Rheometry*, Chapman & Hall (1975).
[9] WHORLOW R W, *Rheological Techniques*, Ellis Horwood (1980).
[10] VAND V, *J Phys Coll Chem* **52** 277, 300, 314 (1948).
[11] MOORE F & DAVIES L J, *Trans Brit Ceramic Soc* **55** 313 (1956).
[12] OKA S, in *Rheology, Theory and Applications*, Vol. 3, Ch. 2, pp. 18–82, in ed. F R Eirich, Academic Press (1960).
[13] TAYLOR G I, *Phil Trans Roy Soc* (*London*) Ser. A **223** 289 (1923).
[14] TAYLOR G I, *Proc Roy Soc* (*London*) Ser. A **157** 546, 565 (1936).
[15] FERRY J D, *Viscoelastic Properties of Polymers*, 2nd edn, John Wiley (1970).
[16] STRIVENS T A, (to be published).
[17] EINSTEIN A, *Ann Physik* **19** (4) 289; (1906) **34** (4) 591 (1911).
[18] GOODWIN J W, in *Specialist Periodical Reports: Colloid Science*, Vol. 2, Ch. 7, The Chemical Society (London) (1975).
[19] RUSSEL W B, in *The Theory of Multiphase Flow*, ed. R Meyer, Wisconsin UP (1982).
[20] RUSSEL W B, *J Rheology* **24** (3) 287 (1980).
[21] MEWIS J, *Adv Coll Interfac Sci* **6** 173 (1976).
[22] GOLDSMITH H L & MASON S G, in *Rheology: Theory and Applications*, ed. F R Eirich, Academic Press (1967).
[23] STRIVENS T A, (Unpublished results).
[24] BARRY B W, *Adv Colloid Interface Sci* **5** 37 (1975).
[25] FIXMAN M J, *Chem Phys* **36** (2) 310 (1962).
[26] SIMHA R, *J Appl Phys* **23** (9) 1020 (1952).
[27] YARON I & GAL-OR B, *Rheol Acta* **11** 241 (1972).
[28] SATHER N F & LEE K J, *Progr Heat Mass Transfer* **6** 575 (1972).
[29] YAMAKAWA H, *Modern Theory of Polymer Solutions*, Harper & Row (1971).
[30] BOHDANECKY M & KOVAR J, *Viscosity of Polymer Solutions*, Elsevier (1982).
[31] MEWIS J, *J Non-Newtonian Fluid Mech* **6** 1–20 (1979).

15

The rheology of paints

T A Strivens

15.1   Introduction

In this chapter, the rheology (flow properties) of paint will be described both for the bulk paint and the film of paint after application, as well as methods for measuring the paint rheological properties in both states. Brief mention will be made also of other areas in the paint industry, where rheology is important, for example in various important stages of paint manufacture, such as pigment dispersion (wetting), transport through pipelines, and mixing operations.

Inevitably, because of the complexity of most practical paint-flow situations, both the theory and the experimental evidence provided to understand paint flow will appear sketchy or incomplete in many areas. However, the importance of understanding and, by consequence, the ability to control, paint rheology remains paramount to the successful utilization of such products. New legislation to control emission to the atmosphere (air pollution) and safety at work (lower and lower maximum levels of solvent vapours which may be toxic or may present fire hazards) is increasing. Such legislative pressure is increasing in industry throughout the world. The consequence of such pressure is to limit the freedom of choice of both paint formulators and users. Two trends which have emerged from such limitations are the move towards higher and higher solids paints (less solvent to be eliminated in the formation of the solid paint film) and towards water-based paints (less toxicity hazard in principle). The inevitable result of such trends is paint with more complex rheological properties than before, both in the bulk and in the film.

15.2   General considerations on paint rheology — paint application processes

The assertion has just been made that paint rheology control is essential to the successful utilization of paint. This assertion can be justified by consideration of some examples. An important example is provided by the application of conventional paints to a surface. Whatever application technique is used, be it spray gun, brush, roller, etc., the process has three stages:

- transfer of paint from bulk container to applicator;
- transfer of paint from applicator to the surface to form a thin, even film; and
- flow-out of film surface, coalescence of polymer particles (emulsion paints), and loss of medium by evaporation.

In each of these stages, the paint rheology has a strong controlling influence on the process.

15.2.1   Transfer stage

In the bulk container, the paint will normally be low in viscosity. This is so that it can be readily utilized in the chosen applicator; for example, it can readily penetrate the spaces between the bristles of a brush or the porous surface of a hand roller where it will be held by capillary/surface tension forces during the transfer to the surface to be painted. Of course if the total weight of paint loaded onto the brush by dipping it into the paint in the bulk container is sufficient to overcome the capillary forces, then the paint will drip or run off the brush (definitely not an attractive property for the user!). If the brush-load, without incurring the dripping penalty, is too low, then only a thin film, or a thicker film over a smaller surface area, will be obtained on brushing out. In the former case, solvent loss may be too rapid, and the consequent increase in film viscosity too quick to allow proper flow-out after application (see below). In the latter case, the painting operation becomes too laborious and time-consuming to be satisfactory. What determines the optimum film thickness? Apart from flow-out properties (see below), the required colour density and covering power (the ability of the paint to mask previous coatings — primers, old paint — or surface texture and colour) are important considerations, as well as the protective properties of the final solid film.

Increased brush-loading may be achieved by increasing the bulk paint viscosity or by introducing rheological structure. Both strategies include the useful bonus of slowing down or eliminating pigment settlement by gravity during storage of the bulk containers. However, increasing the bulk viscosity carries the penalty of increasing the mechanical effort required to spread the paint into a film (an important consideration with hand-operated applicators). To be satisfactory, rheological structure must break down quickly under the relatively low shear stress or strain of dipping a brush into the paint or the higher stresses imposed by a hand roller on paint in a tray. This breakdown facilitates the loading/penetration processes, but then the structure must reform quickly to prevent dripping, running, etc.

Turning to industrial application techniques such as spraying or roller-coating, the considerations are similar. Thus, in the case of spraying, the viscosity of the bulk paint must be low enough to allow the paint to be pumped through the fine jet of the spray gun with minimal application of pressure. Usually, such paints are thinned from higher solids bulk immediately before use, so settlement is less of a problem. The mechanism of droplet formation is reasonably well understood for simple liquids [1,2], but the influence of such factors as the presence of pigment or polymer particles in suspension or dispersion and the presence of polymer in solution, on the droplet size and size distribution in the spray is largely unknown. However, it can be expected that lowering the bulk viscosity will decrease the droplet size, possibly increasing the danger of wastage by small droplets being carried past the object to be sprayed by the air currents originating from the spray gun. Equally, evaporation of medium from the large surface area of the droplets will be influenced by droplet size and consequent total droplet surface area, and so will influence the initial rheology of the film built up by impaction and coalescence of spray droplets at the surface of the object being sprayed. In the case of industrial roller-coating processes, the paint must be considerably thicker, being able to flow under gravity or low pumping energy to the surface of the application roller, where it may be spread into an even layer by the action of the doctor blade or by another roller, etc. In this situation, the mechanical work required to cause the paint to flow is much less important. However, the paint must be viscous enough to prevent it running off or being thrown off the roller by centrifugal force.

The important feature of both these processes is the very high fluid flow rate and operating speeds, respectively, and the consequent high stresses and strain rates of deformation applied to the paint. However, it should be noted that the paint remains in the spray gun jet for such a short time (or in the 'nip' between the rollers) that a steady state is never attained and, therefore, only transient (for high frequency oscillatory) measurement methods are likely to produce relevant rheological parameters. Such methods require complex equipment and techniques, particularly at the high stress and deformation rates attained in the application process. Schurz [3] quotes shear rates of $100\,000\,s^{-1}$ applied for 1 ms in high speed roller-coaters. Such high values also have a further possible consequence if polymer is present in solution in the paint formulation. At such values, the presence of polymer in solution at a concentration practicable for modern paint formulations and at a molecular weight above about 10 000 can lead to the development of a high extensional viscosity component, both in the fluid jet from a spray gun nozzle and in the splitting film at the rear of a roller coater nip. Glass [4] has shown that the extensional viscosity of 'thickened' water-based emulsion paints influences such application properties as tracking, spattering, etc., during hand-roller application of such paints. It is reasonable to expect that the appearance of such a high extensional viscosity can interfere with the process of filament or jet rupture to form spray droplets. By Trouton's law, the extensional viscosity of a simple liquid is three times the shear viscosity. However, the presence of a few tenths of a percent of high molecular weight polymer in solution in this liquid can raise the extensional viscosity to as much as ten thousand times the shear viscosity [4].

15.2.2   Film formation

The loading and transfer of paint by a hand applicator, such as a paint brush, from the bulk container to the surface to be painted, is followed by regular movement of the hand applicator over the surface to transfer the load of paint from the applicator to the surface and spread it out in an even layer. During this process, hand pressure on the applicator causes shearing and compression of the brush bristles or fibres or of the rubber foam or fibrous mat typically covering the surface of a hand roller. Such shear and compression pushes paint out of the interstices in the applicator. The flow processes involved are very complex and probably impossible to analyse quantitatively. However, attempts have been made to measure the corresponding shear rates appoximately for brush application either directly or by correlation with subjective assessments of brushability of Newtonian paints. Ranges of

$15–30\,s^{-1}$ for brush dipping and $2500–10000\,s^{-1}$ for brush spreading are quoted in the literature [5]. Kuge [6] has attempted to measure the forces exerted during brushing and to relate these to the rheological properties of paint, using simple theoretical equations.

15.2.3   Flow-out of paint film

Over 30 years ago, the suggestion was made that the characteristic irregularities produced on the paint film surface by brushing, the surface striations produced when paint is applied by drawdown bars, and (by implication) also those produced by roller-coating application, have a common origin in hydrodynamic instabilities produced by the unstable flow resulting from the paint film being forced to split between the substrate and the receding applicator edge [5, p. 553, 7]. While the physical process is understood, it is not always possible to quantify it. Theoretical treatments of the effect have been attempted by Pearson [8] and more recently by Savage [9].

In a previous section, the possible importance of extensional viscosity in the application transfer process from rollers was mentioned. Glass [4] made measurements of extensional viscosity on water-based emulsion paints, thickened with water-'soluble' polymers, such as various cellulose derivatives, acrylamide/acrylic acid copolymers, and polyethylene oxide polymers, and he tried to relate the phenomena of paint spatter and surface tracking to the extensional viscosity measurements. Unfortunately, because of the limitations of the spinning-fibre technique that he used for measuring extensional viscosity, he was forced to use concentrations and molecular weights of the thickening polymers, which gave unacceptable amounts of web formation, filament formation, and surface irregularity when the paint formulations were applied by roller. He did also measure both the elasticity recovery [10] and the normal force under steady shear conditions. The former was done using a technique first described by Dodge [11] in which the sample is first sheared at high shear rate $(2600\,s^{-1})$ and then instantaneously the elasticity is measured as a function of time on cessation of shearing, using a low-frequency oscillatory shear deformation of frequency $0.3\,Hz$ and maximum shear rate of $0.07\,s^{-1}$. Such measurements may be readily achieved, using an instrument such as the Weissenberg Rheogoniometer. Because of the systems Glass measured for extensional viscosity, whilst acknowledging the importance of elasticity recovery, he tended to ascribe most of the effects he studied to elongational viscosity. However, when this elongational viscosity value was low or moderate, he found the elastic recovery rate correlated well with roll track formation and flow-out.

The critical features of paint application processes are that the paint is first sheared at very high shear rate and then, secondly, forced to split cohesively, either at the spray gun nozzle exit or else on the substrate between the receding edge of the applicator and the layer adhering to the substrate. Both processes take times of only a small fraction of a second. Myers [12] has argued that the high shear rate of the application process will completely destroy any structure present in the paint before application, and that, although the paint film now lacks any rigidity, two factors prevent the normal liquid cohesive splitting by viscous flow:

1   Rapid separation of the applicator from the substrate leads to tensions which cannot be alleviated by transverse flow of the fluid.

2   Most paint compositions contain some polymer in solution which can contribute an appreciable elasticity to the solution.

So, while the usual split of the liquid into filaments by cavitation takes place, the elastic component of the paint, combined with the speed of the separation, both result in less necking and longer temporal persistence of the filaments than would be expected for a simple viscous liquid. The presence of pigment or polymer particles, present in suspension in the paint, may well assist the nucleation of the cavitation process (see Trevenna [13]). Such an explanation implicitly rules out extensional viscosity (not mentioned by Myers), although it may be equally as important as paint elasticity under some application conditions; and, as Walters [14, p. 219] has pointed out, extensional viscosity is not *a priori* a function of shear viscosity. Indeed, some recent work with dilute high-molecular-weight polymer solutions [15] suggests that sharp increases in extensional viscosity at certain extensional deformation rates are due to the polymer molecules becoming fully extended in the deformation field, which is in sharp contrast to the oscillatory average coil configuration probably adopted by such molecules in the oscillatory shear field, customarily used for measuring shear elasticity modulus. Walters [14, p. 233] cites an example of a 100 ppm aqueous polyacrylamide solution of shear viscosity 1.4 cP exhibiting an extensional viscosity of 90 P.

Thus, although something is understood about the physics of how surface irregularities arise during the application and formation of wet paint films, there is no agreement about what rheological properties of the paint are relevant to the formation of such irregularities, although there is no dispute about the relevance of such properties.

When one turns to consider the processes of surface levelling, which affect such important practical properties as colour uniformity, hiding power, etc., as well as the more major flow faults such as sagging and slumping, there appears to be an equal lack of understanding of relevant paint rheological parameters. Although the work of Glass [10] and Dodge [11] has clearly indicated a connection between shear elasticity recovery rate and surface irregularity flow-out (levelling), the existing published theories and experimental work on levelling and sagging still regard paint films as Newtonian or pseudoplastic fluids (possibly with a yield point), thus representing no essential advance over the pioneering work of Orchard on the levelling of Newtonian liquids, published over 30 years ago [16]. Indeed, an edition [5] of a highly regarded textbook manages to treat the whole of paint flow and pigment dispersion processes without once mentioning shear elasticity or extensional viscosity of paint (cf. [17]). There is, however, abundant evidence of viscoelastic behaviour in paints as well as pigment dispersions or millbases [18, 19].

The situation is further complicated by the effects of solvent evaporation. Not only does this affect the rheology of the paint film, but it also affects the surface tension at the wet-film/air interface. Surface tension and gravity forces provide the shear stresses which drive the levelling and sagging processes. The results of evaporation will be to increase solution polymer concentration and to produce cooling at the film surface. Both of these effects will (unless they occur uniformly all over the film surface) lead to a tangential surface shearing force (the Levich–Aris force). Overdiep [20] has recently argued that the hydrostatic pressure gradient in a paint film due to surface tension is insufficient to explain levelling results, as argued by Smith *et al.* [7] and those that have followed or developed their treatment. He attempts to demonstrate both theoretically and experimentally, using solvent-borne alkyd paints, that, while surface tension tends to produce a flat surface, irrespective of the substrate surface profile underneath, the surface tension gradient developing over the wet paint film surface tends to produce a uniform paint film thickness, i.e. the surface profile of the paint film mirrors exactly the surface profile of the substrate underneath.

In addition, solvent evaporation leads to gradients in solvent content through the film as well as across the surface, and, consequently, to density gradients. Both density and surface tension gradients could contribute to circulatory patterns being set up in the wet paint film. Such circulatory systems may well contribute to the disorientation or randomization of aluminium flake orientation in metallic automotive topcoats, although most publications consider only film viscosity as the controlling factor [21]. In extreme cases, they may lead to the formation of Bénard cell patterns, such as are more commonly seen at the surface of boiling or rapidly evaporating bulk liquid samples. Such effects as a cause of surface irregularities of paint or polymer films have been considered by Anand et al. [22] and Higgins and Scriven [22] amongst others.

While such considerations allow estimates to be made of the forces operative in controlling the rheology of the applied paint film, it is important to realize the potential complexity of the film's rheology because of its physical and chemical *composition*. From what has been said in the previous chapter, it can be expected that not only will the film material be viscoelastic but also, even at the low stress values involved, highly non-linear and with time-dependent effects deriving from the high shear during application. Something of the complexity of rheological behaviour of bulk systems of analogous composition may be gauged by referring to the published experimental work over many years of Onogi and his co-workers on polymer particle dispersions in polymer solutions and melts [24]. In view of this, attempts to model the levelling behaviour of paint films by considering the material as a Newtonian or pseudoplastic liquid seem to be too simplistic, as do simple rheological measurements based on such concepts. Not only that, the presence of concentration gradients through the film thickness is likely to mean that the rheology will vary through the depth of the film. By comparison, the effect of density gradients in the film is likely to be minor.

A comprehensive review by Kornum and Raaschou Nielsen [25] assesses the balance of factors involved in the levelling and other flow processes of wet paint films, particularly with reference to surface defects. They state the operative forces in levelling to be in the range 3–5 Pa (30–50 dyne cm^{-2}), and in sagging to be about 0.8 Pa (8 dyne cm^{-2}) at the surface of a typical paint film. They also quote estimates of shear rates for levelling processes in paint films of 0.001–0.5 s^{-1}. Because it is shear stress, resulting from gravitational and surface tension forces that controls the flow in levelling, sagging, etc., estimates of the shear rate are irrelevant to the consideration of the flow processes; and, similarly, when trying to measure the relation of paint rheology to flow properties, it is desirable to use controlled stress instruments, rather than the more conventional controlled shear rate instruments.

15.2.4 Desirable rheology for paint application

We can now summarize the desirable rheology for paint application. Initially, the paint must lose its structure at rest and become low in viscosity to facilitate transfer by, or through, the paint applicator. Because of the high shear rates and short timescales involved in the transfer process, both elasticity and extensional flow processes may modify the pattern of surface irregularities on the paint film. These would be expected to arise anyway from unstable hydrodynamic flows produced by cohesive splitting at either the spray gun jet exit or at the interface between the adherent paint film on the substrate and the material on the receding edge of an applicator, such as a roller, moving over the substrate being painted.

The paint must now remain low in viscosity for a sufficient time for the surface irregularities to flow out to an acceptable extent (dependent on whether, for example, gloss or protective properties by evenness of film thickness is the prime property required). However, while the viscosity is low, the paint will flow on vertical surfaces under the influence of gravity. If the thickness (film depth) builds up too much, the effect (sagging) may become noticeable (i.e. offensive!) to the observer, whilst lumps of the thickened paint may slide on the substrate in an irregular fashion to give rise to the effects of curtaining, slumping, etc. In either case, a definite time after cessation of application is required before such effects become noticeable.

Thus the initial low-viscosity period must be followed by a sharp rise in viscosity, resulting from loss of solvent by evaporation, or by a rapid recovery of elasticity (rheological structure) destroyed by the shearing of the transfer process. In either case, the effect is the same, the drying film is virtually immobilized, and the sagging process ceases before it becomes noticeable. The loss of solvent may be affected by differential volatility as well as solvency of components in the solvent mixture forming the paint medium or diluent towards polymeric components in the paint. These solvency effects, in turn, will control the viscosity rise of the drying paint film as it loses solvent. Losses by solvent evaporation will produce cooling of the drying paint film surface (latent heat of evaporation), particularly in the case of fast-evaporating solvents. This cooling may also influence film viscosity.

15.3   Experimental methods for measuring paint rheology for application and flow-out after application

15.3.1   Paint rheology for application

A number of general instruments for rheological measurements, described in the previous chapter, may be utilized for studying paint application and flow-out properties. However, within the limits discussed in the previous sections, many paints may be considered as quasi-Newtonian liquids. Also there is a need for simple instruments for quality control and user viscosity adjustment before application. This need has been largely satisfied by specialized instruments, developed within the paint industry.

This group of simple instruments consists principally of various flow-cups of different designs, and simplified rotational instruments, such as the ICI cone and plate instrument. Flow cups are used in other industries, e.g. the petroleum industry. The principle of the instrument is illustrated in Fig. 15.1. A known volume of paint is held within a vertical cylindrical cup, whose bottom has a short capillary of controlled length and diameter. The paint is released to flow through the hole in the bottom of the cup (usually by the operator removing his finger!), and the time for

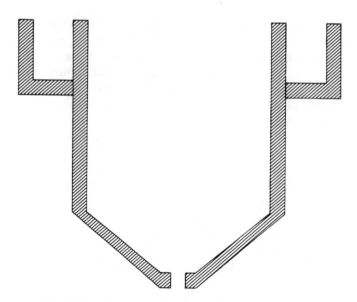

Fig. 15.1 — Section of a flow-cup (approx. full scale).

the liquid to flow out of the cup is measured with a stop-watch. The flow end-point is normally taken as the point at which the continuous liquid jet breaks up into drops. A number of points emerge in considering this type of measurement. Firstly, because the liquid height varies during the test, the force (due to gravity) driving the liquid through the capillary also varies. So, if the paint is non-Newtonian, the viscosity result may be very misleading. Secondly, the capillaries are always short, so stable flow conditions within the capillary are not obtained, and this, as well as entry and exit errors, may also affect the result, particularly if the material is slightly structured (elastic). Thirdly, the presence of abrasive particles in the paint may lead to wear of the metal capillary, and so flow-cups should be checked frequently with Newtonian liquids of known viscosity. The cup bottoms may be conical or flat; the former is likely to reduce entry errors. Finally, tests are normally run at ambient temperatures, but it is preferable to control the paint temperature carefully before and during the test to produce accurate and comparable results. This type of test should be used only with near-Newtonian paints. The various types of flow-cup have been reviewed in detail by McElvie [26].

A different type of instrument is exemplified by the ICI cone and plate instrument, first described by Monk [27, 28]. This instrument consists of a fixed lower plate, thermostatted at $25.0 \pm 0.1\,°C$ by a frigistor system, and an upper cone driven by a motor at 900 rev/min via a torsion spring. The operating shear rate is $10000\,s^{-1}$ and the measurement range is 0–5 poise with a reading accuracy (by pointer moving over a scale) of better than 0.1 poise. The cone is truncated to reduce wear and to avoid particles jamming in the gap. Only a small sample of paint is required (less than 1 ml), and the instrument is quick and easy to use. Clearly, this instrument is superior to the flow-cup in that the sample temperature is carefully controlled and the sample viscosity is measured under conditions (high shear rate) relevant to application conditions. However, the instrument still provides only a single point measurement, and consequently gives no indication of the paint rheology at rest or during film flow-out. To assess this, a measurement at low shear rates or stresses is necessary.

Other methods are available for measuring the application viscosity of paints, and these have been reviewed by McGuigan [29]. A particular problem group of paints are thixotropic or highly structured paints. Some idea of the complexity of rheological behaviour of such materials can be gained from the discussion in the previous chapter, but finding simple test methods for quality control is very difficult. Measuring of such paints at rest is relatively easy — some sort of vibrational method, as discussed in the next section under rheology during storage, may be used — but it is extremely difficult to measure breakdown and recovery of structure accurately (the crucial factor in determining application and film flow-out properties of such materials). An excellent review of the methods available has been given by Walton [30]. In the present author's experience, the only satisfactory method is to destroy all the structure by shearing the sample at high shear rate for a sufficient time (this also eliminates errors due to rheological 'history' produced by the sample-handling and loading into the rheometer), followed by monitoring the recovery of structure with time. This can be done by reducing the shear rate to a very low value, and recording apparent viscosity as a function of time; or better still, using a small oscillatory stress or strain, and measuring both dynamic viscosity and elasticity modulus as a function of time (see also [11]).

In the first edition of this book, the low shear viscometer (LSV) range of instruments, developed by ICI to do this type of measurement, was described. Since then, the availability of computer-controlled rheometers, in particular, stress controlled instruments from all the major manufacturers, has made the experimental realization of such complicated test routines very easy, at least in principle. It should be borne in mind that in designing such routines to simulate application conditions and post-application flow-out, the instruments will often be operating at their high and low limits respectively. This means that interpretation of results must make proper allowances for the limitations of the instruments when operating at the extremes of their ranges.

Standard methods of viscosity measurement of paints put out by official standards organizations tend to specify flow-cups, in spite of the increasing tendency of paints to show viscosity values dependent on applied shear rate or stress. Thus the British Standards Institute (BSI), the Standards Association of Australia (SAA), and the International Standards Organization (ISO) all have flow-cup specifications [26, 31, 32]. The SAA has in addition a guide to test methods for rheological properties [33], measurement of viscosity by cone and plate instruments [34], and consistency measurements by Stormer viscometer [35], Rotothinner [36] and rotational viscometer [37]. American specifications include use of the ICI cone and plate viscometer and the Sturmer viscometer. The Deutshes Institut fur Normung DIN 53214 (1982) method describes the determination of 'rheograms and viscosities' of paints and varnishes, using rotational viscometers, as do French and Czech standards [38, 39].

15.3.2   Paint film flow-out (levelling and sagging)

15.3.2.1   Direct measurements of rheology during flow-out
To summarize the main problems involved in direct measurement of paint film rheology during flow-out (levelling):

- The rheology of the paint film material is extremely complex (not only viscoelastic, but extremely non-linear).
- The rheology can change rapidly with time, owing to compositional changes (solvency balance changes, solids changes, etc. as solvent is lost, as well as rheological structure recovery).
- Small volumes, possibly also inhomogeneous in composition, through the film depth.

The emphasis must therefore be on rapid methods (short individual determination time and, therefore, high repeat rate) and maximizing the information obtained from each determination (for example, by deconvoluting the response curve from an instantaneous stress application to obtain data over a wide frequency range). For this reason we will start by surveying impact and high frequency oscillatory methods before proceeding to a consideration of other methods.

Impact method (bouncing ball)
Impact methods have been used widely in many different forms for testing polymers in the form of cylindrical or disk specimens and for testing the mechanical properties of solid (cured) paint films during their service life (see Chapter 16). However, their use for studying paint film flow-out and curing processes appears to have been overlooked, apart from a brief reference by Snow [40] to the bouncing ball method. For bulk polymer samples, a similar method has been briefly analysed by Flom [41] and, in more detail, by Tillett [42], Jenckel and Klein [43], and Pao [44].

The present author [45] has confirmed the original findings reported by Snow. A typical set of results is shown in Fig. 15.2, where rebound height is plotted as a function of paint drying time for a 0.5 cm diameter (0.5 g weight) steel ball dropped on to a 12 mm ($\frac{1}{2}$ in) thick glass slab coated with the paint under test. As the paint film dries, the viscosity increases owing to solvent loss; and, consequently, the energy dissipated by the film during the ball impact with the glass increases also, and the ball rebound height decreases initially with drying time. However, when curing starts, either by autoxidation or by 'lacquer-type' drying, as in the example shown, the film develops some elasticity and the rebound height increases again. While it is possible to derive a simple theory relating rebound height to film viscosity, based on ball momentum and energy losses, a number of important factors are neglected in this simple treatment. It is thus better to calibrate with Newtonian standard oils and use an empirical formula to relate the rebound height to viscosity. Among these factors is the hydrodynamic force, which will prevent the ball ever actually touching the substrate surface, as well as liquid elasticity effects at the short times of impact (a few tenths of a millisecond).

In spite of these drawbacks, the method is simple and easy to use. Slabs of glass, metal, or even wood, together with a range of balls of different size or density and a graduated 'fall' tube for measuring the rebound height, are the only apparatus required. By temperature programming a metal slab it ought to be possible to use the technique to study the curing of thermoset systems in a similar fashion to that used by Gordon and Grieveson [46] for polymers. By contrast with the rolling-ball technique (to be described below), this technique does have the capability of measuring the development of film elasticity during curing.

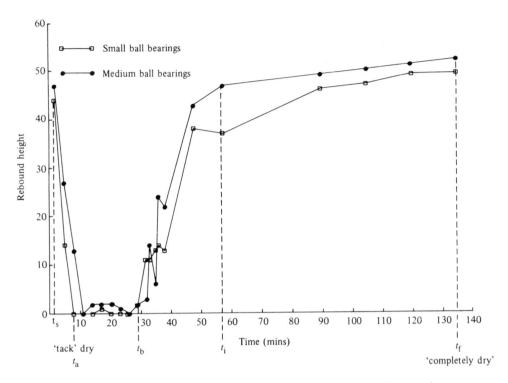

Fig. 15.2 — Time *vs* rebound height for a 200 μm refinish paint film on glass.
t_a = end of viscosity increase period, t_b = start of gel formation, t_i = hard film formation,
t_f = end of experiment.

High frequency (impedance methods)
The mechanical impedance of an elastic shear wave propagating through a medium
is changed by the presence of a viscoelastic layer at the surface of the medium. If
the elastic wave is completely damped in this layer, the change in characteristic
impedance can be related to the rheological parameters of the layer material. This
is difficult to achieve with many paint films, but nevertheless, the method can be
used to follow changes in the paint-film rheology during drying and curing.

In practice, pulses of high frequency oscillations are generated by means of a suitably excited piezoelectric crystal attached to the support medium. After propagation through the support, the attenuated pulses are again transformed into electrical
signals by a piezoelectric crystal attached to the support. The phase angle and
attenuation of the received pulses are measured, and often changes in the values of
either, or both, of these quantities are used to compare rate changes in drying and
curing of different paint films. Either the pulses are directed at the surface of the
support at a shallow angle, and the reflected pulses detected by a separate receiver
crystal, as in the apparatus (Fig. 15.3) of Myers [12, 47–50], or the train of pulses is
reflected normally from a face of the support along the same path so that it can be
detected by the same crystal as in the torsional quartz rod method of Mewis [51].
In either case, the paint to be studied is coated onto the support.

Although it is not explicitly stated, the present author has the distinct impression

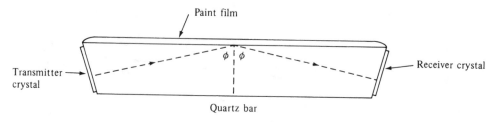

Fig. 15.3 — Myers' impedance technique.

that these methods are limited by the sensitivity of the equipment to a limited range of viscoelastic parameter values, and that, therefore, in many paint systems, the technique can be applied to measure film properties during only a limited part of the total drying/curing process of the paint film. Furthermore, the adhesion of the drying paint film to the support material can be expected to have a strong (and possibly, unknown) effect on the results [50]. Myers [12] used frequencies of 2–100 MHz, gated as 4 μs pulses, and measured attentuation from the signals collected as a series of exponentially decaying echoes. Mewis [51] used somewhat lower frequencies around 100 kHz for his torsional-rod method.

Rolling-ball method
The rolling-ball method is, like the bouncing-ball method, simple in conception and execution. A coated panel is inclined at a convenient angle, and the time taken for a small steel ball to roll over a measured distance on the coated surface is recorded as a function of drying/curing time. Alternatively, the distance rolled by the ball as a function of time is recorded. The method was introduced by Wolff and Zeidler [52], who used it to study the effect of different solvents and plasticizers on the viscosity of nitrocellulose finishes during drying. Quach and Hansen [53] used it to correlate viscosity with surface flow-out in emulsion paints; Taylor and Foster [54] used it to study the reactivity of stoving enamels in the temperature range 100–140 °C and Göring et al. [55] used it to study viscosity changes in high solids and electrodeposited paints as well as fillers. The latter authors [55] have derived a simple theory, based on force balance considerations, which the present author has found to be inadequate to explain results.

Recently, van der Berg and de Vries from the TNO Institute in the Netherlands have described an ingenious automated version of this technique. The coated panel under study is mounted on a revolving inclined table mounted in a temperature-controlled box. A sphere is placed on the edge of the coated panel, and this is illuminated by a light source which also falls on an array of photocells. The output from these photocells is used to control the table rotation speed accurately, so that the ball remains stationary in relation to the light beam. The authors have improved the theory without giving details, but admit that factors such as variations of surface tension and consequent wetting of the ball, paint pick-up during rolling, and the flow pattern around the sphere (bow wave and wake) have not been properly treated. Even so, the results obtained are impressive, although they relate only quantitatively to viscosity changes.

Again, parallel work on rolling friction on solid polymers as well as mechanical test methods for hardness of paint films, such as the Sward Rocker (see Chapter 16), would give the impression that elasticity values could also be derived as with the bouncing-ball technique, but so far this has not been done. Unlike the bouncing-ball, this would require a modification of the experimental technique.

ICI (relaxation) low-shear viscometer
This instrument was designed to simulate the conditions of high shear rate, followed by flow under low (decreasing) stress forces, which obtain when, for example, paint is applied by a brush. A description of the operating principles of the instrument has been given by Colclough, Smith & Wright [56]. The theory of this instrument and its variants has been given by Strivens [57].

The principle of the instrument is illustrated in Fig. 15.4. It is based on a torsion pendulum. If the pendulum is twisted through a small angle and released, the twist in the coil spring suspension unwinds, driving the bob of the pendulum back towards its initial (equilibrium) position. The inertia of the bob now carries the bob past its initial position, re-twisting the spring in an equal and opposite sense. The bob motion then reverses, and the back-and-forth oscillation continues *ad infinitum* in the absence of energy-dissipative processes (Fig. 15.4b) with an amplitude and frequency of oscillation determined by the mass of the bob (inertia) and the material elasticity modulus of the spring (and its physical dimensions). In a real situation, energy losses always occur, therefore the oscillation amplitude decreases (decays) with time, i.e. the oscillation is damped (Fig. 15.4c). If the bob is placed in contact with a viscous liquid during its oscillations, the oscillation will be still further damped. If the viscosity of the damping liquid is increased, the damping will be increased still further until, even at quite low viscosity, the bob no longer oscillates and the deflection decreases smoothly with time from its first maximum deflection position towards its initial (equilibrium) position. The system is said to be over-damped (Fig. 15.4d), and it is this condition that is used in the instrument.

In the instrument, the liquid (paint sample) is held between parallel plates (S in Fig. 15.5), the lower one being fixed and thermostatted, the upper one with its supporting rod (F) and measurement vane (L) forming the bob of the pendulum system. The supporting rod passes through an air bearing (D), which locates it and prevents any motion except the required torsional movement with minimal frictional loss. The measurement vane forms the moving part of an air condenser, which forms part of a tuned quartz oscillator circuit, so that the slightest movement of the vane produces an off-balance voltage, which may be used to measure deflection. At the top of the rod is the coil spring (K), whose outer end is secured to the frame of the instrument head. The whole assembly may be raised and lowered on a vertical column (C) by means of the spring-loaded arm at the side of the instrument.

In operation, in the initial state the upper plate is deflected electromechanically through an angle of 5°, the sample is inserted, allowed to equilibrate, and then the start button pressed. This releases the plate, allowing it to move back towards its zero angular deflection position under the influence of the spring and against the resistance offered by the sample. The resulting deflection–time curve can be recorded over a period of about 30 s before the plate is automatically reset to its initial deflection, ready for the next run. The instrument automatically measures and stores three values of the deflection (θ) at 3, 10, and 25 s after the start (time t), as well as measuring and storing two values of viscosity (by measuring the average slope $d\theta/dt$ over deflection ranges 80–60% and 40–30% of initial deflection). These values, as well as the instantaneous deflection values during the run, can be registered on the digital display of the instrument.

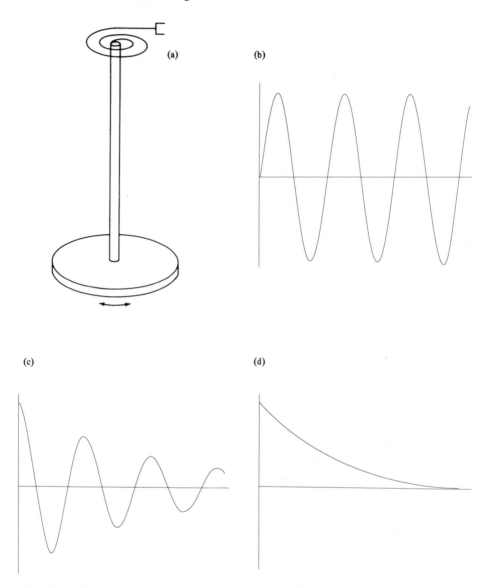

Fig. 15.4 — The principle of the low-shear viscometer (LSV): (b) Sinusoidal oscillatory curve; (c) Damped oscillatory curve; (d) overdamped oscillatory curve.

By means of a mechanical clutch, the upper plate can be rotated at high speed (900 rev/min, corresponding to a maximum (rim) shear rate of $2500\,s^{-1}$) for a few seconds to simulate the high shear experienced by the sample during brush-out. Thus families of 'relaxation' curves can be generated from the sample at rest, immediately after shear and at timed intervals afterwards to study structure recovery in thixotropic paints. As will be seen from Fig. 15.6(b), these families of curves correlate very well with the results of subjective assessments of paint flow-out, based

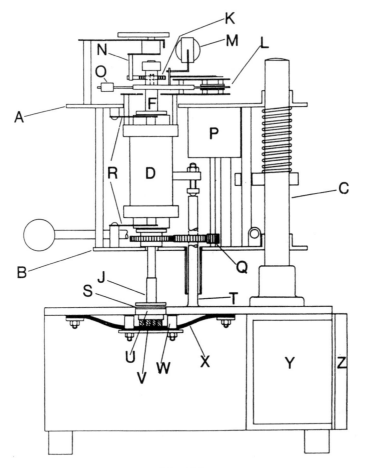

Fig. 15.5

on large-scale brush-outs by skilled painters. This is probably due to thixotropic structure recovery having a stronger influence on the initial stages of brushmark flow-out than solvent loss, at least in the paints tested.

By using the theory derived by Orchard for brushmark flow-out and the equation of motion of the torsional oscillator, Colclough et al. [56] were able to choose suitable coil springs to give the right level of initial stress (3–5 Pa). Furthermore, as the spring unwinds and the upper plate moves towards zero deflection, the driving force is reduced, i.e. the stress exerted on the sample decreases, in an analogous fashion to what happens in practice, where the driving forces produced by surface tension are proportional to the wavelength and amplitude of the brushmarks, and so, as the brushmarks flow out, the driving force is reduced.

Two variant instruments have also been developed [57]. In one of these, the coil spring is wound up at controlled speed through an angle of 180°, thus applying to the sample a stress which increases linearly with time. During this wind-up process, the instantaneous upper plate deflection is recorded as a function of time on a chart

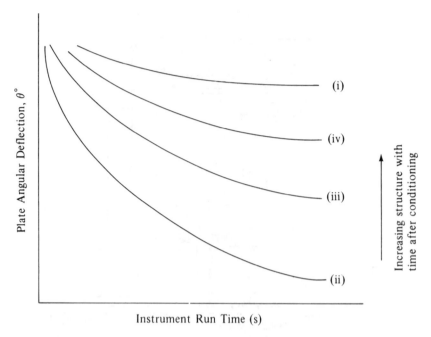

(i)   Initial condition of paint, before subjecting the sample to a predetermined shearing regime.
(ii)   Sample sheared at $2500\,s^{-1}$ for 4 s; then measurement made immediately upon cessation of shear.
(iii) Sample sheared at $2500\,s^{-1}$ for 4 s; 1 min allowed to elapse before making measurement.
(iv)   Sample sheared at $2500\,s^{-1}$ for 4 s; 5 min allowed to elapse before making measurement.

Fig. 15.6(a) — Typical pattern of results, indicating method of use of the low shear viscometer (LSV).

recorder. A Newtonian liquid shows a deflection which varies as the square of the time from start. However, if there is weak elastic structure in the paint, the initial deflection–time plot will be linear rather than parabolic, with slope proportional to elasticity; then, at some critical stress, the structure starts to break down and the deflection–time plot changes towards that of the Newtonian liquid. Thus with this instrument the kinetics of structure breakdown can be studied.

In the second variant, the coil spring is oscillated by a reciprocating drive of variable amplitude and speed. By this means, an oscillatory (sine-wave) stress of variable frequency and amplitude may be applied to the sample. By measuring the phase angle and the amplitude of the upper plate motion in relation to the drive motion, it is possible to derive the dynamic viscosity and elasticity modulus of the sample, as a function of frequency (0.02–3 Hz range) and stress amplitude (four discrete values 0.6, 1.5, 5, and 10 Pa). Studies of the viscoelasticity of concentrated suspensions, using this instrument, have been published by Strivens [58, 59]. Because this instrument is fitted with the high speed shear facility as described above for the relaxation LSV instrument, the kinetics of structure build-up can be followed in more detail than with that instrument, and, as both viscosity and elasticity may be measured at frequent intervals as a function of reformation time, there is potentially more information available than is the case for the relaxation instrument. The

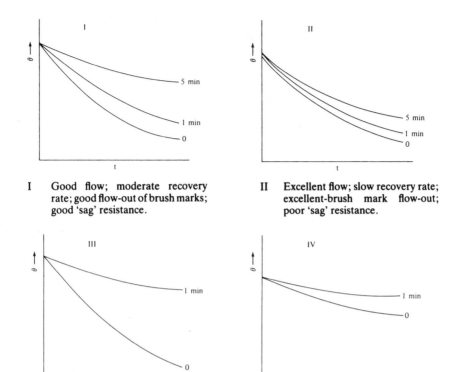

I Good flow; moderate recovery rate; good flow-out of brush marks; good 'sag' resistance.

II Excellent flow; slow recovery rate; excellent-brush mark flow-out; poor 'sag' resistance.

III Good breakdown with rapid recovery; after ~30 s brush marks cease to flow-out, therefore only moderate performance, but good 'sag' resistance.

IV Little breakdown; poor flow; poor brush marking; excellent 'sag' resistance.

Fig. 15.6(b) — Typical curves for paints having different brushing characteristics and flow-out.

relaxation instrument is easy and simple to use and suitable as a development or quality control instrument, whilst the oscillatory instrument is meant to be a research instrument.

The major defect of the instrument just described is the limit placed by the coil spring design on the accessible range of paint rheology. Modern controlled stress instruments do not rely on direct mechanical coupling, as the LSV instruments, while the response measurements (resultant rotational velocity or angular deflection) are based on more sensitive and accurate methods, such as optical gratings. With the advent of much more complex and highly structured (thixotropic) paints, the greater versatility of modern computer-controlled instruments is essential to understanding the rheology of such paints.

Sled method
By sandwiching the paint film between its substrate and a thin plate (say, a small microscope slide), the simplest shear geometry is obtained. Attaching a weight to the slide via a pulley, and recording the motion of the slide as a function of time, allows determination of the paint film viscosity.

Such a simple technique has been described by Kornum [51], among others. Practical difficulties with maintenance of shear geometry and edge effects, which can be foreseen, are discussed.

Torsional pendulum (torsional braid)
Whilst these methods can be used for studying early stages of curing of paint, while the paint is still liquid or semi-solid they are more properly described in the next chapter, as being more suitable for studying fully cured paint films.

Other methods
Within the limitations of this chapter, it has been possible to describe only a few of the vast number of techniques available. Inevitably, the author's choice has been dictated by personal experience working within a desire to concentrate on methods that yield absolute values of rheological parameters, and, apart from those, to concentrate on methods commonly used in the industry.

A more comprehensive survey of viscometers of all types used in the paint industry has been given by McGuigan [22].

15.3.2.2   Direct measurement of film flow-out
As an adjunct to the measurement of film rheology during flow-out, the direct measurement of the flow-out or other flow phenomena, such as sagging, has also been attempted. To an extent, this has been in an effort to confirm, or otherwise, predictions based on the observed rheology.

If an even coating of paint can be spread on a flat substrate, and, by some means, a series of regular striations can be produced on the coating surface, resembling brushmarks, but rather more regular, then it should be possible to observe the course of flow-out directly with drying time and to make measurements of amplitude and wavelength directly as a function of time. Such measurements have been made by using light interference methods, by Wapler [62]; while Kheshgi and Scriven [63] have used Moiré topography. Wapler has compared his results with the theoretical predictions of Biermann [64]. Kheshgi and Scriven, in addition to studying levelling, has also used their technique to study surface profiles in liquid flowing down a slope, liquid curtains falling vertically from a slot, and disturbances in surface profiles due to encounters with surface-active particles. Whilst less sensitive than interference methods, Moiré fringe techniques are more flexible in that they cover a wide range of displacements (from several micrometres to several centimetres) over which their sensitivity is good. Klarskov [65] has described a dual beam optical interference method for measuring the levelling of wet paint films.

By contrast, there are a number of technological methods for measuring both levelling and gross flow defects, such as sagging. These are comprehensively surveyed by McGuigan [29]. The usual pattern of sag test applicator consists of a bar-type spreader whose trailing edge is cut in a series of slots of different depth. Moving the spreader across a glass plate leaves a series of parallel stripes of paint of different thicknesses. The test plate is then placed in a rack usually inclined at an angle of 60°, and the edge profile is observed visually during drying. Sagging resistance is assessed by the minimum film thickness, whose edge profile deforms significantly during the period of observation. Such an instrument has been described by Schaeffer [66] amongst others. Again, Biermann [65] has attempted theoretically to predict the onset of sagging, curtaining, and similar phenomena.

15.4   Paint rheology during manufacture and storage

Four areas, where rheology control or measurement can be useful in paint manufacture, can be listed:

- milling of pigments (preparation of millbases);
- transport of paint or intermediates used in paint production through pipelines, etc;
- mixing operations; and
- storage.

15.4.1   Pigment millbase production

Pigments are delivered to the paint manufacturer in the form of (more or less) dry powders which, for various reasons, contain numbers of pigment particle agglomerates and aggregates. The objectives of the pigment-dispersion process (ball milling, triple rolling, sand milling, etc.) are:

- to break down pigment particle agglomerates and aggregates; and
- to ensure complete wetting of the pigment surface by the paint medium, in particular by surface-active agents present in the medium, which will prevent subsequent flocculation of the pigment.

The development of the required optical properties of the final paint film will depend strongly on the quality of the pigment dispersion, so it is important to optimize the conditions of pigment dispersion.

While the basic science of the pigment-dispersion process is well understood, the application of such scientific principles to optimize dispersion conditions is often, in practice, rather more difficult to achieve. Parfitt [67] has reviewed this basic science, while Kaluza [68] has reviewed the causes and effects of pigment flocculation in paints. Tsutsui and Ikeda [69] have reviewed methods of evaluating particle size and distribution and degree of pigment dispersion. Patton [5, ch. 7–12] has dealt with the practical aspects of pigment millbase formulation and assessment. Vernardakis [70] has recently reviewed all aspects of pigment dispersion.

Starting from the basics of dispersion rheology, a number of principles can be derived. Firstly, the effectiveness of the milling will depend on the amount of (mechanical) energy dissipated. To maximize this dissipation, the overall viscosity of the dry pigment/medium mixture should be as high as possible, i.e. the pigment content must be high. The limit to this is an effective volume fraction of around 0.64 where critical packing of pigment particles occurs and the mixture rapidly becomes like a solid. In practice, the onset of dilatancy, with accompanying solid-like behaviour and fracture, may lower this limit still further, as will the presence of anisotropic pigment particles, agglomerates, or aggregates.

The initial effective volume occupied by the pigment particles will be larger than the total effective volume of the individual pigment particles, owing to volume occluded within the pigment particle aggregates and agglomerates. As these are broken down by the dispersion process, the pigment effective volume is reduced, so the viscosity of the pigment/medium mix falls. Also, some of the mechanical energy is dissipated as heat, so unless adequate cooling is used, this reduces the viscosity still further. The net result of this viscosity decrease is to reduce the dissipation of mechanical energy in the dispersion, and thus to reduce the effectiveness of the dispersion process. Thus the attainment of constant viscosity, as a function of dispersion time, could be used as an indicator of the end of the effective part of the dispersion process. However, changes in particle size and size distribution could still take place without significantly altering the effective volume and, hence the dispersion rheology. To measure such changes, methods other than rheology are more appropriate; for example, fineness of grind gauges, as discussed by Tsutsui and Ikeda [69].

A number of authors have used rheological techniques to study aspects of the pigment-dispersion process. Oesterle [71] has used such techniques to study the course of pigment dispersion, as well as the effectiveness of various dispersion machines. McKay [72] has studied the effectiveness of various dispersants in organic pigment dispersion processes; Heertjes and Smits [73] have used rheological measurements to study the dispersion of titanium dioxide in alkyd media, particularly in relation to alkyd molecular weight and the effect of the presence of saturated fatty acid molecules. Strivens [18], Mewis and Strivens [74], and Zosel [75] have studied the viscoelastic properties of pigment dispersions, the former during the course of ballmilling, the latter to assess the nature of the pigment particle interactions and, by implication, the effectiveness of the dispersant and the dispersion process used. Strivens used an oscillatory stress amplitude controlled rheometer (see Section 15.3.2.1), and Zosel used his own design of creep apparatus. McKay [76] has studied of the influence of the crystal morphology of organic pigments on the rheology of their millbases, while Priel and Torriano [77] have described two methods of characterizing the flow properties which they use to predict their performance.

15.4.2 Pipe flow
The early literature on the flow of pastes and paints in pipelines has been reviewed by Weltmann [78]. Patton [5, ch. 6, pp. 156–183] has also devoted a chapter to the subject.

In general, the problem is to calculate the pressure required to move a given paint or intermediate through a pipe at a required flow rate. Although, either in laminar or turbulent flow, this can be done reasonably accurately for Newtonian liquids, non-Newtonian liquids present more of a problem. By measuring apparent viscosity as a function of shear rate over a relevant range of values, using a rotational viscometer and fitting some empirical viscosity-shear rate equation such as the Casson or Bingham equation, approximate predictions of required pressure can be made which may suffice for the purpose of engineering design. However, time-dependency effects (thixotropy) may vitiate some of these calculations, particularly if flow rates are low. Also, strong interactions in the material, giving rise to strong elasticity, as well as dilatancy, owing to high disperse solids, may also lead to unacceptably high initial pressure values being required to start flow of the material. In this case, measurements with a pipeline rheometer (similar to a capillary viscometer, but with a wider bore) may be more useful.

15.4.3 Mixing
In this area, rheological characterization of the materials to be mixed has an important role to play in designing plant and studying the efficiency of mixing processes.

Something of this role may be judged by consulting some of the more recent general textbooks on this subject, for example, that of Oldshue [79].

15.4.4 Storage

The measurement of changes in rheological structure during storage presents a particularly delicate problem, as exemplified by thixotropic paints. If a can of such a paint is opened and a sample withdrawn and placed in a rheometer, some of the structure will be broken down (cf. the dipping of a paint brush into the can). So what one is trying to measure is being altered by the sampling process, and it would thus be desirable to measure the structure of the paint in its container.

An instrument to achieve this, known as OSCAR (acronym for oscillating can rheometer), was developed by ICI (Paints Division) some years ago. Its principle of operation, briefly described by Strivens [57], is similar to that of the oscillating variant of the low-shear viscometer (LSV, discussed in Section 15.3.2.1), also developed by ICI. A circular table on top of the instrument is caused to oscillate around its centre, by means of a reciprocating drive operating through a coil spring at a closely-controlled frequency (close to 10 Hz). The amplitude and phase angle of the table motion are measured in relation to that of the drive by a similar condenser system to that of the LSV. If a solid cylindrical can or container is placed on the table, the phase angle difference between the table and the drive is zero, while the ratio of the amplitudes has a certain value, dependent on the container mass. This is taken as unity on the instrument scale, and all subsequent measurements are referred to this value for a given container mass. When a can containing a viscoelastic material, such as the thixotropic paint under test, is placed on the table, the amplitude ratio and the phase angle difference alter, because energy is now being dissipated, owing to the viscoelastic property of the can contents. These alterations can now be used to calculate the dynamic viscosity and elasticity of the paint. In practice, because of the complexity of the mathematics of the cylindrical geometry used, this is done numerically, and a graph is produced to allow the results to be read off directly from the instrument's readings. Variations in can weights can be accommodated by means of a potentiometer dial on the front of the instrument. Because the shear wave dies away quickly from the can side, and, as the maximum deflection angle is 5° for a 250 ml can of diameter 9 cm, the maximum shear strain is small and the bulk of the sample is virtually undisturbed.

Measurements can be made very rapidly. Storage samples in cans are weighed and placed on the oscillating table, the weight is dialled on the potentiometer, and normally within half a minute, the digital display of the instrument shows steady readings of phase angle difference and amplitude ratio. Using these readings, values of dynamic viscosity and elasticity modulus may be read off from the standard charts provided. Changes in rheological structure show as changes mainly in elasticity and, to a lesser extent, in viscosity. The instrument may also be used to detect hard settlement, as well as to follow the kinetics of structure formation in bulk gelation processes.

References

[1] RICHARDSON E G, in *Flow Properties of Disperse Systems*. Ch VI, pp. 266–98, ed Hermans J J, North-Holland (1953).

[2] GREEN H L, *Loc cit* [1] Ch VII pp. 299–322.

[3] SCHURZ J, *EUCEPA 20th Conf Budapest*, Paper 11/13 (1982), *World Surface Coatings Abstr*, 83/8306.

[4] GLASS J E, *J Coatings Technol* **50** (641) 56 (1978).

[5] PATTON T C, *Paint flow and pigment dispersion: A Rheological Approach to Coating and Ink Technology.* 2nd edn, Wiley-Interscience, pp. 365–7 (1979).

[6] KUGE Y, *J Coatings Technol* **55** (701) 59 (1983).

[7] SMITH N D P, ORCHARD S E, & RHIND-TUTT A J, *J Oil Colour Chem Assoc* **44** 618–33, part pp. 621–3 (1961).

[8] PEARSON J R A, *J Fluid Mech*, **7** 481 (1960).

[9] SAVAGE M D, *J Fluid Mechanics* **80** 743 (1977).

[10] GLASS J E, *J Oil Col Chem Assoc* **58** 169 (1975).

[11] DODGE J S, *J Paint Technol* **44** (564) 72 (1972).

[12] MYERS R R, *J Polym Sci, Pt C* (35) 3 (1971).

[13] TREVENNA D H, *J Phys, D: Appl Phys*, **17** 2139 (1984).

[14] WALTERS K, *Rheometry*, Chapman & Hall, (1975).

[15] (a) KELLER, (b) *Loc cit* [13] p. 233.

[16] ORCHARD S E, *J Appl Sci Res*, A11, 451 (1962).

[17] WU S, *ACS, Division of Org Coatings & Plastics Chem, papers*, **37** (2) 315, 323 (1977).

[18] STRIVENS T A (unpublished results).

[19] AMARI T & WATANABE K, *Polym Eng Rev* **3** (2–4) 277 (1983).

[20] OVERDIEP W S, in *Physicochemical Hydrodynamics. V G Levich Festschrift*, Vol II, ed Spalding D B, Advance Publications p. 683 (1985).

[21] See, for example, Wojtkowiak J J, *J Coatings Technol* **51** (658) 111 (1980).

[22] ANAND J N & BALWINSKI R Z, *J Coll Interfac Sci*, **31** (2) 196 (1969); Anand J N, *Ibid*, 203; Anand J N & Karam H J, *Ibid*, 208.

[23] HIGGINS B G & SCRIVEN L E, *Ind Eng Chem, Fundamentals* **18** (3) 208 (1979).

[24] MATSUMOTO T, SEGAWA Y, WARASHINA Y, & ONOGI S, *Trans Soc Rheol* **17** (1) 47 (1970).

[25] KORNUM L O & RAASCHOU NIELSEN H K, *Prog Org Coatings* **8** 275 (1980).

[26] MCELVIE A N, *Prog Org Coatings* **6** 49 (1978).

[27] MONK C J H, *J Oil Colour Chem Assoc* **49** 543 (1965).

[28] Manufactured by Research Equipment (London) Ltd, 64 Wellington Rd, Hampton Hill, Middlesex, UK.

[29] MCGUIGAN J P, in *Paint Testing Manual* (*ASTM Special Technical Publication* No 500), 13th edn, ch 3.2. pp. 181–212, ed Sward G G, American Society for Testing of Materials (1972).

[30] WALTON A J, *Paint Manufacture*, **47** (5) 13–7 (1977); *J Oil Colour Chem Assoc* **44** 16–22, 24 (1961).

[31] Standards Association of Australia, Paints and related materials: methods of test, Method 214.2, Flow cup.

[32] International Standards Organization, 'Paints and varnishes: determination of flow times by use of flow cups' ISO 2431, 1993: Paints and varnishes, Vol 1, General test methods. ISO (Switzerland), pp. 83–92 (1994).

[33] Standards Association of Australia, Paints and related materials: methods of test, Method 214.0, Rheological properties: Guide to test methods, AS 1580.214.0 (1990).

[34] Standards Association of Australia, Paints and related materials: methods of test, Method 214.3: viscosity (cone and plate), AS1580.214.3(1993).

[35] Standards Association of Australia, Paints and related materials: methods of test, Method 214.1: consistency, Stormer viscometer, AS1580.214.1 (1990).

[36] Standards Association of Australia, Paints and related materials: methods of test, Method 214.4: consistency, Rotothinner, AS1580.214.4 (1990).

[37] Standards Association of Australia, Paints and related materials: methods of test, Method 214.5: consistency, rotational viscometer, AS1580.214.5 (1990).

[38] Association Francaise de Normalisation, NFT 30–029 (1980).

[39] *Urad pro Normalisaci* CSN67 3016 (1981).

[40] SNOW C I, *Official Digest* (392) 907 (1957).

[41] FLOM D G, *J Appld Physi* **31** 306 (1960).

[42] TILLETT J P A, *Proc Phys Soc* **B67** 677 (1954).

[43] JENCKEL E & KLEIN E Z, *Naturf* **7a** 619 (1952).

[44] PAO Y H, *J Appld Phys* **26** 1082 (1955).

[45] STRIVENS T A, (unpublished results).

[46] GORDON M & GRIEVESON B M, *J Polym Sci* **29** 9 (1958).

[47] MYERS R R, *Official Digest* **33** (439) 940 (1961).

[48] MYERS R R & SCHULTZ R K, *Official Digest* **34** (451) 801 (1962).

[49] MYERS R R & SCHULTZ R K, *J Appl Polym Sci* **8** 755 (1964).

[50] MYERS R R, KLIMEK J & KNAUSS C J, *J Paint Technol*, **38** (500) 479 (1966).

[51] MEWIS J, *FATIPEC IX Congr Brussels Proc* pt 3, pp. 120–4 (1968).

[52] WOLFF H & ZEIDLER G, *Paint Varnish Produ Manager* **15** (8) 7 (1936).

[53] QUACH A & HANSEN C M, *J Paint Technol,* **46** (592) 40 (1974).
[54] TAYLOR R & FOSTER H J, *J Oil Colour Chem Assoc* **54** 1030 (1971).
[55] GÖRING W, DINGERDISSEN N & HARTMANN C, *Farbe Lack* **83** (4) 270 (1977).
[56] COLCLOUGH M L, SMITH N D P & WRIGHT T A, *J Oil Colour Chem Assoc* **63** 183 (1980).
[57] STRIVENS T A, *Quart Rep Paint Research Assocn* (79/4) 11 (1979).
[58] STRIVENS T A, *Colloid Polym Sci* **261**, 74 (1983).
[59] STRIVENS T A, *Colloids Surfaces* **18** 395 (1986).
[60] STRIVENS T A (in preparation).
[61] KORNUM L O, *FATIPEC Congr XIV Budapest Proc* pp. 329–36 (1978).
[62] WAPLER D, *Farbe Lack* **81** (8) 717; (9) 822; (10) 924 (1975).
[63] KHESHGI H S & SCRIVEN L E, *Chem Eng Sci* **38** (4) 525 (1983).
[64] BIERMANN M, *Rheol Acta* **7** (2) 138 (1968).
[65] KLARSKOV M, *Farbe Lack* **36** (3) 53 (1990).
[66] SCHAEFFER B, *Official Digest* **34** (453) 1110 (1962).
[67] PARFITT G D, *FATIPEC Congr XIV Budapest Proc* pp. 107–117 (1978).
[68] KALUZA U, *Progr Org Coatings* **10** 289 (1982).
[69] TSUTSUI K & IKEDA S, *Prog Org Coatings* **10** 235 (1982).
[70] VERNARDAKIS T G, in *Coatings Technology Handbook*, pp. 529–550, ed Satas D, Marcel Dekker (1991).
[71] OESTERLE K M, *FATIPEC Congr XIV Budapest Proc* pp. 329–36 (1978).
[72] MCKAY R B, *FATIPEC Congr XIII Cannes Proc* pp. 428–434 (1976).
[73] HEERTJES P M & SMITS C I, *Powder Technol* **17** 197 (1977).
[74] MEWIS J & STRIVENS T A, Paper given at *8th Int Congr of Chem Eng, Chem Equipment & Apparatus (CHISA)*, Prague (1984).
[75] ZOSEL A, *Rheol Acta* **21** 72 (1982).
[76] MCKAY R B, *Prog Org Coatings*, **22** 211 (1993).
[77] PRIEL S & TORRIANO G, *Rheology* **1** 223 (1991).
[78] WELTMANN R N, in *Rheology: Theory and Applications*, Vol III Ch 6, pp. 236–40, ed Eirich F R (1960).
[79] OLDSHUE J Y, *Fluid Mixing Technology*, McGraw-Hill (1983).

16

Mechanical properties of paints and coatings

T A Strivens

16.1 Introduction

The mechanical properties of paint and coatings are of paramount importance in maintaining the important protective as well as to a lesser extent the decorative functions of such treatments during their service life. Paint films are subject to a great variety of mechanical forces and deformations. Thus they may suffer large forces concentrated over a small surface area for very short times, as in the impact of stones, gravel, etc. on car body paints, or they may suffer a succession of slow cycles of deformation, as happens to decorative gloss paints on wood, for example, on house window frames, as the wood expands and contracts in response to changes in atmospheric moisture and temperature. Such forces and deformations can be large: of the order of gigapascals per unit area in the impact case, or 10–15% strain in the case of wood expansion and contraction (such deformation is also anisotropic, owing to the grain structure of the wood). So it is the ultimate mechanical properties of paint films, which is of most practical importance, i.e. the stress or strain that leads to plastic yield (irreversible deformation) or failure by cracking of the film.

Not only do paint films suffer a wide variety of mechanical stresses and strains during their service life, but also their mechanical properties change during it. This affects the ultimate mechanical properties such as yield and fracture, and in the end determines how long a coating film can preserve its physical integrity and can fulfil its protective role satisfactorily. The continual exposure to air and water condensation (dew) on the paint surface leads to steady leaching of low molecular weight species, such as retained solvent, plasticizer, or low molecular weight polymeric species as well as degradative reaction products which might otherwise soften the coating and increase its resistance to brittle failure (cracking). Equally, exposure to oxygen in the air and to light (particularly the ultraviolet radiation content) can lead to various photolytic reactions, which may generate free radicals and peroxides which can increase the degree of crosslinking of the film and, consequently, its brittleness. Ultimately, the coating film fails by cracking, either in the film matrix or, less frequently, at an interface, i.e. adhesive failure. The permeation of water into the film may be beneficial because it often 'softens' or plasticizes the film. However, if this is combined with permeation of oxygen or anions, this may lead to mechanical failure by blistering or accumulation of solid corrosion products at a coating/metal interface.

Most paints and coatings are based on organic polymers. They are thus visco-elastic and highly non-linear in their response to mechanical stress or strain. Not only this, but the mechanical properties of the matrix polymer is substantially modified by the presence of dispersed polymer particles, as well as phase-separated (insoluble or incompatible) components of the polymer matrix which separate during application or curing, pigment and extender particles, etc. The role played by pigment or polymer particles in modifying the mechanical properties of the coating is analogous to the role played by fillers and disperse polymer phases in determining the mechanical properties of bulk polymer composite systems. However, there is little evidence that this concept has been grasped, and that any attempt has been made to apply the considerable body of ideas and theory developed for polymer composites to interpreting the mechanical properties of paint films.

In addition, most coating systems are multilayered. Although the polymer matrix may be constant, the pigment nature and content, etc., will usually vary from layer to layer. Not only that, but the processes of solvent loss and spatially non-uniform crosslinking during cure may lead to gradients or heterogeneities in mechanical properties, both through the thickness and across the surface area of the film. Crosslinking and solvent loss processes usually lead to shrinkage of the polymer matrix with a consequent build-up in internal stress within the film. The presence of such internal stresses will modify the mechanical properties of the coating on its substrate. This will in turn affect the decision as to whether to test attached or detached coating samples; and indeed, in extreme cases, could totally invalidate the results of the latter.

Enough will have been said to give an idea of the complexity and importance of coating mechanical properties as well as the importance of monitoring such changes during actual or simulated environmental exposure tests, such as accelerated weathering.

The chapter will be organized as follows. After a brief general survey of the viscoelasticity and ultimate mechanical properties of polymers, consideration will be given to methods of determining such properties, particularly in relation to coating specimens attached to their substrates. After a survey of practical mechanical test methods for coatings and their interpretation in terms of more fundamental properties, the chapter will close with a section on the application of acoustic emission to monitoring changes in the mechanical properties during weathering.

16.2 Viscoelastic properties of polymers

In Chapters 14 and 15 of this book, the concepts and experimental techniques for measuring viscoelastic liquids have been outlined, and reference should be made to the first of those chapters for an explanation of the rheological terms used here. Many accounts of polymer viscoelasticity are available; books by Ferry [1], Nielsen [2], and McCrum et al. [3], as well as review articles by Graessley [4], Hedvig [5], etc. These should be consulted for more detailed accounts than will be given here.

The characteristic property of most polymers of sufficiently high molecular weight or degree of crosslinking is that they are elastic solids at room temperature. If a constant mechanical stress (creep experiment) or mechanical strain (stress relaxation experiment) is applied to a specimen of a viscoelastic solid polymer, the response is predominantly elastic. In other words, there is insufficient time for the polymer molecules or their component units to move in relation to each other. Consequently they move by change of configuration in response to the applied mechanical disturbance, thus stretching and changing the original bond lengths and angles. When the mechanical disturbance is removed, this stretching and distortion result in the molecules recovering their original configurations. Thus mechanical energy has been stored and released (elastic response). A similar oscillatory energy storage and release process occurs if an oscillatory or since wave (dynamic experiment) mechanical stress or strain is applied to the specimen, provided that the frequency is high enough (i.e. the timescale of the oscillation is short enough).

Now, if the timescale is lengthened (or equivalently, the frequency of an oscillatory experiment is reduced), there is now sufficient time for a significant number of the polymer molecules to rearrange their positions in relation to each other, and for component units to achieve new equilibrium bond length and angle positions. Thus, when the mechanical disturbance is removed, there is no driving force to return the polymer molecules from their new to their original positions, consequently a significant amount of energy has been dissipated. Although there may not have been a change in sample shape, there still will have been irreversible deformation on a microscopic scale. Continuing this process will lead to irreversible sample deformation. Thus, even a sheet of glass or a glass fibre will bend or stretch irreversibly under a load, given sufficient time.

If the sample temperature is increased, then these irreversible rearrangements are facilitated by the extra translational, rotational, and vibrational energies possessed by the polymer molecules. In gross terms, the sample changes as temperature is raised from a glassy material through a rubbery intermediate state (particularly if the polymer molecules are linked by chemical bonds (crosslinking)) through to a viscous liquid. An equivalent change from predominantly energy storage response to energy dissipative response is observed, if the frequency is lowered or the time scale is increased. This equivalence was recognized and formulated as the Williams–Landel–Ferry (WLF) time–temperature superposition principle [6]. This WLF principle has great experimental importance. Most polymers have wide molecular weight distributions, and, as a result, a wide range of characteristic time scales for rearrangements. The transition in mechanical properties can thus cover many decades of frequency or time. As many instruments for determining mechanical properties cover only comparatively small ranges of frequency or time, the proper characterization of such transitions in mechanical properties (the 'glass transition temperature' or T_g) presents problems, unless a number of instruments with different frequency or time ranges are available. However, if one instrument is used and the sample measured at a number of different temperatures, the data can be reduced, using the WLF principle, to one standard temperature so as to cover a much wider frequency range than the instrument range. Of course, particularly in the case of coatings, due consideration must be given to the effect of temperature alone on the specimens (solvent or plasticizer loss, additional crosslinking, or thermal degradation) before applying such a procedure to determine T_g. This determination, and particularly its change during environmental exposure, is of great practical importance for assessing and predicting coatings performance.

If sufficient mechanical stress or strain is applied over a time scale where the response is predominantly elastic, these irreversible rearrangements may be forced to take place. The behaviour of the sample is now non-linear, the measured response is no longer proportional to the mechanical input, and, taken far enough, mechanical yield and fracture on a macroscopic scale are observed. Thus the ultimate properties of yield and brittle fracture are also time- (frequency) and temperature-dependent. Put another way, as most ultimate property determinations are made by deforming the specimen at a controlled strain rate and measuring the instantaneous force as a function of strain, the results of such measurements are a function of both strain rate and temperature.

The ease or difficulty of such irreversible rearrangements are determined by a number of factors. The chain flexibility of the polymer molecules is of great importance, thus a flexible hydrocarbon polymer such as polyisobutylene will rearrange more easily than a stiff polymer chain with bulky side chains or rigid ring structures in the side chains or main chain, e.g. polystyrene, celluloses.

Other important factors are the presence, number, and flexibility-length relationships of chemical bonded linking molecules between the polymer molecules (thus a vulcanized rubber is 'harder' than the corresponding unvulcanized rubber). Regular arrangements of polymer chains in crystalline materials increases the T_g considerably, as does the presence of other intermolecular forces due to hydrogen bonding, ionic interactions, etc. The presence of low molecular weight species, such as plasticizer molecules, or bulky flexible side chains, as in alkyl methacrylates, increases the volume between polymer molecules, thus making rearrangement easier and lowering the T_g. Also, the polymer molecular weight plays an important role in determining T_g. Put crudely, the longer the polymer chains, the more entangled they become and the higher the T_g. For most polymers the T_g does not increase much over a molecular weight of 20000. Many coatings polymers have been formulated precisely in the range 5000 and 20000 in order to achieve low viscosity solutions suitable for optimum application processes, so the increase of T_g, as the coating dries or crosslinks, is crucial to its performance.

The presentation of viscoelastic data for a solid is analogous to that for a viscoelastic liquid already described in Chapter 14. Thus an apparent modulus (or compliance) is plotted as a function of time with temperature as a parameter for controlled stress or strain experiments. Equally, the real and imaginary components of the modulus (usually referred to as the elasticity and the loss moduli respectively) are plotted as a function of frequency in dynamic experiments, again with temperature as a parameter. However, many commerical instruments operate at a single fixed or approximately constant frequency, and, in this case, it is more usual to plot the real component of the modulus and the ratio of the imaginary to the real components of the modulus (the 'loss tangent', written as $\tan \delta$) as a function of temperature. Most of these instruments operate by some form of temperature scan, and T_g is determined by locating the temperature at which $\tan \delta$ reaches a peak value. However, again, perception of the T_g value must take due account of the effects of temperature scan rate.

In other types of dynamic experiments, in particular a.c. impedance and dielectric measurements, much use is made of Argand diagrams, where the imaginary component is plotted as a function of the real component. Smooth or distorted semicircular or circular arc plots are obtained with points along the plot corresponding to each frequency value used. The distortion of the arc or the limitation of the arc to less than 180° (part of a semicircle) is related to the presence of more than one characteristic time or the existence of a distribution of characteristic times. The perception of these parameters is perhaps easier on the Argand diagram (known as a Cole–Cole plot in dielectric, and a Nyquist plot in impedance, work) than in the more conventional form of presentation. No doubt the generally more limited frequency range available for dynamic mechanical measurements has restricted its use. However, as Havriliak and Negami [7, 8] have demonstrated, it can be very powerful, particularly when comparisons are required with other dynamic data on polymers, in particular, dielectric data. More recently, Laout and co-workers [9–11] have made extensive use of Argand diagrams in presenting their dynamic mechanical measurements for drying alkyd and emulsion paints. They have fitted a simple equation to these results and have shown by comparison with the results of electron microscopy and other physical techniques, that the equation parameters are related to the degree of crosslinking and integration of the films respectively. Toussaint and Dhont [12] have also recently reviewed the theory of these Argand type plots in relation to dynamic mechanical measurements on paint films.

    Examples of these modes of presentation are given as idealized curves in Fig. 16.1(a) and (b). Figure 16.2 illustrates some experimental results, derived from some of the author's own unpublished data on a polyester film. (δ'' and δ' are related to, but are not identical with, G'' and G'.)

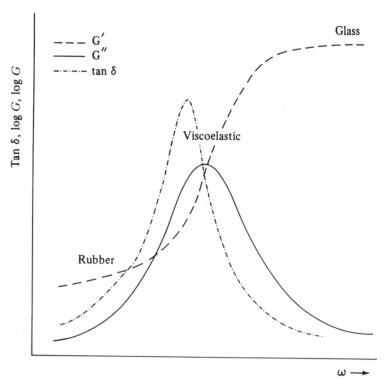

Fig. 16.1(a) — Idealized representation of dynamic mechanical properties of a solid (crosslinked) polymer.

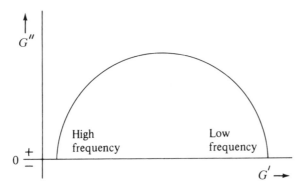

Fig. 16.1(b) — Idealized Argand diagram for mechanical properties of a viscoelastic solid.

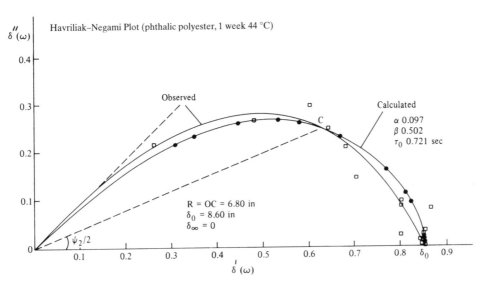

Fig. 16.2 — Argand plot of dynamic mechanical analysis (DMA) measurements on a polyester film (cf. Fig. 16.5).

16.3   Ultimate mechanical properties of polymers

Again, as with the linear viscoelastic behaviour of polymers, only a very brief intro-duction can be given, and more detailed accounts must be sought in books and review articles [13–16]. In the context of ultimate properties of polymers, failure of two principal types is considered, namely, ductile failure (yielding) or brittle failure (cracking). Figure 16.3(a) and (b) show idealized stress–strain curves for the two types of failure.

If a force is applied for a short enough time to a sample, for example, by a blow or impact, then there is insufficient time for viscous deformation to take place, i.e. for the polymer molecules or molecular units to move relative to each other and obtain new equilibrium positions. Consequently, the polymer reacts like an elastic glass (Fig. 16.3(a)) and, provided that the stress does not increase beyond a certain limit, brittle fracture does not take place. In principle, it is possible to calculate this

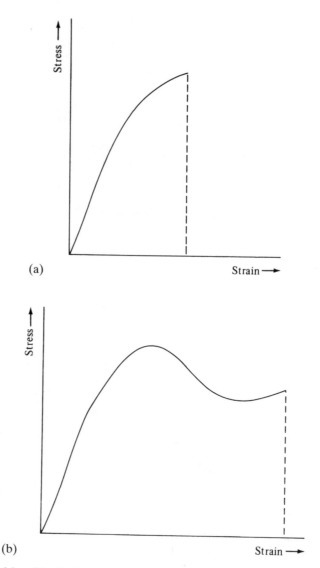

Fig. 16.3 — Idealized stress-strain curves for (a) brittle and (b) plastic yield
failure of solids.

limit (brittle strength) from a knowledge of bond strengths and intermolecular
forces (see, for example, [17]). In practice, values several orders of magnitude lower
than the theoretical are nearly always found. This is generally attributed to structural irregularities in the sample, which serve to concentrate the stress locally. Such
structural irregularities include cracks and fissures as well as imbedded particles of
foreign bodies or degraded material. The Griffith theory [18] of the strength of
brittle solids, which rests on the basic assumption that cracks are responsible for
strengths lower than theoretical, has gained considerable acceptance. However,
although this simple theory has considerable drawbacks and has since been much
modified (see, for example, Andrews' book [14, pp. 123–131]), the basic assumption
is undoubtedly sound and experimentally proven [19].

If, however, sufficient time is allowed and viscous deformation takes place, then the stress passes through a maximum (yield stress) before falling, as the polymer sample yields and deforms before fracture finally occurs. This type of failure is referred to as 'ductile failure'. Attempts have been made to predict yield stress using theories based on Eyring type equations [13].

The importance of time and temperature in determining whether brittle or ductile failure occurs is evident. Smith & Stedry [20] were able to verify the applicability of the WLF principle to ultimate properties as well. Toussaint [21] has discussed both theory and experimental work relating to the effects of pigment content on the ultimate mechanical properties of paint films.

16.4   Experimental methods for determining mechanical properties of coatings

16.4.1   Introduction
In the previous section a general view of polymer viscoelastic properties and their interpretation has been given. A number of reviews are available relevant to, or covering, both experimental methods for determining viscoelastic properties of coatings films and interpretation of the results. As many of these have appeared in the journal *Progress in Organic Coatings* and other readily accessible sources, it is proposed in this section to cover the methods and results rather briefly. The reviews should be consulted for more detail and for a fuller list of published source material.

16.4.2   Transient methods (creep, stress relaxation)
Transient methods have been reviewed by Zosel [22]. Most of the methods use free film specimens and tensile elongation using commercial tensile testing equipment. In creep tests, a weight is applied to the specimen, and the resulting deformation is measured as a function of time. If the weight is abruptly removed, the elastic recovery of the specimen may also be measured. Specific apparatus for making this type of measurement has described by de Jong and van Westen [23] and Zosel [24].

In stress relaxation experiments, the specimen is stretched rapidly, again in tensile mode to a predetermined extent, and the stress is recorded as a function of time. A typical apparatus for this type of test is described by Pierce [25].

So far these types of test use free film specimens. Because of internal stress factors, it is often debatable how relevant such results are to the mechanical properties of the coating/substrate system. For this reason, and for convenience and relevance in testing coatings during weathering and other environmental exposure tests, it is convenient to be able to test relatively small specimens of coatings *in situ* on the substrate.

The ICI pneumatic microindentation apparatus was designed for this purpose. The apparatus was first described by Monk and Wright [26], and is commercially available. The displacement of a loaded needle as it sinks into the painted film alters the gap between a flat plate (the flapper) and a small nozzle fed with a restricted air supply. The changes in pressure resulting from these movements are amplified pneumatically and recorded on an air-pressure strip chart recorder. A painted specimen is cut or punched out from a coated panel. The needle is tipped by a steel or sapphire ball (0.01–0.25 mm radius) and the specimen is mounted on a temperature-controlled platform under the needle. The platform temperature is controlled by a Frigistor unit. This unit is a Peltier effect device, which can bring the specimen to any temperature within the range −20 °C to +90 °C in a few minutes. The maximum deflection on the recorder corresponds to 6 µm indenter movement. The recorder chart can be read to about 0.1 µm, and measurement accuracy is about double this value. Examples of the curves obtained with this instrument for alkyd films as a function of temperature and weathering time are given in Monk and Wright's paper [26]. Interpretation of the shapes of the curves in terms of the coatings' predominant mechanical characteristics, i.e. whether plastic, rubbery, viscoelastic, or glassy, are also given. The theoretical interpretation is a little more difficult. It can be done by applying the basic Hertz theory of spherical indentation [27] as modified by Lee and Radok [28] and Radok [29] for viscoelastic materials, to obtain a compliance as a function of time and temperature in the usual way. However, it should be borne in mind that the stress field beneath the indenter is extremely non-linear and varies with the depth of indentation, as well as the load applied. Also, if the film thickness is too low or the indentation depth too great, this stress field may interact with the substrate and modify the indentation–time curve. Morris [30] has discussed these aspects of the instrument's results critically. In spite of this, the instrument is very useful: disk specimens about one inch in diameter can be cut at intervals during the course of panel weathering, or the indentation temperature can be changed systematically for each of a series of different locations on the same disk and a steady or fixed time delay indentation depth measured as a function of temperature to give a glass transition (T_g) value. Either procedure allows the change in mechanical properties during weathering to be accurately assessed.

16.4.3   Dynamic methods

The subject of dynamic mechanical testing of paint films has been reviewed in considerable detail by Ikeda [31]. This review covers not only experimental methods but also the interpretation of results, the calculation of coating properties from those of the coating plus support (particularly if supported coating specimens are used), the effects of formulation variables (in particular, pigmentation), and the correlation of these results with impact strength.

Four basic methods are available:

1   free oscillating torsion pendulum methods;
2   resonance methods in forced vibration;
3   forced non-resonance vibration methods (longitudinal deformation); and
4   ultrasonic impedance methods.

Method (1) is normally a low-frequency single-value method, operating around 1 Hz. Methods (2) and (3) normally operate in the range 1 Hz to 10 kHz with single frequency values only for the resonance methods. Method (4), as its name suggests, utilizes a high frequency range (a few hundred kHz to ten MHz).

16.4.3.1   Free oscillating torsion pendulum method

The basic arrangement resembles the rotary pendulum of some clocks: the coating specimen forms the suspension, and a weight is hung from it. The weight is given a slight horizontal twist and released. The weight then vibrates back and forth, twisting the coating specimen (usually in the form of a narrow strip) back and forth about its vertical axis. Energy dissipation in the specimen and its suspension leads to a decrease in the amplitude of successive 'twists', while the mass of the weight (more accurately, the moment of inertia of the pendulum bob) and the elasticity of the specimen (or suspension) determine the frequency of the twisting oscillation. Thus, by measuring the frequency and the amplitude decreases of successive oscillations the shear elasticity and loss moduli (G' and G'') can be found.

The fragility or ductility of many coatings specimens as free films make it difficult to use this single frequency measurement over a sufficiently wide range of temperatures, whilst the tension in the film, because of the weight, may affect the results. Two solutions to this problem exist: (a) either the pendulum is inverted and the bob weight counterbalanced so as to minimize the tensional force applied to the specimen (shown schematically in Fig. 16.4), or (b) the coating is applied to a metal foil or woven glass fibre braid which forms the suspension of the simple pendulum arrangement and bears the weight of the bob. If a metal foil is used, the moduli of the coating can still be derived if those of the foil are also determined. However, a serious problem in accurate determination of the moduli is the geometry factor for this type of deformation, as the equation for the modulus contains the difference of cubes of the thicknesses for the coating and foil, and the foil alone (Inone and Kobatake [31]). As the thickness is often difficult to determine with sufficient accuracy and may vary with temperature, owing to curing, solvent loss, etc., this can represent a serious drawback. A less obvious disadvantage is that the operating frequency also changes with temperature (because of changes in the coating elasticity, which may be significant even if the coating is supported). These difficulties may be overcome by using forced extensional or flexural vibration, as described in the following sections.

In spite of these drawbacks, the use of glass braids as a support has achieved some popularity, particularly as method of studying cure reactions of polymer and coating compositions, under the name of 'torsional braid analysis' with commercial equipment available [32, 33]. With the irregular geometry of the impregnated braid, it is not possible to derive the moduli of the coating, but for following curing reactions it appears to be satisfactory. Here again, if temperature scanning is used (as is tempting with modern computer control techniques available), the results may be affected by the temperature scan rate. Far more serious, in the author's view, is the question of wetting and adhesion of the liquid and solid coatings respectively, to the fibres of the braid, which may affect the results. Also, there are scattered

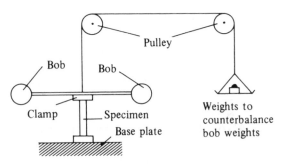

Fig. 16.4 — Schematic diagram of the inverted torsion pendulum.

reports of spurious loss peaks due to such causes as rubbing of the braid fibres together.

16.4.3.2   Resonance methods in forced oscillation

The resonance technique consists of applying a sinusoidal force to the coating specimen and measuring the amplitude of the specimen's response as a function of frequency. The resonance frequency is identified as the frequency corresponding to the maximum or 'peak' amplitude. By measuring this parameter and the 'peak width', the moduli can be calculated. The peak width is defined as the frequency range over which the amplitude has a value equal to the peak amplitude divided by the square root of two. It is sometimes possible to identify higher frequency resonance peaks due to more complex deformation modes, but these are generally smaller in peak amplitude than the normal or fundamental deformation mode.

The method is simple to use. A reed-shaped piece of free coating film or coated substrate is clamped firmly to a vibrator at one of its ends and made to vibrate laterally. The resulting swing amplitude of the free end can be measured by means of a microscope equipped with an eyepiece graticule. Firm clamping is essential, or spurious resonances will occur. Alternatively, if a strip of thin steel shim (for example, razor blade steel) is coated and suspended or supported by fine threads, flexural oscillations can be excited electromagnetically.

In the author's experience, these methods all suffer from the same disadvantage: that, unless the coatings are hard or glassy (i.e. energy dissipation is low), it is difficult to identify the resonance frequency peak precisely enough for a free film, or it will not differ significantly from the value of the substrate alone if a supported film is used.

Again, as with the free oscillation torsional pendulum methods of the previous section, the method is a single-frequency method, and anything which alters the coating elasticity modulus, e.g. temperature, or progress of cure, will shift the resonance frequency. The thickness measurement is slightly less critical for free films — it enters as a square term; however, for supported coating specimens the equations are more complex and involve linear, square, and cubic terms in the coating/substrate thickness ratio [34, 35]. To be valid, the theory also assumes perfect adhesion between coating and substrate.

16.4.3.3   Forced non-resonance vibration methods

Better than the methods described above is measurement of the amplitude as a function of frequency, as with the resonance method, and the phase difference between the applied force or strain and the response waveform. This method is the exact analogue for the solid films of the method used for determining the viscoelastic properties of liquids and paints described in Chapters 14 and 15. It allows a range of frequencies to be covered at constant temperature, supplemented by use of the WLF principle, or a range of temperature at constant frequency. The former is a more accurate way of determining T_g, while the latter is a more critical way of studying curing processes. Such methods are collectively referred to as 'dynamic mechanical analysis' (DMA). It is significant that one of the leading commercial models of DMA equipment controls both frequency and temperature. Gearing [36] has published illustrative results obtained with this equipment on paint samples (using flexural deformation).

Potentially, the frequency range of the technique is large (10^{-5} and 10^4 Hz). However, a number of factors limit the frequency range accessible with a particular piece of equipment. Specific limiting factors are the range of frequencies that can be generated, the sensitivity of the response measuring equipment, etc. A general limiting factor for the upper frequency limit is that above the resonance frequency, the response becomes more and more dominated by the inertia of the moving parts of the apparatus, while the contribution of the sample inertia increases also with frequency. Thus the frequency range can be extended only by making the moving parts as light as possible. By using longitudinal (tensile) extension of strip samples, the coating thickness measurement becomes less critical, as it now enters as a linear term.

Some years ago the author built a dynamic test apparatus for paint-free and supported film samples, in which he tried to incorporate some of these principles [39]. The coating sample strip was clamped between two small aluminium screw clamps, each carried on two light suspensions, firmly attached to a rigid steel cage. The driver suspension consisted of a loudspeaker coil, attached by a light yoke to the clamp. The yoke was supported by a cruciform arrangement of fine piano wires to allow the yoke to move only vertically. The magnet surrounding the coil was firmly attached to the base of the apparatus. The other clamp was carried on a cruciform phosphor bronze leaf spring system, at the centre of which was mounted the core of a linear displacement transducer (LVDT). The coil of the LVDT and the leaf spring system were attached firmly to a plate that could be raised and lowered by a fine-pitch screw mounted in the top of the apparatus, so that a minimum constant tension could be maintained in the sample. The apparatus was surrounded by a temperature-controlled air or inert gas enclosure. Samples of model alkyd coatings were studied to establish the effects of chemical structure and weathering on the dynamic mechanical properties of these coatings. Some typical results are shown in Fig. 16.5. Even with very soft films on aluminium foil substrates, the apparatus sensitivity was good enough to determine the modulus values of the coatings. Similar equipment has been reviewed by Whorlow [38].

Other geometries, such as torsional or flexing, can also be used with forced, non-resonance measurements. In particular, it is possible to produce a dynamic micro-indenter [39]. However, there may be problems if the film surface is not highly elastic or soft enough for adhesion or contact to be maintained over a constant surface area, during the oscillation cycle. However, this does represent an attractive way of measuring film viscoelasticity when the coating is *in situ* on its substrate. The oscillation amplitude should be kept small to minimize these effects and to prevent substrate interference. More recently, an interesting modification of the thermal mechanical analysis technique (TMA) for this purpose has been described by Reichert and Dönnebrink [40].

16.4.3.4   Ultrasonic impedance methods

These methods applied to studies of coating cure have been developed by Myers [41]. Myers describes the development and use of the technique.

The principle of the technique is shown schematically in the previous chapter, Fig. 15.3. A shear wave generated in a plane layer passes across the interface with another layer in proportion to how closely the shear mechanical impedances are matched. If a sensor is mounted to detect the reflected energy, the echo suffers an attenuation directly proportional to the impedance of the substrate. A trapezoidal

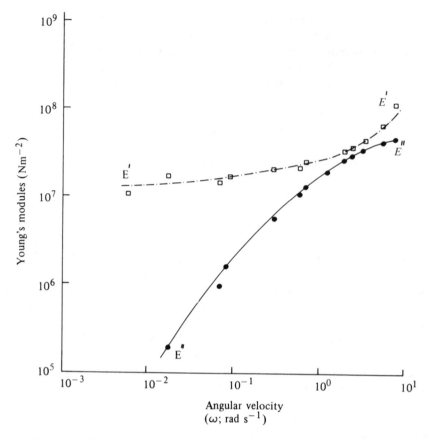

Fig. 16.5 — Phthalic polyester (1 week exposure, 44 °C test temperature).

section substrate forms the support for the coating under test. Fused quartz of known impedance was found to be a suitable substrate for propagation of shear waves into an adhering film, and 11° was the optimum angle of incidence [42]. Steel bars can also be used satisfactorily [43].

Four microsecond gated pulses were generated at frequencies ranging from 2.0–100 MHz, applied to a piezoelectric transducer attached to one sloping face of the bar, propagated and reflected off the coated surface, and collected as a series of exponentially decreasing pulses, and the attenuation was measured and recorded. However, from this measurement only the real component of the complex shear modulus (G') can be derived. The imaginary component (G'') requires measurement of the phase relationships at the interface. However, G' is relatively insensitive to any measurement parameter except attenuation [44], provided that the phase angle on reflection is finite.

In practice, a sigmoidal shaped curve is obtained when attenuation is monitored during the liquid to solid(!) conversion during cure and plotted against curing time [45].

Using a similar principle, but utilizing trains of torsional oscillatory pulses down a fused quartz rod, Mewis [46] described an apparatus operating at a frequency of 100–150 kHz. Again, the coating to be tested was applied to the rod surface. In this apparatus, both attenuation and reflected phase angle were measured, so both components of the shear modulus could be calculated. As Mewis [47] points out in a later review, the frequency range really determines the sensitivity of the technique to different portions of the drying process: thus, his own apparatus is really more suited to monitoring the initial stage of the drying or curing process, whereas the higher frequency equipment is better for the later stages of the process.

However, the electronic equipment required for this technique is considerable, and it is debatable whether rather simpler techniques such as the bouncing-ball (described in Chapter 15), or torsional braid analysis, described in previous sections of the present chapter, may not be just as effective for monitoring cure or drying processes.

16.4.4   Measurement of ultimate (failure) properties

The ultimate properties of a coating may be measured in the same way as for polymers, i.e. by tensile extension at a controlled rate of a rectangular free film specimen of the coating in a commercial instrument, and by the plotting of stress–strain curves at several different rates. Temperature and humidity should be controlled during such tests. The use of these measurements has been reviewed by Evans [48]. Methods for attached films are less frequently described, apart from the technological tests to be described shortly, where test conditions may not be adequately controlled or definable. However, there is recent evidence of more interest in instrumenting and more critically controlling experimental conditions in impact tests. Simpson [49] has designed a cyclical flexure machine, in which coatings applied to metal shims can be tested before and during accelerated weathering tests. Failure is assessed visually by the appearance of surface cracking. Such tests can be used to predict durability or to monitor changes in mechanical properties during weathering. A rather more sensitive technique for detecting failure by cracking (and which will detect subsurface cohesive or adhesive failure, as well) is provided by the acoustic emission technique, which will be described in the last section of this chapter. This technique may be used in a number of deformation geometries, and the coatings are applied to metal foil, shear, or sheet substrates.

The major problem with using free film specimens for failure property tests, apart from the internal stress factors discussed above, is the difficulty of obtaining perfect free film specimens, as the presence of edge defects, for example, will seriously affect the accuracy and repeatability of the measured test parameters. In general, it is better to design and use tests on coatings *in situ* on their substrates.

16.5   Discussion of experimental methods

The choice of experimental method from the array of test methods available is difficult to specify; apart from obvious factors, such as time and equipment available, and expense, the purpose of the tests is the most important determining factor. However, a few general principles of choice can be listed:

1   If it is experimentally practicable, a test on a coating *in situ* on its normal substrate is always preferable to a free film test.
2   Mechanical properties always depend on temperature, and the time scale of the test (frequency, time, strain rate, etc.), and frequently on the humidity. It is there-

fore absolutely essential to control the first two closely, and desirable to control the humidity also during the test. Failure to do so represents a major inadequacy and source of unreliability in most of the technological tests used in the industry for assessing mechanical properties (to be described in the next section).

3 If an assessment or prediction of coating durability is required, then ultimate properties will have to be measured.

However, if the purpose is to study drying/curing processes or the effect of chemical or physical variations in the coating structure (formulation variables, changes during weathering, etc.), then there is little doubt that the transient or dynamic tests using small forces or deformations provide the most information and are the most easy to interpret in terms of the basic physics and chemistry of the coating system. They are also often more reliable because accidental defects in the coatings are less critical under the small force or deformation conditions used.

16.6 Technological tests for mechanical properties

There is a wide range of hardness, flexibility, and other types of test used in the coatings industry for measuring the mechanical properties of a coating. As these have been very adequately described in reviews [50–52], only brief comments will be given here, with a more detailed account of some of the better-defined tests, which give a better chance of relating their results to fundamental viscoelastic properties.

16.6.1 Hardness tests

As Sato [52] has pointed out, hardness is a very imprecise term. It is in common-sense terms the rigidity of the substance, i.e. its resistance to deformation by externally applied force; but, as has already been discussed, it is not a unique property, as it will depend both on the magnitude of the force applied and the rate at which is it applied. There are three basic types of hardness measurement: indentation hardness; scratch hardness; and pendulum hardness.

16.6.1.1   Indentation hardness

Indentation hardness testers have been developed from the Brinell or Rockwell hardness testers used in the rubber and plastics industry. To minimize substrate interference, penetration depths must be small. ASTM D1474 specifies the Tukon indentation tester for determining the Knoop indentation hardness of coatings. The diamond indenter has a carefully cut tip of narrow rhombohedral shape, and the coating, on a solid block of glass or metal, is deformed by the indenter under a load of 25 g for 18 ± 2 s. After releasing the load, the long diagonal of the indentation is measured and the Knoop hardness number (KHN) calculated. Mercurio [53] plotted KHN and Young's modulus with temperature for poly(methyl methacrylate) coatings, and demonstrated that they both were of similar shape and showed identical T_g values (110°C). Inone [54], using his own equipment, showed that the indentation hardness for a number of thermoplastic coatings was proportional to the Young's modulus of the coating. Probably, the most sensitive of these instruments is the micro-indenter, developed at ICI (Paints Division) and described earlier in this chapter. An objection to using pointed conical or prismatic indenters is that these lead to discontinuous stress fields in the coating, i.e. the stress becomes infinite in the vicinity of points and edges. The spherical indenter of the ICI instrument is not open to this objection, whilst the maximum penetration depth of 6μm minimizes substrate interference effects.

Most workers with these instruments take the penetration depth of a fixed time after penetration commences. However, it should be borne in mind that with a spherical indenter it is possible to use the penetration-time curve to evaluate the shear compliance-time relationships at each temperature, as in more conventional transient measurements.

16.16.1.2   Scratch hardness

Scratch hardness tests vary from the more or less qualitative pencil scratch hardness tests, commonly used in the industry, to the use of pointed loaded indenters drawn across films at a constant rate, and measurement of the groove width (see, for example [54, p. 150]). Mercurio [53] pointed out that the pencil hardness of the film is related to the elongation at break, i.e. the coating is broken only when the maximum stress, due to the pencil or indenter scratching, exceeds the tensile strength of the coating film.

16.6.1.3   Pendulum hardness

Pendulum hardness relates to an indenter performing a reciprocating rolling motion on a horizontal coating. The motion is thus an example of a free oscillating pendulum with the amplitude of oscillation decaying with time. The initial driving force is provided by the initial deflection and the resultant rotational moment, provided by the force of gravity, when the pendulum is released. The energy of motion is dissipated by the coating, thus the frequency of oscillation and the decrement of the oscillation are related to the viscoelastic properties of the coating, when these are compared to the values for the uncoated glass or metal block used as the substrate. Pendulum hardness may be expressed in a number of ways: the time for the amplitude to decrease to half (or some other fixed fraction) of its original value, or as time in seconds, or as a number of swings, or relatively, as a percentage of the corresponding time, measured on a standard glass plate. The assumption that the hardness measured is inversely proportional to the damping capacity of the coating is, however, false, as Inone and Ito [55] have demonstrated. It is thus misleading to compare coatings with different viscoelastic properties by hardness alone, although the comparison is valid if the viscoelastic properties are similar.

Two basic kinds of pendulum are available: those popular in Europe (the Koenig and Persoz pendulums), which use an indenter as the pivot for the pendulum, and those like the Sward rocker, popular in Japan and the USA, where a circular cage, containing an internal pendulum, rolls back and forth across the coated substrate. The advantages and disadvantages of the two types have been listed by Sato. The pivot type is superior in accuracy and reproducibility, and has a smaller area of contact with the film.

In terms of the basic physics, the pendulum hardness test has been analysed by Persoz [56] and Inone and Ikeda [57] for the pivot type, and Baker et al. [58], Roberts and Steels [59], and Pierce et al. [60] for the Sward rocker.

Comparisons of the temperature-dependence of the oscillatory decrement of free torsional oscillatory measurements and pivot pendulum decrement were made by Sato and Inone [61, 62] for melamine/alkyd coating systems. Good agreement was obtained between the two methods. However, it must be remembered that the damping will depend on the pivot surface area in contact with the film, and that this will increase as the film softens with temperature and the indenter pivot sinks further into the film. In spite of this, good results can be obtained for the viscoelastic properties, and the simplicity of technique makes it attractive. Sato [52] has surveyed all this work in some detail.

16.6.2   Flexibility tests

Flexibility tests are of two basic types: the bend test, in which painted panels are bent around mandrels, and the Erichsen test in which the panel metal is deformed by a large hemispherical-ended indenter.

In the bend test it is customary for thin coated metal panels to be bent around mandrels of varying diameter with coating on the outside of the end in the panel. The object of the test is to discover the smallest diameter mandrel at which cracking occurs in the coating. Clearly, it is important to control the temperature and the bending rate to obtain comparable results.

In the other types of test, the Erichsen film tester, a hemispherical indenter is forced into the panel (again from the metal side), and the indentation depth at which the film starts to crack is measured. Again, it is important to control both temperature and deformation rate.

Both tests could be made more sensitive by using a conventional tensile tester to control (and, if required, vary) the deformation rate, and by using an acoustic emission detector as a more sensitive detector of cracking than visual observation. That this is possible for the bend test has been shown by Strivens and Bahra [63], using a three-point bend test adapter with fixed mandrel size, and an industrial tensile tester. Although not so sensitive as tests done in the conventional tensile test mode with acoustic emission, it is still possible to differentiate between coatings of different performance.

16.6.3   Impact tests

It appears to be difficult at present to predict impact resistance from the viscoelastic properties of the coating, owing to the complicated stress and strain profiles produced and the lack of reliable data concerning the mechanical properties at very short times. Timoshenko and Goodier [64] have derived a theory of impact which allows the magnitude of the most significant impact parameters to be estimated. Using this theory, Zorll [65] estimated the impact time to be some tens of microseconds. This is in agreement with experimental figures. For example, Kirby [66] found a figure of $18 \pm 3\,\mu s$ at room temperature for an 8 mm steel ball striking a thick glass block at a terminal velocity of $1.70\,m\,s^{-1}$, whilst Calvit [67] found a figure of $100\,\mu s$ for a block of PMMA below its T_g (5 mm ball, $0.70\,m\,s^{-1}$ impact velocity).

A particular requirement for automotive coatings is resistance to gravel impact. Zosel [68] built an apparatus to stimulate gravel impact, where a spherical steel indenter of low mass is forced against the panel at velocities of between zero and $25\,m\,s^{-1}$. The minimum velocity at which damage becomes visible at the impact point and the area A of the impact mark, are measured. From these measurements there appears to be a temperature range around the coating T_g for which impact resistance appears to be optimum. Similar conclusions follow from the results of

Bender [69, 70] comparing Gravelometer results with the viscoelastic properties of coatings.

Intuitively, it would be expected that the impact behaviour will also depend critically on the shape of the impacting object and its angle of approach to the panel. This has been verified by, among others, Breinsberger and Koppelmann [71] who compared conical and spherical impactors. They again verified the conclusions of Zosel and Bender regarding the relation of coating T_g to optimum impact resistance.

Also important in determining impact resistance are the mechanical properties of the individual layers of the total coating system and their relationship to each other, as the work of Bender demonstrates. Technological tests rely on dropping graded objects down pipes onto panels: for example, dropping 1/4 in (6 mm) hexagonal nuts down a 4.5 m-long pipe (Brit. Std. BS AU 148); or carefully graded gravel or steel shot is blown against the panel by a strong air blast, as in the Gravelometer apparatus.

16.7   Acoustic emission

The use of acoustic emission to detect cracks in engineering structures under stress, such as North Sea oil platforms, high-pressure, vessels, aircraft wings, etc., has been commonplace for a number of years. However, it is a novel application to study the acoustic emission of coatings under stress. This technique, pioneered by ICI (Paints Division), has proved to be of considerable use in monitoring and even, in some cases, predicting the durability of coatings during environmental testing, such as natural weathering, accelerated weathering, salt spray corrosion tests, etc. In addition, it has proved very useful in evaluating the effects of formulation variables on the ultimate mechanical properties of coatings, and in evaluating these properties for the individual layers of a coating system, as well as elucidating the ways in which these properties interact in producing the total properties of the full system. The technique is, in principle, extremely simple. Any sudden microscopic movement in a body, e.g. crack formation and propagation, may give rise to acoustic emission. For example, strain is concentrated at the growing tip of a crack. As this crack propagates, strain energy is released in two main forms: as thermal energy and as acoustic energy. The acoustic energy radiates as a deformation wave from the source, and is refracted and reflected by solid inclusions and interfaces until it reaches the surface of the body. Here the surface waves may be detected by sensitive detectors: usually piezoelectric or capacitive transducers. The amplified signal from the transducer is then analysed. Familiar examples of acoustic emission, at frequencies and intensities audible to the human ear, are the cracking of ice on a pond or the creaking of the treads of wooden stairs under the weight of a human body.

The paint is coated onto one side of a metal foil strip, and this is then inserted in the jaws of a tensile tester; the transducer is attached and the sample stretched. The noise emitted is analysed, and some noise characteristic is plotted as a function of total strain. Although tensile testing is more usual, there is no reason why bending or any other form of deformation may not be used. The only essential is that there should be no spurious noise generated by slippage between the specimen and the instrument's clamps. This is why, in practice, slow deformation rates are used. Apart from this source of noise, there is no need to shield the apparatus acoustically,

Strivens and co-workers [72–74] have found narrow-band resonant piezoelectric devices (resonant frequency around 150kHz) to be satisfactory for this purpose. Good acoustic coupling between the transducer surface and the specimen is essential to maximize detector sensitivity: this is readily achieved by means of a thin connecting layer of silicone grease.

The methods of analysis available for characterizing the acoustic emission are numerous. Because of the simultaneous occurrence of many noise sources, often of different kinds, as well as the modification of the waveforms both by propagation through the body of the specimen and by the response characteristics of the detector itself, it is very difficult with acoustic emission from paint specimens to analyse the complex signal forms to obtain information about the original signal source. There is also too little theory or experimental work with 'model' systems relating waveform characteristics to source mechanism. Thus complicated frequency analysis or amplitude analysis techniques are not generally useful, although amplitude analysis can be revealing if the failure mechanism changes drastically, for example, if there is a change from micro- to large-scale cracking at a particular strain value. Strivens and co-workers have concentrated on using simple analysis techniques, such as plots of 'ring-down' or event counts against total strain to characterize the coating, and have then used these on a comparative basis, for example, to monitor changes in the failure properties of coatings during weathering, to study the effects of changes in formulation or chemical structure on the ultimate mechanical properties of the coatings, etc. This simple utilization of the technique has proved extremely useful. Not only can the strong effect of moisture in alkyd films be clearly seen (Fig. 16.6), but also, in most cases, adhesion failure can be seen as abrupt, step-

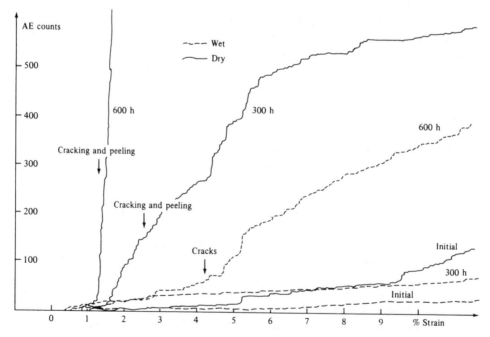

Fig. 16.6 — Effect of humidity and accelerated weathering on noise emission with deformation (exterior alkyd gain system).

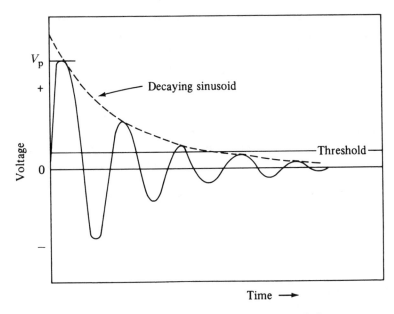

Fig. 16.7 — Method of counting in acoustic emission.

wise discontinuities in the plots, whilst cracking tends to produce uniform smooth changes.

The principle of 'ring-down', and event counting, can be seen in Fig. 16.7. If the amplified voltage output of the transducer, corresponding to a single event, is idealized as a sinusoidal decay curve, then in ring-down counting, one count is registered every time the voltage rises above a threshold voltage (this is imposed to stop random electrical noise affecting the analysis). In event counting, a preset delay is imposed after the first count before another can be registered. By proper choice of the delay time, comparison of ring-down and event counts will give a crude estimate of the average amplitude of the signals. A cumulative plot of total counts against strain is then produced for each.

Further details and examples of results will be found in the publications already cited.

16.8 Recent developments

A number of reviews of methods for the determination of hardness and viscoelastic properties of coatings have been published, as well as standards for the determination of hardness. There has, naturally enough, been an emphasis on making such methods more sensitive and automated, particularly with reference to indentation tests. Finally, acoustic emission has continued to find applications in determining the adhesion of coatings to metallic substrates.

Among the reviews, Molenaar and Hoeflaak [75] have surveyed micro-indentation and methods of determining viscoelastic properties. Berlin and Hartmann [76] have studied the repeatability and resolution of pendulum and micro-indentation tests. Krauss and Sickfeld [77] have reviewed hardness measurements, while Euverard [78] has covered the specific methods of pendulum, rolling ball and triple penetration tests.

Among the novel methods of determining the mechanical properties of coatings, perhaps the most interesting is the optical observation of thermal expansion, diffusion, and acoustic waves, resulting from the impact of a single laser shot on the film by Rogers and Nelson [79]. The dynamic moduli can be obtained after single measurements lasting for a millisecond, so it is possible to follow curing processes. Also, thermal diffusivities can be obtained. The acoustic responses are sensitive to the bond strength of the film to the substrate, thus providing the possibility of determining this parameter by a non-invasive method. Neumaier [80] has claimed a new method of measuring elasticity and hardness, based on the indentation test. Piffard [81] has described a new automated penetration apparatus, which allows measurement of hardness as a function of depth in the coating. Oosterbroek and de Vries [82] and Zorll [83] have studied aspects of impact and coatings dynamic properties. Weiler [84] has described a Vickers-type indentation apparatus, which has ultra low load application (0.025–1 N) and a direct displacement measuring device to replace the measurements by eye, previously required by the method.

Hardness measurement standard methods have been published in France (Persoz pendulum) and by the ASTM in the USA, using the Konig and Persoz pendulums, as well as the Sward Rocker [85–88].

Finally, the use of acoustic emission for testing coatings properties has been reviewed by Rawlings [89], and Mielke [90], Kempe *et al.* [91], and Mosle and Hessling [92] have specifically studied the application of acoustic emission techniques to the adhesion of coatings to metal substrates.

References

[1] FERRY J D, *Viscoelastic Properties of Polymers* 2nd edn, J Wiley (1970).
[2] NIELSEN L E, *Mechanical Properties of Polymers*, Reinhold (1962).
[3] MCCRUM N G, READ B E & WILLIAMS, G, *Anelastic and Dielectric Effects in Polymeric Solids*, J Wiley, (1967).
[4] GRAESSLEY, W W, *Adv Polym Sci* **16** 1 (1974).
[5] HEDVIG P, *J Polym Sci, Macromol Rev* **15** 375 (1980).
[6] WILLIAMS M L, LANDEL R F & FERRY J D, *J Amer Chem Soc* **77** 3701 (1955).
[7] HAVRILIAK S & NEGAMI S, *Brit J Appl Phys* Ser 2 **2** 1301–1315 (1969).
[8] HAVRILIAK S & NEGAMI S, *Polymer* **8** 161–210 (1967).
[9] LAOUT J C & DUPERRAY B, *XIVe Congrès AFTPV, Aix-les-Bains* 143 (1981).
[10] LAOUT J C, *XVIth FATIPEC Congress, Liège* I, 165 (1982).
[11] DUPERRAY B, LAOUT J C, & MANSOT J L, *Double liaison* **330** 61 (1983).
[12] TOUSSAINT A & DHONT L, *Prog Org Coatings* **11** (2) 139 (1983).
[13] WARD I M, *Mechanical Properties of Solid Polymers*, Wiley-Interscience (1981).
[14] ANDREWS E H, *Fracture in Polymers* Oliver & Boyd (1968).
[15] KINLOCH A J & YOUNG R J, *Fracture Behaviour of Polymers* Applied Science (1983).
[16] KINLOCH A J, *Adv Polym Sci* **72** 46 (1985).
[17] CUTHRELL R E, in *Polymer Networks: Structural and Mechanical Properties*, eds Chompff A J & Newman S, p. 121 Plenum Press (1958).
[18] GRIFFITH A A, *Phil Trans Roy Soc (London)* **A221** 163 (1920/1).
[19] BERRY J P, *J Polym Sci* **50** 313 (1961).
[20] SMITH T L & STEDRY P J, *J Appl Phys* **31** 1982 (1960).
[21] TOUSSAINT A, *Proc 5th Int Conf on Organic Coatings Sci and Tech, Athens* (1979) 489.
[22] ZOSEL A, *Prog Org Coatings* **8** (1) 47 (1980).
[23] DE JONG J & VAN WESTEN G C, *J Oil Colour Chem Assoc* **55** 989 (1972).
[24] ZOSEL A, *Farbe Lack* **82** 115 (1976).
[25] PIERCE P E, in *Treatise on Coatings*, Vol 2, Pt I, Ch 4, eds Myers R R & Long J S, p. 112 Marcel Dekker, New York (1969).

[26] MONK C J H & WRIGHT T A, *J Oil Colour Chem Assoc* **48** 520 (1965).
[27] HERTZ H, *J Reine Angew Math* **92** 156 (1881).
[28] LEE E H & RADOK J R M, *Proc 9th Internat Congr Appld Maths* **5** 321 (1957).
[29] RADOK J R M, *Quart Appl Maths* **15** 198 (1957).
[30] MORRIS R L J, *J Oil Colour Chem Assocn* **56** 555 (1973).
[31] IKEDA K, *Kolloid-Z* **160** 44 (1958).
[32] ROLLER M B & GILLAM J K, *J Coatings Technol* **50** (636) 57 (1978).
[33] ROLLER M B, *Metal Finishing* **78** (3) 53; (4), 28 (1980).
[34] VAN HOORN H & BRUIN P, *Paint Varnish Prodn* **49** (9) 47 (1959).
[35] YAMASAKI R S, *Official Digest* **35** 992 (1963).
[36] GEARING J W E, *Polym Paint Colour J* **172** (4081) 687 (1982).
[37] STRIVENS T A (unpublished results).
[38] WHORLOW R W, in *Rheological Techniques*, Ch 5., Ellis Horwood (1980).
[39] SMITH N D P & WRIGHT T A (unpublished results).
[40] REICHERT K-H W & DÖNNEBRINK G, *Farbe Lack* **86** (7) 591 (1980).
[41] MYERS R R, *J Polymer Sci*, C **35** 3 (1971).
[42] MASON W P, BAKER W O, MCSKIMMIN H J & HEISS J H, *Phys Rev* **75** 936 (1949).
[43] PECK G (private communication).
[44] MYERS R R & KNAUSS C J, *Polymer Colloids*, Plenum Press (1971).
[45] MYERS R R, KLIMEK J & KNAUSS C J, *J Paint Technol* **38** 479 (1966).
[46] MEWIS J, *FATIPEC IX, Brussels*, 3–120 (1968).
[47] MEWIS J, in *Rheometry: Industrial Applications* ed. Walters K Ch 6, p. 323 John Wiley (1980).
[48] EVANS R M, in *Treatise on Coatings*, Vol 2, Pt 1, Ch 5, eds Myers A R & Long J S Marcel Dekker (1969).
[49] SIMPSON L A, *FATIPEC XVI, Liège* **1** 33 (1982).
[50] CORCORAN E M, in *Paint Testing Manual* 13th edn, ASTM Special Publication No 500, Pt 5, Ch 1, Hardness p. 181 ed. Sward G G (1972).
[51] SCHURR G G, in *Paint Testing Manual*, 13th edn, ASTM special Publication No 500 Pt 5, Ch 4, Flexibility, p. 333, ed Sward G G (1972).
[52] SATO K, *Prog Org Coatings* **8** (1) 1 (1980).
[53] MERCURIO A, *Official Digest* **33** 987 (1961).
[54] INONE Y, *Seni Gakkaishi* **6** 147 (1950).
[55] INONE Y & ITO Y, *Shikizai Kyokaishi* **27** 37 (1954).
[56] PERSOZ B, *Peint, Pigm Vernis* **21** 194 (1945).
[57] INONE Y & IKEDA K, *Kobunshi Kagaku* **11** 409 (1954).
[58] BAKER D J, ELLEMAN A J & MCKELVIE A N, *Official Digest* **22** 1048 (1950).
[59] ROBERTS J & STEELS M A, *J Appl Polym Sci* **10** 1343 (1966).
[60] PIERCE P E, HOLSWORTH R M & BOERIO F J, *J Paint Technol* **39** 593 (1967).
[61] SATO K & INONE Y, *Kogyo Kagaku Zasshi* **60** 1179 (1957).
[62] SATO K & INONE Y, *Kobunshi Kagaku* **15** 421 (1958).
[63] STRIVENS T A & BAHRA M S (unpublished results).
[64] TIMOSHENKO S & GOODIER J N, *Theory of Elasticity* p. 383 McGraw-Hill (1951).
[65] ZORLL U, *FATIPEC IX Congr Brussels*, p. 3 (1968).
[66] KIRBY P L, *Brit J Appl Phys* **7** 227 (1956).
[67] CALVIT H H, *J Mech Phys Solids* **15** 141–150 (1967).
[68] ZOSEL A, *Farbe Lack* **83** 9 (1977).
[69] BENDER H S, *J Appl Polym Sci* **13** 1253 (1969).
[70] BENDER H S, *J Paint Technol* **43** (552) 51 (1971).
[71] BREINSBERGER J & KOPPELMANN J, *Farbe Lack* **88** (11) 916 (1982).
[72] STRIVENS T A & RAWLINGS R D, *J Oil Colour Chem Assoc* **63** 412 (1980).
[73] STRIVENS T A & BAHRA M S, *J Oil Colour Chem Assoc* **66** (11) 341 (1983).
[74] STRIVENS T A, BAHRA M S & WILLIAMS-WYNN D E A, *J Oil Colour Chem Assocn* **67** (5) 113; (6) 143 (1984).
[75] MOLENAAR F & HOEFLAAK M, *Proc XXth FATIPEC Conf, Nice* (1990), 159–167.
[76] BERLIN H & HARTMANN K, *Ind-Lack-Betrieb* **60** (9) 298 (1992).
[77] KRAUSS H & SICKFELD J, *Ind-Lack-Betrieb* **58** (1) 12 (1990).
[78] EUVERARD M R, *J Coatings Technol* **66** (829) 55 (1994).
[79] ROGERS J A & NELSON K A, ACS Abstr of papers, 205th meeting, Denver 1993, Vol II, Divn of Phys Chem, Abstr 65.
[80] NEUMAIER P, *Ind-Lack-Betrieb* **61** (1) 21 (1993).
[81] PIFFARD D, *Surfaces* **1992** (228) 55.
[82] OOSTERBROEK M & DE VRIES O T, *J Oil Colour Chem Assoc* **73** (10) 398 (1990).
[83] ZORLL U, *Mettalloberflaeche* **43** (2) 65 (1989).
[84] WEILER W W, *J Test Eval*, **18** (4) 229 (1990).

[85] *Standard Methods for Hardness of Organic Coatings by Pendulum damping Tests*, ASTM D4366-92, 1993, Annual Book of ASTM standards, Vol 06.01, pp. 605–607.
[86] *Standard Test Methods for Indentation Hardness of Organic Coatings*, ASTM D1474-92, 1993, Annual Book of ASTM Standards, Vol 06.01, 224–228.
[87] *Standard Test Methods for Film Hardness by Pencil Test*, ASTM D3363-92a, 1993, Annual Book of ASTM Standards, Vol 06.01, 454–455.
[88] *Standard Test Method for Determining Hardness of Organic Coatings with Sward Type Hardness Rocker*, ASTM D2143–93, 1994 Annual Book of ASTM Standards, Vol 06.01, 269–271.
[89] RAWLINGS R D, in *Surface Coatings: 2*, pp. 71–105, eds Wilson A D, Nicholson J V, & Prosser H J, Elsevier Applied Science (1988).
[90] MIELKE W, *Proc XIXth FATIPEC Congress, Aachen*, 1988, 2, 971–986.
[91] KEMPE M, KRELLE G, IMHOF D & WELLER G, *Plaste Kautschuk* **37** (11).
[92] MOSLE H G & HESSLING C, *Farbe Lack* **96** (10) 780 (1990).

17

Appearance qualities of paint — Basic concepts

T R Bullett

17.1 Introduction

One of the most useful properties of paint is an almost infinite capacity to modify the appearance of a substrate. Camouflage and ornamentation have always been two of the main themes in the use of paint. Craftsmen employed great pains and much ingenuity to reproduce, with paint, the appearance of naturally occurring materials; graining and marbling to simulate wood and polished marble were major subjects in the days when painters served long apprenticeships. The term 'enamel' came into the paint industry when paintmakers sought to reproduce the appearance, and hardness, of vitreous enamelled jewellery; they were so successful that vitreous enamellers now insist that the paint products must be qualified as 'enamel paint'. But paint is also a medium in its own right which can be prepared and manipulated to yield smooth surfaces, ranging from highly glossy to full matt, textures of many kinds, and a vast range of colour effects.

This chapter will be concerned, largely, with the basic physics underlying the more important appearance qualities, namely gloss, opacity, and colour. While reflection, scattering, and absorption of light are subject to the laws of physics, appearance is also a function of the observers, including their physiology and, in many instances, their psychology.

17.2 Physics of reflection by paint/air interfaces

17.2.1 Plane surfaces

When a beam of light reaches an interface between two materials of different optical density a proportion of the light is reflected, the remainder travelling on, with change of direction (refraction), into the second material (Fig. 17.1). The proportion reflected depends on the refractive indices of the two media and on the angle of incidence. Quantitative description of reflection is complicated by the fact that

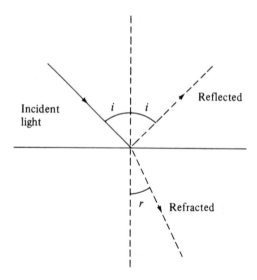

Fig. 17.1 — Reflection and refraction at air/paint interface.

light polarized in the plane of the surface is reflected more easily than light polarized perpendicularly; this is somewhat analogous to the way that a flat stone will bounce off water if thrown horizontally but will penetrate and sink if its long axis is vertical. Mathematically [1], reflectivity for light polarized in the plane of the surface is:

$$R_s = \left[\frac{\sin(r-i)}{\sin(r+i)}\right]^2 \tag{17.1}$$

and for light polarized at right angles to this plane is:

$$R_p = \left[\frac{\tan(r-i)}{\tan(r+i)}\right]^2 \tag{17.2}$$

Figure 17.2 shows the variations of R_s and R_p with angle of incidence for a refractive index of 1.5 for the second material, a typical if slightly low figure for a paint medium.

For unpolarized light, reflectivity is the average of R_s and R_p and increases steadily, for $n_2 = 1.5$, from 0.04 (4%) at normal incidence to 1 (100%) at grazing incidence ($i = 90°$).

Two further consequences of equations (17.1) and (17.2) are of interest. Firstly, when $r + i = 90°$, $\tan(r + i)$ becomes infinite, and thus, $R_p = 0$, which means that the reflected light is completely polarized in the plane of the surface. For $r + i = 90°$, $\sin r = \cos i$, so that from Snell's expression, $n = \sin i/\sin r = \tan i$ for this condition. This is the basis for the Brewster angle method for measuring refractive index in which the angle of incidence is found for which the reflected beam can be completely cut off by a polarizing filter rotated to the correct orientation. The technique requires high-quality apparatus for precision, but is useful for determining the refractive index of black glass (used for gloss standards) or of resin or varnish films on black glass. It can also be used with some loss of sensitivity for glossy paint films.

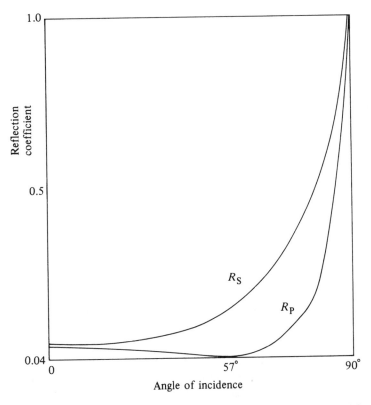

Fig. 17.2 — Variation of R_S and R_P with angle of incidence ($n_2 = 1.50$).

Secondly, for normal incidence ($i = 0$) reflectivity reduces to a simple expression:

$$R = \left(\frac{n_1 - n_2}{n_1 + n_2} \right)^2 \qquad (17.3)$$

where n_1 and n_2 are the refractive indices of the first and second materials. Figure 17.3 shows how R increases with refractive index ratio n_2/n_1 over the range 1.2 to 2.0. It will be seen that the intensity of specular reflection increases sharply with refractive index, so that much brighter reflection is possible from a paint or varnish based on a high refractive index resin (e.g. a phenolic resin) than from one based on a low refractive index resin (e.g. polyvinyl acetate). These observations are very significant to both the practical operation of glossmeters and the formulation of high-gloss paints.

17.2.2 Effects of surface texture

The discussion in Section 17.2.1 assumes an optically plane interface. When the surface is distorted, for example by uneven shrinkage over pigment particles or by residual texture from irregular application, individual facets of the surface present different angles to the incident beam. The reflected light thus becomes spread over a wider range of angles, and clear mirror-like reflection is destroyed. The scale of texture necessary to break up specular reflection is related to the wavelength of

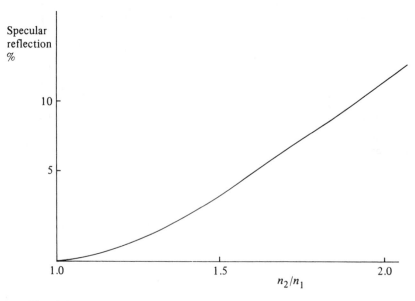

Fig. 17.3 — Increase in specular reflection with refractive index ratio.

light and to the angle of incidence. For normal incidence and for angles up to about 45°, surface roughness on a scale and depth equal to the wavelength of light (0.4–0.7 μm) is sufficient to give at least a veiling effect on specular reflection; for grazing incidence much larger texture is necessary to destroy low angle sheen. Thus, when the surface of a gloss paint begins to erode on weathering, the first effect is loss of gloss when viewed at high angles to the surface, whilst it is not until the film has begun to craze or micro crack that all grazing incidence sheen is lost. Also a fully matt paint film can be produced only by incorporating particles that are coarse, relative to the wavelength of light. Typically, particles of 10–15 μm diameter are necessary in thick films.

Appearance variation is not a simple scale from fully glossy to matt. Large-scale ripples of low amplitude, such as residues of brushmarks, give visible disturbance of the specular image if the peak-to-trough amplitude exceeds 0.5 μm. Disturbance of the surface by large flat particles under the surface can give a diffused specular reflection, resulting in a pearly appearance.

17.3   Light scattering and absorption by paint films

17.3.1   General

Light refracted into a paint film is partly absorbed by the medium (resin or varnish), but mainly encounters particles where it is scattered, absorbed, or transmitted in various proportions. Light scattering by a pigment particle depends on its size, relative to the wavelength of the light, and its refractive index ratio to that of the medium. Absorption by a pigment particle depends on the path length of light through the particle and on the extinction coefficient (absorbance) of the pigmentary material for the particular wavelength. Most of the theoretical treatments of

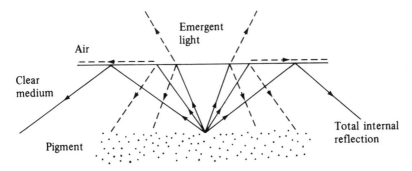

Fig. 17.4 — Internal reflection at paint/air interface.

light reflection by paint film, such as that of Kubelka & Munk [2], assume that repeated scattering by large numbers of particles sets up a completely diffused flux of light, such as is encountered in a cloud or thick mist. It is important to realize that such treatments are imperfect when applied to thin or lightly pigmented films where the necessary degree of scattering cannot occur.

17.3.2   Reflection at interfaces

Light transmitted through a paint film is partly absorbed by the substrate and partly reflected; a proportion of the reflected light eventually re-emerges, so that the film appears lighter over a more reflective substrate.

A full mathematical treatment of these phenomena is complicated even for the simplified case of a glossy paint film of uniform thickness over a smooth substrate. This is because a considerable proportion (usually over one half) of the light scattered back, diffusely, from a pigment layer is internally reflected at the paint/air interface (Fig. 17.4), and thus is attenuated again by absorption by pigment or substrate before reaching the interface for a second time. A similar situation occurs at the paint/substrate interface where light is inter-reflected between the substrate and the pigmented layer [3]. Figure 17.5 indicates some of the infinite series of inter-reflections that occur until all the light has been absorbed or has emerged finally from the film. Full mathematical analysis of this situation has not been attempted.

The effect of the air/paint interface can be calculated fairly simply by summing the geometric progressions of inter-reflections. The result obtained is:

$$R_T = R_E + \frac{(1 - R_E)(1 - R_I)R_P}{(1 - R_I R_P)} \tag{17.4}$$

where R_T is the reflectivity of the glossy paint film,
R_E the external reflection coefficient of the interface,
R_I the internal reflection coefficient of the interface,
R_P the reflectivity of the pigmented layer over its substrate.

Table 17.1 indicates how R_T varies with R_P for $R_E = 0.05$ and $R_I = 0.55$, typical values for 45° illumination on to a film with a medium of refractive index 1.5. It will be seen that the effect of the interface is to reduce reflection markedly, especially for the medium range of reflectivity. A consequence of this relationship is that most

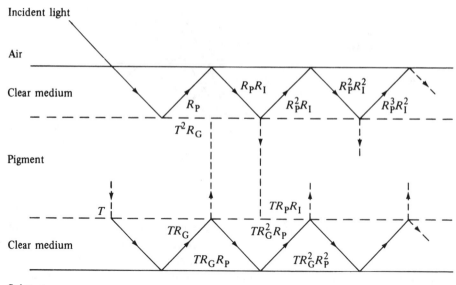

Fig. 17.5 — Inter-reflections at surface and substrate.

Table 17.1 — Effect of reflections at air/paint interface on reflectance of glossy paint films (assumptions: $R_E = 0.05$, $R_I = 0.55$).

Reflectivity of pigmented layer (R_p)	External reflectance	
	Total	Specular excluded
0.05	0.072	0.022
0.10	0.095	0.045
0.20	0.148	0.098
0.30	0.204	0.154
0.40	0.269	0.219
0.50	0.345	0.295
0.60	0.433	0.383
0.70	0.537	0.487
0.80	0.661	0.611
0.90	0.812	0.762
0.95	0.901	0.851
1.0	1.0	0.950

non-glossy materials (including matt paint films) are darkened appreciably if varnished over. For surfaces that are very dark the effect of varnishing depends on whether the observer picks up the externally reflected (specular) light; if he or she does, as, for example, when the surface is indirectly illuminated by reflection from a cloudy sky or from walls, the original colour will appear lighter. This apparent lightening of dark colours is often greater than would be suggested by Table 17.1 because for completely uniform illumination from all directions R_E rises to a value of about 0.1.

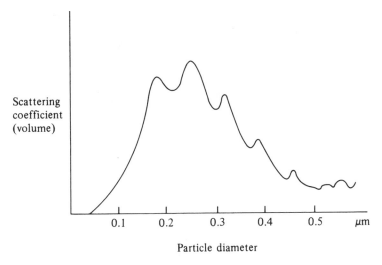

Fig. 17.6 — Scattering *vs* particle size for single spheres (calculated for rutile TiO$_2$ in linseed oil).

One further consequence of reflection at interfaces is important in considering the optical properties of paint films. A detached film or a film applied to transparent foil has an air/paint interface on its underside as well as on the side which is illuminated. This second interface also has an internal reflection coefficient of about 0.55 for diffuse light. Thus a detached film of non-hiding thickness will show the same reflectivity as it would have if painted over a light grey substrate; if laid loosely over a black-and-white checkerboard the contrast will be much reduced from that of the same film painted over the board.

17.3.3   Scattering by white pigments

White pigments are made from transparent, almost colourless materials used in paint as fine particles. The relationship between particle size and light-scattering was investigated by Mie in 1908 [4] who showed that maximum scattering per unit amount of material occurred for a particle diameter rather less than the wavelength of light. Figure 17.6 indicates the variation of scattering with particle size for uniform spheres of various diameters. Strictly, this curve refers to single scattering, that is light scattered only once by each particle, but in practice the optimum size is not greatly modified in paint films, except at very high levels of pigmentation where scattering by more closely spaced particles is considerably reduced. Commercial white pigments are developed to the particle diameter that gives maximum scattering of green light (for maximum film opacity); this is about 0.25 µm for rutile titanium dioxide. Particles of this size are less efficient in scattering yellow or red light, so that thin white paint films show distinctly orange transmission.

Work by the rutile pigment industry has assessed the extent to which scattering coefficients are reduced at high pigment concentrations. For particles of the optimum size of scattering at low concentrations the scattering per particle is approximately halved at a PVC of 30%. For this particle size, increasing concentration of pigment beyond 30% gives no further gain in opacity; indeed, opacity may actually fall over a range of concentrations where the gain in scattering from an increased number of particles is less than the loss from closer packing. At very high PVC there is insufficient medium to fill the spaces between pigment particles, so that air/pigment interfaces give dry hiding and increased opacity. There is some evidence that at and above 30% PVC rather larger pigment particles, say 0.4 µm instead of 0.25 µm, give better opacity. It has also been shown, by Stieg [5] and others, that replacement of a proportion of the rutile pigment particles by small extender particles of low refractive index (such as fine silica or calcium carbonate) greatly increases opacity at high PVC.

These findings are consistent with the idea that at high pigment concentrations scattering can be considered as being from the edges of holes between particles rather than from isolated particles. In theory a scattering system based on cavities of low refractive index in a continuum of higher refractive index could be as effective as the reverse system; indeed the opacity of much biological material, for example dry bones, derives from cavities. Many attempts have been made to utilize cavities for developing opacity in paints. A simple method is to emulsify an immiscible liquid, such as white spirit into an aqueous gelatin solution; on drying and eventual evaporation of the white spirit an opaque white film, filled with minute spherical cavities, is left. Other more sophisticated techniques have given similar results with better film properties [6–8]. Possibly the most successful industrial development has been the production of conglomerate pigment particles consisting of small lumps of resin incorporating both small cavities and small particles of high refractive index pigments [9]. Theory also suggests that if particles of high refractive index material of optimum size were coated with a shell of low refractive index, an opacifying pigment less sensitive to concentration or to quality of dispersion would be obtained [10].

17.3.4   Absorption by pigments

All pigments absorb radiation of some wavelengths, but for white pigments the absorption becomes strong only in the ultraviolet region. Black pigments absorb at all visible light wavelengths but may be transparent in the infrared, a property of importance for use in camouflage paints. Most coloured pigments absorb strongly for parts of the visible light spectrum but are transparent for others. In a film where coloured pigment is mixed with white scattering particles the total absorption and hence the depth of colour depends on the particle size of the coloured pigment. Provided that the particles are fully dispersed, absorption increases steadily as particle size is reduced. This is because the cross-section of each particle is proportional to d^2, where d is the particle diameter, and the number of particles per unit volume is proportional to $1/d^3$. Hence the total cross-section offered to the light is proportional to $1/d$. While the path through a particle is still long enough for a large proportion of the incident light to be absorbed, colour strength is approximately inversely proportional to particle diameter; when absorption per particle is much less, the gain in strength for further reduction is smaller. Figure 17.7 shows data for K/S (see Section 17.4.2) reported by Carr [11] for paints containing organic pigments after varying periods of grinding replotted against $1/d$. For the strongly absorbing phthalocyanine blue the proportionality between K and

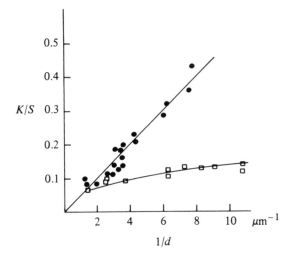

Fig. 17.7 — Absorption coefficient *vs* reciprocal particle size for organic pigments (data of Carr [11]). ● Phthalocyanine Blue; □ Pigment Green B (both measured as 1:12.5 reductions with titanium dioxide).

$1/d$ holds down to a particle size of $0.15\,\mu m$, but for the weaker Pigment Green B it does not.

17.4   Colour of pigment mixtures and pigmented films

17.4.1   Perception of colour

A full account of colour vision would be inappropriate in this manual. Readers are referred to other textbooks [12–16]. For present purposes it is sufficient to recognize that in daylight the human eye analyses the light of wavelengths $0.4–0.75\,\mu m$ in terms of three primary sensations (approximately to blue, green, and red), and that the colour perceived results from the balance of these sensations. Analysis of colour matching results with lights of known spectral energy distribution showed that each of the primary sensations was caused, with varying effectiveness, by light over a wide band of wavelengths. Figure 17.8 illustrates the distribution of this sensitivity in terms of the three primaries X, Y, and Z of the internationally recognised CIE system. It will be noted that the red sensation, X, is stimulated by light of $0.43–0.45\,\mu m$, which is the violet part of the spectrum, as well as by light of $0.55–0.65\,\mu m$, which is the yellow, orange, and red part of the spectrum. These 'mixture curves' were obtained from observations by a group of observers, all judged to have normal colour vision, working under standardized conditions. The perception of colour by any single individual is likely to differ somewhat from this average performance, and is certainly influenced by surrounding colours, stimuli to which the eyes have just been exposed, and other factors. However, most practical colour measurement is based on the mixture curves as shown in Fig. 17.8, which were obtained with colorimeters where the coloured patch subtended a 2° angle at the eye or, on slightly different data, obtained with a 10° field.

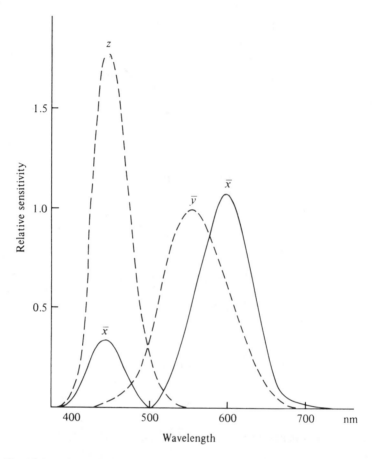

Fig. 17.8 — CIE spectral sensitivity distributions: equal energy stimulus.

From a set of response curves such as those in Fig. 17.8 it is possible to calculate for any energy distribution of incident light (E_λ) and distribution of reflection (R_λ) what the relative sensations will be. Mathematically,

$$X = \int R_\lambda E_\lambda \bar{x}_\lambda d\lambda$$
$$Y = \int R_\lambda E_\lambda \bar{y}_\lambda d\lambda \qquad (17.5)$$
$$Z = \int R_\lambda E_\lambda \bar{z}_\lambda d\lambda$$

In practice it is usual to tabulate values of the energy distribution of the illuminant (E_λ) and $\bar{x}_\lambda$, $\bar{y}_\lambda$, and $\bar{z}_\lambda$ at say 10 nm wavelength intervals, and to calculate X, Y, and Z by summing the products with R_λ. If the wavelength intervals are large, some accuracy is lost, particularly when materials with sharp absorption peaks are measured.

The relative values of X, Y, and Z correlate with the depth of colour perceived, which is termed 'chromaticity' in the original CIE system. Chromaticity is expressed as x and y where

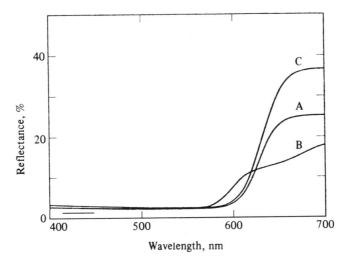

Fig. 17.9 — Reflection curves of metameric matching colours.

$$x = \frac{X}{X + Y + Z} \tag{17.6}$$

and

$$y = \frac{Y}{X + Y + Z}$$

The absolute values of X, Y, and Z correlate with the brightness of the colour, that is the proportion of the incident light reflected. The CIE primaries were chosen such that the Y value corresponds to the amount of light reflected, X and Y simply indicating the extent of variation from neutral white or grey. Thus Y, x, and y constitute a complete specification of the colour of a surface under a particular illumination. If the illumination is changed, say from north sky daylight to tungsten filament lamplight, Y, x, and y will all change, the extent of change depending on the particular reflection curve of the surface. Because the response curves to the three primary stimuli (Fig. 17.8) are broad it is possible for two widely different spectral reflection curves to give the same integrated values of X, Y, and Z for one particular illuminant. Such a colour match is destroyed by a change in illuminant because the new tristimulus values under the second illuminant will no longer be identical. Two colours of this type are termed 'metameric', and the phenomenon of matching under one illuminant but not under another is termed 'metamerism'. A colour which exhibits a marked change in hue with change in illuminant is termed 'dichroic'. Figure 17.9 shows reflection curves for three colour chips prepared by the Munsell Color Co.; A and B match under Macbeth 7500°K daylight lamp; B and C match under a cool white fluorescent lamp; none match under tungsten lamps.

17.4.2 Colour of pigmented films

The objective quantity determining the colour of a paint film is the amount of the incident light reflected at each wavelength in the visible spectrum, or to be precise, the amount of that light received by the observer. The first successful attempt to relate this reflection to light scattering and absorption in the film was that of Kubelka & Munk [2, 17]. They made the assumption that a thin horizontal slice of a film (Fig. 17.10) would

- scatter light equally backwards and forwards in proportion to a scattering coefficient S and the slice volume, and
- absorb light in proportion to an absorption coefficient K and the slice volume.

From these simple assumptions the flux of light transmitted or reflected by the film can be integrated. For a film of finite thickness, h, the resultant expression for R is complicated:

$$R_P = \frac{S\left[1 - e^{-2h\sqrt{K(K+2S)}}\right]}{K + S + \sqrt{K(K+2S)} - K + S - \sqrt{K(K+2S)}e^{-2h\sqrt{K(K+2S)}}} \tag{17.7}$$

For films of infinite thickness this simplifies to

$$R_P = \frac{S}{K + S + \sqrt{K(K+2S)}} \tag{17.8}$$

which may be transformed to,

$$\frac{K}{S} = \frac{(1 - R_P)^2}{2R_P} \tag{17.9}$$

The ratio K/S thus determines the reflection from a pigmented layer of sufficient thickness to give complete hiding. Because of reflections at the air/paint interface, not considered in the original Kubelka–Munk treatment, R_P is not the same as the actual reflection from the paint film (see Section 17.3.2 above). However, in dealing with the colour of pigmented films it is convenient to calculate R_P at each wavelength and then to make appropriate corrections to give R_λ.

The second development in the theory of colour of pigmented films was the assumption by Duncan [18] that absorption and scattering coefficients of different pigments could be added in proportion to the concentrations of the pigments.

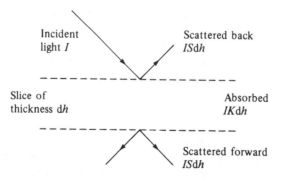

Fig. 17.10 — Kubelka–Munk absorption and scattering in slice of paint film.

Mathematically,

$$\frac{K}{S} = \frac{C_1 K_1 + C_2 K_2 + \dots}{C_1 S_1 + C_2 S_2 + \dots} \tag{17.10}$$

This assumption, which amounts to saying that the presence of other pigments in the film does not affect the scattering or absorbing power of a pigment particle, proved, like several similar rules in physics, e.g. Dalton's law of partial pressures, to be a useful working rule. It enables R_P, and thus R, to be calculated for any mixture of pigments once the appropriate absorption and scattering coefficients are known for the individual pigments.

17.4.3   Colour prediction

The prediction of the mixture of pigments that will produce a desired colour in a paint film, obtained by computation, has largely replaced trial and error formulation by mixing pigments on a slab. The first colour prediction programs were for the simplified case of films of hiding thickness, but as more powerful computers became available calculation has been extended to films of non-hiding thickness over backgrounds of specified reflectivity. These more complex calculations also deal with absolute concentrations of pigments rather than the relative proportions predicted by the earlier computations. The accuracy of the predictions must depend on the degree to which the underlying Kubelka–Munk theory applies to scattering and absorption in the film, although shortcomings of the theory can be partly confounded by using K and S values determined for each pigment, in the media used in the paint, and at concentrations close to those to be used in the final mixed system.

Many different colour prediction programs have been produced. They can be based on calculating the amounts of selected pigments that will give a close match to the colour values X, Y, Z of the desired colour, in which case metameric solutions are possible, or on a close match to a series of R_λ values. For simple tristimulus matching a unique solution can be found only for three tinting pigments (more can given an infinite number of solutions); spectrum matching at more than three wavelengths accommodates more variables, but as more pigments are introduced the calculations become much more lengthy and more difficult to resolve to the optimum solution. One compromise is to calculate formulations for several selected sets of tinting pigments, and to choose between the results on the basis of cost, light-fastness, or other known properties and optical factors such as degree of metamerism, which can be computed easily.

A useful spin-off of the methods for predicting formulations is that the effect on the colour of the film of small changes in the concentrations of each pigment can be calculated without difficulty; indeed, this is often a necessary part of the method of successive approximations used to arrive at a solution of the main problem. From these tinting differentials or tinting factors it is a matter of simple algebra to calculate the additions necessary to correct a batch initially off colour by small amounts ΔX, ΔY, and ΔZ. Again, these corrections will be accurate only if the paint system behaves in accordance with the Kubelka–Munk and Duncan assumptions. Major disturbances such as flotation of one pigment of flocculation in a mixed system cannot be accommodated.

17.5   Changes in paint films

17.5.1   General

Discussion of factors controlling appearance is not complete without considering the changes that occur during film formation and subsequently. For solution type paints the main change is a rapid shrinkage during the drying stage and, often, a continuing slow shrinkage as residual voltatile material is lost over a period of days or months. Oxidizing media also shrink slowly (by as much as 10% over 3 months) owing to scission and loss of volatile breakdown products; these processes also result in a rise in refractive index of the residual binder, thus reducing scattering by white pigment particles. Some basic pigments such as zinc oxide and white lead react with acidic products of oxidation, so that metal soaps build up around the pigment particles. This process, combined with general shrinkage, results in roughening of the surface with severe loss of gloss if the particles or clusters of particles are large; the 'hazing' of zinc oxide paints is an example.

Latex paints form films by aggregation and partial coalescence of the polymer globules as water is lost from the film. This process involves the loss of light scattering from the latex particles and concentration with some clustering of the pigment particles. The result is again reduction in light scattering, leading to reduced opacity of the dried film and increase in depth of colour, because tinting pigments are less affected by the changes. It is difficult to achieve high gloss from latex paints because pigment tends to cluster between groups of coalesced polymer globules, giving some surface roughness (see Fig. 17.11). One method used to preserve more uniform pigment dispersion is to incorporate a proportion of water-soluble, film-forming material that can hold the particles apart and plasticize the film shrinkage process.

17.5.2   Opacity changes in films

For the reasons discussed in Section 17.5.1, and others, white or lightly coloured paint films can show very considerable changes in opacity as the films dry. There are two main factors.

17.5.2.1   Changes in refractive index

Scattering is proportional to

$$\frac{(n_1 - n_2)^2}{(n_1 + n_2)^2}$$

where n_1 is the refractive index of the pigment and n_2 that of the medium around it. When a long oil alkyd resin thinned with white spirit dries and oxidizes, n_2 may

Fig. 17.11 — Diagrammatic structure of poor quality latex paint film. ○○ Partially coalesced polymer particles; ●● Clusters of pigment particles.

rise from perhaps 1.46 to 1.53. For a pigment of refractive index 2.7 (rutile titanium dioxide) the reduction in scattering is 14%. For a latex paint where the medium is thinned to at least 60% water ($n = 1.34$) the reduction in scattering may be even greater. It is important to realize that a reverse effect occurs with a lightly bound latex paint or a distemper when after film formation some of the pigment is effectively in air. A rutile particle in air will, theoretically, scatter 158% more than one in a medium of refractive index 1.5; in practice not all this gain is achieved because in the dry film the particles are not completely separated, but there is a major gain in opacity. This phenomenon is sometimes termed 'dry hiding'; it explains why a cheap, underbound emulsion paint film can be more opaque than an expensive, fully bound paint containing more prime pigment.

17.5.2.2 Changes in concentration

Pigments scatter most effectively when each particle is at a considerable distance from its neighbour. Experiments have shown that scattering coefficients, calculated on the amount of pigment, remain reasonably constant up to a volume concentration of 5%, and then fall rapidly (Fig. 17.12) [19–21]. It follows that the increase of concentration that occurs as solvent evaporates from a film, where the pigment concentration in the dried film exceeds 5%, must result in some decrease in opacity. This effect is often intensified by some flocculation of the pigment or by crowding into regions of the film, resulting from Bénard cell movements. The overall result is to produce a greater fall in opacity than that due to refractive index changes [22].

17.5.3 Colour changes

Reduction in scattering for the reasons discussed in Section 17.5.2 usually results in significant increase in the depth of colour as a paint film dries; the exception is where pigmentation is high enough to give dry hiding, including also application to a

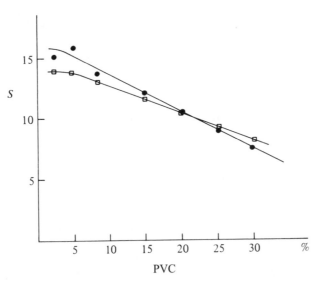

Fig. 17.12 — Scattering coefficient (calculated on volume of pigment)/PVC. Rutile TiO$_2$ in alkyd resin. ⊙ 0.34 µm particle size. ☐ 0.44 µm particle size.

porous substrate where binder can be lost by penetration. For most paints the colour change on drying is reproducible, so that it is possible to control the colour of the dry film by adjusting the wet paint colour. This makes possible fully automated colour control during manufacture of repeat batches of paint, by blending liquid tinting paints to give the required wet bulk colour.

17.6   Fluorescence and phosphorescence

Discussion so far in this chapter has been concerned solely with reflection of incident light or its absorption by films and substrates. Consideration of optical properties of coatings is incomplete without reference to changes in colour of reflected light that can occur with some materials, and stimulation of emission of light by other forms of energy.

Fluorescence is the phenomenon of absorption of radiation by a material followed by release of the energy absorbed as radiation of longer wavelength (that is, as quanta of lower energy). The energy release may be virtually instantaneous or phased over a considerable period, depending on the probability of return of atoms from the excited state, caused by absorption, to their normal condition. Substantially delayed energy release is usually called phosphorescence. The latter term is also used, loosely, to cover light emission due to chemical changes (chemiluminescence) and light emission resulting from radioactivity.

Fluorescence is significant in coatings because some materials transform near ultraviolet radiation which is present in daylight into visible light emission, or blue light (for example) into green. The result is that the light of certain wavelengths leaving the surface may exceed in intensity the incident light of that wavelength. Daylight fluorescent paints and inks exploit this effect for display purposes and hazard warnings. Fluorescence complicates colour measurement considerably, for two main reasons. First, measurement must allow for the proportion of near ultraviolet radiation present in the source under which the coating may be viewed; later revisions of CIE standardized illuminants allow for this. Second, it cannot be assumed that the energy reflected from a surface will be light of the same wavelength as the incident light, so that it is necessary to illuminate with the complete spectrum and then to analyse the light reflected rather than to illuminate with narrow wavelength bands in succession. Phosphorescence presents even greater problems of measurement, but fortunately is confined to a small proportion of materials used only for very specialized purposes.

17.7   Colour appreciation

17.7.1   General

It is a truism that colour exists only in the mind of the observer; nobody can ever know how another person perceives colour, only what he or she accepts as the same colour or judges to be different. Experimental evidence does, however, indicate the extent to which people agree and the factors that govern the average person's judgement. It is also possible to infer some relationships between perception of colour and the physical factors of wavelength distribution and intensity that influence perception (see Section 17.4.1). Some of these relationships and some of the variations among individuals will be discussed; for more detailed consideration further reading is recommended.

17.7.2   Anomalous colour vision

At least 10% of human males, but a considerably smaller proportion of females, show colour perception markedly different from the normal described in Section 17.4.1. The higher male figure is because most anomalous colour vision results from a sex-linked recessive genetic deficiency that is transmitted through females but rarely shown by them. The anomalies may be of several types, the most common being protanomaly (ascribed to reduced sensitivity of the red receptor) and deuteranomaly (ascribed to reduced ability to separate the responses of the red and green receptors). The differences are easily shown by an anomaloscope in which a mixture of red and green lights is used to match a spectral yellow; the protanomalous observer needs more red than a normal observer, and a deuteranomalous observer more green. A quick check can also be made with Ishihara test plates [23] which consist of patterns of variously coloured spots that either a normal or an anomalous observer, respectively, may recognize as clear numbers or patterns or 'Colorules' — slide rules with which highly metameric coloured patterns are matched [24, 25]. Much more rarely, observers are found with no colour discrimination (presumably seeing only shades of grey) or with tritanopia, that is a reduced blue response so that some blues and greens and some pinks and yellows are confused.

Severe colour vision anomaly is a serious deficiency for anyone engaged in a colour-using industry so that it is desirable to screen workers for any sensitive department. But the widespread incidence of anomalous vision also means that metameric matches are likely to be unacceptable to a significant proportion of observers, which is a strong argument in favour of complete spectral matching.

Even the normal observer analyses differently light received on different parts of the retina, the light-sensitive coating on the back of the eye. This is because of anatomical variations; near the central, most sensitive area the cone receptors are closely packed to give the highest spatial discrimination, but further away the cones are interspersed with rod receptors that operate largely in dim light and are responsible for twilight vision characteristics (when yellows become dark and blues relatively light). When under normal daylight an observer focuses on an object the image of which covers only the foveal area the response approximates to that of the tritanope. In extreme cases metameric matches that are acceptable when viewed closely, i.e. by a large area of the retina, will be utterly rejected when seen from a few yards away; often with such matches, when the two patterns are placed in contact, a red spot can be seen on one pattern as the eye scans the boundary, balanced by a blue green or greyish spot on the other.

17.7.3   Fatigue and contrast

The photochemical changes in the retina that result from light-absorption are communicated to the brain by electrical nerve pulses. This complex mechanism succeeds surprisingly well in conveying information on changes in intensity and wavelength distribution, but fatigues, at some stage, under extreme conditions or where extreme contrasts occur. This results in reduced discrimination when very bright and saturated colours are present in the visual field, and anomalies in colour perception, especially when contrasting colours are presented. These factors have some bearing on visual colour matching. Thus, samples are best judged against standards with a background similar in lightness to that of the standard, and while a considered judgement may be made on light or unsaturated colours, for bright oranges and reds a snap judgement on a match is likely to be more discriminating.

17.7.4   Appreciation of colour differences

The CIE system of colour measurement in terms of X, Y, and Z can be looked upon as establishing a position in a colour solid in which X, Y, and Z are measured along rectilinear axes. The distance between two colours represented by X_1, Y_1, Z_1 and X_2, Y_2, Z_2 is thus:

$$\sqrt{\left[(X_1 - X_2)^2 + (Y_1 - Y_2)^2 + (Z_1 - Z_2)^2\right]} \tag{17.11}$$

Unfortunately, equal distances in this colour solid do not represent equally perceptible colour differences. This was shown by MacAdam [26] who studied the variation of colour matches, and thus of just perceptible colour differences, for different colours. Figure 17.13 shows MacAdam ellipses for colours of constant lightness but

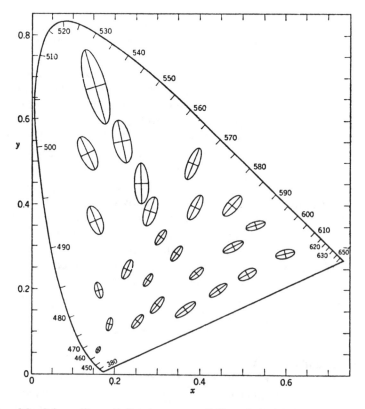

Fig. 17.13 — MacAdam ellipses indicating preceptibility of chromaticity differences on the x/y diagram. The ellipses correspond in size to 10 times the standard deviation of matches to a colour represented by the central point.

varying x and y. Clearly, distances in this x/y plot exaggerate colour differences in the green region of the field, and minimize those in the blue and orange/brown. Many attempts have been made to transform the XYZ colour space into one giving more uniform correspondence with visual appreciation. A perfect answer is not possible because visual appreciation varies with field size and colour of surrounding field and to some extent illumination levels, but much better approximations have been found. Thus the response of the eye to lightness change, represented by the luminance factor Y in the CIE system, is not a linear relationship; that is, if Y is 100 for white, the grey visually equidistant from black and white will be considerably darker than that for which Y is 50. Again, x and y can be transformed by linear equations to new coordinates α and β which convert the MacAdam ellipses into something nearer circles of more uniform size. Of the many transformations proposed the $L^*\ a^*\ b^*$ system published by the CIE in 1976 has proved to be one of the most useful in practice [16, 27], and its use is standardized in ISO 7724 Paints and Varnishes — Colorimetry. This system is complicated, but basically employs cube root functions of X, Y, and Z for all light colours, and linear functions for dark colours. The colour difference is calculated as $\sqrt{(\Delta L^{*2} + \Delta a^{*2} + \Delta b^{*2})}$. It is recommended that for near-white specimens the colour difference should be described in terms of ΔL^*, Δa^*, and Δb^*, but for colours the difference can be analysed into ΔL, a chroma difference ΔC^* (representing a difference in depth of colour), and a hue difference ΔH^*.

17.7.5 Effects of illumination

The effect of variation in the energy distribution of light sources on perceived colour has already been indicated. These effects are of great importance in all specification and measurement; even the opacity of a white paint film will be greater when measured with a light containing a high proportion of blue radiation than with one where red predominates. International specifications of illuminants have generally been related to sunlight or to a tungsten-filament lamp. The CIE initially standardized Illuminant A, corresponding to a tungsten lamp at a colour temperature of 2854 °K, Illuminant B, corresponding to sunlight, and Illuminant C, corresponding to day-light. Either B or C was used generally for colorimetry until it was found that measurements on rutile titanium dioxide paints were not in accord with visual experience. This led to more careful examination of response curves and to the adoption of new standardized illuminants with greater short wavelength contents. The CIE standardized D_{65} based on natural daylight at a colour temperature of 6500 °K is now the most widely used; its energy distribution is tabulated from 320–780 nm (see Table 17.2) and compared with those of Illuminants A and C.

17.8 Further reading

Judd and Wyszecki [13] is probably the best general review for the paint technologist; the earlier 1962 edition by Judd alone is shorter and more easily read. Billmeyer and Saltzman [14] is a clearly presented state of the art summary with a useful appendix on more specialized publications. McLaren [16] is particularly valuable on colouring properties of dyes and pigments. Wright [12] written by a pioneer of colour physics is possibly the best source book on colour vision. Wyszecki and Stiles

Table 17.2 — Energy distribution of strandard illuminants (normalized at $E_{560} = 100$).

Wavelength (nm)	Relative spectral energy		
	D65	Illuminant C (daylight)	Illuminant A (tungsten)
320	20.2	0.01	
340	39.9	2.6	
360	46.6	12.3	
380	50.0	31.3	9.8
400	82.8	60.1	14.7
420	93.4	93.2	21.0
440	104.9	115.4	28.7
460	117.8	116.9	37.8
480	115.9	117.7	48.2
500	109.4	106.5	59.9
520	104.8	92.0	72.5
540	104.4	97.0	86.0
560	100	100	100
580	95.8	92.9	114.4
600	90.0	85.2	129.0
620	87.7	83.7	143.6
640	83.7	83.4	158.0
660	80.2	83.5	172.0
680	78.3	79.8	185.4
700	71.6	72.5	198.3
720	61.6	64.9	210.4
740	75.1	58.4	221.7
760	46.4	55.2	232.1
780	63.4	56.1	241.7

[15] has the most comprehensive treatment of colour perception and the mathematics of presentation of colour space and calculation of colour differences, but is perhaps a book for the specialist rather than the general technologist. Hunter [28] is recommended for general background on perception and measurement of appearance properties.

References

[1] HEAVENS O S, *Optical Properties of Thin Solid Films*, Butterworth (1955).
[2] KUBELKA P & MUNK F, *Z Tech Physik* **12** 593 (1931).
[3] ROSS W D, Theoretical computation of light scattering power, *J Paint Technol.* **43** 50 (1971).
[4] MIE G, *Ann Physik* **25** 377 (1908).
[5] STEIG F B, *Official Digest* **31** 52 (1959).
[6] SIENER J A & GERHART H L, *XI Fatipec Congress 1972, Proceedings* 127.
[7] PPG Industries, US patent 3 669 729.
[8] CHALMERS J R & WOODBRIDGE R J, Air and polymer extended paints, *Paint R.A. 5th International Conference, P.R.A. Progress Report* **3** 16 (1983).
[9] KERSHAW R W, A new class of pigments, *Australian OCCA Proceedings and News* **8** no 84 (1971).
[10] LONEY S T, Scattering of light by white pigment particles, *Paint R.A. Technical Paper* no 213 (1960).
[11] CARR W, *J Oil Colour Chem Assoc* **65** 373 (1982).
[12] WRIGHT W D, *The Measurement of Colour* 4th edn, Adam Hilger (1969).
[13] JUDD D B & WYSZECKI G, *Color in Business, Science and Industry* 3rd edn, John Wiley (1975).
[14] BILLMEYER F W & SALTZMAN M, *Principles of Color Technology*, 2nd edn, John Wiley (1981).
[15] WYSZECKI G & STILES W S, *Color Sciences — Concepts and Methods*, 2nd edn, John Wiley (1982).
[16] MCLAREN K, *The Colour Science of Dyes and Pigments*, Adam Hilger (1983).
[17] KUBELKA P, *J Opt Soc Amer* **38** 448 (1948).

[18] DUNCAN D R, *J Oil Colour Chem Assoc* **32** 296 (1949).
[19] VIAL F, On the dependency of scattering coefficients on PVC, *IX Fatipec Congress* 1968 1–40.
[20] Tioxide Plc, *Opacity with tioxide pigments*, BTP Leaflet 107.
[21] STEIG F B, *Official Digest* **29** 439 (1957).
[22] BULLETT T R & HOLBROW G L, *J Oil Colour Chem Assoc* **40**, 991 (1957).
[23] ISHIHARA S, *Test for Colour Blindness*, Tokyo, Kanehara Shuppan (1973).
[24] Glenn Colorule — obtainable from AATCC, Research Triangle Park, North Carolina, USA.
[25] Davidson & Hemmendinger Color Rule — obtainable from Munsell Color Co., Newburgh, NY, USA and Kohlmorgen (UK) Ltd, Bridgewater House, Sale M33 1EQ.
[26] MACADAM D, *J Opt Soc Amer* **32** 247 (1942).
[27] ROBERTSON A R, *Color Res Appl* **2** 7 (1977).
[28] HUNTER R S, *The Measurement of Appearance*, John Wiley (1975).

18

Specification and control of appearance

T R Bullett

18.1 Gloss

18.1.1 Classification

The gloss of paint films is classified according to the degree to which they exhibit specular reflection. The broad descriptions are:

- full gloss — showing clear specular reflection at all angles of view;
- semi-gloss — showing specular reflection when viewed at low angles to the surface but only a hazy reflection at higher angles;
- eggshell — showing hazy reflection for all angles of view with, possibly, clear specular reflection near grazing incidence;
- flat (matt) — showing no specular reflection even at grazing incidence, and, for full matt, no preferential reflection around the specular angle.

The term 'oil gloss' is sometimes used for a level between full gloss and semi-gloss, and represents the typical appearance of old-fashioned oil paints when first applied. In BS 2015:1965 *Glossary of Paint Terms* [1] 'eggshell' is further divided into 'eggshell gloss' and 'eggshell matt'. 'Silk' or 'satin' are also used, particularly for emulsion paints, for the eggshell range.

Gloss is not a simple property that can be assigned a value on a linear scale. There are several different characteristics that must be considered:

- Intensity of specular reflection or brightness or reflection at or close to the specular angle.
- Distinctness of images, that is the detail that can be resolved in a pattern reflected in the surface.
- Grazing incidence sheen, that is preferential reflection near the specular angle for light at near-grazing incidence.

Even these three characteristics are not sufficient to differentiate between all surfaces that are visually different, particularly in the semi-gloss range where it is necessary to analyse the distribution of reflected light over a wide range of angles, and

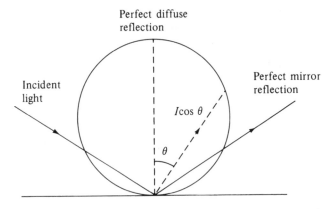

Fig. 18.1 — Distribution of reflected light from perfect mirror and perfect diffuser.

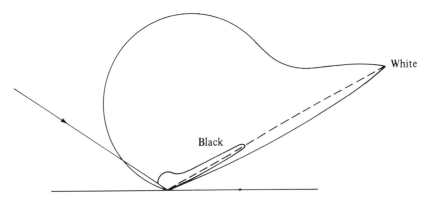

Fig. 18.2 — Distribution of reflected light from semi-gloss black and white films.

in the matt range where a rough surface causes preferential back reflection, so that some films may appear much brighter when viewed from the direction of incidence than from the usual direction of specular reflection.

To understand the reasons for these differences in appearance it is useful to consider the distribution of light reflected from some typical films. In Fig. 18.1 and 18.2 illumination by a very narrow beam at 60° is assumed. For a perfect mirror all the light is reflected at the specular angle. For a perfect diffuser the intensity of reflected light varies as the cosine of the angle with the normal, giving the circle shown in Fig. 18.1; for an extended surface this is equivalent to looking equally bright from any direction of observation. A semi-gloss white paint shows a circular plot of reflection, from the underlying pigment, with a spike of reflection around the specular angle; the narrowness of the spike and its height at the specular angle are a measure of the gloss of the paint. A semi-gloss paint shows an almost insignificant circular plot with a spike again varying in shape with the gloss (Fig. 18.2).

18.1.2   Specular reflection gloss measurement
A great deal of attention has been given to the measurement of gloss by specular reflection. For many years the standard method in the UK was that detailed in BS

3900:Part D2:1967 in which specular reflection at 45° is compared with that from a glossy black glass standard. This method has now given place, largely, to BS 3900:Part D5:1980 [2], which is essentially the international standard ISO 2813–1978, in turn based on a much older ASTM standard [3]. Part D5 also involves comparison with reflection from a black glass standard, but angles of incidence of 20°, 60°, or 85° are used according to the gloss levels involved and the reasons for measurement. A high angle to the surface, e.g. 20° incidence, gives a sensitive measure for high gloss paints, but for following loss of gloss on weathering 45° or 60° gives a more open scale; 85° gloss is useful only for measurement of sheen of eggshell or near-matt paints.

From Fig. 18.2 it is clear that the amount of reflected light recorded as specular reflection from a semi-gloss paint will depend on the spread of angles around the specular picked up by the glossmeter. Figure 18.3, reproduced from BS 3900:Part D5, illustrates how this is controlled in the method. Both the range of angles in the incident beam, which depends on the size of the lamp filament in relation to the focal length of the collimating lens, and the range of angles picked up, which depends on the angular size of the field stop, have to be controlled. Increasing either of these angles reduces discrimination for high gloss and opens the scale at the low gloss end [4].

There are several other points to remember:

1   The intensity of light reflected is dependent on the refractive index as well as on the planarity of the surface (see equation (17.3) Chapter 17). For this reason the refractive index of the black glass used for comparison has to be clearly specified. Additionally, a paint based on a medium of high refractive index will give a higher specular reflection value than a paint of equal surface planarity based on a medium of lower refractive index. Table 18.1 indicates the refractive index of some typical media and the specular reflection from perfectly plane surfaces,

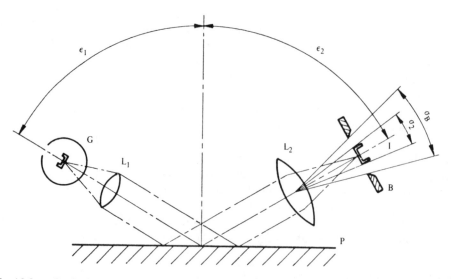

Fig. 18.3 — Optical specification of 60° glossmeter BS 3900:Part D5. G = lamp; L_1 and L_2 = lens; B = receptor field stop; P = paint film; $\varepsilon_1 = \varepsilon_2 = 60 \pm 0.2°$; σ_B = receptor aperture angle = $4.4 \pm 0.1°$; σ_2 = source image angle = $0.75 \pm 0.25°$; I = image of filament.

Table 18.1 — Refractive indices of paint media and comparative materials with relative specular reflection intensity.

Material	n_{D25}	Specular reflection (cf. BS 3900: D5 Standard = 100)
Glass standard (BS 3900:D2)	1.523	88.1
Glass standard (BS 3900:D5)	1.567	100
Polytetrafluorethylene (PTFE)	1.35	45.5
Polyvinylidene fluoride (PVF$_2$)	1.42	61.7
Polybutyl acrylate	1.466	73.2
Polyvinyl acetate	1.466	73.2
Polyester resins	1.523–1.54	88.1–92.6
Long oil alkyd resin films	1.53–1.55	89.9–94.7
Epoxy resins	1.55–1.60	94.7–109.1
Chlorinated rubber	1.55	94.7
Polyvinylidene chloride	1.60–1.63	109.1–117.6
Phenolic resins (unmodified)	1.66–1.70	126.1–137.8

compared with the black glass standard of BS 3900:D5 as 100. The human eye also tends to be impressed by the higher lustre of reflection from a high refractive index surface (diamonds are distinguished from glass!), but imperfect surfaces may be wrongly rated by glossmeters because of refractive index differences.

2   Directional defects such as residual brush marks strongly affect gloss measured across the direction of the defects, but have much less effect on measurements along the direction.

3   Substrates must be very flat because of the small tolerances in angles of incidence and reflection. It is usual, therefore to spread films for measurement on glass plates. When transparent films are measured the glass must be either opaque or the underside must be coated to avoid a second specular reflection from the underlying glass/air interface; merely placing the glass plate over a black surface is not sufficient.

18.1.3   Distinctness of image gloss

There are several shortcomings to the specular reflection method for assessing gloss of very glossy surfaces. Anomalies due to refractive index differences have already been mentioned; these can result in a rise in specular reflection with film age for alkyd gloss paints, which is not correlated with any improvement of glossiness. A second problem is that accurate specular reflection measurements can be made only on a plane substrate; the method is inapplicable on the curved surfaces of a car body. Perhaps even more important is that specular reflection glossmeters to the ASTM and BS specifications do not enable significant differences in gloss resulting from improved dispersion of very fine pigments to be measured; these differences involve the clarity of mirror reflection without substantial change in the amount of light reflected in a cone of 1° or 2° about the specular angle. These various shortcomings are avoided when gloss is assessed in terms of distinctness of image.

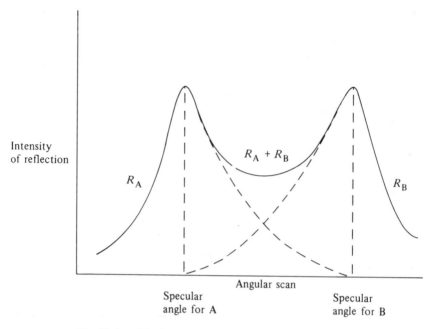

Fig. 18.4 — Distinctness of images from point sources.

Direct visual methods for assessing distinctness of image have been based on assessing the finest patterns of lines or figures that can be clearly resolved under controlled conditions [5]. The technique can be sensitive but is subject to variations of judgement between individuals common to all such sensory methods. Nevertheless, such methods are most useful for assessment of gloss achieved in production, for example on car bodies. Many attempts have also been made to measure the physical equivalent of distinctness of image. This is essentially the sharpness of the specular spike as indicated in Fig. 18.2. In Fig. 18.4 the specular reflection from two equally bright points in the object at slightly different angles to the surface is indicated. The two points will be distinct in the image if the minimum in the $R_A + R_B$ curve is significantly below the heights of the peaks corresponding to reflection from A and B. Direct measurement of this discrimination using a twin source requires very precise equipment. A rather more straightforward method is to scan one reflected beam with a photometer that is swung across the image plane of the lens and to record, for example, the angular width between peak and half-peak recordings. A cruder technique that was tentatively introduced by ASTM was to compare specular gloss readings obtained using a very small and a larger field lens aperture [6]; this technique eliminates refractive index anomalies in the medium range of gloss but is insufficiently sensitive, at least with relatively inexpensive equipment, for the highgloss range.

18.1.4   Assessment of sheen
Specular reflection at low angles to the surface is the source of glare on table tops under bright light, and of stray light in optical instruments; and it is often undesirable in interior decoration, where it shows up unevenness of walls and ceilings. One

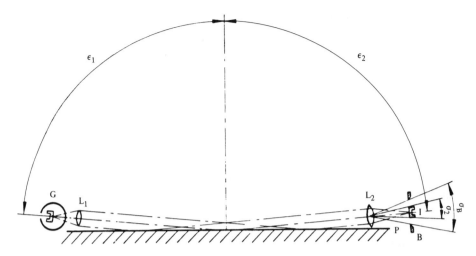

Fig. 18.5 — Optical specification of 85° glossmeter BS 3900:Part D5. G = lamp; L_1 and L_2 = lens; B = receptor field stop; P = paint film; $\varepsilon_1 = \varepsilon_2 = 85 \pm 0.1''$; σ_B = receptor aperture angle = $4.0 \pm 0.3''$; σ_2 = source image angle = $0.75 \pm 0.25''$; I = image of filament.

method for assessing this so-called sheen effect is to measure specular reflection at a near-grazing angle, e.g. 85° to the normal [2]. Results obtained by this method are sensitive to planarity of test specimens and to correct alignment of the glossmeter, but the technique can be used as a go/no-go method to exclude finishes showing significant sheen even if its value for exact measurement is doubtful. Visual methods might seem simpler and, therefore, more attractive. From about 1940 to 1980 the standard UK method, described in its final form in BS 3900:Part D3:1975 [7] (withdrawn June 1983), was to assess the highest angle from the surface at which an image of a pattern of black and white bars, reflected in the surface, could be resolved. The technique gave reproducible results for the same observer; but, inevitably, there were frequent differences of opinion as to whether a pattern was just visible or not. Eventually, despite many attempts to improve the test apparatus and to standardize viewing conditions more closely, the method was dropped by the committee responsible for BS 3900 methods, being replaced, effectively, by the 85° specular reflection method of BS 3900:Part D5:1980 (Fig. 18.5).

18.2 Opacity of paint films

18.2.1 General
One of the practically important characteristics of an applied paint film is the uniformity of colour over the surface. Unless films are applied at a thickness sufficient to give complete opacity, which is rarely the case than for very dark or metallic paints, colour variation will result either for thickness variation or from variation in colour of the substrate. Most measurements of paint film opacity or hiding power are evaluated in terms of ability to obscure the colour variation of the substrate, and are expressed as contrast ratio, e.g. reflection over black divided by reflection over white; contrast ratio may be expressed either as a fraction or a percentage. For

strongly coloured paints the variation in colour due to film thickness variation over a uniform substrate may, in practice, be more significant. Examples are the effects of heavy brush marks and the effects of retraction from sharp edges, which are often evident on painted mouldings. These effects are linked with the flow properties of the paint and are probably best assessed by practical trials on standardized substrates. Their significance can be reduced in practical painting by appropriate choice of undercoats. One practical method for assessment of film thickness dependence for emulsion paints is to paint a large blackboard with one coat overall, a second coat over the lower half, and a third over the right-hand half; the solidity of areas carrying one, two, and three coats can then be compared. Tests of this kind, carried out by experienced painters, enable the relative hiding characteristics of different paints to be compared under a simulation of practical application.

18.2.2   Contrast ratio determination

The most widely used method for assessing the opacity of paints is to apply a film of controlled thickness over a background consisting partly of black areas and partly of white. Initially, checkerboard patterns such as Morest charts were used, and the contrast over black and white areas was judged visually. This subjective judgement is not reproducible, and has largely given place to photometric measurements. There are several aspects of contrast ratio assessment that merit discussion; some are practical points involved in the measurement, some are theoretical, and some involve interpretation of the results obtained in relation to practical painting.

18.2.2.1   Measurement of contrast ratio

Meaningful results are only obtained on uniform films of known thickness. Drawdowns using doctor blades on flat black-and-white glass panels can be effective. An alternative is to spread the paint on transparent plastic foil, e.g. polyester of uniform thickness, and to measure with the painted foil in optical contact with glass plates (a liquid of similar refractive index to that of the foil gives optical contact) [8]. With both methods the thickness of the paint film can be found from its weight per unit area. White paints are much more transparent to red light than to blue, so that to achieve correspondence with visual assessment the combination of light source and recording system must approximate in spectral response to that of the eye in daylight. Failure to achieve correct spectral response is one of the chief causes of error in contrast ratio measurements.

18.2.2.2   Theoretical basis of contrast ratio

If the paint film is considered to have reflectivity R_0 over a non-reflecting background and R over a background with reflectivity R_G, and the transmission of the film is T, then it can be shown that:

$$R = R_0 + \frac{T^2 R_G}{1 - R_0 R_G} \tag{18.1}$$

This expression is not exact because it does not allow for the dependence of T on the distribution in angle of the incident light, or the fact that R_G may be modified by painting over, which eliminates a glass/air interface on a glass standard. However, the expression is useful when considering the effect of variation in R_G on contrast. Table 18.2 shows the variation of R with R_G for two typical cases: a white paint

Table 18.2 — Effect of luminance factor of substrate on reflectance of white and light grey paint films

| Substrate reflectance | (a) White paint ($R_0 = 0.7$, $T = 0.25$) | | (b) Grey paint ($R_0 = 0.5$, $T = 0.1$) | |
	R	Contrast ratio (R_0/R)	R	Contrast ratio (R_0/R)
0	0.7	1	0.5	1
0.1	0.707	0.99	0.501	0.998
0.2	0.715	0.98	0.502	0.996
0.4	0.735	0.95	0.505	0.990
0.6	0.772	0.91	0.509	0.982
0.8	0.814	0.86	0.513	0.975
0.9	0.852	0.82	0.516	0.969
1.0	0.908	0.77	0.520	0.961

film ($R_0 = 0.7$, $T = 0.25$), and a tinted paint of good opacity ($R_0 = 0.5$, $T = 0.1$). In both cases R is seen to be much more sensitive to change of R_G for light substrates than for dark substrates. The conclusion from these calculations is that to make reproducible contrast ratio measurements, close specification of the black substrate is not necessary, but the reflectivity of the white substrate must be very closely controlled.

The calculations in Table 18.2 also show how much absorption in the film contributes to hiding; in case (a) only one-twentieth of the incident light was assumed to be absorbed in one passage through the film, but for case (b) 40% was absorbed.

18.2.2.3   Relation to practical painting

The argument in Section 18.2.2.2 might seem to suggest that it is much easier to hide variations in a dark substrate than those in a lighter one. However, when the variations in R that can arise from film thickness variation are also considered, the use of dark substrates becomes less attractive; the optimum substrate (or undercoat) colour is generally close to that of a thick film of the coat to be applied.

With dark-coloured paints, which are usually much more opaque than lighter paints, contrast ratio measurements of little value; with these paints failure to hide the substrate is usually due to thin streaks in the film that relate more to flow properties and application defects than to intrinsic poor opacity.

For white paints, possibly the most useful parameter is the degree of hiding that will be obtained for application at an average spreading rate. ISO 3906–1980 published by BSI as BS 3900:Part D6:1982 [9] describes a method for determining contrast ratio corresponding to a spreading rate of $20 \, \text{m}^2 \text{l}^{-1}$ (i.e. wet film thickness $50 \, \mu\text{m}$, reasonably typical of brush application). The method involves measurement on six films spanning the desired spreading rate and interpolation.

An alternative approach is to determine the film thickness or weight that will give a specified degree of hiding, by plotting contrast ratio against film weight and interpolating or extrapolating. When this method was first developed the required level specified was a contrast ratio of 0.98. This figure was chosen because a 2% difference in luminance factor was taken to be the smallest contrast that the eye could recognize clearly. Such a high contrast ratio is probably too extreme a requirement for normal painting because, in practice, paint is rarely applied over a black and white substrate. There are also serious experimental difficulties in the accurate determination of high contrast ratios because of inevitable errors in the measurement of luminance factors. For both these reasons it is preferable to work with a contrast ratio of 0.95 over black and white as equivalent in practice to satisfactory hiding for a white paint. To be acceptable a white paint should achieve this figure for an application rate not below $10\,m^2\,l^{-1}$, i.e. in two coats each of $20\,m^2\,l^{-1}$. Calculation shows that for a white paint with almost negligible absorption a contrast ratio of 0.95 in two coats corresponds to approximately 0.85 for a single coat. Thus a figure of 0.85 attained in the BS 3900:D6 method implies satisfactory hiding in two coats, uniformly applied.

18.2.3   Paint film opacity and scattering coefficients

Opacity can be related to absorption and scattering using Kubelka–Munk theory. This is the basis of the method developed in the USA for assessing the hiding power of paints, which was standardized as ASTM D2805 and in Germany as DIN 53162. The technique was adopted as an international test method by ISO and published as ISO 6504/1–1983, and by BSI as BS 3900:Part D7:1983 [10]. The working form of the Kubelka–Munk relationships used is:

$$R = \frac{1 - R_G(a - b \coth bSt)}{a + b \coth bSt - R_G}\tag{18.2}$$

where

$$a = \frac{1}{2}\left(\frac{R_\infty + 1}{R_\infty}\right), \quad b = a - R_\infty,$$

R is the reflectance of a paint film of thickness t applied over a substrate of reflectance R_G,

S is the scattering coefficient per μm,

t is the film thickness in μm,

R_∞ is the reflectance of a film so thick that further increase in thickness does not increase the reflectance, and

a and b are parameters, mathematically related to R_∞, introduced for simplification of the working equation.

It will be seen that the reflectance over a black substrate ($R_G = 0$) reduces to

$$R_B = \frac{1}{a + b \coth bSt}\tag{18.3}$$

Equation (18.3) can be rearranged to give

$$St = \frac{1}{b}\operatorname{arcoth}\left(\frac{1 - aR_B}{bR_B}\right)\tag{18.4}$$

To simplify calculation the product St is usually determined from published graphs relating St to R_B and R_∞. Once St has been determined it is then possible to calculate the film thickness or spreading rate corresponding to a contrast ratio of 0.98.

The method requires only the determination of R_∞, which is most easily done for light-coloured paints by applying a thick film over an opal glass or white ceramic tile, and the measurement of R_B for a film of known thickness. For the R_B measurement, films can be supplied directly, by doctor blade, to black glass. Alternatively, they can be spread on polyester foil and laid over black glass, a few drops of white spirit being first applied to the glass to give optical contact. Despite the complexity of the theory and the apparently unwieldly form of equations (18.2), (18.3), and (18.4), the method when used with all the aids to computation given in BS 3900:Part D7 is not unduly difficult to use.

For the simple case where absorption in the film can be neglected in relation to scattering, a much simpler expression for R_B can be derived:

$$R_B = \frac{St}{St+1}$$

which rearranged gives

$$St = \frac{R}{1-R} \tag{18.5}$$

Hence a plot of $R/(1-R)$ against t should give a straight line plot of slope S. Figure 18.6 shows results of measurements at various wavelengths for films of a white paint applied to a black glass substrate. It will be seen that at the longer wavelengths the

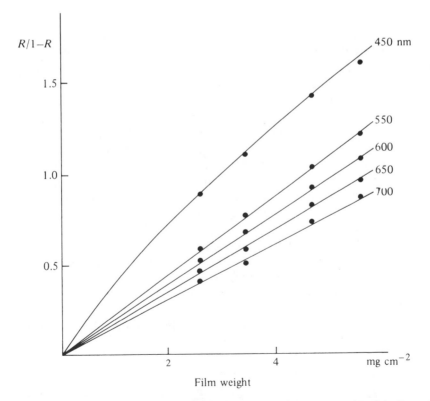

Fig. 18.6 — Scattering coefficients from plot of $R/(1-R)$ vs film weight (ZnO in linseed oil 20% PVC).

linear relation holds, and that S increases with decreasing wavelength. At 450 nm, where absorption by the pigment cannot be neglected, linearity is lost.

18.3   Specification and control of colour

18.3.1   General

Colour can be controlled purely by reference to standard colour cards by visual judgements of matching. It can be argued that ultimately the eye is the only true arbiter, but this begs many questions, of which the most difficult is, possibly, 'Whose eye?' Apart from the genetic abnormalities discussed earlier it is known that colour vision changes with age owing to build up of yellow macular pigmentation in the eye, and that the standard observer is a statistical abstraction. There is thus a strong argument for basing all colour control on physical measurements. Equally, these measurements and their interpretation must be related closely to the responses of visual observers. Colour control in practice, accordingly, has two branches, visual and instrumental.

18.3.2   Visual colour control

18.3.2.1   Colour systems and colour standards

Many attempts have been made to set up comprehensive visual colour systems. The most universal is the *Munsell Book of Color* first published in 1929 [11]. The complete system has 40 pages, each of a different hue running around the spectrum to red and on through purple back to violet (PB in the Munsell notation). The colours on each page are arranged in rows of equal *Value* (corresponding to Y value) and in columns of equal *Chroma* (corresponding to saturation or depth of colour) (see Fig. 18.7). Each colour has three references corresponding to hue, value, and chroma, e.g. 5YR/5/10 is a saturated orange. A wide range of Munsell colours are available as small chips (either glossy or matt), but the numbering system allows for interpolation or extrapolation. The Munsell system has also been standardized by reflection measurements and some smoothing of spacing in the original system, so that a Munsell book can be used for visualization of tristimulus values. Bearing in mind that the human eye can distinguish at least half a million colours, under optimum viewing conditions, it is not surprising that while the Munsell system enables a colour to be specified approximately, it does not replace the use of individual colour cards for precise specification. A further reason for the use of colour cards for industrial purposes is that visual colour matching is difficult to standardize unless the gloss and texture of the surfaces to be compared is also similar. Thus one approach to control of colour of successive batches of paint is to match a first master batch very carefully to a standard, and then to use colour cards prepared from this master batch as working standards.

18.3.2.2   Visual colour matching

Apart from selection of observers with normal or average colour vision the most important factor in visual judgements of matches is the illumination. It is only necessary to look at the colour change of blue or purple flowers in a shaft of sunlight

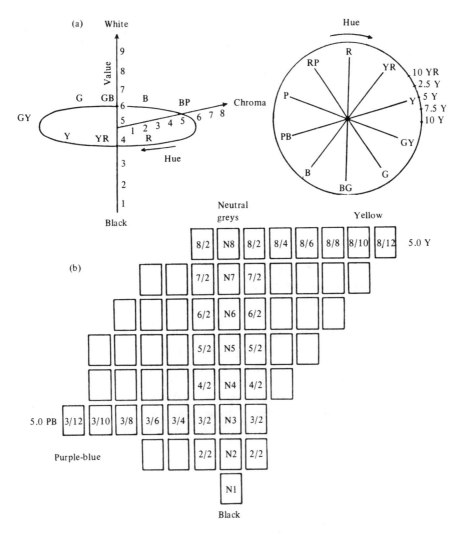

Fig. 18.7 — Arrangement in Munsell book of color.

to realize how greatly colour can change with illumination. The old method in paint factories was to arrange the colour-matching bench under north sky daylight, possibly the most constant natural source. Most matching is now done in booths with carefully selected fluorescent lamps and controlled conditions of viewing. Standard conditions for visual comparison of the colour of paints are laid down in ISO 3668–1976, reproduced as BS 3900:Part D1:1978 [12]. This standard covers both daylight matching under north sky daylight of at least 2000 lux intensity, and artificial light (D 65) matching with illumination between 1000 and 4000 lux. A background of a neutral grey of about 15% luminance factor (Munsell N4 to N5) is recommended for general use, but for whites and near-whites a higher level of 30% (Munsell N6) is preferred. Specimens to be compared are best positioned with a long touching or overlapping edge, and should be viewed from a distance of about 500 mm. Metamerism should be checked by, for example switching to a tungsten lamp or another source of radically different spectral distribution. Often, difficulty in deciding on the quality of a match is an indication of metamerism because, where there are marked differences in spectral reflection curves, the visual response from the foveal region of the retina indicates a mismatch, while that from the surrounding areas may correspond to matching. In extreme cases an observer may actually see a reddish area on one side of the touching line and a blue green area on the other when strongly metameric colours are compared.

One problem with visual colour standards is that they may fade or otherwise change in colour. Regular checking by instrumental means is imperative to guard against drift in standards. Working standards that show significant change must be replaced.

18.3.3   Instrumental colour control

18.3.3.1   General

There are two basic methods for measuring the colours of surfaces. The first is to simulate the supposed analysis made by the eye in terms of responses to three stimuli (as discussed in Section 17.4.1). This technique is known as 'tristimulus colorimetry', and it sets out to measure X, Y, and Z directly. The second method is to determine reflectance (R) for each wavelength band in turn across the range of the spectrum to which the eye is sensitive, and then to calculate the visual responses by summing products of R and the standard values for distribution of the sensitivity of the three colour responses ($\bar{x}$, $\bar{y}$, and $\bar{z}$). The tristimulus method has theoretical advantages where the materials to be measured are fluorescent, but there are serious practical problems in assuming that a tristimulus colorimeter exactly matches human vision, that is, in eliminating colour blindness from the instrument.

18.3.3.2   Tristimulus colorimetry

A tristimulus colorimeter has three main elements:

- a source of illumination, usually a lamp operated at a constant voltage;
- a set of three combinations of filters used to modify the energy distribution of the incident or, better, the reflected light;
- a photoelectric detector that converts the reflected light intensity into an electrical output.

The general requirement is that the product of energy distribution from the source (E_λ), filter transmission (F_λ), and detector sensitivity (S_λ) shall match the product of the spectral distribution of the sensitivity of the eye and the energy distribution of the illuminant (e.g. D 65) to which the tristimulus values are to be referred. A perfect match is impossible, but the best tristimulus colorimeters achieve a reasonable compromise. Usually no attempt is made to match the two peaks of the $\bar{x}$ distribution; instead, the shape of the short wavelength peak is assumed to be close to that of the $\bar{z}$ distribution, and the measured X value (X_M) is increased by an equivalent proportion of the measured Z value, e.g. the recorded tristimulus values may be:

$$X = X_M + 0.18Z_M$$

$$Y = Y_M$$

$$Z = Z_M$$

Measurements made on a tristimulus colorimeter are normally comparative, the instrument being standardized on glass or ceramic standards. Because the correct responses are not always attained, or maintained during use of the instrument, for best accuracy standardization should be carried out using calibrated standards of similar colours to the materials to be measured. This 'hitching post' technique enables reasonably accurate tristimulus values to be obtained even when the colorimeter is demonstrably colour blind. However, tristimulus colorimeters are most useful for quick comparison of near-matching colours. When, as on most modern instruments, the electrical output is digitalized, colour differences can be automatically computed in L, a, b or L, C, and H units, for use in quality control systems. As with all digitalized recording, it is necessary to remember that the accuracy of results depends on the input rather than the computer; if the spectral response of the lamp/filter/photodetector system is wrong, the recorded colour differences are most probably wrong, despite the precision of the computation and the number of figures in the printout.

18.3.3.3   Spectrophotometry

For precise measurement of colour in absolute terms it is advisable to use a spectrophotometer, that is, to measure the reflectance for each wavelength in turn and then to calculate tristimulus values. The advantage over tristimulus colorimetry is that sufficient information is obtained to calculate colour values for any illuminant and that metamerism is automatically detected. The disadvantages are that high quality spectrophotometers are very expensive and that measurements take longer (although this disadvantage has been greatly reduced by instrument development). As with colorimeters, a built-in or add-on computer can be used to process readings to give tristimulus values under a range of illuminants, colour differences from standards, and variance of colour between repeat specimens or over parts of a surface.

In a spectrophotometer the light is usually split into a spectrum by a prism or a diffraction grating before each wavelength band is selected in turn for measurement. Instruments have also been developed in which narrow bands are selected by interference filters. If fluorescent materials are to be measured, the specimen must be illuminated with the complete spectrum and the reflected light split up for analysis [13]. The spectral resolution of the instrument depends on the narrowness of the bands utilized for each successive measurement. For most paint work a 10 nm bandwidth gives sufficient resolution, but where there are sharp-edged absorption bands, as with some dyestuffs, sharper resolution is desirable. In theory, a spectrophotometer could be set up to compare reflected light directly with incident light, but it is more usual to calibrate against an opal glass standard that has been calibrated by an internationally recognized laboratory. Checks must also be made on the optical zero, e.g. by measurements with a black light trap, because dust or other problems can give rise to stray light in an instrument, which would give false readings.

18.3.3.4   Illumination and viewing conditions

With both types of colour-measuring equipment, results obtained and their correlation with visual observation depend on illumination and viewing angles; the effects are greatest for dark, glossy specimens. Colorimetry of paints and varnishes is covered by an international standard ISO 7724 (1984) which is reproduced in BS

Table 18.3 — ISO standard illumination and viewing conditions. Paints and varnishes: colorimetry (in BS3900 Part D8 Table 4)

| Measurement conditions | | Designation |
Illumination	Viewing	(abbreviation)
directional 45° ± 5°	directional 0 ± 10°	45°/normal (45/0)
directional 0 ± 10°	directional 45° ± 5°	normal/45° (0/45)
hemispherical integrating sphere	directional 8° ± 2°	diffuse/8° (d/8)
hemispherical integrating sphere with gloss trap	directional 8° ± 2°	diffuse/8° (d/8) specular reflection excluded
directional 8° ± 2°	hemispherical integrating sphere	8°/diffuse (8/d)
directional 8° ± 2°	hemispherical integrating sphere with gloss trap	8° diffuse (8/d) specular reflection excluded

Note: In the last four conditions 8° is preferred to 0° because with normal illumination or viewing there may be difficulties with inter-reflections between glossy specimens and the illuminating or viewing optics; specular reflection from glossy specimens can largely be trapped with an 8° angle.

3900 (Part D8–10 1986). This standard describes six different illuminations and viewing conditions (see Table 18.3). Of these the first, 45/0, and fourth, d/8, specular reflection excluded, probably correspond most closely to visual examination in a colour matching cabinet. However, the third and fifth conditions where the specular reflection is included have the advantage of minimizing the effect of gloss differences on colour measurements, and, of course, give results that correspond to colour seen under a cloudy sky or by indirect illumination in a room.

Many colour-measuring instruments now incorporate integrating spheres for illumination or viewing. The sphere is coated internally with matt white paint, which must have a uniform reflectivity throughout the visible spectrum (pure barium sulphate is the usual pigment for the topcoat). There are ports for the specimen to be measured, to admit the incident beam, and for viewing; the specimen port should represent only a small proportion of the total surface area, since otherwise repeated reflections from the specimen may increase the saturation of the colour in the sphere. By careful design of the sphere geometry, removable caps or light-trapping cones can be inserted to eliminate specular components and thus enable measurements to be made of diffuse reflection only or of total reflection.

18.3.3.5   Colour tolerances

Under optimum conditions the eye can detect extremely small colour differences, about 1% in Y value or luminance factor and about 0.1 in ΔE on the CIE $L\,a\,b$ system [14]. Industrially, it is unpractical to control colour to within such limits even when sufficiently precise methods for measurement are available. For routine control, therefore, acceptable limits of colour tolerance should be set. The best correlation with visual judgements is possibly given by setting limits on the differences in psychometric lightness L^*, chroma, C_{ab}^*, and hue H_{ab}^* as defined in the CIELAB 1976 formulae. These limits can be calculated back to allowable variations in X, Y, and Z for any particular colour, but with dedicated computer facilities available on most colour-measuring equipment it is simplest to work directly in L, C, and H.

Some investigations have shown, however, that acceptability is not always simply related to perceptibility of colour difference [14]. Thus a customer may tolerate departure of a neutral grey paint towards blue but not towards yellow.

18.4 Colour control in paint manufacture

Reference has already been made to calculation of colour of pigment mixtures and computerized formulation and colour correction. Such systems work to maximum advantage if close quality control is kept on all pigments and media used [15]. Metameric batches are avoided if the number of tinting pigments used is kept to a minimum, usually no more than three, and if these pigments have reproducible colouring properties. Given adequate control, most paints can be prepared from single-colour bases and tinting paints by volumetric blending. This is also the logical approach to retail paint supply where a large number of colours are demanded, for example for car refinishing. Such mixing systems are widely used in retail outlets in Scandinavia and North America, although they have not been so generally accepted in the United Kingdom.

Acknowledgements

The author wishes to thank the staff of the Paint Research Association Library for general assistance, and Mr David Cain of the Paint Research Association for help with the section of colour measurement.

Extracts from BS 3900 are reproduced by permission of the British Standards Institution. Complete copies of the document can be obtained from BSI at Linford Wood, Milton Keynes MK14 6LE, UK.

References

[1] BS 2015:1965: *Glossary of Paint Terms*.
[2] BS 3900:Part D5:1980 *Measurement of specular gloss of non-metallic paint films at 20°, 60° and 85°*.
[3] ASTM D523 — 1939 revised as D523 — 67.
[4] BULLETT T R & TILLEARD D L, *J Oil Colour Chem Assoc* **36** 545 (1953).
[5] HUNTER R S, Gloss evaluation of materials ASTM Bulletin No. 186 48 (1952).
[6] ASTM D 1471 *Test for 2-parameter 60° specular gloss*.
[7] BS 3900:Part D3:1975 *Assessment of sheen* (withdrawn June 1983).
[8] BS 3900:Part D4:1974 *Comparison of contrast ratio (hiding power) of paints of the same type and colour*.
[9] BS 3900:Part D6:1982 *Determination of contrast ratio (opacity) of light-coloured paints at a fixed spreading rate, using polyester foil*.
[10] BS 3900:Part D7:1983 *Determination of hiding power of white and light-coloured paints by the Kubelka-Munk method*.
[11] MUNSELL A H, *Munsell Book of Color*, Munsell Color Co., Baltimore, Mary-land, USA.
[12] BS 3900:Part D1:1978 *Visual comparison of the colour of paints*.
[13] BILLMEYER F W, *Colorimetry of fluorescent specimens. A state of the art review*, Technical Report NBS — GCR 79–185. National Bureau of Standards Washington DC (1979).
[14] TILLEARD D L, *J Oil Colour Chem Assoc* **40** 952 (1957).
[15] JOHNSTON R M, *J Paint Technol* **41** 415 (1969).

For general reading on colour measurement and control see textbooks, references [12–16] in Chapter 17 and on gloss Hunter ref. [28] Chapter 17.

19

Durability testing

R Lambourne

19.1 Introduction

Durability may be defined as the capacity of a paint to endure; that is, to remain unchanged by environment and events. The 'events' we are concerned with are those that impose stresses and strains on the paint system that may be of short duration (e.g. impact) or of long duration (e.g. slow expansion or contraction of the film and substrate). Effects of environmental conditions have an enormous effect on durability, and test methods for developing and monitoring the performance of paint systems are always designed to simulate conditions of usage. They are usually designed to accelerate the degradative processes to which paints are subjected. The need for this acceleration of the degradation processes is to provide early warning of paint failure. Another aspect of durability is the capacity of the paint to withstand abuse under conditions of use, and in this case the requirements for different markets dictates different methods of test. Broadly, these tests are of two types: chemical resistance tests and physical or mechanical tests for all types of paint. The durability of a paint system is often dependent on the nature of the substrate, and this must be taken into account when designing appropriately realistic test methods.

The extent and range of tests that may be applied vary according to a number of circumstances. In the industrial paint markets the paint manufacturer may have to meet specifications laid down by the user. Test method specifications have been drawn up by a number of national and international organizations such as the American Society for Testing Materials (ASTM), the British Standards Institution (BSI), and the International Standards Organization (ISO). In the United Kingdom other organizations have developed and are responsible for maintaining certain standard test specifications. Examples of these organizations are the Society of Motor Manufacturers and Traders (SMMT) and the Ministry of Defence (MOD DEF Specifications). Establishments such as the Building Research Station and the Paint Research Association have contributed to the development of test methods and influence the standardization of test methods through representation on technical panels of the national and international standards organizations.

19.1.1   Why do paints fail?

The reasons for paint failure are legion. Nevertheless some reasons for failure are readily identifiable, and attempts can be made to combat them. Architectural paints based upon autoxidizable binders have the seeds of degradation within them. The oxidation process does not stop when the film has dried. Oxidation proceeds, giving an increasingly crosslinked film. Oxidation products are lost from the film, the net effect being embrittlement ultimately to the stage that changes in the substrate cannot be accommodated. The final effect is the cracking and flaking of the film. The adequacy of durability of modern exterior gloss paint is due to a careful choice of binder which aims to keep the oxidizability of the film to the minimum, maximizes the extensibility of the film while maintaining adequate hardness, among other considerations.

Failure of an industrial finish, e.g. on a washing machine, may not be a problem during the lifetime of the appliance, unless the paint is unsatisfactory with respect to detergent or alkali resistance. In this case very different criteria are applicable, and the methods of test reflect these differences. With motor vehicles not only does the paint system have to retain an attractive appearance on exposure to ultraviolet radiation and weathering generally, but it must be resistant to stone chipping. If stone chipping does occur the system should be designed to resist the spread of rust from the damaged site. These examples serve to indicate the extent of the problem of maintaining adequate standards of durability. Formulation of paints for different markets to meet more or less exacting requirements has been discussed in Chapters 9–13. It will be clear that the requirements to be met include both chemical resistance and optimum mechanical properties. Failure of paints will be due to either of these factors or to a combination of them. We shall now discuss an extensive range of tests that have been devised to give an indication of the probable performance of a paint appropriate to its everyday use.

19.1.2   Methodology

Durability testing is carried out for a number of reasons. It may be used as a quality control on standard products, or it may be applied to new products under development. It is done to establish that certain criteria with respect to performance are met. This means that the paint manufacturer may have to test very large numbers of samples, usually applied to test panels which also are required to meet some standard which will be specified. Because of the large number of individual samples it is often not possible to carry out the tests with statistically significant sample numbers in any one comparable series. At best, sample panels may be duplicated, and it is essential to include a standard or control sample of established performance within each series. This standard is used as a basis for comparison. If the standard composition performs less well than expected the whole series must be suspect and should be repeated. This can happen when some unusual circumstance has arisen such as overheating or contamination of a test solution has occurred, which might otherwise have gone unheeded.

19.1.2.1   British Standard specifications covering the testing of paints

To ensure that there is a consensus throughout the paint industry, and for the convenience of paint users, the methods employed in carrying out the testing of paints need to be standardized. In the UK the British Standards Institution is the body that in conjunction with representatives of industry draws up and publishes test specifications.

The main British Standard relating to paints is BS 3900 which is issued in ten parts, dealing with all aspects of the testing of liquid paints, film formation, film properties, and durability testing. These are now closely related to international standards (ISO) and the national standards of many countries. The following list indicates briefly the contents of the various parts of the specification:

Part

O   General introduction (last revised in 1989)
A   Physical properties of liquid paints
B   Chemical testing
C   Film formation
D   Optical properties
E   Mechanical properties
F   Durability testing
G   Environmental testing (resistance properties)
H   Paint and varnish coating defects
J   Coating powders (sampling, size distribution, gel time, etc

In the context of this chapter, Parts E and F are most relevant. These parts are sub-divided as follows:

E	Mechanical Properties	Last revised
1	Bend test (cylindrical mandrel)	1970
2	Scratch test	1992
3	Impact test (falling weight) — now E13	1993
4	Cupping (deformation) test	1976
5	Pendulum damping test (includes both the Konig and Persoz pendulums)	1973
6	Cross-cut adhesion test	1992
7	Impact test (falling ball)	1974
8	Impact test (pendulum)	1974
9	Bucholz indentation test	1976
10	Pull-off adhesion test (now superseded by EN 24624)	1993
11	Bend test (conical mandrel)	
12	Indentation test (spherical or pyramidal)	1986
13	as E3	

F	Durability	
1	Alkali resistance of plaster primer (now withdrawn)	
2	Resistance to humidity under condensation conditions	1973
3	Artificial weathering (enclosed carbon arc) (now obsolescent) — An addendum was added in 1978 entitled 'Notes on guidance on the operation of artificial weathering apparatus'	1971
4	Continuous salt spray test	1968
5	Determination of lightfastness	1972

6	Natural weathering tests	1976
7	water resistance (immersion)	1972
8	Humid atmospheres containing SO_2	1993
9	Humidity test (continuous condensation)	1982
10	Resistance to cathodic disbonding of marine coatings	1985
11	Cathodic disbonding of coatings for land based (buried) structures	1985
12	Neutral salt spray test	
13	Filiform corrosion of steel	1986

19.1.3 Test sample preparation

The type of panel is frequently specified. It is common to use mild steel panels ($6\,in \times 4\,in$, $150\,mm \times 100\,mm$). For UK goverment specifications these panels must conform to BS specification 1449, which designates EN2A deep drawing quality. For government specifications the panels are prepared and the paint applied by an appropriate method as given in DEF Standard 1053. Essentially this means that one should ensure that the panel is free from surface imperfections, such as rolling marks, scores, and corrosion. It should be degreased thoroughly with trichloroethylene and dried. The panel is then abraded on the test side(s) with 180 grade silicon carbide paper and wiped with petroleum spirit (SBP No. 3) to remove any contaminants. The material to be tested is then applied in accordance with the appropriate product or test specification, including any necessary pretreatment of the panel and the application of primer and/or undercoat. Care should be taken that at no time between degreasing and painting, the prepared surfaces are touched by hand or otherwise contaminated.

The coated panels are air-dried or stoved, as required. In some cases the paint system extends to both sides of the panel. In others only one side is coated, and the back of the panel may be protected (e.g. against corrosion) by a different material. This 'backing' may be a quick air-drying paint or in some cases a wax coating. Usually the backing will be specified in the test method.

In testing products under development it is the practice to have only one experimental paint as part of the system. For example, if a finishing coat is being tested it should be applied over a standard pretreatment, primer, and/or undercoat as the case may be. Similarly if an experimental primer is under test it should be tested in combination with standard pretreatment, undercoat, and finishing coats.

Mild steel panels are not used exclusively, and in some tests aluminium or tinplate panels may be preferred; wood, hardboard, chipboard (particle board), asbestos cement, and glass may be used for some tests, particularly in connection with mechanical testing (see Section 19.3) and weathering.

19.2 Chemical resistance testing

It is not proposed to reproduce in this book detailed specific test methods. Those that are commonly adopted and are derived from (or identical to) specifications drawn up by one of the standards institutions will be identified by their reference numbers. They will be described in sufficient detail for the reader to appreciate the applications which they are designed to meet, but the reader should consult the original specification if intending to make use of them. Unless one adheres strictly to specifications it is unlikely that reproducible results will be obtained, and it is likely that a user may reject experimental data unless it has been obtained by laid-down methods. In many cases tests are developed by an individual paint manufac-turer to suit his or her needs, or in conjunction with a user. In any of these cases, the methods of test may form part of a contractual agreement between the paint manufacturer and the customer.

19.2.1   Specific tests

19.2.1.1   Water resistance
The purpose of this test is to assess the resistance of a surface coating to immersion in distilled water. It is sometimes referred to as 'blister resistance'. The method is set out in SMMT 57 (Standards for the British Automobile Industry). It is applied to a wide range of industrial products.

The test is carried out in a thermostatically controlled water bath equipped for mechanical stirring. A rectangular laboratory water bath 151 capacity is suitable; the water is heated electrically to $38\,°C \pm 0.25\,°C$. The panels, prepared as described in Section 19.1.3, are supported in panel racks made of material inert to water (e.g. Perspex). The panels are packed in pairs, back-to-back vertically in the panel racks. The racks are placed across the tank so that the water, which is circulated by a pro-peller situated at one end, can pass across the face of the panels. After 24 hours' immersion the panels are removed from the tank and gently wiped dry with a dry soft cloth. They are examined immediately for blistering and for loss of gloss. Blis-tering within 12 mm of the edge of the panel is usually disregarded. After exami-nation the panels are replaced in the bath and the immersion continued until the specification limit is reached, usually 7 days at least. The panels are examined every 24 hours up to this point. At the end of the test the panels are removed from the bath, and after examination are allowed to dry at ambient temperature so that they may be subjected to other tests as desired. Most commonly they are examined for adhesive failure, and it can usually be established whether failure has taken place between the coating and the substrate or is an intercoat failure.

Blistering is commonly assessed by using photographic standards published in ASTM Standard method D714-56. The photographs enable the classification of blis-ters by size and number or density. Thus each blister size is categorized by four levels of density, designated 'few', 'medium', 'medium dense', and 'dense'. Loss of gloss is also estimated. Where significant blistering has taken place it is not practicable to do this instrumentally so that the gloss is estimated on a numerical scale in com-parison with a control that has not been subject to test. A 1–10 rating is used where 10 = excellent, i.e. no loss of gloss, to 1 = complete loss of gloss.

19.2.1.2   Moisture resistance: BS 3900 part F2
This is similar to DEF-1053 Method 25. This test differs from the water resistance test in that it is concerned with behaviour of paints under conditions of tempera-ture and humidity cycling such as may be encountered on motor vehicles or other coated metal objects in many climates. It is a test frequently specified in govern-ment contracts. In this test the humidity cabinet itself is specified. It consists of a

closed cabinet in which the relative humidity is maintained at approximately 100% and in which the temperature cycles between 42 and 48 °C to ensure condensation on the panels. Heating is by immersion heater in a water reservoir at the bottom of the cabinet. The air temperature cycles continuously between the two extremes in 60 ± 5 minutes. The air circulation (by fan) is such that the air temperatures at any two points in the air space do not differ by more than 1.0 °C at any given moment.

Test panels, prepared as for the water immersion test, are examined visually every 24 hours for signs of deterioration. When they are replaced in the humidity cabinet they are put in different positions to minimize any effects that might be due to their position in the cabinet. At the end of the test they are removed and allowed to stand at room temperature for 24 hours and then examined for loss of adhesion, change of colour, or embrittlement. In some cases, a trip of paint may be removed from the panel with a non-corrosive paint stripper and the metal substrate examined for corrosion.

19.2.1.3   *Resistance to salt spray*
Salt spray tests are probably the most common tests applicable to corrosion resistance, and the most controversial. It is well established that salts such as sodium chloride can cause rapid corrosion of ferrous substrates, and it is useful to have information on the behaviour of a particular system in protecting such substrate from corrosion both with intact and damaged paint films. Controversy arises largely from the interpretation of the data because of the poor reproducibility of the tests. However, they are well established, and, despite the problem of reproducibility, are quite a useful guide to performance in the absence of longer term corrosion data. They are thus unlikely to be discarded. They are considered to be unrealistic by some workers because of the degree of acceleration of the corrosion process that they achieve and the variability of the extent of 'damage' that is inflicted in some of the tests.

Two tests are in common use: the continuous salt spray test and the intermittent.

The continuous salt spray test
This test was originally developed in the UK as a government specification (DEF-1053 Method No. 24). It was developed subsequently into the British Standards Institution Method BS3900: Part F4: — 1968 (Salt-Spray). The test is of the 'pass' or 'fail' type, whereby the coating is subjected to treatment for a specified time and then examined for failure. Care must be exercised in interpreting the results of the test; it is not intended to be used as an accelerated test for normal weathering.

The panels are prepared by the method described in BS 3900: EN 605 (1992). This is essentially the same as described earlier in Section 19.1.3. The back and the edges of the panel are coated with a good protective air-drying material. The panel(s) is aged for 24 hours before starting the test. If the panel is aged for a longer period the time of ageing should be recorded since this might influence the results. Using a scalpel, the coating is cut through to the metal, starting 1 inch (25 mm) from the top of the panel and finishing 1 inch (25 mm) from the bottom. The cut should be parallel to the longer side of the panel, and it is important that the surface of the metal should be scored. The test is carried out in a chemically inert container (e.g. glass or plastic) with a close-fitting lid. A salt mist is produced by spraying a synthetic sea-water solution through an atomizer. The panels are supported on non-metallic racks so that their faces are approximately 15° to the vertical. The spray is so arranged that it does not impinge directly onto the panel surfaces. The solution which drains from the test panels is not recirculated. Panels are examined after 48 hours, 1, 2, and 3 weeks. They are rinsed in running tap water and dried with absorbent paper and examined immediately for blistering, adhesion, and corrosion from the cut. Blistering is assessed as previously described, with reference to photographic standards. In this case, however, particular attention is paid to the extent that blistering occurs in the vicinity of the cut. It is reported as extending 'under $\frac{1}{8}$ inch' (3 mm), 'under $\frac{1}{4}$ inch' (4 mm), or 'under $\frac{1}{2}$ inch' (12.5 mm). Loss of adhesion is estimated by testing the ease of removal of a coating with a finger nail, and how far in from the cut the poor adhesion extends. The observations are reports as 'no change', 'slight loss of adhesion', or 'bad loss of adhesion' as appropriate. For reasons of reproducibility a standard salt solution prepared from 'analytical quality' reagents is used. The test solution composition is specified as follows:

Sodium chloride (as NaCl)	26.5 g
Magnesium chloride (as $MgCl_2$)	2.4 g
Magnesium sulphate (as $MgSO_4$)	3.3 g
Potassium chloride (as KCl)	0.73 g
Sodium hydrogen carbonate (as $NaHCO_3$)	0.20 g
Sodium bromide (as NaBr)	0.28 g
Calcium chloride (as $CaCl_2$)	1.1 g
Distilled water	to 1000 ml

It is specified that the calcium chloride should be added last and that the temperature of the solution during the test should be $20 \pm 2\,°C$.

The intermittent salt spray test
This test is similar to the continuous test except that the mist is produced each day for 8 periods of 10 minutes at intervals of 50 minutes. It is carried out for 5 consecutive days and then 'rested' for 2 days. It is normally confined to government contract specifications and is clearly not as severe as the continuous salt spray method.

A number of variants of these tests are in use. They have often arisen through the development of test methods by major paint users such as the motor manufacturers. In some of these test the panels are scored in a different way to that specific in the BS standard. They are scored diagonally, as in the diagram:

It is claimed that the loss of adhesion is easier to detect (and is probably manifested at an earlier stage) at the centre of the ×. It is also common practice to specify the type of surgical scalpel and the force that should be exerted in scoring the panel. The latter is done by placing the panel on a laboratory balance and pressing down to achieve 1 kg force whilst scoring. Panels prepared in this way and showing different degrees of corrosion are shown in Plates 19.1a and 19.1b.

In recent years a number of cyclic corrosion tests (CCTs) have been developed to meet specific requirements. Most of these CCTs are automated, but can be performed manually. They are generally carried out using commercially available equipment specially designed for the purpose. Thus, CCT units are available, for example, from the manufacturers listed at the end of the chapter as supplying accelerated weathering machines.

Some common CCTs are the 'Prohesion' test, which was developed in the UK for industrial maintenance coatings applications; a corrosion/weathering cycle, which also involves irradiation with UV light; and other CCT cycles (using higher concentrations of NaCl) designed to test automotive paint systems. The Prohesion test is reputed to be a good test for filiform corrosion. The test electrolyte solution used (0.05% sodium chloride + 0.035% ammonium sulphate, pH 5.0–5.4) is more dilute than traditional salt spray (normally 5% NaCl only). The cycle comprises 1 hour salt fog followed by an hour drying out period [1].

19.2.1.4 Alkali and detergent resistance
At least three alkali resistance tests are in common use. These are based upon the use of trisodium orthophosphate, soda ash (anhydrous sodium carbonate), and

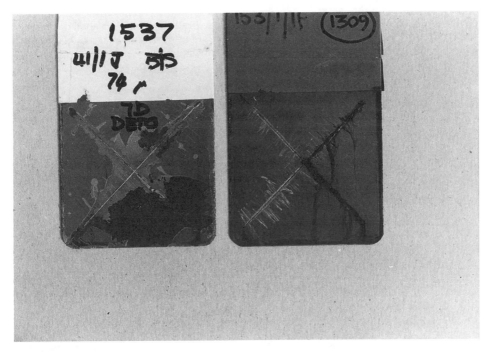

Plate 19.1a — Salt spray tests using mild steel panels. The left-hand panel illustrates a film that has poor adhesion, resulting in ready detachment from the substrate and severe rusting under the film. The right-hand panel shows a film that has good adhesion. Rusting has occurred at the score mark, and while staining is apparent, the rusting does not extend under the film which remains firmly attached despite attempts to scrape it away from the score mark.

Plate 19.1b — Salt spray tests using aluminium panels. The left-hand panel illustrates a coil coating composition that gives excellent protection to the aluminium substrate. The right-hand panel shows a typical industrial paint, not specifically designed for good corrosion resistance, which illustrates gross failure resulting from blistering and adhesive failure from the region of damage.

sodium hydroxide respectively. They are commonly applied to industrial finishes such as may be used on domestic appliances such as washing machines. In addition a detergent test may also be used for these types of product. These alkalis represent different degrees of aggressiveness. The panels are prepared as described previously for all three tests. It is usual for the panels to be coated on both sides (and edges), but if only one side is coated the back and edges can be sealed with molten paraffin wax (mp 49–60 °C). When it is intended to coat both sides of the panel with the paint system it is convenient to have a small hole drilled near one end of the panel so that it can be suspended in the oven during stoving, or air-drying. In tests which have been carried out in ICI Paints Division laboratories for many years the concentration of solute and temperature of test are as follows:

trisodium phosphate	10% w/w in distilled water;	75 °C
soda ash	10% w/w in distilled water;	65 °C
sodium hydroxide	5% w/w in distilled water;	25 °C

In these tests the panels are partly immersed in the solutions, to a depth of 3–4 inches (75–100 mm). As in previously described tests, they are supported in the tank in non-metallic racks such that they are close to the vertical. The effect of partial immersion is to provide an opportunity to make direct comparisons of the effect of the solution on the paint system. It is essential if comparisons with an accepted standard are to be made to include a control in the series under test. The panels are removed for examination after the first 4 hours' immersion and then subsequently at 24 hour intervals as appropriate. After removal from the tank the panels are rinsed with tap-water, dried with a chamois leather, and examined for blistering, softening, loss of gloss, and erosion. They are then allowed to dry at room temperature for 1 hour and examined for cracking and peeling. They are then re-immersed in the solution and re-examined after 24 hours (and at subsequent intervals), using the same procedure. Blistering is assessed as described previously for water resistance, using the ASTM method. Hardness or degree of softening may be assessed by comparing the pencil hardness (Section 19.3.2) of the immersed portion with that of the non-immersed portion of the paint system. Adhesion loss is normally evident by 'lifting' and 'shrivelling' of the coating. The film may be easily lifted off with the finger-nail. If a more accurate method of assessing loss of adhesion is required the 'cross-hatch' method (Section 19.3.5) may be applied. Loss of gloss, discoloration, or staining may all be reported on a crude scale as 'none', 'slight', or 'bad'.

The detergent test is carried out in a similar manner to the alkali tests. In the author's laboratories a 5% solution of 'Deepio' (ex Procter and Gamble) is used for the test. The test is carried out at $74 \pm 0.5\,°C$. The period of test is shorter than the alkali tests, the panels being examined after 1, 2, 4, 6, 24, and 48 hours from the start of the test. The same criteria for the assessment of deterioration of the paint are adopted as for the alkali tests. These tests are abitrary, but nevertheless they can give information that is relevant to the performance of the paint in practice. Once accepted as standard methods they are seldom changed, although the proliferation of tests is indicative of other criteria being sought by paint manufacturers and industrial users alike. In most cases the tests which are used are derived by collaboration between paint manufacturer and user.

19.2.1.5   Solvent resistance

Solvent resistance may be tested for very different reasons. Tests of resistance to petrol and diesel fuel are carried out on compositions that may be expected to encounter contact or intermittent splashing with these liquids, e.g. motor vehicle finishes, storage tanks, etc. The use of polar solvents such as ketones is often used to assess the degree of cure of a cross-linkable composition.

In the case of the petrol resistance test it is preferable to use a synthetic petrol of known composition, because of the wide variation in the composition of the commercial petrols and their incorporation of materials such as antiknock additives. A typical synthetic petrol has the following composition;

> SBP petroleum spirit No. 3   40%
> SBP petroleum spirit No. 4   40%
> Industrial methylated spirits   20%
>   (74° over proof)

For the diesel fuel test, derv is suitable. It should have an aniline point between $60\,°C$ and $70\,°C$ as measured by the Institute of Petroleum Method No. 2.

For solvent resistance, methyl isobutyl ketone is recommended. The panels are prepared as previously described and are immersed to a depth of 4 inches (100 mm) in a tank of the appropriate liquid, while being maintained in a near-vertical position. All of these tests are carried out at ambient temperature. The panels are removed for examination 1, 2, 4, 6, 24, and 48 hours from the start of the test. In the case of the volatile solvents they are examined immediately after removal for blistering, hardness, adhesion, and discoloration, and again after allowing 5 minutes for the evaporation of the solvent. The reason for this procedure is to detect the initial degree of softening and to observe the recovery of hardness on evaporation, if it occurs. In the case of the diesel fuel, evaporation does not take place significantly in the time allowed, and the panels are allowed to drain for one minute before testing.

This type of test may, of course, use any solvent, depending on the type of information sought. In general the test will be useful to assess the performance of a particular system in use, the solvent chosen representing a real contaminant for the cured system which may be the solvent itself or solutions of other materials, e.g. adhesives or mastics. Alternatively the test may serve to give an indication of the degree of cure, in which case it is common to use the solvent or solvent mixture that carried the original paint.

In addition to immersion testing, solvent resistance may be assessed by a solvent–rub test. In this case the surface of the panel is rubbed with a piece of soft cloth moistened with the solvent. Solvent resistance may be measured by the number of rubs necessary to disintegrate the film or rub it through to the next coat. Alternatively a specified number of rubs may be carried out and the appearance of the film assessed against an accepted scale, say 0–5, where 0 indicates that the solvent has had no effect and 5 indicates complete failure of the film either by disruption or dissolution; in between these extremes, varying degrees of loss of gloss and film attack will be observed.

19.2.1.6   Resistance to staining

It is particularly important that kitchen equipment such as refrigerators and washing machines should be coated with materials that are resistant to staining by a wide range of household products. It is thus necessary to test new finishes and compare them with existing standards before introducing them. Panels are prepared as described in Section 19.1.3. In the case of these tests it is, however, much more common to use 12 in × 4 in (300 × 100 mm²) panels because of the number of contaminants used.

A typical range of contaminants might include lipstick, red wax crayon, black and brown shoe polishes, mustard pickle, tomato ketchup, and a slice of lemon. The contaminants are applied to the surface of the flat panel and are individually covered with watch glasses. Contact time is normally 24 hours, after which the contaminants are wiped off with a clean soft cloth. The extent of staining is assessed visually and reported accordingly. It is essential that a standard paint is included for comparison to establish a point of reference.

19.3   Testing mechanical properties of paints

The significance of the mechanical properties of paint films and their measurement has been discussed in detail in Chapter 16. The testing, on a routine basis, of specific mechanical properties, in many cases related to the end use of the paint, has resulted in the establishment of standard tests. A selection of some commonly used tests will be discussed here.

Although the paint film performance is the primary consideration, several of the standard tests involve damage to the substrate as well as the paint. Ideally it should be possible to cause considerable substrate deformation before paint failure by cracking, peeling, and loss of gloss occur. However, in practice it is necessary to compromise between hardness, flexibility, and extensibility in order to achieve adequate performance. In these cases some arbitrary limits of acceptance are laid down. In this section we shall deal with resistance to impact, scratching, and bending; adhesion measurement; and indentation. Although these tests are primarily of the 'pass' or 'fail' type according to predetermined accepted standards and are applied to new (or at least non-weathered) films, they may also be applied to weathered films. In this case they may be used as an adjunct to accelerated or natural weathering.

19.3.1   Resistance to impact
Three types of impact test are in use, a pendulum test, a falling weight impact test and the Erichsen indentation test.

19.3.1.1   Pendulum test
This method was developed as DEF-1053 Method 17(b). It calls for the use of a special piece of test equipment which is also specified in the method. The apparatus is available from Sheen Instruments (Sales) Ltd.

The apparatus consists of two swinging arms, each supporting a piece of steel tubing coated with the paint system under test. One of the test pieces is pivoted on a horizontal axis, while the other piece is held horizontally. The mechanism is such that when the arms are swung towards each other, the tubes strike and the pivoted test piece slides across the other. The damage to the coating is then assessed visually. This is one of the few tests that does not use a flat steel panel. The test pieces are tubes, $1\frac{1}{2}$ in (37 mm) external and $1\frac{1}{8}$ in (28 mm) internal diameter, 5 in (125 mm) long. The normal methods of sample preparation, using the sequence of operations as described previously for flat panels (Section 19.1.3) are used. Whilst the test is simple to carry out, some experience is needed in interpreting the results. The extent of damage at the point of impact is assessed in terms of loss of adhesion.

19.3.1.2   Falling weight test
This test was devised to determine the degree to which a paint on a metal surface could accommodate rapid deformation. The panel on which the surface coat has been applied is deformed rapidly by allowing a weighted indenter to fall on it from a fixed height. The impact may be onto the coated surface or onto the back of the panel (reverse impact) according to the specification requirements for the material. The test is published as BS 3900: Part E3 (Impact). The apparatus is available from Sheen Instruments (Sales) Ltd. It consists of a steel block which slides vertically between two guides. Mounted under the block is a tool holder in which is fixed an indenter. The block and tool fall under gravity onto two die blocks with a hole in the centre. The test panel clamped between the die blocks is impacted by the tool. The depth of the indentation is varied by inserting washers of known thickness between the indenter and the tool holder. The test panels are prepared as described previously. However, in this test it is important to use films of standard thickness, 0.001 in (0.025 mm), unless otherwise specified. The thickness of the steel panel must also be constant. In this case the panel is somewhat smaller than those used for chemical resistance testing. Panels 4 in × 2 in (100 × 50 mm^2) are specified having a

thickness of 0.0495 in or 1.257 mm. Failure of the paint is shown by cracking and by loss of adhesion at the deformed portion of the panel.

19.3.1.3   The Erichsen indentation test

This test is concerned with the degree to which a paint is able to accommodate slow deformation of the substrate. A panel coated with the paint system under test is deformed slowly by forcing a ball-shaped indenter into the uncoated side. The distance the indenter has moved when the coating first shows signs of cracking is taken as a measure of its deformation characteristics. Two machines were developed for carrying out the test: the Erichsen Lacquer Testing Machines, Model 229/E (Manual) and 225/E (Mechanical).

Both instruments required a panel 70 mm wide to fit the instrument. The panels, prepared from standard panels as previously described, were cut to size after coating and just before testing. The panel was clamped painted side upwards by a circular boss against a fixed face. A hole in the face allows a 20 mm diameter hemispherical indenter to be moved into the back of the panel. In Model 229/E a manually operated wheel was used to exert hydraulic pressure on a piston connected to the indenter. A self-illuminating microscope placed at the centre of the boss enables the operator to observe the deformation taking place and to stop the indenter when the first break occurs in the coating. A dial gauge micrometer continuously shows the depth of the indentation. In Model 225/E the piston is similarly hydraulically operated but actuated by an electric motor instead of by hand. The motor drives the indenter at 0.2 mm per second into the back of the panel. These models have now been superseded by models 200 and 202.

Adhesional failure is shown by a clear-cut circular break around the apex of the deformation or radial breaks often combined with circular breaks. In such cases the coating is unable to accommodate the dimensional changes in the substrate. This type of break generally occurs quite suddenly. Cohesional failure is indicated by finely divided cracks. In this case the coating adheres well to the substrate, but the cohesion of the film breaks down as the deformation is increased. This does not occur in as clear-cut a way as with adhesional failure, and careful observation is necessary. With the type of steel used for test purposes, fracture of the panel occurs between 9 and 10 mm. Failures of the test film may be due to such fracture, and the damaged area must therefore be examined very carefully (see Plate 19.2).

19.3.2   Hardness testing

It may seem inappropriate to consider hardness testing in the context of durability, since the general mechanical properties of paint films are covered in Chapter 16. However, here we shall cover only some of the standard methods of test that have been developed. Firstly, they can be considered in relation to adequacy of cure, which will have a profound effect on ultimate durability. Also, in some cases changes in hardness as a result of weathering give an important insight into the rate and nature of the weathering process.

A number of tests have been developed over the years to give information on the hardness of coatings. Hardness is very difficult to define in absolute terms, except to say that it is a composite function of the mechanical properties of a material that is related to resistance to deformation. This definition is, however, too simple, since

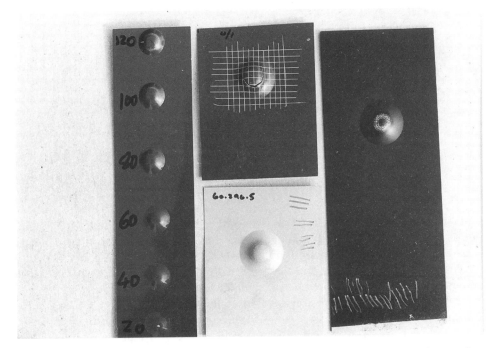

Plate 19.2 — Typical Erichsen indenter test results, including a precross-hatched example, which also shows failure of the metal substrate.

materials can be brittle, plastic, elastic, etc., and it will be perceived that two materials although deformed to the same extent under load may differ in their behaviour on removal of the load. For example, one material may be permanently deformed, the other not. These two materials will have undergone plastic deformation in the first case and elastic deformation in the second. The paint technologist has to take a more pragmatic view of hardness, and for this reason, simple, standard methods of measurement are adopted. Since one is concerned with measuring the properties of a thin film on a variety of substrates it is important to recognize that the substrate can influence the apparent hardness of the film. It is common therefore to make hardness measurements on rigid substrates such as glass or steel and to accept that one may be compressing the test material between the substrate and the probe, or whatever device may be used.

It may be necessary to measure hardness on the production line, so that simple tests and the use of portable equipment will be required in some cases. Carrying out hardness testing in the laboratory, under standard conditions, with carefully prepared samples is always to be preferred whenever this is possible.

Four methods of determining the hardness of paint films, (a) Pencil, (b) Sward Rocker, (c) Tukon tester, and (d) ICI Indenter, are in common use. Pendulum and indentation tests have been described in Chapter 16. In the context of durability studies it is worth noting that the Sward and Tukon test methods are of little value on eroded films. The ICI Indenter can, however, give useful information on almost any type of film.

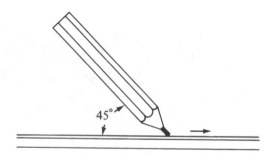

Fig. 19.1 — Pencil hardness testing.

Pencil hardness is mainly used on fresh rather than weathered films. It is one of the simplest methods of measuring and recording hardness (albeit on an arbitrary scale). It uses a range of pencils of different hardness as a basis for comparison. As geologists use the Mohr scale of hardness, with a well-defined range of 'standard' minerals, so it is possible to relate in a similar way paint film hardness to conventional pencil hardness, using a series of specially prepared pencil points of varying hardness.

The pencils are prepared, as indicated in Fig. 19.1, such that approximately $\frac{1}{4}$ in (6 mm) of lead is exposed from the wood. The lead is cylindrical in shape with the end squared off with a fine grade abrasive paper (e.g. 400 grade). In use the pencil lead is pushed firmly along the surface of the paint film at 45° to the surface, using the maximum force, i.e. just less than sufficient to break off the lead. The hardness of the film is equated with that of the pencil that just *fails* to damage the coating. With practice a high degree of reproducibility can be achieved. A good-quality set of pencils ranging from 6B to 9H should be used. (Eagle Turquoise 'chemi-sealed' pencils are recommended.) It is not necessary to work through the whole range of pencils once one is familiar with the typical characteristics of the paint under test. Thus for checking low-bake refinishes only the four pencils ranging from 2B to F are likely to be required. Atmospheric conditions (temperature and moisture) can affect surface hardness, and ideally all films should have been prepared, aged, and tested under exactly the same conditions.

The original Wolff–Wiborn method upon which the above description is based differs only in specifying the hardness of the *softest* pencil that just marks the surface.

Better reproducibility can be achieved by using a pencil-holding device marketed under the name Wolff–Wiborn Scratch Hardness Tester, Model 291, available from Erichsen GmbH.

19.3.3 Scratch resistance
The purpose of this test is to assess the ability of a surface coating to withstand scratching. A needle with a spherical steel point (of specified diameter), carrying a predetermined weight, is lowered onto the paint film and drawn across the surface at a set speed.

The method has been used in the paint industry (and in government specifications) for many years. The test may be operated in either of two ways, as a 'pass or fail' test using a specific weight for the material under test, or (in assessing new materials) with an increasing load until failure occurs. The method was standardized originally in BS 3900: Part E32 (Scratch), itself based upon DEF-1053 Method No. 14.

The apparatus consists essentially of a horizontal sliding panel to which the test panel is clamped, coated face upwards. The sliding panel is moved beneath the needle point at a speed of 3–4 cm per second. The 'needle' itself consists of a shank to which has been soldered a 1 mm diameter grade A1 steel ball. It is fixed into a holder at the end of a counterpoised arm, which is kept horizontal by adjusting the length of the needle in the holder. Weights are placed over the needle in the holder and the panels set in motion. The forward travel lowers the needle gently onto the surface. At least 6 cm of scratch is required for the test.

Although manual and motordriven apparatus are available for the test, the latter is preferred since more reproducible results can be obtained. (The apparatus is available from Sheen Instruments (Sales) Ltd.)

The panels are prepared as previously described. They may be of steel, tinplate, or aluminium, 5 in × 2 in ($125 × 50$ mm^2) in size.

The test panel is removed from the panel after testing and examined visually.

19.3.4   Bend tests

The ability of a paint to undergo bending (on a suitable substrate) has called for the development of several related bend tests. This type of test is applied to materials that may be expected to undergo bending as a result of the method of fabrication of the articles to which they are applied. Thus the tests find greatest use in connection with industrial finishes.

Most commonly, the paint system to be tested is applied to a suitable panel which can be bent through 180° around a cylindrical mandrel of known diameter. The coating is then examined for cracking or loss of adhesion. Although usually regarded as a test of flexibility, the test is a composite one which encompasses adhesion, flexibility, and extensibility. One form of the test (developed originally as DEF-1053 Method No. 13 and subsequently as BS 3900 Part E1) uses a 'hinge' into which the mandrel is incorporated. The hinge consists of two rectangular metal flaps hinged at the extremities of one of their shorter sides with a long pin. The mandrel is fitted around this pin, such that it rotates freely around it. It is so positioned that when the hinge is open there is sufficient clearance to insert the test panel between the mandrel and the flaps. Each mandrel is fitted in its own hinge so that it is necessary to have a complete range of hinges to apply the test in increasing severity as the diameter of the mandrel is reduced. The normal range of mandrels is $\frac{1}{8}$ in, $\frac{3}{16}$ in, $\frac{1}{4}$ in, $\frac{3}{8}$ in, $\frac{1}{2}$ in, $\frac{3}{4}$ in, and 1 in (3, 5, 6, 9, 12, 18 and 25 mm) diameter. It is important that the test panel meets certain dimensional requirements. For government specifications 4 in × 2 in ($100 × 50$ mm^2) aluminium panels conforming to BS 3900: Part A3 are used. In non-specification testing tinplate or aluminium panels may be used, but they should not be thicker than 0.012 in (0.3 mm).

The hinge is used in the following manner: the panel is inserted into the hinge with the coated side outward from the direction of bending. The hinge is closed evenly, without jerking, in not less than one second and not greater than $1\frac{1}{2}$ seconds, bending the panel through 180°. The coating is examined immediately after bending and before removing it from the hinge for evidence of cracking or loss of adhesion.

Cracking closer to the edge of the panel than $\frac{1}{4}$ in (6 mm) is ignored. The test should be carried out in a constant temperature room (e.g. at 25 °C), allowing at least two hours for the panels and hinges to come to temperature.

This extremely simple test can be very informative since it may be related to conditions of use of many of the compositions which it may be used to test. It can be developed in different ways to give information of the acceptability or otherwise of materials that may be subject to environmental changes after a coated substrate has been formed. Thus compositions that pass the test at a given mandrel diameter at 25 °C may crack and lose adhesion when they are heated (subject to a second stoving sequence) or subject to excessive cooling, when the coating becomes glassy.

A variation of the previously described test uses rod mandrels which are supported in a steel cradle. The method is used for test panels which are too thick to fit into the hinges described above. However, the thicker the substrate, the smaller the range of mandrels that can be successfully employed. Thus 26 gauge (0.018 in; 0.45 mm) mild steel can be successfully bent over all mandrels, whereas 20 gauge (0.039 in; 1 mm) mild steel can be bent only over mandrels greater than $\frac{5}{16}$ in (7.5 mm) diameter.

19.3.5   Adhesion: the cross-hatch test

Adhesion, as has been established in Chapter 16, is a very difficult property to measure. Nevertheless it is a very important property of a paint system, and some useful empirical information can be obtained by the cross-hatch test. In this test a die with a number of close-set parallel blades is pressed into the test successively in two directions at right angles to each other. The second pressing is superimposed on the first, giving a pattern of squares. A strip of self-adhesive tape stuck over the pattern is removed sharply, and the adhesion of the film is assessed from the amount of the coating removed.

In the test applied in the laboratories of ICI Paints Division the die, made of hardened carbon steel, consists of nine parallel blades $\frac{1}{16}$ in (1.5 mm) apart and 1 in (25 mm) long. The tip of each blade has a radius of 0.003 in–0.004 in (0.08–0.1 mm), and the shoulders of each make an angle of 60° with one another. The die is applied to the coating with an hydraulic press at 2000 p.s.i. ($14 \times 10^6\,\mathrm{N\,m^{-2}}$). The tape is applied over the pattern left by the die and pressed firmly down, using a soft rubber eraser. It is left in contact for 10 seconds and then stripped rapidly by pulling the tape back on itself at an angle of approximately 120°. The adhesion is reported as good if there is little or no removal of the coating; moderate, where there is some removal, small particles of coating still left adhering in the middle of all or most of the squares; poor, if almost complete removal of the coating occurs. In this test it is important to use a 'non-release' cellulose type of tape. A suitable type is Sellotape NR Cellulose 1101.

If a suitable die and hydraulic press are not available, a variant of the test which uses a scalpel to cut an equivalent pattern into the coating can give useful if not reproducible data. In this case one should attempt to score through the film to the substrate, using a surgical scalpel. In some cases of poor adhesion some detachment of the film may occur during the scoring and before the attachment of the adhesive tape. In these circumstances it is preferable to score the film more carefully in another place so as to make comparable assessments between different coatings.

19.4 Accelerated weathering

The durability of modern paint systems is such that on exposure to natural weathering they may show little signs of deterioration for periods well in excess of a year. Indeed users of coil coatings may expect guarantees that the paint system will not break down in less than ten years. In formulating these types of coating, the more durable they are the more difficult the testing becomes. Means of accelerating the processes have been sought for many years, and some standard methods have been developed. However, a major problem arises in assessing how reliable a particular test method may be. Incorporation of known standards against which experimental paints may be compared helps to overcome this problem to a certain extent. It does not follow that a good correlation will be obtained between accelerated and natural weathering in all cases, and results from accelerated tests must always be treated with extreme caution.

The deterioration of organic coatings on exposure to the elements is due to the effects of radiation (particularly ultraviolet radiation), moisture, and temperature. The degradation processes are the result of chemical change (oxidation) and the effects of mechanical stresses. The accelerated weathering methods seek to intensify these effects so that film breakdown occurs in a fraction of the time that it would do naturally. To achieve this a number of weathering machines have been designed in which radiation/moisture cycles are maintained to achieve perceptible change within periods of up to 2000 hours exposure.

The type of weathering machine and the test cycle may be specified by the customer, and this is commonly the case with large customers such as the motor car manufacturers. Specifications have been drawn up by bodies such as the BSI and the ASTM. These specifications may be the basis of tests agreed between paint manufacturer and customer, but in many cases the customer may have designed his own tests to which the suppliers' product must conform.

Several types of weathering machine have been developed, differing principally in the source (and hence the spectral distribution) of the radiation employed in the test and whether or not a moisture cycle is required. Water may be introduced into the test procedure in several ways. It may simply be sprayed (usually intermittently) onto the test panels, introduced as a fog (using a very fine spray) or as the result of creating within the machine conditions under which condensation can occur on the panels (humidity cycles). In the test cycles, intensity of radiation, spectral output of the radiation source, and use of a moisture/humidity cycle can be chosen to simulate atmospheric conditions that may be encountered throughout the world. Extreme conditions, representing more or less aggressive environments may thus be selected. The manufacturers of the machines in some cases provide a natural weathering service to their customers, with exposure sites (e.g. in Florida or Arizona) where intense natural UV irradiation climatic conditions occur enable correlative data for artificial and natural weathering to be collected.

Three types of radiation source are in common use, carbon arc, xenon arc, and UV-emitting fluorescent tubes. These differ in spectral output and have different effects on the rates of degradation of paint films exposed to them. In general, the higher the intensity of irradiation in the range 300–400 nm the more 'aggressive' the test. However, the relationship between test method and its relevance to natural weathering of a given system is very complex. This has led to the development of machines which can be employed using various test cycles, conforming in most cases to one or other of the specifications of the national standards organization of a particular country. Most manufacturers of weathering machines produce a range of models specifically tailored to the needs of specific or related technologies, e.g. paints and inks, plastics, and textiles. Some will be designed to accelerate colour changes (yellowing or fading), others (including paint applications) will be concerned with the overall integrity of the film, but nevertheless may also call for the monitoring of gloss and colour changes. The diversity of test machines is such that it is not possible to give more than a cursory account, in the following section, of some that are commercially available and are in everyday use in paint-testing laboratories around the world. Some of these machines represent the culmination of developments that have taken place over the past three decades, resulting in greater reliability and refinement. A list of the principal manufacturers of weathering machines (and other test equipment) and their agents in the UK is given in Section 19.6.

19.4.1   Weathering machines — carbon arc source

19.4.1.1   The Marr Carbon Arc weathering machine
The Marr machine was one of the first accelerated weathering machines to be manufactured (by J B Marr & Co.) in the UK (see Plates 19.3 and 19.4). It is now manufactured by Westlairds Ltd, who also offer a maintenance service for this machine. It was designed to conform to BS 3900, Part F3, when the following cycle is used:

> 4 hours wet — atomizers on and fan off
> 2 hours dry — atomizers off and fan on
> 10 hours wet — atomizers on and fan off
> 2 hours dry — atomizers off and fan on
> 5 hours wet — atomizers on and fan off
> 1 hour stopped for servicing and assessments.

The Marr consists of a circular tank with the radiation source, a 1600 watt enclosed carbon arc, suspended at the centre. The painted panels (standard 6 in × 4 in (150 × 100 mm^2) prepared as described in Section 19.1.3) are secured on the periphery of an inner concentric circular supporting ring which rotates slowly during the humidity/radiation cycle. The cycle may be varied to meet different needs, but usually includes intermittent periods of wet and dry.

The Marr is still in use in many laboratories, because Bs 3900 continues to be specified by paint users. However, especially in the automotive paint market (and particularly outside the UK), other machines have been developed to meet more exacting criteria. The Marr Arc machine also has applications outside the paint industry.

19.4.1.2   The Atlas CDMCA Enclosed Carbon Arc Weather-ometer®
The CDMCA (manufactured by the Atlas Electric Devices Co) was first introduced in the USA in 1918. Since then it has evolved into a modern automated unit that satisfies most international weathering test methods which specify the enclosed carbon arc light source. The machine has undergone considerable refinement, such that temperature and humidity conditions, test duration, and black panel temperature are monitored and controlled automatically. The black panel temperature

Plate 19.3 — The Marr Carbon Arc Weathering Machine.

approximates to the maximum temperature a specimen may reach under a given set of conditions. The machine employs two light sources and can be operated in both the normal twin arc or single arc exposure modes. The CDMCA uses solid and cored electrodes burning within the semi-sealed atmosphere of a borosilicate globe. The globe filters out portions of the low UV energy not present in daylight. A microprocessor-based program controls the light and dark cycles. Plate 19.5 illustrates the dual carbon arc system of the CDMCA.

19.4.1.3   The Suga 'Sunshine Long Life Weather Meter': model WEL-SUN-DC
This machine, manufactured by the Suga Test Instruments Co. in Japan, is of the enclosed chamber type, with radiation source and water spray; temperature

Plate 19.4 — Interior of the Marr, showing test panels in place.

Plate 19.5 — Dual enclosed carbon arc lamps employed in the CDMC Weather-ometer.

Plate 19.6(a) — The Atlas Ci65A Weather-Ometer.

and humidity being controlled by an integral microcomputer. It is illustrated in Plate 19.6(b). The machine features unique 'ultra long life carbon arc lamps' (developed by the Suga Company) which can be run continuously for 78 hours. The spectral output of the lamps approximates closely to that of sunlight over the UV range. Stable arc and radiation energy output are achieved by using a specially designed servo-amplifier system and automatic voltage regulator. The DC model offers a dew cycle system which simulates a night and day condition, a feature that many paint-testing laboratories consider to be essential. Models that do not offer this facility are also available, but are more likely to be used when colour changes (e.g. fading) are under investigation. The machine is also available equipped with a 6.0 kW water-cooled xenon long life are lamp (WEL-6X-HC), but this machine does not have the dew cycle facility.

Plate 19.6(b) — The Suga Super Xenon Model SC700-W.

19.4.2   Weathering machines — xenon arc source

19.4.2.1   The Atlas Ci65A Weather-Ometer® controlled irradiance system
Radiant energy in the Ci65A is provided by a single water cooled 6500 W xenon arc lamp whose filtered spectral output closely simulates natural daylight. The machine (illustrated in Plate 19.6(a)) incorporates a system of specially developed interchangeable filters that can be used in various combinations to enable the selection of spectral distributions to meet specific end-use conditions found in different environments. Thus, six different filter combinations, involving the use of borosilicate glass, quartz, soda lime glass, Type-'S' borosilicate glass and IR-absorbing glass in the inner and outer glasses of the lamp assembly. Most commonly in paint durability testing, borosilicate glass is used for both inner and outer glasses. For light fastness testing quartz is used for both glasses. The Ci65A is one of the

Plate 19.7 — The 6500 watt xenon arc system employed in the Ci65.

more sophisticated machines of its kind (see Plate 19.7). Other models that are in use, in some cases with lower rated power sources (3500 or 4500 W lamps) of essentially the same design are available (these include the Ci35A and Ci3000 Weather-Ometers).

19.4.2.2   The Heraeus Xenotest® 1200 weathering machine

This machine (manufactured by Heraeus Instruments GmbH) is of the closed chamber modular type, which is available with a number of optional features, depending on the testing schedules that it is required to perform. The LM version is the most sophisticated, incorporating means of measuring and controlling irradiance and black surface temperature, chamber temperature (up to 70 °C), relative humidity, 'rain function' (water spray up to 8.51h^{-1}), and air volume control. The machine may be operated in static or turning modes (i.e. in the latter case the sample holders rotate during the weathering cycle effectively doubling the surface area of the samples exposed). The machine can house 25 test panels with a total surface area of 10000 cm^2 in the static mode. The power input to the lamp ranges from 2.8 to 6 kW and the spectral output of the lamp is guaranteed to remain constant for at least 1500 hours. A smaller machine, the Xenotest 150S (which conforms to BS 3960), is also available for paint testing. It employs a long arc xenon lamp with a rating of 1.1–1.5 kW, but only accommodates ten test panels.

19.4.2.3   The Suga Super Xenon Weather Meter SC700-W

Another machine of the closed chamber type, the SC700-W, employs a 7 kW water-cooled xenon lamp, with an irradiation control system designed to monitor and control (automatically) the irradiation at the specimen plane. The lamp delivers radiant energy 60–180 W m^{-2} in the UV region (300–400 nm). It is provided with a black panel temperature control system which measures and controls temperature in the specimen plane, considered to be an important feature, since variations in specimen temperature during an irradiation cycle can significantly affect the reproducibility of the test results. The machine does not appear to be available in a form that permits humidity cycling or the use of a water spray to simulate the effects of rain.

19.4.3   Static panel weathering testers

When paints are exposed to sunlight they are affected mostly by the UV light (about 5% of the sun's total spectral output). If one is interested primarily in the effects of UV on a coating, then tests can be carried out using UV sources such as low wattage fluorescent lamps which emit almost all of their energy in the UV region. They can be employed within compact cabinets in which the panels are static, thus providing a relatively inexpensive test facility. The test may also include a humidity (dew) cycle, which brings it closer to the reality of natural weathering and the more complex simulation of conditions that is the basis of the machines described above. These tests tend to be more aggressive than the weathering cycles previously described, but do not correlate as closely with natural weathering data. However, they can give useful information on the onset of breakdown of a paint film in comparatively short times and therefore tend to favoured by some paint users (particularly the car manufacturers), who need to have information to enable them to make decisions, e.g. with respect to colour changes on their models, every year.

The Atlas 'UVCON®', the Q-Panel 'Q-U-V' Accelerated Weathering Tester, and the Suga 'Dewpanel Light Control Weather Meter', model DPWL-5R, are described briefly below.

19.4.3.1   The Atlas UVCON

The UVCON employs eight 40 watt fluorescent lamps mounted horizontally in two banks within a cabinet with access doors on both sides. The test panels (26 or 75 × 300 mm or 20 or 100 × 300 mm) are secured in racks such that the paint films face a bank of four tubes, with their backs exposed to a flow of air saturated with water vapour from a reservoir at the bottom of the cabinet. During the dark cycle condensation takes place on the surface of the panels, providing uniform wetting. The temperatures of the UV cycle and condensation cycle can be set independently, the black panel temperature is monitored constantly and displayed on a digital meter. To maintain constancy of irradiation the lamps are subject to a rotation and replacement programme. Three types of UV fluorescent tubes are available with UV outputs covering the spectral ranges 295–340 nm, 295–265 nm, and 295–410 nm. The most commonly employed lamps for paint testing with this tester (under the most aggressive conditions) are the F40 UVB type with an output covering the range 295–340 nm. These lamps are specified in test method SAE J2020 and recommended in ASTM G 53.

19.4.3.2   The Q-Panel Q-U-V Accelerated Weathering Tester

The Q-U-V was the first weathering machine of its kind, being introduced in the USA in 1969, and is now widely accepted for paint testing around the world. It combines simplicity of operation with reliability. The standard model is designed to take either 75×300 mm (25) or 75×150 mm (51) panels, although other options are available on request. A variety of UV lamps are available for use in the tester, depending on the nature of the test cycle to be used. The manufacturers recommend UVB-313 which has a higher output in the UV-B region than the older FS 40 type, which was originally employed in the Q-U-V. The recommended lamp life is 1600 hours, at which time the lamps have lost about 40% of their original output. Irradiance is controlled by replacing the two oldest of the Q-U-V's eight lamps and rotating the remaining lamps in a specified manner. Lamp maintenance is required only every four to six weeks.

The latest model of the Q-U V features an automatic system of irradiance control, called the 'Solar Eye®'. The Solar Eye is a precision light control system which enables the user to select a level of irradiance which is then maintained automatically. It monitors the UV intensity continuously using four sensors at the sample plane. A four channel feedback loop maintains the programmed irradiance level by compensating for any variability in lamp output by adjusting the power to the lamps. Q-Panel claims that the new system gives more reproducible results, increases the usable life of lamps, and reduces lamp maintenance by eliminating the need for lamp rotation,

19.4.3.3   The Suga Dewpanel Light Control Weather Meter model DWPL-5R

This machine is similar to those described above. It uses eight 40 watt lamps and employs a humidity/condensation cycle. The choice of lamp spectral output is essentially the same as previously described for this type of tester. The model 5R uses an automatic light control system which monitors the light energy at the surface of the test panels and displays the energy level in mW cm^{-2}. The energy level is maintained at a constant value automatically. This eliminates the need for lamp rotation. A manually operated machine (Model 5M) is available at lower cost. Both machines satisfy the requirements of ASTM G-53 and JIS D0205 specifications.

19.4.4   Assessment methods: definition of criteria

Paints subjected to accelerated (or natural) weathering are assessed for changes in certain properties such as gloss and colour, and for specific types of breakdown under the following headings:

(a) chalking, (b) bronzing, (c) cracking and checking,
(d) blistering, (e) flaking, (f) rusting and corrosion of substrate,
(g) corrosion from a cut, (h) erosion, (i) water spotting,
(j) water marking, (k) dirt retention and ingrained dirt,
(l) mould growth, (m) chalking into film, (n) efflorescence.

These terms are defined in BS 2015 (1965), *Glossary of Paint Terms*.

For the most part the assessment of change and breakdown is done visually, assisted in some cases by the use of a low-power magnifying glass or microscope. Occasionally, gloss or colour measurements are made instrumentally, but in many cases this is not essential. Comparisons are made with unexposed panels or in some cases (particularly with accelerated tests) with parts of the panel that have been covered as a result of the method of securing the panels in the machine. An arbitrary 0–10 scale is adopted in most cases. Thus in assessing *gloss* a rating of 0 signifies no loss of gloss, and rating 10, corresponding to a completely matt surface, signifies a total loss of gloss. In making judgements of gloss it is common to assess the sharpness of an image reflected by the paint film. A 'gloss box' has been developed to improve the method [2], commercial versions of which are now available.

Colour change can take many forms, and often the type of change is as important as the degree. A colour may become darker or faded, duller or cleaner. Fading may be real or apparent. For example, chalking (q.v.) gives an appearance of fading, but in this case the original colour can be restored by cleaning to remove chalking in the surface layer. 'Duller' would indicate an increase in greyness, i.e. a loss of purity of colour, not necessarily associated with fading. 'Cleaner' indicates the development of a purer colour and is the opposite of 'duller'. Specific changes in colour may be recorded as such, e.g. redder or bluer than standard.

Gloss and colour assessments may be made before and after washing, i.e. wiping with a water-wet cloth, leathering off, and drying.

Chalking is defined as 'the formation of a friable powdery coating on the surface of the paint film caused by the disintegration of the binding medium due to disruptive factors during weathering'. Chalking can occur in any colour, but is more usually associated with pastel shades where the white component gives it prominence. In deeper shades, particularly blues and maroons, the chalking exhibits a lustrous effect and is described as *bronzing*. Chalking can be rated easily and with a reasonable degree of accuracy by the 'finger stroke' method. The end of the index finger is drawn with light pressure along the film and the amount of 'chalk' removed related to photographic standards. Bronzing is assessed in the same way. *Cracking* is specifically a case of film breakdown in which the cracks penetrate at least one coat and may propagate through the whole paint system. *Checking* is a lesser form of cracking where the cracks do not penetrate the topcoat. Three types of cracking and checking are defined;

- micro (mc): confirmed by examination under a low-power binocular microscope at ×16 or ×32;
- minute: fine cracks, but clearly visible to the naked eye;
- cracking or checking, where the effect is larger and immediately obvious.

Cracking and checking form patterns which can be classified according to their type. These are as follows:

- Fine, or linear with approximately parallel cracks.
- Pattern, where the angle between cracks is random and the cracks can join up to form a continuous network.
- Crowsfoot, where cracks radiate from a point after the manner of a bird's foot.
- Hairline and grain cracking specifically in wood finishes.

All of these defects are rated according to photographic standards, as are the majority of those other defects listed. The use of photographic standards for *blistering* has

Plate 19.8 — Blistering of a decorative gloss paint on softwood.

already been noted, in connection with chemical resistance testing. An example of blistering of a gloss paint on soft wood is shown in Plate 19.8.

The defects of *flaking*, *rusting*, and *corrosion* from a cut are likewise judged against photographic standards for the type of finishes involved. Erosion is akin to chalking, but is a more severe defect which might lead ultimately to exposure of the underlying surface. It may be due to the action of rain drops, the abrasive action of wind borne particles of grit, or a combination of these effects. *Water spotting* and *water marking* are different phenomena. The former is the spotty appearance of a paint film caused by drops of water on the surface which remain after the water has evaporated. The spots usually appear to be lighter in colour than the surrounding paint. They may or may not be permanent. Water marking, on the other hand, is due to a distortion of the paint film by water droplets. No colour change or whitening occurs. Both of these defects can be related to the photographic standards for blister as an indication of abundance and size. Dirt retention, ingrained dirt, and mould growth are more likely to occur on natural weathering than under accelerated weathering conditions for obvious reasons. Ingrained dirt and mould growth can sometimes be confused, and it is often necessary to resort to microscopic examination to distinguish them. Panels are assessed after wiping one half of the panel lengthwise, with a damp soft cloth, taking care not to remove or disturb the dirt on the remaining half. The panel may then be replaced for a further period of weathering. The panel is assessed for colour changes against an unexposed standard. Mould growth can readily be distinguished from dirt by using a ×16 binocular microscope. It is characterized by spidery tentacles radiating from a dark nucleus. Even after washing the remains of the tentacles can still be recognized, and it is common to observe a regrowth of mould on the washed area when the panel is re-exposed.

'Chalking into film' is one of the common defects. It is an effect that looks like chalking, but which cannot be removed by washing. Efflorescence is the development of a powdery deposit on the surface of brick and cement due to the migration of water-soluble salts through to the surface and their subsequent deposition by the evaporation of water. In some cases the deposit may be formed on the top of any paint film present, but normally the paint film is detached and broken by the efflorescence under the coat.

19.5   Natural weathering

By exposing paint systems to the elements one might assume that one would collect the most reliable data on paint performance. This is true only up to a point, since there are so many variables involved, and it is extremely important to take into account all significant variables. Because natural weathering is a slower degradative process than can be achieved artificially it is particularly important to design the experimental series with care as any fault in the experimental design may not become apparent for one or several years.

Good documentation of the exposure series is called for because it may extend over many years and there may be changes in staff who are likely to be assessing the panels, and the staff who may be seeking information from the series. The geographical location of the exposure site may be important. If the paint is to be used in an industrial environment (e.g. structural steelwork) it is probably best to weather the test panels in an industrial rather than rural environment. On the other hand it might be useful to use some sites that are known to give rise to more rapid breakdown than the location of anticipated use. Thus many technologists regard the North American sites in Florida and Texas favourably in this way, and regard them as offering 'accelerated' weathering compared to sites in the UK. This is because of the higher UV radiation levels impinging on the panels in these areas coupled, with, in some cases, large variations in temperature and humidity. Typical 45° south facing panel racks are shown in Plate 19.9.

19.5.1   Choice of test pieces

Not only does the nature of the substrate affect the performance of panels in a test series, but the physical shape and method of construction can also do so. The substrates most commonly used for tests are wood, mild steel, aluminium, and some forms of masonry material, e.g. asbestos cement panels. The choice will depend upon the market for which the paint system has been developed. Thus architectural paints may be tested on wood, metal, or masonry, whereas automotive finishes and refinishes will be tested only on metal panels. Exceptions do occur in some cases, for example with compositions specifically designed for application to plastic body shells, where it is important to use the appropriate substrate. Wood is probably the most variable of all substrates. It is anisotropic and subject to change as a result of variations in temperature and humidity, and is particularly sensitive to the ingress of liquid water.

Plate 19.9 — Panel racks for exterior exposure of panels.

With the exception of wood, most products are tested on flat panels of a convenient size. They are assessed, usually at quarterly or half-yearly intervals, for periods of up to five years. Some compositions, e.g. coil coatings, may be exposed for longer periods because of their expected long life as protective coatings. The methods of assessment are akin to those described in Section 19.4.

For many years it was the practice to collect weathering data for decorative paints on flat panels, 12 in × 6 in (300 × 150 mm^2), of best quality British Columbian Pine. This type of panel was chosen because of the good reproducibility of data that could be obtained, and it is less demanding than other more variable substrates. In recent years, however, a number of research establishments and paint manufacturers have attempted to devise test methods that are more demanding of the coating and yet more realistic of 'in use' performance. For example, Whiteley (Building Research Station) has developed test methods based upon a test house, in which glazed window frames are used as test pieces. The orientation of the test house is such that maximum exposure to direct sunlight is obtained. The interior temperature and humidity are cycled such that condensation of water occurs on the interior of the glass surfaces and this runs down onto the (interior) uncoated wooden frame test pieces. This test method simulates the worst conditions of use for decorative paint systems in housing, i.e. in kitchens and bathrooms. This is a rather expensive way of collecting data, and other workers have attempted to achieve the same objective in alternative ways. Thus the Paint Research Association devised weathering cabinets that accommodated test pieces that were similar to those used by Whiteley, but in which the glass was replaced by a rectangular piece of marine plywood. Softwood frames were used as before.

Both of these methods simulate the problem of maintaining film integrity over joints between different pieces of wood with grain running in different directions at the junction between individual parts of the frame. The differences in dimensional change on the absorption or loss of water are considerable. Across the grain of a softwood, water absorption may cause a dimensional increase of about 8%. Along the grain, under the same conditions, only about 1% dimensional change may be observed. These dimensional changes can cause the opening or closing of the joint at the junction between the two pieces. Thus the paint film is under considerable stress at the joint, and this is one of the main causes of breakdown, at or around joints in standard softwood joinery. In window frames the lower two joints and the lower rail are the first sites for paint failure because of these considerations. The formulation of paint systems to meet these demanding situations has been discussed in Chapter 9.

A simpler method for simulating these effects has been devised by ICI Paints Division. It too recognized the need to induce effects due to water absorption through the end-grain of joinery sections. Because of its simplicity and hence low cost we shall consider this method in more detail.

19.5.2   ICI joinery section test method

This method seeks to obtain weathering data on paints, using a realistic approach designed to incorporate some of the most demanding conditions that such systems are likely to encounter. It was arrived at after careful consideration of the factors most likely to contribute to the rapid breakdown of paint systems. The most important factor was found to be the ingress of water through the end grain, usually through joints, e.g. at the junction of pieces of wood forming a window frame. Also, the prolonged contact with water that arises when liquid water collects at joints or between frames an sills was found to have an important bearing on paint performance. The test piece (see Plate 19.10) is thus designed with a 'water trap' which the joinery section abuts. The end of the joinery section abutting the water trap is unpainted. The test pieces, mounted on a backing panel, are exposed (in the northern hemisphere) at 45° facing south, so that they have the maximum incident UV and actinic light energy impinging upon them. Rainfall on the test piece drains into the water trap which is therefore the last part of the test piece to dry out. The test has been found to be reliable and cost effective. It should, however, be realized that there are many variables that can influence reproducibility of results in weathering, particularly on wooden substrates.

19.5.2.1   Selection of panels for a given series

It should be emphasized that the nature of the wood has a major influence on the performance of the coating, which in many cases will surpass effects arising from the coatings themselves. This must be allowed for in the design of the experiment. For reference purposes and later correlation, the ratio of heartwood to sapwood should be recorded. This may be estimated using a stain test [3]. The grain orientation may also be recorded by pressing a representative specimen of the end grain of the wood onto an ink pad and then transferring an imprint to a piece of paper. Because weathering is a process which leads to an inherent distribution of results, some replication will be necessary if valid comparisons are to be made between different systems. The degree of replication is dependent on the magnitude of the dif-

Plate 19.10 — ICI joinery section test, showing part of a series of experimental paints after 2 years' exposure.

ference being sought and the standard deviation within the series. This will vary between series, but for independent systems it is unlikely that anything less than fourfold replication will be sufficient. By 'independent system' is meant any two systems which have no intrinsic relationship to each other, i.e. there are no known common factors. If the systems have a relationship, e.g. a PVC ladder, then statistical methods can be used to analyse the results, and the degree of replication may be reduced.

In a small series it will be possible to cut all the test pieces from a single piece of wood. This will minimize the variability, though as a precaution the pieces should be taken in random order. It must be remembered, however, that the results of the exposure will be relative, and care must be taken in drawing any absolute comparisons with other series. The total credibility of the series can be increased by replication on other test pieces cut from another piece of joinery section. For a large series the number of joinery sections (test pieces) is taken from two or more pieces of wood. The test pieces are pooled and used in random order. Alternatively, a note may be kept of the source of each test piece and the original lengths of wood treated as an experimental variable.

19.5.2.2   Preparation of joinery section test pieces

The test pieces ($7\frac{1}{2}$ in (185 mm) long) are cut from lengths of deal joinery section and are numbered consecutively. The joinery section is cut to the profile indicated in Plate 19.10 from average quality (HMS grade, i.e. Swedish Brack Group 1, Fifth Quality), shake-free lengths of red deal, excluding the first eight growth rings. Sections that are damaged, have poor sharp edges, dead knots, bad knotting at sharp edges, or any other major defect are not used. The top end of the section is sealed with a single coat of aluminium primer sealer and then finished with the system under test. The system to be tested is applied to all surfaces of the test piece with the exception of the bottom end which is to be in contact with the water trap. At least the minimum recommended drying time should be allowed between coats. The backing panel (12 in × 3 in (300×75 mm^2)) and water trap (3 in × 3 in (75×75 mm^2)) are constructed out of flat panels withdrawn from previous series. The backing panel and water trap are screwed to the test piece. Both are painted with standard exterior paint but are not subject to assessment. A coat of gloss paint is given to the upper face of the water trap which will be in contact with the unpainted end given of the joinery section.

19.5.2.3   Examination and assessment procedure

The joinery sections are evaluated by considering two distinct regions of the test piece, the bottom 2 in (50 mm) adjacent to the water trap and the top $5\frac{1}{2}$ in (140 mm). The latter is assessed in the same way as plane (flat) panels for all of the appropriate defects listed in Section 19.4. The region close to the water trap is examined specifically for signs of cracking and flaking since it is to be expected that these defects will appear more rapidly in this region. The joinery sections are part washed at examination times, on the raised 1 in wide $7\frac{1}{2}$ in long (25×185 mm^2) strip and on the adjacent side. The criteria of assessment are as previously discussed (Section 19.4.4).

19.5.3   Recording of weathering data

The recording of weathering data takes place over such long periods, and may be so voluminous and diverse, that it must be done systematically. A systematic approach to origination and data-recording calls for the adoption of standard methods both of testing and reporting, so that data retrieval is facilitated. This type of data storage and retrieval is readily computerized, but it is important that any system adopted is sufficiently flexible to take into account variables that have not yet been defined, yet may assume considerable importance sometime in the distant future. The system should also include a detailed record of the compositon under test. Typically the weathering data should, in addition to the results obtained during the period of test, contain at least the following information:

- series identification; originator; purpose of the series;
- date of initial exposure; site, substrate; product type;
- system, indicating number of coats of each paint;
- pigmentation; binder; minor compositional variants, if important to the series;
- application conditions (e.g. temperature, humidity, film thickness, and recoat time).

It is also useful to record weather conditions so that if any abnormal conditions are encountered that might affect particular series they will not be overlooked.

19.5.4   'Natural' versus 'artificial' durability testing

The purpose of testing durability is to establish the probable performance of a product that is designed to meet a particular need. In most cases paints are sufficiently durable to survive many months or even years of exposure to the elements before showing the first signs of breakdown. This period may in the absence of other data preclude the marketing of a product until confidence in its performance has been established. If this period can be shortened, the response of the development group can be more rapid and potential problems identified and overcome. The benefits of an accelerated test procedure can very readily be seen in terms of a reduction in development costs and possibly a greater lead time over competition for the manufacturer developing a new product.

Acceleration of durability testing can be carried out in several ways. If the main causes of degradation an failure are due to ultraviolet irradiation, heat, and moisture, it may be convenient to submit a product in a part of the world that will provide more intense irradiation and higher temperatures for longer periods than can be expected in the UK. If mould growth is a problem, regions more conductive to mould growth, i.e. where the temperature and humidity are high, such as in Malaysia, may be ideal for testing a given product. More often than not, however, the researcher may wish to accelerate degradation by a factor far greater than can be achieved by using a natural tropical exposure site, and he or she will resort to the type of equipment described in Section 19.4. In doing so he or she may risk the possibility that the more aggressive test will bear little relationship to the performance of the test paint in reality. In these cases it is often assumed, with some justification, that if an experimental paint fails before a comparable standard paint of known performance it is unlikely to perform better than the standard paint in practice. If the experimental paint performs better than the standard on the accelerated test there is no guarantee that it will do so in practice. In general the test conditions will be programmed to simulate as closely as possible the type of exposure to which the paint may be exposed. The comparative distribution of irradiance for sunlight and for various artificial sources has been quoted by Scott [4] and are shown in Table 19.1. The 6500 xenon source, with a borosilicate glass inner and outer filter, provides irradiation with a distribution closest to that of sunlight. The open carbon arc has slightly more than double the radiation in the 340–400 nm band than sunlight. Fluorescent tubes have a very much more intense output in the region below 400 nm, 97%, compared to 6.1% in-sunlight. It is to be expected that this intense UV output will be much more aggressive, and this is borne out in practice. This makes data from weathering testers like the QUV and the UVCON much more difficult to interpret than those obtained from machines utilizing less aggressive radiation sources.

Table 19.1 — Comparative distribution of irradiance (from Scott [4])

Wavelength (nm)	Sunlight %	6500 W Xenon %	Carbon arc (open flame) %	Fluorescent source %
300	0.01	0.01	0.5	14.0
300–340	1.6	1.5	2.5	70.0
340–400	4.5	5.0	11.0	13.0
Total to 400	6.1	6.5	14.0	97.0
400–750	48.0	51.5	34.0	3.0
750	46.0	42.0	52.0	0.0
Total to 750	94.0	93.5	86.0	3.0

Nevertheless, some major paint users such as the motor manufacturers may demand a certain standard of performance under these particularly aggressive conditions, although they may not correlate well with condition of use.

Scott [34] has pointed out that laboratory testing may be designed to accelerate the natural processes of degradation or to simulate a wide range of environmental conditions. The former represents the traditional use of weathering machines that has led to standard sets of conditions, i.e. specifications of performance according to predetermined cycles. The latter more recent approach is more concerned with simulating, under controlled laboratory conditions, the natural environment in any part of the world.

Sophisticated weathering machines now available are capable of being programmed to provide environmental cycles of radiation, temperature, wet/dry cycling, and atmospheric pollutants such as NO_2, SO_2, and ozone. This type of equipment enables a more rigorous approach to the study of the effect of the environment on paint degradation.

19.6   Suppliers of accelerated weathering test equipment

- Atlas Electrical Devices Co., 4114 N. Ravenswood Avenue, Chicago, Illinois 60613, USA.
- UK Agent: Alplas Technology, 11 Kings Meadow, Ferry Hinksey Road, Oxford OX2 0DP, UK.
- Heraeus Equipment Ltd, Unit 9, Wates Way, Brentwood CM15 9TB, UK.
- Suga Test Instruments Co. Ltd, 4-14 Shinjuku 5-Chome, Shinjuku-Ku, Tokyo, Japan.

UK Agent: Westlairds Ltd.

- Westlairds Ltd, North Green, The Green, Datchet, Slough SL3 9JH, UK.
- The Q-Panel Company, 26200 First Street, Cleveland, Ohio 44145, USA.
  European Branch: Express Trading Estate, Farnworth, Bolton BL4 9TP, UK.

Other manufacturers or suppliers of paint testing equipment:

- Pearson Panke Equipment Ltd, 1–3 Hale Grove Gardens, London NW7 3LR, UK.

UK Agents for Erichsen testing equipment.

- Research Equipment (London) Ltd, 72 Wellington Road, Twickenham, Middlesex TW2 5NX, UK.

Acknowledgements

The author is indebted to a number of suppliers of paint testing equipment who provided information, particularly on accelerated weathering machines, for the revision of this chapter. In this context special thanks are due to Mr A L Batty (of Westlairds) and Dr Shigeo Sua (of Suga Test Instruments); Bill Bethell (of Q-Panel Co); and Mrs Ulrike Sellen (of Alplas Technology) who provided a wealth of useful information. Thanks are also due to the author's former colleagues at ICI Paints, notably M Bahra, C W A Bromley, A Doroszkowski, and D Farrow, for supplying the test panels from which plates illustrating various aspects of paint failure under test were prepared.

References

[1] Technical Bulletin F-8144, Q-Panel Co, 1993.
[2] *J Oil Colour Chem Assoc* **47** (11) 864 (1964).
[3] *J Wood Sci* **5** (6) 21–4 (1971).
[4] scott j i, *J Oil Colour Chem Assoc* **65** 182 (1982).

20

Computers and modelling in paint and resin formulating

J Bentley

20.1   Introduction

With personal computers now used in all working environments, this chapter presents both an introduction to, and an overview of, the many computer aids available to the paint technologist [1]. Its aim is to make the reader aware of the full range of possibilities and opportunities, with the emphasis on that which can be available at the desktop. This is done with some trepidation, because it is a changing area and while equipment and utility software costs fall, computing power, scientific software development and communications through access to local, inter- and intra-networks, advance at a rapid rate.

Simpler software, and mathematical and modelling techniques are all covered. While many aspects of paint and resin formulating and manufacturing are included, the emphasis is on resin preparation and properties, where the author has both written programs, and has actively used many of the modelling techniques described. Other topics include information technology, formulation, prediction, experimental design, manufacturing and quality assurance. For a number of these topics, only an outline is given, though in all cases sources for further study are referenced. In many sections, one or more sources of commercially available software are mentioned. These may only be examples from a more extensive range available.

General source books are included in a bibliography, and a number have software included or available on companion discs. Those wishing to pursue any topic are urged to consult the references, a number containing the methodology or the modelling equations, while a few have usable program listings; others reference or review commercial software sources. Readers may gain benefit from approaching some authors directly, especially those working academically, since they may be seeking sponsorship, have industrial collaboration 'clubs', or be willing to work co-operatively with an industrial partner.

20.2   Software in the laboratory

It is assumed that the reader is familiar with the basic PC and the standard software for word processing, spreadsheets, databases and presentations, etc. However there are extensions to these programs available to aid chemists, and there is also general scientific software aimed at the scientific community and chemists in particular. Most basically, scientific and technical spelling checkers are available to function with popular word processors, and useful collections of clip art of chemical apparatus for presentations can also be found. As discussed later, spreadsheets are frequently used for formulation development and even simple modelling (See Bibliography, Section 20.4 and Section 20.6.2)

The prominent development for chemists is now the availability of so-called chemical office suites (ChemOffice, CambridgeSoft and ChemWindow, Bio-Rad). These can offer structural drawing and property database packages, with optionally NMR and property prediction, and coupling to simpler molecular modelling modules, and additionally Internet linkages to chemical suppliers and reaction information.

Finally, in many laboratories which have analytical or test laboratory accreditation, and where it is a requirement to record electrical equipment safety checking and hazardous material inventories specialist database software is available.

20.3   Information technology and knowledge based systems

The Internet is now a primary information source [2], and there are many sites with items of potential interest (paint manufacturers, raw material suppliers, university departments, trade and research organisations, journal contents (ACS), complete journals (RSC), etc.). As starting points, the Paint Research Association Coatings Superstore may be found at http://www.pra.org.uk and the European Coatings Net at http://www.coatings.de. Both of these provide links to many other coatings related sites.

Other information is available by the direct on-line accessing and searching of remote databases, which now include both abstracts (*Chemical Abstracts*, ACS; *World Surface Coatings Abstracts* or WSCAs, PRA), and often complete journal contents [3]. A specific area where valuable databases are available is the area of safety, health and environment (SHE), and these can assist for seeking both substance data and for legislation information [4]; on the Internet, http://www.chemfinder.com is an outstanding source. Patent sourcing can also come from either of the above abstract sources, or from the Internet, where the European Coatings Net has an expanding and searchable patents section. American patents can be searched at http://www.ibm.com/patents.

Some 'public domain' software is available, and there are 'shareware' distributors who include scientific and technical programs (e.g. The Public Domain Software Library). Again the Internet is a source of all kinds of software, and commercial software can also be found there for download, often at substantial discount from store prices.

As an alternative to searching remotely, much information is now readily available on CD-ROM, and information sources published in this medium are expanding rapidly. Bench chemists can now have many of their journals, abstracts

(WSCAs), chemical indexes (e.g. Merck Index, CRC/Chapman & Hall) and catalogues, safety data sheets and chemical data search facilities (Aldrich) at hand on their desk top computer. In the analytical laboratory, the analyst/spectroscopist has searchable peak libraries available for most spectroscopic techniques.

For representing structures both in documents and in database searching, a range of established data formats are now available to chemists. For drawing alone, the range of packages includes Chem-Draw (Cambridge Soft) and WIMP 95 (Aldrich). Chem-Draw now has physical property data (real where available and estimated from group contribution techniques where not) immediately available for drawn structures. Common formats for 3-D and X-ray crystallographic data, and for spectroscopic data are used by specialists in these areas. An introduction, giving a useful list of web sites has been noted [5].

The organisation of data searching methods incorporating 'intelligence' is well developed such that expert or knowledge-based systems are available for interrogation on various topics [6]. Simpler expert systems are principally knowledge providers, with the user asking questions and responding to choices, enabling a specific data item to be searched. Examples of expert systems include various coating selectors [7, 8] and equipment selection for items such as mixing equipment [9]. Writing expert systems may require the use of special declarative programming languages such as Lisp or Prolog, particularly suitable for processing sets of logical rules.

Information management also includes laboratory information management systems (LIMS), principally of use in such locations as analytical and raw material testing laboratories, aiding in sample organization, test scheduling and result distribution [10].

20.4   Modelling and mathematical techniques

Modelling is the discipline applied to finding an explanation for the chemical and physical phenomena controlling processes, and expressing the result in some form of mathematical expression. A mathematical model as encountered, may hence be one or a set of equations used to describe a property, a part of a process or a complete system. The model can be used to calculate (or predict) characteristics related to the state of the system. Useful models, if at all complex, will be made more accessible as a computer program incorporating frequently encountered data parameters in a database to save repetitive entry. Outputs will be screen displayed or printed, in tabular or graphical form as appropriate. The elements of a computer model are shown in Fig. 20.1.

Models may be empirical or mechanistic, or sometimes semi-empirical. Models of real processes will always be a simplification of reality bounded within a known area; empirical models should always be used within that known area, but even with good mechanistic models, some caution in employing extrapolative prediction is necessary. Notwithstanding, models are powerful tools to indicate areas of interest for subsequent investigation, and can explain the effects of previously unexpected or unobserved phenomena [11]. In practice many models are semi-empirical, and these are often surprisingly the most useful and developed. Used properly, and kept up to date in their databases, models are both teaching aids and repositories of acquired knowledge, complementing other appropriate techniques.

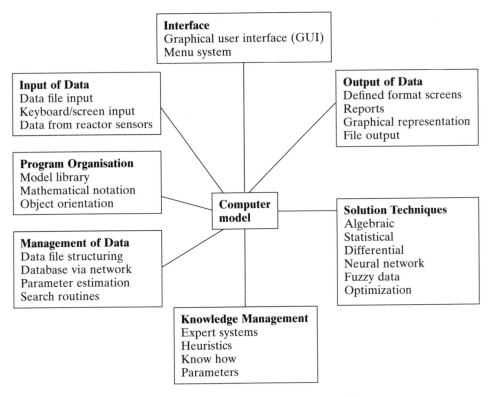

Fig. 20.1 — Elements of a computer model.

Good experimental design allied to the appropriate statistical analysis techniques is a basic requirement for both model assembly and parameter acquisition, and uses are discussed later.

Group contribution techniques may be simple additive functions to provide adequate estimates of properties, or may be more complex, and many have a sound theoretical basis. Molecular weight is an additive property by definition. The use of group contribution techniques is well recognized in polymer property estimation, and discussed further in Section 20.8.

Predictive models may be assembled by a variety of techniques. The simplest use basic algebraic equations, and ready examples are the Fox equation for T_g and the Arrhenius equation for initiator breakdown, the latter extended by simple integration to cover typical practical situations (see later).

Where analysis is more difficult, extended methods may be required. Many polymer properties are those of populations and require statistical techniques in analysis. Many processes are stochastic in being made up of events, each of which occurs with a certain probability, rather than deterministic in being absolutely determined by some physical consideration. These statistical probabilities may be assigned by consideration of energies, of reaction rates or some previous statistical counting activity; they may not always be easily accessible. To model usefully where mechanisms, and probabilities or rates are reasonably certain, a variety of techniques can be used. These include the use of statistical (probability) analysis, use of Monte Carlo techniques, and by the solving of differential equations (deterministic). A discussion of the use of all of these techniques for polymer kinetic modelling may be found in the literature [12, 13].

Statistical methods were evident early in the formation of branching and gelation theory for step growth systems; the developments begun with Flory's theory are used in gelation prediction for alkyds and polyesters, and in gel and post-gel network calculations. Both are discussed later.

Monte Carlo modelling is so-called because of its use of random numbers, such as those occurring in roulette games or from throwing dice. By generating these numbers within a given range, and testing them against event probabilities, a large range of processes may be simulated. Monte Carlo modelling may be considered a hybrid of statistical and kinetic models since it has elements of both. Computer use is essential, since the technique requires complete randomness, and the generation of enough 'events' to reduce 'shimmer' to acceptable levels, so that the distributions produced are suitably smooth and predictions within acceptable tolerances. This versatile technique has been used for applications ranging from parameter estimation [14, 15], and kinetic simulation [16], to addition polymerization composition prediction and network modelling, and paint film structure simulation, as discussed later.

The most rigorous approach used, where a number of kinetic events are present, involves the assembly of a system of differential rate equations, which when solved against time, provide full reaction progress information. This combined with statistical techniques such as 'Method of Moments' is the basis of the full kinetic models for addition polymerization described later.

Those writing computer modelling programs will find a range of languages and techniques available to them. Most satisfactory is likely to be the use of a compiled procedural programming language such as Basic, Fortran, Pascal or C. Powerful libraries of mathematical routines are available to enhance their use (IMSL, Visual Numerics; NAG, NAG Ltd). While prejudices and preferences exist, versions of all of these are available with structured programming features and with optimising compilers, making the choice essentially personal to the user. Nevertheless, C++ is currently the chosen 'professional' language, because of its portability, availability of advanced features and libraries, and possession of efficient compilers. If programming for the personal computer Windows (Microsoft) graphical user interface, Microsoft Visual Basic or Visual C++, or Borlands Delphi will be probable choices.

Simpler mathematical 'programming' can be carried out within the capabilities of spreadsheets (123, Lotus; Excel, Microsoft), that have graphical output available, and possess extensive macro languages. This is, however, suitable in the author's opinion, only for individual use, and should be discouraged on grounds of maintenance and portability difficulties for all but the simplest model. Alternatively there are mathematical and simulation programs working from a graphical or flow chart user interface, such as MathCad (Mathsoft Inc.) and VisSim (Visual Solutions). The former has good chemical data and formulae modules available.

The modelling process can be shown diagrammatically as in Fig. 20.2.

Models may be used analytically, or predictively, where analytical use shows the consequences of given ingredient charges or process conditions. Predictive calculation, to determine the conditions necessary to achieve a required goal, will always be possible, but the difficulty in deriving a predictive model will depend on model complexity. Simple algebraic equations may be rearranged into reverse form, or simple solving of multiple equations may be possible; however, other situations may

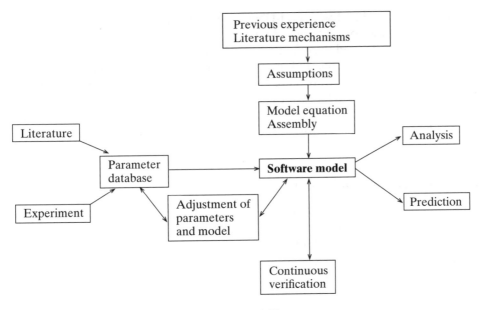

Fig. 20.2 — The modelling process.

require iterative calculation, or the use of optimizing routines, of which the SIMPLEX search method is useful and robust (Nelder & Meade [17]). All formulating programs will be more valuable if they include costs, and in particular facilities for cost optimization.

A number of specialized techniques have also emerged and these include chemometrics, neural networks and fuzzy logic. Knowledge-based and expert systems have been referred to under information technology. Fractal analysis has now found application in such diverse areas as colloidal and dendritic structure investigation, and in surface profiling [18].

Chemometrics is not a new technique; the term is used for the application of a combination of chemistry, and mathematical and statistical techniques in processing data particularly from spectroscopic analysis [19]. Powerful packages are available for multivariate calibration, prediction and experimental design in this context (Unscrambler, CAMO). The now extensive use of computers in analytical instrument control and data analysis is not further discussed here. However, as modelling techniques, it should be noted that simulation of certain spectra is possible, and as examples, proton NMR (gNMR, Cherwell Scientific), C_{13} NMR [20] and mass spectroscopy (MasSim, SCS International) models are available.

Neural network techniques attempt to mimic human brain processes, and can be used in areas where statistical analysis might have been a first choice; most appropriate are areas where a pragmatic approach is satisfactory, or indeed the only one possible. A particular and powerful use is in applications involving pattern recognition. Application can also be very useful where mechanisms are unknown or too complex, and where extensive data are available. They are more accommodating to noisy data than statistical packages. However, prediction should be kept within the bounds of the training region. Neural networks can be used with chemometrics packages, and modules are now offered as an alternative or adjunct to statistical packages (Unscrambler has Neural-UNSC as an add-on module). A powerful application is also in the complete process design environment. Commercial software is now available to be used in conjunction with experiment and simulation in understanding and optimization and in formulating [21, 22] (CAD/Chem Custom Formulation System, AI Ware Inc). The outstanding feature of these programs is their ability to 'learn', and for this reason there is a visible and growing use in process control applications [23, 24, 25] (see also later in this chapter). Fuzzy logic is also applied here, and the papers cited [26] should be referred to.

20.5   Molecular modelling

Molecular modelling is included here, though as a technique it may have limitations for the practising polymer chemist. Molecular modelling determines the conformations and interactions of materials from calculations at a molecular level. Desktop computer programs are available which can optimize the conformational information on simpler molecules and provide useful 3D visualizations (Molecules-3D, Molecular Arts Corporation; Chem-X, Chemical Design). By contrast, almost all calculations concerning polymers require larger amounts of computing power so that graphic workstations, rather than personal desk computers have been required. Also applications to real situations (solvation, interaction with surfaces) are complex in calculation and in interpretation and hence in this instance molecular modelling methods are useful more as tools suited to those engaged in theoretical studies. However, the potential is enormous as the examples below will show.

Molecular modelling encompasses a range of molecular orbital calculations, and may be either *ab-initio* or 'semi-empirical' in approach; the methods used for calculations for larger molecules (polymers) will be found discussed in reference [27]. Visible developments that have come to the author's attention, of potential use for the polymer chemist include calculation of:

- Q and e values for chain addition polymerization [28],
- propagation and transfer rate constants in polymerization [29],
- partition coefficients [30], (used for determining monomer partition in emulsion polymerization),
- rates of free radical initiator breakdown [31],
- bond energies for allylic hydrogen substitution [32] (in evaluation of allyl ethers as potential autoxidative oligomers),
- T_g of acrylate and methacrylate polymers [33],
- thermal deblocking rates of masked isocyanates [34],
- autoignition temperatures [35],
- polyol structure effects on polyester durability [36].

Fullest use is developing for conformational structures of polymers and those interested may refer to a comprehensive example [37].

More immediately applicable data from molecular modelling of use to paint technologists has been obtained in the area of solvents and solubility. For example methods for gaining UNIFAC predictions of vapour liquid equilibria (VLE) [38] have been developed. Hansen solubility parameters for surfactants, and for cellulose derivatives used as thickeners, have been determined using POLYGRAF (Molecular Simulations) software [39].

20.6   Resin formulating and processes

20.6.1   Step growth polymerization

Many features of resin formulating and processing can be simulated, and final resin properties estimated or predicted. These features vary from individual facets, to full systems and their process kinetics; some of these calculations are essential to safe and effective formulation and manufacture.

For step growth (condensation) polymerization detailed stoichiometry calculation is part of formulating and has already been described in Chapter 2; both calculation prior to, and control during preparation, are essential to ensure that required molecular weights and the required end-group concentrations are attained, and at the same time avoid gelation. While calculating basic factors and stoichiometry (including molar ratios, oil length, excess hydroxyl content) may be simple, reverse calculation and calculating the likelihood of gelation is more complex [40].

Various equations have been produced as gelation theory has developed, by for example Flory and Stockmeyer, based on statistical methods (see Section 2.5.5). These provide the means to calculate a trial formulation giving the extent of reaction at gelation, most usefully expressed as the acid value at gelation (AV gel); in many formulations this will be around five units below that desired in the final resin. Formulating variables will be the hydroxyl content, expressed as % excess OH or hydroxyl value, or the average functionality K, adjusted in order to alter the AV gel.

As well as calculation of water of reaction and final yield, following on from gelation calculation it will be possible to calculate the molecular weight averages M_n and M_w at given degrees of polymerization. While the M_n calculation is simple, the M_w calculation is more complex [41]. (Users should note the corrected form of the calculation in the second paper cited.)

Reaction is normally followed in the practical situation by the progress of acid value (AV) which decreases as polymerization proceeds. Following calculation of molecular weights, branching functions and functionality of reactive groups for the average molecule may be calculated as very useful indicators of practical performance.

Calculation may also be required to account for multi-stage processing, where following partial condensation, further reaction with for example a di-isocyanate, epoxy resin or dicarboxylic anhydride may be required to produce urethane oil, epoxy capped low AV resin, or high AV water soluble resins respectively.

Those formulating alkyds (and polyesters) regularly are likely to use computer programs to calculate both stoichiometry and gelation parameters, incorporating the above concepts. While programs of high functionality have been written [42], writing one's own programs is quite feasible, and simpler programs are available from raw material suppliers [43]. Model schemes are being developed by Ray to cope comprehensively with all classes of condensation polymerization [44].

Compared to chain growth polymerization (see section 20.6.2), there has been surprisingly little investigation of species distributions in step growth copolymerization, especially since unequal reactivity of species is well recognized. Presumably this is due to the absence of any scheme comparable to the reactivity ratios used in the former. A few pre-gelation studies have been carried out, however, and Monte Carlo investigation [45] is here again a powerful tool for investigating these features, and has been used for examining monomer sequence distribution [46, 47]. The author has found a simple model, following O'Driscoll's methodology [47] useful for example for modelling distributions of species after monoglyceride equilibra-tion, and for modelling linear polyester preparation.

While the *process kinetics* of polyester and alkyd step growth processes is not generally modelled, some specific examples exist, for example for unsaturated poly-ester [48]. Monte Carlo simulation has shown significant potential for insight into the detailed features of this type of polymerization, and can include an overview of the process kinetics (see also section 20.7).

A number of instances of detailed modelling of epoxy curing reactions however exist, in some cases involving computer use [49].

20.6.2   Chain growth polymerization

Many features of chain growth polymers and polymerization can be modelled. In formulating, calculated T_g will have been a principal consideration, modelled using the Fox or Johnston equations. The Johnston equation requires knowledge both of the diad distribution of the polymer, and most importantly the T_g of the strict alter-nating copolymer, as well as that of the respective homopolymers. The T_g calcula-tion is easily carried out predictively. In the case of emulsion polymerization this may possibly be with data from a more complex partial polymerization model such as that developed by Guillot [50].

$$\textit{Fox Equation} \quad \frac{1}{T_g} = \frac{W_1}{T_{g1}} + \frac{W_2}{T_{g2}} + \text{etc.}$$

$$\textit{Johnston Equation} \quad \frac{1}{T_g} = \frac{W_{11}}{T_{g11}} + \frac{W_{22}}{T_{g22}} + \frac{W_{12}}{T_{g12}}$$

where, W_1 etc is the weight fraction of monomer 1 in the copolymer and T_{g1} its T_g. W_{11} etc is the weight fraction of diad 11 in the copolymer, and T_{g11} the homopoly-mer T_g. W_{12} is the weight fraction of diad 12 in the copolymer, and T_{g12} the alternat-ing copolymer T_g.

A point for care in the calculation of a copolymer T_g is to ensure that homopoly-mer T_gs applied have been determined on high molecular weight material. T_g is chain length (Xn) dependent and a number of complex relationships have been developed [51, 52]. O'Driscoll has recently shown that a simpler empirical model [53],

$$T_g = T_{g\,\text{inf}} - \frac{k}{Xn^{2/3}}$$

where $T_{g\text{inf}} = T_g$ at infinite molecular weight and k = polymer-dependent constant, is applicable to a number of useful polymers for which he gives values of k. The author has used this to illustrate molecular weight effects to users of Fox calculations who would otherwise have been uncritical of calculation results. While T_g is decreased by branching compared to the linear molecule, it is of course raised in many prac-tical polymer applications by crosslinking [54] (see section 20.8).

Sequence structures are readily calculated and reactivity ratios are available and generally adequately reliable. Harwood *et al.* [55] and more recently Cheng *et al.* [56] have published models for normal polymerization. The importance of sequence distribution is not always fully appreciated; favourable distributions, for example, minimize the effects of hydrolysis and neighbouring group reactions. It is also self-evident that in most instances, functional groups should be well spaced along the polymer chains, otherwise properties such as hardness and solvent resistance following crosslinking will be affected. Cumulative diad and triad content calculations can be useful for T_g and NMR investigations respectively. Using a sequence model and the Johnston equation, it is possible to predict thermogram shapes from the polymer prepared from a given process [57].

The Arrhenius equation for initiator breakdown

$$k = A\ e^{-E/RT}$$

and the equation for initiator concentration

$$[I] = [I]_0\ \exp(-kt)$$

where A = pre-exponential constant, E = Activation Energy, where t = elapsed time from t_0, K = breakdown rate, T = temperature

can provide a useful radical production rate model in process design, and manufacturers' literature provides starting data. This is readily extended to modelling radical production rates in practical situations, which typically involve a combination of shot, feeding and holding regimes, sometimes complicated by changing temperature. This is illustrated in Fig. 20.3. These figures show the typical decay in radical production rate following a shot of initiator and the slow rise to a steady rate of radical production from a feed of initiator. Only when an initial shot and a feed are combined, can a steady rate be achieved throughout a process. This analysis is particularly useful in the 'mopping up' of unreacted monomer near the end of

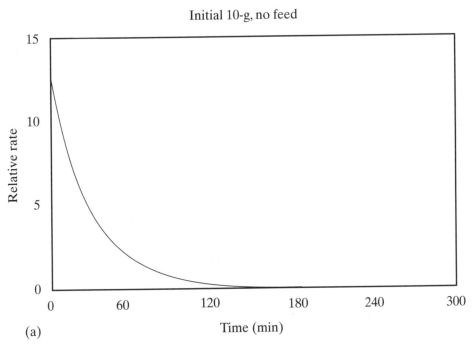

(a)

Fig. 20.3 — Radical production rates with AIBN, $t = 90\,°C$.

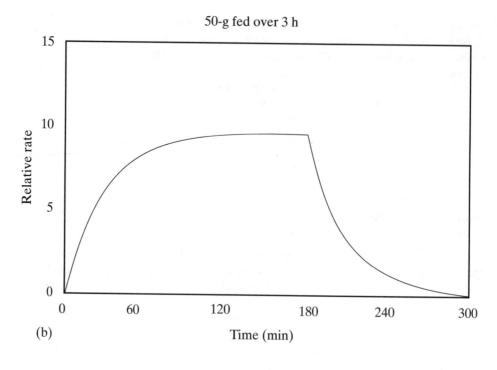

(b)

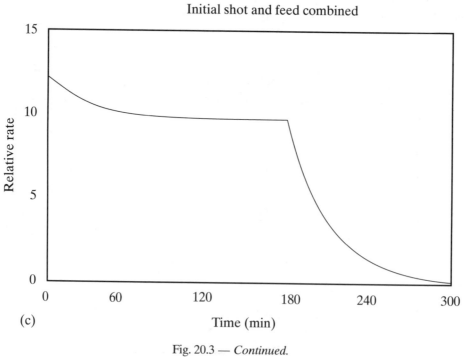

(c)

Fig. 20.3 — *Continued.*

a process, when a radical concentration appropriate to the amount of free monomer actually present may be determined.

Use of this approach for redox initiation systems is not easily achieved because of lack of data for most systems in actual use.

Consideration of the heat of reaction, extended to the heat produced or required in the overall process, and the in-process heat balance, are important for safety, scale-up, and plant and process design. The need for care can be illustrated by translating the relatively similar molar heats of polymerization of most acrylic monomers, and the heats of evaporation of xylene and water, into heats per unit mass for these materials (Table 20.1).

With practical knowledge of the process, including feed and distillation rates (if any), useful process heat balances can be estimated. It can be vital to apply this calculation at the formulation stage, especially for 'one-shot' methods, when the process can be designed around a maximum temperature or temperature rise, with the formulation at optimum solids and initial monomer concentration.

The factors in considering an overall heat balance are shown in Table 20.2. It is easy to calculate that with no applied cooling, a typical vinyl acetate/2-ethyl hexyl acrylate latex seed stage will exotherm by 22 °C. Similarly, with the assumption that monomer is consumed at the same rate that it is fed, in the preparation of a methyl

Table 20.1 — Typical heats of reaction and evaporation

Reaction	Heat per mol	Heat per kg
ROH + R'COOH > ester	+6 to +12 KJ mol^{-1}	+60 kJ kg^{-1}
Anhydride + ROH > $\frac{1}{2}$ ester*	−42 KJ mol^{-1}	−200 kJ kg^{-1}
Epoxide + acid > Mono ester**	−77 KJ/epoxide	−170 kJ kg^{-1}
Epoxide + anhydride > polyester***	−108 KJ/epoxide	−310 kJ kg^{-1}
R-NCO + ROH > urethane****	−37 KJ/NCO	−230 kJ kg^{-1}
Addition monomer > polymer	−50 to −90 KJ mol^{-1}	−200 to −1500 kJ kg^{-1}
Heat of vaporization xylol at 250 °C	+32 KJ mol^{-1}	+302 kJ/kg^{-1}
Water at 250 °C	+32 KJ mol^{-1}	+1780 kJ kg^{-1}

* Phthalic anhydride with butanol.
** Epikote 828 (Shell) with linseed oil fatty acid.
*** Phthalic anhydride with Cardura E10 (Shell).
**** TDI and butanol

Table 20.2 — Factors in estimating process heat balances

Thermal feature	Reaction no reflux	Reaction with reflux	Reaction process Alkyd fusion process	Alkyd solvent process	Typical acrylic process
Heat of reaction	✓	✓	✓	✓	✓
Heat loss from vessel	✓	✓	✓	✓	✓
Heat-up of incoming feeds					✓
Solvent distillation (latent heat)		✓		✓	✓
Reheat of returning solvent		✓		✓	✓
Heat loss from vapour stack		✓		✓	✓
Water distillation (latent heat)			✓	✓	

methacrylate copolymer by a typical solution polymerization process on a 10000 kg scale, a net heat production rate of 125 kW can be calculated. In the absence of applied heating or cooling, this could sustain solvent reflux at 1300 L/hr.

Data for many commercial monomers are not available, though in the author's experience, *molar* heats of polymerization may be satisfactorily estimated by considering chemical class. Thus methacrylates fall in one band of values, acrylates in another. Certain heats of reaction may be calculated from heats of formation. One unresolved issue is the effect of *copolymerization* in addition polymerization [58] and it is unclear whether this effect is significant.

Chain growth copolymerization processes are readily modelled from both reactivity ratios and from Q and e values by straight calculation [59] and by matrix methods [60]. Problems with the Q and e scheme have already been referred to (see 2.7.2) where the 'Revised Patterns of Reactivity Scheme' may well be a more satisfactory alternative. Carrying out these calculations predictively to show charge conditions required to achieve a given composition is also straightforward. An example of this is as follows. Figure 20.4 illustrates the modelling of the 'Instantaneous Copolymer Composition' as conversion to polymer precedes, for an initial monomer composition of methyl methacrylate (MMA), vinyl acetate (VA) and butyl acrylate (BA) of composition MMA/VA/BA 45/25/30 by weight. This shows that MMA is consumed fastest, it's concentration in the copolymer steadily reducing as conversion proceeds. By contrast, BA content initially rises but peaks at around 80% conversion and then falls; VA content rises steadily throughout and the final polymer is 100% VA in composition. As shown, polymer most rich in each of the three monomers in turn is formed at different intervals during polymerization. Calculation shows that a constant copolymer composition could be achieved during conversion by feeding MMA/BA 42.5/27.5 at a steady rate into an initial charge of

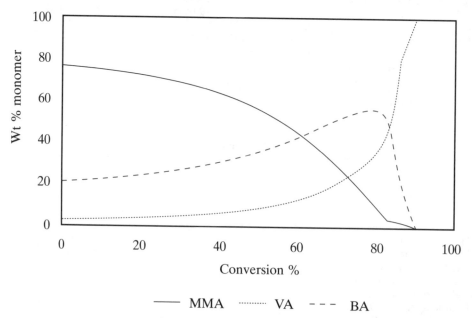

Fig. 20.4 — Instantaneous copolymer composition from polymerising MMA/VA/BA mixture.

MMA/VA/BA 2.5/25/2.5, providing polymerization also occurred smoothly during the feed period. (The simulation is not shown, since the respective graph would simply have three horizontal lines). This technique is effective in situations where as here, one monomer (VA) is low in reactivity and can only be fully copolymerised by being present all in the initial charge; only the faster reacting monomers are fed.

Simpler programs have been published [61] for manipulating reactivity ratios, and Q and e values; a spreadsheet method has recently been disclosed [62]. A program is available for the determination of reactivity ratios from experimental data by the more rigorous 'error in variables' method [63], giving more valid results than the regression methods still used.

Monte Carlo methods have been used by O'Driscoll and others, to simulate both irreversible and reversible polymerization [64], and these can give copolymer composition and sequence distributions [65], and even molecular weight distributions [66]. These latter can include useful analysis of such processes as random branching by transfer to polymer [67]. The methodology assuming the irreversible terminal mechanism in binary polymerization for illustration, involves the calculation of probabilities as follows

$$P_{ij} = \frac{1}{1 + (r_{ij} * x_i / x_j)}$$

where p_{ij} is the probability of a chain ending in M_i adding monomer M_j, and $p_{ii} + p_{ij}$ = 1. x_i and x_j are the mol fractions of monomer present in the feed.

A unique feature of this technique is the ability to analyse a process for compositional heterogeneity, and to calculate, for example, the number and weight of polymer chains not containing functional monomer, which is illustrated below. In this example which might not appear atypical in formulating higher solids systems, taking none and one-functional polymer together, it is seen that over 10% of the polymer cannot fully contribute to crosslinking. Some 4% of material (that which is non-reactive) will remain solvent extractable, even with all functional groups present having been linked into the network (see also the example of Miller–Macosko analysis given later). This analysis also allows molecular weight distributions to be modelled and Fig. 20.5 shows the result of both simulations.

Kinetic modelling of addition polymerization can give the fullest data on process conversion and molecular weight and property development in a time frame, and the kinetic processes are relatively well-defined [68]. However, uncertainties still exist in some areas such as in details of copolymerization mechanism, in solvent effects and where reversible reactions occur. The range of parameters required is large, even for the simplest process. Schemes exist to assemble a full range of models for addition polymerization processes [69].

Kinetic model assembly can be illustrated from the following simplified scheme:

Initiator decomposition $\quad I \xrightarrow{k_d} 2\,R_0{}^*$

Initiation $\quad\quad\quad\quad R_0{}^* + M \xrightarrow{k_i} R_1{}^*$

Propagation $\quad\quad\quad R_n{}^* + M \xrightarrow{k_p} R_{n+1}{}^*$

Termination $\quad\quad\quad R_n{}^* + R_m{}^* \xrightarrow{k_t} P_{n+m}$ or $P_n + P_m$

The simplified set of equations for concentration of initiator, monomer and free radicals then becomes

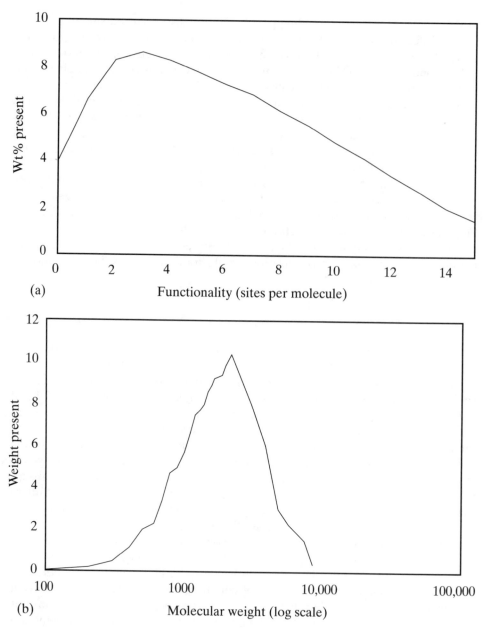

Fig. 20.5(a) — Distribution of functional species. 5(b) — Molecular weight distribution of one-functional polymer, M_n 1038, M_w 1944. System containing 10% by wt. of functional monomer overall. Total polymer M_n 5014, M_w 9960. F_{av} of chains = 7.7 (number average).

$$\frac{d[I]}{dt} = -k_d[I] + F_I$$

$$\frac{d[M]}{dt} = -k_p[M][R] + F_M$$

$$\frac{d[R]}{dt} = 2k_d[I] - k_t[R]^2$$

where F_I and F_M are input flows of initiator and monomer respectively.

The above has ignored the necessary volume term, but has included flow terms for initiator and monomer, so that the reader will see that this is immediately applicable to many practical situations. In practice, chain transfer terms (to solvent, polymer, chain transfer agent) are required, plus consideration of the two modes of termination.

To match results more seen in practice, it is necessary to expand the above to include most importantly diffusion effects on propagation and termination, as well as initiator efficiency and inhibitor effects. Copolymerization with a number of monomers expands the number of equations, as does adding calculation for molecular weight and sequence distribution. To appreciate the complexity of the calculation scheme fully, the reader is directed to the literature [70, 71].

A large literature exists describing mechanisms and models [72, 73, 74], though available models may currently be for narrowly defined compositions or processes. Only full kinetic models can predict the molecular weight averages (M_n and M_w). Where complete process models are available, these can be extremely powerful both for the experimenter, and for the scale-up/production engineer. Experimental time for new process and process change investigation will be dramatically reduced, and what-if simulation can be carried out to investigate safe plant loading and simulate such occurrences as transfer and cooling failures. Even basic kinetic models can be used to great advantage in graduate teaching [75], and in new process examination [76].

Inspection shows, however, that data is scarce for systems actually used in the paint industry, and for many of the reactive monomers and transfer agents used, little or no data exists; data is poor or missing for almost all monomers at the higher temperatures used in practice. Similarly it is doubtful whether current kinetic schemes with assumptions of chain length independence of kinetic constants, and long chain approximations are valid for the lower molecular weight materials (oligomers) now of interest for high solids paint applications. With one exception [11], most models disclosed are only apparently effective with a simple monomer system and probably not useful within the coatings industry.

20.6.3 Emulsion polymerization

Modelling of emulsion polymerization adds the complication of reaction in two phases, with the need to define the volumes and compositions of those phases, the particle size of the disperse phase, and the equilibria and transport occurring between the two phases. Again, a range of kinetic models exists for specific systems [77, 78]. Here while the mechanistic schemes are rapidly becoming more refined and applicable (uncertainties being around particle nucleation and issues such as radical transport), a number of parameters will remain experimentally inaccessible or have

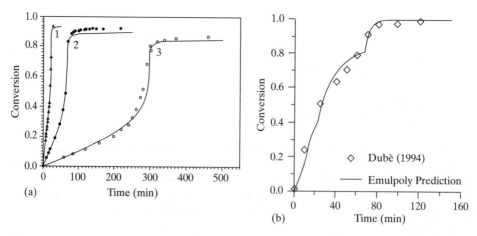

Fig. 20.6 — Comparison of kinetic model prediction *vs* experimental results (simulations shown by permission of A. Penlidis, University of Waterloo). (a) Experimental and simulation results for the bulk polymerisation of methyl methacrylate. Continuous line, simulation; discrete points are experimental (Balke S T & Hamielec A E, *J App Poly Sci* **17** 905 (1973). (b) Experimental and simulation results for the emulsion polymerisation of BA/MMA/VA. Continuous line, simulation; discrete points are experimental (Dubé [171]).

to be adjusted to a much greater extent than for bulk/solution models [70]. While general models remain imprecise, individual specific models exist, useful for the experimenter and for process modelling. Complex models are now emerging [79]. Only kinetic models can predict molecular weight and particle size completely.

Where the full kinetic models are unavailable, simpler partition models predicting microstructure, composition, and even molecular weight, have been developed, initially by Guillot [50, 80], and more recently by German [81], and Storti [82]. Because of the phases present (three initially, if including monomer droplets), copolymer composition calculation must be preceded by calculating phase composition, and these models use partition coefficients, monomer solubility and maximum swelling ratios for this purpose.

Monte Carlo modelling has also been applied by Tobita to molecular weight simulation [83, 84], and to microgel formation [85, 86] in emulsion polymerization. Models for the prediction of particle morphology are now well developed [87]. Both the Guillot model described above and kinetic models [88] have been found useful in morphology prediction.

The current possibilities of kinetic modelling for both solution and emulsion polymerization are illustrated by results by Penlidis and co-workers [89] in Fig. 20.6.

20.6.4   Control of polymerization

While complete automated control of polymerization reactors to achieve user requirements (closed loop) may be an ultimate goal, formulators regularly apply formulating 'policies' which are in effect open-loop control [90], in developing formulations. Hence composition, molecular weight and conversion control can all be achieved by applying such policies, and predictive techniques which will be used as computer programs can be applied to a number of these. Thus the standard practice of 'starved feed' polymerization, where short half-life initiators are increasingly used has been thoroughly analyzed [91]. Composition control (as illustrated earlier) is relatively straightforward in solution polymerization [70, 92], where reactivity ratios are accurately known, and control can be extended throughout the process. The minimum requirement is that the system has been characterized for standing monomer levels during polymerization. Policies for optimizing the completion of conversion, and for molecular weight control have been proposed [93, 94].

In emulsion polymerization, much has been achieved in composition control, which has been applied using partition models [80]. Some form of reaction monitoring has been found advisable for the most satisfactory results, with gas chromatography being the technique favoured by most workers. Storti in applying his model-based control used ultrasound [95]. Asua has worked with co- and ter-polymer models in composition control [96]. The current status of process modelling and control in emulsion polymerization has been recently reviewed [97].

20.7   Resin scale-up and manufacture

As already mentioned, for exothermic processes, heat balance calculations may readily be carried out, and will be a first reference for safety considerations. These calculations can be sufficiently reliable to be used for both the overall process and for stages in the process where steady conditions can be expected. Table 20.2 showed the factors that need consideration in calculating a heat balance; using these, first estimates of heating/cooling loads and cooling reserves can be determined. The major goal for attempting fuller kinetic modelling has been to load plant productively and to control exothermic processes fully [11].

Pilot plants now demand full data logging in order to provide kinetic and thermal data and may have partial or full computer control, and allow safety features to be tested [98, 99]. The least that should be provided is the monitoring necessary to determine the heat balance (summation of heat supplied by heating surfaces, removed by coil and condensor, etc.) at all times, and to record this data. Ideally experimenters should also have access to controlled and logged reactors at the laboratory stage and commercial laboratory reaction calorimeters are available (RC1, Mettler-Toledo).

Computer control of full-scale resin plant is now normal and provides material transfer, agitation rate, heating/cooling and process sequence control, plus many safety features. All types of plant have now had computer control fully implemented including alkyd [100], latex [101, 102] and multi-purpose resin plant [103].

As well as its use for process analysis, and its predictive use in formulation and scale-up, a further use of kinetic modelling is linkage into control systems to enable closed loop process control [104]. This brings in issues including the need for *in-situ* sensors for chemical and polymer properties where the requirements and status were first assessed by Penlidis and have more recently been re-assessed [105].

The scheme for extended reactor control to ensure consistency of properties such as MW and composition using state estimation is illustrated in the block diagram in Fig. 20.7. A typical state estimator would include a Kalman filter incorporating a simple model of the process.

Significant progress is being made both in the development of models suitable for process optimization and in control algorithm development [106]; dynamic neural network models have recently been explored [107]. Calculation relating to optimization is well advanced, and a large literature exists, and the interested reader is referred to the review cited [108].

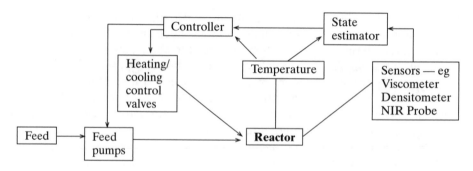

Fig. 20.7 — Block diagram of extended reactor control system.

Working on the larger scale involves concern with heat transfer. Asua has extended his emulsion polymerization partition models to deal with situations where heat removal capability is limited [109], and has also critically reviewed the broader situation with extended emulsion polymerization reactor control [110].

Chemists and chemical engineers involved in scale-up and problem solving on the full scale will at some stage become concerned with mixing considerations, and flow simulation by computational fluid dynamics (CFD) is well-established [111]. Relating reaction kinetics and mixing is an ultimate modelling goal, and given the development still required in the basic kinetic models, it is not surprising that progress has been slow. Study of micromixing and effects on polymerization have been published [112] and developments are evident for addition polymerization reactions, where analysis has shown how compositional distribution will be affected [113]; in semi-batch processes, the effect on molecular weight distributions has been modelled [114]. Those concerned with emulsion polymerization on the larger scale will be fully aware how critical heat and mass transfer, mixing, control mechanisms and methods of feed addition are in emulsion reactors [115]. The pressing need of those scaling up disperse polymers is for help in controlling or containing particle coagulation, manifested variously as size distribution broadening, seed and bit formation, or at the extreme polymer deposition and fouling on the reactor walls. While micro particle coagulation is recognised to be part of the nucleation process, coagulation processes occurring in the later stages of feed processes, through incomplete surfactant coverage or through shear instability are neither wanted nor in many cases can they be fully controlled. Detailed phenomenological studies have been made of coagulation and mixing [116], and some progress is evident in understanding and modelling coagulation [117]. An overview and industrial perspective has recently been provided [118].

20.8   Polymer properties, curing and network properties

Polymer property prediction, which in the future is likely to gain much from molecular modelling methods, currently relies principally on a range of group contribution techniques. This methodology assumes that component groups contribute on a mainly additive basis and can have a sound theoretical basis where the groups are chosen sensibly. In the case of a polymer, this means taking structures such as (CH_2), (COO) etc. as sub groups, and in most cases treating backbone (main chain) and side chain (branch) constituents separately. Thus characteristics such as T_g, density, refractive index, surface tension and solubility parameter may be estimated [119]; this ability has been incorporated into a screening technique for candidate structures with a required set of properties [120]. Chem-Draw has a property module included (see Section 20.3); Polymer Design Tools (DTW Associates) claims to estimate 30 physical properties of polymers. While many group contribution methods give useful results, they can often lack precision and should only be used where no other data are available. For safety, results should be inspected against known experimental values from similar structures and possibly then adjusted. Where more than one method is available, the most accurate should be chosen [121].

Most paint polymer films only attain their properties from a curing or crosslinking reaction, whereby MW is increased until an infinite 3D structural network is formed; the crosslinking process and analysis of the final structure has had much attention. Nevertheless, mechanical property measurements on thin paint films are not easy, and films are invariably inhomogeneous and often multiphase. Thus neither the modelling nor the characterisation of the network can be precise at the present time, though this area is vital to the properties of coatings.

Models for network analysis have grown out of Flory's statistical theory, through Gordon's probability functions, to Dusek and Eichinger's non-lattice Monte Carlo approach. Duseks approach has most recently been applied to the modelling of the formation and crosslinking of star oligomers [122]. The most attention has been given to the recursive approach of Miller and Macosko [123, 124], and simple source code is available in some of these references. A number of authors have shown results of investigations linked to this technique with examples for urethanes and M/F resin curing [125]. The Miller models are comprehensive and embrace both the pre- and post-gelation situations, and can include various reactant combinations and reactivities. Post-gel properties include extensive analysis of the network structure for number and type of crosslinks, and pendant and sol fractions. Miller has extended the technique by linking a kinetic model of pre-gel functional polymer production by addition polymerization to the post-gel situation [126]. The formalism is illustrated in Fig. 20.8. Example output graphs are shown in Fig. 20.9.

The example shows the random crosslinking of long chain growth polymer chains incorporating a comonomer with a reactive group, which is capable of reaction with another similar group. The copolymer in this case had a negative binomial molecular weight distribution. As applied here, this methodology assumes equal reactivity of all sites, an absence of substitution effects and that condensation by-products are absent or of negligible weight.

As an alternative to the rather complex Miller/Macosko method, Hill has developed a modification of the Scanlan equation for the calculation of crosslink density [127]. Unlike Miller/Macosko (Fig. 20.9b), this equation however has more limited application, and does not account for sol fraction.

The calculated information on the network of effective strands (or between elastically effective network junctions) at full conversion can be used with simple relationships to get a first estimate of the T_g change due to crosslinking by the relationships of either DiMarzio or Nielsen [128].

$$\Delta T_g = \frac{T_{g0} \times K \times M \times \chi/\gamma}{1 - K \times M \times \chi/\gamma}$$

$$\Delta T_g = \frac{3.9 \times 10^4}{M_c}$$

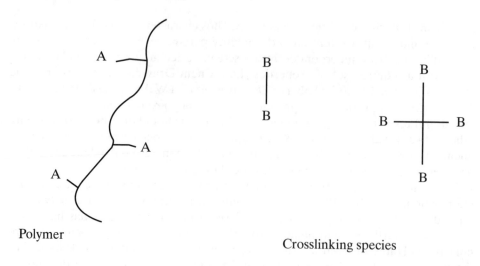

Polymer

Crosslinking species

Reaction constituents

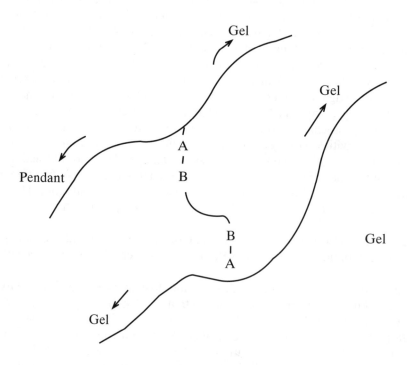

Elastically effective 'node' with pendant branch

Fig. 20.8 — Miller–Macosko formalism showing species involved and reacted gel.

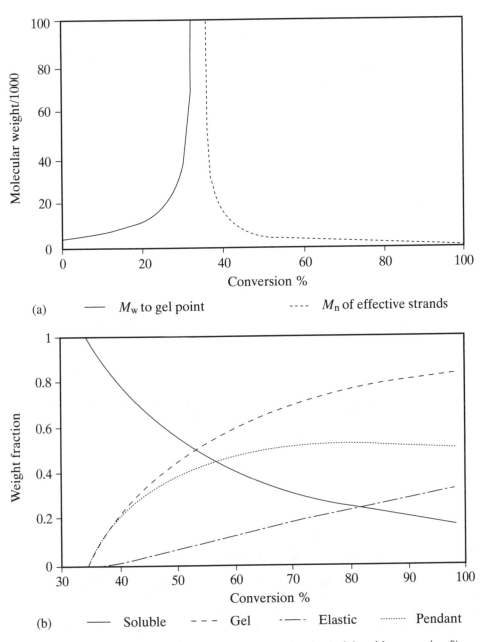

(a)   —— M_w to gel point   ---- M_n of effective strands

(b)   —— Soluble   --- Gel   ·—· Elastic   ······· Pendant

Fig. 20.9(a) — Pre- and post-gelation changes in molecular weight with conversion %.
(b) Post-gelation soluble and gel fractions, and concentration of pendant and
elastic chains. Non-functional monomer molecular weight 80, functional
monomer molecular weight 100, 5% present by wt. Degree of
polymerization = 40.

DiMarzio's Simplified Relationship
where $T_{g0} = T_g$ of uncrosslinked polymer

Nielsen's Relationship
M_c = No average molecular weight between network junctions

χ = no of crosslinks per gram
M = molecular weight of residue
γ = no of flexible bonds per residue
K = constant, aprox. 1.3×10^{-23}

Monte Carlo simulation of network formation is another approach, which can be successfully applied [129] to a range of multicomponent systems with both competitive and sequential reactions. Obvious application is to step growth polymerization, where polyurethane and epoxy/amine reactions are amongst those that may be modelled [130]. Commercial software to model pre- and post-gel polymerization is available such as DryAdd (Oxford Materials). Monte Carlo models of this type are effective in modelling rate effects and can include the effects of feeding reactants and of temperature variation. DryAdd can also model chain growth poly merization and can be used with CAD/Chem (mentioned earlier) as a preparation/property prediction package [131]. This type of model is valuable in exploring species and sequence distributions, as referred to earlier. Other approaches have been applied to this very complex area and have been reported [132]. Network formation analysis by Monte Carlo methods in emulsion polymerization crosslinking has already been mentioned.

As described above, both Miller/Macosko and Monte Carlo methods are powerful in analysis of reaction processes. Current limitations of network models are, however, in their treatment of inter- versus intra-molecular reactions, and these can only as yet be treated approximately [133]. In practical situations, steric and geometric factors can dominate; these along with diffusion effects make kinetic prediction difficult. Hence, prediction of conversion at gelation and gelation *times* are not achieved accurately from first principles. However for reaction of step growth monomers to intermediate molecular weight, and for lower levels of crosslinking of well-defined prepolymers these analyses can give very useful insight.

20.9   Solvents and solubility properties

Paint systems comprise polymer, pigment and solvent, the solvent system typically being a mixture; release of that mixture is a normal part of the film formation process. Typically, it will have been necessary to prepare the polymer component in at least part of that solvent mixture. Calculating the properties of solvent mixtures has inherent difficulties due to non-ideal mixing behaviour of solvents; it is most difficult if water is one of the solvents. In considering full paint systems, the polymer content contributes added complications; solvent–polymer interaction occurs, and in later stages of drying, the presence of polymer and the occurrence of crosslinking reactions, raise system viscosity, restricting molecular mobility and slowing evaporation. At the limit, if the polymer is glassy, and the system then becomes glassy, solvent loss ceases and solvent can be retained in the coating.

Modelling and computer prediction of straight solvent evaporation behaviour owes its development to chemical engineers' need for accurate VLE data for calculations on distillation processes, and to developments in UNIFAC theory [134].

Hence prediction of solvent evaporation, with the ancillary calculation of mixed flash points, viscosity, solvency power, surface tension, etc., is well-developed [135]. Paint manufacturers will find that solvent suppliers have computer models to which they may gain access [1, 136–139]. Prediction of some properties such as autoignition temperature however may seem elusive, but here molecular modelling has been used with some success [36]. The capability of current evaporation models is illustrated in Fig. 20.10. In this model (BP Solve version 3.1, BP Chemicals), polymer solvent interaction related to evaporation is modelled very simply. The model, however, incorporates excellent provision to adjust for the effects of application including spraying under a variety of conditions, and onto different substrates [140].

Solubility prediction remains in the domain of solubility parameters (see Chapter 4) where the Hansen 3D approach is most used [141, 142]. BP Solve 3.1 now also includes a good range of commercial resin data, and presents Hansen plots with solubility spheres, following calculation of the mixed solvent parameters. Where not determined practically, group contribution methods are used extensively for determining solubility parameters [118]. Hoy has described his method, which is applicable to both solvents and polymers; he has illustrated its use as a design tool in formulating water-borne coatings [143]. Complex models deal more rigorously with polymer/solvent interaction. Wallström has developed a solubility and evaporation prediction model (dG, EnPro ApS) which can deal with the presence of polymer, and even handle multiphase systems [144].

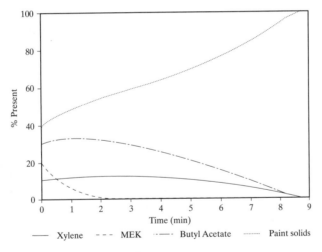

Calculated properties of initial solvent mixture

Hildebrand solubility parameter		9.006	Density (kg/l @ 20C)	0.851
Hydrogen bonding index		8.264	Viscosity (cP @ 20C)	0.607
Fractional polarity		0.235	Vapour pressure (mbar @ 15.0C)	38.5
Hansen solubility parameter	δ_D	7.98	Mixture VOC (g/l)	599.4
	δ_P	2.49	Flash point °C	4
	δ_H	2.56		

Fig. 20.10 — Modelling of evaporation of polymer solution containing xylene, methyl ethyl ketone and butyl acetate at 15° following application by dipping (by permission of BP Chemicals).

Prediction of polymer solubility in water-based systems is a goal in which progress is being made [145]. The solubility parameter approach has been used in the study of phase separation of polymers, which has been exploited in the development of self-stratifying coatings [146].

An area of current interest is the efficiency of coalescing aids, which is related to their location. Toussaint has disclosed a method for predicting the distribution coefficient of these aids in disperse systems [147]. Partitioning has also been simulated using UNIFAC, where a recent example has shown how Texanol (Eastman) partitions almost totally into the polymer phase, while butoxyethanol is more strongly partitioned into the water phase [148].

20.10   Paint formulation, manufacture and use, and coating performance

Just as in resin formulating, manipulation of weights and volumes, and calculation of PVC, extender and drier ratios, etc., merit the use of computer programs in coating formulation. The mathematics is straightforward but a typical paint formulation requires the handling of solutions and ingredients (e.g. driers) whose active component is less than 100%. Techniques of handling this using a computer have been discussed [149] and commercial software is available for formulation [150, 151]. Formu-Tools (DTW Associates) is an example [152]. Aids are available for more complex situations such as predicting the effect of the inclusion of opaque polymer beads, whose effect is only obtained by water loss from their core on drying [153, 154].

Modelling and property prediction in overall coatings formulation is not easy, and use of experimental and statistical design and analysis is essential practice. This has been used for evaluating total performance, and predicting service life [155, 156]. Neural network modelling has recently been applied to formulation optimization [157]. Specific areas such as the effect of rheology modifiers have been studied [158]. With coatings being multiphase, and many properties dependent on ratios of pigment/extender/polymer, and on the spacing of the first two particulate components in the polymer continuous portion, a number of notable attempts have been made at modelling packing, and aggregation and cluster effects, and associated colloidal behaviour by Monte Carlo techniques [159–162]. Film formation via the coalescence of disperse polymer has also been extensively studied and models produced; this topic is however beyond the scope of this chapter.

Colour measurement and analysis and in particular the mathematics of colour are not capable of coverage here, but are of course integral in paint manufacture. Development in light scattering theory continues to be reported [163], and the use of computational methods in colour matching is fully discussed in references [164] (see Chapter 19).

Attempts continue to correlate weathering performance to various aspects of composition and exposure, and durability testing is covered elsewhere (see Chapter 19). In the narrow context of computer modelling, a recent paper has shown the simulation of surface erosion related to variations in pigment clustering [165].

20.11   Experimental analysis, design and quality control

This section has been left to last because while experimental design and analysis are obvious mathematical and computational applications, it was not the intention to discuss details of applying them in this chapter. It should however be understood that all paint and resin experimentation and testing must benefit from the application of good design and statistical analysis of results. Doubts are sometimes raised because of the simple and subjective nature of much testing, but this endorses rather than diminishes the reasons for good practice. Readers interested in the overall methodology are referred to the many good guides available [166]. Sophisticated software aids are now available for graphing, curve fitting and for full statistical analysis (Statgraphics, Statistical Graphics Corporation; SYSTAT, SPSS).

Numerous application examples exist in paint formulation [167, 168] and in polymer process investigation [169, 170]. Good examples are evident of application of refined technique in, for example, polymerization kinetic studies [171] and parameter investigation [172]. The obvious benefit of applying this methodology is the ability to optimize performance and properties, not the least for cost, which is often the ultimate criterion for commercial viability for a formulation.

The requirements of quality system registration (ISO 9002) mean that most paint producers will use statistical process control (SPC) in manufacture, in order to demonstrate good control of their production processes [173]. SPC requires the collection of process intermediate, and production and product test data and then the calculation of statistical distribution parameters, control limits and capability indices. The methodology includes the plotting and analysis of Pareto, Cusum and Shewhart charts and scatter diagrams [174, 175]. The application of SPC to aspects of resin manufacture can be more difficult and give conflict in interpretation [176]. Value from applying these methods includes aid in specification setting and in the ability to monitor incoming raw materials effectively, as well as in the pinpointing of production disturbances and their cause. Computer programs are available to assist in calculations, of which Quality Analyst (North West Analytical) is an example.

Bibliography

Chemistry and computer techniques

ZIELINSKI T J & SWIFT M L, *Using Computers in Chemistry and Chemical Education*, ACS 1997. ISBN 0 8412 3465 5.

DIAMOND D & HANRALTY V C A, *Spreadsheet Applications in Chemistry using Microsoft Excel*, Wiley 1997. ISBN 0 471 14987 2.

JURS P C, *Computer Software Applications in Chemistry*, 2nd ed, Wiley Interscience 1996 ISBN 0 471 10587 2.

GRANT G H & GRIFFITHS W G, *Computational Chemistry*, O.U.P. 1995. (An introduction to techniques of molecular modelling). ISBN 0 198 55740 X.

EBERT K, EDERER H & ISENHAUER T L, *Computer Applications in Chemistry* VCH 1989. ISBN 3 527 27807 9.

Polymers and polymerization

PROVDER T (ed), *Computer Applications in Applied Polymer Science II, Automation, Modelling and Simulation*, ACS Symposium Series No. 404, American Chemical Society 1989. ISBN 0 841 21662 2.

BRUINS W, MOTOC I & O'DRISCOLL K F, *Monte Carlo Applications in Polymer Science* (Lecture Notes in Chemistry), Springer Verlag 1981.

MATTSON J S, MARK H B & MACDONALD H C (eds), *Computers in Polymer Science*, Marcel Dekker 1977.

LOWRY G G (ed), *Markov Chains and Monte Carlo Calculation in Polymer Science*, Marcel Dekker 1969.

Polymer properties
BICERANO J (ed), *Prediction of Polymer Properties*, Marcel Dekker 1996. ISBN 0 8247 99781 3.
GELIN B R, *Molecular Modeling of Polymer Structures and Properties*, Hanser 1995. ISBN 3 446 16553 3.
COLBOURN E A (ed), *Computer Simulation of Polymers*, Longman 1994. ISBN 0 582 08374 5.
BICERANO J (ed), *Computational Modeling of Polymers*, Marcel Dekker 1993. ISBN 0 8247 9119 3.
VAN KREVELAN D W, *Properties of Polymers. Their Estimation and Correlation with Chemical Structure*, 3rd Edition Elsevier 1990. ISBN 0 444 88160 3.
KAELBLE D H, *Computer Aided Design of Polymers and Composites*, Marcel Dekker 1985. ISBN 0 824 77288 1.

Paint formulation
WOODBRIDGE R, *Principles of Paint Formulation*, Blackie 1991. ISBN 0 216 93144 4.

Chemical and reactor engineering
MCGREAVY C (ed), *Polymer Reactor Engineering*, Blackie 1994 ISBN 0 751 40083 1.
ROSS G, *Computer Programming Examples for Chemical Engineers*, Elsevier 1987.
RAMON R, *Chemical Process Computations*, Elsevier 1985 ISBN 0 85334 341 1.

Solvents
FREDENSLUND T, GMEHLING J & RASMUSSEN P, *Vapour Liquid Equilibria using UNIFAC*, Elsevier 1977.

Mathematics
DANBY J M A, *Computer Modeling: From Sports to Space Flight . . . From Order to Chaos*, Willmann-Bell 1997. ISBN 0 943396 51 4.
ZUPAN J & GASTEIGER J, *Neural Networks for Chemists*, VCH 1993. ISBN 3 527 28603 9.
PRESS W H, TEUKOLSKY S A, VETTERLING W T & FLANNERY B P, *Numerical Recipes*, 1992–1996. (2nd edition with separate editions in C, Fortran and Fortran 90.) ISBN 0 521 43108 5, 0 521 43064 X, 0 521 57439 0.
PRESS W H, FLANNERY B P, TEUKOLSKY S A, VETTERIMOS W T, *Numerical Recipes*, Cambridge UP 1986. (First edition in Fortran with supplements for Basic and Pascal).

References

[1] REISCH M S, *Chem. Eng. News* 18 October 1993, 34.
[2] BACHRACH S M, *The Internet: a guide for chemists*, ACS Computer Applications in Chemistry Books, 1996. ISBN 0 841 23224 5.
[3] JOTISCHKY H, KENNEDY R, MORGAN N & WATSON J, *J Oil Col Chem Assoc* **71** (6) 167 (1988).
[4] PANTRY S, *Health & Safety at Work* **16** (1) 15 (1994); Anon., *Environ. Business*, Supplement January 1993.
[5] LANCASHIRE R J, *Spectroscopy Europe* **9** (2) 22 (1997).
[6] SCOTT A, *Europ Coatings J* **1991** (3) 146.
[7] BASSNER S L, *Am. Paint & Coatings J.* **77** (7) 46 (1992).
[8] MATHUIS A, SWIFT K & ROBINSON P, *Finishing* **17** (10) 26 (1993).
[9] KOIRANEN T, KRASLAWSKI A & NYSTRÖM L, *Ind Eng Chem* **33** 1756 *(1994)*; *Ind Eng Chem* **34** *3059 (1995)*.
[10] NAKAGAWA A S, *LIMS: Implementation and Management*, RSC London, 1994. ISBN 0 851 86824 X
[11] WU D T, *Chemtech* January 1987 26; p170 in *Computer Applications in Applied Polymer Science* Provder T (Ed.) *ACS Symposium Series* **313** (1986).
[12] LAURENCE J L, GALVAN R & TIRREL M V, Chapter 3 in *Polymer Reactor Engineering* in McGreavy C, Blackie (1994).

[13] JURS P C, *Computer Software Applications in Chemistry*, Chapter 8, John Wiley (1996).
[14] DUEVER T A & REILLY P M, *Can J Chem Eng* **68** 1040 (1990).
[15] DUEVER T A, REILLY P M & O'DRISCOLL K F, *J Poly Sci A Poly Chem* **26** 965 (1988).
[16] PLATOWSKI K & REICHERT K-H, *Polymer* **40** 1057 (1999).
[17] JURS P C, *Computer Software Applications in Chemistry*, Chapter 9, John Wiley (1986).
[18] KUNZ B & PROVDER T, *Progr Org Coatings* **27** 219 (1996).
[19] ADAMS M J, *Chemometrics in Analytical Spectroscopy*, RSC Analytical Spectroscopy Monographs 1995; OTTO M, *Chemometrics*, Wiley-UCH (1999). ISBN 3 527 29628 X.
[20] CHENG H N & BENNETT M A, *Analytica Chimica Acta* **242** 43 (1991).
[21] ROWE R C, *Manufacturing Chemist* October 1996, 21.
[22] VERDUIN W H, *Tappi* **77** 100 (1994).
[23] HUNT K & SBARBARO D, *Process Eng* May 1991 59.
[24] PSICHOGIOS D C & UNGAR L B, *Ind Eng Chem Res* **30** 2564 (1991).
[25] SEMINO D, MANNING N & BRAMBILLA A, *DECHEMA Monograph* **131** 693 (1995).
[26] ROFFEL S & CHIN P A, *Hydrocarbon Processing* June 1991 47.
[27] YEH E B & FRIED J R, *Chemtech* **23** (2) 22, ibid. **23** (2) 35 (1993); Chapters 1, 3, 4 in Colbourn E A. *Computer Simulation of Polymers*.
[28] ROGERS S C, MACKRODT W C & DAVIS T P, *Polymer* **35** (6) 1258 (1994).
[29] GILBERT R G et al, *Macromolecules* **28** 8771 (1995); ibid. **32** 5175 (1998); *Macromol. Symp.* **111**, 147 (1996).
[30] BODOR N, GABANYI Z & WONG C-K, *JACS* **111** (11) 3783 (1989).
[31] WOLF R A, *ACS Symposium Series* **404** 416 (1989).
[32] HARRIS S H, GOOD A C, GOOD R & GUO S-H, *Water-Borne, Higher-Solids, and Powder Coatings Symposium 1994, New Orleans* 386.
[33] TAN T M M & RODE B M, *Macromol Theory Simul* **5** 467 (1996).
[34] MURAMATSU I et al, *Progr Org Coatings* **22** 279 (1993).
[35] EGOLF L M & JURS P C, *Ind Eng Chem* **32** 1798 (1992).
[36] SULLIVAN C J & COOPER C F, *J Coatings Tech* **67** (847) 53 (1995).
[37] SUN H, *Macromolecules* **28** 701 (1995).
[38] WU H S & SANDLER S I, *Ind Eng Chem Res* **30** 889 (1991).
[39] KAVASSALIS T A, CHOI P & RUDIN A, *Molecular Simulation* **11** (2–4) 229 (1993); *Ind Eng Chem Res* **33** 3154 (1994).
[40] MISEV T A, *J Coatings Tech* **61** (772) 49 (1989); *Progr Org Coatings* **21** 79 (1992).
[41] STOCKMEYER W H, *J Poly Sci* **91** (1952); ibid. **11**, 424 (1954).
[42] NELEN P J C, *17th FATIPEC Congress* 283 (1984).
[43] *Technical Report No GTSR-74, AMOCO Chemicals Company; Technical Information Leaflet No. 0005 T10, Perstorp Polyols*.
[44] APPERT T, JACOBSEN L & RAY W H, *DECHEMA Monograph* **127** 189 (1992).
[45] PÉTIAND R et al, *Polymer* **33** (23) 5057 (1992); ibid. **36** (22) 4309 (1995).
[46] RAVI N, *ACS PMSE* **74** 270 (1996).
[47] JOHNSON A F & O'DRISCOLL K F, *Eur Polym J* **20** (10) 979 (1984).
[48] LEHTONEN J, SALMI T, IMMONEN K, PAATERO E & NYHOLM P, *Ind Eng Chem Res* **35** 3951 (1996).
[49] ASPIN I P et al, Surf Coat Int, **1998** (2) 68; Mauri A N et al, *Macromolecules* **30** 1616 (1997).
[50] GUILLOT J, p221 in *Polymer Reaction Engineering*, Reichert K H & Geisler W (eds), Hüthig & Wepf, 1986. ISBN 3857391170.
[51] BEEVERS R V & WHITE E F T, *Trans Farad Soc* **56** 117 (1960).
[52] DIBENEDETTO A T & DILAUDRO L, *J Poly Sci Poly Phys Ed* **27** 1405 (1989).
[53] O'DRISCOLL K F, *Macromolecules* **24** 4479 (1991).
[54] FOX T G & LOSHAEK S, *J Poly Sci* **15** 371 (1955); Loshaek S ibid. **15** 391 (1955).
[55] HARWOOD H J, CHEN T K & LIN F T, *ACS Symposium Series* **247** 197 (1984).
[56] CHENG H N, TAM S B, & KASEHAGEN L J, *Macromolecules* **25** (14) 3779 (1992).
[57] GUILLOT J, p295 in *Polymer Reaction Engineering* Reichert K H & Geisler W (eds), VCH 1989. ISBN 3 527 28015 4; *Macromol Symp* **92** 223 (1995).
[58] ALFREY T & LEWIS C, *J Poly Sci* **4** 221 (1949).
[59] CHOI K Y, *J App Poly Sci* **37** 1429 (1989); Seiner J A, *J Poly Sci* **A3** 2401 (1965).
[60] ROLAND M T & CHENG H N, *Macromolecules* **24** (8) 2015 (1991).
[61] RUDIN A, ROUNSEFELL T D'A & PITTMAN C U, Chapters 4 and 6 in *Computers in Polymer Science*, Mattson J S et al.
[62] REICH L S, PATEL S H & KHORRAMIAN B, *J App Poly Sci* **66** 891 (1997).
[63] POLIC A L, DUEVERT A & PENLIDIS A, *J Poly Sci Poly Chem Ed* **36** 813 (1998).
[64] O'DRISCOLL K F, *J Coatings Tech* **55** 57 (1983) and Chapter 3 in *Computers in Polymer Science*, Mattson J S et al.
[65] MOAD G et al, *Macromolecules* **20** 675 (1987).
[66] MOAD G et al, *Makromol Chem, Macromol Symp*, **51** 127 (1991); Tobita H, *Macromolecules* **28** 5119 (1995); ibid. **28** 5128 (1996).

[67] TOBITA H, *Macromol Theory Simul* **5** 129 *(1996)*.
[68] RAY W H, *Can J Chem Eng* **69** 626 (1991); Chapters 2, 12, 15, 16 in *Comprehensive Polymer Science* Volume 3.
[69] RAY W H, p105 in *Polymer Reaction Engineering* ed Reichert K H & Geisler W (eds), VCH 1989. ISBN 3 527 28015 4.
[70] HAMIELEC A E, MACGREGOR J F & PENLIDIS A, *Makromol. Chem Macromol. Symp* **10/11** 555 (1987).
[71] DUBÉ M A, SOARES B P, PENLIDIS A & HAMIELEC A E, *Ind & Eng Chem Research,* **36** 937 *(1997)*.
[72] GAO J & PENLIDIS A, *J Macromol Sci Revs* **C36** (2), 201 (1996).
[73] KUINDERSMA M E, PENLIDIS A & O'DRISCOLL K F, *ISCRE 11, July 8–11 1990, Toronto.*
[74] DUBE M A, PENLIDIS A & O'DRISCOLL K F, *Chem Eng Sci* **45** 2785 (1990).
[75] KUINDERSMA M A, PENLIDIS A & O'DRISCOLL K F, *ASEE Chem Eng Division Annual Conference Proceedings New Orleans 1991* 1464.
[76] CORNER T, *Advances in Polymer Science* **62** 95 (1984).
[77] DOUGHERTY E P, *J App Poly Sci* **32** 3051 (1986).
[78] RICHARDS J R, CONGALIDIS J P & GILBERT R G, *J App Poly Sci* **37** 2727 (1989).
[79] GILBERT R G *et al, Polymer* **39** 7099 (1998); VAN HERK A M & GERMAN A L, *Macromol Theory Simul* **7** 557 (1998).
[80] KONG X Z, PICHOT C, GUILLOT J & CAVAILEE J Y, *ACS Symposium Series* **492** 163 *(1992)*.
[81] VAN DOREMAELE G H J, VAN HERK A M, GERMAN A L, *Polym Internat* **27** 95 (1992).
[82] STORTI G, CANEGALLO S, CANU P & MORBIDELLI M, *DECHEMA Monograph* **127** 379 (1992); J. App. Poly. Sci. 54(12) 1899 & 1919 (1994).
[83] TOBITA H, TAKADA Y, & NOMURA M, *J Poly Sci Part A Poly Chem* **33** 441 (1995).
[84] TOBITA H, *Macromolecules* **28** 5128 (1995).
[85] TOBITA H & YAMAMOTA K, *Macromolecules* **27** 3389 (1994).
[86] TOBITA H, *DECHEMA Monograph* **131** 3 *(1995)*.
[87] DURANT Y G & SUNDBERG D C, *ACS Sympsium Series* **663** 44 (1997) ISBN 0 8412 3501 5; Gonzalez-Ortiz L J & Asua J M, *Macromolecules* **29** 4520 (1996).
[88] MESTACH D E & LOOS F, *XXIV FATIPEC 1998* B-91.
[89] KUINDERSMA E M, MASc Thesis, University of Waterloo 1992; Wilson T S, MASc Thesis, University of Waterloo 1994.
[90] MORBIDELLI M & STORTI G, p349 in *Polymeric Dispersions: Principles and Applications*, Asua J M (Ed.), Kluwer, 1997. ISBN 0 7923 4549 5.
[91] O'DRISCOLL K F & BURCZYK A F, *Polym React Eng J* **1** (1) 111 (1992–3).
[92] HAMIELEC A E, MACGREGOR J F, p21 in *Polymer Reaction Engineering*, Reichert K H & Geisler W (eds), Hanser, 1983. ISBN 3 446 13951 6.
[93] O'DRISCOLL K F, PONNUSWAMY S R, & PENLIDIS A, *ACS Symposium Series* **404** 321 *(1989)*.
[94] TAYLOR T, GONZALES V & JENSEN K F, p261 in *Polymer Reaction Engineering*, Reichert K H & Geisler W (eds), Hüthig & Wepf, 1986. ISBN 3 857 39117 0.
[95] STORTI G & MORBIDELLI M, *J App Poly Sci* **57** 1333 (1995); *DECHEMA Monograph* **131** 149 (1995).
[96] ASUA J M *et al, Polym. Internat* **30** 455 (1993).
[97] SCHORK, F J, Chapter 10 in *Emulsion Polymerisation and Emulsion Polymers*, Lovell P A & El-Aasser M S (eds), Wiley, 1997. ISBN 0 471 96746 7.
[98] BENTLEY J & BARKER S L, *ACS Symposium Series* **404** 454 (1989).
[99] BUDDE U & REICHERT K H, p140 in *Polymer Reaction Engineering* ed Reichert K H & Geisler W (eds), VCH 1989. ISBN 3 527 28015 4; Moritz H U. ibid. p248.
[100] ANON, *Polym Paint Col J* **173** 786 (1983).
[101] FELADY R, *Färg och Lack Scandinavia* 1980 (4) 15; ibid. 1980 (5) 85. (in Swedish).
[102] JOHN D, *Control Systems* January 1988, 28.
[103] Anon, *The Chemical Engineer* May 1983 (388) 85.
[104] PENLIDIS A, *Canadian J Chem Eng* **72**. 385 (1994).
[105] CHIEN D C H & PENLIDIS A, *J Macromol Sci-Rev Macromol Chem Phys* **C30** (1) 1 (1990); Hergeth W-D, p267 in *Polymeric Dispersions: Principles and Applications*, Asua J M (Ed.).
[106] PENLIDIS A, PONNUSWAMY S R, KIPARISSIDES C & O'DRISCOLL K F, *Chem Eng J* **50** 95 (1992).
[107] SEMINO D, MANNING N & BRAMBILLA A, *DECHEMA Monograph* **131** 693 (1995).
[108] TIEU D, CLUETT W R & PENLIDIS A, *Polym React Eng J* **2** (3) 275 (1994).
[109] ASUA J M *et al, Ind Eng Chem Res* **30** (6) 1343 (1991); *Angew Makromol. Chem* **194** 47 *(1992)*.
[110] LEIZA J R & ASUA J M, p363 in *Polymeric Dispersions: Principles and Applications*, Asua J M (Ed.).
[111] MCGREAVY C & NAUMAN E B, Chapter 6 in McGreavy C. *Polymer Reactor Engineering*, Blackie (1994).
[112] SCHMIDT-NAAKE G, ZESSING J & ENGELMANN U, *DECHEMA Monograph* **131** 424 (1995).
[113] MECKLENBURGH J C, *Can J Chem Eng* **48** (6) 279 (1970); Attiqullah A L, *Europ Poly J* **29** (12) 1581 (1993).
[114] TOSUN G, *AIChF J* **38** (3) 425 (1992).

[115] POEHLEIN G W, p305 in *Polymeric Dispersions: Principles and Applications*, Asua J M (Ed.); Soares J B P & Hamielec A E, p289 ibid; KLOSTERMAN R *et al, DECHEMA Monograph* **134** 295 (1998).

[116] KIEHLBAUCH M W & SCHATZ D D, *DECHEMA Monograph* **127** 93 (1992).

[117] MAYER M J J, MEULDIJK J & THOENES D, *J App Poly Sci* **59** 83 (1996); STORTI G *et al, DECHEMA Monograph* **134** 219 (1998).

[118] PENCZEK P *et al, XXIV FATIPEC 1998* B-107.

[119] VAN KREVELAN D W, *Properties of Polymers* Elsevier 3rd Edition; van Krevelan D W, Chapter 1 in Bicerano J (ed), *Computational Modeling of Polymers*; Perry R W & Green D, *Perry's Chemical Engineers' Handbook* McGraw-Hill 7th Edition. 1997 ISBN 0-07-115448-5.

[120] TASAKI K & RAVI N, *ACS PMSE* **74** 110 (1996).

[121] WU D T, *Int. Waterborne High Solids & Powder Coatings Symp. 1998*, New Orleans, p423.

[122] HUYBRECHTS & DUSEK K, *Surface Coatings Int.* 1998 (3) 117; ibid. 1998 (4) 172; ibid. 1998 (5) 235.

[123] MILLER D R & MACOSKO C W, *Macromolecules* **9** (2) 199 & 206 (1976); *J Poly Sci B Polym Phys* **26** 1 (1988).

[124] BAUER D R, *J Coatings Tech* **60** (758) 53 (1988); *ACS Symposium Series* **404** 190 (1989); Eichinger B E & Akgiray O, Chapter 8 in Colbourn E A. *Computer Simulation of Polymers*, Longman (1994).

[125] CLAYBOURN M & READING M, *J App Poly Sci* **44** 565 (1992); Bauer D R, *Progr Org Coat* **14** (3) 193 (1986).

[126] MILLER D R, *Makromol Chem Macromol Symp* **30** 57 (1989).

[127] HILL L W, *Progr Org Coat* **31** 235 (1997).

[128] DIMARZIO E A, J Res Natl Bur Stand Sect A **68A** 611 (1964); Nielsen L E, *J Macromol Sci-Revs Macromol Chem* **C3** 69 (1969).

[129] CHENG K-C & CHIU W-Y, *Macromolecules* **27** 3406 (1994).

[130] READING M, *23rd NATAS Conference 1994* USA; Colbourn E A, *Polym Paint Col J* **187** (3) 13 (1997).

[131] COLBOURN E, *Paint & Ink Int.* Mar/Apr 1997, 2; PRA 5th Nurnberg Congress Paper 12 (1999).

[132] TERMONIA Y, *Macromolecules* **22** 3633 (1989); Tobita H, *Macromolecules* **26** 5427 (1993).

[133] DUTTON S, STEPTO R F T & TAYLOR D J R, *ACS PMSE* **74** 172 (1996).

[134] FREDENSLUND T *et al, Vapour Liquid Equilibria using UNIFAC.*

[135] WU D T & KLEIN J A, *FATIPEC XX Nice 1990* 367.

[136] BEERS N C M, *Europ Coatings J* 1993 (3) 147.

[137] LYONS D, *J Oil Colour Chem Ass* **73** (2) 82 (1990); Kelsey J R, Lyons D & Sutley S R, *J Oil Colour Chem Ass* **74** (8) 278 (1991).

[138] VANCE R G, MORRIS N H & OLSON C M, *J Coatings Tech* **63** (802) 47 (1991).

[139] GOFF A, *Mod Paint & Coatings* May 1993 8.

[140] Alston D, Paper 14a, *PRA 8th Asia Pacific Conf.* Bangkok June 1998; EDDOWES D, *Polym Paint Col J* **188** (4410) 38 (1998).

[141] CARR C, *Paint Oil Col J* **181** (4278) 112 (1991).

[142] VAN DYK J W, FRISCH H L & WU D T, *Ind Eng Chem Prod Res Dev* **24** 473 (1985).

[143] HOY K L, *ACS Polymer Materials Science Engineering* **51** 556 (1984).

[144] RASMUSSEN D & WALLSTRÖM E, *J Oil Colour Chem Ass* **77** (3) 87 (1994); ibid. **77** (8) 323 (1994).

[145] NIELSEN C & WALLSTRÖM E, *XXIII Fatipec*, Brussels June 1996, C57.

[146] CARR C, *Progr Org Coatings* **28** 219 (1996); Misev T A, *FATIPEC XX Nice 1990* **101**; *J Coatings Tech* **63** (795) 23 (1991).

[147] TOUSSAINT A & DE WILDE M, *Progr Org Coatings* **30** 173 (1997).

[148] NIELSEN C & WALLSTRÖM E, *Färg och Lack Scandinavia* 1996 (6) 5 (in Danish).

[149] CUTRONE L, Chapter 6 in Woodbridge R, *Principles of Paint Formulation.*

[150] ANON, *Paint & Coatings Industry* Apr 1998 32.

[151] STOCK B S, *J Oil Colour Chem Assoc* **76** (3) 104 (1993); Anon, *Paint & Coatings Industry* Nov 1992 26.

[152] WU D T, *Modern Paint & Coatings* June 1996 34.

[153] WILLIAMS J R & FASANO D M, chapter 7 in *Handbook of Coatings Additives Vol 2* Calbo L J (ed) Marcel Dekker 1992. ISBN 0 8247 8716 2.

[154] MYNTTINEN R, *Färg och Lack Scandinavia* 1988 (10) 237 (in Swedish).

[155] CAREW J A, *J Oil Colour Chem Assoc* **77** (12) 515 (1994).

[156] SIMAKASKI M & HEGEDUS C R, *J Coatings Tech* **65** (817) 51 (1993).

[157] TUSAR L, TUSAR M & LESKOVSEK N, *Surface Coatings Internat* **1995** (10) 427.

[158] MAVER T L, *J Coatings Tech* **64** (812) 45 (1992).

[159] TEMPERLEY J, WESTWOOD M J, HORNBY M R & SIMPSON L A, *J Coatings Tech* **64** (809) 33 (1992).

[160] MOON M, *Trends in Polymer Science* **5** (3) 80 (1997).

[161] NOLAN G T & KAVANAGH P E, *J Coatings Tech* **67** (850) 37 (1995).

[162] BROWN R et al, Paper 8, *PRA 6th Annual Asia Pacific Conf.* Hong Kong *1996; Progr Org Coatings* **30** 185 & 195 (1997).

[163] SCOTT R M, *Lab Microcomputer* **12** (2) 55 (1993).

[164] BODEN D, PAINT & COATINGS INDUSTRY Feb 1998 60; Rich D C. *J Coatings Tech* **67** (840) 53 (1995); Hunston M & Carmody R, *Polym Paint Colour J* **184** (4348) 186 (1994).

[165] HUNT F Y, GALLER M & MARTIN J W, *J Coatings Tech* **70** (880) 45 (1998).
[166] MONTGOMERY D C, *Design of Experiments*, Wiley (1991)
[167] THOMAS T J P, *J Oil Colour Chem. Assoc* **76** (7) 272 (1993).
[168] BASSNER S L, *Amer Paint & Coatings J* **77** (6) 46 (1992).
[169] SCOTT P J, PENLIDIS A & REMPELL G L, *J Poly Sci A Poly Chem Ed* **31** (13) 2205 (1993).
[170] BURKE A L, DUEVER T A & PENLIDIS A, *Macromolecules* **27** 386 (1994).
[171] DUBÉ M A & PENLIDIS A, *Polymer* **36** (3) 587 (1995); *Macromol Chem Phys* **196** 1101 (1995); *Polym, Internat* **37** 235 (1995).
[172] BURKE A L, DUEVER T A & PENLIDIS A, *J Poly Sci A Poly Chem* **31** 3065 (1993).
[173] BARBRA A & ANTON C, *Europ. Coatings J.* 1993 (3) 161.
[174] BELL A, *Surface Coatings Austral.* Nov 1995, 19.
[175] MONTGOMERY D C, *Introduction to Statistical Quality Control*, Wiley, 1997. ISBN 0 471 30353 4.
[176] NEOGI D & SCHLAGS C E, *Ind Eng Chem Res* **37** 3971 (1998).

Trade marks and copyright of all software cited is acknowledged to the organization whose name follows mention of that software.

21

Health and safety in the coatings industry

G R Hayward

21.1   Introduction

In recent years there has been a considerable change in the coatings industry. Now the formulator has to satisfy many requirements in addition to that of producing a varnish, paint, adhesive, or ink that satisfies the stringent performance characteristics required by the customer. Today all companies and staff have to be fully aware of the health, safety, and environmental factors involved in producing a product from the initial formulation and development, manufacture, marketing, and technical backup. Every stage in a coatings development now involves the technologist to answer four basic questions:

- What? What materials are going to be used in this product? Are the substances and intermediates hazardous and what are the risks? This question will then trigger many further queries which will be discussed later in this chapter.
- Where? Where is this coating going to be manufactured? Is special plant or precautions necessary?
- Who? Who is going to manufacture this coating? Do the personnel involved with manufacture require additional training or equipment?
- How? How is this coating to be used? How is the coating going to be applied and cured? Is the coating for industrial or public use? This can restrict the use of some materials and may require re-formulation.

When reviewing any aspect of safety involving coatings one must first decide whether a hazard exists and what is the risk.

When one considers any aspect of safety the possible hazard that exists with a particular substance or process is considered. A hazard may be biological, chemical, or physical, including fire and explosion, the consequences of which can cause damage, injury, or loss. Having considered the hazards with a substance or process the actual risk needs to be evaluated, e.g. the risk from handling a very toxic chemical delivered in drum quantities is high unless very stringent safety precautions are taken.

The same substance used in a closed circuit system with no manual handling and where personnel do not come into physical contact with the substance is minimal, thus for the hazard of the substance is the same but the risk is low.

21.2   Raw materials and intermediates

The majority of starting materials, intermediates, additives, and solvents used for the manufacture of coatings present some degree of hazard. Information in the form of a safety data sheet (SDS) giving details or the health, safety, fire, and explosion and environmental hazards associated with material has to be presented to the user either with or preferably before delivery. This information is necessary for all materials considered to be dangerous and must be supplied, even for small laboratory samples. Currently the safety data sheet only has to be supplied with materials considered dangerous and for industrial use; however, it is expected that data sheet will eventually be required for materials not considered to present any risk and for public as well as professional users. Full details of the contents of safety data sheets and the necessary details which have to be given by agreed legislation in Europe will be given in Section 21.7.

A brief description of the hazards and potential risks associated with coating manufacturing materials, many of which are transferred to the final coating, is given below. The possible risks associated with these substances and products may be subdivided into two main groups: physicochemical properties and health effects.

21.2.1   Physicochemical properties

21.2.1.1   Flammability
Many of the solvents, intermediates, pigments, and additives have the potential to cause fire or explosions if stored or handled incorrectly. The possible risk from fire or explosion is greatest from the volatile solvents used in coatings but pigments and fillers may also give rise to dust explosions.

21.2.1.2   Solvents
The solvents used in coatings and intermediates usually present more possible hazards and risks to the users than other materials used in coating formulations. The major problems occur with flammability of solvents, although some do have acute and chronic toxicity effects which will be considered later.

Many solvents and intermediates are considered to be highly flammable, flammable, or combustible. Fires are usually caused by the creation of a flammable atmosphere involving the substance mixing with air under atmospheric conditions in proportions sufficient that it may be ignited by sparks, arcs, or high temperatures.

Organic solvents are usually defined by their ease of flammability which is determined by a physical property known as the flash point. This is the minimum temperature at which a material gives off sufficient vapour to form a flammable atmosphere above the surface of the liquid within the apparatus used for the determination. Many different standard methods can be used for the determination of flash point; usually when a flash point is quoted it is suffixed by an abbreviation indicating the method used for the determination.

Solvents also possess characteristics known as lower and upper flammable limits (LFL and UFL). The LFL is the percentage volume of a material mixed with unit volume of air below which the mixture is too weak to sustain the propagation of flame. The UFL is the percentage volume of a material mixed with unit volume of air above which the mixture is too rich to sustain the propagation of flame.

Another important parameter when solvents and other combustible materials are considered is the autoignition temperature. This is the lowest temperature to which, under ideal conditions, a substance has to be heated to initiate self-sustained combustion independently of any ignition source. Autoignition is also not an absolute property but varies with the method of determination and conditions.

Combustible substances, although not flammable, cause fire by undergoing an exothermic reaction with air when ignited.

Typical physical properties for some of the common solvents used in coatings are given in Table 21.1. Inspection of these values shows that all solvents have very specific flammability characteristics. For example there is no relationship between flash point and autoignition temperature; acetone has a low flash point but a very high autoignition temperature, whereas white spirit has a reasonably high flash point but one of the lowest autoignition temperatures. Even a reasonably stable substance such as ethylene glycol is flammable over the widest range of concentrations in air.

Another often overlooked property of some solvents is their ability to generate static electricity by flow or movement. This generation cannot be prevented but precautions can be taken to ensure that incendive sparks are not generated.

Liquids of low conductivity, e.g. toluene and xylene, can become charged when handled. Charges can be induced by flowing through pipes, strainers or filters; spraying, splashing, stirring or mixing, free fall, and in many other ways.

The distance a liquid is allowed to free fall or the velocity at which it is pumped or speed of stirring are all factors that can influence the generation of static.

Table 21.1 — Flammability properties of common solvents

Substance	Flash point (°C)	Boiling point (°C)	LFL (% v/v)	UFL (% v/v)	Autoignition temperature (°C)
Acetone	−17	56	2.1	13	538
Amyl alcohol	42	128–133	1.2	12.4	325
Butyl alcohol	33	117–118	1.4	11.3	340
Butyl glycol	68	167–173	1.1	10.6	240
Butyl glycol acetate	88	186–194	0.5	3.7	386
Diacetone alcohol	54	150–173	1.8	6.9	600
Diethyl ether	−45	34	1.9	36	160
Ethyl acetate	−5	76–78	2.2	11	425
Ethyl alcohol	12	78	3.3	19	365
Ethyl glycol acetate	54	145–165	1.7	10.1	390
Methoxy propanol	34	117–125	1.9	13.1	286
Methyl ethyl ketone	−7	80	1.8	11.5	514
Ethylene glycol	116	193–205	3	28	400
Toluene	4	111	1.2	8	550
Trichloroethylene	None	87	8	10.5	410
White spirit	36–41	152–198	0.6	8	230
Xylene	25	138–142	1.1	6.6	490

Various precautions are necessary to prevent the generation of static electricity. The most effective method is to ensure that all conducting parts of the plant and equipment are bonded to earth.

21.2.1.3 Pigments, fillers, and other substances

Many other chemicals in addition to solvents possess flammable characteristics. Some of the chemicals used to manufacture resins and intermediates are equally as flammable as many solvents; even the solid materials have the potential to cause dust explosions during handling. Details of a typical selection of these are given in Table 21.2.

Many powders are insulating and can easily become charged during handling. Those that are conducting can become charged if they are pneumatically conveyed or form a dust cloud during charging. Some powders are difficult to ignite but ignition becomes easier as the particle size is reduced. The range of concentrations over which powders mixed with air can propagate flame is defined in a similar way to solvents except that with powders they are called lower or upper explosible limits.

With powders, again the essential precaution to ensure that sparks are not generated is for all metal or conducting components associated with the process to be bonded to earth.

21.2.1.4 Explosive and oxidizing properties

Some of the raw materials used in resins and coatings have explosive or oxidizing properties. Explosive substances or preparations may be in solid, liquid, or paste form. They react exothermically, evolving gases which under specific conditions may detonate, burst into flame or explode when heated. Fortunately very few of the materials used in the coatings industry have explosive characteristics.

Table 21.2 — Flammability properties of common chemicals

Substance	Flash point (°C)	Melting point (°C)	Boiling point (°C)	LFL (% v/v)	UFL (% v/v)	Autoignition temperature (°C)
Acetic acid (glacial)	40	17	118	5.4	16.1	485
Acrylic acid	54	14	141	2.9	8.0	390
Acrylonitrile	0	−82	78	3	17	480
Adipic acid	196	151	338	—	—	420
Benzyl alcohol	101	−15	206	1.7	15	428
n-Butylamine	−12	−50	78	1.7	9.8	312
Castor oil	229	−12	313	—	—	453
Dibutyl phthalate	157	−35	340	—	—	400
Diethylamine	<−18	−50	57	1.8	10.1	312
Diethylene glycol	124	−8	244	—	—	225
Formaldehyde	750	−118	−19	7	73	424
Formic acid	69	8.4	101	18	57	520
Glycerol	160	18	290	—	—	393
Morpholine	38	−3	128	1.4	13.1	310
Phenol	79	41	182	1.5	9.0	605
α-Pinene	33	−55	155	0.8	—	255
Styrene	32	−31	146	1.1	6.1	490
Vinyl acetate	−8	−100	73	2.6	13.4	385

Some of the organic peroxides and other substances used to initiate free radical reactions in the manufacture of acrylic resins or to cure unsaturated polyesters or acrylic systems have the potential to be explosive and are classified as explosive substances. For this reason the majority of these materials are sold mixed with inert solids or liquids, e.g. plasticizers to reduce the risk. Typical of this class of substances are dibenzoyl peroxide and cyclohexanone peroxide.

21.2.1.5   Oxidizing substances

These are substances which react exothermically when in contact with other substances, particularly flammable substances.

21.2.2   Health effects

Although the risk of fire or explosion is a potential risk associated with many raw materials, intermediates, and finished coatings, many possess health risks which are often overlooked to some degree, as they may be longer term and not less tangible than fire which is instant and recognized by everyone. The health effects may involve the toxicity of the substance or preparation: it may be classified as being very toxic, toxic, or harmful. Sometimes the material may have additional corrosive, irritating, or sensitizing properties. The possible risks involved during use may be due to inhalation, ingestion, or contact with the skin or the eyes.

Simple descriptions of these health effects are given below. These may help when the terms are encountered in assessing the hazard of any materials or preparations (coatings) encountered.

The terms very toxic, toxic, and harmful are applied to substances or preparations which may cause death or acute or chronic damage to health by inhalation, or when swallowed, or by skin absorption, Acute effects are those that develop rapidly as opposed to chronic effects which usually the result of long and continuous exposure.

Corrosive is the term applied to substances that destroy living tissues on contact; they also corrode and attack many metals rapidly. Irritant materials are those that, on contact with the skin or mucous membrane, cause inflammation. Sensitization is a property that is becoming increasingly recognized as a potential heath risk. It is the ability of certain substance or preparations when inhaled or when they penetrate the skin to cause a reaction by hypersensitization such that further exposure causes characteristic adverse effects. The quantities of a material and the degree of exposure necessary to produce a sensitization response is very small and the effects, similar to a severe asthma attack, can be life-threatening.

Toxic substances or preparations may have carcinogenic, mutagenic, or reproductive effects.

21.3   Occupational exposure

During manufacture, application, and cure of surface coatings, personnel may be exposed to a variety of substances which may give off vapours, fumes, dusts, or fibres. Under certain conditions these may have a harmful effect on health, and these are regarded as hazardous substances. If the exposure to these substances is not properly controlled they may cause ill health in a number of ways.

These substances may cause harm by excessive amounts being taken into the body by breathing vapours or fumes, absorbed though intact skin, swallowing, or acting directly at some point of contact, e.g. skin or eyes. The effect of these substances on the body may be rapid or the symptoms of exposure may not appear until a long time after the initial exposure.

It is very important to know the level of these substances that is considered safe under normal conditions of use so that people working with these substances may be suitably protected. Many countries have introduced level of exposure values which, although they may vary in terms and the application of these values, are generally similar.

21.3.1 Threshold limit values and biological exposure indices

The initial concept of limiting exposures for chemical substances and physical agents was introduced in the USA by the American Conference of Governmental Industrial Hygienists (ACGIH). These limits are known as threshold limit values (TLVs) and biological exposure indices (BEIs). In general the values assigned to substances is very similar to those applied in other countries. One difference is the use of a threshold limit value — ceiling (TLV-C) term which is the concentration that should not be exceeded during any part of the working day and the use of A1:confirmed human carcinogen and A2:suspected human carcinogen classifications.

21.3.2 MAK values

Another range of exposure values which may be encountered are the German MAK values. These are issued as part of the TRGS 900 standard. MAK values use the following codes:

- G — MAK total dust;
- IIb — substances for which no MAK values can be listed;
- III — identified as a cancer-causing agent;
- A1 — capable of producing malignant tumours as shown by experience with humans;
- A2 — unmistakably carcinogenic in animal experiments only;
- B — justifiably suspected of having carcinogenic potential.

In addition there are MAK peak exposure limitation categories as given in Table 21.3. The values given in Table 21.3 are similar to the short-term exposure limits applied in the UK and US but are more specific.

Table 21.3 — MAK peak exposure limitation categories

Category	Peak	Duration	Frequency per shift
I	$2 \times$ MAK	5 min momentary value	8
II.1	$2 \times$ MAK	30 min average value	4
II.2	$5 \times$ MAK	30 min average value	2
III	$10 \times$ MAK	30 min average value	1
IV	$2 \times$ MAK	60 min momentary value	3
V	$2 \times$ MAK	10 min momentary value	4

There are also TRK values, which are shift mean values for an 8 hour work-day, 40 hour work week in accordance with TRGS 900 Manufacturer.

21.3.3   UK occupational exposure limits

In the UK the values for the control of exposure are known as occupational exposure limits (OELs). These will be discussed in detail, bearing in mind that most of the terms, conditions, and units are applicable to the majority of controls in other countries. The limits are set annually by the Health and Safety Executive (HSE) and published in EH40 [1].

Two types of occupational exposure limit for hazardous substances are used in the UK, occupational exposure standard (OES) and maximum exposure limit (MEL). Both limits are for the concentration of hazardous substances in air, averaged over a specific period of time referred to as a time weighted average (TWA). The exposure is assessed over two periods of time: long term, 8h, and short term, 15 min.

An OES is set at a level which is not expected to damage the health of workers exposed to it by inhalation day after day.

MELs are assigned to substances that may cause more serious health effects such as cancer or asthma and for which safe levels of exposure cannot be determined or for substances for which safe levels of exposure exist but control to these levels is not practicable. Both levels are set on the recommendations of the Health and Safety Commission's Advisory Committee on Toxic Substances (ACTS) and the Working Group on the Assessment of Toxic Chemicals (WATCH).

The HSE also publisher a complete range of documents giving precise details on precautions required when handling harmful materials (the EH [2] series) and methods for monitoring specific substances (the MDHS [3] series of publications).

21.3.4   Units of measurement

Occupational exposure limits for volatile substances are usually expressed in parts per million by volume in air (ppm) and milligrams per cubic metre ($mg\,m^{-3}$). For dusts and fume, the limits are quoted in $mg\,m^{-3}$; this normally is applied to the total inhalable fraction unless the respirable fraction is indicated. Man made fibres are expressed as $mg\,m^{-3}$ or fibres per millimetre of air (fibres ml^{-1}).

In addition to giving the exposure level in the specified units, some substances may have the Sk annotation which indicates that the substance may be absorbed through intact skin and/or the Sen notation where the substance may give rise to sensitization problems.

The document EH40 also contains biological monitoring guidance values (BMGVs), a list of substances under review, the method for calculating a limit for mixtures of hydrocarbon solvent, and additive substances. Appendices cover specific instructions for substances such as lead, asbestos, and substances defined as carcinogens.

21.3.5   Community exposure limits

There are two types of exposure limits being introduced into the EC European Communities: indicative limit values (ILVs) which give exposure levels deemed to ensure protection from occupational ill-health (analogous to the UK OESs) and binding limit values, for substances for which no health based limits can be set (similar to the UK MELs). To date only 27 substances have been assigned ILVs. The process of assigning limits in the EU is very slow owing to the complicated legislative procedures required in arriving at an agreed limit.

21.4   Provision of information

An essential and legal obligation for all involved in the manufacture, supply, and use of coating materials is the provision of information which provided by means of an SDS and the appropriate labelling for supply and transport. In previous years information on the hazards of using a raw material, intermediate, or finished product had only to be supplied on request. These data sheets were in variable formats and the information supplied was often sparse. In addition, some countries had their own regulations regarding provision of information.

In Europe with the introduction of the Dangerous Substances Directive 67/548/EEC which has had 8 amendments and 22 adaptations to technical progress, the Dangerous Preparations Directive 88/379/EEC with 4 adaptations to technical progress, and the Safety Data Sheets Directive 91/155/EEC with amendment 93/112/EC, the need to examine and investigate any dangerous substances and preparations thoroughly has become a legal obligation.

All these EC Directives have to be incorporated in national laws within the required time scale. In the UK this implementation has been by means of statutory instruments (SIs) which are laid before parliament which agrees a date for the legislation coming into force. This legislation is currently incorporated into UK law by means of The Chemicals (Hazard Information and Packaging for Supply) CHIP Regulations 1994 and amendments. This is a comprehensive document which explains the application of the regulation to any substance or preparation for supply and introduces the various concepts involved.

As previously stated all these regulations are continually being revised and amended. In the UK the current CHIP version is CHIP 97, which implemented the 22nd Adaptation to Technical Progress (ATP) of the Dangerous Substances Directive 67/548/EEC, the 8th Amendment to the DSD, the 4th ATP to the Dangerous Preparations Directive 88/379/EEC and the 2nd ATP of the 14th Amendment to the Marketing and Use Directive 76/769/EEC (in part).

21.5   The approved supply list

This is the information approved for the classification and labelling of substances and preparations for supply, issued by the Health and Safety Commission and updated as the EC adopts the various amendments to technical Progress of Council Directive 67/548/EEC. The current version is the third edition and incorporates the EC agreed risks and safety phrases and labelling symbols required at specified concentrations. This is the essential starting point for assessing the hazards and risks of any substance or preparation: if the starting substances are all listed then the evaluation for safety purposes is relatively simple. How the hazards of unlisted substances is dealt with will be discussed later in this chapter.

Table 21.4(a)

Substance name	Index No.	CAS No.
Acrylic acid	607-061-00-8	79-10-7
Ethylene glycol monobutyl ether(butyl glycol)	603-014-00-0	111-76-2
Phthalic anhydride	607-009-00-4	85-44-9

Table 21.4(b)

Index No.	Classification	Hazard detail EC No.	Specific concentration limits
607-061-00-8	R10 C; R34	C; R10-34 S(1/2)26-36-45 201-177-9	≥25% Conc C; R34 ≥2–<25% Xi; R36/38
603-014-00-0	Xn; R20/21/22 Xi; R37	Xn; R20/21/22 S (2) 24/25 203-905-0	≥20% Conc Xn; R20/21/22-37 ≥12.5–<20% Xn; R20/21/22
607-009-00-4	Xi; R36/37/38	Xi; R 36/37/38 S (2) 201-607-5	≥5% Xi; R36/37/28

Examples of how substances are entered in the Approved Supply List are given in Table 21.4.

21.5.1 Numbering system

The index number is the main identification of the substance which incorporates the atomic number of the most characteristic element or class for organic substances, a consecutive number for the substance, when the substance was placed on the market, and a check digit.

The EC number is from the European Inventory of Existing Commercial Chemical Substances (EINECS) and the Chemical Abstracts Service (CAS) number is to assist identification.

21.5.2 Classification system

The basis of the classification of substances and preparations for hazardous properties is to establish agreed risk and safety phrases; these are given for completeness in Appendix 1. When a specific risk hazard is assigned to a material, this also is indicated for major risks by the following abbreviations E explosive; O oxidizing; F+ extremely flammable; F flammable; T+ very toxic; T toxic; Xn harmful; Xi irritant; C corrosive; and N dangerous for the environment.

21.5.3 Determination of classification

Given below are the factors which are used to decide the classification of substances and preparations dangerous for supply not transport.

21.5.3.1   Explosive

Solid, liquid, pasty, or gelatinous substances and preparations which may also react exothermically without atmospheric oxygen, thereby quickly evolving gases, and which under defined test conditions detonate, quickly deflagrate or upon heating explode when partially confined.

21.5.3.2   Oxidizing

Substances and preparations which give rise to a highly exothermic reaction in contact with other substances, particularly flammable substances.

21.5.3.3   Extremely flammable

Liquid substances and preparations having an extremely low flash point ($\leq 0\,°C$), and a low boiling point ($\leq 35\,°C$), and gaseous substances and preparations that are flammable in contact with air at ambient temperature and pressure.

21.5.3.4   Highly flammable

- Substances and preparations which may become hot and finally catch fire in contact with air at ambient temperature without any application of energy.
- Solid substances and preparations which may readily catch fire after brief contact with a source of ignition and which continue to burn or to be consumed after removal of the source of ignition.
- Liquid substances and preparations having a very low flash point ($\leq 21\,°C$).
- Substances and preparations which, in contact with water or damp air, evolve highly flammable gases in dangerous quantities.

21.5.3.5   Flammable

Liquid substances and preparations having a low flash point ($\geq 21 - \leq 55\,°C$).

21.5.3.6   Very toxic

Substances and preparations which in very low quantities cause death or acute or chronic damage to health when inhaled, swallowed or absorbed via the skin.

21.5.3.7   Toxic

Substances and preparations which in low quantities cause death or acute or chronic damage to health when inhaled, swallowed or absorbed via the skin.

21.5.3.8   Harmful

Substances and preparations which, may cause death or acute or chronic damage to health when inhaled, swallowed, or absorbed via the skin.

21.5.3.9   Corrosive

Substances and preparations which may, on contact with living tissues, destroy them.

21.5.3.10   Irritant

Non-corrosive substances and preparations which, through immediate, prolonged or repeated contact with the skin or mucous membrane, may cause inflammation.

21.5.3.11   Sensitizing

Substances and preparations which, if they are inhaled or if they penetrate the skin, are capable of eliciting a reaction by hypersensitization such that on further exposure to the substance or preparation, characteristic adverse effects are produced.

21.5.3.12   Carcinogenic
Substances and preparations which, if they are inhaled or ingested or if they penetrate the skin, may induce cancer or increase its incidence.

21.5.3.13   Mutagenic
Substances and preparations which, if they are inhaled or ingested or if they penetrate the skin, may induce heritable genetic defects or increase their incidence.

21.5.3.14   Toxic for reproduction
Substances and preparations which, if they are inhaled or ingested or if they penetrate the skin, may produce, or increase the incidence of, non-heritable adverse effects in the progeny and/or an impairment of male or female reproductive functions.

21.5.3.15   Dangerous for the environment
Substances which, were they to enter the environment, would present or may present an immediate or delayed danger for one or more components of the environment.

For labelling purposes these abbreviations are depicted as pictograms, a black symbol on an orange background as shown in Fig. 21.1.

21.5.3.16   Carcinogenic, mutagenic, or toxic to reproduction substances
Some substances are additionally classified if they have agreed carcinogenic, mutagenic or teratogenic properties.

Carcinogenic substances
- Category 1 — substances known to be carcinogenic to man. There is sufficient evidence to establish a causal association between human exposure to the substance and the development of cancer.
- Category 2 — substances which should be regarded as if they are carcinogenic to man. There is sufficient evidence to provide a strong presumption that human exposure may result in the development of cancer.
- Category 3 — substances which cause concern for man owing to possible carcinogenic effects but available information is not adequate for making a satisfactory assessment.

Categories 1 and 2 are classified as toxic and assigned the toxic symbol with the R 45 phrase; if the risk is only from inhalation then the R 49 phrase is used. Category 3 is classified as harmful and assigned Xn harmful symbol with the R 40 risk phrase.

Mutagenic substances
- Category 1 — substances known to be mutagenic to man. There is sufficient evidence to establish a causal association between human exposure to the substance and heritable genetic damage.
- Category 2 — substances which should be regarded as if they are mutagenic to man. There is sufficient evidence to provide a strong presumption that human exposure may result in the development heritable genetic damage.

Hazard	Symbol letter	Symbol
Toxic	T	
Harmful	Xn	
Corrosive	C	
Irritant	Xi	
Dangerous for the environment	N	

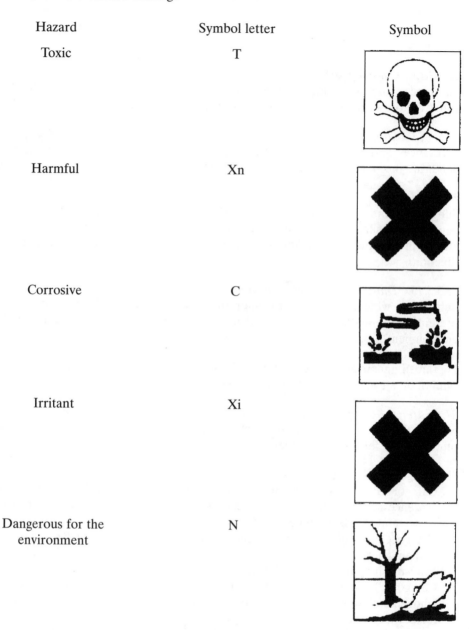

Fig. 21.1 — Hazard and symbols for substances and preparations for supply.

- Category 3 — substances which cause concern for man owing to possible mutagenic effects. There is evidence from studies but the available information is not adequate for making a satisfactory assessment.

Categories 1 and 2 are classified as toxic and assigned the T toxic symbol with the R 46 phrase. Category 3 is classified as harmful and assigned Xn harmful classification and symbol with the R 40 risk phrase.

Hazard	Symbol letter	Symbol
Explosive	E	
Oxidizing	O	
Extremely flammable	F+	
Highly flammable	F	
Very toxic	T+	

Fig. 21.1 — *Continued.*

Substances toxic to reproduction
- Category 1
  — Substances known to impair fertility in humans. There is sufficient evidence to establish a causal relationship between human exposure and impaired fertility.
  — Substances known to cause development toxicity in humans. There is sufficient evidence to establish a causal relationship between human exposure and subsequent developmental toxic effects in the progeny.

- Category 2
  - Substances which should be regarded as if they impair fertility in humans. There is sufficient evidence to provide a strong presumption that human exposure may result in impaired fertility.
  - Substances which should be regarded as if they cause developmental toxicity in humans. There is sufficient evidence to provide a strong presumption that human exposure may result in developmental toxicity.
- Category 3
  - Substances which cause concern for human fertility.
  - Substances which cause concern for humans owing to possible developmental toxicity effects.

Category 1 substances that impair fertility are assigned the T, toxic, classification and symbol and R 60 phrase. Substances that cause developmental toxicity are assigned the T, toxic, classification and symbol and R 61 phrase.

Category 2 substances that should be regarded as if they impair fertility are assigned the T, toxic, classification and symbol and R 60 phrase. Substances that should be regarded as if they cause developmental toxicity are assigned the T, toxic, classification and symbol and R 61 phrase.

Category 3 substances which cause concern for human fertility are assigned the Xn, harmful, classification and symbol and R 62 phrase. Substances which cause concern for humans owing to possible developmental toxicity effects are assigned the Xn, harmful, classification and symbol and R 63 phrase.

21.6   Hazard details

These are essentially the risk and safety phrases which are necessary for the production of an SDS and label.

Concentration limits require specific indications of the required hazard and safety phrases at that concentration. For single substances it this is a fairly simple task; however, for preparations containing many substances and intermediates the assessment is either by using a calculation method or by toxicity testing the actual preparation or mixture of substances.

The classification as toxic or harmful from the results of toxicity tests is carried out using the parameters given in Table 21.5 laid down under the EC Directives and given in.

Table 21.5

Classification	LD_{50} absorbed orally rat ($mg\,kg^{-1}$)	LD_{50} absorbed percutaneously in rat or rabbit ($mg\,kg^{-1}$)	LC_{50} absorbed by inhalation in rat, $mg\,l^{-1}$ (4h)	
Very toxic	≤25	≤50	≤0.25*	≤0.5[†]
Toxic	≥25–≤200	≥50–≤400	≥0.25–≤1.0	0.5–≤2
Harmful	≥200–≤2000	≥400–≤2000	≥1–≤5	≥2–≤20

* For aerosols or particulates.
[†] For gases or vapours.

For substances or preparations which are corrosive, irritant, or sensitizing other tests with limits are used. Environmental classification has also similar limits applied to the testing of fish, Daphnia and algae. On the basis of these results the agreed risk phrase(s) are assigned (see phrases R 50–59 in Appendix 1). It should be noted that the environmental classification of preparations is not required under EC law at present.

21.7   Safety data sheets

The provision of SDSs for substances and preparations in Europe is covered by Directive 93/112/EC. This lists the following obligatory headings which must be included in an SDS and the recommended order in which they appear on the data sheet.

1   *Identification of the substance/preparation and company.* This should be the approved name for the substance or preparation and the full name, address, and telephone number of the supplying company. An emergency telephone number should also be included.
2   *Composition/information on ingredients.* Here sufficient information should be given to identify the risks together with the concentration.
3   *Hazard identification.* This should be a brief indication of the hazard(s).
4   *First-aid measures.* A description of the first-aid measures and if there is a need for immediate attention.
5   *Fire-fighting measures.* Detailed here should be the required extinguishing media, hazards arising from combustion of the product, unsuitable extinguishing media, and special protection for fire-fighters.
6   *Accidental release measures.* Give details of any personal precautions required, conditions to avoid, and clean-up methods.
7   *Handling and storage.* Here guidance on safe storage and handling are given, paying attention to extraction of fumes, risk from fire, etc.
8   *Exposure controls/personal protection.* This section included the exposure limits as previously discussed and any personal protection required and the type of equipment necessary.
9   *Physical and chemical properties.* These should be given where available and determined where essential for safety or classification by the approved test method. The following data should be given where applicable:
   (a)   appearance — physical state and colour
   (b)   odour — brief description
   (c)   pH value
   (d)   boiling point or range
   (e)   melting point or range
   (f)   flash point
   (g)   flammability
   (h)   autoflammability
   (i)   explosive properties
   (j)   oxidizing properties
   (k)   vapour pressure
   (l)   solubility — water, solvents, oils

(m)   density

(n)   partition coefficient — *n*-octanol/water

(o)   other data considered important

10   *Stability and reactivity.* The stability of the substance and any conditions which must be avoided, e.g. temperature, metal, water.

11   *Toxicological information.* This section should include precise details of the health effects, symptoms from the different routes of exposure, and any delayed or immediate effects from short- or long-term exposure.

12   *Ecological information.* Any short- or long-term effects on aquatic and soil organisms, plants and animals are included in this section, also the mobility, degradability, and accumulation aspects.

13   *Disposal information.* Conditions for safe and approved methods of disposal should be given taking into account any national measures.

14   *Transport information.* This is any important section and covers labelling and conveyance by land, rail, air, and sea transport. It should be noted that the data are different from any labelling required for supply and totally different regulations apply.

15   *Regulatory information.* This section involves the classification and danger symbols previously discussed in Section 21.5.2.

16   *Other information.* This includes any specific information of safety or health importance, any restrictions on use, sources of data, and any specialized training required.

All SDSs must be dated with the date of publication or revision and include details of the revision and revision number. Any revision of the safety data should be brought to the attention of any recipient of the product within the previous 12 months.

Similar health and safety information is provided in the USA, details of which are included in the Federal Register under Part 1910 of title 29 of Code of Federal Regulations (CFR). The material safety data sheet (MSDS) basically has 12 section but provides similar data; it does not cover the transport or supply aspects of the European sheet. However, it must be stated if the material is listed in the National Toxicology Programme (NTP) Annual Report on Carcinogens or has been found a potential carcinogen in the International Agency for Research on Cancer (IARC) monographs or by the Occupational Safety and Health Administration (OSHA).

Additional information involves the use of a Hazardous Material Information System which numerically classifies the health hazard, fire hazard, reactivity, and personal protection requirements. The National Fire Protection Association also has a similar system which classifies the flammability, health, reactivity, and any specific hazard numerically.

21.8   Labelling of substances and preparations

There is a legal obligation to label appropriately all substances and preparations which are considered dangerous. This is an essential part of the various provision of information and packaging regulations embracing evaluation of hazards, provision of safety data sheets, and labelling.

Two types of labelling may be necessary: one for supply (use) and another for conveyance (transport), although under some circumstances a combined supply and carriage label is acceptable. Labelling for conveyance requires different classification limits and test methods and will be discussed further in Section 21.9. Labelling for supply applies to all substances or preparations which are classified as dangerous for supply by EC Directives or by self-classification methods previously discussed.

Classification is also carried out by specific industry groups, where groups of experts agree on labelling and handling requirements for materials not classified or under consideration. A example of this is the classification of dibutyl and diethyl hexyl phthalates as harmful and dangerous for the environment with appropriate risk and safety phrases by the CEFIC, Plasticizer Sector Group.

Other groups have suggested labelling for alkyl phenol ethoxylates, epoxy resin, cement, etc. Sometimes, associations or federations suggest additional labelling for products which specifically used or manufactured by their members. A typical example of this is the labelling for decorative paints recommended by the British Coatings Federation.

The specified dimensions of the label in relation to package size are:

Capacity of package	Dimensions of label
not exceeding 3 litres	at least 52 × 74 millimetres (if possible)
>3 litres not exceeding 50 litres	at least 74 × 105 millimetres
exceeding 50 litres not exceeding 500 litres	at least 105 × 148 millimetres
exceeding 500 litres	at least 148 × 210 millimetres

The details required on the label are slightly different for substances and preparations.

The following particulars are required for substances (Fig 21.2):

- Name and full address with telephone number of the company or person responsible for supplying the substance.
- Name of substance or internationally recognized name.
- Indication of danger and corresponding symbol(s).
- Risk phrases in full.
- Safety phrases in full.
- EC number for substances listed in Annex 1 with the words EC label.

For preparations (see Fig. 21.3):

- Name and full address with telephone number of the company or person responsible for supplying the preparation.
- Trade name or other designation of the preparation; if this contains dangerous substances these should be named if above certain levels.
- Indication of danger and corresponding symbol(s).
- Risk phrases in full.
- Safety phrases in full.
- If the preparation is for supply to the general public, an indication of the nominal quantity.

For both substances and preparation it is suggested that four risk and four safety phrases are used on the label, although more appear to be used in many cases.

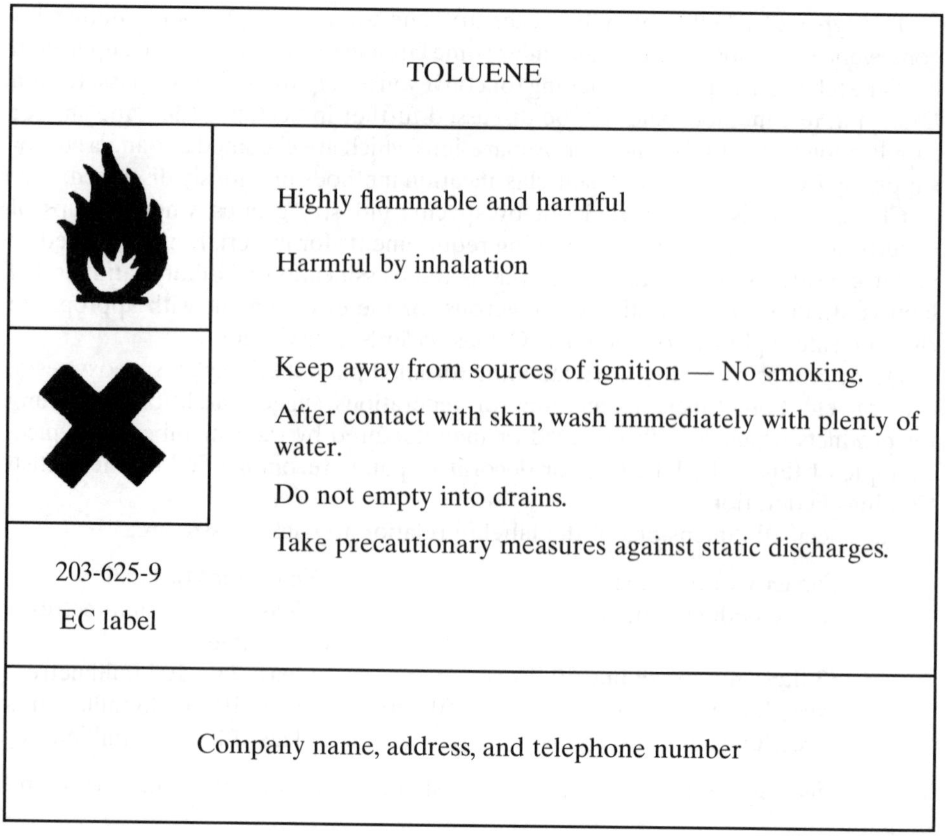

TOLUENE

	Highly flammable and harmful Harmful by inhalation
	Keep away from sources of ignition — No smoking. After contact with skin, wash immediately with plenty of water. Do not empty into drains. Take precautionary measures against static discharges.
203-625-9 EC label	

Company name, address, and telephone number

Fig. 21.2 — Example of a label for the supply of a substance.

21.8.1 Derogations

There are some derogations which apply to the labelling of small quantities of dangerous materials. If the substance or preparation is not explosive, very toxic, toxic, or sensitizing and the amount is so small that there is virtually no risk then it need not be labelled.

If the package contains less than 125 ml and the material is not explosive, very toxic, toxic, corrosive, or extremely flammable or if sold to the general public; classified as harmful, there is no need to show the risk and safety phrases.

Safety phrases may be put on a separate sheet if the container is too small.

21.8.2 Special labelling phrases

There are also special provisions for certain preparations whether or not they are dangerous for supply, many of which have implications for the coatings industry.

1 Paints and varnishes containing lead.
 (a) Labels of paints and varnishes containing lead in quantities exceeding 0.15% (wt of metal in total wt of preparation) — 'Contains lead.

POLYESTER RESIN
contains STYRENE

Flammable and harmful

Harmful by inhalation
Irritating to eyes and skin

Keep out of reach of children*.

Do not breathe vapour or spray.

1 Litre*

Company name, address, and telephone number

---

* Norminal quantity and S 2 safety phrase added if preparation is supplied to the general public.

Fig. 21.3 — Example of a label for the supply of a preparation.

Should not be used on surfaces that are liable to be chewed or sucked by children'.

(b) If the package contains less than 125 ml — 'Warning. Contains lead'.

2 Cyanoacrylate adhesives
(a) The immediate package containing cyanoacrylate shall be labelled —
'Cyanoacrylate.
Danger.
Bonds skin and eyes in seconds.
Keep out of reach of children'.
(b) Appropriate safety advice must accompany the package.

3 Preparations containing isocyanates; the label for preparations containing isocyanates (monomers, oligomers, prepolymers, etc., or as mixtures) shall state — 'Contains isocyanates. See information supplied by the manufacturer'.

4 Preparations containing epoxy constituents: if the preparation contains epoxy constituents with average molecular weight ≤700 the label should state — 'Contains epoxy constituents. See information supplied by the manufacturer'.

21.8.4   Additional requirements
The following additional packaging and labelling requirements may apply to some products offer for sale to the **general public**.

21.8.4.1   Child-resistant packages
*   Child-resistance fastenings to ISO 8317 shall be used for all substances or preparations classified as very toxic, toxic, or corrosive.
*   Those containing ≥3% methanol or ≥1% dichloromethane.
*   Liquid preparations containing ≥10% aliphatic or aromatic hydrocarbons (there are viscosity derogations).

21.8.4.2   Tactile warning label
A product which requires labelling as very toxic, toxic, corrosive, harmful, extremely flammable, or highly flammable shall not be supplied unless it carries a tactile warning of danger in accordance with EN Standard 272 (Fig. 21.4).

21.8.4.3   CHIP 96/97
As previously stated CHIP 96 introduced about 700 substances into the Approved Supply List (ASL) and a new labelling phrase 'Restricted to professional users' for certain carcinogens, mutagens, reproductive toxicants, and some chlorinated solvents, when present at a level exceeding 0.1%. CHIP 97 was implemented into UK law on 31 May 1998 and implements the 8th Amendment the DSD, 22nd Adaptation to the DSD, the 4th Adaptation to the DPD and the 2nd Adaptation to the Marketing and Use Directive. Essentially this adds 52 new entries to the Approved Supply List, amends many existing entries, introduces a code of practice on test methods, and extends CHIP to cover LPG and other gas cylinders.

It introduces a new aspiration hazard phrase R 65 'Harmful, may cause lung damage if swallowed', for various specified hydrocarbons in concentrations greater than 10%. This phrase is only applicable to low viscosity products and there is a viscosity range which exempt products from using this phrase. In the UK, the HSE issued an exemption clause which enabled the use of this phrase before its formal adoption, and thus preventing unnecessary expense.

The current version is CHIP 98, introduced in December 1998. This introduced a carcinogenic classification for some forms of man-made mineral fibres (MMMF); it deals only with specific man-made vitreous fibres (MMVF) which are a subset

Fig. 21.4 — Tactile warning of danger

of MMMF. It also classifies refractory ceramic fibres of specific fibre length as carcinogenic, cat. 2. A new edition of the ASL was also issued at this time implementing Commission Directive 97/69/EC, adopting to technical progress for the 23rd time Council Directive 67/548/EEC on the laws, regulations and administration provisions relating to the classification, packaging and labelling of dangerous substances.

This labelling criteria with symbols and standard phrases is reasonably uniform throughout Europe. In the USA the Code of Federal Regulations requires that dangerous products are correctly labelled, but pictograms and standardization of risk and safety phrases are not included. The supplier has to ensure that any warnings are in line with OSHA requirements for regulated substances.

21.9   Classification and labelling for transport (conveyance)

This is yet another hurdle with which the manufacturer of coatings or ancillary products has to comply. The regulations that apply to the transport of dangerous goods are similar in some respects to those required for the supply of a substance or preparation.

When considering the transport of a substance or preparation the following aspects have to be considered: (a) mode of transport, (b) destination, (c) pack size, type, and outer pack and, (d) tremcard (transport emergency card) or other requirements.

In order to transport packages or bulk containers of substances, preparations, or items classified as dangerous for transport various regulations have to be complied with, depending upon the different modes of transport that may be encountered in transit:

- For road transport: ADR (the European Agreement concerning the International Carriage of Dangerous Goods by Road).
- For rail transport: RID (International Regulations concerning the Carriage of Dangerous Goods by Rail).
- For sea transport: IMDG (International Maritime Dangerous Goods Code).
- For transport: IATA (International Air Transport Association — Dangerous Goods Regulations).
- United Nations: Recommendations by the Committee of Experts on the Transport of Dangerous Goods (the UN Orange Book).

All these regulations classify materials in a broadly similar way for hazard, but differ in various important areas depending upon the mode of travel.

All the publications referred to above give specific details for the substance or material concerned; again there are variations, although attempts are being made to harmonize the regulations for different modes of transport. Basically road, rail, sea, and air publications all give, for the material concerned, a proper shipping name (in capitals), the United Nations number, the classification code, subsidiary hazard number, and packing group. In addition whether the material may be transported in bulk or tanks, restrictions on package size and materials and methods of manufacture, hazard identification number, emergency action code, special provisions, storage position, restrictions on aircraft type, e.g. cargo or passenger.

21.9.1 Definition of hazard involved

The system used to define the actual hazard of a substance or preparation gives each type of hazard a class number. This is further subdivided into packing groups 1, 2 or 3, ADR/RID using letter a, b, or c, and IMO, a decimal system 3.1, 3.2, or 3.3.

Class 1:   Explosive
Class 2:   2.1 Flammable gases
            2.2 Non-flammable non-toxic gas
            2.3 Toxic gas
Class 3:   Flammable liquids
Class 4:   4.1: Flammable solids
            4.2: Substances liable to spontaneous combustion
            4.3: Substances which emit flammable gas on contact with water
Class 5:   5.1: Oxidizing substances
            5.2: Organic peroxide
Class 6:   6.1: Toxic substances
            6.2: Infectious substances
Class 7:   Radioactive substances
Class 8:   Corrosive substances
Class 9:   Miscellaneous dangerous substances

In addition to these main hazard classifications, subsidiary hazard codes and labelling is used where there is a secondary hazard which involves lower risk.

21.9.2 Labelling requirements

In order to transport a material it must in addition to other requirements be correctly labelled. These labels are known as danger signs and cover the hazard class numbers discussed in Section 21.9.1. The labels are in the shape of a diamond of minimum 100 mm side length and include a symbol indicating the hazard, the hazard classification code on a background colour which may be red, green, yellow, blue, or white depending upon the classification. Some examples of typical hazard labels are given in Table 21.6. There is also a subsidiary hazard code with accompanying subsidiary hazard sign as shown in Table 21.7.

For transport involving movement by sea an additional label is used to designate that the substance has been classified as a marine pollutant (Fig. 21.5). Marine pollutants are classified as either severe marine pollutants or marine pollutants. In addition to any single substances classified, any solution or mixture containing 10% of a listed marine pollutant or 1% or a severe marine pollutant require this label to be used.

A package containing dangerous goods must be labelled with the following particulars:

- the designation of the goods;
- the UN number;
- the danger sign;
- the subsidiary hazard sign, if any;
- the name and address or telephone number, or both of the consignor or other person who can give expert advice on the dangers created by the substance.

Table 21.6 — Danger signs

1 Classification code	2 Classification	3 Danger sign
5.1	Oxidizing substance	
5.2	Organic peroxide	
6.1	Toxic substance	
6.2	Infectious substance	

21.9.3   Approved Carriage List

A considerable amount of the information required to comply with the transport regulations for road and rail may be obtained by consulting the Approved Carriage List which states the information approved for the carriage of dangerous goods by road and rail. In addition the publication, *Approved Requirements and Test Methods for the Classification and Packaging of Dangerous Goods for Carriage*, gives full

Table 21.7 — Subsidiary hazard signs

1 Subsidiary hazard code	2 Subsidiary hazard	3 Subsidiary hazard sign
1	Liable to explosion	
2.1	Danger of fire (flammable gas)	
3	Danger of fire (flammable liquid)	
4.1	Danger of fire (flammable solid)	

details of how materials which are not listed in the Approved Carriage List may be tested and evaluated and hence classified for transport.

The Approved Carriage List also contains details of the hazard identification number for each listed substance or class of materials. This consists of either two or three figures indicating the following hazards:

- 2 — emission of gas due to pressure or reaction,
- 3 — flammability of liquids, gas, or self-heating liquids,
- 4 — flammability of solids or self-heating solids,

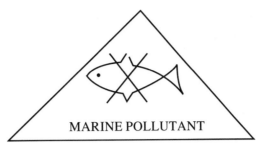

MARINE POLLUTANT

Fig. 21.5

- 5 — oxidizing or (fire-intensifying) materials,
- 6 — toxicity,
- 7 — radioactivity,
- 8 — corrosivity,
- 9 — risk of spontaneous reaction.

The prefix X is used to indicate that the material will react dangerously with water. Doubling the figure denotes an increased hazard. A combination of figures is used to denote more than one hazard or a subsidiary hazard. A typical example of this numbering system for a flammable liquid is: 30 [Flammable liquid, flash point between 23 °C and 61 °C]; 33 [Highly flammable liquid flash point below 23 °C], and X338 [Highly flammable, corrosive liquid which reacts dangerously with water].

Also listed in the Approved Carriage List are emergency action codes, also known as the Hazchem scheme. This essentially give details of the action necessary in case of fire and was introduced by the fire brigade. Under this system flammable liquids are either 3YE or 3Y. The 3 indicates that foam should be used as the extinguishing agent, the Y that it should be contained and breathing apparatus used, the E that, evacuation should be considered. A full explanation of this coding system is given in Fig. 21.6.

21.9.4   Supply and conveyance requirements

Classification and labelling for transport involve different considerations from classification and labelling for supply. Thus substances classified as toxic and flammable for supply may only be regarded as flammable for transport.

Transport classification uses three classes for toxic substances and has no harmful classification as used for supply. Flash point values as an indication of flammability also cover a different range, for transport liquids with a flash point between 23 and 61 °C are classified as flammable liquids. Below 23 °C the packing group used indicates the flammability of the product. For supply a liquid is classified as flammable if the flash point is 21–55 °C, below 21 °C the classification is highly flammable or extremely flammable.

For both supply and conveyance, products of above a specified viscosity are not classified as flammable.

The transport of dangerous goods in packages, bulk containers, or tanker quantities is a very complicated subject; given here is just an indication of the complexity and problems involved in ensuring that a material is transported in the legally agreed way.

Fire Extinguishants

Code 1: Water jet
Code 2: Fog
Code 3: Foam
Code 4: Dry agent

P	v	FULL	DILUTE
R			Spillage may be washed to drain with large quantities of water. However, due care must be undertaken to avoid unneccessary pollution of water courses
S	v	BA	
S		BA for FIRE only	
T		BA	
T		BA for FIRE only	
W	v	FULL	CONTAIN
X			Prevent, by any means available, spillage from entering drains and water courses
Y	v	BA	
Y		BA for FIRE only	
Z		BA	
Z		BA for FIRE only	

E	Consider Evacuation

Consider evacuation but depending on the nature of the Incident, it may be safer to keep people indoors

V	Can be violently or even explosively reactive

FULL	Full body protective clothing with BA

BA	Breathing apparatus plus protective gloves

Fig. 21.6 — Hazchem scheme: emergercy action codes.

21.10   Control of hazardous substances

An essential aspect in the manufacture and use of coatings is the protection of the personnel involved. One aspect of this is the control of substances hazardous to health which in the UK is known as the Control of Substances Hazardous to Health (COSHH) Regulations. These regulations, together with the Approved Codes of Practice (ACOP) define substances that are hazardous to health and specify how employers must protect workers from risks to their health from hazardous substances. Basically the regulations:

- require employers to assess, prevent, or adequately control the risks to health caused by hazards in the work-place;
- require that employees make proper use of control measures;
- simplify existing law and make compliance easier;

- set clear objectives to be achieved;
- set out a framework for preventing or controlling exposure to most substances hazards to health;
- provide a framework for implementing existing and future EC Directives on worker protection.

Employers must

- assess the risks from exposure and precautions necessary to prevent exposure or introduce adequate controls;
- introduce appropriate control measures to prevent or control any risk;
- ensure that these control measures are used;
- ensure that these control measures are properly maintained and periodically examined;
- monitor worker exposure where necessary;
- carry out health surveillance where necessary;
- inform, instruct, and train employees about risk and precautions to be taken.

The regulations apply

- to all substances or preparations classified as dangerous in the approved lists or which have been self-classified as very toxic, toxic, harmful, corrosive, or irritant;
- to any substance which has a maximum exposure limit (MEL) or occupational exposure standard (OES) as specified in EH40;
- to dusts of any kind when present in air in concentrations greater than $10\,mg\,m^{-3}$ over an 8 hour time weighted average of total inhalable dust or $4\,mg\,m^{-3}$ of respirable dust.

The hazard involved is the potential of a substance to cause harm. The risk is the likelihood that it will harm personnel in the actual circumstances of use. This risk not only depends upon the hazard presented by the substance but also on:

- how it is used;
- how it is controlled;
- who is exposed — to what amount and for how long a period of time;
- what process or operation is being carried out.

This is a brief description of the regulations that must be observed in the workplace and reinforces the need for accurate information in the form of manufacturers' information, data sheets, and exposure limits. Armed with this data the management must carry out an initial assessment to detemine which dangerous substances are handled, measure exposure levels if necessary, and evaluate any protective equipment used or necessary. From this assessment it must be decided how any exposure which is above agreed limits may be controlled. The use of personal protective equipment is only regarded as a short-term answer. Where reasonably practicable the following possibilities should be considered:

- changing the process so that the substance is no longer used;
- replacement by a safer alternative;
- enclosing the process;
- using a ring main system with no exposure to the substance.

Typical examples of areas where assessment is necessary are:

- In coating manufacture
  - delivery and charging of raw materials and intermediates
  - storage of materials (warehouse)
  - grinding and mixing processes
  - tinting and testing operations
  - straining and filtration
  - application testing and approval
  - filling out and packaging
- Application and use
  - testing for quality, etc.
  - storage of coatings and intermediates
  - application, spray, dip, brush, etc.
  - cure, oven stoving, UV cure, acid cure, etc.
  - any air drying emissions
  - waste paint storage and disposal

As more attention is paid to hazardous materials and environmental emissions so the paint user is requesting more help with assessments and monitoring of levels. In coatings applications these vary from solvents emissions from spray application and stoving to formaldehyde and other harmful materials evolved during cure. Even the water from spray booths can contain harmful products which have to be monitored and safely disposed of.

21.11   New substances regulations

The need to provide information about new substances was incorporated in the EC Directive amending for the seventh time Directive 67/548/EEC, commonly known as the Dangerous Substances Directive on the Approximation of the Laws, Regulations and Administrative Provisions relating to the Classification, Packaging and Labelling of Dangerous Substances. The object is to protect people and the environment from the possible ill-effects of new substances. A new substance is a substance *not* on the lists published by the Commission of the European Communities. This list is known as the European Inventory of Existing Commercial Chemical Substances (EINECS).

Requirements for the notification and assessment of existing substances is set out in the EC Existing Substances Regulation.

The degree of notification and assessment requirements depend upon the quantity of new substance to be supplied. Suppliers are required to inform the Competent Authority of their intention to place a new substance on the market and to produce a dossier of information about the substance. The Competent Authority is required to carry out a risk assessment of the substance and make recommendations.

The amount of information required on the substance is set out in the regulations which lists the various tests required in the Annexes. This starts at 10–100 kg marketed, e.g. placed on the EC market annually. For one tonne or more what is known as a 'base set' dossier is required with an approximate cost for evaluation and testing of £50000–90000. The lowest quantity involves costs of about £4000 but if quantities of the order of 100–1000 tonnes per annum are involved the costs can become excessive; for the largest amount this can be greater than £500000.

These regulations seriously restrict the development of new products unless an almost guaranteed market niche has been established early in a product's conception. It undoubtedly discourages innovation and affects a companys ability to compete internationally: some countries have lower notification costs, while others impose even more stringent testing requirements.

Another Directive recently introduced is the Biocidal Products Directive which will affect formulators of some types of coatings. This will require a biocide to be authorized for use within member states and be registered. Again core data sets and evaluation are involved. It is suggested that for a formulated biocidal paint that the company may need to obtain authorization before placing the product on the market.

21.12   Food contact coatings

21.12.1   Introduction

There has been concern about the possible problems that may occur when coatings are in contact with foodstuffs over many years and progress towards a European legislative document for coatings has been slow.

Initially any restrictions on coatings for food contact and the raw materials and intermediates involved were based on the American FDA (Food and Drug Administration) 175.300 regulations, which listed monomers, catalysts, additives, and pigments that could be safely used as the food-contact surface of articles intended to for use in producing, manufacturing, packing, processing, preparing, treating, packaging, transporting, or holding food. Also, 176.170/180 covers paper and board requirements.

In Europe, the Commission representative dealing with food contact materials is Dr L Rossi, principal administrator for DG III. The first European Directive was adopted in 1976. This is known as the Framework Directive 76/893/EEC on the approximation of the laws of the member states relating to materials and articles intended to come into contact with foodstuffs. Initially a task force was set up which produced a summary of existing different legislation and practice in Europe at that time. The scope was limited to coatings for metal substrates intended or liable to be used as food contact.

The second framework Directive 89/109/EEC superseded the 76 Directive and included the provision for the development and adoption of Directives for specific classes of materials. These include:

- plastics, including varnishes and coatings;
- regenerated cellulose;
- elastomers and rubbers;
- paper and board;
- ceramics;
- glass;
- metal and alloys;
- wood, including cork;
- textile products;
- paraffin waxes and micro-crystalline waxes.

21.12.2   The plastics directive

The plastics *only* part was introduced as The Plastics Food Contact Directive, this was Directive 90/128/EEC relating to plastics materials and articles intended to come into contact with foodstuffs.

This Directive deals with plastics materials only and does not include coatings, and has been implemented into national laws.

21.12.3   Synoptic documents

These are EC Commission documents which summarize the status of the substances listed and gives the current evaluations by the Scientific Committee for Food (SCF). Synoptic documents have been issued over many years, the current one, Synoptic Document No. 7, 15 May 1994 with supplement to July 1998, is titled Draft of Provisional List of Monomers and Additives used in the Manufacture of Plastics and *Coatings* intended to come into contact with Foodstuffs'. 'Coatings' have now been added to the list which initially was for plastics only.

However, the document states that it is a provisional and incomplete list of monomers and additives for food contact plastics as well as coatings authorized or used in the member states of the European Communities. It is stressed that this is a working document and not a legal or binding document. Updated lists are issued as different substances are evaluated by the SCF or Working Groups.

Substances are submitted from industry often through trade organizations, these are examined from the toxicological standpoint by the SCF, each substance is then classified into one of ten lists, numbered 0–9, as detailed below.

- List 0 — Substances, e.g. foods, which may be used in the production of plastic materials and articles, e.g. food ingredients and certain substances known from intermediate metabolism in man and for which an acceptable daily intake (ADI) need not be established for this substances.
- List 1 — Substances, e.g. food additives, for which an ADI, a temporary ADI (t-ADI), a maximum tolerable daily intake (MTDI), a provisional maximum tolerable daily intake (PMTDI), or the classification 'acceptable' has been established by this committee.
- List 2 — Substances for which a TDI or t-TDI has been established by this Committee.
- List 3 — Substances for which an ADI or TDI could not be established but where the present use could be accepted.
- List 4
  - Section A (for monomers): Substances for which an ADI or TDI could not be established but which could be used for if the substance migrating into foods or food simulants is not detectable by an agreed sensitive method.
  - Section B: Substances for which an ADI or TDI could not be established but which could be used if the levels of monomer residues in materials or articles intended to come into contact with foodstuffs are reduced as much as possible.
- List 4 (for additives) — Substances for which an ADI or TDI could not be established but which could be used for if the substance migrating into foods or food simulants is not detectable by an agreed sensitive method.
- List 5 — Substances which should not be used.

- List 6 — Substances for which there exists suspicions about their toxicity and for which data are lacking or are insufficient. The allocations of substances to this list are mainly based upon similarity of structure of chemical substances already evaluated or known to have functional groups that indicate carcinogenic or other severe toxic properties.
  - — Section A: Substances suspected to have carcinogenic properties. These substances should not be detectable in food or in food simulants by an approved sensitive method for each substance.
  - — Section B: Substances suspected to have toxic properties (other than carcinogenic). Restrictions may be indicated.
- List 7 — Substances for which some toxicological data exists, but for which an ADI or TDI could not be established. The required additional information should be furnished.
- List 8 — Substances for which no or only scanty and inadequate data were available.
- List 9 — Substances and groups that could not be evaluated owing to lack of specifications (substances) or to lack of adequate descriptions (groups of substances). Groups of substances should be replaced, where possible by individual substances actually in use. Polymers for which the data on identity specifies in SCF guidelines are not available.
- List W — 'waiting list'. Substances not yet included in the existing positive lists of member states. Although these substances appear in the synoptic documents, they are not susceptible to be included in the Community lists, lacking the data requested by the Committee.

This document was issued with 'Practical Guide No. 1', a 140 page document giving details of how to submit substances for assessment and the criteria applied by the SCF.

The regulations for plastics control the levels of substances migrating from the plastic material to the foodstuffs. The overall migration limit is set at $60\,mg\,kg^{-1}$ of foodstuffs or $10\,mg$ released from $1\,dm^2$ of the surface area of the plastic material.

The measurement of the total amounts of substances which have migrated into the foodstuffs is not practicable and so the testing is carried out on the plastic using simulants. Four food simulants are specified in Directives 82.711/EEC and 93/8/EEC. These are distilled water, 3% w/v aqueous acetic acid, 15% w/v aqueous ethanol, and olive oil.

Directives list simulants for different classes of foods; however, in some cases levels migrating into some simulants are significantly higher than the actual food and the levels recorded are reduced by factors of 2–5 depending upon the foodstuff involved.

These migration tests on plastic are carried out under standard test conditions at specified temperatures and time.

The Plastics Directive 90/128/EEC contains lists of monomers and other starting substances divided into two sections: Section A, permanent list, and Section B, temporary list, substances for which insufficient toxicological data are available. Substances may be transferred to Section A on submission of data and approval by the SCF. Section B is due to be removed in 2001; after this date substances remaining in Section B will not be permitted for use.

Substances in Section A with TDIs or ADIs of $1\,mg\,kg^{-1}$ body weight or less have been assigned specific migration limits.

To date there have been four amendments to Directive 90/128/EEC and one amendment 93/8/EEC to Directive 82/711/EEC on migration testing.

The 5th amendment to 90/128/EC which includes further additions to the incomplete list of additives and details of monomer added or transferred to Section A has yet to be agreed. This will prohibit from 1 July 1999 the manufacture or importation of plastic materials and articles which do not comply with this Directive.

In addition to agreeing the migration limits, considerable discussion also takes place on many other aspects of food contact for example functional barriers, daily intakes, levels of no effect.

It is proposed that the 6th Amendment to 90/128/EEC will include a more realistic assessment of exposure to take account of food type (fatty or aqueous), plastic utilization and possible thickness (diffusion).

21.12.4   Council of Europe

The Council of Europe is made up of representatives from Austria, Belgium, Denmark, Finland, France, Germany, Greece, Ireland, Italy, Luxembourg, the Netherlands, Norway, Portugal, Spain, Switzerland, and UK. The Council of Europe provides Resolutions compiled by the representatives from the above countries these are submitted to the EC Commission.

The Council of Europe has issued Resolutions on Surface Coatings Intended to Come into Contact with Foodstuffs, the current one, AP(96)5, was adopted by the Committee of Ministers in October 1996. This document is divided into two sections, List 1, Monomer and Additives lists 0–4, and List 2, Monomers and Additives lists 6–8, classified in accordance with SCF recommendations.

21.12.5   Conclusions

Many amendments to the Plastic Directive still have to be agreed and the position of many plastics additives clarified. When a positive list for additives is established, attention will specifically turn to coatings. Future directives on coatings may be some way off. If the Council of Europe proposals are accepted by the European Community, then progress will be faster. Although current legislation still only applies to coatings for plastics, many companies use the various EC and CoE (Council of Europe) documents for guidance when formulating coatings for food contact, as the FDA regulations were before involvement with the EC. As stressed in the previous section on new substances, once a positive list is established for food contact coatings, very few new substances will be added unless there are guarantees of extensive use in the coatings for food markets owing to the high cost of testing necessary to providing sufficient toxicological data to the EU Commission.

21.13   Major accident hazards

The consequences of major accidents occurring as a result of industrial activities has concerned the EC over many years, examples being the Flixborough and Seveso incidents. These helped to develop what has become known as the Seveso

Directives 82/501/EEC of 24 June 1982 on major accident hazards. In the UK this was implemented by the Control of Industrial Major Accident Hazards (CIMAH) regulations and subsequent amendments.

The aim of these regulations was to prevent major accidents resulting from industrial activities involving the production, processing, or storage of certain dangerous substances, and to limit their consequences for people and the environment.

Other major accidents have given rise to further concerns about the hazards which arises when dangerous sites and dwellings are in close proximity.

The EC Fourth Action programme on the Environment in 1987 called for the need for more effective implementation of Directive 82/501/EEC, and if necessary a widening of scope, together with a greater exchange of information between member states.

A resolution in Council on 16 October 1989 called on the Commission to exercise greater controls on land-use planning for new installations and for development around existing sites.

The Fifth Action programme on the Environment and a subsequent Council resolution in February 1993, also called for better risk and accident-management. It was decided that the 'Seveso' directives should be revised and supplemented in order to ensure high levels of protection throughout the EC.

Having carried out a fundamental review of the so-called 'Seveso' directives the EC Environmental Working Group having reviewed the existing directives published a formal proposal early in 1994 and a common position was reached on the proposal in June 1996. On 9 December 1996 the Council of the European Union formally adopted the proposals as Council Directive 96/82/EC. The new Directive is known as the Control Of Major Accident Hazards (COMAH) Directive, also known as the Sereso II Directive. This Directive came into force on 3 February 1997. Member states have 24 months to implement the necessary legislation into national law.

The CIMAH Directives and corresponding national legislation will remain in force until replaced by the COMAH legislation.

21.13.1 COMAH

The Directive 96/82/EC applies to establishments where dangerous substances are present on site in quantities equal to or in excess of threshold levels listed in Annex I Parts 1 and 2.

The new COMAH Directive contains a number of significant differences to the CIMAH directives:

- Exemptions to explosives and chemical risks at nuclear plants will no longer apply.
- The distinction between 'process' and 'storage' will also no longer apply.
- Includes a new category 'dangerous for the environment'.
- Now extended to cover land-use planning for the first time.
- The use of generic categories such as toxic or highly flammable greatly reduces the 178 substances, previously listed.
- Management duties, although largely in line with UK practice, have been expanded.

Requirements now include:

(a) Notification within prescribed time limits (Article 6).
(b) Written major accident prevention policies (Article 7).
(c) Safety reports (Article 9).
(d) Safety audits.
(e) On and off-site emergency plans (where Article 9 applies)
(f) Testing of emergency plans.

- The Competent Authority's duties are extended to include 'inspection' to ensure that the Directive's requirements are being met and 'Monitoring' the way major accident policies are put into effect.
- Account must be taken of any major hazard establishment in land-use policies. The UK already does and has for some time.
- COMAH also requires the Commission to collect and evaluate data on mechanisms for regulating dangerous substances through pipelines and take any necessary action.

Below are further brief details of the Directive with some of the substances and categories of dangerous substances include.

Part 1 of Annex 1 gives specific substances listed by chemical name, together with the qualifying quantities for the application of the various articles in the directive.

e.g.

Dangerous substances	Qualifying quantity (tonnes) for the application of	
	Articles 6 and 7	Article 9
Toluene diisocyanate	10	100

Part 2 give quantities for substances and preparations not specifically named in Part I

Categories of dangerous substances	Qualifying quantity (tonnes) of dangerous substances as delivered in Article 3(4), for the application of	
	Article 6 and 7	Article 9
Very toxic	5	20
Toxic	50	200
Oxidizing	50	200
Explosive (where the substance or preparation falls within the definition given in Note 2 (a))	50	200
Explosive (where the substance or preparation falls within the definition given in Note 2 (b))	10	50
Flammable (where the substance or preparation falls within the definition given in Note 3 (a))	5000	50000
Highly flammable (where the substance or preparation falls within the definition given in Note 3(b) (1)	50	200

Categories of dangerous substances	Qualifying quantity (tonnes) of dangerous substances as delivered in Article 3(4), for the application of	
	Article 6 and 7	Article 9
Highly flammable liquids (where the substance or preparation falls within the definition given in Note 3 (b) (2)	5000	50 000
Extremely flammable (where the substance or preparation falls within the definition given in Note 3 (c))	10	50
Dangerous for the environment in combination with risk phrases:		
(i) R50: 'Very toxic to aquatic organisms'	200	500
(ii) R51: 'Toxic to aquatic organisms'; and R53: 'May cause long-term adverse effects in the aquatic environment'	500	2000
Any classification not covered by those given above in combination with risk phrases:		
(i) R 14: 'Reacts violently with water' (including R14/15)	100	500
R 29: 'in contact with water, liberates toxic gas	50	200

Annex I of this directive also defines of the various categories, e.g. highly flammable, toxic.

Annexes II to VI give requirements for data and information, management systems, and organization, emergency plans, information for the public, and criteria for notification of an accident.

21.14   Environmental protection

This is another group of legislative controls which have faced the coatings manufacturer and user in recent years as the result of UK and EC legislation. The object is reduce pollution and to ensure protection of the environment. A dual regime has been established in the UK Integrated Pollution Control (IPC) which regulates releases to air, water, and land, and Air Pollution Control which only regulates releases to air.

Both coatings manufacture and application are included in the schedule and in some cases there is an overlap where the coatings manufacturer or user can fall into the IPC category if the output of special waste is over a specified amount or if certain specific chemicals are involved. Resin manufacturers fall into the Part 1 (IPC) group because of the use of chemicals such as vinyl chloride and anhydrides.

Basically authorization is given for the process by the environmental agency or the local authority, these are given on the basis that operators employ BATNEEC, the best available technique not entailing excessive cost to prevent or minimize pollution or render any emissions harmless. Industry is required to upgrade the process or operation within a specified time scale. Various guidance notes are issued for each

Table 21.8

Process or operation	Guidance note
Adhesives	
Application of adhesives	PG 6/32
Manufacture of adhesives	PG 6/10
Automotive	
OEM finishing	PG 6/20
Refinishing	PG 6/34
Components	PG 6/23
Coatings	
Metal and plastics	PG 6/23
Wood	PG 6/33
Coatings manufacture	PG 6/10
Coil coatings, application	PG 6/13
Drum coating	PG 6/15
Film coatings, application	PG 6/14
Metal spraying	PG 6/35
Packaging	
Printing of flexible packaging	PG 6/17
Metal printing and coating of small drums	PG 6/7
Paper coating processes	PG 6/18
Powder coatings	
Application	PG 6/31
Manufacture	PG 6/9
Printing ink manufacture	PG 6/16
Printing	PG 6/11
Wood preserving	PG 6/3

industrial process which list emission limits, monotoring requirements, operational controls, and agreed upgrading measures with times.

Most companies that manufacture, use, or apply resins, coatings, inks, adhesives have to register with the appropriate authority, obtain authorization for the process, produce an initial survey of emission sources, meet interim operating conditions, agree long-term upgrading policies, monitor emissions, keep records, and notify appropriate authorities if emissions are outside agreed operating parameters.

Authorization applications have to be accompanied by the required fee per component or process. There is an additional substantial variation fee and an annual subsistence fee. For a company operating many processes or those involving many components a considerable amount of money is involved, not only in obtaining authorization and continuing the process but with additional costs for ensuring continues compliance with the authorization.

Given in Table 21.8 are some of the process guidance notes issued by the environmental agency which must be complied with if an operation or process is undertaken. This is not a comprehensive list but those of interest to coating manufacturers and applicators.

21.14.1   Integrated pollution prevention and control (IPPC)

The EC IPPC Directive was finally adopted in 1996. This Directive is very similar in some respects to the UK IPC regulations. It introduces a permit system for installations covered in the directive and sets thresholds for solvent consumption, for coating, printing and surface treatment of >150 kg h^{-1} or >200 tonnes per year.

Pollution not only covers emission of substances but vibration, noise, heat, and energy.

The list of emission limit values include VOCs, organotin compounds, biocides, materials discharged into water, carcinogenic and mutagenic emissions to air, etc.

This directive, which has been under discussion for many years, will come into operation in October 1999 and must be fully implemented by companies by the year 2007.

Another directive which has been agreed and a common position reached is the Council Directive on the limitation of emissions of volatile organic compounds due to the use of organic solvents in certain activities and installations (known informally as The Solvents Directive). The directive will control emission limits, abatement, reduction and verification of compliance for many coating processes and installations. It is similar to the UK, IPC system of authorization of the process with compliance to agreed emission limits. Existing installations will have to meet the requirements for new installations, but within the transition period of the IPPC Directive, that is 30 October 2007.

21.15   Conclusions

This chapter is a snap-shot of some of the legislative problems and restrictions which face the coating manufacturer, supplier, or applicator. Virtually every topic warrants a large chapter or book in its own right. In all cases the appropriate Directive, Statutory Instrument, and Guidance Notes should be consulted. In most instances there have been numerous adaptations and amendments to the original directive. In others they have been embodied into existing directives.

Other reference areas and legislation should be considered, although most are EC Directives and have to be implemented into UK law by means of a Statutory Instrument (SI) agreed by the UK Parliament:

- The Marketing and Use of Dangerous Substances 76/769/EEC
- Child-resistant Fastenings and Tactile Danger Warnings 83/467/EEC
- The Safety and Health Framework Directive 89/391/EEC
- The Workplace Directive 89/654/EEC
- The Work Equipment Directive 89/655/EEC
- The Personal Protective Equipment Directive 89/686/EEC
- The Manual Handling Directive 90/269/EEC
- The Visual Display Equipment Directive 90/270/EEC
- Safety Data Sheets Directive 91/155/EEC
- The Cadmium Directive 91/338/EEC
- Child-resistant Fastenings for Dangerous Preparations 91/442/EEC
- Hazardous Waste Directive 91/689/EEC
- The Carcinogens at Work Directive 90/394/EEC
- The Asbestos at Work Directive 91/382/EEC
- Exposure at Work: Limit Values Directive 91/322/EEC
- Safety Signs Directive 92/58/EEC
- Risk Assessment Directive 93/67/EEC
- Aerosol Directive 94/1/EC
- Explosive Atmospheres Directive 94/9/EC

- Transport of Dangerous Goods by Road Directive 94/55/EC
- Young Workers Directive 94/93/EC
- Eco Label Criteria for Paints & Varnishes 96/13/EC
- Safety Advisors for Transport Directive 96/35/EC
- RID Directive 96/49/EC

References

[1] HSE Guidance Note Environmental hygiene EH 40 — Occupational exposure Limits 1998 (revised annually) ISBN 0-7176-1474-3.
[2] Enviromental hygiene series of publications.
[3] MDHS Medical series of health and safety publications.

Appendix 1: Risk phrases

Indication of particular risks

R 1: Explosive when dry
   2: Risk of explosion by shock, friction, fire, or other sources of ignition
   3: Extreme risk of explosion by shock, friction, fire, or other sources of ignition
   4: Forms very sensitive explosive metallic compounds
   5: Heating may cause an explosion
   6: Explosive with or without contact with air
   7: May cause fire
   8: Contact with combustible material may cause fire
   9: Explosive when mixed with combustible material
  10: Flammable
  11: Highly flammable
  12: Extremely flammable
  14: Reacts violently with water
  15: Contact with water liberates extremely flammable gases
  16: Explosive when mixed with oxidizing substances
  17: Spontaneously flammable in air
  18: In use may form flammable/explosive vapour–air mixture
  19: May form explosive peroxides
  20: Harmful by inhalation
  21: Harmful in contact with skin
  22: Harmful if swallowed
  23: Toxic by inhalation
  24: Toxic in contact with skin
  25: Toxic if swallowed
  26: Very toxic by inhalation
  27: Very toxic in contact with skin
  28: Very toxic if swallowed
  29: Contact with water liberates toxic gas
  30: Can become highly flammable in use
  31: Contact with acids liberates toxic gas
  32: Contact with acids liberates very toxic gas
  33: Danger of cumulative effects
  34: Causes burns
  35: Causes severe burns
  39: Danger of very serious irreversible effects
  40. Possible risk of irreversible effects
  41: Risk of serious damage to eyes
  42: May cause sensitization by inhalation
  43: May cause sensitization by skin contact

44: Risk of explosion if heated under confinement
45: May cause cancer
46: May cause heritable genetic damage
48: Danger of serious damage to health by prolonged exposure
49: May cause cancer by inhalation
50: Very toxic to aquatic organisms
51: Toxic to aquatic organisms
52: Harmful to aquatic organisms
53: May cause long-term adverse effects in the aquatic environment
54: Toxic to flora
55: Toxic to fauna
56: Toxic to soil organisms
57: Toxic to bees
58: May cause long-term adverse effects in the environment
59: Dangerous for the ozone layer
60: May impair fertility
61: May cause harm to the unborn child
62: Possible risk of impaired fertility
63: Possible risk of harm to the unborn child
64: May cause harm to breast fed babies

Combined risk phrases

14/15:	Reacts violently with water, liberating extremely flammable gases
15/29:	Contact with water liberates toxic, extremely flammable gas
20/21:	Harmful by inhalation and in contact with skin
20/21/22:	Harmful by inhalation, in contact with skin and if swallowed
20/22:	Harmful by inhalation and if swallowed
21/22:	Harmful in contact with skin and if swallowed
23/24:	Toxic by inhalation and in contact with skin
23/24/25:	Toxic by inhalation, in contact with skin, and if swallowed
23/25:	Toxic by inhalation and if swallowed
24/25:	Toxic in contact with skin and if swallowed
26/27:	Very toxic by inhalation and in contact with skin
26/27/28:	Very toxic by inhalation, in contact with skin and if swallowed
26/28:	Very toxic by inhalation and if swallowed
27/28:	Very toxic in contact with skin and if swallowed
36/27:	Irritating to eyes and respiratory system
36/37/38:	Irritating to eyes, respiratory system, and skin
36/38:	Irritating to eyes and skin
37/38:	Irritating to respiratory system and skin
39/23:	Toxic: danger of very serious irreversible effects through inhalation
39/23/24:	Toxic: danger of very serious irreversible effects through inhalation and in contact with skin
39/23/24/25:	Toxic: danger of very serious irreversible effects through inhalation, in contact with skin, and if swallowed
39/23/24:	Toxic: danger of very serious irreversible effects through inhalation and if swallowed
39/24:	Toxic: danger of very serious irreversible effects in contact with skin
39/24/25:	Toxic: danger of very serious irreversible effects in contact with skin and if swallowed
39/25:	Toxic: danger of very serious irreversible effects if swallowed
39/26:	Very toxic: danger of very serious irreversible effects through inhalation
39/26/27:	Very toxic: danger of very serious irreversible effects through inhalation and in contact with skin
39/26/27/28:	Very toxic: danger of very serious irreversible effects through inhalation, in contact with skin, and if swallowed
39/26/28:	Very toxic: danger of very serious irreversible effects through inhalation and if swallowed

39/27:	Very toxic: danger of very serious irreversible effects in contact with skin
39/27/28:	Very toxic: danger of very serious irreversible effects in contact with skin and if swallowed
39/28:	Very toxic: danger of very serious irreversible effects if swallowed
40/20:	Harmful: possible risk of irreversible effects through inhalation
40/20/21:	Harmful: possible risk of irreversible effects through inhalation and in contact with skin
40/20/21/22:	Harmful: possible risk of irreversible effects through inhalation, in contact with skin, and if swallowed
40/20/22:	Harmful: possible risk of irreversible effects through inhalation and if swallowed
40/22:	Harmful: possible risk of irreversible effects if swallowed
40/21:	Harmful: possible risk of irreversible effects in contact with skin
40/21/22:	Harmful: possible risk of irreversible effects in contact with skin and if swallowed
42/43:	May cause sensitization by inhalation and skin contact
48/20:	Harmful: danger of serious damage to health by prolonged exposure through inhalation
48/20/21:	Harmful: danger of serious damage to health by prolonged exposure through inhalation and in contact with skin
48/20/21/22:	Harmful: danger of serious damage to health by prolonged exposure through inhalation, in contact with the skin and if swallowed
48/20/22:	Harmful: danger of serious damage to health by prolonged exposure through inhalation and if swallowed
48/21:	Harmful: danger of serious damage to health by prolonged exposure in contact with skin
48/21/22:	Harmful: danger of serious damage to health by prolonged exposure in contact with skin and if swallowed
48/22:	Harmful: danger of serious damage to health by prolonged exposure if swallowed
48/23:	Toxic: danger of serious damage to health by prolonged exposure through inhalation
48/23/24:	Toxic: danger of serious damage to health by prolonged exposure through inhalation and in contact with skin
48/23/24/25:	Toxic: danger of serious damage to health by prolonged exposure through inhalation, in contact with skin, and if swallowed
48/23/25:	Toxic: danger of serious damage to health through inhalation and if swallowed
48/24:	Toxic: danger of serious damage to health by prolonged exposure in contact with skin
48/24/25:	Toxic: danger of serious damage to health by prolonged exposure in contact with skin and if swallowed
48/25:	Toxic: danger of serious damage to health by prolonged exposure if swallowed
50/53:	Very toxic to aquatic organisms, may cause long-term effects in the aquatic environment
51/53:	Toxic to aquatic organisms, may cause long-term adverse effects in the aquatic environment
52/53:	Harmful to aquatic organisms, may cause long-term adverse effects in the aquatic environment

Appendix 2: Safety phrases

Safety precautions

S 1: Keep locked up
   2: Keep out of reach of children

3: Keep in a cool place
4: Keep away from living quarters
5: Keep contents under ... (appropriate liquid to be specified by the manufacture)
6: Keep under ... (inert gas to be specified by the manufacturer)
7: Keep container tightly closed
8: Keep container dry
9: Keep container in a well-ventilated place
12: Do not keep the container sealed
13: Keep away from food, drink, and animal feeding stuffs
14: Keep away from ... (incompatible materials to be indicated by the manufacturer)
15: Keep away from heat
16: Keep away from sources of ignition — No smoking
17: Keep away from combustible material
18: Handle and open container with care
20: When using do not eat or drink
21: When using do not smoke
22: Do not breathe dust
23: Do not breathe gas/fumes/vapour/spray (appropriate wording to be specified by the manufacturer)
24: Avoid contact with the skin
25: Avoid contact with the eyes
26: In case of contact with eyes, rinse immediately with plenty of water and seek medical advice
27: Take off immediately all contaminated clothing
28: After contact with skin, wash immediately with plenty of ... (to be specified by the manufacturer)
29: Do not empty into drains
30: Never add water to this product
33: Take precautionary measures against static discharges
35: This material and its container must be disposed of in a safe way
36: Wear suitable protective clothing
37: Wear suitable gloves
38: In case of insufficient ventilation, wear suitable respiratory equipment
39: Wear eye/face protection
40: To clean the floor and all objects contaminated by this material use ... (to be specified by the manufacturer)
41: In case of fire and/or explosion do not breath fumes
42: During film ignition/spraying wear suitable respiratory equipment (appropriate wording to be specified)
43: In case of fire, use (indicate in the space the precise type of fire-fighting equipment. If water increases the risk add — Never use water)
45: In case of accident or if you feel unwell, seek medical advice immediately (show label where possible)
46: If swallowed seek medical advice immediately and show this container or label
47: Keep at temperature not exceeding ... °C (to be specified by the manufacturer)
48: Keep wetted with ... (appropriate material to be specified by the manufacturer)
49: Keep only in the original container
50: Do not mix with ... (to be specified by the manufacturer)
51: Use only in well-ventilated areas
52: Not recommended for interior use on large surface areas
53: Avoid exposure — obtain special instruction before use
56: Dispose of this material and its container to hazardous or special waste collection point
57: Use appropriate containment to avoid environmental contamination
59: Refer to manufacturer/supplier for information on recovery/recycling
60: This material and/or its container must be disposed of as hazardous waste
61: Avoid release to the environment. Refer to special instructions/safety data sheet
62: If swallowed, do not induce vomiting: seek medical advice immediately and show this container or label

Combined safety precautions

1/2:	Keep locked up and out of the reach of children
3/9/14:	Keep in a cool, well-ventilated place away from . . . (incompatible materials to be indicated by manufacturer)
3/9/14/49:	Keep only in the original container in a cool well-ventilated place away from . . . (incompatible materials to be indicated by the manufacturer)
3/9/49:	Keep only in the original container in a cool well-ventilated place
3/14:	Keep in a cool place away from . . . (incompatible materials to be indicated by the manufacturer)
3/7:	Keep container tightly closed in a cool place
7/8:	Keep container tightly closed and dry
7/9:	Keep container tightly closed and in a well-ventilated place
7/47:	Keep container tightly closed and at a temperature not exceeding . . . °C (to be specified by manufacturer)
20/21:	When using do not eat, drink or smoke
24/25:	Avoid contact with skin and eyes
29/56:	Do not empty into drains, dispose of this material and its container to hazardous or special waste collection point
36/37:	Wear suitable protective clothing and gloves
36/37/39:	Wear suitable protective clothing, gloves, and eye/face protection
36/39:	Wear suitable protective clothing and eye/face protection
36/39:	Wear suitable gloves and eye/face protection
47/49:	Keep only in the original container at temperature not exceeding

Index